Modern Birkhäuser Classics

Many of the original research and survey monographs, as well as textbooks, in pure and applied mathematics published by Birkhäuser in recent decades have been groundbreaking and have come to be regarded as foundational to the subject. Through the MBC Series, a select number of these modern classics, entirely uncorrected, are being re-released in paperback (and as eBooks) to ensure that these treasures remain accessible to new generations of students, scholars, and researchers.

Advanced Calculus

A Differential Forms Approach

Harold M. Edwards

Reprint of the 1994 Edition

Harold M. Edwards
Courant Institute
New York University
New York, NY, USA

ISSN 2197-1803 ISSN 2197-1811 (electronic)
ISBN 978-0-8176-8411-2 ISBN 978-0-8176-8412-9 (eBook)
DOI 10.1007/978-0-8176-8412-9
Springer New York Heidelberg Dordrecht London

Library of Congress Control Number: 2013953495

Printed on acid-free paper

Springer is part of Springer Science+Business Media (www.birkhauser-science.com)

Harold M. Edwards

Advanced Calculus

A Differential Forms Approach

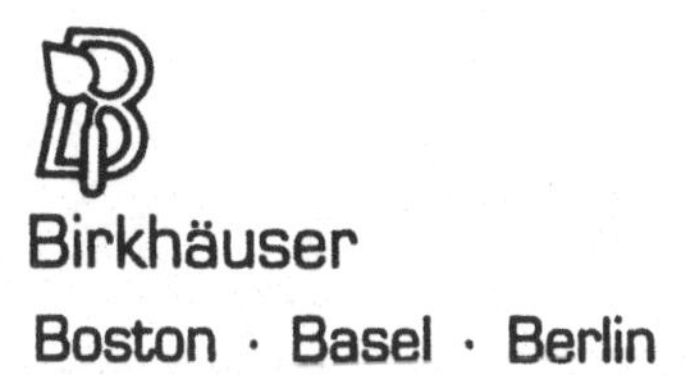
Birkhäuser
Boston · Basel · Berlin

Harold M. Edwards
Courant Institute
New York University
New York, NY 10012

Library of Congress Cataloging In-Publication Data

Edwards, Harold M.
Advanced calculus : a differential forms approach / Harold M. Edwards. -- [3rd ed.]
p. cm.
Includes index.
ISBN 0-8176-3707-9 (alk. paper)
1. Calculus. I. Title.
QA303.E24 1993 93-20657
515--dc20 CIP

Printed on acid-free paper

Reprinted 1994 with corrections from the original Houghton Mifflin edition.

Birkhäuser

ISBN 0-8176-3707-9
ISBN 3-7643-3707-9

Printed and bound by Quinn-Woodbine, Woodbine, NJ
Printed in the USA

9 8 7 6 5

To my students—and especially to those who never stopped asking questions.

Preface to the 1994 Edition

My first book had a perilous childhood. With this new edition, I hope it has reached a secure middle age.

The book was born in 1969 as an "innovative textbook"—a breed everyone claims to want but which usually goes straight to the orphanage. My original plan had been to write a small supplementary textbook on differential forms, but overly optimistic publishers talked me out of this modest intention and into the wholly unrealistic objective (especially unrealistic for an unknown 30-year-old author) of writing a full-scale advanced calculus course that would revolutionize the way advanced calculus was taught and sell lots of books in the process.

I have never regretted the effort that I expended in the pursuit of this hopeless dream—only that the book was published as a textbook and marketed as a textbook, with the result that the case for differential forms that it tried to make was hardly heard. It received a favorable telegraphic review of a few lines in the *American Mathematical Monthly*, and that was it. The only other way a potential reader could learn of the book's existence was to read an advertisement or to encounter one of the publisher's salesmen. Ironically, my subsequent books—*Riemann's Zeta Function, Fermat's Last Theorem* and *Galois Theory*—sold many more copies than the original edition of *Advanced Calculus*, even though they were written with no commercial motive at all and were directed to a narrower group of readers.

When the original publisher gave up on the book, it was republished, with corrections, by the Krieger Publishing Company. This edition enjoyed modest but steady sales for over a decade. With that edition exhausted and with Krieger having decided not to do a new printing, I am enormously gratified by Birkhäuser Boston's decision to add this title to their fine list, to restore it to its original, easy-to-read size, and to direct it to an appropriate audience. It is at their suggestion that the subtitle "A Differential Forms Approach" has been added.

I wrote the book because I believed that differential forms provided the most natural and enlightening approach to the calculus of several variables. With the exception of Chapter 9, which is a bow to the topics in the calculus of one variable that are traditionally covered in advanced calculus courses, the book is permeated with the use of differential forms.

Colleagues have sometimes expressed the opinion that the book is too difficult for the average student of advanced calculus, and is suited only to honors students. I disagree. I believe these colleagues *think* the book is difficult because it requires that they, as teachers, rethink the material and accustom themselves to a new point of view. For students, who have no prejudices to overcome, I can see no way in which the book is more difficult than others. On the contrary, my intention was to create a course in which the students would learn some useful methods that would stand them in good stead, even if the subtleties of uniform convergence or the rigorous definitions of surface integrals in 3-space eluded them. Differential forms are extremely useful and calculation with them is easy. In linear algebra, in implicit differentiation, in applying the method of Lagrange multipliers, and above all in applying the generalized Stokes theorem $\int_{\partial S} \omega = \int_S d\omega$ (also known as the fundamental theorem of calculus) the use of differential forms provides the student with a tool of undeniable usefulness. To learn it requires a fraction of the work needed to learn the notation of div, grad, and curl that is often taught, and it applies in any number of dimensions, whereas div, grad, and curl apply only in three dimensions.

Admittedly, the book contains far too much material for a one year course, and if a teacher feels obliged to cover everything, this book will be seen as too hard. Some topics, like the derivation of the famous equation $E = mc^2$ or the rigorous development of the theory of Lebesgue integration as a limiting case of Riemann integration, were included because I felt I had something to say about them which would be of interest to a serious student or to an honors class that wanted to attack them. Teachers and students alike should regard them as extras, not requirements.

My thanks to Professor Creighton Buck for allowing us to reuse his kind introduction to the 1980 edition, to Sheldon Axler for his very flattering review of the book in the *American Mathematical Monthly* of December 1982, and to Birkhäuser for producing this third edition.

Harold Edwards
New York 1993

Introduction

It is always exciting to teach the first quarter (or first semester) of a calculus course—especially to students who have not become blasé from previous exposure in high school. The sheer power of the new tool, supported by the philosophical impact of Newton's vision, and spiced by the magic of notational abracadabra, carries the course for you. The tedium begins when one must supply all the details of rigor and technique. At this point, one becomes envious of the physicist who frosts his elementary course by references to quarks, gluons and black holes, thereby giving his students the illusion of contact with the frontiers of research in physics.

In mathematics, it is much harder to bring recent research into an introductory course. This is the achievement of the author of the text before you. He has taken one of the jewels of modern mathematics—the theory of differential forms—and made this far reaching generalization of the fundamental theorem of calculus the basis for a second course in calculus. Moreover, he has made it pedagogically accessible by basing his approach on physical intuition and applications. Nor are conventional topics omitted, as with several texts that have attempted the same task. (A quick glance at the Index will make this more evident than the Table of Contents.)

Of course this is an unorthodox text! However, it is a far more honest attempt to present the essence of modern calculus than many texts that emphasize mathematical abstraction and the formalism of rigor and logic. It starts from the calculus of Leibniz and the Bernouillis, and moves smoothly to that of Cartan.

This is an exciting and challenging text for students (and a teacher) who are willing to follow Frost's advice and "take the road less travelled by."

R. Creighton Buck
January 1980

Preface

There is a widespread misconception that math books must be read from beginning to end and that no chapter can be read until the preceding chapter has been thoroughly understood. This book is not meant to be read in such a constricted way. On the contrary, I would like to encourage as much browsing, skipping, and backtracking as possible. For this reason I have included a synopsis, I have tried to keep the cross-references to a minimum, and I have avoided highly specialized notation and terminology. Of course the various subjects covered are closely interrelated, and a full appreciation of one section often depends on an understanding of some other section. Nonetheless, I would hope that any chapter of the book could be read with some understanding and profit independently of the others. If you come to a statement which you don't understand in the middle of a passage which makes relatively good sense, I would urge you to push right on. The point should clarify itself in due time, and, in any case, it is best to read the whole section first before trying to fill in the details. That is the most important thing I have to say in this preface. The rest of what I have to say is said, as clearly as I could say it, in the book itself. If you learn anywhere near as much from reading it as I have learned from writing it, then we will both be very pleased.

New York 1969

Contents

Synopsis

There are four major topics covered in this book: convergence, the algebra of forms, the implicit function theorem, and the fundamental theorem of calculus.

Of the four, convergence is the most important, as well as the most difficult. It is first considered in §2.3 where it arises in connection with the definition of definite integrals as limits of sums. Here, and throughout the book, convergence is defined in terms of the Cauchy Criterion (see Appendix 1). The convergence of definite integrals is considered again in §6.2 and §6.3. In Chapter 7 the idea of convergence occurs in connection with processes of successive approximation; this is a particularly simple type of convergence and Chapter 7 is a good introduction to the general idea of convergence. Of course any limit involves convergence, so in this sense the idea of convergence is also encountered in connection with the definition of partial derivatives (§2.4 and §5.2); in these sections, however, stress is not laid on the limit concept *per se*. It is in Chapter 9 that the notions of limit and convergence are treated in earnest. This entire chapter is devoted to these subjects, beginning with real numbers and proceeding to more subtle topics such as uniform continuity, interchange of limits, and Lebesgue integration.

The algebra of forms is the most elementary of the four topics listed above, but it is the one with which the reader is least likely to have some previous acquaintance. For this reason Chapter 1 is devoted to an introduction of the notation, the elementary operations, and, most important, the motivating ideas of the algebra of forms. This introductory chapter should be covered as quickly as possible; all the important ideas it contains are repeated in more detail later in the book. In Chapter 2 the algebra of forms is extended to non-constant forms (§2.1, §2.4). In Chapter 4 the algebra of constant forms in considered again, from the beginning, defining terms and avoiding appeals to geometrical intuition in proofs. In the same way Chapter 5 develops the algebra of (non-constant) forms from the beginning. Finally, it is shown in Chapter 6 (especially §6.2) that the algebra of forms corresponds exactly to the geometrical ideas which originally motivated it in Chapter 1. Several important applications of the algebra of forms are given; these include the theory of determinants and Cramer's rule (§4.3, §4.4), the theory of maxima and minima with the method of Lagrange multipliers (§5.4), and integrability conditions for differential equations (§8.6).

The third of the topics listed above, the implicit function theorem, is a topic whose importance is too frequently overlooked in calculus courses. Not only is it the theorem on which the use of calculus to find maxima and minima is based (§5.4), but it is also the essential ingredient in the definition of surface integrals. More generally, the implicit function theorem is essential to the definition of any definite integral in which the domain of integration is a k-dimensional manifold contained in a space of more than k dimensions (see §2.4, §2.5, §6.3, §6.4, §6.5). The implicit function theorem is first stated (§4.1) for affine functions, in which case it is little more than the solution of m equations in n unknowns by the techniques of high school algebra. The general (non-affine) theorem is almost as simple to state and apply (§5.1), but it is considerably more difficult to prove. The proof, which is by the method of successive approximations, is given in §7.1. Other more practical methods of solving m equations in n unknowns are discussed later in Chapter 7, including practical methods of solving affine equations (§7.2).

The last of the four topics, the fundamental theorem of calculus, is the subject of Chapter 3. Included under the heading of the "fundamental theorem" is its generalization to higher dimensions

$$\int_S d\omega = \int_{\partial S} \omega$$

which is known as Stokes' theorem. The complete statement and proof of Stokes' theorem (§6.5) requires most of the theory of the first six chapters and can be regarded as one of the primary motivations for this theory.

In broad outline, the first three chapters are almost entirely introductory. The next three chapters are the core of the calculus of several variables, covering linear algebra, differential calculus, and integral calculus in that order. For the most part Chapters 4–6 do not rely on Chapters 1–3 except to provide motivation for the abstract theory. Chapter 7 is almost entirely independent of the other chapters and can be read either before or after them. Chapter 8 is an assortment of applications; most of these applications can be understood on the basis of the informal introduction of Chapters 1–3 and do not require the more rigorous abstract theory of Chapters 4–6. Finally, Chapter 9 is almost entirely independent of the others. Only a small amount of adjustment would be required if this chapter were studied first, and many teachers may prefer to order the topics in this way.

[illegible] calculus of several variables, systems of linear alge[illegible] ordinary differential equations, and integral calculus in the [illegible] order. For the most part Chapters 4 to 6 do not rely on Chapters [illegible] except to provide motivation for the abstract theory. Chapter 7 is almost entirely independent of [illegible] other chapters and can be read either before or [illegible] Chapter [illegible] is a [illegible] of applications. [illegible] of these applications can be understood on the basis of [illegible] introduction of Chapters 1–3 and [illegible] abstract theory of [illegible] Chapter 6 is almost entirely inde[illegible]pendent of the other chapters [illegible] small amount of adjustment would be required if the chapter were read first, and many teachers [illegible] the topics in this [illegible]

constant forms

chapter 1

The purposes of this chapter are to introduce the notation and the algebraic operations of constant forms, and to illustrate the sorts of mathematical ideas which are described by forms. This notation and these algebraic operations are used throughout Chapters 2–8. Readers who are perplexed rather than enlightened by such primitive physical notions as work and flow should skip the discussions involving these ideas.

1.1

One-Forms

Suppose that a particle is being moved in the presence of a force field. Each displacement of the particle requires a certain amount of work, positive if the force opposes the displacement, negative if the force aids the displacement. Thus work can be considered as a rule or function assigning numbers (the amount of work) to displacements of the particle. To make the situation as simple as possible it will be assumed that the force field is *constant*; that is, it will be assumed that the direction and magnitude of the force do not change, either from one point in space to another or from one time to another. Moreover, only straight line displacements will be considered, to begin with, so that work can be considered as a function

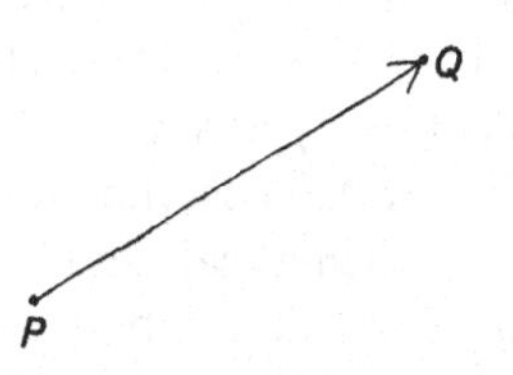

assigning numbers (the amount of work) to directed line segments PQ (the displacement).

Let A be the amount of work required for the displacement of the particle from $P = (0, 0, 0)$ to $Q = (1, 0, 0)$. Since the force field is constant, the amount of work required to go from any point $P = (x_0, y_0, z_0)$ to the point $Q = (x_0+1, y_0, z_0)$ with the x-coordinate increased by one is also equal to A; that is, A is the amount of work required for a unit displacement in the x-direction. Moreover, the amount of work required for a displacement of h units in the x-direction, from $P = (x_0, y_0, z_0)$ to $Q = (x_0+h, y_0, z_0)$, is hA; that is, going h times as far requires h times as much work. This holds for all values of h including, for example, $h = -1$, in which case the statement is that $-A$ units of work are required to go back from $P = (1, 0, 0)$ to $Q = (0, 0, 0)$.

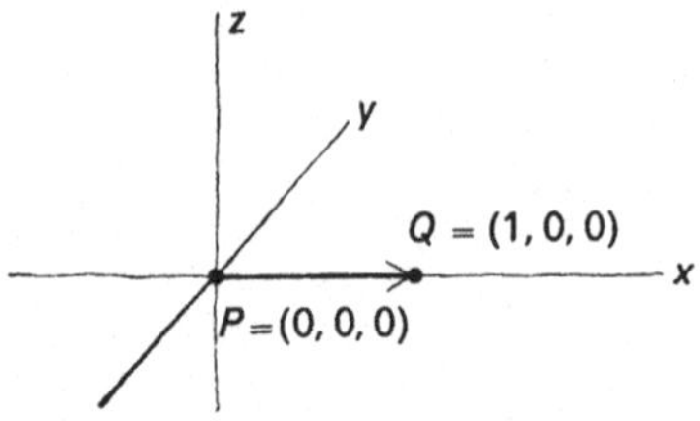

z

y

$Q = (1, 0, 0)$

x

$P = (0, 0, 0)$

Thus the one number A suffices to determine the amount of work required for any displacement in the x-direction. Similarly, if B, C are the amounts of work required for unit displacements in the y- and z-directions respectively, then the amount of work required for any displacement in the y- or z-direction can be found. Since every displacement is a superposition of displacements in the coordinate directions, the amount of work for any displacement can be found by addition. For example, displacement from $P = (3, 2, 4)$ to $Q = (4, 5, 1)$ involves a displacement of 1 unit in the x-direction, 3 units in the y-direction, and -3 units in the z-direction, a total amount of work equal to $A \cdot 1 + B \cdot 3 + C \cdot (-3)$. In general, displacement from (x_1, y_1, z_1) to (x_2, y_2, z_2) requires an amount of work equal to $A(x_2-x_1) + B(y_2-y_1) + C(z_2-z_1)$.

In summary, work is a function assigning numbers to directed line segments, which is of the form

$$\text{work} = A\,dx + B\,dy + C\,dz \tag{1}$$

where A, B, C are numbers and where dx, dy, dz are *functions** assigning to directed line segments PQ the corresponding change in x, change in y, change in z, respectively. A function of the type $A\,dx + B\,dy + C\,dz$ is called a *one-form* (1-form). Such functions occur in many contexts, both mathematical and physical, of which work in a constant force field is only one example.

**The reader has doubtless seen the symbol dx used in other ways, particularly in integrals and as a differential. In this book dx will always mean a function from directed line segments to numbers. The connection with other uses of the symbol will be explained.*

Another example of a one-form arises in connection with a constant planar flow. Imagine a thin layer of fluid uniformly distributed over a plane and flowing at a

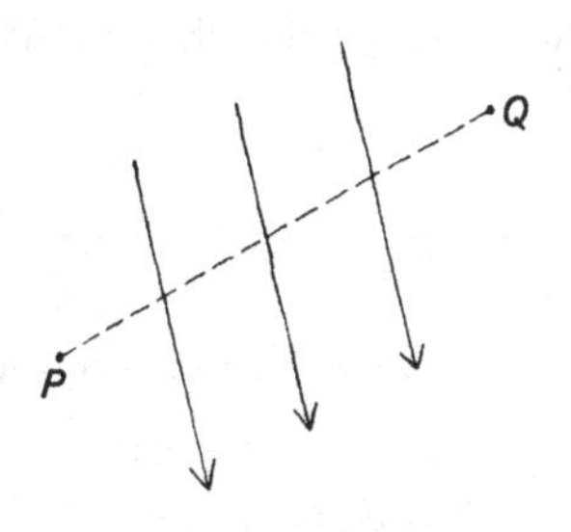

constant rate, that is, flowing in such a way that the density and velocity of the fluid do not change either from one point of the plane to another or from one time to another. (With these assumptions the 'fluid' could in fact be a sheet of tin sliding across the plane.) Given any line segment in the plane, one can measure the amount of fluid which crosses it in unit time. In this way there is associated with the flow a rule or function assigning numbers (mass per unit time) to line segments. This function 'flow across' is not a 1-form, but it becomes a 1-form once the *direction* of the flow is taken into account as follows: The flow across a line segment can be in either of two directions which can be described as 'left-to-right' or 'right-to-left' when left and right are determined by an observer standing on the plane at P and facing toward Q. The direction 'left-to-right' will be designated as the positive direction and the function 'flow across' will be defined to be the amount of fluid which crosses PQ in unit time if it crosses from left to right, and minus the mass per unit time if it crosses from right to left. Thus in particular the flow across QP is *minus* the flow across PQ.

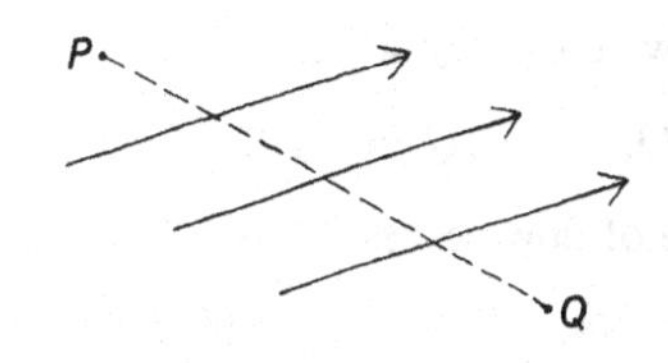

The function defined in this way is a 1-form, that is, it can be written

(2) $$\text{flow across} = A\,dx + B\,dy$$

where A, B are numbers and where dx, dy are the functions 'change in x' and 'change in y' assigning numbers to directed line segments in the xy-plane. The argument that flow across is a 1-form can be stated as follows: If two line segments are parallel and oriented in the same direction then the ratio of the flow across these two segments is equal to the ratio of their lengths. In particular, the flow across a segment in the x-direction is $A\,dx$ where A is the flow across the segment from (0, 0) to (1, 0), and the flow across a segment in the y-direction is $B\,dy$ where B is the flow across the segment from (0, 0) to (0, 1). To find the flow across an arbitrary line segment from (x_1, y_1) to (x_2, y_2) consider the right triangle whose hypotenuse is the given segment and whose sides are parallel to the coordinate axes. The flow across one side is $A\,dx$ and the flow across the other side is $B\,dy$. But the constancy of the flow implies that the amount of fluid inside the triangle is constant, hence the flow into the triangle across the hypotenuse is equal to the flow out of the triangle across the sides; hence, taking orientations

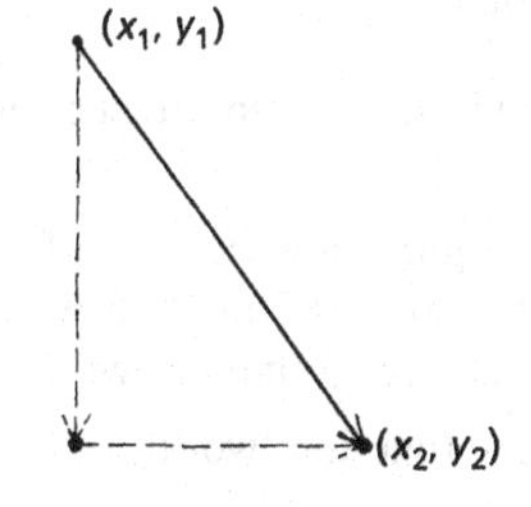

into account, the flow across the hypotenuse is $A\,dx + B\,dy$.

Exercises

1 Evaluate the 1-form $2\,dx + 3\,dy + 5\,dz$ on the line segment PQ where

(a) $P = (0, 1, 0)$ $\quad Q = (0, 0, 1)$
(b) $P = (3, 12, 4)$ $\quad Q = (11, 14, -7)$
(c) $P = (-1, 3, -5)$ $\quad Q = (3, -1, -7)$.

2 (a) Given that the rate of flow (of a constant flow in the plane) across the segment PQ is

3 when $P = (2, 1)$, $\quad Q = (3, 1)$
1 when $P = (-3, 2)$, $\quad Q = (-3, 3)$

find the rate of flow across PQ when

$$P = (3, 4), \qquad Q = (0, 0).$$

(b) Given that the rate of flow across PQ is

5 when $P = (4, 2)$, $\quad Q = (6, 3)$
2 when $P = (-2, 1)$, $\quad Q = (1, 3)$

find the 1-form $A\,dx + B\,dy$ describing the flow.

3 (a) If the flow across is given by the 1-form $3\,dx - 2\,dy$, find several segments across which the flow is zero, then find the most general segment across which the flow is zero. Sketch the lines of flow. In which direction is the flow along these lines (according to the orientation convention stated in the text)?
(b) Answer the same questions for the case in which flow across is given by $dx + dy$.
(c) If flow across is given by $A\,dx + B\,dy$, what are the lines of flow? Across which points of the plane would a particle of the fluid which was at $(0, 0)$ at time 0 pass at subsequent times?

4 If the force field is constant and if displacement of a given particle

from $(0, 0, 0)$ to $(4, 0, 0)$ requires 3 units of work
from $(1, -1, 0)$ to $(1, 1, 0)$ requires 2 units of work, and
from $(0, 0, 0)$ to $(3, 0, 2)$ requires 5 units of work

find the 1-form describing the function 'work'.

5 If work is given by the 1-form $3\,dx + 4\,dy - dz$, find all points which can be reached from the origin $(0, 0, 0)$ without work. Describe the set of these points geometrically. Describe the direction of the force geometrically.

6 If $3\,dx + 2\,dy$ describes
 (a) flow across, draw an arrow indicating the direction of flow;
 (b) work, draw an arrow indicating the direction of the force.

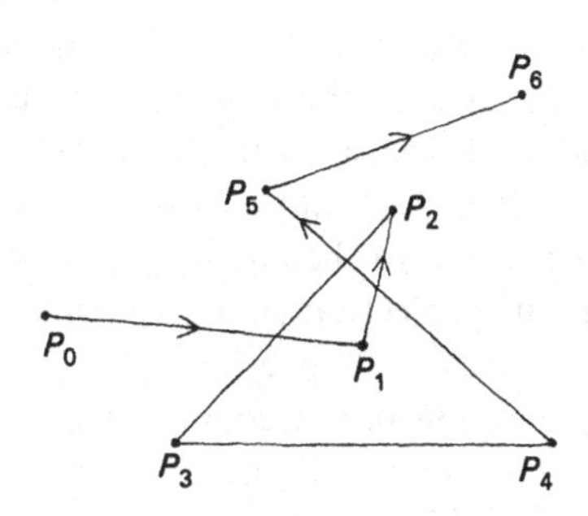

7 Show that if the work required for straight line displacements is given by the 1-form $A\,dx + B\,dy + C\,dz$ then the work required for a displacement along a polygonal curve $P_0P_1P_2 \ldots P_n$ is the same as the work required for displacement along the straight line P_0P_n. (This shows that there is no loss of generality in considering only straight line displacements in a constant force field.)

1.2

Two-Forms

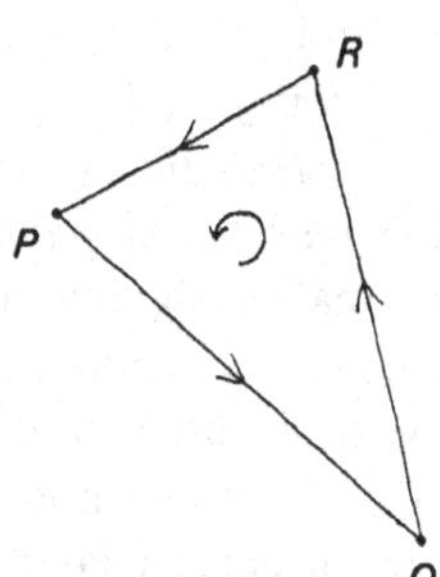

PQR or *QRP* or *RPQ*
not *RQP* or *QPR* or *PRQ*

Consider a constant flow in *xyz*-space. The function 'flow across', which in the case of a planar flow assigned numbers to oriented curves, will in this case assign numbers to oriented surfaces. For the sake of simplicity, only the flow across triangles will be considered, that is, flow across will be considered as a function assigning numbers to oriented triangles in space. An orientation of such a triangle, which is needed to establish the sign of the flow across it, can be indicated by drawing a rotational arrow ↺ or ↻ in the center of the triangle. More formally, the orientation can be described by naming the vertices in a particular order, thereby indicating one of the two possible rotational directions in which the bounding curve can be traced. The convention for determining the sign of the flow across an oriented triangle in space is to say that the flow is positive if it is in the direction indicated by the thumb of the right hand when the fingers are curled in the rotational direction which orients the triangle. This convention, like the analogous convention for planar flows, is in itself unimportant; all that is important is the fact that the sign of the flow across an oriented triangle can be defined in a consistent way.

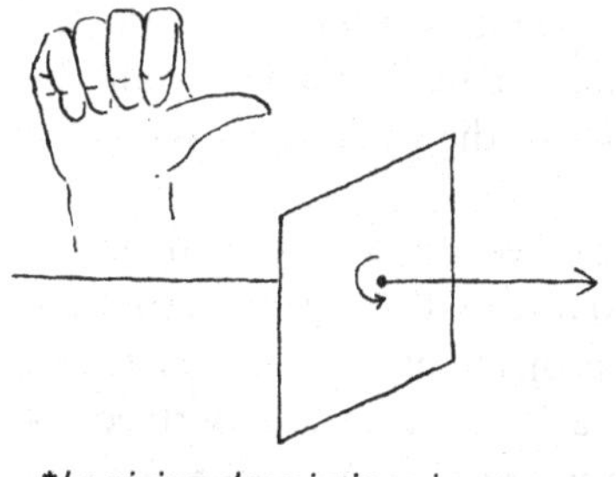

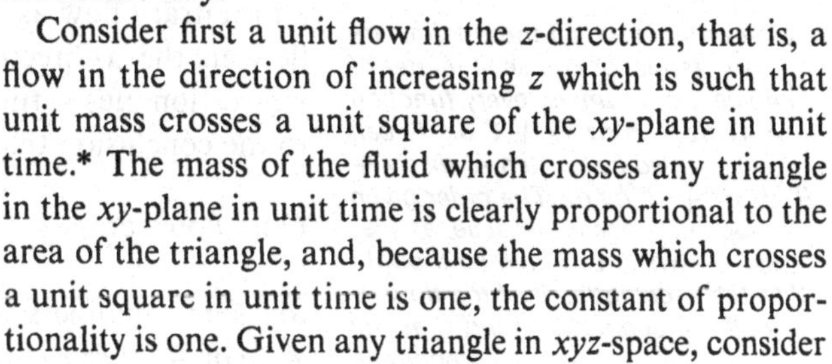

Consider first a unit flow in the *z*-direction, that is, a flow in the direction of increasing *z* which is such that unit mass crosses a unit square of the *xy*-plane in unit time.* The mass of the fluid which crosses any triangle in the *xy*-plane in unit time is clearly proportional to the area of the triangle, and, because the mass which crosses a unit square in unit time is one, the constant of proportionality is one. Given any triangle in *xyz*-space, consider

**In giving descriptions in xyz-space the direction of increasing x will be taken to be east (left to right), that of increasing y to be north (bottom to top of the page), and that of increasing z to be up (from the page toward the viewer). Thus the flow will be called 'upward' rather than 'in the direction of increasing z', and the rotational direction (0, 0, 0) → (1, 0, 0) → (0, 1, 0) → (0, 0, 0) will be called 'counterclockwise'.*

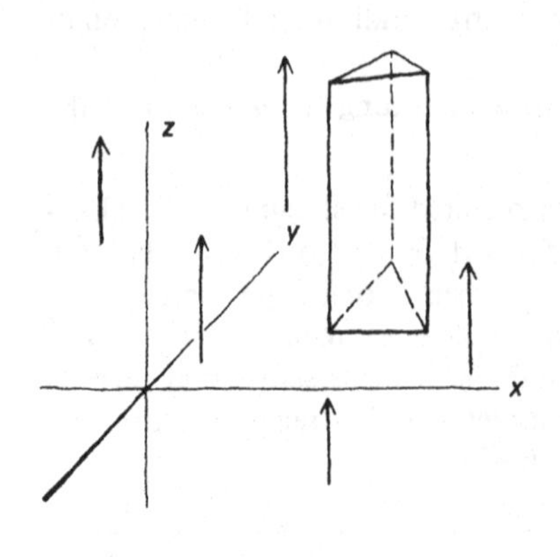

the prism generated by lines parallel to the z-axis through points of the triangle. It is clear that the mass of the fluid which crosses any triangle formed by intersecting this prism with a plane is the same as the mass of the fluid which crosses the original triangle, simply because there is no flow across the sides and what goes in the bottom must come out the top. Thus the amount of flow across any triangle in xyz-space is equal to the amount of flow across the triangle which is its projection on the xy-plane. This in turn is equal to the area of its projection on the xy-plane. Finally, taking the signs into account, the flow across an oriented triangle is equal to the area of its projection on the xy-plane if this projection is oriented counterclockwise, minus the area of its projection if the projection is oriented clockwise. This conclusion can be abbreviated by defining the oriented area of an oriented triangle in the xy-plane to be the area if its orientation is counterclockwise, and to be minus the area if its orientation is clockwise. Then, *for a unit flow in the z-direction, the flow across any oriented triangle is equal to the oriented area of its projection onto the xy-plane.*

The function 'oriented area', assigning numbers to oriented triangles in the xy-plane, is denoted $dx\,dy$.* The positive rotational direction designated by $dx\,dy$ can be described independently of pictures and notions such as 'clockwise' by saying that an oriented triangle described by going from the origin to a point on the positive x-axis to a point on the positive y-axis and back to the origin is positively oriented. Then it is natural to use $dy\,dx$ to denote $-dx\,dy$, that is, to denote the function 'oriented area' in which the orientation of a triangle from the origin to the positive y-axis to the positive x-axis to the origin is positive.

This notation derives from the formula for the area of a rectangle—the x-dimension times the y-dimension—although for other polygons the area is not given by a product. The entire symbol dx dy should be thought of as representing the function 'oriented area' which is not, in general, the product of dx and dy.

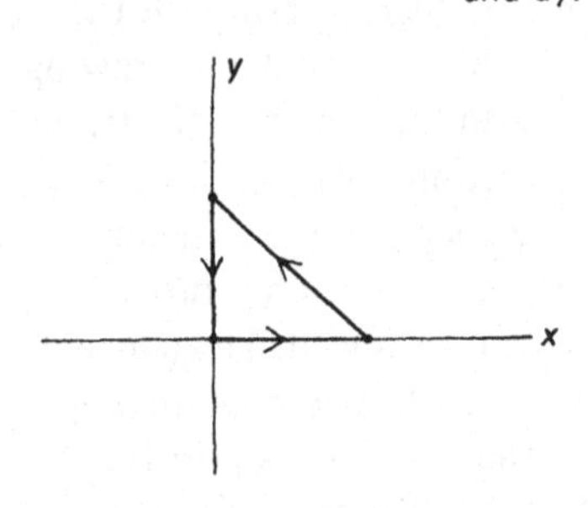

Thus unit flow in the z-direction is described by the function $dx\,dy$ assigning to oriented triangles in space the oriented area of their projections on the xy-plane. Similarly, unit flow in the x-direction is described by $dy\,dz$ and unit flow in the y-direction by $dz\,dx$.† Writing an arbitrary flow as a superposition of A times the unit flow in the x-direction plus B times unit flow in the y-direction plus C times unit flow in the z-direction leads to the conclusion that for an arbitrary constant flow

$$\text{(1)} \quad \text{flow across} = A\,dy\,dz + B\,dz\,dx + C\,dx\,dy$$

for some numbers A, B, C. A function of the type $A\,dy\,dz + B\,dz\,dx + C\,dx\,dy$‡ is called a *two-form*

†*Note the reverse order. This is dictated by the orientation convention stated above, as the diagram on the opposite page indicates. An alternative derivation, not depending on a diagram, will be given later (Exercise 6, §1.5).*

‡*The order in which these terms are written is, of course, a matter of choice. For example, every function A dy dz + B dz dx + C dx dy can also be written A′ dx dy + B′ dx dz + C′ dy dz. The order given above is chosen because, as was just shown, A can then be thought of as the x-component of the flow, B as the y-component, and C as the z-component.*

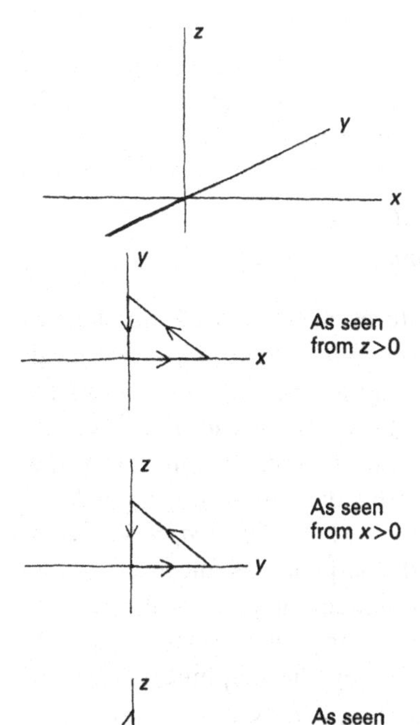

(2-form). The conclusion (1) is therefore that the function 'flow across' associated with a constant flow in xyz-space is a 2-form.

A 2-form $A\,dy\,dz + B\,dz\,dx + C\,dx\,dy$ is thus a function assigning numbers to oriented triangles in xyz-space. Of course it also assigns numbers to other oriented polygons, such as rectangles and parallelograms, because the notion of area is just as meaningful for polygons of arbitrarily many sides as it is for triangles. Moreover, a 2-form assigns numbers to oriented polygonal surfaces (just as a 1-form assigns numbers to oriented polygonal curves), the value of the 2-form on the entire surface being defined to be the sum of its values on the individual pieces of the surface. The values of 2-forms on oriented triangles will be emphasized because these are the simplest oriented polygonal surfaces to describe, but the reader should continue to think of a 2-form as a function assigning numbers to arbitrary polygonal surfaces.

The actual evaluation of 2-forms on particular polygons is less trivial than the evaluation of 1-forms. The technique of evaluation is described in the next section.

Exercises

1 Find the value of the 2-form $dx\,dy + 3\,dx\,dz$ on the oriented triangle with vertices

$$(0, 0, 0), \qquad (1, 2, 3), \qquad (1, 4, 0)$$

in that order. [Draw pictures of the projection of the triangle in the relevant coordinate planes.]

2 Find the value of the 2-form $dy\,dz + dz\,dx + dx\,dy$ on the oriented triangle with vertices

$$(1, 1, 2), \qquad (3, 5, -1), \qquad (4, 2, 1)$$

in that order.

3 Give a necessary and sufficient condition for the oriented area of the triangle $(0, 0)$, (x_1, y_1), (x_2, y_2) to be positive. [The equation $y_1x - x_1y = 0$ is the equation of the line through $(0, 0)$ and (x_1, y_1). The sign of the number $y_1x - x_1y$ tells which side of this line the point (x, y) lies on.]

4 Give a necessary and sufficient condition for the oriented area of the triangle (x_1, y_1), (x_2, y_2), (x_3, y_3) to be positive. [Use Exercise 3.] Show that interchanging two vertices changes the sign of the oriented area.

1.3

The Evaluation of Two-Forms Pullbacks

The algebraic rules which govern computations with forms all stem from the following fact: Let

$$
(1) \qquad \begin{aligned} x &= au + bv + c \\ y &= a'u + b'v + c' \end{aligned}
$$

be a function assigning to each point of the uv-plane a point of the xy-plane (where a, b, c, a', b', c' are fixed numbers). A function from the uv-plane to the xy-plane is also called a *mapping** or a *map* instead of a function, and a mapping of the simple form (1) (in which the expressions for x, y in terms of u, v are polynomials of the first degree in u, v) is called an *affine mapping.* Given an oriented polygon in the uv-plane, its image under the affine mapping (1) is an oriented polygon in the xy-plane. For example, the oriented triangle with vertices (u_0, v_0), (u_1, v_1), (u_2, v_2) is carried by the mapping (1) to the oriented triangle (x_0, y_0), (x_1, y_1), (x_2, y_2) where

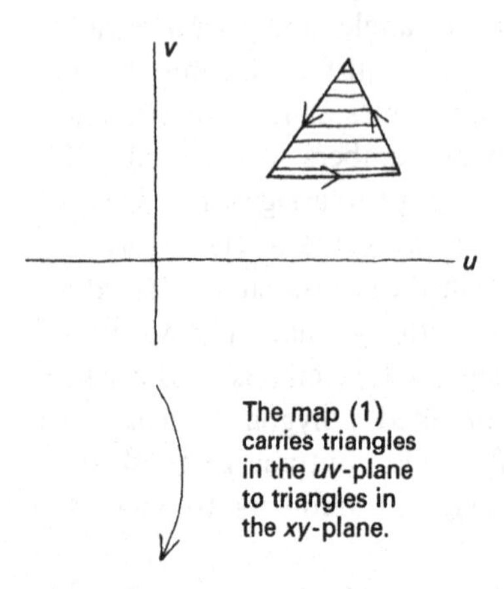

The map (1) carries triangles in the uv-plane to triangles in the xy-plane.

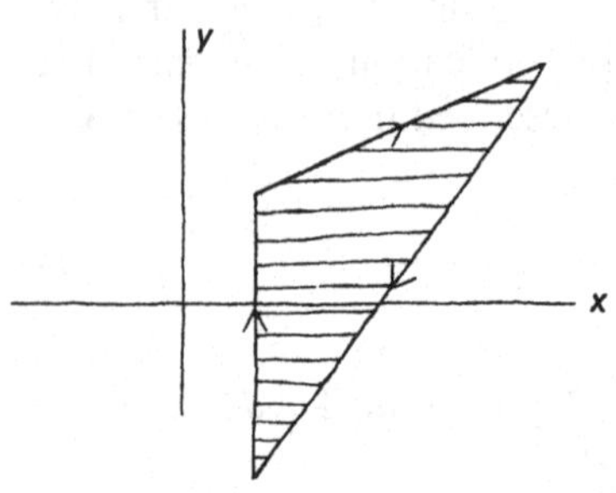

$$
\begin{aligned} x_0 &= au_0 + bv_0 + c \\ y_0 &= a'u_0 + b'v_0 + c' \end{aligned}
$$

and similarly for x_1, y_1, x_2, y_2. It will be shown that the oriented area of the image of any oriented polygon under the map (1) is $ab' - a'b$ times the oriented area of the polygon itself. That is, the map (1) 'multiplies oriented areas by $ab' - a'b$', a fact which is conveniently summarized by the formula

$$
(2) \qquad dx\, dy = (ab' - a'b)\, du\, dv.
$$

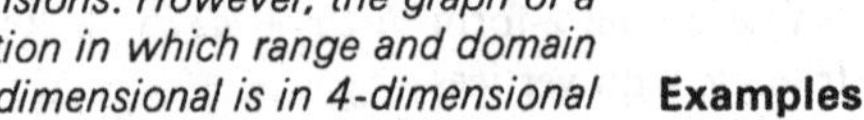

**A function $y = f(x)$, in which range and domain are 1-dimensional, can be pictured geometrically by means of its graph, a curve in two dimensions. However, the graph of a function in which range and domain are 2-dimensional is in 4-dimensional space. Since geometrical space is only 3-dimensional, the graph therefore cannot be visualized. However, such functions can be visualized as mappings of one plane onto another, which is the origin of this term (see Exercise 7).*

Examples

The affine mapping

$$
\begin{aligned} x &= -v \\ y &= u \end{aligned}
$$

carries $(0, 0)$ to $(0, 0)$, $(1, 0)$ to $(0, 1)$, $(0, 1)$ to $(-1, 0)$, $(3, 2)$ to $(-2, 3)$, etc., and can be visualized as a rotation of 90° in the counterclockwise direction. For this mapping $ab' - a'b$ is $0 \cdot 0 - (1)(-1) = 1$, hence $dx\, dy = du\, dv$, that is, the oriented area of the image of any polygon is the same as its oriented area. In short, a rotation of 90° does not change oriented areas.

The affine mapping

$$x = -u$$
$$y = v$$

can be visualized as a *reflection*. In this case $ab' - a'b = -1$; hence, $dx\,dy = -du\,dv$. This says that the reflection leaves areas unchanged but reverses orientations, a fact which is clear geometrically.

The affine mapping

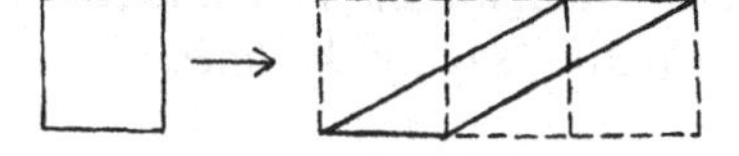

$$x = u + 3v$$
$$y = v$$

can be visualized as a *shear*. In this case $dx\,dy = du\,dv$, i.e. areas and orientations are unchanged by the mapping, which agrees with geometrical intuition.

The affine mapping

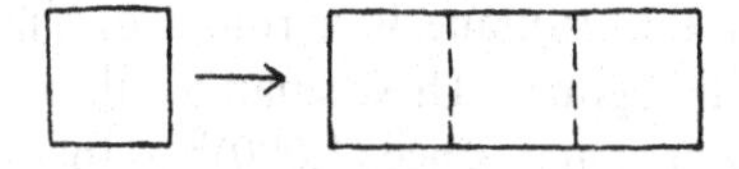

$$x = 3u$$
$$y = v$$

changes lengths in the horizontal direction by a scale factor of 3 and leaves lengths in the vertical direction unchanged. Here $dx\,dy = 3\,du\,dv$; that is, orientations are unchanged and areas are trebled.

An affine mapping

$$x = u + c$$
$$y = v + c'$$

is a translation. In this case $dx\,dy = du\,dv$, that is, oriented areas are unchanged.

The affine mapping

$$x = u$$
$$y = 0$$

collapses the entire uv-plane to the x-axis. Thus the image of any polygon is a 'polygon' of zero area. This agrees with the formula $dx\,dy = 0 \cdot du\,dv$ given by (2).

A rigorous proof of formula (2) would require a rigorous definition of area as a double integral. Informally, however, the plausibility of the formula is easily deduced from the intuitive idea of area as follows:

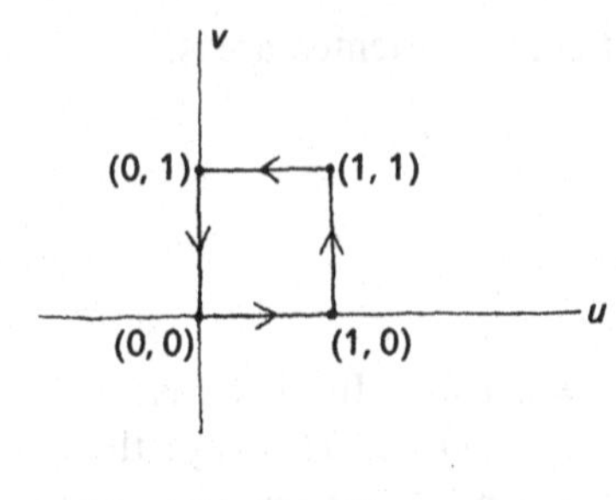

The map (1) carries the unit square in the *uv*-plane to a parallelogram in the *xy*-plane.

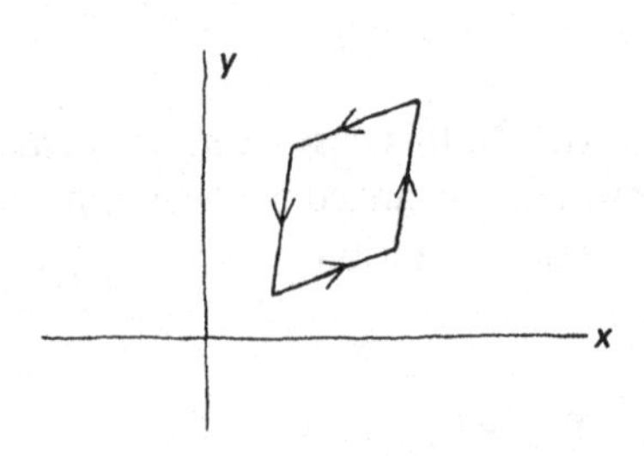

It is clear that for any given affine mapping (1) there is some number, say A, such that $dx\,dy = A\,du\,dv$—that is, such that the oriented area of the image in the *xy*-plane of any oriented polygon in the *uv*-plane is A times its oriented area. To reach this conclusion, consider first the unit square in the *uv*-plane with vertices $(0, 0)$, $(1, 0)$, $(1, 1)$, $(0, 1)$ in that order. Its image under the map (1) is an oriented parallelogram in the *xy*-plane. Let A be the oriented area of this parallelogram. It is to be shown that if PQR is any oriented triangle in the *uv*-plane then the oriented area of the image of PQR is A times the oriented area of PQR itself. Imagine the *uv*-plane to be ruled off into very small squares, say by the lines

$$u = \frac{\text{integer}}{1{,}000} \qquad v = \frac{\text{integer}}{1{,}000}$$

dividing it into squares which are $(1{,}000)^{-1}$ on a side. Then the area of PQR is roughly equal to $(1{,}000)^{-2}$ times the number of squares which lie in PQR.* Now the image of this ruling of the *uv*-plane is a ruling of the *xy*-plane into small parallelograms. These small parallelograms all have the same area and, since $(1{,}000)^2$ of them make up the parallelogram which is the image of the unit square, this area must be $|A| \cdot (1{,}000)^{-2}$, where $|A|$ is the absolute value of A. The image of the triangle PQR contains roughly (area PQR) $\times\ (1{,}000)^2$ of these small parallelograms, each of area $|A| \times (1{,}000)^{-2}$; hence the area of the image of PQR is $|A|$ times the area of PQR. It is easy to see that if $A > 0$ then the map preserves all orientations (carries clockwise triangles to clockwise triangles and counterclockwise to counterclockwise) whereas if $A < 0$ then the map reverses all orientations.† Thus the oriented area of the image of PQR is A times the oriented area of PQR as was to be shown.

**This is true only roughly because this number is defined only roughly—some squares neither lie in PQR nor lie outside PQR but lie part way in and part way out. The number of such borderline cases is insignificant relative to the number of squares inside.*

†If $A = 0$ then the map (1) collapses the uv-plane to a single line in the xy-plane. Therefore all polygons are collapsed to figures with zero area, and $dx\,dy = A\,du\,dv = 0$ holds.

Now it must be shown that A is given by the formula $A = ab' - a'b$. This formula is certainly correct for the simple affine maps considered in the examples above. The general case will follow from these particular cases if it is shown that

(i) if (2) is true of two affine maps then it is true of their composition, and
(ii) every affine mapping is a composition of reflections, rotations by 90°, shears, translations, and multiplications of the coordinate directions by scale factors.

Statement (ii) is plausible geometrically and is not difficult to prove algebraically (Exercise 8). Statement (i) can be verified as follows:

Let two affine maps

$$u = \alpha r + \beta s + \gamma \qquad x = au + bv + c$$
$$v = \alpha' r + \beta' s + \gamma' \qquad y = a'u + b'v + c'$$

be given. Assuming that the first map multiplies oriented areas by $\alpha\beta' - \alpha'\beta$ and that the second multiplies oriented areas by $ab' - a'b$, it follows that the composed map multiplies oriented areas by $(\alpha\beta'-\alpha'\beta)(ab'-a'b)$. On the other hand, the composed map is given explicitly as a map of the rs-plane to the xy-plane by

$$x = a(\alpha r+\beta s+\gamma) + b(\alpha' r+\beta' s+\gamma') + c$$
$$y = a'(\alpha r+\beta s+\gamma) + b'(\alpha' r+\beta' s+\gamma') + c'$$

that is

$$x = (a\alpha+b\alpha')r + (a\beta+b\beta')s + (a\gamma+b\gamma'+c)$$
$$y = (a'\alpha+b'\alpha')r + (a'\beta+b'\beta')s + (a'\gamma+b'\gamma'+c').$$

Thus the statement to be proved is

$$(ab'-a'b)(\alpha\beta'-\alpha'\beta) = (a\alpha+b\alpha')(a'\beta+b'\beta') - (a'\alpha+b'\alpha')(a\beta+b\beta')$$

which is easily verified.

There is no need to memorize formula (2) because it can be derived immediately from the formal algebraic rules*

*The letter d represents the operation, here purely formal and algebraic, of taking a differential.

$$du\,du = 0,\ dv\,dv = 0,\ du\,dv = -dv\,du$$
$$d(au+bv+c) = a\,du + b\,dv$$
$$d(a'u+b'v+c') = a'\,du + b'\,dv$$

which give

$$\begin{aligned} dx\,dy &= (a\,du+b\,dv)(a'\,du+b'dv) \\ &= aa'\,du\,du + ab'\,du\,dv + ba'\,dv\,du + bb'\,dv\,dv \\ &= (ab'-a'b)\,du\,dv. \end{aligned}$$

The 2-form $(ab'-a'b)\,du\,dv$ is called 'the pullback of the 2-form $dx\,dy$ under the affine map (1)'. The name 'pullback' derives from the fact that the map goes from the uv-plane to the xy-plane while the 2-form $dx\,dy$ on the xy-plane 'pulls back' to a 2-form on the uv-plane;

that is, the pullback goes in the direction (xy to uv) opposite to the direction of the map itself (uv to xy).

The pullback should be considered to be defined by the algebraic rules $du\,du = 0$, $du\,dv = -dv\,du$, $d(au+bv+c) = a\,du + b\,dv$, etc., and not by the formula (2). Then the same rules serve to define the pullback of a 2-form $A\,dy\,dz + B\,dz\,dx + C\,dx\,dy$ under an affine map

$$\begin{aligned} x &= au + bv + c \\ y &= a'u + b'v + c' \\ z &= a''u + b''v + c'' \end{aligned}$$

of the uv-plane to xyz-space. (A 2-form in xyz pulls back under the map to give a 2-form in uv.) One merely performs the substitution and applies the algebraic rules to obtain

$$\begin{aligned} &A(a'\,du+b'\,dv)(a''\,du+b''\,dv) + B(a''\,du+b''\,dv)(a\,du+b\,dv) + C(a\,du+b\,dv)(a'\,du+b'\,dv) \\ &\quad = Aa'a''\,du\,du + Aa'b''\,du\,dv + \cdots \\ &\quad = [A(a'b''-a''b')+B(a''b-ab'')+C(ab'-a'b)]\,du\,dv. \end{aligned}$$

The justification of these formal rules of computation—which appear quite mysterious at first—is completely pragmatic. They are simple to apply and they give very quickly the solution to the problem of *evaluating 2-forms*. For example: In order to find the value of the 2-form $dy\,dz - 2\,dx\,dy$ on the oriented triangle whose vertices are $(1, 0, 1)$, $(2, 4, 1)$, $(-1, 2, 0)$ one can first write this triangle as the image of the oriented triangle $(0, 0)$, $(1, 0)$, $(0, 1)$ in the uv-plane under the affine mapping

$$\begin{aligned} x &= 1 + u - 2v \\ y &= 4u + 2v \\ z &= 1 - v. \end{aligned}$$

The value of $dy\,dz$ on this triangle is the oriented area of its projection on the yz-plane, that is, the oriented area of the image of $(0, 0)$, $(1, 0)$, $(0, 1)$ under

$$\begin{aligned} y &= 4u + 2v \\ z &= 1 - v \end{aligned}$$

which is the value of $(4\,du+2\,dv)(-dv) = -4\,du\,dv$ on $(0, 0)$, $(1, 0)$, $(0, 1)$. Since the oriented area $du\,dv$ of this triangle is $\frac{1}{2}$, the answer is -2. Similarly the value of $-2\,dx\,dy$ is found by

$$\begin{aligned} -2\,dx\,dy &= -2(du-2\,dv)(4\,du+2\,dv) \\ &= -2(2\,du\,dv-8\,dv\,du) \\ &= -20\,du\,dv \end{aligned}$$

hence the value on the triangle is -10. Altogether the value of $dy\,dz - 2\,dx\,dy$ on the given triangle is therefore -12. The computation is best done all at once by writing

$$\begin{aligned} dy\,dz - 2\,dx\,dy &= (4\,du+2\,dv)(-dv) - 2(du-2\,dv)(4\,du+2\,dv) \\ &= -4\,du\,dv - 4\,du\,dv + 16\,dv\,du \\ &= -24\,du\,dv \end{aligned}$$

so that the value on the triangle (0, 0), (1, 0), (0, 1) is -12.

The evaluation of an arbitrary 2-form $A\,dy\,dz + B\,dz\,dx + C\,dx\,dy$ on an arbitrary oriented triangle (x_0, y_0, z_0), (x_1, y_1, z_1), (x_2, y_2, z_2) can be accomplished in the same way using the computational rules for finding pullbacks. In the following chapters it is these computational rules $du\,du = 0$, $du\,dv = -dv\,du$, and $d(au+bv+c) = a\,du + b\,dv$, which are of primary importance. Their use in the evaluation of 2-forms is only one of their many applications.

Exercises

1 Redo Exercises 1, 2, 3 of §1.2 using the techniques of this section.

2 Find the oriented area of the triangle PQR (oriented by this order of the vertices) in each of the following cases by (a) drawing the triangle and using geometry, and (b) using the techniques of this section.

$$\begin{array}{lll} P = (0,0) & Q = (1,2) & R = (2,0) \\ P = (1,1) & Q = (3,1) & R = (2,3) \\ P = (0,0) & Q = (2,1) & R = (1,2) \end{array}$$

3 Evaluate the 2-form $3\,dy\,dz + 2\,dx\,dy$ on the triangle PQR where

(a) $P = (3,1,4)$ $Q = (-2,1,4)$ $R = (1,4,1)$
(b) $P = (0,0,0)$ $Q = (1,2,1)$ $R = (-1,7,-1)$

and on the parallelogram $PQRS$ where

$P = (1,5,3)$, $Q = (2,7,6)$, $R = (8,12,10)$, $S = (7,10,7)$.

4 (a) Find the formula for the oriented area of a triangle whose vertices are

$$P = (0, 0) \qquad Q = (x_1, y_1) \qquad R = (x_2, y_2)$$

in that order.

(b) Find the oriented area of the triangle PQR when

$$P = (x_0, y_0) \qquad Q = (x_1, y_1) \qquad R = (x_2, y_2).$$

Write the answer as the sum of three similar terms.

(c) For a unit flow in the z-direction find the total flow *into* the tetrahedron with vertices

$$P = (x_0, y_0, 0),\ Q = (x_1, y_1, 0),\ R = (x_2, y_2, 0),\ S = (0, 0, 1),$$

that is, find the flow across each of the four sides, orienting them appropriately, and add. Relate the answer to (b).

(d) If S is a point *inside* the triangle PQR then

$$\text{area }(PQR) = \text{area }(PQS) + \text{area }(QRS) + \text{area }(RPS).$$

Is the same true for S outside the triangle? What if oriented area is used instead of area? Draw pictures illustrating a few examples.

5 (a) Continuing Exercise 4, find the formula for the oriented area of a quadrilateral with vertices

$$P = (x_0, y_0),\ Q = (x_1, y_1),\ R = (x_2, y_2),\ S = (x_3, y_3)$$

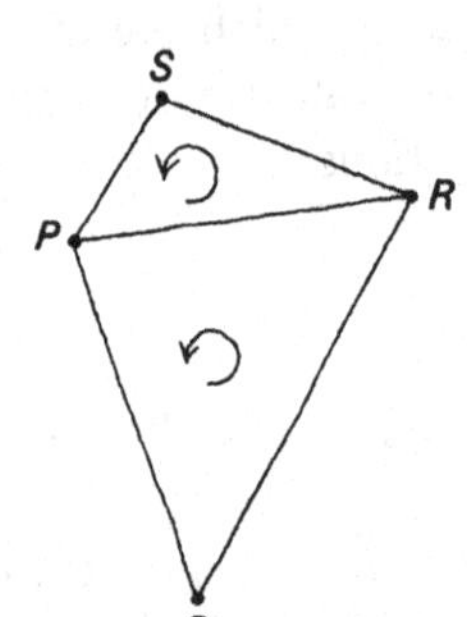

[Write the quadrilateral as a pair of oriented triangles joined along a side.]

(b) Show that the formula for the oriented area of an n-gon is a sum of n similar terms, one for each side of the n-gon.

(c) A closed, oriented, polygonal surface in space is a set of oriented polygons with the property that the boundary cancels; that is, every oriented line segmented PQ which occurs in the boundary of one polygon occurs with the opposite orientation QP as a part of the boundary of another polygon (in the same way that a closed oriented polygonal curve is a collection of oriented line segments with the property that every point which is the beginning point of one line segment is the end point of another segment), or, more precisely, PQ occurs with the same multiplicity as QP. Using (b), show that the total value of $dx\,dy$ (unit flow in the z-direction) on any closed, oriented polygonal surface is zero.

(d) Generalize (c) to show that the total value of any 2-form on any closed, oriented polygonal surface is zero.

(e) Generalizing Exercise 7, §1.1, show that the flow

across an oriented polygonal surface depends only on the oriented curve which is its boundary.

6 Given that flow across is described by the 2-form $3\,dy\,dz - 7\,dz\,dx + 11\,dx\,dy$, what, based on the derivation of §2.2, would be the flow vector; that is, what is the direction of flow? Verify your guess by showing that the flow across any parallelogram, one side of which has this direction, is zero.

7 Let

$$u = 2 + 3x - y$$
$$v = 1 - 4x + y$$

be an affine mapping of the xy-plane to the uv-plane.

(a) Give a picture of this 'mapping' by drawing the lines $u = \ldots, -2, -1, 0, 1, 2, \ldots$ and $v = \ldots, -2, -1, 0, 1, 2, \ldots$ in the xy-plane.

(b) In the same way draw the lines $x =$ const., $y =$ const. in the uv-plane.

(c) Draw a few triangles in the xy-plane and show their images in the uv-plane.

(d) Judging from the drawings, would you say that the mapping preserves orientations or reverses them?

(e) Find the pullback of $du\,dv$. Relate the answer to part (d).

8 Show that every affine map can be written as a composition of the simple types listed in the text. [The map (1) can be so written if every such map in which $c = c' = 0$ can be so written. The map $x = v$, $y = u$ can be so written. The map (1) with $c = c' = 0$ can be so written if either of the maps

$$\begin{aligned} x &= bu + av \\ y &= b'u + a'v \end{aligned} \qquad \text{or} \qquad \begin{aligned} x &= a'u + b'v \\ y &= au + bv \end{aligned}$$

can be so written. The map $x = 0$, $y = 0$ can be so written. Therefore one can assume $a \neq 0$. Therefore one can assume $a = 1$. Therefore one can assume $a' = 0$. Therefore one can assume $b = 0$. Thus it suffices to show that the map $x = u$, $y = b'v$ can be so written, which is obviously the case.]

1.4

Three-Forms

A *three-form* (3-form) on xyz-space is an expression of the form $A\,dx\,dy\,dz$ where A is a number. The *pullback* of a 3-form $A\,dx\,dy\,dz$ under an affine map

$$\begin{aligned} x &= au + bv + cw + e \\ y &= a'u + b'v + c'w + e' \\ z &= a''u + b''v + c''w + e'' \end{aligned} \tag{1}$$

of uvw-space to xyz-space is defined to be the 3-form $B\,du\,dv\,dw$ found by performing the substitutions

$$\begin{aligned} dx &= a\,du + b\,dv + c\,dw \\ dy &= a'\,du + b'\,dv + c'\,dw \\ dz &= a''\,du + b''\,dv + c''\,dw \end{aligned}$$

and multiplying out, using the distributive law of multiplication and the formal rules $du\,du = 0$, $du\,dv = -dv\,du$, etc., to reduce the resulting expression to the form $B\,du\,dv\,dw$.

When carried out, the substitution gives

$$\begin{aligned} A\,dx\,dy\,dz &= A(a\,du+b\,dv+c\,dw)(a'\,du+b'\,dv+c'\,dw)(a''\,du+b''\,dv+c''\,dw) \\ &= Aaa'a''\,du\,du\,du + Aaa'b''\,du\,du\,dv + \cdots \text{ (27 terms)} \\ &= Aab'c''\,du\,dv\,dw + Aac'b''\,du\,dw\,dv + \cdots \text{ (6 terms)} \\ &= A(ab'c''+bc'a''+ca'b''-cb'a''-ba'c''-ac'b'')\,du\,dv\,dw. \end{aligned}$$

However, in most cases it is as easy to carry out the substitution directly as it is to use this formula.

The 3-form $dx\,dy\,dz$ can be interpreted as the function 'oriented volume' assigning numbers to oriented three-dimensional figures in xyz-space in the same way that $dx\,dy$ is the oriented area of two-dimensional figures in the xy-plane and in the same way that dx is the oriented length of intervals of the x-axis. The principal fact about 3-forms which is needed to establish the plausibility of the interpretation of $dx\,dy\,dz$ as oriented volume is the following purely algebraic statement.

Theorem

Let two affine maps

$$\begin{aligned} u &= \alpha r + \beta s + \gamma t + \zeta \\ v &= \alpha' r + \beta' s + \gamma' t + \zeta' \\ w &= \alpha'' r + \beta'' s + \gamma'' t + \zeta'' \end{aligned} \qquad \text{and} \qquad \begin{aligned} x &= au + bv + cw + e \\ y &= a'u + b'v + c'w + e' \\ z &= a''u + b''v + c''w + e'' \end{aligned}$$

be given. Then the pullback of $dx\,dy\,dz$ under the composed map (of rst-space to xyz-space) is equal to the pullback under the first map (of rst-space to uvw-space) of the pullback under the second map (of uvw-space to xyz-space) of $dx\,dy\,dz$. In short, the pullback under a composed map is equal to the pullback of the pullback.

Computationally the theorem states that the 3-form $dx\,dy\,dz$ can be 'expressed in terms of rst' either by first expressing xyz directly in terms of rst and then forming the pullback or by first expressing $dx\,dy\,dz$ as a 3-form in uvw and then expressing this 3-form as a 3-form in rst.

This theorem can be proved, in the same way that the analogous theorem for 2-forms was proved in §1.3, by writing out the computations explicitly in terms of the 18 coefficients $a, b, c, a', \ldots, \beta'', \gamma''$ and showing that the two 3-forms in $dr\,ds\,dt$ which result are algebraically identical. These computations are quite long and can be avoided by examining more carefully the nature of the algebraic rules by which pullbacks are found. A simple proof based on such an examination of the algebraic rules is given in §4.2. For the moment it is the practical application of the computational rules to specific examples which should be emphasized. The proof of the theorem will therefore be postponed to Chapter 4.

Orientations agree

In order to show that $dx\,dy\,dz$ can be interpreted as oriented volume it is necessary to have an intuitive idea of how 3-dimensional figures can be oriented. To see how this is done it is useful to reformulate the idea of the orientation of 2-dimensional figures as follows: An orientation of a plane can be specified by giving three non-collinear points $P_0P_1P_2$. Two orientations $P_0P_1P_2$, $P_0'P_1'P_2'$ are said to agree if the points $P_0P_1P_2$ can be moved to $P_0'P_1'P_2'$ in such a way that throughout the motion the three points remain non-collinear. Otherwise the orientations are said to be opposite. Then it is geometrically plausible that the orientations $P_0P_1P_2$ and $P_1P_0P_2$ are opposite (do not agree) and that every orientation agrees either with $P_0P_1P_2$ or with $P_1P_0P_2$. Thus all orientations $P_0'P_1'P_2'$ are divided into two classes by $P_0P_1P_2$—those which agree with $P_0P_1P_2$ and those which agree with $P_1P_0P_2$. In the xy-plane these classes are called clockwise and counterclockwise—the counterclockwise orientations being those which agree with the orientation (0, 0), (1, 0), (0, 1).

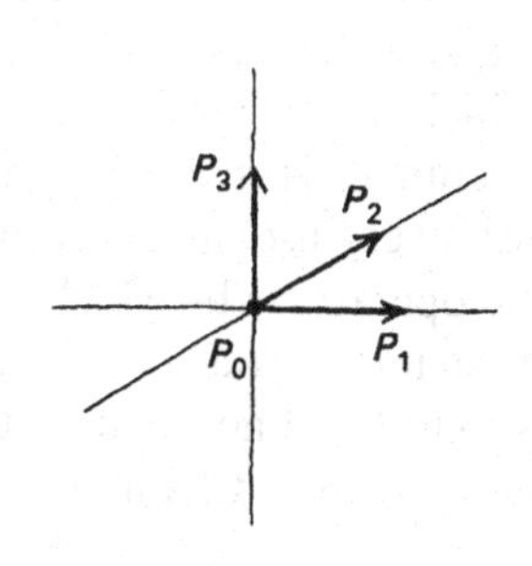

In the same way, an orientation of space can be described by giving four non-coplanar points $P_0P_1P_2P_3$. Two orientations $P_0P_1P_2P_3$ and $P_0'P_1'P_2'P_3'$ agree if the points of one can be moved to the points of the other keeping them non-coplanar all the while. All orientations fall into two classes such that two orientations in the same class agree. In xyz-space these classes are called left-handed and right-handed, an orientation being called right-handed if it agrees with the orientation (0, 0, 0), (1, 0, 0), (0, 1, 0), (0, 0, 1).*

*Because when coordinate axes are drawn as indicated in §1.2 the orientation (0, 0, 0), (1, 0, 0), (0, 1, 0), (0, 0, 1) agrees with the orientation $P_0P_1P_2P_3$ where P_0 is the base of the thumb, P_1 the tip of the thumb, P_2 the tip of the index finger, and P_3 the tip of the third finger of a right hand held in the natural position so that these points are non-coplanar.

With this definition of orientation the 3-form $dx\,dy\,dz$ can be described as the function 'oriented volume' assigning to oriented solids in xyz-space the number which is the volume of the solid if its orientation is right-handed and which is minus the volume if its orientation is left-handed. If the solid consists of several oriented pieces then the oriented volume of the whole is defined to be the sum of the oriented volumes of the pieces.

The notion of 'pullback' can then be described as follows: Let $A\,dx\,dy\,dz$ be given and let a map (1) be given. The map (1) carries oriented solids in uvw-space to oriented solids in xyz-space. The 3-form $A\,dx\,dy\,dz$ assigns numbers to oriented solids in xyz-space—namely, A times the oriented volume. The *composition* of these two operations assigns numbers to oriented solids in uvw-space. This new rule assigning numbers to oriented solids in uvw-space is clearly proportional to oriented volume; that is, it is of the form $B\,du\,dv\,dw$ for some number B. This geometrically defined 3-form $B\,du\,dv\,dw$ is in fact identical with the *pullback* of $A\,dx\,dy\,dz$ under the map (1) as defined algebraically above.

As in §1.3, the plausibility of the assertion that the 3-form $B\,du\,dv\,dw$ can be found by applying the algebraic rules which define the pullback operation, can be established by decomposing the given affine map into a sequence of simple operations for which the algebraic rules clearly give the correct answer—operations such as rotation by 90° around a coordinate axis, reflections in coordinate planes, shears, scale factors, and translations. The rigorous proof, which must await the rigorous definition of 'oriented volume', is given in Chapter 6. Meanwhile this fact gives a very useful conceptual interpretation of the algebraic operations by which pullbacks are defined. This conceptual interpretation will frequently be used to explain ideas, but it will not be used to define concepts or to prove theorems until Chapter 6.

Exercises

1 Show that the pullback of $dx\,dy\,dz$ under the composition of the maps

$$\begin{aligned} u &= r + 3s & \qquad x &= 2u + v \\ v &= \quad\; 2s + t & \qquad y &= 3u + v \\ w &= r - s + t & \qquad z &= u + v + w \end{aligned}$$

is equal to the pullback of the pullback; that is, verify the theorem for this case.

2 Do the same for the composition of the maps

$$\begin{aligned} u &= 10r - 7s + t & x &= 2u + v + w \\ v &= 3r + 5s - t & y &= 13u - v + w \\ w &= r + s + 2t & z &= -u + 7v - 2w. \end{aligned}$$

3 For each of the following triples of points in the plane determine whether they are collinear or determine a counterclockwise orientation or determine a clockwise orientation.

(a) $(0, 0), (2, 1), (4, 3)$ (b) $(3, 4), (7, -5), (1, 2)$
(c) $(-1, 3), (3, 5), (7, 7)$ (d) $(7, 1), (7, 2), (-1, 3)$.

[Write each triple as the image of the standard triple $(0, 0)$, $(1, 0)$, $(0, 1)$ under an affine map and find the effect of the map on oriented areas. Verify the answers by drawing pictures.]

4 For each of the following quadruples of points in space, determine whether they are coplanar or determine a right-handed orientation or determine a left-handed orientation.

(a) $(0, 0, 0), (2, 1, 4), (3, 4, 7), (2, 2, 9)$
(b) $(3, -1, -2), (-1, 0, 1), (0, 1, 1), (4, 2, -2)$
(c) $(2, 1, 4), (7, 9, 6), (3, 5, 6), (2, 4, 6)$.

5 Show that the oriented volume of the tetrahedron $(0, 0, 0)$, $(1, 0, 0)$, $(0, 1, 0)$, $(0, 0, 1)$ is the same as that of the tetrahedron $(0, 0, 0)$, $(1, 0, 0)$, $(1, 1, 0)$, $(1, 1, 1)$. (A tetrahedron is oriented by its vertices in the obvious way.) Show that this volume is $1/6$ by showing that the unit cube $\{0 \le x \le 1, 0 \le y \le 1, 0 \le z \le 1\}$ can be divided into 6 such tetrahedra.

6 Find the oriented volume of the tetrahedra in Exercise 4.

7 How is the orientation of a line described?

1.5

Summary

A 1-form in the three variables x, y, z is an expression of the form $A\,dx + B\,dy + C\,dz$ where A, B, C are numbers and dx, dy, dz are symbols. A 2-form in x, y, z is an expression of the form $A\,dy\,dz + B\,dz\,dx + C\,dx\,dy$ in which A, B, C are numbers and $dy\,dz$, $dz\,dx$, $dx\,dy$ are symbols. A 3-form in x, y, z is an expression of the form $A\,dx\,dy\,dz$ where A is a number and $dx\,dy\,dz$ is a symbol.

The symbol dx should be thought of as representing 'oriented length of the projection on the x-axis', a function assigning numbers to oriented curves in space. The symbols dy, dz have analogous interpretations. The symbol $dx\,dy$ should be thought of as representing 'oriented area of the projection on the xy-plane', a function assigning numbers to oriented surfaces in space. The symbols $dy\,dz$, $dz\,dx$ have analogous interpretations. The symbol $dx\,dy\,dz$ should be thought of as representing 'oriented volume', a function assigning numbers to oriented solids in space.

Given a k-form ($k=1, 2$, or 3) in the variables x, y, z and given an affine mapping

$$\begin{aligned} x &= au + bv + cw + e \\ y &= a'u + b'v + c'w + e' \\ z &= a''u + b''v + c''w + e'' \end{aligned} \tag{1}$$

there is a k-form in u, v, w called the *pullback* of the given k-form under the given affine map, defined by the computational rules $dx = d(au + bv + cw + e) = a\,du + b\,dv + c\,dw$, $du\,du = 0$, $du\,dv = -dv\,du$, etc.

In the same way the computational rules define the pullback of a k-form in m variables under an affine mapping which expresses the m variables in terms of n other variables. Only cases in which m, n are less than or equal to 3 and k is less than or equal to m were considered, but the method of computation extends immediately.

Geometrically, the pullback operation can be interpreted as a composed function. For example, if $A\,dy\,dz + B\,dz\,dx + C\,dx\,dy$ is a 2-form on xyz-space and if (1) is an affine map of uvw-space to xyz-space then the rule 'evaluate $A\,dy\,dz + B\,dz\,dx + C\,dx\,dy$ on the image under (1)' assigns numbers to oriented surfaces in uvw-space. This function is a 2-form on uvw-space, namely, the 2-form obtained from the given 2-form and the given affine map by forming the pullback. Other pullbacks have analogous interpretations. The connection between this geometrical interpretation of pullback and its actual algebraic definition has been indicated by plausibility arguments. A rigorous statement and proof are given in Chapter 6.

In addition to the functions 'oriented length', 'oriented area', and 'oriented volume', examples of forms are provided by the functions 'work required for displacements in a constant force field', which is a 1-form, and 'rate of flow of a constantly flowing fluid across oriented surfaces in space', which is a 2-form.

Exercises

1 Under the affine mapping

$$\begin{aligned} x &= 2 + 3u + 4v \\ y &= 1 + 2u - 3v \\ z &= 7 - 5u + 2v \end{aligned}$$

find the pullbacks of dx, $3\,dx + 2\,dy - 2\,dz$, $dx + dy + dz$, $3\,dx\,dy$, $8\,dy\,dz + 3\,dz\,dx + dx\,dy$.

2 Under the affine mapping

$$\begin{aligned} u &= 7 - 3x + 4y + 12z \\ v &= \quad\;\; x + y + z \end{aligned}$$

find the pullbacks of $2\,du + 3\,dv$, $du + 3\,dv$, $du\,dv$, $4\,du\,dv$, $3\,du\,dv$.

3 If work in a constant force field is given by $3\,dx - 2\,dy + 2\,dz$ and if $x = 3t$, $y = t$, $z = 4 + 3t$ is the position (x, y, z) of the particle at time t, how much work must be done during the time interval $0 \leq t \leq 3$? During $0 \leq t \leq 2$? During $-10 \leq t \leq -8$? Is the function 'work per time interval' a 1-form in this case? Describe this as a pullback.

4 A 1-form in 3 variables has 3 components, a 2-form 3 components, and a 3-form 1 component. How many components does a 1-form in 4 variables have? A 2-form? A 3-form? A 4-form? How many components does a k-form in n variables have? [How many 'k-dimensional coordinate planes' are there in an n-dimensional space?]

5 A natural way to describe a constant flow in the plane is by saying that in time t the point (x, y) moves to $(x + At, y + Bt)$ where A, B are the x- and y-components of the constant flow. Assuming the fluid has unit density find the 1-form which describes this planar flow. [The fluid crossing a given line segment in unit time is contained in a certain parallelogram.]

6 Let $(x + tA, y + tB, z + tC)$ describe a flow in space. As in Exercise 5, assume the fluid has unit density and find the 2-form which describes this flow.

integrals

chapter 2

2.1

Non-Constant Forms

In Chapter 1 a constant force field is described by a 1-form

$$A\,dx + B\,dy + C\,dz,$$

where

A = work required for unit displacement in x-direction
B = work required for unit displacement in y-direction
C = work required for unit displacement in z-direction.

In a force field in which the force depends on the location (x, y, z) the quantities A, B, C depend on x, y, z, that is, $A = A(x, y, z)$, $B = B(x, y, z)$, $C = C(x, y, z)$. The expression

$$(1)\quad A(x, y, z)\,dx + B(x, y, z)\,dy + C(x, y, z)\,dz$$

can then be regarded as assigning to each point (x, y, z) of space the 1-form which describes the force field at that point.

Similarly, a non-constant flow is described by an expression of the form

$$(2)\quad A(x, y, z)\,dy\,dz + B(x, y, z)\,dz\,dx + C(x, y, z)\,dx\,dy$$

assigning to each point (x, y, z) of space the 2-form which describes the flow at that point.

Henceforth an expression such as (1) will be called a *1-form on xyz-space*, and what was called a 1-form in Chapter 1 will be called a *constant* 1-form (a 1-form in which the functions A, B, C are constant). Similarly, a *2-form on xyz-space* will mean an expression of the form (2), a *1-form on the xy-plane* will mean an expression of the form

$$A(x, y)\, dx + B(x, y)\, dy,$$

a *3-form on xyz-space* will mean an expression of the form

$$A(x, y, z)\, dx\, dy\, dz,$$

and so forth. If there is danger of confusion the term 'variable' will be used in parentheses—e.g. (variable) 1-form, (variable) 2-form—to distinguish 1-forms from constant 1-forms, but henceforth '1-form' will *always* mean '(variable) 1-form'.*

The mathematical terminology is unfortunately ambiguous, the term 'form' referring both to (variable) forms and to constant forms. [The terminology of physics makes a very useful distinction between vectors and vector fields which is exactly the distinction between constant forms and forms. A (variable) form is a certain kind of tensor field (namely a field of alternating covariant tensors) and physicists may prefer to think of forms as alternating covariant tensor fields.] Since constant forms are the exception rather than the rule, it is only reasonable to use the shortest term possible ('form') for the idea which occurs most frequently and to use the longer term ('constant form') for the exceptional cases.

Exercises

1 *The central force field.* Newton's law of gravitational attraction states that the force exerted by a massive body (the sun) fixed at the origin $(0, 0, 0)$ on a particle in space is directed toward the sun and has magnitude proportional to the inverse square of the distance to the sun. Show that this means that the force field is described by the 1-form

$$\frac{kx}{r^3}\, dx + \frac{ky}{r^3}\, dy + \frac{kz}{r^3}\, dz$$

where k is a positive constant and $r = r(x, y, z) = \sqrt{x^2 + y^2 + z^2}$; that is, where

$$A(x, y, z) = kx/(x^2 + y^2 + z^2)^{3/2}, \text{ etc.}$$

[Show that for each fixed point (x, y, z) this 1-form represents a constant force which has the right direction (see Exercise 6, §1.1). Then use the fact that the magnitude of a force is measured by the amount of work required per unit displacement in the direction opposing the force.]

2 *Flow from a source (planar flow).* Find the 1-form describing a planar flow from a source at $(0, 0)$, assuming the flow is outward at all points (x, y) and has magnitude inversely proportional to the radius $r = r(x, y) = \sqrt{x^2 + y^2}$ (so that the flow is radially symmetric and the total flow across a circle of radius r about the origin is independent of r; that is, so there is no source between two circles of different radius). The magnitude of a flow is measured by the rate of flow across a line perpendicular to the direction of flow. Sketch the flow vectors.

3 *Flow from a source* (*spatial flow*). Find the 2-form describing flow in space from a source at (0, 0, 0), again assuming that the flow is *outward* at all points with a magnitude depending only on r, and assuming that there are no sources between spheres about the origin. [The surface area of the sphere of radius r is $4\pi r^2$.]

4 *Linear flows*. A linear flow is described by a (variable) 0-form assigning numbers to oriented points. Find the 0-form describing flow from a source at 0 on the line. How would a 0-form be described in general? What is a constant 0-form?

2.2 *Integration*

If the force field is not constant, then finding the amount of work required for a given displacement requires a process of *integration*. The essential idea is that the 1-form

$$(1) \quad A(x, y, z)\, dx + B(x, y, z)\, dy + C(x, y, z)\, dz$$

which describes the force field gives the approximate amount of work required for small displacements. The amount of work required for a displacement which is not small can then be described as a limit of sums of values of (1).

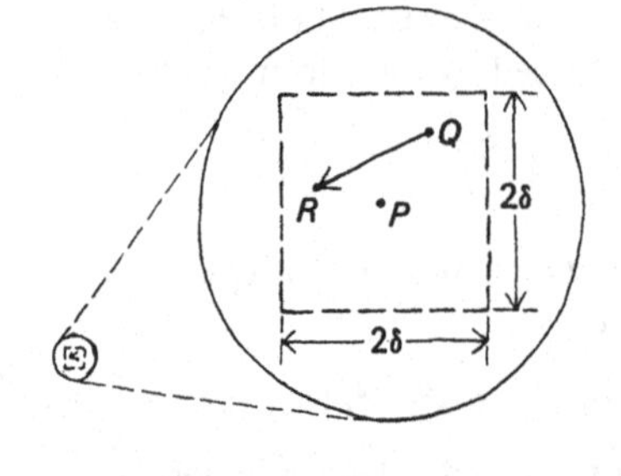

It will be assumed that *the force field depends continuously on* (x, y, z); that is, it will be assumed that for any given point $P = (\bar{x}, \bar{y}, \bar{z})$, the values of the functions A, B, C at points near P differ only slightly from their values at P. This means that throughout a small neighborhood of P, say the cube $\{|x - \bar{x}| < \delta, |y - \bar{y}| < \delta, |z - \bar{z}| < \delta\}$, the force field (1) is practically equal to the constant force field

$$A(\bar{x}, \bar{y}, \bar{z})\, dx + B(\bar{x}, \bar{y}, \bar{z})\, dy + C(\bar{x}, \bar{y}, \bar{z})\, dz$$

so that the work required for a displacement QR inside this neighborhood is practically equal to the value of this constant 1-form on the oriented line segment QR.

This relationship between the 1-form (1) and work is expressed by writing

$$(2) \quad \textit{work required for small displacements} \sim A(x, y, z)\, dx + B(x, y, z)\, dy + C(x, y, z)\, dz.$$

[Read ~ as 'is approximately equal to'.] Thus, if the points $P = (\bar{x}, \bar{y}, \bar{z})$, $Q = (x_1, y_1, z_1)$, $R = (x_2, y_2, z_2)$ are close together, then the amount of work required to go from Q to R is approximately equal to

$$A(\bar{x}, \bar{y}, \bar{z})(x_2 - x_1) + B(\bar{x}, \bar{y}, \bar{z})(y_2 - y_1) + C(\bar{x}, \bar{y}, \bar{z})(z_2 - z_1),$$

and the closer together they are, the better the approximation.*

It is precisely in this context, of course, that dx, dy, dz are thought of as being 'infinitesimals'. The point of view taken here, however, is that dx, dy, dz are functions assigning numbers to directed line segments. What is 'infinitesimal', then, is the line segment on which they are evaluated. Instead of saying that (2) holds for small line segments, with the approximation improving for shorter line segments, it is often said simply that work required for 'infinitesimal' displacements = $A(x, y, z)\,dx + B(x, y, z)\,dy + C(x, y, z)\,dz$.

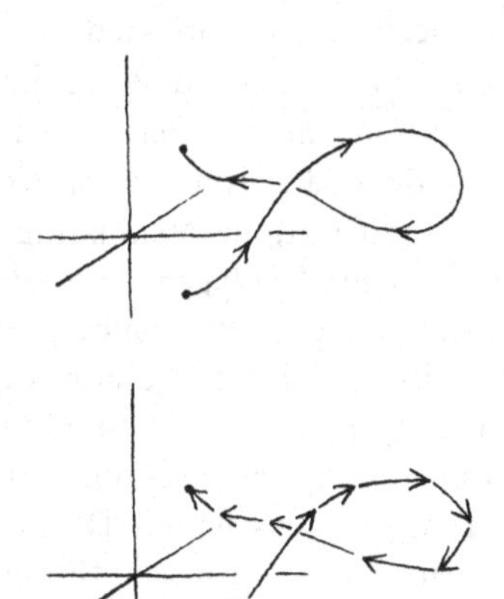

If S is any oriented curve, then an approximation to the amount of work required for the displacement S is found as follows: Approximate S by an oriented polygonal curve consisting of short straight-line displacements. The approximate amount of work required for each of these is found from (2), and the amount required for S is approximately equal to the sum of these values. The number found in this way is called an approximating sum; there are two approximations involved in the process: first, the approximation of the curve S by a polygonal curve, and second, the approximation of the amount of work required for each segment of the polygonal curve by (2). Since both approximations can be improved by taking the polygonal approximation to S to consist of more and shorter segments, it would be expected that the approximating sum could be made arbitrarily close to the true value by refining the approximation in this way. That is, the amount of work required for the displacement S should be equal to the *limiting value* of the approximating sums as the approximating curve is taken to consist of more and shorter segments fitting the curve S more and more closely. This limiting value (if it exists) is called the *integral* of the 1-form (1) over the curve S and is denoted

$$\int_S (A(x, y, z)\,dx + B(x, y, z)\,dy + C(x, y, z)\,dz)$$

or simply

$$\int_S (A\,dx + B\,dy + C\,dz),$$

where it is understood that A, B, C are functions of x, y, z.

Similarly, let a flow in space be described by a (variable) 2-form $A\,dy\,dz + B\,dz\,dx + C\,dx\,dy$. Then, again assuming that A, B, C are continuous, the flow across a small polygon near $P = (\bar{x}, \bar{y}, \bar{z})$ is approximately equal to the value on this polygon of the constant 2-form

$$A(\bar{x}, \bar{y}, \bar{z})\, dy\, dz + B(\bar{x}, \bar{y}, \bar{z})\, dz\, dx + C(\bar{x}, \bar{y}, \bar{z})\, dx\, dy.$$

This is summarized by saying that

(3) *flow across small polygons* $\sim A(x, y, z)\, dy\, dz + B(x, y, z)\, dz\, dx + C(x, y, z)\, dx\, dy.$

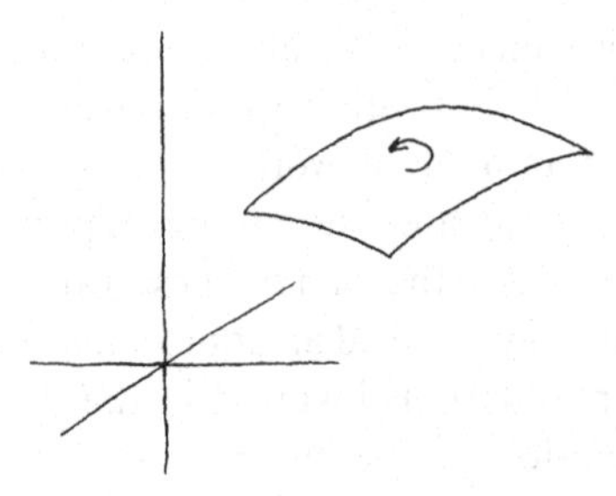

Then for an arbitrary oriented surface S the flow across S is equal to the limit of approximating sums obtained by constructing a polygonal approximation to S in which all polygons are small, using (3) to find the approximate rate of flow across each polygon, and adding. The limit of the approximating sums (if it exists) is called the integral of the 2-form over S and is denoted

$$\int_S (A\, dy\, dz + B\, dz\, dx + C\, dx\, dy).$$

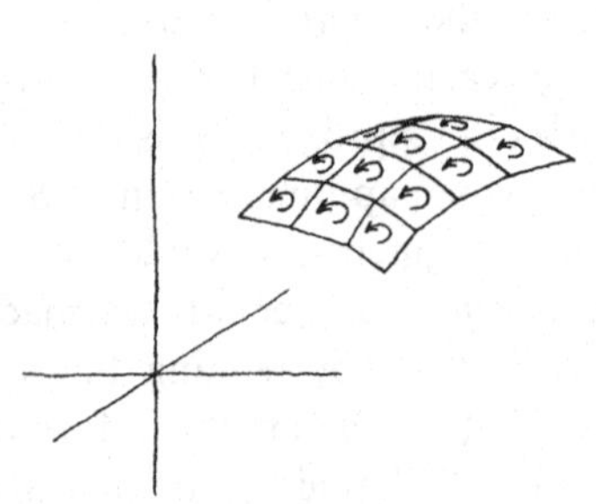

In general, an integral is formed from an *integrand*, which is a 1-form, 2-form, or 3-form, and a *domain of integration* which is, respectively, an oriented curve, a surface, or a solid. The integral is defined as the limit of approximating sums, and an approximating sum is formed by taking a finely divided polygonal approximation to the domain of integration, 'evaluating' the integrand on each small oriented polygon, and adding. The integrand is 'evaluated' on a small oriented polygon by choosing a point P in the vicinity of the polygon, by evaluating the functions A, B, etc. at P to obtain a constant form, and by evaluating the constant form on the polygon in the usual way (as in Chapter 1).

At this point two questions arise: How can this definition of 'integral' be made precise? How can integrals be evaluated in specific cases?

It is difficult to decide which of these questions should be considered first. On the one hand, it is hard to comprehend a complicated abstraction such as 'integral' without concrete numerical examples; but, on the other hand, it is hard to understand the numerical evaluation of an integral without having a precise definition of what the integral is. Yet, to consider both questions at the same time would confuse the distinction between the *definition* of integrals (as limits of sums) and the *method of evaluating* integrals (using the Fundamental Theorem of Calculus). This confusion is one of the greatest obstacles to understanding calculus and should be avoided at all costs. Therefore, all consideration of the evaluation of integrals is postponed to Chapter 3, and the remainder of this chapter is devoted solely to the question of the definition and elementary properties of integrals.

Exercises

1 Let $A\,dx + B\,dy + C\,dz = (1/r^3)[x\,dx + y\,dy + z\,dz]$ be the central force field with $k = 1$ (see Exercise 1, §2.1), and let S be the line segment from (1, 0, 0) to (2, 0, 0). Find the approximating sum to $\int_S (A\,dx + B\,dy + C\,dz)$ formed by dividing S into 10 segments of equal length and using the midpoint of each interval to evaluate $A\,dx + B\,dy + C\,dz$. [Express the answer as a number times a sum of reciprocals of integers; actual numerical evaluation of the sum is difficult.] Call this number $\sum(10)$. Similarly, let $\sum(n)$ be the approximating sum formed by dividing S into n equal segments (n = positive integer), and evaluating at midpoints. Express $\sum(n)$ as a number times a sum of reciprocals of integers. Suppose that a computing machine has been programmed to compute $\sum(n)$ for any n, the value being rounded to two decimal places. Find an upper bound for the magnitude of the difference $|\sum(10) - \sum(20)|$. [Each term of $\sum(10)$ corresponds to two terms of $\sum(20)$. To get an upper bound on the magnitude of the difference $|(1/x_1^2) - (1/x_2^2)|$ it suffices to observe that the slope of the chord of the graph of $1/x^2$ from $(x_1, (1/x_1^2))$ to $(x_2, (1/x_2^2))$ $(1 \le x_1 \le x_2)$ is greater (both are negative) than the slope of the tangent at (1, 1) which gives

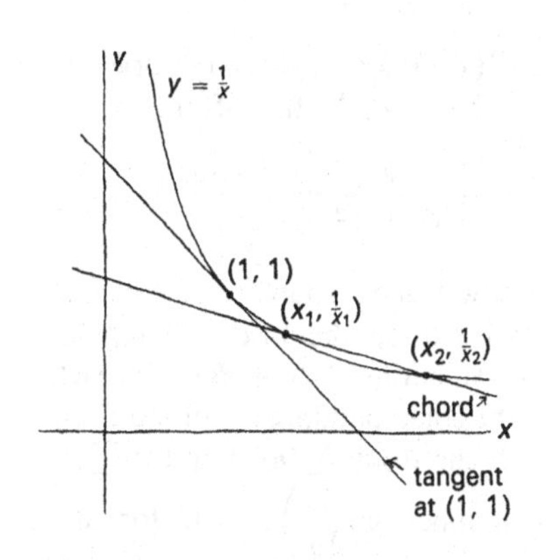

$$\frac{1}{x_1^2} - \frac{1}{x_2^2} \le 2(x_2 - x_1).]$$

If $\sum(10)$ and $\sum(20)$ are both rounded to two decimal places, how great can the difference of the resulting numbers be? Find an integer N such that $|\sum(N) - \sum(mN)| < .005$ for all integers m. Show that moreover $|\sum(n) - \sum(mn)| < .005$ for all $n \ge N$ and all m. Show that $|\sum(N) - \sum(n)| < .01$ for all $n \ge N$. Conclude that the number produced by the computer for any $n \ge N$, no matter how large, will differ from the number it produces for N only by ± 1 in the last decimal place. This is what it means to say that the number $\sum(N)$ rounded to two decimal places represents the limiting value (the integral) with an accuracy of two decimal places.

2 *Computation with decimals and decimal approximations.* A number a (the approximate value) is said to represent a number t (the true value) with an accuracy of 3 decimal places if $|a - t| < .001$. Show that:

(a) If a represents t with an accuracy of 3 decimal places, if a_3 is a rounded to 3 decimal places, and if t_3 is t rounded to 3 decimal places, then a_3 differs from t_3 by at most ± 1 in the last (third) place.

(b) No matter how close a is to t, a_3 may still differ from t_3 by ± 1 in the last place.

3 Form approximating sums to $\int_S \frac{1}{x}\,dx$ in the same way as in Exercise 1; that is, divide the interval, S, into n equal segments and evaluate at midpoints. Suppose the computer has been programmed to find this number $\sum(n)$ rounded to three decimal places. Find an N such that $\sum(N)$ represents all $\sum(n)$ for $n \geq N$ (and hence represents the limiting value) with an accuracy of three decimal places.

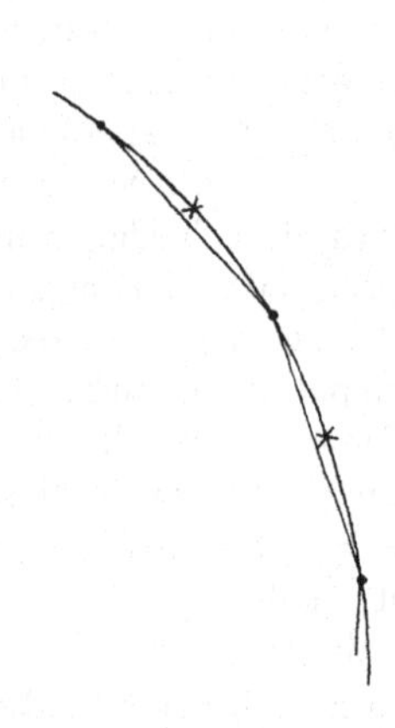

4 Let S be the circle $\{(x, y) : x^2 + y^2 = 1\}$ oriented counterclockwise, and let $A\,dx + B\,dy$ be the 1-form

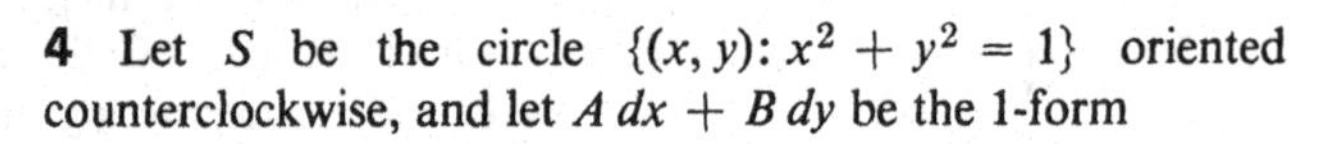

$$\frac{x}{x^2 + y^2}\,dy - \frac{y}{x^2 + y^2}\,dx$$

of Exercise 2, §2.1, giving flow from a source at the origin. Estimate $\int_S (A\,dx + B\,dy)$ by using the inscribed regular n-gon to approximate S and evaluating $A\,dx + B\,dy$ for each segment at the midpoint of the corresponding arc of the circle (because this is easiest). Call the result $\sum(n)$. Express $\sum(n)$ explicitly in terms of the number $\sin\left(\frac{\pi}{n}\right)$. [The formula $\sin(x + y) - \sin(x - y) = 2 \cos x \sin y$ and the analogous formula for $\cos(x + y) - \cos(x - y)$ are used.] The formula

$$\lim_{x \to 0} [(\sin x)/x] = 1$$

can be used to evaluate the limit as $n \to \infty$.

5 Let $x^3y^2z\,dx\,dy$ be a 2-form describing a flow in space and let S be the rectangle $\{(x, y, z) : 0 \leq x \leq 1,\ 0 \leq y \leq 2,\ z = 1\}$ oriented counterclockwise. Let $\sum(n, m)$ be the approximating sum to $\int_S x^3y^2z\,dx\,dy$ obtained by dividing the x interval $0 \leq x \leq 1$ into n equal parts, dividing the y interval $0 \leq y \leq 2$ into m equal parts, orienting each of the mn rectangles counterclockwise, and evaluating the 2-form at the midpoint of each rectangle. Find an N such that $\sum(N, N)$ represents $\sum(n, m)$ with four-place accuracy for $n > N$, $m > N$.

2.3

Definition of Certain Simple Integrals. Convergence and the Cauchy Criterion

The greatest difficulty in giving a precise formulation of the informal definition of §2.2 lies in describing precisely what is meant by a 'finely divided polygonal approximation to the domain of integration'. This section is devoted to giving a precise formulation of the definition of §2.2 for integrals in which the domain of integration is a simple domain for which a 'finely divided polygonal approximation' can be described explicitly.

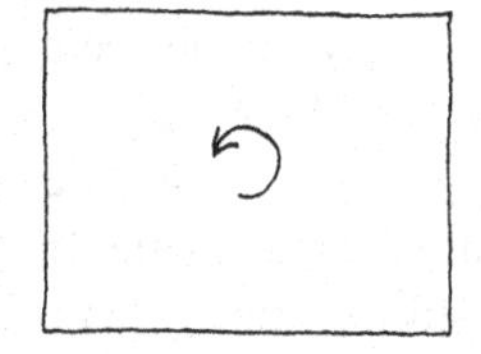

Consider first the case of an integral $\int_R A\,dx\,dy$ in which $A\,dx\,dy$ is a (variable) 2-form on the xy-plane and in which the domain of integration R is a rectangle

$$R = \{a \leq x \leq b, c \leq y \leq d\}$$

in the xy-plane oriented counterclockwise. [If R is oriented clockwise then $\int_R A\,dx\,dy$ is defined to be $-\int_{-R} A\,dx\,dy$ where $-R$ denotes R with the opposite orientation. This is in accord with the usual definition* $\int_b^a f(x)\,dx = -\int_a^b f(x)\,dx$ of integrals over intervals $\{a \leq x \leq b\}$ which are oriented from right to left.] In this case a 'finely divided polygonal approximation' to the domain of integration R can be obtained simply by drawing lines $x =$ const., $y =$ const. to divide R into subrectangles and by orienting each subrectangle counterclockwise in agreement with the orientation of R. An approximating sum corresponding to this 'finely divided approximation to R' is obtained by 'evaluating' $A(x, y)\,dx\,dy$ on each subrectangle and adding over all rectangles. To form such an approximating sum it is necessary to choose:

**The notation $\int_a^b f(x)\,dx$ denotes, of course, the integral of the 1-form $f(x)\,dx$ over the interval $\{a \leq x \leq b\}$ oriented from a to b. Unfortunately there is no such convenient notation for indicating orientations of 2-dimensional integrals.*

lines $x = x_i$, where $a = x_0 < x_1 < x_2 < \cdots < x_{m-1} < x_m = b$
lines $y = y_j$, where $c = y_0 < y_1 < y_2 < \cdots < y_{n-1} < y_n = d$
points P_{ij}, one in each of the mn rectangles R_{ij}

$$= \{(x, y): x_{i-1} \leq x \leq x_i, y_{j-1} \leq y \leq y_j\}.$$

The approximating sum is then

$$(1) \quad \sum_{i,j} [A(P_{ij})][\text{oriented area of } R_{ij}] = \sum_{i=1}^{m} \sum_{j=1}^{n} A(P_{ij})(x_i - x_{i-1})(y_j - y_{j-1}).$$

The choices x_i, y_j, P_{ij} will be denoted collectively by the letter α and the corresponding sum (1) by $\sum(\alpha)$.†

†This sum is a number once A, R are given and all the choices α are made. It is helpful to imagine that a computer has been programmed to compute this number $\sum(\alpha)$ whenever it is provided with the choices α. See Appendix 1.

To say that the approximating sums $\sum(\alpha)$ approach a limit as the approximation is refined means essentially that the choices α do not significantly affect the result $\sum(\alpha)$, provided only that the polygons R_{ij} on which the approximating sum $\sum(\alpha)$ is based are all small. Let $|\alpha|$ be the largest dimension of any of the rectangles specified by α, that is, $|\alpha| = \max(x_1 - x_0, x_2 - x_1, \ldots, x_m - x_{m-1}, y_1 - y_0, \ldots, y_n - y_{n-1})$. $|\alpha|$ is called the *mesh size* of α. For the limit to exist means that if the mesh size $|\alpha|$ is small the resulting sum $\sum(\alpha)$ is insensitive to the choices to a very high degree; that is, another approximating sum $\sum(\alpha')$ in which the mesh size is similarly small will differ very little from $\sum(\alpha)$. Specifically, convergence of the sums $\sum(\alpha)$ is defined by the Cauchy Convergence Criterion:

The integral $\int_R A\,dx\,dy$ is said to *converge* if it is true that given any margin for error ϵ there is a mesh size δ such that

> any two approximating sums $\sum(\alpha)$, $\sum(\alpha')$ in which the mesh sizes are both less than δ differ by less than the prescribed margin for error ϵ,

that is,

$$|\alpha| < \delta, |\alpha'| < \delta \text{ imply } |\textstyle\sum(\alpha) - \sum(\alpha')| < \epsilon.$$

If this is the case then the *limiting value* can be defined to be the number* which is determined to within any margin of error in the obvious way. For example, if it is desired to find the limiting value with an accuracy of five decimal places (see Exercise 2, §2.2) set $\epsilon = .00001$, let δ be the corresponding mesh size, choose *any* α with $|\alpha| < \delta$, form $\sum(\alpha)$, and round to five decimal places. The number which results is determined, except for ± 1 in the last place, solely by the 2-form $A\,dx\,dy$ and the domain of integration R. If a different set of choices α is used, no matter how different they may be and, in particular, no matter how much finer a subdivision of R they may involve, the result will be the same except for at most ± 1 in the last place. The limiting value defined in this way is called the *integral* of $A\,dx\,dy$ over R and is denoted $\int_R A\,dx\,dy$.

It is an axiom of the real number system that a number is defined once such a process for computing it to any prescribed degree of accuracy has been given (see §9.1).

If this is not the case, that is, if the integral does not converge, then it is said that the integral *diverges* and that the limiting value $\int_R A\,dx\,dy$ *does not exist.*

This completes the definition of the integral of a 2-form $A(x, y)\,dx\,dy$ over an oriented rectangle R. The integral $\int_R A\,dx\,dy$ either is a *number*, defined above, or it does not exist. Only minor modifications are necessary to define the integral of a 3-form over an oriented rectangular parallelopiped or the integral of a 1-form over an oriented interval. The integral of a 2-form over a more general oriented domain D of the plane—for example, over the disk $D = \{(x, y): x^2 + y^2 \leq 1\}$ oriented counterclockwise—can be defined by the simple trick of taking a rectangle R containing the domain D, setting the integrand equal to zero outside D, and proceeding as before. Assuming then that D is an oriented domain of the plane which can be enclosed in a rectangle R, the integral of a 2-form $A\,dx\,dy$ over D is defined by setting

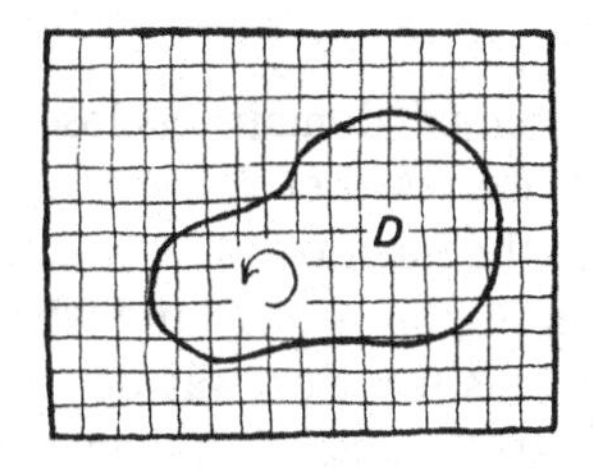

$$A_D(x, y) = \begin{cases} A(x, y) & \text{if } (x, y) \text{ is in } D \\ 0 & \text{if } (x, y) \text{ is not in } D \end{cases}$$

and defining

$$\int_D A(x, y)\,dx\,dy = \int_R A_D(x, y)\,dx\,dy,$$

where the orientation of R is chosen to agree with that of D.* The integral of a 3-form over any oriented domain in space which can be enclosed in a rectangular parallelopiped, and the integral of a 1-form over any oriented domain of the line which can be enclosed in an interval, are defined by the same trick. Such domains are called *bounded* domains (that is, they are domains which stay within certain finite bounds). In summary, the integral of a k-form over a bounded, oriented domain of k-dimensional space has been defined for $k = 1, 2, 3$.

However, the definition begs a substantial part of the question, namely, *does the integral converge*? Generally speaking, one can say that in all reasonable cases the answer is yes. In particular the answer is yes in all cases which are considered in the following chapters, namely those cases in which the bounded domain D is *a region enclosed by a finite number of curves*† and the function A is *bounded*‡ and *continuous*§ on D.

An outline of the proof of the fact that these conditions on A and D guarantee the convergence of the integral is given below. This proof involves some rather difficult mathematical arguments, and one should not expect to understand it completely on the first or second reading.

It is easy to show that convergence or divergence of the integral, as well as the limiting value in the case of convergence, is independent of the choice of the enclosing rectangle R (see Exercise 6).

†Meaning 'reasonable' curves, that is, differentiable curves. The detailed statement of the theorem is given in Chapter 6.

‡This means that there is a number M such that $|A(P)| \leq M$ at all points P of D.

§This means that given a point $(\bar{x}, \bar{y})$ of D and a margin for error ϵ there is a distance δ such that the value of A varies by less than ϵ on the square $\{|x - \bar{x}| < \delta, |y - \bar{y}| < \delta\}$; that is, given $(\bar{x}, \bar{y})$ in D and $\epsilon > 0$ there is a $\delta > 0$ such that $|A(x, y) - A(\bar{x}, \bar{y})| < \epsilon$ whenever $|x - \bar{x}| < \delta$, $|y - \bar{y}| < \delta$. Intuitively it means that if (x, y) is near $(\bar{x}, \bar{y})$ then $A(x, y)$ is near $A(\bar{x}, \bar{y})$.

However, there is no better way to grasp the meaning of convergence of integrals than to study the proof of this theorem, particularly in concrete cases such as those of Exercises 1, 2, 3 at the end of this section, and Exercises 1, 3, 4, 5 of the preceding section.

Outline of proof

Let D be a bounded, oriented domain enclosed by a finite number of smooth curves. Let R be a rectangle containing D and let A be the function on R which is the given continuous function at points of D and which is 0 at points of R not in D (this function was denoted A_D above). The orientation of R, which is taken to agree (counterclockwise or clockwise) with that of D, can be assumed to be counterclockwise, since this affects only the sign of the result. It is to be shown that for such a function A on R the sums $\sum(\alpha)$ defined by (1) converge.

For the sums $\sum(\alpha)$ to converge means that $\sum(\alpha')$ can be made to differ arbitrarily little from $\sum(\alpha)$ by making $|\alpha|$ and $|\alpha'|$ small. In particular, if α' differs from α only in the choice of the points P'_{ij} and not in the choice of rectangles—that is, if $R'_{ij} = R_{ij}$—then

$$\left|\sum(\alpha) - \sum(\alpha')\right| = \left|\sum_{i,j} [A(P'_{ij}) - A(P_{ij})] \text{ area } (R_{ij})\right|$$

can be made small by making $|\alpha| = |\alpha'|$ small. Letting S denote the subdivision of R into subrectangles R_{ij} common to α and α', this implies that the *maximum* value can be made small, that is,

$$(2) \quad U(S) = \sum_{i,j} \left[\max_{P,P' \text{ in } R_{ij}} \{A(P') - A(P)\} \right] [\text{area } (R_{ij})]$$

can be made small by making the mesh size $|S|$ of S small. (Since the maximum is ≥ 0, all terms are positive and the absolute value signs may be dropped.) Specifically, if the integral converges, then given any margin for error ϵ there is a mesh size δ such that the number $U(S)$ defined by (2) is less than ϵ whenever the mesh size of the subdivision S is less than δ, i.e. $|S| < \delta$, implies $U(S) < \epsilon$. In brief, if the integral converges, then $U(S) \to 0$ as $|S| \to 0$.

The converse of this statement is also true; that is, *if* $U(S) \to 0$ as $|S| \to 0$ *then* the integral converges. One need only note that if α is based on a subdivision S and if α' is based on a subdivision S' which is a refinement* of S then it is still true that $|\sum(\alpha) - \sum(\alpha')| \leq U(S)$.

*A subdivision S' is said to be a refinement of a subdivision S if it is obtained from S by adding more lines $x = const.$, $y = const.$

(Exercise 4.) Then if α, α' are any two sets of data there is a third set of data, α'', based on a refinement of both S and S' (merely take S'' to include all lines $x =$ const., $y =$ const. specified by either S or S'), and hence

$$\begin{aligned} |\textstyle\sum(\alpha) - \sum(\alpha')| &= |\textstyle\sum(\alpha) - \sum(\alpha'') + \sum(\alpha'') - \sum(\alpha')| \\ &\leq |\textstyle\sum(\alpha) - \sum(\alpha'')| + |\sum(\alpha'') - \sum(\alpha')| \\ &\leq U(S) + U(S'). \end{aligned}$$

If $U(S) \to 0$ as $|S| \to 0$ this can be made small by making $|S|$ and $|S'|$ both small; that is, given ϵ, there is a mesh size such that $|S| < \delta$ and $|S'| < \delta$ implies $|\sum(\alpha) - \sum(\alpha')| < \epsilon$, and therefore the integral converges. Thus *the integral converges if and only if* $U(S) \to 0$ as $|S| \to 0$.

This important conclusion is perhaps more comprehensible when it is formulated as follows: The number $U(S)$ represents the 'uncertainty' of an approximating sum $\sum(\alpha)$ to $\int_D A\,dx\,dy$ based on the subdivision S. Any approximating sum based on any refinement of S differs from any approximating sum based on S by at most $U(S)$. Thus further refinement changes the result by at most the 'uncertainty' $U(S)$. The integral converges if and only if this uncertainty can be made small by making the mesh size small.

The problem, therefore, is to show that the assumptions on A and D are sufficient to guarantee that $U(S) \to 0$ as $|S| \to 0$. In order to do this, it is useful to decompose the sum (2) which defines $U(S)$ into two parts $U(S) = U_1(S) + U_2(S)$ as follows: A term of $U(S)$ corresponding to a rectangle R_{ij} of S is counted in the first sum $U_1(S)$ if R_{ij} is contained entirely in D, and counted in the second sum $U_2(S)$ if R_{ij} lies partly in D and partly outside D. (If R_{ij} lies entirely outside D then A is identically zero on R_{ij} and the corresponding term in $U(S)$ is zero.) It will be shown that the numbers $U_1(S)$ and $U_2(S)$ are both small when $|S|$ is small (but for quite different reasons).

The sum $U_2(S)$ is small because the total area of the rectangles R_{ij} which lie partly inside and partly outside D is small. More specifically, each term is at most $(A(P') - A(P))[\text{area }(R_{ij})] \leq 2M[\text{area }(R_{ij})]$ where M is a bound on the bounded function A, so the total $U_2(S)$ is at most $2M$ times the total area of such R_{ij}. It is intuitively clear that if the boundary of D consists of a finite number of reasonable curves then the total area of such rectangles R_{ij} can be made arbitrarily small by making the mesh size small. The rigorous proof of this statement must await a precise definition of 'reasonable' curves. (See Chapter 6 for the proof. For specific examples this statement can be proved directly—see Exercise 1.)

The sum $U_1(S)$ is small because $|A(P') - A(P)|$ is small (when P' is near P) by the continuity of A. Specifically, it can be shown that for every $\epsilon > 0$ there is a $\delta > 0$ such that $|A(P) - A(P')| < \epsilon$ whenever $|P - P'| < \delta$ (meaning that both the x-coordinates and the y-coordinates of P, P' differ by less than δ). In other words A is *uniformly* continuous on D (see §9.3.). Then, if $|S| < \delta$, it follows that $|U_1(S)|$ is at most ϵ times the total area of the rectangles R_{ij} inside D. Since ϵ is arbitrarily small and the total area of the rectangles in D is bounded (by the area of a rectangle containing D, for instance) it follows that $U_1(S)$ can be made small by making $|S|$ small. (Again, the rigorous proof is postponed to Chapter 6. In specific examples it can be proved directly that $U_1(S) \to 0$ as $|S| \to 0$; see, for example, Exercise 1, §2.2.)

This completes the outline of the proof that the integral $\int_D A\, dx\, dy$ converges.

Exercises

1 *Computation of π.* The number π is defined to be the area of the circle (disk) of radius 1, that is,

$$\pi = \int_D dx\, dy$$

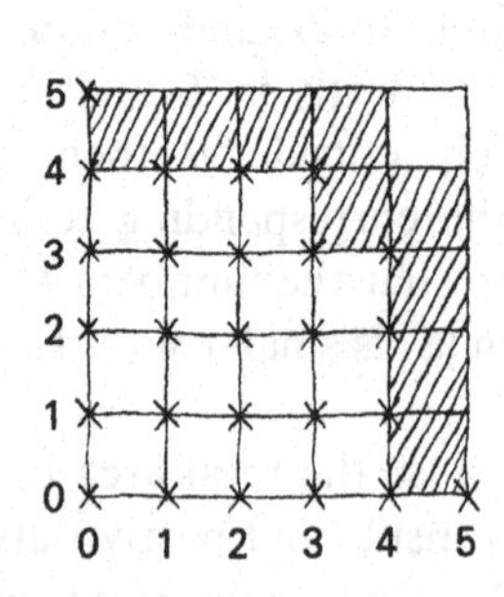

$U_5 = 9$

$C_5 = 15$

$A_5 = 3.12$

where D is the disk $\{(x, y): x^2 + y^2 \le 1\}$. Given an integer n, draw a square grid of n^2 squares and $(n + 1)^2$ vertices, labeling the vertices with integer coordinates (p, q), $0 \le p \le n$, $0 \le q \le n$. Mark the vertices (p, q) for which $p^2 + q^2 \le n^2$ with an X. Shade all squares for which the inner vertex (p, q) has an X but for which the outer vertex $(p + 1, q + 1)$ does not. Let

U_n = # of shaded squares (U = uncertain)

C_n = # of squares whose vertices all have X's (C = certain)

$$A_n = 4\left(\frac{1}{n^2}\cdot C_n + \frac{1}{n^2}\cdot\frac{1}{2}\cdot U_n\right) \quad (A = \text{approximation})$$

Find A_{10}. Find A_{20}. Show that any approximating sum to the integral $\int_D dx\,dy$ which is based on the subdivision by lines $x = \pm\frac{p}{n}$, $y = \pm\frac{q}{n}$ or on any refinement of this subdivision differs from A_n by at most $2n^{-2}U_n$. What accuracy (how many decimal places) does this estimate guarantee for the approximation $A_{10} \sim \pi$? What is the actual accuracy? What accuracy does it guarantee for A_{20}? What is the actual accuracy? Find a formula for U_n. [Count the crossings of the lines x = const. and y = const. separately.] How large would n have to be for this estimate to guarantee two-place accuracy? Note that the approximations are in fact more accurate than this estimate of the error would indicate. Explain this.

2 Many mathematicians, notably Karl Friedrich Gauss (1777–1855), have investigated the number N_r of points $(\pm p, \pm q)$ with integer coordinates contained in the circle of radius r (including points on the circle).

(a) Find N_5, $N_{\sqrt{7}}$.

(b) Show that N_r/r^2 is an approximating sum to $\pi = \int_D dx\,dy$. [Subdivide the plane by lines $x = \pm\frac{n}{r} + \frac{1}{2r}$, $y = \pm\frac{m}{r} + \frac{1}{2r}$ and evaluate at midpoints of squares.]

(c) Use the argument of Exercise 1 to prove that the number of squares which lie on the boundary of D is $\leq 8(r + \frac{1}{2})$.

(d) Show that any approximating sum based on any refinement of the subdivision of (b) differs from N_r/r^2 by less than $8(r + \frac{1}{2})/r^2$, and hence that the *limiting value* π differs from N_r/r^2 by less than $8(r + \frac{1}{2})/r^2$, i.e.

$$|\pi r^2 - N_r| < 8(r + \tfrac{1}{2}).$$

As was seen in the preceding problem, this estimate of the error is much too large; Gauss conjectured that the error is of the order of magnitude of $\sqrt{r}$, that is, that there is a number M such that $|\pi r^2 - N_r| < M\sqrt{r}$, but this conjecture has never been proved or disproved.*

**(Note added in 1980) G. H. Hardy (Collected Papers, Vol. 2, p. 290) did disprove $|\pi r^2 - N_r| < M\sqrt{r}$. He conjectured that $< Mr^{(1/2)+\epsilon}$ was correct, and this has never been disproved. The statement about Gauss conjecturing $< M\sqrt{r}$ is not historically reliable.*

3 Show that the integral $\int_D dx\,dy$ defining π converges. [This is of course a special case of the theorem proved in the text, so it is a question of extracting the necessary parts of that proof. Take a fine subdivision by lines $x = \pm\frac{p}{n}$, $y = \pm\frac{q}{n}$, take $B_1, \ldots, B_N$ to be the squares which lie on the boundary, estimate their area, and estimate for *any* subdivision S the total area of the squares which touch one of the B_i in terms of the mesh size $|S|$. Show that this total area, and hence $U(S)$, can be made small by making $|S|$ small.]

4 Show that if $\sum(\alpha)$ is an approximating sum based on the subdivision S and if $\sum(\alpha')$ is any approximating sum based on a subdivision S' which is a refinement of S, that is, for which every rectangle R'_{ij} of S' is contained in or equal to some rectangle of S, then

$$\left|\sum(\alpha) - \sum(\alpha')\right| \leq U(S)$$
$$= \sum_{\substack{\text{rectangles} \\ R_{ij} \text{ of } S}} \left[\max_{P, P' \text{ in } R_{ij}} \{A(P) - A(P')\}\right][\text{area } R_{ij}]$$

[Lump together all terms of $\sum(\alpha')$ which correspond to the same rectangle R_{ij} of S.]

5 Show that if an integral converges and if L is the limiting value then any approximating sum $\sum(\alpha)$ based on the subdivision S is within $U(S)$ of the limiting value $|L - \sum(\alpha)| \leq U(S)$. [This observation was used in Exercise 2. It suffices to show that $|L - \sum(\alpha)| \leq U(S) + \epsilon$ for all $\epsilon > 0$. Note that by definition of L, $|L - \sum(\alpha')|$ can be made less than any $\epsilon > 0$ by making $|S'|$ small where S' is the subdivision of α'. Take S' to be a refinement of S.]

6 *Irrelevance of the enclosing rectangle.* Suppose that A_D is a bounded function which is zero outside the domain D, and suppose that the rectangles R, R' both contain D. Show that $\int_R A_D\,dx\,dy = \int_{R'} A_D\,dx\,dy$. [First show that one may as well assume that $R' \subset R$ and, in fact, that $D = R'$. Then since every approximating sum to $\int_{R'}$ is also an approximating sum to $\int_R$ it is easy to show:

$$\int_R \text{ converges implies } \int_{R'} \text{ converges;}$$

if both converge the limits are equal.

The only statement remaining is

$$\int_{R'} \text{ converges implies } \int_R \text{ converges.}$$

This is proved by showing that the rectangles of a subdivision S of R which lie on the boundary of R' make an insignificant contribution to $U(S)$.]

7 The definition of 'integral' given in the text was first given by Bernhard Riemann (1826–1866) in whose honor this notion of integration is called 'Riemann integration' and the approximating sums are called 'Riemann sums'. Riemann gave the following necessary and sufficient condition for the convergence of the integral $\int_R A\,dx\,dy$ where R is a rectangle and A is an *arbitrary* function on R:

Riemann's criterion. For any positive number σ and any subdivision S of R, let $s(S, \sigma)$ be the total area of those rectangles of S for which the variation $U_{ij} > \sigma$, i.e. in which there are points P, P' with $A(P) - A(P') > \sigma$. Then the integral $\int_R A\,dx\,dy$ converges if and only if

(1) A is bounded on R and
(2) for every $\sigma > 0$ the total area $s(S, \sigma)$ of those rectangles on which the variation is $> \sigma$ can be made small by making the mesh size $|S|$ small; that is, given $\sigma > 0$ and $\epsilon > 0$ there is a $\delta > 0$ such that $s(S, \sigma) < \epsilon$ whenever $|S| < \delta$. In short, for every $\sigma > 0$, $s(S, \sigma) \to 0$ as $|S| \to 0$.

Prove Riemann's criterion. [From the argument of the text it suffices to show that $U(S) \to 0$ as $|S| \to 0$ if and only if (1) and (2) hold. To prove 'only if', show that if A is unbounded then for any fixed S the maximum $U(S)$ is in fact infinite; whereas if (2) is false for some σ then $U(S) > \sigma \cdot s(S, \sigma)$ does not go to zero as $|S| \to 0$. To prove 'if', split $U(S)$ into two sums, one consisting of terms in which $U_{ij} \leq \sigma$ and the other in which $U_{ij} > \sigma$, make the first sum small by making σ small, and make the second sum small by using (1) and (2).]

8 In the text the condition 'the integral diverges' is defined to mean the negation of the statement 'the integral converges'. State precisely what this condition says about the approximating sums $\sum(\alpha)$; in other words, reformulate the *denial* of the Cauchy Criterion as a positive statement. Show that $\int_R A\,dx\,dy$ diverges when R is a rectangle and when $A(x, y)$ is the function

$$A(x, y) = \begin{cases} 1 & \text{if } x, y \text{ are both rational numbers} \\ 0 & \text{if one or both coordinates are irrational.} \end{cases}$$

If L is a number with the property "Given $\epsilon > 0$ and given $\delta > 0$ there exists an approximating sum $\sum(\alpha)$ to $\int_R A\,dx\,dy$ with $|L - \sum(\alpha)| < \epsilon$ and $|\alpha| < \delta$" what values can L have? [Use the fact that every interval contains both rational and irrational numbers to show that L can be any number between 0 and the oriented area of R.]

9 Suppose that a, b, c, d are four numbers such that the lines $ax + by = \text{const.}$ and $cx + dy = \text{const.}$ are not parallel. Then a bounded domain can also be subdivided using lines $ax + by = \text{const.}$ and $cx + dy = \text{const.}$ Describe how to form an approximating sum to $\int_D A\,dx\,dy$ based on such a subdivision. Define the *mesh size* of such a subdivision to be largest dimension (in x- or y-direction) of any parallelogram of the subdivision. Sketch a proof of the fact that if D is a rectangle and if A is continuous then these approximating sums converge. The proof that their limit is $\int_D A\,dx\,dy$ is examined in Exercise 10.

10 Given an arbitrary subdivision of the plane into polygons, describe how to form an approximating sum to $\int_D A\,dx\,dy$ based on the subdivision. Define 'mesh size'. Sketch a proof of the fact that if D is a rectangle and if A is continuous then these approximating sums converge. Show that their limit is $\int_D A\,dx\,dy$. Why is it necessary to restrict to polygonal subdivisions? The approximating sums of the text were based on rectangular subdivisions because they are easiest to describe (being defined simply by the numbers x_i, y_j) and because it is quite difficult to define precisely what is meant by 'an arbitrary subdivision of the plane into polygons'.

2.4

Integrals and Pullbacks

In §2.2 the integral of a k-form over an oriented k-dimensional domain was described as a limit of approximating sums, an approximating sum being formed by constructing a finely divided oriented polygonal domain which approximates the domain of integration, 'evaluating' the k-form on each polygon of this polygonal domain, and adding; the 'evaluation' involves first evaluating the k-form at some point near the polygon to obtain a constant k-form, and then proceeding as in Chapter 1.

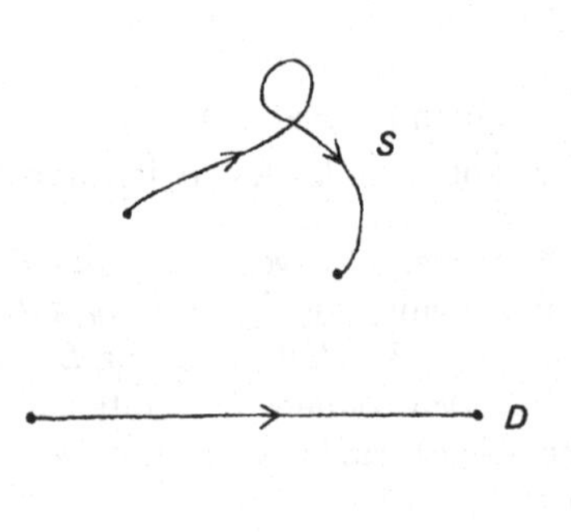

In §2.3 this description was made the basis of a rigorous definition for cases in which the domain of integration is a rectangle (interval, rectangular parallelopiped) by describing explicitly what is meant by a 'finely divided polygonal approximation' to a rectangle. This definition was extended to arbitrary bounded domains in k-space by the simple trick of enclosing such a domain in a k-dimensional rectangle, setting the integrand equal to zero outside the domain, and proceeding as before.

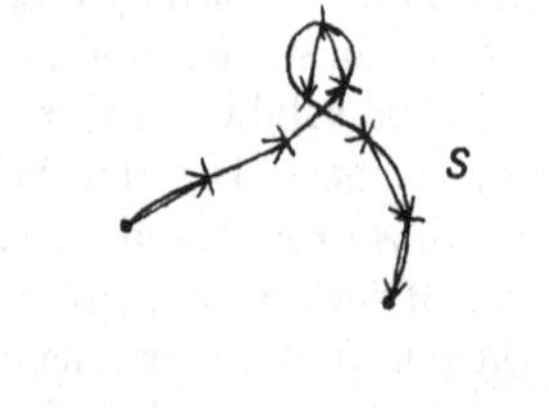

The definition of §2.3 does not apply, however, to integrals over curves in the plane, over curves in space, or over surfaces in space. Such integrals will be defined in this section by assuming that the domain of integration can be described *parametrically* and by defining the integral as an integral over the parameter space.

Let the curve S in the xy-plane represent a displacement of a particle. Then S can be described by a pair of functions $(x(t), y(t))$ giving the coordinates of the particle as functions of time, these functions being defined for the time interval $D = \{a \le t \le b\}$ during which the displacement S occurs. A curve represented in such a way is called a *curve defined by a parameter*, as opposed to a *curve defined by an equation*. (For example, the curve $\{(\cos t, \sin t)\colon 0 \leq t \leq 2\pi\}$ is a curve defined by a parameter, whereas the set $\{(x, y)\colon x^2 + y^2 = 1\}$ is a curve defined by an equation.) Then a polygonal approximation to S can be constructed by giving a subdivision $a = t_0 < t_1 < t_2 \ldots < t_{n-1} < t_n = b$ of the parameterizing time interval D and drawing the polygonal curve from $(x(t_0), y(t_0))$ to $(x(t_1), y(t_1))$ to ... to $(x(t_n), y(t_n))$. The corresponding sum approximating $\int_S (A\,dx + B\,dy)$ is $\sum_{i=1}^{n} (A\,\Delta x_i + B\,\Delta y_i)$ where $\Delta x_i = x(t_i) - x(t_{i-1})$, $\Delta y_i = y(t_i) - y(t_{i-1})$ and where the functions A, B are evaluated at some point in the neighborhood of the line segment from $(x(t_{i-1}), y(t_{i-1}))$ to $(x(t_i), y(t_i))$—at the point $(x(\hat{t}_i), y(\hat{t}_i))$ where $\hat{t}_i$ is a point in the ith interval $t_{i-1} \leq \hat{t}_i \leq t_i$, for instance. Thus $\int_S (A\,dx + B\,dy)$ is approximately

$$
\begin{aligned}
&\sum_{i=1}^{n} [A(x(\hat{t}_i), y(\hat{t}_i))\,\Delta x_i + B(x(\hat{t}_i), y(\hat{t}_i))\,\Delta y_i] \\
&\qquad = \sum_{i=1}^{n} \left[A\,\frac{\Delta x_i}{\Delta t_i} + B\,\frac{\Delta y_i}{\Delta t_i}\right] \Delta t_i
\end{aligned}
\tag{1}
$$

where $\Delta t_i = t_i - t_{i-1}$ = value of dt on the directed interval $t_{i-1}t_i$. But the sum on the right is approximately the integral of the 1-form

$$\left[A(x(t), y(t))\,\frac{dx}{dt}(t) + B(x(t), y(t))\,\frac{dy}{dt}(t)\right] dt$$

over the oriented interval from a to b. This integral has been defined and converges provided that the functions $A(x, y)$, $B(x, y)$ are continuous, which is assumed already, and provided that the functions $\frac{dx}{dt}(t)$, $\frac{dy}{dt}(t)$ exist and are continuous, which will be assumed henceforth. Therefore $\int_S A\,dx + B\,dy$ can be *defined* to be $\int_D \left[A\,\frac{dx}{dt} + B\,\frac{dy}{dt}\right] dt$ and equation (1) shows that the resulting number corresponds to the intuitive description of $\int_S (A\,dx + B\,dy)$ given in §2.2.

A slightly different way of arriving at the same result is to interpret $A\,dx + B\,dy$ as representing 'work' and to ask how much work is done during a short time interval. This is approximately

$$\left[A(x(\hat{t}_i), y(\hat{t}_i))\frac{\Delta x_i}{\Delta t_i} + B(x(\hat{t}_i), y(\hat{t}_i))\frac{\Delta y_i}{\Delta t_i}\right]\Delta t_i$$

and the shorter the time interval the better the approximation; that is,

$$\begin{matrix}\text{work done during} \\ \text{short time intervals}\end{matrix} \sim \left[A\frac{dx}{dt} + B\frac{dy}{dt}\right]dt.$$

The work done during a time interval which is not short is then found by integration

$$\begin{aligned}\int_S (A\,dx + B\,dy) &= \text{work done during the time interval } D \\ &= \int_D \left[A\frac{dx}{dt} + B\frac{dy}{dt}\right]dt.\end{aligned}$$

Seen in this light, what is involved is a pullback operation in which a 1-form on the t-line is obtained from the 1 form $A\,dx + B\,dy$ on the xy-plane and a map of the t-line to the xy-plane. The 1-form $\left(A\frac{dx}{dt} + B\frac{dy}{dt}\right)dt$ is called the pullback of the (variable) 1-form under the (non-affine) map $x = x(t)$, $y = y(t)$.

The justification for defining

$$\int_S (A\,dx + B\,dy + C\,dz) = \int_D \left[A\frac{dx}{dt} + B\frac{dy}{dt} + C\frac{dz}{dt}\right]dt,$$

when S is a parametric curve $\{(x(t), y(t), z(t))\colon t \text{ in } D\}$ in which the functions x, y, z have continuous derivatives is exactly the same. The 1-form

$$\left[A\frac{dx}{dt} + B\frac{dy}{dt} + C\frac{dz}{dt}\right]dt$$

on the line is called the pullback of the (variable) 1-form $A\,dx + B\,dy + C\,dz$ under the (non-affine) map $x = x(t)$, $y = y(t)$, $z = z(t)$.

Similarly, let S be a surface which is given in the form

$$S = \{(x(u, v), y(u, v), z(u, v)\colon (u, v) \text{ in } D\}$$

where $x(u, v)$, $y(u, v)$, $z(u, v)$ are given functions of two variables u, v defined on a domain D of the uv-plane.

Such a surface is called a surface defined by parameters, as opposed to a surface defined by an equation. (For example, the surface

$$S = \left\{(\cos\theta\cos\varphi, \sin\theta\cos\varphi, \sin\varphi): 0 \leq \theta \leq 2\pi, -\frac{\pi}{2} \leq \varphi \leq \frac{\pi}{2}\right\}$$

is a surface defined by parameters, whereas the set $\{(x, y, z): x^2 + y^2 + z^2 = 1\}$ is a surface defined by an equation.) The integral of a 2-form over such a surface

$$\int_S [A\,dy\,dz + B\,dz\,dx + C\,dx\,dy]$$

S will be *defined* to be the integral of a pulled-back 2-form over the parameter space D. This 2-form on D is to assign to a small polygon in D the approximate value of $A\,dy\,dz + B\,dz\,dx + C\,dx\,dy$ on its image.

To find such a 2-form on D, consider a particular point $(\bar{u}, \bar{v})$ in D. Defining the partial derivatives, as usual, by

$$\frac{\partial x}{\partial u}(\bar{u}, \bar{v}) = \lim_{u\to\bar{u}} \frac{x(u, \bar{v}) - x(\bar{u}, \bar{v})}{u - \bar{u}}$$

$$\frac{\partial x}{\partial v}(\bar{u}, \bar{v}) = \lim_{v\to\bar{v}} \frac{x(\bar{u}, v) - x(\bar{u}, \bar{v})}{v - \bar{v}}$$

it is to be expected that the approximation

$$x(u, v) \sim x(\bar{u}, \bar{v}) + \frac{\partial x}{\partial u}(\bar{u}, \bar{v})(u - \bar{u}) + \frac{\partial x}{\partial v}(\bar{u}, \bar{v})(v - \bar{v})$$

holds for (u, v) near $(\bar{u}, \bar{v})$; that is, the given function $x(u, v)$ is well approximated by the affine function on the right. The given functions $y(u, v)$ and $z(u, v)$ can be approximated in the same way, leading to the conclusion that the map $x = x(u, v)$, $y = y(u, v)$, $z = z(u, v)$ defining the surface is well approximated near $(\bar{u}, \bar{v})$ by the affine map

$$\begin{aligned} x &= x(\bar{u}, \bar{v}) + \frac{\partial x}{\partial u}(\bar{u}, \bar{v})(u - \bar{u}) + \frac{\partial x}{\partial v}(\bar{u}, \bar{v})(v - \bar{v}) \\ y &= y(\bar{u}, \bar{v}) + \frac{\partial y}{\partial u}(\bar{u}, \bar{v})(u - \bar{u}) + \frac{\partial y}{\partial v}(\bar{u}, \bar{v})(v - \bar{v}) \qquad (2) \\ z &= z(\bar{u}, \bar{v}) + \frac{\partial z}{\partial u}(\bar{u}, \bar{v})(u - \bar{u}) + \frac{\partial z}{\partial v}(\bar{u}, \bar{v})(v - \bar{v}). \end{aligned}$$

The image of a small polygon in D under the actual map is nearly its image under the affine map (2), and the value of $A\,dy\,dz + B\,dz\,dx + C\,dx\,dy$ on its image is therefore nearly the value of the pullback of

$$(3)\quad A(\bar{x},\bar{y},\bar{z})\,dy\,dz + B(\bar{x},\bar{y},\bar{z})\,dz\,dx + C(\bar{x},\bar{y},\bar{z})\,dx\,dy$$

on the polygon itself (where $\bar{x} = x(\bar{u},\bar{v})$, etc.). It is reasonable to define the pullback of $A\,dy\,dz + B\,dz\,dx + C\,dx\,dy$ under the map $x = x(u,v)$, $y = y(u,v)$, $z = z(u,v)$ to be the 2-form in uv whose value at any point $(\bar{u},\bar{v})$ is the pullback of the constant form (3) under the affine map (2).

This (variable) 2-form in uv is easily computed. One need only set

$$\begin{aligned} dx &= \frac{\partial x}{\partial u}\,du + \frac{\partial x}{\partial v}\,dv \\ dy &= \frac{\partial y}{\partial u}\,du + \frac{\partial y}{\partial v}\,dv \\ dz &= \frac{\partial z}{\partial u}\,du + \frac{\partial z}{\partial v}\,dv \\ x &= x(u,v) \\ y &= y(u,v) \\ z &= z(u,v), \end{aligned}$$

substitute these expressions in $A(x,y,z)\,dy\,dz + B(x,y,z)\,dz\,dx + C(x,y,z)\,dx\,dy$, and simplify using the usual rules $du\,du = 0$, $du\,dv = -dv\,du$. For example, the pullback of $x\,dy\,dz - y^2\,dx\,dy$ under the map

$$\begin{aligned} x &= e^u + v \\ y &= u + 2v \\ z &= \cos u \end{aligned}$$

is

$$\begin{aligned} &(e^u + v)(du + 2\,dv)(-\sin u\,du) - (u + 2v)^2(e^u\,du + dv)(du + 2\,dv) \\ &\qquad = [(e^u + v)2\sin u + (u + 2v)^2(1 - 2e^u)]\,du\,dv. \end{aligned}$$

Note that the pullbacks of 1-forms above were found by the same method: The pullback of $A(x,y)\,dx + B(x,y)\,dy$ was found merely by performing the substitutions $x = x(t)$, $y = y(t)$, $dx = \frac{dx}{dt}\,dt$, $dy = \frac{dy}{dt}\,dt$ to 'express $A\,dx + B\,dy$ in terms of t'.

Thus the pullback of a (variable) k-form under a (not

necessarily affine) map can be defined by these computational rules

$$(4)\quad dx = \frac{\partial x}{\partial u}\,du + \frac{\partial x}{\partial v}\,dv,\ du\,du = 0,\ du\,dv = -dv\,du, \text{etc.,}$$

and it is reasonable to interpret the resulting k-form in uv as the k-form whose value on a small k-dimensional figure in uv-space is equal to the value of the given k-form on the image of this figure under the given map.

Using the definition of pullback by the computational rules (4), one can define the integral of a k-form over a k-dimensional domain which is defined by parameters to be the integral of the pullback over the parameterizing domain D. [Assuming that the partial derivatives of the given map exist and are continuous, the pullback is a continuous k-form on a k-dimensional space; hence the integral is defined and converges whenever the parameterizing domain D is reasonable in the sense of §2.3.]

To prove that this definition of integrals over curves and surfaces has all the desired properties is a rather long task (see Chapter 6). What is important for the moment is the computation of pullbacks and an intuitive understanding of the relation of this operation to the notion of integration as described in §2.2.

Exercises

1 Find the pullbacks of

(a) $x\,dy\,dz$ under $x = \cos uv$, $y = \sin uv$, $z = uv^2$
(b) $xy\,dz\,dx$ under $x = u \cos v$, $y = u + v$, $z = u \sin v$
(c) $z^3\,dx\,dy$ under $x = e^u + v$, $y = e^u - v$, $z = 2$

2 If $x\,dy + y\,dx$ gives the work required for small displacements in the plane, and if $(x, y) = (\sqrt{t}, t)$ for $t > 0$ gives position as a function of time, find rate of work as a function of time (=work done/time elapsed as elapsed time $\to 0$). Graph the motion by drawing the curve along which the particle moves and labeling points with the time at which they are passed. Write the amount of work done between times $t = 1$ and $t = 4$ as an integral.

3 The mapping

$$x = 2u/(u^2 + v^2 + 1)$$
$$y = 2v/(u^2 + v^2 + 1)$$
$$z = (u^2 + v^2 - 1)/(u^2 + v^2 + 1)$$

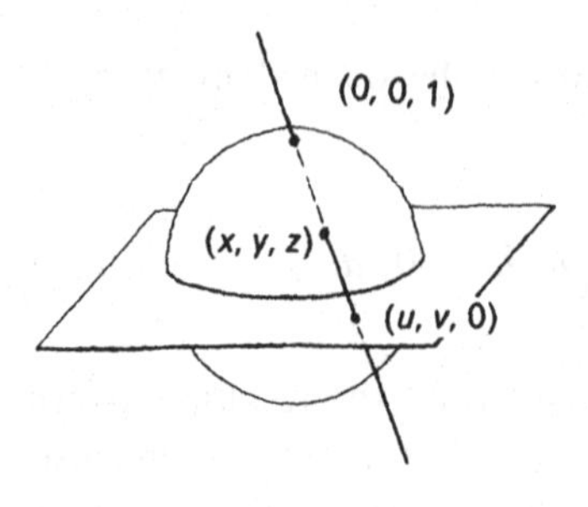

arises from the *stereographic projection* of the sphere $x^2 + y^2 + z^2 = 1$ onto the plane $z = 0$. Specifically, the point (x, y, z) given by these formulas is the unique point of the sphere which lies on the line through $(u, v, 0)$ and $(0, 0, 1)$. Check this fact and show that every point of the sphere except $(0, 0, 1)$ corresponds to exactly one point of the plane. Find the pullback under this map of the 2-form $x\,dy\,dz + y\,dz\,dx + z\,dx\,dy$. [The computation is long but the answer is simple.]

4 Find the pullback of $x\,dy\,dz + y\,dz\,dx + z\,dx\,dy$ under the map

$$x = \cos\theta\cos\varphi,\ y = \sin\theta\cos\varphi,\ z = \sin\varphi$$

giving spherical coordinates on $x^2 + y^2 + z^2 = 1$.

5 Write the integral of Exercise 4, §2.2, as an integral over $\{0 \le \theta \le 2\pi\}$.

6 Find the pullback of $dx\,dy$ under the map

$$x = r\cos\theta,\ y = r\sin\theta$$

giving polar coordinates on the xy-plane. What is the approximate area of a ring-shaped region $\{r_1^2 \le x^2 + y^2 \le r_2^2\}$? What are the orientations when $r < 0$?

2.5

Independence of Parameter

When the domain of integration is described parametrically the integral can be defined, as in the previous section, to be the integral of the pullback of the integrand over the parameterizing domain. However, it is frequently necessary—as will be seen in Chapter 3—to deal with integrals over oriented domains which are not described in this way, to deal, for example, with integrals of 2-forms over the sphere $S = \{x^2 + y^2 + z^2 = 1\}$. If the sphere is oriented by the convention 'counterclockwise as seen from the outside' (a mathematical description of this convention is given below) then the intuitive description of the integral given in §2.2 is applicable but the exact definition of §2.4 is not. The solution of this problem is simply to parameterize the domain of integration, but this raises some very difficult questions:

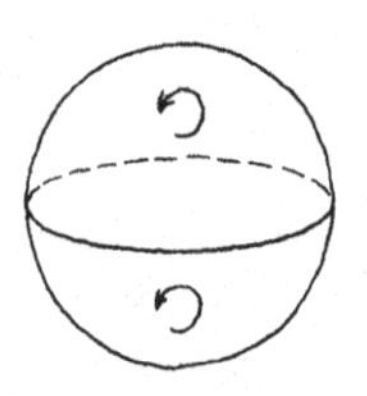

(i) When can a domain be parameterized?
(ii) What, precisely, does it mean to say that a parametric domain is a parameterization of a given domain?

(iii) If a domain is parameterized in two different ways, is the integral the same?

Rigorous answers to these questions will not be given until Chapter 6. In general, the first two questions are not important in practice, and the answer to the third question, which is very important in practice, is 'yes' under very broad assumptions.

For example, consider some parameterizations of the sphere $\{x^2 + y^2 + z^2 = 1\}$. A very common one is given by spherical coordinates

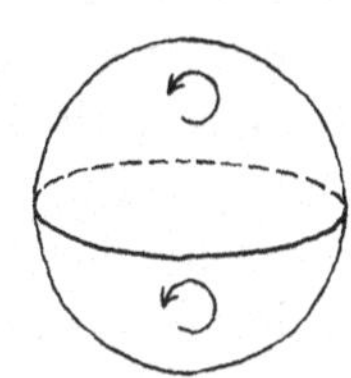

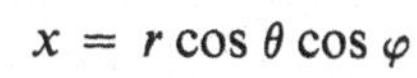

$$
\begin{aligned}
x &= r\cos\theta\cos\varphi \\
y &= r\sin\theta\cos\varphi \\
z &= r\sin\varphi
\end{aligned}
$$

on the rectangle $r = 1,\ 0 \le \theta \le 2\pi,\ -\frac{\pi}{2} \le \varphi \le \frac{\pi}{2}$.

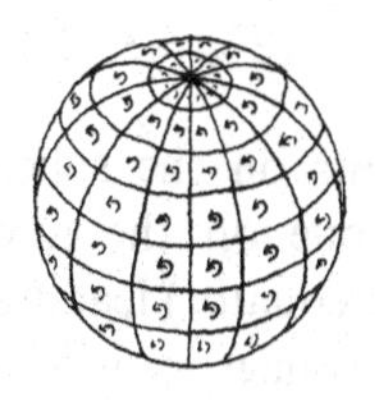

Denoting this rectangle by D and the sphere by S, a mapping from D to S has been given which can be seen geometrically to cover all of S; in fact, the lines θ = const. are the meridians of longitude, φ = const. are the parallels of latitude, and these coordinates are the usual ones for locating points on the earth's surface.

To integrate a 2-form over this parametric surface it is necessary to orient the parameterizing domain $D = \left\{r = 1, 0 \le \theta \le 2\pi, -\frac{\pi}{2} \le \varphi \le \frac{\pi}{2}\right\}$ and this is to be done in such a way that it gives the orientation 'counterclockwise as seen from the outside' to the sphere. A few pictures will suffice to show that the correct orientation of D is 'counterclockwise' when θ, φ are drawn as shown, but the same result can be reached without the aid of pictures as follows: An orientation of the $\theta\varphi$-plane can be described by specifying either $d\theta\, d\varphi$ or $d\varphi\, d\theta$ and saying that a triangle is positively oriented if the value of the specified 2-form on it is positive; thus $d\theta\, d\varphi \leftrightarrow$ counterclockwise and $d\varphi\, d\theta \leftrightarrow$ clockwise. Similarly, an orientation of the sphere can be described by specifying a non-zero 2-form and by saying that a small triangle on the sphere is positively oriented if the value of the specified 2-form is positive. In this particular example the 2-form $x\, dy\, dz + y\, dz\, dx + z\, dx\, dy$ describes the given orientation of the sphere; it is $dx\, dy$ at $(0, 0, 1)$, $dy\, dz$ at $(1, 0, 0)$, $-dx\, dy$ at $(0, 0, -1)$, $dz\, dx$ at $(0, 1, 0)$, etc.—all of which can be seen to describe the orientation 'counterclockwise as seen from the

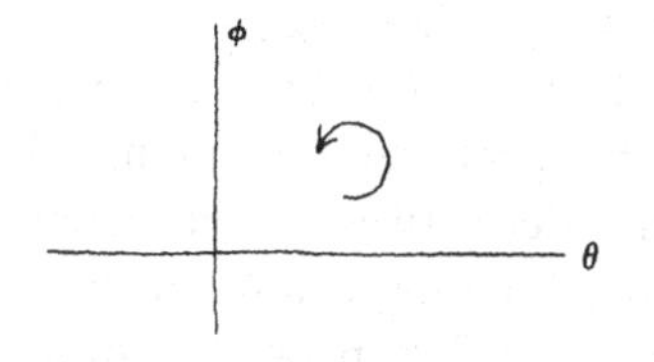

outside' near the points in question. The pullback of $x\,dy\,dz + y\,dz\,dx + z\,dx\,dy$ to $\theta\varphi$ therefore will show how the parameter space should be oriented; this pullback is of the form $f(\theta, \varphi)\,d\theta\,d\varphi$ where $f(\theta, \varphi)$ is a function whose values in the rectangle

$$D = \left\{0 \leq \theta \leq 2\pi, -\frac{\pi}{2} \leq \varphi \leq \frac{\pi}{2}\right\}$$

are all positive (see Exercise 4, §2.4). Thus the orientation of the parameterized sphere by $d\theta\,d\varphi$ agrees with the orientation of the sphere by $x\,dy\,dz + y\,dz\,dx + z\,dx\,dy$ which is the orientation described verbally by the phrase 'counterclockwise as seen from outside'.

A second method of parameterizing the sphere is to take x, y as coordinates, that is, to project onto the xy-plane. Each point inside the disk

$$D = \{(x, y): x^2 + y^2 \leq 1\}$$

corresponds to two points on the sphere, one on the upper hemisphere and one on the lower hemisphere (except for points on the equator) leading to the parameterization of S by *two* parametric surfaces

$$S^+ = \{(x, y, \sqrt{1 - x^2 - y^2}): x^2 + y^2 \leq 1\}$$
$$S^- = \{(x, y, -\sqrt{1 - x^2 - y^2}): x^2 + y^2 \leq 1\}.$$

Since the orientation of S is described by $dx\,dy$ at $(0, 0, 1)$ and by $dy\,dx$ at $(0, 0, -1)$ these parametric surfaces should be oriented by orienting the domain $D = \{(x, y): x^2 + y^2 \leq 1\}$ using $dx\,dy$ for S^+ and $dy\,dx$ for S^-. This method of parameterizing S has the disadvantage that the parameterizing mappings do *not* have continuous partial derivatives at the equator so that the pullbacks of a 2-form on S are not defined at points of the boundary of D. The parameterization can nonetheless be used to find the integral $\int_S (A\,dy\,dz + B\,dz\,dx + C\,dx\,dy)$ of a 2-form over S by integrating over

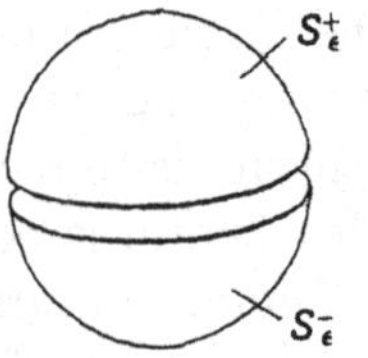

$$S_\epsilon^+ = \{(x, y, \sqrt{1 - x^2 - y^2}): x^2 + y^2 \leq 1 - \epsilon\}$$

oriented $dx\,dy$ and

$$S_\epsilon^- = \{(x, y, -\sqrt{1 - x^2 - y^2}): x^2 + y^2 \leq 1 - \epsilon\}$$

oriented $dy\,dx$, adding the results, and letting $\epsilon \to 0$.

A third parameterization of the sphere is given by the stereographic projection

$$\left\{\left(\frac{2u}{u^2+v^2+1}, \frac{2v}{u^2+v^2+1}, \frac{u^2+v^2-1}{u^2+v^2+1}\right) : \text{all } u, v\right\}$$

(see Exercise 3, §2.4). The reader will easily find the orientation of the uv-plane which corresponds to the given orientation of S (Exercise 5). This parameterization has the disadvantage that the point $(0, 0, 1)$ is omitted. It can nonetheless be used to find the integral of a 2-form over S by integrating the pullback over $u^2 + v^2 \leq A$ and letting $A \to \infty$. Alternatively, the stereographic projection can be used to parameterize the lower hemisphere $\{u^2 + v^2 \leq 1\}$ and its mirror image used to parameterize the upper hemisphere.

These various parametric representations lead to various methods of computing 'the number $\int_S (A\,dy\,dz + B\,dz\,dx + C\,dx\,dy)$'. The fact that all of them result in the same value *requires proof* because no definition of 'the number $\int_S (A\,dy\,dz + B\,dz\,dx + C\,dx\,dy)$' has been given other than these methods of computing it. This fact, that integrals over domains can be computed using any parameterization of the domain, is called the principle of independence of parameter.

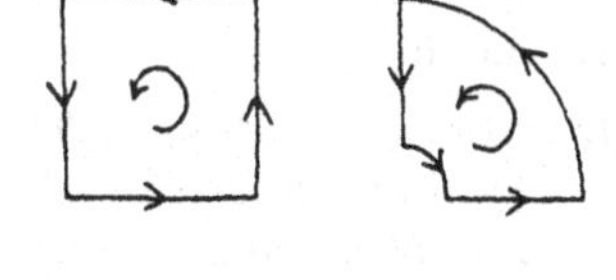

Another example of the principle of independence of parameter is contained in the rule for conversion to polar coordinates in a double integral. For example, if D is the disk $D = \{(x, y): x^2 + y^2 \leq 1\}$ oriented counterclockwise, then D is parameterized in polar coordinates

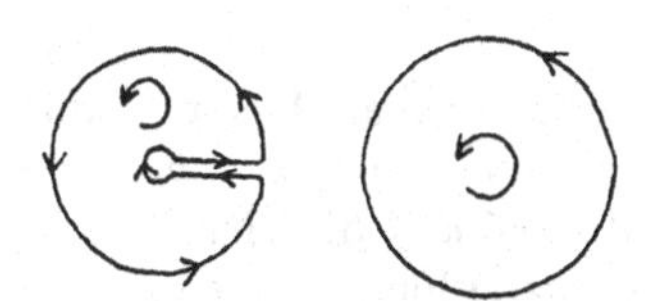

$$x = r\cos\theta$$
$$y = r\sin\theta$$

by the rectangle $0 \leq r \leq 1$, $0 \leq \theta \leq 2\pi$ oriented counterclockwise. Therefore the integral of $A(x, y)\,dx\,dy$ over the disk is equal to the integral of the pullback

$$A(r\cos\theta, r\sin\theta)(\cos\theta\,dr - r\sin\theta\,d\theta)(\sin\theta\,dr + r\cos\theta\,d\theta) = A(r\cos\theta, r\sin\theta)r\,dr\,d\theta$$

over the rectangle. The orientation $dx\,dy$ corresponds to the orientation $dr\,d\theta$ because $r > 0$.

Similarly, the integral of a 1-form over an oriented curve is independent of the choice of parameter. For example, if S is the curve $x^2 = y$ between the points

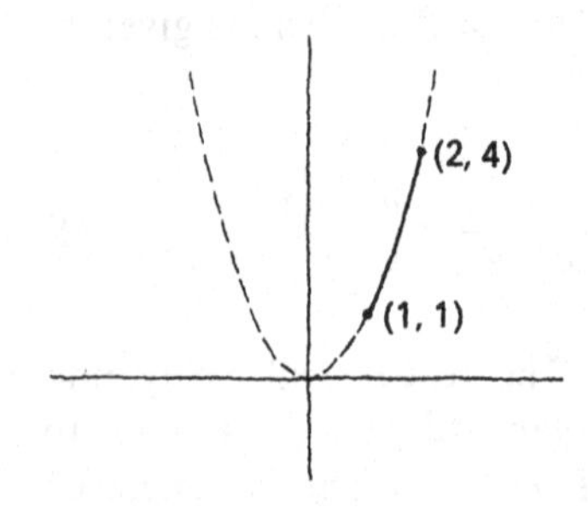

(1, 1) and (2, 4) oriented from the first toward the second, then the integral of $A\,dx + B\,dy$ over S can be computed using either the parameterization $S = \{(x, x^2): 1 \leq x \leq 2\}$ or the parameterization $S = \{(\sqrt{y}, y): 1 \leq y \leq 4\}$ oriented by dx and dy respectively.

Plausible as these statements are in specific cases, it is rather difficult to give a precise definition of the statement that two parameterized surfaces are parameterizations of the same surface, and more difficult still to prove rigorously that integrals are independent of parameter. Until these subjects are dealt with carefully (in Chapter 6) the notion of the integral of a form over an oriented surface is *not defined* until a particular parameterization of the oriented surface is given. However, this is strictly a technical difficulty; integrals *are* independent of parameter, and the informal description of integrals in §2.2 is nearer to their true meaning than is the precise definition of integrals over parameterized surfaces given in the preceding section.

Exercises

1 Parameterize the surface $x + y + z = 1$. Orient the parameterization so as to agree with the orientation given on the original surface by $dy\,dz$; by $dz\,dx$; by $dx\,dy$. Sketch the given surface showing its orientation.

2 Find the pullbacks of $dy\,dz$, $dz\,dx$, and $dx\,dy$ under each of the parameterizations of the sphere considered—spherical coordinates, projection on the xy-plane, and stereographic projection.

3 Let $\int_a^b f(x)\,dx$ be a given integral. Let $x = x(u)$ be a 'parameterization' of the interval $\{a \leq x \leq b\}$ by a new parameter u on an interval $\{\alpha \leq u \leq \beta\}$. That is, let $x(u)$ be a differentiable function establishing a one-to-one correspondence between points of the interval $\{\alpha \leq u \leq \beta\}$ and points of the interval $\{a \leq x \leq b\}$. State the principle of independence of parameter in this case. Pay particular attention to the orientation. Apply this to the integral $\int_a^b x^n\,dx$ $(a > 0, b > 0)$ when $x = e^u$.

4 In order to find the correct orientation of spherical coordinates $\theta\varphi$ it is not necessary to compute the pullback of $x\,dy\,dz + y\,dz\,dx + z\,dx\,dy$ at all points, since the pullback at any one point is sufficient to determine the sign. The point $(x, y, z) = (1, 0, 0)$, $(\theta, \varphi) = (0, 0)$ is particularly simple. Find the pullback at this point.

5 Find the correct orientation of the parameterization by stereographic coordinates. [Find the pullback of $x\,dy\,dz + y\,dz\,dx + z\,dx\,dy$ at the point $(0, 0, -1)$.]

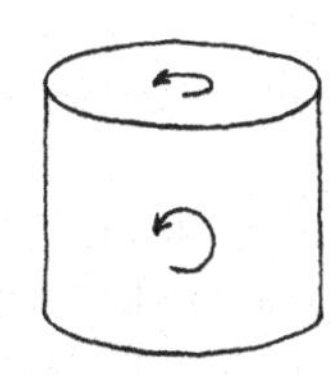

6 Parameterize the three pieces of the boundary of the cylinder $\{(x, y, z): x^2 + y^2 \leq 1, -1 \leq z \leq 1\}$. Orient each of the three pieces by the rule 'counterclockwise as seen from the outside'.

7 Parameterize the surface obtained by rotating the circle $(x - 2)^2 + z^2 = 1, y = 0$ about the z-axis. Orient the parameter space to agree with the orientation of the given surface by the rule 'counterclockwise as seen from the outside'. [First parameterize the given circle and rotate about the z-axis using cylindrical coordinates (r, θ, z). Convert to (x, y, z) coordinates using $x = r\cos\theta$, $y = r\sin\theta$, $z = z$.] This surface is called a *torus*.

2.6 Summary. Basic Properties of Integrals

Chapter 1 was devoted to constant k-forms and to their evaluation on simple k-dimensional figures such as oriented line segments, triangles, parallelograms, and cubes. This chapter has been devoted to (variable) k-forms and to their evaluation on oriented k-dimensional domains. The 'value' of a k-form on an oriented k-dimensional domain has been defined as a limit of sums, that is, as an *integral.*

The precise definition of the notion of 'integral' is difficult for two reasons—first, because it involves the notion of 'limit', and second, because it involves the notion of 'oriented k-dimensional domain'. The notion of 'limit' is defined by the Cauchy Convergence Criterion, which was discussed in detail in §2.3. The problem of defining 'oriented k-dimensional domain' is more difficult and has been avoided entirely by restricting consideration to *specific* domains such as rectangles, disks, spheres, etc., and to domains which are parameterized by such domains.*

**The precise definition of 'k-dimensional domain', which is given in Chapters 5 and 6, depends essentially on the Implicit Function Theorem.*

The following properties of integrals are all immediate consequences of the definition of integrals as limits of sums. They are stated specifically for integrals of the form $\int_R A(x, y)\,dx\,dy$ where R is an oriented rectangle in the xy-plane and $A(x, y)$ is a continuous function defined at all points of R, but they all have analogs which are true for integrals of k-forms on k-dimensional domains in the general cases described in §2.2:

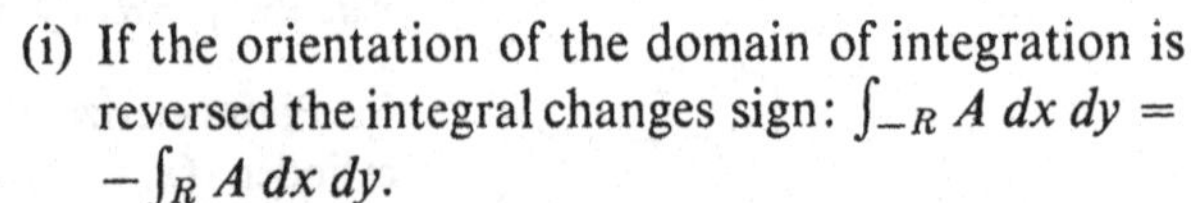

(i) If the orientation of the domain of integration is reversed the integral changes sign: $\int_{-R} A\,dx\,dy = -\int_R A\,dx\,dy$.

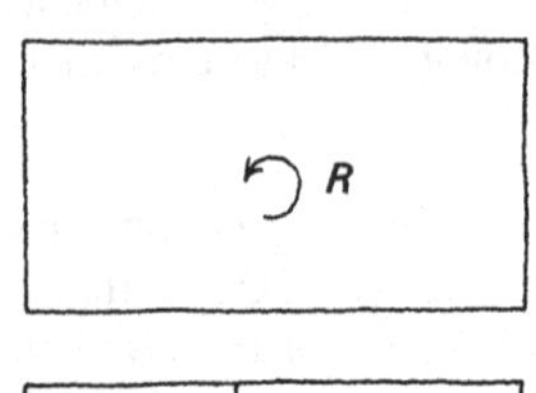

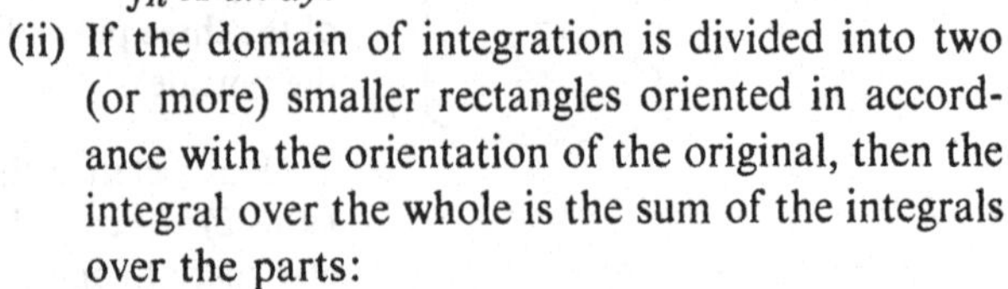

(ii) If the domain of integration is divided into two (or more) smaller rectangles oriented in accordance with the orientation of the original, then the integral over the whole is the sum of the integrals over the parts:

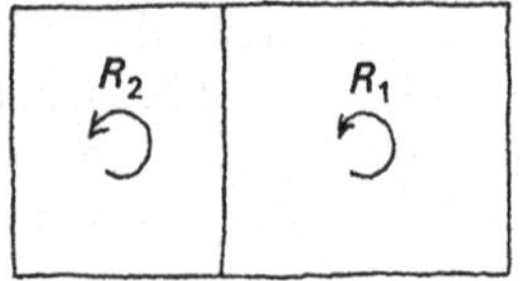

$$\int_{R_1+R_2} A\,dx\,dy = \int_{R_1} A\,dx\,dy + \int_{R_2} A\,dx\,dy.$$

(iii) If the integrand is multiplied by a constant the integral is multiplied by the same constant: $\int_R cA(x, y)\,dx\,dy = c\int_R A(x, y)dx\,dy$.

(iv) If the integrand is a sum of two (or more) terms then the integral of the sum is the sum of the integrals: $\int_R (A_1 + A_2)\,dx\,dy = \int_R A_1\,dx\,dy + \int_R A_2\,dx\,dy$.

If the rectangle R is subdivided into a very large number of very small pieces, then, by (ii), $\int_R A\,dx\,dy$ is the sum of the integrals over the individual pieces. The integral over a very small rectangle is roughly the value of A on the rectangle times the oriented area of the rectangle. This is only roughly true because, of course, A does not have *a* value on the rectangle but many values. However, the assumption that A is continuous is the assumption that A is nearly constant on sufficiently small rectangles, so that the integral over such a rectangle is nearly 'the' value of A times the oriented area. Specifically, the definition of continuity of a function easily implies the following:

(v) Given a point $(\bar{x}, \bar{y})$ and given $\epsilon > 0$ there is a $\delta > 0$ such that if R is any rectangle containing $(\bar{x}, \bar{y})$ and contained in the square $\{|x - \bar{x}| < \delta, |y - \bar{y}| < \delta\}$, then $\int_R A\,dx\,dy$ differs from $A(\bar{x}, \bar{y})$ times the oriented area of R by less than ϵ times the area of R

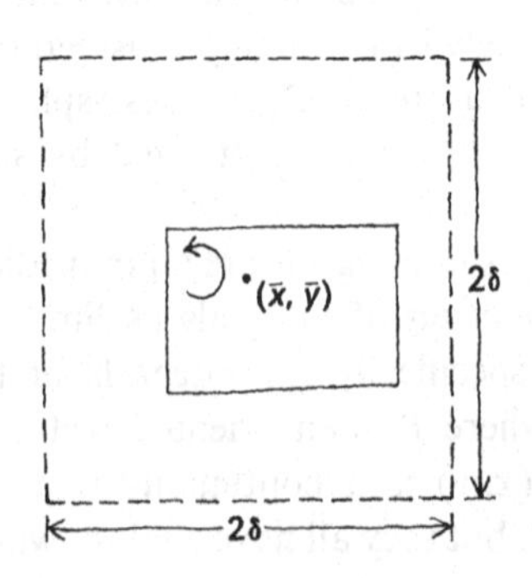

$$\left|\int_R A\,dx\,dy - A(\bar{x}, \bar{y})\int_R dx\,dy\right| < \epsilon \cdot \left|\int_R dx\,dy\right|.$$

A useful way to remember this statement is by means

of the formula

$$\lim_{R \to P} \frac{\int_R A\, dx\, dy}{\int_R dx\, dy} = A(P)$$

where the rectangle R is thought of as shrinking down to the point P. This statement about integrals over small rectangles, together with the subdivision property (ii), is the substance of the intuitive idea of 'integral' as it is described in §2.2.

Another type of formula which is frequently useful is the formula for a double integral as an iterated integral

$$(1) \qquad \int_R A(x, y)\, dx\, dy = \int_c^d \left[\int_a^b A(x, y)\, dx \right] dy$$

where $A(x, y)$ is a continuous function on the rectangle $R = \{a \leq x \leq b, c \leq y \leq d\}$, where R is oriented counterclockwise, and where $\int_a^b A(x, y)\, dx$ is considered as a function of y. For the proof of this formula see Exercise 2.

The integral of a form over a domain which is described parametrically is defined in terms of the pullback of the form under the parameterizing map. The pullback operation, which is a simple generalization of the pullback operation for constant forms under affine maps, is studied further in Chapter 5.

Exercises

1 Prove the properties (i)–(v) of $\int_R A\, dx\, dy$.

2 Prove that if $A(x, y)$ is continuous on $R = \{a \leq x \leq b, c \leq y \leq d\}$ then the integral on the right side of (1) converges and the formula (1) holds. [Use the fact, stated at the end of §2.3, that for every $\epsilon > 0$ there is a $\delta > 0$ such that $|A(x, y) - A(\bar{x}, \bar{y})| < \epsilon$ whenever (x, y), $(\bar{x}, \bar{y})$ are points of R such that $|x - \bar{x}| < \delta$, $|y - \bar{y}| < \delta$. (This is Theorem 2 of §9.4.) Then if $\sum(\alpha)$ is an approximating sum to the integral on the right side of (1) based on a subdivision of $\{c \leq y \leq d\}$ finer than δ, it follows that $\sum(\alpha)$ differs by less than $\epsilon(b - a)(d - c)$ from an approximating sum to $\int_R A\, dx\, dy$, which in turn differs by less than $\epsilon(b - a)(d - c)$ from $\int_R A\, dx\, dy$. Thus $\sum(\alpha) \to \int_R A\, dx\, dy$ as was to be shown.]

integration and differentiation

chapter 3

3.1

The Fundamental Theorem of Calculus

The evaluation of integrals in elementary calculus is accomplished by the Fundamental Theorem of Calculus, which can be stated as follows:

I. Let $F(t)$ be a function for which the derivative $F'(t)$ exists and is a continuous function for t in the interval $\{a \le t \le b\}$. Then

$$\int_a^b F'(t)\,dt = F(b) - F(a). \tag{1}$$

II. Let $f(t)$ be a continuous function on $a \le t \le b$. Then there exists a differentiable function $F(t)$ on $a \le t \le b$ such that $f(t) = F'(t)$.

Part I says that in order to evaluate a given integral *it suffices to write the integrand as a derivative* so that the desired integral is on the left side of equation (1) and a known number is on the right. For example, to compute the integral $\int_1^2 (1/t^2)\,dt$ of Exercise 1, §2.1, it suffices to write the integrand as

$$\frac{1}{t^2} = \frac{d}{dt}\left[-\frac{1}{t}\right]$$

so that (1) says

$$\int_1^2 \frac{d}{dt}\left[-\frac{1}{t}\right] dt = \left(-\frac{1}{2}\right) - \left(-\frac{1}{1}\right)$$

$$\int_1^2 \frac{1}{t^2}\, dt = \frac{1}{2}.$$

Part II says that theoretically this procedure always works, that is, theoretically any continuous integrand can be written as a derivative. Anyone who has been confronted with an integrand such as

$$f(t) = \frac{t}{\sin^3 t}$$

without the aid of a table of integrals, or

$$f(t) = \frac{1}{\sqrt{1 - k^2 \sin^2 t}}$$

with or without a table of integrals, knows how deceptive this statement is. In point of fact, II says little more than that *the (definite) integral of a continuous function over an interval converges*, which was already proved in §2.3 (see the proof of II below).

In the use of part I to evaluate integrals the right side is assumed known and the left side is thereby evaluated; the equation (1) can also be useful when read the other way around, i.e. when knowledge of the left side is used to draw conclusions about the right. For example, it says that if $F'(t)$ is identically zero then $F(b) - F(a) = 0$; that is, a *function whose derivative is zero is constant.* More generally, one value $F(a)$ of the function and all values $F'(t)$ of its derivative suffice to determine all other values

$$F(b) = \int_a^b F'(t)\, dt + F(a)$$

of the function.

Proof of I

The idea of the theorem is the following: Let $a = t_0 < t_1 < t_2 < \cdots < t_n = b$ be a subdivision of the given interval into small subintervals. Then

$$\begin{aligned}F(b) - F(a) &= [F(b) - F(t_{n-1})] + [F(t_{n-1}) - F(t_{n-2})] + \cdots \\ &\quad + [F(t_2) - F(t_1)] + [F(t_1) - F(a)] \\ &= \sum \Delta F = \sum \frac{\Delta F}{\Delta t} \Delta t\end{aligned}$$

where $\sum$ denotes a sum over all subintervals of the subdivision, and where for each subinterval $\{t_{i-1} \le t \le t_i\}$ the symbol ΔF denotes $F(t_i) - F(t_{i-1})$ and Δt denotes $t_i - t_{i-1}$. By the definition of 'derivative', the numbers $\frac{\Delta F}{\Delta t}$ are nearly $F'(t)$ for t in the interval; hence, by the definition of 'integral', the sum $\sum \frac{\Delta F}{\Delta t} \Delta t$ is nearly $\int_a^b F'(t)\,dt$. Since the sum $\sum \frac{\Delta F}{\Delta t} \Delta t$ is exactly equal to $F(b) - F(a)$, this is the statement to be proved.

To make this rough argument into a proof of I, one must estimate the error in the approximations

$$\int_{t_{i-1}}^{t_i} F'(t)\,dt \sim F'(t)\,\Delta t \sim \frac{\Delta F}{\Delta t} \Delta t = \Delta F.$$

In doing this it is helpful to divide the difference between $\int_{t_{i-1}}^{t_i} F'(t)\,dt$ and ΔF by Δt and to estimate

$$\frac{1}{\Delta t}\left\{\int_{t_{i-1}}^{t_i} F'(t)\,dt - \Delta F\right\}, \tag{2}$$

which can be thought of as the *average difference per unit length* between the numbers $\int_{t_{i-1}}^{t_i} F'(t)\,dt$ and $\Delta F = F(t_i) - F(t_{i-1})$. Assuming that the theorem is true, this average difference per unit length is of course zero for all subintervals $\{t_{i-1} \le t \le t_i\}$ and this is the statement to be proved. If the interval is further subdivided, then the maximum of this average, like any average, can only increase; that is, the average on at least one of the smaller intervals is as large as the average over the whole interval. since the limit of (2) as $\Delta t \to 0$ is $F'(t) - F'(t) = 0$ this observation will suffice to prove the theorem.

Specifically, for any r, s in the interval $a \le r < s \le b$ let $\mathcal{E}_{rs}$ denote

$$\mathcal{E}_{rs} = \frac{1}{s-r}\left\{\int_r^s F'(t)\,dt - [F(s) - F(r)]\right\}.$$

If c is the midpoint between a and b then

$$\begin{aligned}\mathcal{E}_{ab} &= \frac{1}{b-a}\left\{\int_a^b F'(t)\,dt - [F(b) - F(a)]\right\}\\ &= \frac{1}{b-a}\left\{\int_a^c F'(t)\,dt + \int_c^b F'(t)\,dt - [F(b) - F(c)] - [F(c) - F(a)]\right\}\\ &= \frac{c-a}{b-a}\mathcal{E}_{ac} + \frac{b-c}{b-a}\mathcal{E}_{cb}\\ &= \tfrac{1}{2}(\mathcal{E}_{ac} + \mathcal{E}_{cb}).\end{aligned}$$

Thus either $|\mathcal{E}_{ac}| \geq |\mathcal{E}_{ab}|$ or $|\mathcal{E}_{cb}| \geq |\mathcal{E}_{ab}|$; that is, the average error is at least as great on (at least) one of the two halves as it is on the whole interval. Dividing this half into halves and repeating the argument shows that there is a quarter of the original interval on which $|\mathcal{E}|$ is at least $|\mathcal{E}_{ab}|$. Continuing this process *ad infinitum* gives a sequence of intervals such that the ith interval is one of the halves of the $(i-1)$st (the first interval is $\{a \leq t \leq b\}$) and such that the average error $\mathcal{E}_i$ per unit length on the ith interval satisfies $|\mathcal{E}_i| \geq |\mathcal{E}_{i-1}|$. As $i \to \infty$ the intervals shrink down to a point,* say T, and $\mathcal{E}_i$ approaches

*The proof of this 'obvious' fact is very subtle. See §9.4 and Appendix 4.

$$\lim_{\Delta t \to 0} \frac{1}{\Delta t}\int F'(t)\,dt - \lim_{\Delta t \to 0}\frac{\Delta F}{\Delta t} = F'(T) - F'(T) = 0$$

by (v) of §2.6 and by the definition of the derivative $F'(T)$. Thus $|\mathcal{E}_i| \geq |\mathcal{E}_{ab}|$ for all i and $\lim_{i\to\infty} \mathcal{E}_i = 0$, which implies $\mathcal{E}_{ab} = 0$. This completes the proof of I.

Proof of II

Given a continuous function $f(t)$ on $a \leq t \leq b$, the integral defines a function

$$F(c) = \int_a^c f(t)\,dt \tag{3}$$

assigning numbers (the integral) to points c in the interval $a \leq c \leq b$. It is to be shown that the function F so defined is differentiable and that its derivative is f. But since

$$\frac{F(t_1) - F(t_0)}{t_1 - t_0} = \frac{1}{t_1 - t_0}\int_{t_0}^{t_1} f(t)\,dt,$$

this follows immediately from (v) of §2.6.

Statement II is confusing to many students because of a misunderstanding about the word 'function'. When one thinks of a function one unconsciously imagines a simple rule such as $F(t) = t^2$ or $F(t) = \sin\sqrt{t}$ which can be evaluated by simple computation, by consultation of a table, or, at worst, by a manageable machine computation. The function defined by (3) need not be a standard function at all, and *a priori* there is no reason to believe that it can be evaluated by any means other than by forming approximating sums and estimating the error as in the preceding chapter, in which case the function F on the right-hand side of (1) is just as difficult to evaluate as the integral on the left. The method "write the integrand as a derivative" is better stated "write the integrand as the derivative of a function *whose evaluation is easier than the evaluation of the given integral by direct computation.*" Only when this is possible does the equation (1) give a means of evaluating the integral.

Exercises *Interpretations of the Fundamental Theorem:*

1 Give physical descriptions of the function F and the 1-form $F'(t)\,dt$ in such a way that equation (1) gives two ways of finding the amount of work required to go from a to b. Compare to Exercise 1, §2.2.

2 Give a physical description of the equation (1) in which $F(t)$ is the flow across points of a flow on the line as in Exercise 4, §2.1. What is $F(b) - F(a)$? Give an interpretation of $F(b) - F(a)$ on small intervals ($\sim F'(t)\,dt$) in terms of the mass of the fluid. Describe equation (1) in terms of this model.

3 Let $F(t)$ be position as a function of time. Give physical descriptions of the function $F'(t)$, the 1-form $F'(t)\,dt$, and the equation (1).

4 Let $f(t)$ be a given function and let $A_{[a,b]}$ be the area under the curve, that is, the area between the graph $(t, f(t))$ of the function and the interval $[a, b]$ of the t-axis, counting area as negative if the curve lies below the t-axis. Fix t_0 and consider $A_{[t_0,b]} = F(b)$ as a function of b. Give a geometrical interpretation of the 1-form $F'(t)\,dt$, and hence of the function $F'(t)$. Describe each side of (1) as an expression for $A_{[a,b]}$. (This is an extremely awkward interpretation of the Fundamental Theorem. See Exercise 12, §3.2.)

Analogy of the principle "to integrate, write the integrand as a derivative" and the principle "to sum, write the summand as a difference."

In order to sum a finite series $f(1) + f(2) + f(3) + \cdots + f(n)$ in which the summand $f(n)$ is a function of n, it suffices to find a function $F(n)$ whose differences $F(n) - F(n-1)$ are equal to $f(n)$ since then $f(1) + f(2) + \cdots + f(n) = [F(1) - F(0)] + [F(2) - F(1)] + \cdots + [F(n) - F(n-1)] = F(n) - F(0)$. Given an $f(n)$ it is usually quite hard (much harder than finding antiderivatives) to find a function $F(n)$ such that $f(n) = F(n) - F(n-1)$. However, one can cheat and *start* with F, find its differences $F(n) - F(n-1)$, and see what series $f(n)$ it enables one to sum. This is the method of the following exercises:

5 Set $F(n) = \cos(nA + B)$ where A, B are fixed numbers. Use the formula for $\cos(\alpha + \beta) - \cos(\alpha - \beta)$ to simplify the difference $F(n) - F(n-1)$. Write the result in the form $C \sin(nA + D)$ and find the formula for $\sin \beta + \sin(\alpha + \beta) + \sin(2\alpha + \beta) + \cdots + \sin(n\alpha + \beta)$.

6 Find the formula for $\cos \beta + \cos(\alpha + \beta) + \cos(2\alpha + \beta) + \cdots + \cos(n\alpha + \beta)$.

7 Set $F(n) = r^n$ and show that this yields the sum of the series $1 + r + r^2 + \cdots + r^{n-1}$ for $r \neq 1$.

8 Setting $F(n) = n^2$ gives the sum of $f(n)$ where $f(n) = 2n + 1$. Since the sum of $1 + 1 + \cdots + 1$ is n this gives the familiar formula for $1 + 2 + 3 + \cdots + n$.

9 Set $F(n) = n^3$ and use the formula of Exercise 8 to find a formula for $1 + 2^2 + 3^2 + \cdots + n^2$.

10 Use the method of 9 to find the sum $1 + 2^3 + 3^3 + \cdots + n^3$.

11 Use the same method to show that the sum $1 + 2^k + 3^k + \cdots + n^k$ can be written in the form $\dfrac{n^{k+1}}{k+1} + \cdots$ where the omitted terms are multiples of $n^k, n^{k-1}, \ldots, n^0$.

12 Let $f(t)$ be a function on the interval $\{a \le t \le b\}$. Then by definition the integral $\int_a^b f(t)\, dt$ is equal to the limit of the sums $\sum f\left(\frac{j}{N}\right)\frac{1}{N}$ where the sum is over all integers j such that $a < j/N < b$, that is, such that $Na < j < Nb$, and where the limit is taken as $N \to \infty$. The formulas of Exercises 5–11 give explicit formulas for $\sum f\left(\frac{j}{N}\right)\frac{1}{N}$ when $f(t) = \sin t$, $f(t) = \cos t$, $f(t) = r^t$, $f(t) = t^n$. Use these formulas to write $\int_a^b f(t)\,dt$ as a limit as $N \to \infty$ in these cases. Evaluate the limit; the formulas

$$\lim_{N\to\infty} \frac{\sin\left(\frac{1}{N}\right)}{\frac{1}{N}} = 1 \qquad \lim_{N\to\infty} \frac{\sqrt[N]{r} - 1}{\frac{1}{N}} = \log r$$

are needed.

13 If the function $F(t)$ is monotone on $\{a \leq t \leq b\}$, i.e. if $y = F(t)$ establishes a one-to-one correspondence between the interval $\{a \leq t \leq b\}$ and an interval of the y-line, then part I of the Fundamental Theorem is a special case of the principle of independence of parameter (§2.5). Describe the relationship between the two.

3.2

The Fundamental Theorem in Two Dimensions

The generalization of the Fundamental Theorem* to two dimensions can have two very different forms, depending on the way that the one-dimensional theorem $\int_a^b F'(t)\,dt = F(b) - F(a)$ is interpreted. On the one hand there is the generalization "the difference $F(Q) - F(P)$ between two values of a function $F(x, y)$ on the xy-plane can be written as the integral of a 1-form over a curve from P to Q" and on the other hand there is the generalization "the integral of a 1-form $A\,dx + B\,dy$ around the boundary of a 2-dimensional region can be written as the integral of a 2-form over the region itself."

*In the remainder of this chapter the 'Fundamental Theorem' will mean just Part I of the Theorem.

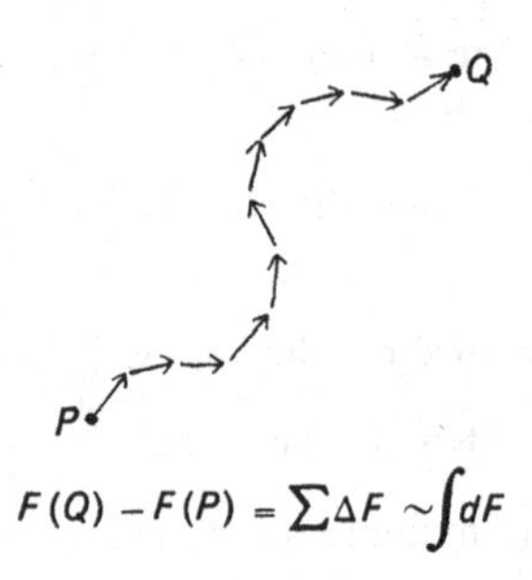

In the first generalization, a function $F(x, y)$ is given and it is claimed that if S is an oriented curve from P to Q in the xy-plane then $F(Q) - F(P)$ can be written as an integral over S, $\int_S (A\,dx + B\,dy)$, of some 1-form $A\,dx + B\,dy$ derived from F. Intuitively, the idea is that the curve S can be well approximated by a polygonal curve $P_0P_1P_2\ldots P_n$, where $P_0 = P$ and $P_n = Q$, in which the line segments $P_{i-1}P_i$ are very short. Then

$$\begin{aligned} F(Q) - F(P) &= [F(P_n) - F(P_{n-1})] + [F(P_{n-1}) - F(P_{n-2})] + \cdots + [F(P_1) - F(P_0)] \\ &= \sum \Delta F, \end{aligned}$$

where the sum is over the line segments $P_{i-1}P_i$ making up the polygonal curve and where $\Delta F = F(P_i) - F(P_{i-1})$. One should be able, for simple functions F, to find a 1-form $dF = A\,dx + B\,dy$ whose values on short line segments are nearly ΔF so that the above sum is nearly an approximating sum of $\int_S dF$. Assuming that there is such a 1-form $dF = A\,dx + B\,dy$, it is easy to see that A must be

$$A(x, y) = \frac{\partial F}{\partial x}(x, y) = \lim_{h \to 0} \frac{F(x + h, y) - F(x, y)}{h}$$

$$\left[F(x + h, y) - F(x, y) = \int_{(x,y)}^{(x+h,y)} (A\,dx + B\,dy) = \int_x^{x+h} A(x, y)\,dx \sim A(x, y) \cdot h\right];$$

similarly, $B = \dfrac{\partial F}{\partial y}$, hence dF must be $\dfrac{\partial F}{\partial x}\,dx + \dfrac{\partial F}{\partial y}\,dy$.

Thus the expected theorem would be

$$(1) \qquad F(Q) - F(P) = \int_S \left(\frac{\partial F}{\partial x}\,dx + \frac{\partial F}{\partial y}\,dy\right)$$

when S is an oriented curve from P to Q.

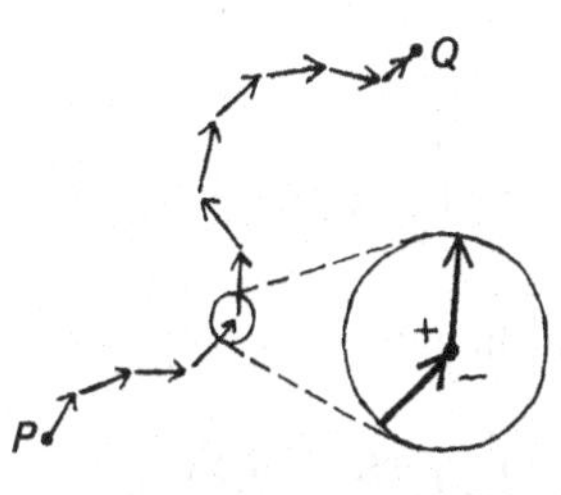

$F(Q) - F(P) = \sum \Delta F$

because of *cancellation at interior points*

In the second generalization, a 1-form $A\,dx + B\,dy$ is given and it is claimed that if S is an oriented 2-dimensional domain in the xy-plane with boundary ∂S oriented accordingly,* then $\int_{\partial S} (A\,dx + B\,dy)$ can be written as an integral over S, $\int_S C\,dx\,dy$, of some 2-form $C\,dx\,dy$ derived from $A\,dx + B\,dy$. Intuitively, the idea is that the domain S can be divided up into a large number of small polygons, say $S_1, S_2, \ldots, S_n$. Then

$$\int_{\partial S} (A\,dx + B\,dy) = \sum_{i=1}^{n} \int_{\partial S_i} (A\,dx + B\,dy)$$

because the interior boundaries cancel, that is, any line segment in the subdivision of S into $S_1, S_2, \ldots, S_n$ is counted twice with opposite orientations in the sum on the right and hence cancels out in the same way that $F(P_i) - F(P_i)$ cancelled out of the sum in the previous case. The idea, then, is that there should be a 2-form $C\,dx\,dy$ such that

$$\int_{\partial S_i} (A\,dx + B\,dy) \sim \int_{S_i} C\,dx\,dy$$

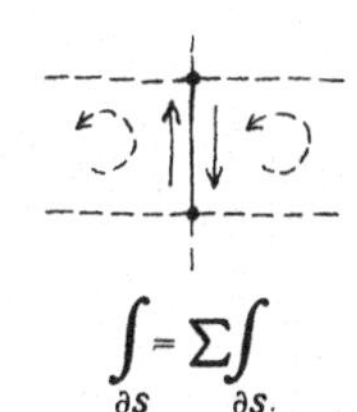

because of *cancellation on interior boundaries*

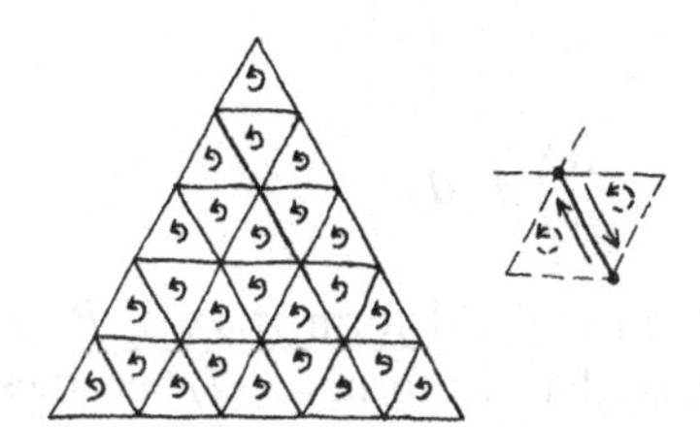

holds for small polygons S_i so that the right side above is nearly $\int_S C\,dx\,dy$. Assuming that there is such a 2-form $C\,dx\,dy$, it is easy to see what it must be; taking S_i to be

*The notation ∂S for 'the boundary of S' is standard and will be used throughout the book. It is not to be confused with the use of the symbol ∂ to denote partial differentiation. Note that an orientation of S gives an orientation to ∂S as well.

a rectangle $\{a \leq x \leq b, c \leq y \leq d\}$ oriented counterclockwise,

$$\int_{\partial S_i} (A\,dx + B\,dy) = \int_a^b A(x, c)\,dx + \int_c^d B(b, y)\,dy + \int_b^a A(x, d)\,dx + \int_d^c B(a, y)\,dy$$

$$= \int_c^d [B(b, y) - B(a, y)]\,dy - \int_a^b [A(x, d) - A(x, c)]\,dx$$

$$= \int_c^d \left[\int_a^b \frac{\partial B}{\partial x}(x, y)\,dx\right] dy - \int_a^b \left[\int_c^d \frac{\partial A}{\partial y}(x, y)\,dy\right] dx$$

(by the Fundamental Theorem)

$$= \int_{S_i} \frac{\partial B}{\partial x}\,dx\,dy - \int_{S_i} \frac{\partial A}{\partial y}\,dx\,dy$$

(by formula (1) of §2.6), and hence

$$\int_{\partial S_i} (A\,dx + B\,dy) = \int_{S_i} \left(\frac{\partial B}{\partial x} - \frac{\partial A}{\partial y}\right) dx\,dy.$$

Since this holds for all rectangles, the desired 2-form must be $C\,dx\,dy = \left(\frac{\partial B}{\partial x} - \frac{\partial A}{\partial y}\right) dx\,dy$. Thus the expected theorem would be

$$(2) \quad \int_{\partial S} (A\,dx + B\,dy) = \int_S \left(\frac{\partial B}{\partial x} - \frac{\partial A}{\partial y}\right) dx\,dy.$$

There is no need to memorize this formula because it is easily derived from the usual rules for computing with differentials:

$$d(A\,dx + B\,dy) = dA\,dx + dB\,dy = \left(\frac{\partial A}{\partial x}dx + \frac{\partial A}{\partial y}dy\right) dx + \left(\frac{\partial B}{\partial x}dx + \frac{\partial B}{\partial y}dy\right) dy$$

$$= \left(\frac{\partial B}{\partial x} - \frac{\partial A}{\partial y}\right) dx\,dy$$

(because $dA = \frac{\partial A}{\partial x}dx + \frac{\partial A}{\partial y}dy$ and because $dx\,dx = 0$, $dx\,dy = -dy\,dx$, $dy\,dy = 0$). The 2-form

$$\left(\frac{\partial B}{\partial x} - \frac{\partial A}{\partial y}\right) dx\,dy$$

is called the *derived 2-form* of the 1-form $A\,dx + B\,dy$, written $d(A\,dx + B\,dy)$, so that the formula (2) becomes

$$(2') \quad \int_{\partial S} (A\,dx + B\,dy) = \int_S d(A\,dx + B\,dy).$$

The formulas (1), (2) can be illustrated by physical ideas. In physics a potential function is a function with the property that $F(Q) - F(P)$ is equal to the amount of work required to go from the point P to the point Q.* Formula (1) says simply that if F is a potential function then the force field is described by the 1-form dF in the manner described in §2.1. [For example, $\frac{\partial F}{\partial x}(\bar{x}, \bar{y})$ is the work required per unit displacement in the x-direction near $(\bar{x}, \bar{y})$; this is minus the x-component of the force.] In the formula (2) it is useful to regard the given 1-form as describing a flow; in this case, however, it is natural to write the 1-form as $A\,dy - B\,dx$ so that A is the x-component of the flow and B the y-component (see Exercise 5, §1.1). Then the formula (2) becomes

*Except that physicists use the opposite sign and speak of the work done by the field rather than the work done by the mover of the particle. See §8.8.

$$\int_{\partial S} (A\,dy - B\,dx) = \int_S \left(\frac{\partial A}{\partial x} + \frac{\partial B}{\partial y}\right) dx\,dy.$$

The 2-form $\left(\frac{\partial A}{\partial x} + \frac{\partial B}{\partial y}\right) dx\,dy$ is also called the *divergence* of the flow represented by $A\,dy - B\,dx$ because it gives the rate at which the fluid is flowing out of small rectangles.

Formulas (1) and (2) require proof, of course, and conditions must be placed on S, F, A, B, etc., in order to ensure that they are true. As for Chapters 1 and 2, rigorous proofs will be postponed to Chapter 6 because of the technical difficulties which they present. However, some discussion of the proofs is useful at this point both because it sheds light on the meaning of the theorems and because it gives an idea of the sorts of difficulties involved in the proofs to come.

The statement of formula (1) assumes that the partial derivatives $\frac{\partial F}{\partial x}, \frac{\partial F}{\partial y}$ exist and are continuous functions of (x, y), and assumes as well that the curve S can be parameterized, $S = \{(x(t), y(t)): a \le t \le b\}$, by functions $x(t)$, $y(t)$ which have continuous derivatives. These assumptions are required in order for the definition of the number $\int_S \left(\frac{\partial F}{\partial x}\,dx + \frac{\partial F}{\partial y}\,dy\right)$ to apply. This number is then defined to be $\int_a^b \left(\frac{\partial F}{\partial x}\frac{dx}{dt} + \frac{\partial F}{\partial y}\frac{dy}{dt}\right) dt$. The number on the left side of (1) is

$$F(Q) - F(P) = F(x(b), y(b)) - F(x(a), y(a)) = \int_a^b \frac{d}{dt}\{F(x(t), y(t))\}\, dt$$

by the Fundamental Theorem. Equation (1) therefore follows from the *Chain Rule of Differentiation* with which the reader may already be familiar:

$$(3) \qquad \frac{d}{dt}\{F(x(t), y(t))\} = \frac{\partial F}{\partial x}\frac{dx}{dt} + \frac{\partial F}{\partial y}\frac{dy}{dt}.$$

This formula of differential calculus, which is proved in §5.3, therefore implies the theorem (1).

Theorem (2) was actually proved above in the case where S is a rectangle. This case is particularly simple because there is a natural way to parameterize the boundary of a rectangle—namely by the coordinate functions—so that the number $\int_{\partial S} (A\,dx + B\,dy)$ can be written in a very explicit form. The essential difficulty in proving (2) for 'arbitrary' domains S is simply that the number $\int_{\partial S} (A\,dx + B\,dy)$ has not been satisfactorily defined (see §2.5). When it is defined in Chapter 6 it will be defined in such a way that the general formula (2) is reduced to the case where S is a rectangle, which was proved above.

Exercises **1** Sketch the following flows. Indicate with a '+' those regions of the plane where the flow is diverging, and with a '−' those where it is converging. Compute the divergence in each case:

(a) dx (b) $y\,dx$ (c) $x\,dx$
(d) $x^2\,dy$ (e) $x\,dx + y\,dy$ (f) $x\,dy - y\,dx$.

2 What would the divergence of flow from a source at the origin (see Exercise 4, §2.2) be expected to be? Check by computation. Since the 1-form describing this flow is not defined at the origin, neither is its divergence; therefore the flow across the boundary of a domain containing the origin cannot be described as the integral of the divergence over the domain. What is the rate of flow across the boundary of an oriented domain which contains the origin? Which does not contain the origin?

3 Show how the function $F(x, y) = \frac{1}{r} = (x^2 + y^2)^{-1/2}$ can be used in finding the amount of work required for displacements in a central force field (see Exercise 1, §2.2).

4 Use the Fundamental Theorem to prove the formula (1) in cases where S is a horizontal line segment parameterized by x or a vertical line segment parameterized by y.

5 If a flow is described by the 1-form $x\,dx + y\,dy$ find the rate of flow across

(a) the line segment from $(0, 0)$ to (x_0, y_0);
(b) the broken line segment from $(0, 0)$ to $(x_0, 0)$ to (x_0, y_0); and
(c) the parabolic arc (t^2x_0, ty_0), $0 \le t \le 1$,

by direct computation. Give a physical explanation of the result.

6 Show that if a flow is described by a (continuous) 1-form of the type $\frac{\partial F}{dx}\,dx + \frac{\partial F}{dy}\,dy$ then the flow across the boundary of any oriented rectangle is zero. Use formula (2) to conclude that if F is a function such that $\frac{\partial}{\partial x}\left(\frac{\partial F}{\partial y}\right)$, and $\frac{\partial}{\partial y}\left(\frac{\partial F}{\partial x}\right)$ both exist and are continuous then

$$\frac{\partial}{\partial x}\left(\frac{\partial F}{\partial y}\right) \equiv \frac{\partial}{\partial y}\left(\frac{\partial F}{\partial x}\right).$$

This theorem is referred to as the *equality of the mixed partials.*

7 Given a 1-form $A\,dx + B\,dy$, the preceding exercise shows that if $A\,dx + B\,dy = \frac{\partial F}{dx}\,dx + \frac{\partial F}{\partial y}\,dy$ for some function F, then $\frac{\partial A}{\partial y} = \frac{\partial B}{\partial x}$. Prove that if A, B are defined at all points of the plane and satisfy $\frac{\partial A}{\partial y} = \frac{\partial B}{\partial x}$ then, conversely, $A\,dx + B\,dy = dF$ for some function F. [Following the proof of part II of the fundamental theorem, *define* $F_1(x, y)$ to be the integral of $A\,dx + B\,dy$ from $(0, 0)$ to $(x, 0)$ to (x, y) and $F_2(x, y)$ to be the integral from $(0, 0)$ to $(0, y)$ to (x, y). Show that $F_1 \equiv F_2$ and that the function they define has the right partial derivatives.]

8 A 1-form $A\,dx + B\,dy$ is called *closed* if $\frac{\partial A}{\partial y} = \frac{\partial B}{\partial x}$. It is called *exact* if there is a function F such that $\frac{\partial F}{\partial x} = A$, $\frac{\partial F}{\partial y} = B$. Exercise 6 shows that every exact form is closed, while Exercise 7 shows that every closed form is exact provided it and its derived form are defined everywhere. Determine which of the following forms are closed, which are exact, and find functions F for those which are exact.

(a) $(x + y)^2\,dx + (x + y)^2\,dy$
(b) $x\,dy + y\,dx$
(c) $x\,dx + y\,dy$
(d) $(ye^{xy}\cos y)\,dx + (xe^{xy}\cos y - e^{xy}\sin y)\,dy$
(e) $(\log xy + 1)\,dx + \left(\dfrac{x}{y}\right)dy$
(f) $\dfrac{x\,dx + y\,dy}{x^2 + y^2}$
(g) $\dfrac{y\,dx - x\,dy}{x^2 + y^2}$

9 A force field $A\,dx + B\,dy$ is called *conservative* if the amount of work required for displacement around any closed path is zero. Show that if $A\,dx + B\,dy$ is conservative, then $\dfrac{\partial A}{\partial y} = \dfrac{\partial B}{\partial x}$. Is the converse true? Show that $A\,dx + B\,dy$ is conservative if and only if there is a function F such that $\dfrac{\partial F}{\partial x} = A$, $\dfrac{\partial F}{\partial y} = B$, that is, if and only if the force field can be described by a potential function F.

10 'Express in polar coordinates' the 1-forms

$$dx,\ dy,\ x\,dy,\ x\,dy - y\,dx,\ \frac{x\,dx + y\,dy}{x^2 + y^2},\ \frac{x\,dy - y\,dx}{x^2 + y^2};$$

that is, find the pullbacks under the map $x = r\cos\theta$, $y = r\sin\theta$. Which of these forms are closed and which are exact? Which of the pullbacks are closed and which are exact? What would you expect the 'expression in polar coordinates' of flow from a source (Exercise 2) to be? Verify the answer.

11 Let $A\,dx + B\,dy$ be a 1-form defined on the disk $D = \{(x, y): x^2 + y^2 \leq 1\}$ oriented counterclockwise, and assume that $\dfrac{\partial A}{\partial x}, \dfrac{\partial A}{\partial y}, \dfrac{\partial B}{\partial x}, \dfrac{\partial B}{\partial y}$ are all defined and continuous on D. Convert $\int_{\partial D}(A\,dx + B\,dy)$ to an integral over the boundary of the rectangle $\{0 \leq \theta \leq 2\pi, 0 \leq r \leq 1\}$ in polar coordinates. Use the formula (2) to convert this to the integral of a 2-form in r and θ over the entire rectangle. Simplify by using the chain rule to write $\dfrac{\partial A}{\partial r}$ in terms of $\dfrac{\partial A}{\partial x}$ and $\dfrac{\partial A}{\partial y}$, etc. Compare the result to $\displaystyle\int_D\left(\frac{\partial B}{\partial x} - \frac{\partial A}{\partial y}\right)dx\,dy$. Conclude that

$$\int_{\partial D} A\,dx + B\,dy = \int_D\left(\frac{\partial B}{\partial x} - \frac{\partial A}{\partial y}\right)dx\,dy$$

is valid when polar coordinates are used to define these integrals.

12 Let S be a curve in the plane parameterized by the coordinate x, $S = \{(x, f(x)): a \leq x \leq b\}$ where f is a differentiable function. Use $d(y\,dx) = dy\,dx = -dx\,dy$ to interpret $\int_S y\,dx$ as an oriented area. Express this integral in terms of x. (This is the interpretation of the integral as 'the area under a curve'. The widespread idea that an integral 'is' the area under a curve is very unfortunate because it completely obscures the meaning of the Fundamental Theorem. An integral 'is' the limit of a sum, and area 'is' a *double* integral.)

13 Approximating the domain S by an n-sided polygon and passing to the limit as $n \to \infty$, show that the formula of Exercise 5(b), §1.3, becomes

$$\int_S dx\,dy = \int_{\partial S} \tfrac{1}{2}(x\,dy - y\,dx)$$

which is a special case of (2). Apply this formula to the case where S is the unit disk by using polar coordinates (Exercise 11).

14 Prove the formula

$$\int_{-\infty}^{\infty} \frac{du}{1 + u^2} = \pi$$

by applying the formula of Exercise 13 to the circle parameterized by stereographic projection (see Exercise 3, §2.4).

3.3 *The Fundamental Theorem in Three Dimensions*

The extension of the ideas of §3.2 to three dimensions is immediate. If $F(x, y, z)$ is a function on xyz-space and if S is an oriented curve from P to Q then S can be approximated by a polygonal curve consisting of short line segments and $F(Q) - F(P)$ can be written as $\sum \Delta F$ by cancellation on interior boundaries, the boundaries in this case being points. Passing to the limit, the formula $F(Q) - F(P) = \sum \Delta F$ becomes

$$F(Q) - F(P) = \int_S dF \tag{1}$$

where dF is a 1-form derived from F. If $A\,dx + B\,dy + C\,dz$ is a 1-form on xyz-space and if S is an oriented surface with boundary curve ∂S then S can be approximated by a polygonal surface consisting of small polygons

S_i and $\int_{\partial S} (A\,dx + B\,dy + C\,dz)$ can be written as $\sum \int_{\partial S_i} (A\,dx + B\,dy + C\,dz)$ by cancellation on the interior boundaries, the boundaries in this case being small curves. Passing to the limit in this formula, it becomes

$$\int_{\partial S} (A\,dx + B\,dy + C\,dz) = \int_S d(A\,dx + B\,dy + C\,dz) \tag{2}$$

where $d(A\,dx + B\,dy + C\,dz)$ is a 2-form derived from $A\,dx + B\,dy + C\,dz$. Finally, if $A\,dy\,dz + B\,dz\,dx + C\,dx\,dy$ is a 2-form on xyz-space and if S is an oriented solid with boundary surface ∂S, then S can be divided into a large number of small polyhedra S_i and $\int_{\partial S} (A\,dy\,dz + B\,dz\,dx + C\,dx\,dy)$ can be rewritten as $\sum \int_{\partial S_i} (A\,dy\,dz + B\,dz\,dx + C\,dx\,dy)$ by cancellation on the interior boundaries, the boundaries in this case being small surfaces. Passing to the limit in this formula it becomes

$$\int_{\partial S} (A\,dy\,dz + B\,dz\,dx + C\,dx\,dy) = \int_S d(A\,dy\,dz + B\,dz\,dx + C\,dx\,dy) \tag{3}$$

where $d(A\,dy\,dz + B\,dz\,dx + C\,dx\,dy)$ is a 3-form derived from $A\,dy\,dz + B\,dz\,dx + C\,dx\,dy$.

In the formula (1) the derived 1-form dF is of course $dF = \frac{\partial F}{\partial x}dx + \frac{\partial F}{\partial y}dy + \frac{\partial F}{\partial z}dz$. The proof of the formula for parameterized curves S reduces immediately, using the Fundamental Theorem, to the Chain Rule of Differentiation

$$\frac{dF}{dt} = \frac{\partial F}{\partial x}\frac{dx}{dt} + \frac{\partial F}{\partial y}\frac{dy}{dt} + \frac{\partial F}{\partial z}\frac{dz}{dt}$$

where $x = x(t)$, $y = y(t)$, $z = z(t)$ and where F is a function of t by composition $F(x(t), y(t), z(t))$. The Chain Rule, and hence (1), is proved in Chapter 5.

The formula (3) presupposes that ∂S is an oriented surface, i.e. that a convention has been established for orienting the bounding surface of an oriented solid. Geometrically, the usual convention for doing this is given by the rule "the boundary of a right-handed solid is oriented to be counterclockwise as seen from the outside and the boundary of a left-handed solid is oriented in the opposite way." Analytically, an orientation of a surface is a rule for deciding whether a triple of nearby points $P_0P_1P_2$ on the surface which are not collinear are 'positively' or 'negatively' oriented. The convention above

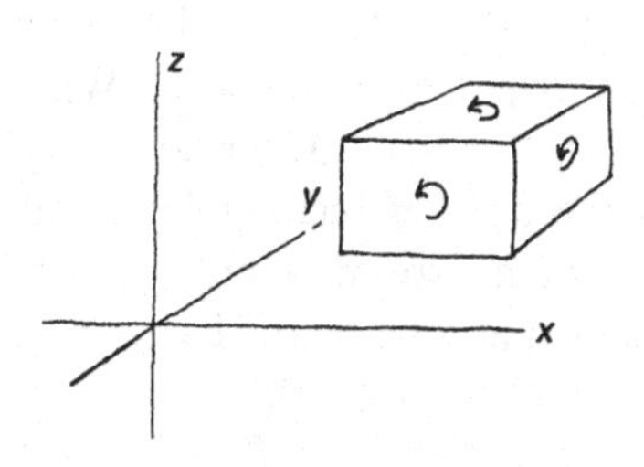

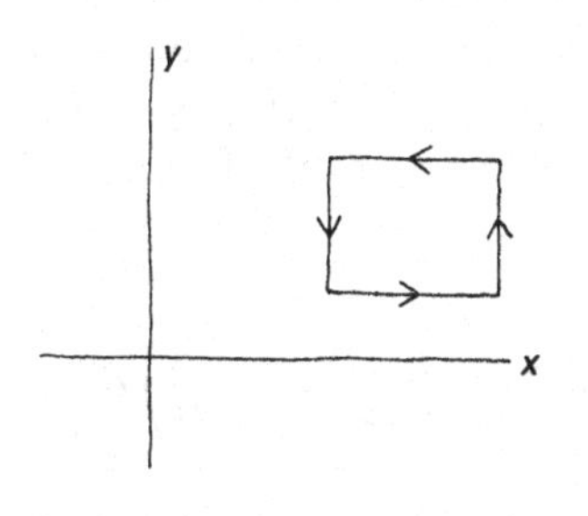

states that $P_0P_1P_2$ is positive if the orientation of $P_0P_4P_1P_2$ agrees with the orientation of S when P_4 is a nearby point *outside* of S.

An alternative statement of this convention for orienting the boundary of an oriented solid is the following: If S is a rectangular parallelepiped $\{a \le x \le b, c \le y \le d, e \le z \le f\}$ oriented by $dx\,dy\,dz$ (right-handed), then the side $x = b$ of S is oriented by $dy\,dz$ (counterclockwise as seen from $x > b$) and the remaining sides are oriented accordingly. Stated in this way the convention is seen as a generalization of the rule for two dimensions: If S is a rectangle $\{a \le x \le b, c \le y \le d\}$ oriented by $dx\,dy$ then the side $x = b$ of S is oriented by dy. (The orientation $dy\,dz$ of a plane $x =$ const. is established by a triple of points $(\bar{x}, \bar{y}, \bar{z})$, $(\bar{x}, \bar{y} + 1, \bar{z})$, $(\bar{x}, \bar{y}, \bar{z} + 1)$ in that order, the orientation dy of a line $x =$ const., $z =$ const. by $(\bar{x}, \bar{y}, \bar{z})$, $(\bar{x}, \bar{y} + 1, \bar{z})$, the orientation $dx\,dy\,dz$ of space by $(\bar{x}, \bar{y}, \bar{z})$, $(\bar{x} + 1, \bar{y}, \bar{z})$, $(\bar{x}, \bar{y} + 1, \bar{z})$, $(\bar{x}, \bar{y}, \bar{z} + 1)$, etc.)

Using this convention, the integral of $A\,dy\,dz$ over the boundary of a right-handed rectangular parallelepiped $\{a \le x \le b, c \le y \le d, e \le z \le f\}$ is the integral of the 2-form $[A(b, y, z) - A(a, y, z)]\,dy\,dz$ over the rectangle $\{c \le y \le d, e \le z \le f\}$ oriented $dy\,dz$. Writing $A(b, y, z) - A(a, y, z) = \int_a^b \frac{\partial A}{\partial x}(x, y, z)\,dx$ by the Fundamental Theorem and applying the 3-dimensional analog of formula (1) of §2.6 gives then

$$\int_{\partial S} A\,dy\,dz = \int_S \frac{\partial A}{\partial x}\,dx\,dy\,dz.$$

(This formula is the real reason for the orientation convention above; see Exercise 2, §3.4.) Similarly

$$\int_{\partial S} B\,dz\,dx = \int_S \frac{\partial B}{\partial y}\,dy\,dz\,dx, \quad \int_{\partial S} C\,dx\,dy = \int_S \frac{\partial C}{\partial z}\,dz\,dx\,dy,$$

and, adding these,

$$\int_{\partial S} (A\,dy\,dz + B\,dz\,dx + C\,dx\,dy) = \int_S \left(\frac{\partial A}{\partial x} + \frac{\partial B}{\partial y} + \frac{\partial C}{\partial z}\right) dx\,dy\,dz.$$

This shows that the derived form in (3) must be defined by

$$(4) \quad d(A\,dy\,dz + B\,dz\,dx + C\,dx\,dy) = \left(\frac{\partial A}{\partial x} + \frac{\partial B}{\partial y} + \frac{\partial C}{\partial z}\right) dx\,dy\,dz.$$

Then *formula* (3) *is proved for the case where S is a rectangular parallelepiped.* To prove (3) for more general oriented solids S the essential difficulty is to define the left-hand side (integration over the boundary). When this is done (in Chapter 6) the formula (3) for 'arbitrary' solids S will follow from the case of rectangular parallelepipeds, which has just been proved. Until then the formula (3) should be accepted as true on the basis of the intuitive argument by which it was derived.

To find the derived 2-form $d(A\,dx + B\,dy + C\,dz)$ which should appear on the right side of (2) it suffices to consider cases in which S is a rectangle in a coordinate direction — $S = \{a \le x \le b,\ c \le y \le d,\ z = \text{const.}\}$ oriented $dx\,dy$, for instance. In this case the formula (2) of §3.2 applies and gives

$$\int_{\partial S} (A\,dx + B\,dy + C\,dz) = \int_{\partial S} (A\,dx + B\,dy) = \int_S \left(\frac{\partial B}{\partial x} - \frac{\partial A}{\partial y}\right) dx\,dy;$$

hence the $dx\,dy$-component of $d(A\,dx + B\,dy + C\,dz)$ must be $\left(\frac{\partial B}{\partial x} - \frac{\partial A}{\partial y}\right) dx\,dy$. Applying the same argument in the zx- and yz-directions gives the other components and hence the formula

$$(5)\quad d(A\,dx + B\,dy + C\,dz) = \left(\frac{\partial C}{\partial y} - \frac{\partial B}{\partial z}\right) dy\,dz + \left(\frac{\partial A}{\partial z} - \frac{\partial C}{\partial x}\right) dz\,dx + \left(\frac{\partial B}{\partial x} - \frac{\partial A}{\partial y}\right) dx\,dy.$$

Note that this formula, like formula (4), need not be memorized because it can be derived by the usual rules. For example, taking B and C to be zero for the sake of simplicity, one obtains

$$\begin{aligned} d(A\,dx) &= dA\,dx \\ &= \left(\frac{\partial A}{\partial x}\,dx + \frac{\partial A}{\partial y}\,dy + \frac{\partial A}{\partial z}\,dz\right) dx \\ &= \frac{\partial A}{\partial z}\,dz\,dx - \frac{\partial A}{\partial y}\,dx\,dy. \end{aligned}$$

Thus in all three formulas (1), (2), (3) the derived forms on the right can be said to be 'found by the usual rules'.

Formula (2) is the most difficult of the three to prove, because in this case *neither* side of the equation has been satisfactorily defined. Taking S to be a surface with a specific parametric description, $S = \{(x(u, v), y(u, v), z(u, v)) \colon (u, v) \text{ in } D\}$, the integral on the right side of (2) becomes the integral of a 2-form in uv over D, while the integral on the left becomes the integral of a 1-form in uv over the bounding curve ∂D. If these two integrals are always equal, then it must be true that the 2-form is the derived form of the 1-form; that is, *the derived form of the pullback must be equal to the pullback of the derived form.* If this is shown to be true, then, just as the formula (1) was reduced to the Fundamental Theorem, the formula (2) will be reduced to the formula (2) of §3.2 (when the S of (2) is parameterized). Taking $B = C = 0$ for the sake of simplicity, the desired formula is that d of the pullback of $A\,dx$ is the pullback of $d(A\,dx)$. This can be proved by writing

$$\begin{aligned} d(A\,dx) &= dA\,dx \\ &= \left(\frac{\partial A}{\partial x}\,dx + \frac{\partial A}{\partial y}\,dy + \frac{\partial A}{\partial z}\,dz\right)dx \\ &= \left[\frac{\partial A}{\partial x}\left(\frac{\partial x}{\partial u}\,du + \frac{\partial x}{\partial v}\,dv\right) + \cdots\right]dx \\ &= \left[\left(\frac{\partial A}{\partial x}\frac{\partial x}{\partial u} + \frac{\partial A}{\partial y}\frac{\partial y}{\partial u} + \frac{\partial A}{\partial z}\frac{\partial z}{\partial u}\right)du + (\cdots)\,dv\right]dx. \end{aligned}$$

Using the Chain Rule

$$\frac{\partial A}{\partial x}\frac{\partial x}{\partial u} + \frac{\partial A}{\partial y}\frac{\partial y}{\partial u} + \frac{\partial A}{\partial z}\frac{\partial z}{\partial u} = \frac{\partial A}{\partial u}, \tag{6}$$

and the analogous formula for $\frac{\partial A}{\partial v}$ this becomes

$$\begin{aligned} d(A\,dx) &= \left[\frac{\partial A}{\partial u}\,du + \frac{\partial A}{\partial v}\,dv\right]\left[\frac{\partial x}{\partial u}\,du + \frac{\partial x}{\partial v}\,dv\right] \\ &= \left(\frac{\partial A}{\partial u}\frac{\partial x}{\partial v} - \frac{\partial A}{\partial v}\frac{\partial x}{\partial u}\right)du\,dv. \end{aligned}$$

On the other hand, the pullback of $A\,dx$ is

$$A(u, v)\left(\frac{\partial x}{\partial u}\,du + \frac{\partial x}{\partial v}\,dv\right)$$

and the derived form of this is

$$\left\{\frac{\partial}{\partial u}\left(A\frac{\partial x}{\partial v}\right) - \frac{\partial}{\partial v}\left(A\frac{\partial x}{\partial u}\right)\right\}du\,dv$$

which is the above plus

$$A\left(\frac{\partial^2 x}{\partial u\,\partial v} - \frac{\partial^2 x}{\partial v\,\partial u}\right)du\,dv.$$

But if the function $x(u, v)$ is assumed to have continuous second partial derivatives then by Exercise 6 of §3.2 it follows that $d(dx) = 0$; that is,

$$\frac{\partial^2 x}{\partial u \partial v} = \frac{\partial^2 x}{\partial v \partial u},$$

and d of the pullback of $A\,dx$ is identical to the pullback of $d(A\,dx)$.

This is the method by which the formula (2) will be proved in Chapter 6. Using the Chain Rule (6) (proved in Chapter 5) it follows that the pullback of the derived form is the derived form of the pullback (when the map is twice continuously differentiable) so that the desired formula is reduced to a previous case (formula (2) of §3.2). This in turn can be reduced to the case where S is a rectangle, which has already been proved above.

Exercises

1 Find the derived 3-form of the 2-form

$$\frac{x\,dy\,dz + y\,dz\,dx + z\,dx\,dy}{(x^2 + y^2 + z^2)^{3/2}}.$$

Interpret the result in terms of Exercise 3, §2.1.

2 The formula (3) implies that the volume of a solid S is equal to

$$\int_{\partial S} \frac{x\,dy\,dz + y\,dz\,dx + z\,dx\,dy}{3}$$

when ∂S is appropriately oriented. Use this formula to find:

(a) the volume of the unit sphere using spherical coordinates.

(b) the volume of the unit sphere using the coordinates of stereographic projection. (Exercise 3, §2.4. Evaluate the resulting double integral by converting to polar coordinates.)

(c) the volume of the torus of Exercise 7, §2.5.

3 Given a 3-dimensional region D in xyz-space, imagine that ∂D is a rigid body (say a metal shell) and that D is filled with a gas under pressure. Then the x-component of the force exerted by the gas on any piece of ∂D is proportional to the oriented area of its projection on the yz-plane. Show that the total x-component of the force on ∂D is zero.

4 Redo the computation of Exercise 11, §3.2 using the method of the text to prove that the pullback of the derived form is the derived form of the pullback.

5 Faraday's Law of Induction says that if the electric force field is given by the 1-form $E_1\,dx + E_2\,dy + E_3\,dz$ and if the magnetic flux is given by the 2-form $H_1\,dy\,dz + H_2\,dz\,dx + H_3\,dx\,dy$ where $E_1, E_2, \ldots, H_3$ are functions of x, y, z, t, then there is a constant k such that

$$\frac{d}{dt}\int_S H = k\int_{\partial S} E$$

for any surface S. Use (2) to state this as an equation involving the partial derivatives of E and H.

6 Any 2-form on space can be pictured in terms of 'lines of force' such that the integral of the 2-form over any surface is equal to the number of lines which cross the surface.* How would flow from a unit source at the origin (Exercise 1) be represented in this way? What is the description of the derived form of the 2-form in terms of lines of force and their endings? How is the fact that $dH = 0$ reflected in the picture of the lines of magnetic force? The electric displacement is represented by a 2-form $E_1\,dy\,dz + E_2\,dz\,dx + E_3\,dx\,dy$ whose derived form is equal to the charge density (a 3-form). Describe this in terms of lines of force.

*See §8.7

7 As in Exercise 8 of the preceding section, a 1-form on space is called *exact* if it can be written in the form

$$\frac{\partial F}{\partial x}dx + \frac{\partial F}{\partial y}dy + \frac{\partial F}{\partial z}dz$$

for some function $F(x, y, z)$. A 1-form $A\,dx + B\,dy + C\,dz$ on space is called *closed* if the derived form

$$\left(\frac{\partial C}{\partial y} - \frac{\partial B}{\partial z}\right)dy\,dz + \left(\frac{\partial A}{\partial z} - \frac{\partial C}{\partial x}\right)dz\,dx + \left(\frac{\partial B}{\partial x} - \frac{\partial A}{\partial y}\right)dx\,dy$$

is zero. Prove that an exact 1-form is always closed

(a) by computation using the equality of mixed partials, and

(b) by arguing directly from the geometrical meaning of the derived form.

8 A 2-form on space is called *exact* if it is the derived form

$$\left(\frac{\partial C}{\partial y} - \frac{\partial B}{\partial z}\right)dy\,dz + \left(\frac{\partial A}{\partial z} - \frac{\partial C}{\partial x}\right)dz\,dx + \left(\frac{\partial B}{\partial x} - \frac{\partial A}{\partial y}\right)dx\,dy$$

of some 1-form $A\,dx + B\,dy + C\,dz$. It is called *closed* if its derived form is zero. Prove that an exact 2-form is always closed

(a) by computation using the equality of mixed partials, and

(b) by arguing directly from the meaning of the derived form.

9 Give an example of a 1-form in three variables which is closed but not exact. Of a 2-form which is closed but not exact. Show that a closed 1-form which is defined at all points of space is exact. [Use the method of Exercise 7, §3.2.]

10 Given a closed 2-form on space define a 1-form on space by saying that the value on a short line segment PQ is the integral of the given 2-form over the triangle OPQ (O = origin). Argue geometrically that this will prove that a closed 2-form defined on all of space is exact. Use this method to show that the 2-form $dy\, dz$ is exact.

3.4

Summary. Stokes' Theorem

All the versions of the Fundamental Theorem stated above can be summarized by this statement: *The integral of a k-form over the boundary of a* $(k + 1)$*-dimensional domain is equal to the integral over the domain itself of the derived* $(k + 1)$*-form found by the rules previously described.* (These are reviewed below.) Intuitively the idea is that the given $(k + 1)$-dimensional domain S can be decomposed into a large number of very small $(k + 1)$-dimensional pieces and the integral over the boundary can therefore (by cancellation on the interior boundaries) be written as the sum of the integrals over the boundaries of the pieces. This sum, which is a sum over all pieces of a very fine subdivision of S, is the sort of sum whose limits define integrals over S. Passing to the limit, the integral over ∂S becomes an integral over S.

Terminology and notation

It is traditional to denote a k-form by the single Greek letter ω (omega). Then the Fundamental Theorem is simply

$$\int_{\partial S} \omega = \int_S d\omega \tag{1}$$

which is called *Stokes' Formula* or *Stokes' Theorem.* The Stokes' Formula (1) includes as a special case the

Fundamental Theorem of Calculus ((1) of §3.1) if a function $F(t)$ is regarded as a '0-form' and if the 'integral' of a 0-form over the boundary of the oriented interval from a to b is defined to be $F(b) - F(a)$.

Stokes' Formula (1) is called by many different names. When $k = 0$ it is called the Fundamental Theorem of Calculus. The case (2) of §3.2 is called Green's Theorem in the Plane. The case (3) of §3.3 is called the Divergence Theorem or Gauss' Theorem. The case (2) of §3.3 is called Stokes' Theorem. As was mentioned in §3.3, the formula is most difficult to prove in the case of Stokes' Theorem (i.e. (2) of §3.3), which is the reason that the general formula (1) takes its name from this case.

A k-form is also called a *differential form*, or an *exterior differential form* to distinguish these forms from quadratic forms, bilinear forms, homogeneous forms, and other sorts of forms which occur in the mathematical vocabulary. The derived $(k + 1)$-form $d\omega$ of a k-form ω is also called the *differential* of ω (particularly in the case $k = 0$) or the *exterior derivative* of ω. The simplest terminology is just to call it $d\omega$ (read 'dee-omega') and this is the terminology which will be used in the remainder of this book.

The rules by which $d\omega$ is defined are the rules

$$dA = \frac{\partial A}{\partial x}\,dx + \frac{\partial A}{\partial y}\,dy + \frac{\partial A}{\partial z}\,dz$$

(where $A = A(x, y, z)$ is a function or '0-form' in three variables)

$$\begin{aligned} d(A\,dx + \cdots) &= dA\,dx + \cdots \\ d(A\,dy\,dz + \cdots) &= dA\,dy\,dz + \cdots \\ dx\,dx &= 0 \\ dx\,dy &= -dy\,dx, \text{ etc.} \end{aligned}$$

A k-form ω is said to be *differentiable* if the coefficient functions A, B, etc. have continuous first derivatives. In the statement of Stokes' Formula (1) above it is tacitly assumed that the k-form ω is differentiable.

The rigorous proof of Stokes' Theorem (1) must await a rigorous definition of the integrals involved. If S is a $(k + 1)$-dimensional rectangle lying in a coordinate direction, then S and all $2k$ pieces of ∂S are explicitly parameterized by the coordinate functions, so the integrals have been rigorously defined; in these cases Stokes' Formula (1) has been rigorously proved above.

Using the Chain Rule of Differentiation it is not difficult (Exercises 7, 8, 9, 10) to prove the formula for simple $(k + 1)$-dimensional polygons S (line segments, triangles, tetrahedra), which implies that the formula holds for polygonal curves, polygonal surfaces, and polyhedra S (when $k = 0, 1, 2$ respectively). Given an 'arbitrary' S and given a finely divided polygonal approximation $\hat{S}$ of S (so that $\partial\hat{S}$ is a finely divided polygonal approximation of ∂S) this implies $\int_{\partial\hat{S}} \omega = \int_{\hat{S}} d\omega$ and hence as $\hat{S} \to S$ it implies $\int_{\partial S} \omega = \int_S d\omega$ provided that the intuitive definitions of these integrals in terms of limits of polygonal approximations (§2.2) are valid.

Exercises

1 Find $d\omega$ for the following forms ω. [Note: It is not necessary to indicate the domain of definition of a form in order to differentiate it—e.g., $d(x\,dy) = dx\,dy$ whether $x\,dy$ is thought of as a form on the xy-plane or on xyz-space.]

(a) $xy\,dz + yz\,dx + zx\,dy$
(b) $x\,dy\,dz + y\,dz\,dx + z\,dx\,dy$
(c) e^{xyz}
(d) $(\cos x)\,dy + (\sin x)\,dz$
(e) $(x + y)^2\,dy + (x + y)^2\,dz$
(f) $\log x$ (g) $\sin x$
(h) x^2 (i) x

2 If R is an oriented line segment then $\int_{\partial R} x$ and $\int_R dx$ are equal *by definition* of dx, and in particular have the same sign. Similarly, the sign of $dx\,dy$ can be defined by the equation $\int_{\partial R} x\,dy = \int_R dx\,dy$; draw the coordinate axes in both possible ways and show that this definition agrees with the one previously given. Once $dy\,dz$ has been defined then the sign of $dx\,dy\,dz$ is determined by the equation $\int_{\partial R} x\,dy\,dz = \int_R dx\,dy\,dz$. Describe the orientation of a rectangular parallelepiped in the coordinate directions which is positive with respect to $dx\,dy\,dz$ by describing the positive orientation of one of its faces.

3 Show that $d(d\omega) = 0$ corresponds geometrically to the statement that a boundary has no boundary. Prove analytically that $d(d\omega) = 0$. [It suffices to show that $d(dA) = 0$ for functions (0-forms) A, which was proved in Exercise 6, §3.2.]

4 The function

$$F(x, y) = \begin{cases} 2xy\dfrac{x^2 - y^2}{x^2 + y^2} & (x, y) \neq (0, 0) \\ 0 & (x, y) = (0, 0) \end{cases}$$

is continuous at $(0, 0)$ because the quotient of two homogeneous polynomials [in this case $2xy(x^2 - y^2)$ and $x^2 + y^2$] in which the denominator is never zero has the limit zero at the origin whenever the degree of the numerator [in this case, 4] is greater than that of the denominator [in this case, 2]. Find $\frac{\partial F}{\partial x}, \frac{\partial F}{\partial y}$ and show that they are continuous functions. Show that $\frac{\partial}{\partial x}\left(\frac{\partial F}{\partial y}\right)$ and $\frac{\partial}{\partial y}\left(\frac{\partial F}{\partial x}\right)$ both exist at the origin but that they are not equal. Why does this not violate 'equality of mixed partials'?

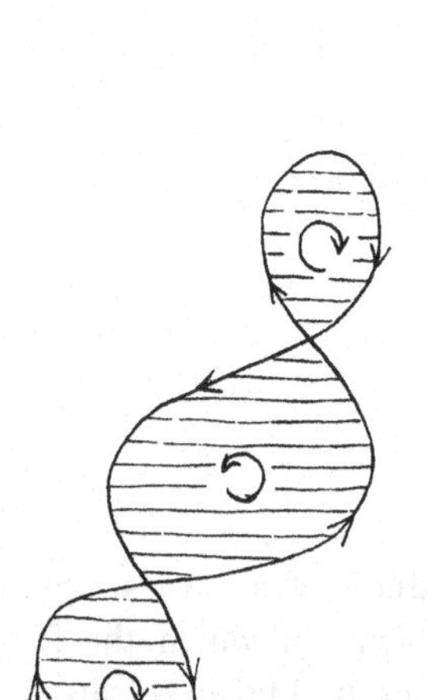

A reasonable closed curve in the plane is the boundary of an oriented domain.

5 Let D be a (reasonable) domain in the plane. If any two points P, Q in D can be joined by curve S, what conclusion can be drawn about a function F, defined throughout D, for which $dF = 0$? Give an example to show that the conclusion does not necessarily hold without the assumption that points can be joined by curves.

6 Let D be a (reasonable) domain in the plane. If D has the property that every (reasonable) closed curve in D is a *boundary*—that is, given an oriented curve there is an oriented domain S such that the given curve is ∂S—what conclusion can be drawn about a 1-form ω defined throughout D for which $d\omega = 0$? Give an example to show that the conclusion does not necessarily hold without the assumption on D.

7 Prove that if S is the oriented triangle with vertices $(0, 0)$, $(1, 0)$, $(0, 1)$ in that order, and, if $A(x, y)$ is continuous, then $\int_S A\, dx\, dy$ can be written as an iterated integral

$$\int_0^1 \left[\int_0^{1-x} A(x, y)\, dy\right] dx \quad \text{or} \quad \int_0^1 \left[\int_0^{1-y} A(x, y)\, dx\right] dy.$$

[Use the method of Exercise 2, §2.6.]

8 Prove that Stokes' Formula $\int_{\partial S} \omega = \int_S d\omega$ holds for ω a differentiable 1-form on the oriented triangle of Exercise 7.

9 Prove that Stokes' Formula $\int_{\partial S} \omega = \int_S d\omega$ holds whenever S is an oriented triangle in xyz-space parameterized, by an affine map, on the triangle $(0, 0)$, $(1, 0)$, $(0, 1)$ in the uv-plane (and whenever ω is a differentiable 1-form on S). [Use the Chain Rule.]

10 Outline a proof that $\int_{\partial S} \omega = \int_S d\omega$ holds for polygonal surfaces S and differentiable 1-forms ω.

linear algebra

chapter 4

4.1

Introduction Linear algebra is the study of a very simple but very important type of problem, of which the following is typical: Certain quantities are known to satisfy relations

$$
(1) \qquad \begin{aligned}
p &= 2S + 4T + 3U - V + 2 \\
q &= S + 2T + 3U + V - 3 \\
r &= 4S + 8T + 10U + 2V + 2.
\end{aligned}
$$

Given values of (p, q, r), find all possible values of (S, T, U, V).

The method of solution is nothing more than the step-by-step elimination of variables. Although certain combinations such as $p - q$ or $r - 4q$ immediately suggest themselves for this particular case, it is better to organize the elimination in a systematic way which will be applicable to all such problems. The organizing principle which will be used is to *move the unknown quantities one by one to the left-hand side*, thus expressing them in terms of the others. In the system (1) one can solve the first equation for V

$$V = -p + 2S + 4T + 3U + 2$$

and use the result to eliminate V from the other two equations

$$
\begin{aligned}
q &= -p + 3S + 6T + 6U - 1 \\
r &= -2p + 8S + 16T + 16U + 6.
\end{aligned}
$$

One of these two remaining equations can be solved for one of the remaining unknowns (S, T, U), say

$$S = \tfrac{1}{3}p + \tfrac{1}{3}q - 2T - 2U + \tfrac{1}{3}$$

and used to eliminate S from the other equations

$$\begin{aligned} V &= -\tfrac{1}{3}p + \tfrac{2}{3}q \qquad - U + 2\tfrac{2}{3} \\ r &= \tfrac{2}{3}p + 2\tfrac{2}{3}q \qquad\qquad + 8\tfrac{2}{3}. \end{aligned}$$

Since the equation for r now contains none of the unknowns the process can go no further. The conclusion is that the given system of equations (1) is equivalent to the system

$$(2) \qquad \begin{aligned} V &= -\tfrac{1}{3}p + \tfrac{2}{3}q \qquad\qquad - U + 2\tfrac{2}{3} \\ S &= \tfrac{1}{3}p + \tfrac{1}{3}q - 2T - 2U + \tfrac{1}{3} \\ r &= \tfrac{2}{3}p + 2\tfrac{2}{3}q \qquad\qquad\qquad + 8\tfrac{2}{3}. \end{aligned}$$

In this form the original problem "given (p, q, r) find all possible values of (S, T, U, V)" can be solved immediately: If (p, q, r) do not satisfy the relation

$$r = \tfrac{2}{3}p + 2\tfrac{2}{3}q + 8\tfrac{2}{3}$$

then there is *no* solution; if they do satisfy this relation then values of T, U can be chosen arbitrarily and V, S determined by the relations

$$\begin{aligned} S &= \tfrac{1}{3}p + \tfrac{1}{3}q - 2T - 2U + \tfrac{1}{3} \\ V &= -\tfrac{1}{3}p + \tfrac{2}{3}q \qquad\qquad - U + 2\tfrac{2}{3}. \end{aligned}$$

This gives all possible solutions (S, T, U, V).

Writing the given equations (1) in the form (2) can therefore be regarded as a solution of the problem of finding all possible (S, T, U, V) given (p, q, r). Similarly, a system of m equations in n unknowns can be solved by rewriting it in a form which gives the values of as many of the unknown quantities as possible in terms of the remaining ones. Thus a system*

$$(1') \qquad y_i = \sum_{j=1}^{n} a_{ij}x_j + b_i \qquad (i = 1, 2, \ldots, m)$$

is solved by the following process: Choose one of the unknowns $x_1, x_2, \ldots, x_n$, solve one of the equations for this unknown, and substitute the result into the remaining equations. For the sake of simplicity assume that the first equation can be solved for x_1 (i.e. assume $a_{11} \neq 0$)

**Notation.* *Although in specific problems it is always preferable to give separate names to the variables, such as (p, q, r) or (S, T, U, V), this is inconvenient in open-ended problems where the number of variables is unspecified. It was once common to write $(x, y, \ldots, z)$ or $(a, b, \ldots, c, \ldots, d)$ to indicate an unspecified number of variables of which x, y were the first two and z the last, or of which a, b were the first two, c some variable in between, and d the last. This notation has obvious drawbacks—e.g. a statement about two variables in between requires separate explanations—and the subscript notation is now prevalent. In this notation the natural numbers 1, 2, 3, . . . are used to index the variables so that $(x_1, x_2, \ldots, x_i, \ldots, x_n)$ denotes an unspecified number n of variables of which a typical one is denoted x_i where i is a natural number $\leq n$. This makes possible the very compact notation (1′) for m equations in n unknowns:*

$$\begin{array}{ccccc} y_1 = a_{11}x_1 + \cdots + a_{1n}x_n + b_1 \\ \vdots \\ y_m = a_{m1}x_1 + \cdots + a_{mn}x_n + b_m. \end{array}$$

and that this is the choice that is made. The result is a new set of equations

$$\begin{aligned} x_1 &= \text{combination of } y_1, x_2, x_3, \ldots, x_n \\ y_2 &= \quad \text{"} \\ &\vdots \\ y_m &= \quad \text{"} \end{aligned}$$

in which one x has been eliminated from the right-hand side in favor of one of the y's. Next, one of the remaining equations with a y on the left is solved for one of the remaining x's on the right and the result substituted into all the other equations. Assuming for the sake of simplicity that it is the second equation which is solved for x_2 the result is a new system

$$\begin{aligned} x_1 &= \text{combination of } y_1, y_2, x_3, \ldots, x_n \\ x_2 &= \quad \text{"} \\ y_3 &= \quad \text{"} \\ y_4 &= \quad \text{"} \\ &\vdots \\ y_m &= \quad \text{"} \end{aligned}$$

and so forth. It may well happen that a point is reached where the next equation cannot be solved for the next unknown (if $a_{11} = 0$ it would not have been possible to solve the first equation for the first unknown) even though all unknowns have not been eliminated from all the remaining equations. In this event it is convenient to rearrange and renumber the remaining equations and/or unknowns so that at the rth stage the equations express $(x_1, x_2, \ldots, x_r, y_{r+1}, \ldots, y_m)$ as combinations of $(y_1, y_2, \ldots, y_r, x_{r+1}, \ldots, x_n)$. The process terminates when none of the remaining equations contains any of the remaining unknowns. At this point the equations are of the form

$$(2') \quad \begin{cases} (2a') \quad x_i = \sum_{j=1}^{r} A_{ij} y_j + \sum_{j=r+1}^{n} B_{ij} x_j + C_i \quad (i = 1, 2, \ldots, r) \\ (2b') \quad y_i = \sum_{j=1}^{r} D_{ij} y_j + E_i \quad (i = r+1, \ldots, m) \end{cases}$$

(where A_{ij}, B_{ij}, C_i, D_{ij}, E_i are numbers) expressing $x_1, \ldots, x_r$ as combinations of $(y_1, \ldots, y_r, x_{r+1}, \ldots, x_n)$ and $y_{r+1}, \ldots, y_m$ as combinations of $(y_1, \ldots, y_r)$. Now

(2′) is a solution of (1′) in that it enables one to find all possible sets of values $(x_1, x_2, \ldots, x_n)$ corresponding to a given set of values $(y_1, y_2, \ldots, y_m)$: If the given $(y_1, y_2, \ldots, y_m)$ do not satisfy the $m - r$ relations (2b′) then *no* set of values $(x_1, x_2, \ldots, x_n)$ can be substituted in the right side of (1′) to obtain the given $(y_1, y_2, \ldots, y_m)$ on the left. On the other hand, if the relations (2b′) are satisfied then there do exist such sets of values $(x_1, x_2, \ldots, x_n)$ and all such sets of values $(x_1, x_2, \ldots, x_n)$ can be obtained by choosing $(x_{r+1}, x_{r+2}, \ldots, x_n)$ arbitrarily and using the equations (2a′) to determine $(x_1, x_2, \ldots, x_r)$.

This simple elimination process is the core of linear algebra and should be examined from as many points of view as possible. In particular, a 'geometrical' formulation is very helpful in understanding the nature of systems of linear equations such as (1) and (1′). The word 'geometrical' is in quotes because most often there are more than three variables, which means that the spaces involved generally have more than three dimensions and hence cannot actually be visualized. Nonetheless, linear algebra makes only very simple statements about these higher dimensional spaces, and it is surprisingly easy to develop an intuitive understanding of these statements as generalizations of statements about planes and lines in space.

A system of two equations in three unknowns, e.g.

$$(1'') \qquad \begin{aligned} u &= 2x + 4y - z + 1 \\ v &= x - y \phantom{{}-z} + 4 \end{aligned}$$

can be visualized as a mapping of space with coordinates (x, y, z) to a plane with coordinates (u, v). The elimination gives

$$\begin{aligned} z &= -u + 2x + 4y + 1 \\ v &= x - y + 4 \end{aligned}$$

and hence

$$(2'') \qquad \begin{aligned} z &= -u - 4v + 6x + 17 \\ y &= - v + x + 4. \end{aligned}$$

Thus there are *no* relations between u and v, and given (u, v) it is always possible to find (x, y, z). This means that the mapping of space to a plane defined by the system (1″) is 'onto',* that is, every point of the uv-plane is

**The word 'onto' has become an adjective in mathematics. A mapping $f: A \to B$ of a set A to a set B is said to be onto if every element of B is f of some element of A.*

the image of some point of xyz-space. For each particular point of the uv-plane, say $(u, v) = (2, 3)$, the set of all points of xyz-space which are mapped to the given point form a line in xyz-space—in this case the line

$$\begin{aligned} z &= -2 - 4 \cdot 3 + 6x + 17 = 6x + 3 \\ y &= -3 + x + 4 = x + 1. \end{aligned}$$

Similarly, the points of xyz-space for which $(u, v) = (1, 1)$ are the points of the line

$$\begin{aligned} z &= 6x + 12 \\ y &= x + 3. \end{aligned}$$

It should be noted that this line is parallel to the previous one and that in fact the equations

$$\begin{aligned} u &= \text{const.} \\ v &= \text{const.} \end{aligned} \tag{3}$$

always describe a line

$$\begin{aligned} z &= 6x + \text{const.} \\ y &= x + \text{const.} \end{aligned}$$

parallel to these lines. Geometrically, the map (1″) describes a map from a 3-dimensional space to a 2-dimensional space*. The 'level surfaces'† of this map—that is, the set of all (x, y, z) for which the map has a given value (u, v)—are a family of parallel lines in xyz-space.

In ordinary usage, 'space' of course means three dimensions. In mathematics the different words line, plane, space are often inconvenient and often it is useful to refer to a line as a 'one-dimensional space' or a plane as a 'two-dimensional space'. Similarly the set of all quadruples of real numbers is called 'four-dimensional space', etc.

†*A set such as (3) where a given function has a given value will be called a 'level surface' regardless of its dimension.*

A system of 3 equations in two unknowns such as

$$\begin{aligned} x &= 3u - v + 2 \\ y &= u + 2v \\ z &= 2u + v + 1 \end{aligned}$$

can be visualized as a mapping of a two-dimensional space to a three-dimensional one. The elimination gives

$$\begin{aligned} v &= -x + 3u + 2 \\ y &= -2x + 7u + 4 \\ z &= -x + 5u + 3 \end{aligned}$$

and then

$$\begin{aligned} v &= -\tfrac{1}{7}x + \tfrac{3}{7}y + \tfrac{2}{7} \\ u &= \tfrac{2}{7}x + \tfrac{1}{7}y - \tfrac{4}{7} \\ z &= \tfrac{3}{7}x + \tfrac{5}{7}y + \tfrac{1}{7}. \end{aligned}$$

In this case a point of xyz-space is the image of some

point in uv-space only if the last relation is satisfied, i.e. only if (x, y, z) lies on the plane

$$(4) \qquad 7z = 3x + 5y + 1.$$

This plane is called the *image* of the map. For each point of the image there is exactly one point (u, v) given by the first two equations; thus the level surfaces of the map are individual points and the mapping is one-to-one.*

*A function $f: A \to B$ is said to be one-to-one if no two points in the domain are carried to the same point of the range, i.e. if $f(a_1) = f(a_2)$ implies $a_1 = a_2$.

The original example (1) can be regarded as a mapping from four-dimensional $STUV$-space to three-dimensional pqr-space. The elimination (2) shows that not every point of pqr-space is the image of a point in the $STUV$-space, but only those points which lie on the plane

$$(4') \qquad 3r = 2p + 8q + 26.$$

This plane is the image of the map (1). For each point (p, q, r) of the image, the first two equations of (2) define a two-dimensional set (parameterized by T, U) where (1) has this value. Thus the level surfaces of (1) are planes in $STUV$-space.

The general system of equations (1′) and its solution (2′) can be described 'geometrically' in a similar way: The equations (2b′) describe the image of the map (1′). It is an r-dimensional subspace of the m-dimensional $y_1y_2 \ldots y_m$-space. The level surfaces of the map (1′) are the $(n - r)$-dimensional subspaces of $x_1x_2 \ldots x_n$-space described by the equations (2a′). Note in particular that the map (1′) is onto if and only if $r = m$, and is one-to-one if and only if $r = n$.

Reverting to a more algebraic terminology, the integer r in (2′) can be described as the *number of independent* y's since (2b′) expresses all of the y's in terms of the r independent values $(y_1, y_2, \ldots, y_r)$. Fixing the values of the y's imposes only r independent conditions on the x's (leaving $n - r$ degrees of freedom) since the equations (2a′) express all of the x's in terms of the y's and in terms of the $n - r$ independent values $(x_{r+1}, x_{r+2}, \ldots, x_n)$. The integer r is called the *rank* of the system (1′).

This discussion leaves unanswered the very important question of *which* sets of r of the y's are independent and *which* sets of $(n - r)$ of the x's are independent when the y's are given. In terms of the elimination process, this is the question: "When can a given set of r variables be eliminated from a given set of r equations?" This question is very conveniently answered in terms of the algebra of forms of Chapter 1. For example: The original mapping (1) has rank $r = 2$. This is reflected in the fact that its image is the 2-dimensional plane (4′). Geometrically this implies that any 3-dimensional solid in $STUV$-space is carried by the map (1) to a figure with no volume (because it lies in a plane); hence, by the geometrical meaning of pullback, one would expect that the pullback of oriented volume $dp\,dq\,dr$ under the map (1) would be zero. This is easily verified:

$$\begin{aligned} dp\,dq\,dr &= [2\,dS + 4\,dT + 3\,dU - dV][dS + 2\,dT + 3\,dU + dV] \\ &\quad \times [4\,dS + 8\,dT + 10\,dU + 2\,dV] \\ &= [(4-4)\,dS\,dT + (6-3)\,dS\,dU + (2+1)\,dS\,dV \\ &\quad + (12-6)\,dT\,dU + (4+2)\,dT\,dV + (3+3)\,dU\,dV] \\ &\quad \times [4\,dS + 8\,dT + 10\,dU + 2\,dV] \\ &= [0 \cdot dS\,dT + 3\,dS\,dU + 3\,dS\,dV + 6\,dT\,dU + 6\,dT\,dV + 6\,dU\,dV] \\ &\quad \times [4\,dS + 8\,dT + 10\,dU + 2\,dV] \\ &= (0 - 24 + 24)\,dS\,dT\,dU + (0 - 24 + 24)\,dS\,dT\,dV \\ &\quad + (6 - 30 + 24)\,dS\,dU\,dV + (12 - 60 + 48)\,dT\,dU\,dV \\ &= 0. \end{aligned}$$

The fact that the first two equations of (1) can be solved for (V, S) as functions of (p, q, T, U) is reflected in the fact that the $dS\,dV$ component of $dp\,dq$ is not zero. Geometrically this can be seen as the statement that the planes $T =$ const. $U =$ const. (coordinatized by S, V) are mapped one-to-one onto the pq-plane (because this map of the SV-plane to the pq-plane multiplies oriented areas by 3 and therefore doesn't collapse the plane to a line or a point); hence T, U, p, q suffice to determine S, V. In the same way, since the pullback of $dp\,dq$ is

$$dp\,dq = 0\,dS\,dT + 3\,dS\,dU + 3\,dS\,dV + 6\,dT\,dU + 6\,dT\,dV + 6\,dU\,dV,$$

it is to be expected that (S, U) can be expressed as functions of (p, q, T, V), that (T, V) can be expressed as functions of (p, q, S, U), etc., except that (S, T) cannot be expressed as functions of (p, q, U, V) because for $U =$ const., $V =$ const., the map $(S, T) \to (p, q)$ multiplies oriented areas by zero, which means that the map collapses planes coordinatized by (S, T) to lines or points. These conclusions are immediately verified by directly solving the relevant equations by step-by-step elimination. Since the coefficients of S, T in p, q are proportional, any combination of p, q which eliminates S must also eliminate T.

Analogous considerations apply to the general system

(1′). If its rank is r then its image is r-dimensional and the projection of the image on any $(r + 1)$-dimensional 'coordinate plane' in y-space has no '$(r + 1)$-dimensional volume'. Thus it is to be expected that the pullback of any $(r + 1)$-form in the y's is zero. On the other hand, the equations (2b′) show that the projection of the image onto the $y_1y_2 \ldots y_r$-coordinate 'plane' is one-to-one and onto; hence there is an r-dimensional figure in x-space whose image in y-space has a non-trivial projection on the $y_1y_2 \ldots y_r$-coordinate plane. Thus it is to be expected that the pullback of $dy_1\, dy_2 \ldots dy_r$ is not zero. More generally, any set of r of the y's can be used to parameterize the image of (1)—that is, all y's can be expressed in terms of these—if and only if the projection of the image on the corresponding r-dimensional coordinate plane is one-to-one and onto, which is true if and only if the pullback of the corresponding r-form is not zero. (In (1) the pullbacks of $dp\, dr$ and $dq\, dr$ are not zero, and, as is predicted by the above, the equation (4′) can be solved for q in terms of p and r or for p in terms of q and r.) Finally, the equations (2a′) show that for any fixed values of $x_{r+1}, x_{r+2}, \ldots, x_n$ the map (1) composed with projection on the $y_1y_2 \ldots y_r$-plane is onto; therefore there is an r-dimensional figure in this plane $x_{r+1} =$ const., $\ldots$, $x_n =$ const. such that the value of $dy_1\, dy_2 \ldots dy_r$ on the image is not zero. Since all r-forms in x except $dx_1\, dx_2 \ldots dx_r$ are zero on this figure (its projection on other coordinate planes is not r-dimensional) this implies that the $dx_1\, dx_2 \ldots dx_r$-component of the pullback of $dy_1\, dy_2 \ldots dy_r$ is not zero. More generally, any set of r of the x's can be eliminated from the equations for $y_1, y_2, \ldots, y_r$ if and only if the corresponding component of the pullback of $dy_1\, dy_2 \ldots dy_r$ is not zero.

In summary: A set of equations of the form

$$(1') \qquad y_i = \sum_{j=1}^{n} a_{ij}x_j + b_j \qquad (i = 1, 2, \ldots, m)$$

can be put in the form

$$(2') \begin{cases} (2a') \quad x_i = \sum_{j=1}^{r} A_{ij}y_j + \sum_{j=r+1}^{n} B_{ij}x_j + C_i \\ \qquad\qquad (i = 1, 2, \ldots, r) \\ (2b') \quad y_i = \sum_{j=1}^{r} D_{ij}y_j + E_i \\ \qquad\qquad (i = r + 1, r + 2, \ldots, m) \end{cases}$$

by (possibly) rearranging the x's and y's and eliminating as many x's as possible from the right-hand side. The integer r is called the *rank* of the system (1′). Algebraically the rank can be thought of as the number of independent y's and geometrically it can be thought of as the dimension of the image of the mapping (1′). These interpretations of the rank make the following assertions plausible: The pullback of any $(r + 1)$-form in the y's, found by the familiar computational rules, is zero. A given set of r of the y's is independent and can be used to find the remaining $(m - r)$ values if and only if the pullback of the corresponding r-form in the y's is not zero. When this is the case, these y's can be used to eliminate a given set of r of the x's if and only if the corresponding term of the pullback is not zero. For example, the form (2′) can be achieved without rearrangement of the x's and y's if and only if the coefficient of $dx_1\, dx_2 \ldots dx_r$ in the pullback of $dy_1\, dy_2 \ldots dy_r$ is not zero.

These conclusions are contained, in a somewhat more concise form, in the Implicit Function Theorem, which is stated and proved in §4.4. The two intervening sections are devoted to developing needed definitions and notation.

Exercises

1 For each of the following systems find one reduction to the form (2′), the value of r, and all possible choices of r independent variables (variables on the right) and r dependent variables (variables on the left) for which such a reduction is possible.

(a) $u = 3x + 2y + 1$
$\quad v = 2x + y - 3$

(b) $x = 2t + 1$
$\quad y = t + 2$
$\quad z = 4t - 3$

(c) $V = x + 2y - z + 7t + 4$

(d) $u = 2x + y - z$
$\quad v = -4x + 3y + 2z - 4$

(e) $a = 3p + q + 4$
$\quad b = 2p - q + 2$
$\quad c = p + q - 1$

(f) $u = 2x + y - z - 4t + 1$
$\quad v = -4x + 3y + 2z - 12t + 1$

(g)
$$\begin{aligned} x &= 2p + q + 3r - 7 \\ y &= 2p + r \\ z &= -7p + 2q + r + 2 \end{aligned}$$

(h)
$$\begin{aligned} u &= 3x + 2y + 8z \\ v &= x - 3y - z \\ w &= 4x + y + 9z \end{aligned}$$

2 Solve the equations of Exercise 7, §1.3, for x, y as functions of u, v. Review this exercise.

3 Consider the triangle in 4-dimensional space whose vertices are $(4, 4, -3, 4)$, $(8, 8, 3, 6)$, $(2, 0, -6, 2)$. Draw its projection on each of the 6 two-dimensional coordinate planes. Find a mapping

$$\begin{aligned} x &= a_{11}u + a_{12}v + b_1 \\ y &= a_{21}u + a_{22}v + b_2 \\ z &= a_{31}u + a_{32}v + b_3 \\ t &= a_{41}u + a_{42}v + b_4 \end{aligned}$$

of the uv-plane to $xyzt$-space which carries the triangle $(0, 0)$, $(1, 0)$, $(1, 1)$ to the given triangle. Can the coefficients a_{ij}, b_i be chosen in more than one way? Which pairs of the equations $xyzt$ can be used to eliminate uv from the right-hand side? Relate the answer to the first part of the exercise. Find the pullbacks of the 2-forms $dx\,dy$, $dx\,dz$, etc. to uv-space.

4 Write the tetrahedron with vertices $(2, 4, 1, -1)$, $(-1, 1, 4, 2)$, $(0, 0, 2, -1)$, $(1, 3, 3, 3)$ as the image under a map

$$\begin{aligned} x &= a_{11}u + a_{12}v + a_{13}w + b_1 \\ y &= a_{21}u + a_{22}v + a_{23}w + b_2 \\ z &= a_{31}u + a_{32}v + a_{33}w + b_3 \\ t &= a_{41}u + a_{42}v + a_{43}w + b_4 \end{aligned}$$

of the tetrahedron in uvw-space with vertices $(0, 0, 0)$, $(1, 0, 0)$, $(1, 1, 0)$, $(1, 1, 1)$. Can this be done in more than one way? Find the pullbacks of each of the 4 basic 3-forms on $xyzt$-space. Which triples of these equations can be used to eliminate uvw from the right-hand side?

5 Show that the equations

$$\begin{aligned} u &= a_{11}x + a_{12}y + a_{13}z + b_1 \\ v &= a_{21}x + a_{22}y + a_{23}z + b_2 \end{aligned}$$

can be solved for x, y as functions of u, v, z if and only if the $dx\,dy$-component of the pullback of $du\,dv$ is not zero. [Form $a_{22}u - a_{12}v$ and $a_{21}u - a_{11}v$. If $a_{11}a_{22} - a_{12}a_{21} \neq 0$ this gives the solution. Otherwise it shows that the map is not onto for fixed z; hence there is no such solution.]

6 Use the method of Exercise 5 to prove a necessary and

sufficient condition for it to be possible to eliminate u, v from the first two equations of the system

$$x = a_{11}u + a_{12}v + b_1$$
$$y = a_{21}u + a_{22}v + b_2$$
$$z = a_{31}u + a_{32}v + b_3.$$

Assuming the condition is met, write the final form (2′) of the solution explicitly in terms of the a's and b's.

7 Under what conditions on the a's and b's is the map of Exercise 5 of rank one? Describe geometrically, i.e. give dimensions of range, domain, image, and level surfaces.

8 Under what conditions on the a's and b's is the map of Exercise 6 of rank one? Describe geometrically.

9 In the main example (1) find the pullbacks of $dq\, dr$ and $dp\, dr$, showing that they are multiples of $dp\, dq$. Derive the same result, along with the specific multiples, using the last of the equations (2).

4.2

Constant k-Forms on n-Space

As the discussion of §4.1 indicates, the algebra of forms is useful in the linear algebra of higher dimensions. This algebra of forms is described very succinctly by the rules

$$(1) \qquad \begin{aligned} dx_i\, dx_i &= 0 \\ dx_i\, dx_j &= -dx_j\, dx_i. \end{aligned}$$

A k-form* in $x_1, x_2, \ldots, x_n$ is described by an expression which is a sum of terms of the form $A\, dx_{i_1}\, dx_{i_2} \ldots dx_{i_k}$ where A is a number and $i_1, i_2, \ldots, i_k$ are integers $1 \le i_j \le n$. The *pullback* of a k-form in $y_1, y_2, \ldots, y_m$ under an affine map

*In this chapter only constant forms will be considered. Therefore, as in Chapter 1, the word 'constant' will be omitted.

$$(2) \qquad y_i = \sum a_{ij}x_j + b_i$$

is the k-form in $x_1, x_2, \ldots, x_n$ obtained by performing the substitution $dy_i = \sum a_{ij}\, dx_j$ and using the rules (1) together with the distributive law of multiplication. In short, the algebraic rules of 1-, 2-, and 3-forms in 3 variables, which were described and used in Chapter 1, will now be applied to cases where there are more than 3 variables.

The basic fact about this algebra of forms is the following statement which, for reasons to be explained in the next section, is known as the Chain Rule.

Chain Rule

Let

$$y_i = \sum_{j=1}^{n} a_{ij}x_j + b_i \qquad (i = 1, 2, \ldots, m)$$

and

$$z_i = \sum_{j=1}^{m} \alpha_{ij}y_j + \beta_i \qquad (i = 1, 2, \ldots, p)$$

be affine maps* from $(x_1, x_2, \ldots, x_n)$ to $(y_1, y_2, \ldots, y_m)$ and from $(y_1, y_2, \ldots, y_m)$ to $(z_1, z_2, \ldots, z_p)$ so that their composition is an affine map from $(x_1, x_2, \ldots, x_n)$ to $(z_1, z_2, \ldots, z_p)$. Then the pullback of a k-form under the composed map is equal to the pullback of the pullback.

*Here, as in Chapter 1, an affine map is a function described by polynomials of the first degree.

This was proved in §1.3 for 2-forms in two variables and was stated, but not proved, for 3-forms in three variables in §1.4. For 1-forms in any number of variables the Chain Rule is proved by direct computation: The composed map is

$$\begin{aligned} z_j &= \alpha_{j1}(a_{11}x_1 + a_{12}x_2 + \cdots + a_{1n}x_n) + \cdots \\ &\quad + \alpha_{jm}(a_{m1}x_1 + a_{m2}x_2 + \cdots + a_{mn}x_n) + \text{const.} \\ &= (\alpha_{j1}a_{11} + \alpha_{j2}a_{21} + \cdots + \alpha_{jm}a_{m1})x_1 + \cdots \\ &\quad + (\alpha_{j1}a_{1n} + \alpha_{j2}a_{2n} + \cdots + \alpha_{jm}a_{mn})x_n + \text{const.} \end{aligned}$$

so the pullback of dz_j under the composed map is by definition

$$\begin{aligned} dz_j = (\alpha_{j1}a_{11} + \cdots + \alpha_{jm}a_{m1})\,dx_1 + \cdots \\ + (\alpha_{j1}a_{1n} + \cdots + \alpha_{jm}a_{mn})\,dx_n \end{aligned}$$

whereas the pullback of the pullback is the pullback of

$$\alpha_{j1}\,dy_1 + \alpha_{j2}\,dy_2 + \cdots + \alpha_{jm}\,dy_m$$

which is

$$\begin{aligned} &\alpha_{j1}(a_{11}\,dx_1 + \cdots + a_{1n}\,dx_n) + \cdots + \alpha_{jm}(a_{m1}\,dx_1 + \cdots + a_{mn}\,dx_n) \\ &\quad = (\alpha_{j1}a_{11} + \cdots + \alpha_{jm}a_{m1})\,dx_1 + \cdots + (\alpha_{j1}a_{1n} + \cdots + \alpha_{jm}a_{mn})\,dx_n. \end{aligned}$$

Thus the pullback of dz_j under the composed map is the pullback of the pullback. But then the same is true for an arbitrary 1-form $\sum A_j\,dz_j$ by superposition of these cases.

The proof of the Chain Rule for k-forms ($k > 1$) requires a more careful examination of the algebraic rules which govern them, and in particular an examination of the rule $dx_i\,dx_j = -dx_j\,dx_i$. What is the meaning of 'equal' in this 'equation'? Clearly if it is to have any meaning at all then it must mean that the same k-form can be represented in different ways; for example, $dx_1\,dx_2\,dx_3 = -dx_1\,dx_3\,dx_2 = dx_3\,dx_1\,dx_2$ must mean that the different expressions $dx_1\,dx_2\,dx_3$ and $dx_3\,dx_1\,dx_2$ *represent the same* 3-*form*. This is formalized by the following definitions.

Definitions

A k-form in the variables $x_1, x_2, \ldots, x_n$ is represented by an expression which is a sum of a finite number of terms of the form $A\,dx_{i_1}\,dx_{i_2} \ldots dx_{i_k}$ in which A is a number and $i_1, i_2, \ldots, i_k$ are integers $1 \leq i_j \leq n$. (Or by a sum of *no* such terms, which is represented by 0.) Two such expressions *represent the same k-form* if one can be obtained from the other by a finite number of applications of the rules

(i) $dx_i\,dx_j = -dx_j\,dx_i$,
(ii) $dx_i\,dx_i = 0$,
(iii) $A\,dx_{i_1}\,dx_{i_2} \ldots dx_{i_k} + B\,dx_{i_1}\,dx_{i_2} \ldots dx_{i_k} = (A + B)\,dx_{i_1}\,dx_{i_2} \ldots dx_{i_k}$,
(iv) the commutative law of addition.

In words: Terms can be arranged in any order (iv). Two terms with the same dx's (identical and in identical order) can be combined (iii). A term which contains two identical factors dx_i adjacent to each other can be stricken (ii). Two adjacent factors $dx_i\,dx_j$ in a term can be interchanged provided the sign of the coefficient A is also changed (i).

Thus, for example,

$$
\begin{aligned}
3\,dx_1\,dx_2\,dx_3 + 2\,dx_1\,dx_3\,dx_1 + 4\,dx_3\,dx_2\,dx_1 &= 3\,dx_1\,dx_2\,dx_3 - 2\,dx_1\,dx_1\,dx_3 - 4\,dx_3\,dx_1\,dx_2 \\
&= 3\,dx_1\,dx_2\,dx_3 + 4\,dx_1\,dx_3\,dx_2 \\
&= 3\,dx_1\,dx_2\,dx_3 - 4\,dx_1\,dx_2\,dx_3 \\
&= -dx_1\,dx_2\,dx_3 \\
&= dx_2\,dx_1\,dx_3
\end{aligned}
$$

etc., where = means 'represent the same 3-form' as defined above. Note that the term 'k-form' itself has not been defined* but that only the terms 'representation of a k-form' and 'represent the same k-form' have been defined. This fine point in the nomenclature has no effect on the way that computations are done, but it does

*Geometrically the k-form $dx_1\,dx_2 \ldots dx_k$ should be imagined as oriented k-dimensional volume of the projection on the $x_1x_2 \ldots x_k$-plane, a function assigning numbers to oriented k-dimensional surfaces in $x_1x_2 \ldots x_n$-space. This intuitive interpretation is, of course, not at all a definition.

simplify the underlying philosophy of the subject considerably. (See Appendix 3.)

The *sum* of two k-forms ω and σ is the k-form $\omega + \sigma$ represented by the expression obtained by adding the terms of a representation of ω to the terms of a representation of σ. This definition is valid because of the obvious fact that if either the representation of ω or the representation of σ is changed to another expression representing the same form, then the expression for the sum represents the same form as before.

The *product* of a k-form ω and an h-form σ is the $(k + h)$-form $\omega\sigma$ represented by the expression obtained by writing representations of ω and σ side by side in that order and multiplying out using the distributive law of multiplication and the rule

$$A\, dx_{i_1}\, dx_{i_2} \ldots dx_{i_k} \cdot B\, dx_{j_1}\, dx_{j_2} \ldots dx_{j_h} = AB\, dx_{i_1}\, dx_{i_2} \ldots dx_{i_k}\, dx_{j_1}\, dx_{j_2} \ldots dx_{j_h}.$$

This definition is valid because of the obvious fact that if either the representation of ω or the representation of σ is changed to another expression representing the same form, then the expression for the product will represent the same form as before. For example

$$(dx_1 + 2\, dx_2)(3\, dx_1\, dx_2 + 4\, dx_2\, dx_3)$$
$$= 3\, dx_1\, dx_1\, dx_2 + 4\, dx_1\, dx_2\, dx_3 + 6\, dx_2\, dx_1\, dx_2 + 8\, dx_2\, dx_2\, dx_3$$

and

$$(2\, dx_2 + dx_1)(3\, dx_1\, dx_2 - 4\, dx_3\, dx_2)$$
$$= 6\, dx_2\, dx_1\, dx_2 - 8\, dx_2\, dx_3\, dx_2 + 3\, dx_1\, dx_1\, dx_2 - 4\, dx_1\, dx_3\, dx_2$$

represent the same 3-form (which is also represented more simply by $4\, dx_1\, dx_2\, dx_3$).

A *zero-form* is a number. If A is a 0-form and $dx_1\, dx_2 \ldots dx_k$ is a k-form then their product is the $(0 + k)$-form $A\, dx_1\, dx_2 \ldots dx_k$ as before.

This completes the definition of the algebraic rules which govern computations with forms.

Clearly $\omega(\sigma_1 + \sigma_2) = \omega\sigma_1 + \omega\sigma_2$, $(\omega_1 + \omega_2)\sigma = \omega_1\sigma + \omega_2\sigma$, $\omega(\sigma\tau) = (\omega\sigma)\tau$. If ω and σ are 1-forms, then $\omega\sigma = -\sigma\omega$ because $dx_i\, dx_j = -dx_j\, dx_i$ (even if $i = j$, in which case both sides are 0). Therefore if ω is a 1-form, $\omega \cdot \omega = -\omega \cdot \omega$, $2\omega \cdot \omega = 0$, which implies $\omega \cdot \omega = 0$. Therefore the rules $\omega \cdot \omega = 0$, $\omega \cdot \sigma = -\sigma \cdot \omega$ apply to 1-forms generally, and not just to the 1-forms

dx_i. (More generally, if ω is a k-form and σ is an h-form, then $\omega\sigma = (-1)^{kh}\sigma\omega$.)

In terms of these algebraic operations on forms, the operation of forming the pullback of a k-form ω in $y_1, y_2, \ldots, y_m$ under an affine map

$$y_i = \sum_{j=1}^{n} a_{ij}x_j + b_i \qquad (i = 1, 2, \ldots, m)$$

is literally an operation of *substitution* of the 1-forms $\sum a_{ij}\,dx_j$ for dy_i in an expression representing ω, the resulting sums and products being regarded as sums and products of 0-forms and 1-forms in the variables $x_1, x_2, \ldots, x_n$. This definition is valid because of the obvious fact that if the expression of ω is replaced by another expression representing the same k-form then the resulting expression for the pullback of ω represents the same k-form (because $\omega \cdot \omega = 0$ and $\omega\sigma = -\sigma\omega$ hold for all 1-forms ω, σ).

Clearly the pullback of a sum is the sum of the pullbacks and the pullback of a product is the product of the pullbacks. Therefore the pullback of a k-form ω can be found by taking any expression of ω as a sum of products of 0-forms and 1-forms, and taking pullbacks of the 1-forms separately (the pullback of a 0-form is itself). Thus the pullback operation is completely determined once its effect on 1-forms is known. But the same is true of the operation 'pullback of the pullback' and, because for 1-forms this is identical to the operation 'pullback under the composed map', the same is true for k-forms.* This proves the Chain Rule.

*This argument can be stated in the language of modern algebra as follows: 'Pullback' is an algebra homomorphism from forms in y to forms in x. Therefore 'pullback of the pullback' is also an algebra homomorphism. To prove it is equal to the algebra homomorphism 'pullback under the composed map' it suffices to show the two are equal for a set of generators for the algebra; hence it suffices to show they are the same for the 1-forms $dz_1, dz_2, \ldots, dz_p$, which is done by direct computation.

In computing with forms it is very useful to allow the expression as much latitude as possible—as was done in the above proof. On the other hand it is also useful, particularly in deciding whether two different expressions represent the same k-form, to have a standard format for the expression of a k-form. This is accomplished as follows.

Definition

An expression representing a k-form in the variables $x_1, x_2, \ldots, x_n$ is said to be in *lexicographic* (dictionary) *order* if it answers to the following description: The expression is a sum of $\binom{n}{k}$ terms, where $\binom{n}{k}$ is the binomial coefficient.† In particular, if $k > n$ then there

†$\binom{n}{k} = \dfrac{n!}{k!(n-k)!}$ if $1 \le k \le n-1$. It is 1 if $k = 0$ or $k = n$ and 0 if $k > n$. It should be imagined here as the number of k-dimensional coordinate planes in n-space.

are no terms and the expression is 0. If $k \leq n$ then the expression is a sum

$$\sum_{1 \leq j_1 < j_2 < \cdots < j_k \leq n} A_{j_1 j_2 \cdots j_k}\, dx_{j_1}\, dx_{j_2} \ldots dx_{j_k}$$

of $\binom{n}{k}$ terms ordered so that the term corresponding to $(j_1, j_2, \ldots, j_k)$ precedes all terms in which j_1 is greater, all terms in which j_1 is the same but j_2 is greater, all terms in which j_1 and j_2 are the same but j_3 is greater, and so forth. In other words, the k-tuples $(j_1, j_2, \ldots, j_k)$ in which $1 \leq j_1 < j_2 < \cdots < j_k \leq n$ (there are $\binom{n}{k}$ such k-tuples) are ordered like words in a dictionary so that $(j_1, \ldots, j_k)$ precedes $(j_1', \ldots, j_k')$ if $j_i < j_i'$ in the first position where they differ. The numbers $A_{j_1 j_2 \cdots j_k}$ may be zero but the terms are written anyway so that there are always $\binom{n}{k}$ terms.

The idea of lexicographic order is easier to illustrate than it is to define in full. For example, for 3-forms in x_1, x_2, x_3, x_4, x_5, the triples in order are

$$(1, 2, 3), (1, 2, 4), (1, 2, 5), (1, 3, 4), (1, 3, 5),$$
$$(1, 4, 5), (2, 3, 4), (2, 3, 5), (2, 4, 5), (3, 4, 5)$$

and an expression in lexicographic order is an expression of the form

$$A_{123}\, dx_1\, dx_2\, dx_3 + A_{124}\, dx_1\, dx_2\, dx_4 + \cdots + A_{245}\, dx_2\, dx_4\, dx_5 + A_{345}\, dx_3\, dx_4\, dx_5 \quad \text{(10 terms).}$$

Theorem

Every k-form can be represented by an expression in lexicographic order. Two expressions in lexicographic order represent the same k-form only if they are identical.

Proof

The first statement of the theorem is the observation that the rules defining 'represent the same k-form' allow one to reduce any expression representing a k-form to one in lexicographic order. This is easily seen to be true (see Exercises 1, 2, 3).

Suppose now that two expressions ω_1 and ω_2 in lexicographic order represent the same k-form but are not identical. Let $dx_{j_1}\,dx_{j_2}\ldots dx_{j_k}$ be a term in which ω_1 and ω_2 differ and let $\bar{\omega}_1, \bar{\omega}_2$ be the pullbacks of ω_1, ω_2 under the affine map

$$x_{j_1} = u_1,\ x_{j_2} = u_2, \ldots, x_{j_k} = u_k$$
$$x_i = 0\ (i \neq j_1, j_2, \ldots, j_k)$$

of $(u_1, u_2, \ldots, u_k)$ to $(x_1, x_2, \ldots, x_n)$. Then $\bar{\omega}_1, \bar{\omega}_2$ are of the form $A_1\,du_1\,du_2\ldots du_k$, $A_2\,du_1\,du_2\ldots du_k$ with $A_1 \neq A_2$, but $\bar{\omega}_1, \bar{\omega}_2$ represent the same k-form in $u_1, u_2, \ldots, u_k$. (If the two expressions represent the same form then so do their pullbacks; for the particular affine map here the proof of this fact is utterly trivial.) This means that it is possible to interpolate expressions $\sigma_0, \sigma_1, \sigma_2, \ldots, \sigma_N$ representing k-forms in $u_1, u_2, \ldots, u_k$ such that σ_0 is identical to $A_1\,du_1\,du_2\ldots du_k$, such that σ_N is identical to $A_2\,du_1\,du_2\ldots du_k$, and such that the step from σ_i to σ_{i+1} involves just one application of just one of the rules

(i) $du_i\,du_j = -du_j\,du_i$,
(ii) $du_i\,du_i = 0$,
(iii) $A\,du_{i_1}\,du_{i_2}\ldots du_{i_k} + B\,du_{i_1}\,du_{i_2}\ldots du_{i_k} = (A + B)\,du_{i_1}\,du_{i_2}\ldots du_{i_k}$,
(iv) the commutative law of addition.

It is to be shown that if $\sigma_0, \sigma_1, \sigma_2, \ldots, \sigma_N$ is such a sequence of expressions then $A_1 = A_2$.

The proof is by induction on k. If $k = 1$ then the expressions σ_i are all sums of multiples of du_1 (there are no other du's) and rules (i) and (ii) do not apply. Let $\sigma_i = C_{i1}\,du_1 + C_{i2}\,du_1 + \cdots + C_{iM}\,du_1$; then the sum $C_{i1} + C_{i2} + \cdots + C_{iM}$ is the same for all i because either of the changes allowed by (iii) and (iv) leaves the sum unchanged. Since this sum is A_1 when $i = 0$ and A_2 when $i = N$, it follows that $A_1 = A_2$ and the case $k = 1$ is proved.

Now suppose the case $k - 1$ is proved and suppose that $\sigma_0, \sigma_1, \ldots, \sigma_N$ are as above. Note first that if any and all terms in the σ_i which contain repeated factors (not necessarily adjacent) are stricken then the new sequence $\sigma_0, \sigma_1, \ldots, \sigma_N$ has the same properties as before except that at a step σ_i to σ_{i+1} where rule (ii) was used the new σ_i and σ_{i+1} are now identical. Therefore it can be assumed at the outset that the given expressions $\sigma_0, \sigma_1, \sigma_2, \ldots, \sigma_N$ contain no terms with repeated factors and that the rule (ii) is never used. Thus each term of each σ_i contains the factor du_k exactly once. Define a sequence of expressions $\hat{\sigma}_0, \hat{\sigma}_1, \ldots, \hat{\sigma}_N$ representing $(k-1)$-forms in $u_1, u_2, \ldots, u_{k-1}$ as follows: In each term of each σ_i strike the factor du_k and multiply the term by ± 1, using $+1$ if the du_k occurs in the kth, $(k-2)$nd, $(k-4)$th, ..., $(k-2\nu)$th place in this product of k factors, and using -1 if it occurs in the $(k-1)$st, $(k-3)$rd, ..., $(k-2\nu+1)$st place. Then $\hat{\sigma}_0 = A_1\, du_1\, du_2 \ldots du_{k-1}$, $\hat{\sigma}_N = A_2\, du_1\, du_2 \ldots du_{k-1}$ and it suffices to show that the expressions $\hat{\sigma}_0, \hat{\sigma}_1, \hat{\sigma}_2, \ldots, \hat{\sigma}_N$ have the property that each can be obtained from the preceding by one of the rules (i)–(iv). If the step from σ_i to σ_{i+1} involves one of the rules (iii) or (iv) then the step from $\hat{\sigma}_i$ to $\hat{\sigma}_{i+1}$ is obtained by the same rule. If the step from σ_i to σ_{i+1} uses rule (i) to interchange two adjacent factors other than du_k then the same applies to the step from $\hat{\sigma}_i$ to $\hat{\sigma}_{i+1}$. Finally, if the step from σ_i to σ_{i+1} interchanges du_k with another du then this term changes sign between σ_i and σ_{i+1} and the position of du_k changes by one, which implies that $\hat{\sigma}_i = \hat{\sigma}_{i+1}$. This completes the proof of the theorem.

Exercises

In the following exercises, sequences of consecutive letters (u, v, w, x, y, z), (p, q, r, s, t), $(\alpha, \beta, \gamma, \delta)$ will be used to denote the variables. Then 'lexicographic order' means quite literally 'alphabetical order' in the everyday sense.

1 Find expressions in lexicographic order which represent the same 4-forms as the following.

(a) $3\, dw\, dx\, du\, dv - 4\, du\, dw\, dv\, dx + 2\, du\, dy\, dw\, dz$
(b) $(3\, du\, dv + 4\, du\, dw)(dy + dz)\, dw$
(c) $(du\, dv + dx\, dy)(du\, dv + dx\, dy)$
(d) $(4\, du\, dv\, dw + 4\, dx\, dy\, dz + 2\, du\, dw\, dy + dv\, dx\, dz)$
$\times (du + dw + dy)$

2 Find expressions in lexicographic order which represent the pullback of each of the 4-forms of Exercise 1 under the affine map

$$\begin{aligned} u &= r + s + 1 \\ v &= r - s \\ w &= 2p + t \\ x &= p - q \\ y &= q - r - 7 \\ z &= p + q + r + s + t. \end{aligned}$$

3 Find expressions in lexicographic order of the pullbacks of the 4-forms of Exercise 2 under the map

$$\begin{aligned} p &= 4\alpha - 3\beta + \gamma \\ q &= 2\alpha - \delta + 14 \\ r &= \alpha + \beta \\ s &= \alpha - \beta \\ t &= 144. \end{aligned}$$

4 Find the composed map $(\alpha, \beta, \gamma, \delta) \to (u, v, w, x, y, z)$ of the maps of Exercises 2 and 3, find the pullbacks of the forms of Exercise 1 by direct computation, and verify the Chain Rule by comparing these answers with the answers to Exercise 3.

5 List all possible orders in which the 1-forms dx, dy, dz, dw can be multiplied to give a 4-form. Which of these represent the same 4-form as $dx\,dy\,dz\,dw$ and which the same as $-dx\,dy\,dz\,dw$?

4.3

Matrix Notation. Jacobians

In most cases, the additive constants b_i in an affine map $y_i = \sum a_{ij}x_j + b_i$ are of minor importance. This is emphasized by writing the affine map in the form

$$(1) \qquad y_i = \sum_{j=1}^{n} a_{ij}x_j + \text{const.} \qquad (i = 1, 2, \ldots, m)$$

The numbers a_{ij}, which are called the *coefficients* of the affine map, can be conveniently exhibited, without the y's and x's, in a rectangular array or *matrix* called the *matrix of coefficients* of the map (1). Thus, for example, the matrix of coefficients of the map

$$\begin{aligned} u &= 3x - y + 2z + 8 \\ v &= x + y + 4z + 14 \end{aligned}$$

is

$$\begin{pmatrix} 3 & -1 & 2 \\ 1 & 1 & 4 \end{pmatrix}.$$

In particular, the number of *columns* of the matrix of coefficients is equal to the dimension of the *domain* of the affine map and the number of *rows* equal to the dimension of the *range*.*

If $f: A \to B$ is a function from the set A to the set B then A is called the domain of f and B the range. Above, the domain is xyz-space and the range is uv-space.

In general a *matrix* is a rectangular array of numbers, and an $m \times n$ *matrix* is a matrix with m rows and n columns. [Thus the matrix above is 2×3.] A matrix is therefore simply a convenient format in which sets of mn numbers can be recorded, which *per se* is of very little interest or importance. What is important, rather, is that certain operations with matrices occur in many contexts and are useful in a great variety of problems. This section is devoted to two of the most important operations with matrices, namely matrix product and exterior power of a matrix. They are in fact nothing new at all, but are simply the operations of composition of affine maps and computation of pullbacks under affine maps written in a new notation.

The first operation, matrix product, is the operation of composition of affine maps. Consider, for example, two affine maps

$$\begin{aligned} u &= 3x - y + 2z + \text{const.} & r &= u + 8v + \text{const.} \\ v &= x + y + 4z + \text{const.} & s &= 7u + 13v + \text{const.} \\ & & t &= -4u + 6v + \text{const.} \end{aligned}$$

with matrices of coefficients

$$M = \begin{pmatrix} 3 & -1 & 2 \\ 1 & 1 & 4 \end{pmatrix} \quad \text{and} \quad N = \begin{pmatrix} 1 & 8 \\ 7 & 13 \\ -4 & 6 \end{pmatrix}$$

respectively. Their composition is given by

$$\begin{aligned} r &= (3x - y + 2z + \text{const.}) + 8(x + y + 4z + \text{const.}) + \text{const.} \\ &= [3 + 8]x + [-1 + 8]y + [2 + 8 \cdot 4]z + \text{const.} \\ &= 11x + 7y + 34z + \text{const.} \\ s &= 7(3x - y + 2z + \text{const.}) + 13(x + y + 4z + \text{const.}) + \text{const.} \\ &= [7 \cdot 3 + 13]x + [7 \cdot (-1) + 13]y + [7 \cdot 2 + 13 \cdot 4]z + \text{const.} \\ &= 34x + 6y + 66z + \text{const.} \\ t &= -4(3x - y + 2z + \text{const.}) + 6(x + y + 4z + \text{const.}) + \text{const.} \\ &= [(-4) \cdot 3 + 6]x + [(-4)(-1) + 6]y + [(-4) \cdot 2 + 6 \cdot 4]z + \text{const.} \\ &= -6x + 10y + 16z + \text{const.} \end{aligned}$$

The matrix of coefficients of the composed map, namely

$$\begin{pmatrix} 11 & 7 & 34 \\ 34 & 6 & 66 \\ -6 & 10 & 16 \end{pmatrix},$$

depends only on the matrices of coefficients M and N of the maps themselves. It is called the *product* (or *com-*

position) of these two matrices. The factors are written in the same reversed order NM as are the factors of a composed map*; thus

*If f and g are functions $f: A \to B$, $g: B \to C$ then the composed function $A \to C$ is denoted $g \circ f$, that is, 'g of f of an element of A'.

$$(2) \qquad \begin{pmatrix} 11 & 7 & 34 \\ 34 & 6 & 66 \\ -6 & 10 & 16 \end{pmatrix} = \begin{pmatrix} 1 & 8 \\ 7 & 13 \\ -4 & 6 \end{pmatrix} \begin{pmatrix} 3 & -1 & 2 \\ 1 & 1 & 4 \end{pmatrix}.$$

More generally, if two affine maps

$$y_i = \sum_{j=1}^{n} a_{ij}x_j + \text{const.} \qquad (i = 1, 2, \ldots, m)$$

and

$$z_i = \sum_{j=1}^{m} b_{ij}y_j + \text{const.} \qquad (i = 1, 2, \ldots, p)$$

are given, with matrices of coefficients

$$M = (a_{ij}) \quad \text{and} \quad N = (b_{ij})$$

respectively, then the matrix of coefficients of the composed map

$$(3) \qquad z_i = \sum_{j=1}^{n} \left[\sum_{k=1}^{m} b_{ik}a_{kj} \right] x_j + \text{const.}$$

can be found from the matrices M and N and is called their *product* NM. In this way the $m \times n$ matrix M combines with the $p \times m$ matrix N to give the $p \times n$ matrix NM. Formula (3) shows that the coefficients of the matrix NM are found from the coefficients of N and M by the rule:

(4) The term in the ith row and jth column of NM is equal to $\sum_{k=1}^{m} b_{ik}a_{kj}$ where b_{ik} is the term in the ith row and kth column of N and a_{kj} is the term in the kth row and jth column of M.

In other words, the term in the ith row and jth column of NM is found from the ith row of N and the jth column of M by multiplying corresponding terms and adding. For example the term 66 in the second row and the third column of the product (2) is found from the row and column

$$(7 \quad 13)\begin{pmatrix} 2 \\ 4 \end{pmatrix}$$

by multiplying corresponding terms and adding $7 \cdot 2 + 13 \cdot 4 = 66$.

Note that the product NM of two matrices N and M is defined if and only if the number of columns of N is equal to the number of rows of M. Thus, for example, the matrices N, M above can be multiplied in the opposite order and the result can be found by the rule (4) to be

$$\begin{pmatrix} 3 & -1 & 2 \\ 1 & 1 & 4 \end{pmatrix} \begin{pmatrix} 1 & 8 \\ 7 & 13 \\ -4 & 6 \end{pmatrix} = \begin{pmatrix} -12 & 23 \\ -8 & 45 \end{pmatrix}.$$

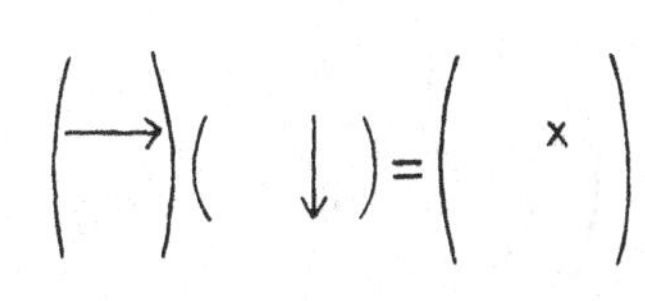

The computation of products of matrices is a simple procedure which can be (and should be) mastered with a little practice. It can be carried out as a two-handed operation, which involves running the index finger of the left hand across a row of the first factor while running the index finger of the right hand down a column of the second, multiplying corresponding entries and keeping a running total (assuming the numbers are simple enough that the arithmetic can be done mentally). Once again, this process merely represents a streamlined computation of the matrix of coefficients of a composed affine map.

The second operation, formation of the *exterior powers** of a matrix, is an abbreviated format for the computation of pullbacks of forms under an affine map. More specifically, the kth exterior power $M^{(k)}$ of a matrix M tells how to find the pullback of k-forms under an affine map whose matrix of coefficients is M. Although it is clear that only the matrix of coefficients $M = (a_{ij})$ is used in finding the pullback of a k-form under an affine map $y_i = \sum a_{ij}x_j +$ const., it is not clear how best to organize the computation. Consider the example of the map

This name is not very descriptive of the operation, since it does not resemble at all the operation of multiplying the matrix by itself—which would be possible only for square matrices—and since there is nothing exterior about it.

$$\begin{aligned} p &= 4S + 4T + 3U - V + 2 \\ q &= S + 2T + 3U + V - 3 \\ r &= 4S + 8T + 10U + 2V + 2 \end{aligned}$$

of §4.1. The matrix of coefficients is

$$M = \begin{pmatrix} 4 & 4 & 3 & -1 \\ 1 & 2 & 3 & 1 \\ 4 & 8 & 10 & 2 \end{pmatrix}$$

and the pullback maps are as follows.

1-forms

A 1-form on pqr-space can be written $A\,dp + B\,dq + C\,dr$. Its pullback is

$$\begin{aligned}&A(4\,dS + 4\,dT + 3\,dU - dV)\\&+ B(dS + 2\,dT + 3\,dU + dV)\\&+ C(4\,dS + 8\,dT + 10\,dU + 2\,dV)\\&\qquad = (4A + B + 4C)\,dS\\&\qquad + (4A + 2B + 8C)\,dT\\&\qquad + (3A + 3B + 10C)\,dU\\&\qquad + (-A + B + 2C)\,dV.\end{aligned}$$

This formula is summarized by the matrix

$$M^{(1)} = \begin{pmatrix} 4 & 1 & 4 \\ 4 & 2 & 8 \\ 3 & 3 & 10 \\ -1 & 1 & 2 \end{pmatrix}$$

which tells how to find the pullback of $A\,dp + B\,dq + C\,dr$ given the numbers A, B, C. Note that $M^{(1)}$ can be obtained from M merely by interchanging rows and columns; that is, the entry in the ith row and jth column of $M^{(1)}$ is the entry in the ith column and jth row of M. $M^{(1)}$ is also called the *transpose* or *adjoint* of M and is also denoted by M^t or M^* as well as by $M^{(1)}$.

2-forms

A 2-form on pqr-space can be written $A\,dp\,dq + B\,dp\,dr + C\,dq\,dr$. (Here, and in the remainder of this section, the lexicographic order is used.) The pullback of $A\,dp\,dq + B\,dp\,dr + C\,dq\,dr$ is

$$\begin{aligned}&A(4\,dS + 4\,dT + 3\,dU - dV)(dS + 2\,dT + 3\,dU + dV)\\&+ B(4\,dS + 4\,dT + 3\,dU - dV)(4\,dS + 8\,dT + 10\,dU + 2\,dV)\\&+ C(dS + 2\,dT + 3\,dU + dV)(4\,dS + 8\,dT + 10\,dU + 2\,dV)\\&\qquad = (4A + 16B + 0 \cdot C)\,dS\,dT\\&\qquad + (9A + 28B - 2C)\,dS\,dU\\&\qquad + (5A + 12B - 2C)\,dS\,dV\\&\qquad + (6A + 16B - 4C)\,dT\,dU\\&\qquad + (6A + 16B - 4C)\,dT\,dV\\&\qquad + (6A + 16B - 4C)\,dU\,dV\end{aligned}$$

where the terms have again been arranged in lexicographic order $dS\,dT, dS\,dU, dS\,dV, dT\,dU, dT\,dV, dU\,dV$.

This formula is summarized by the matrix

$$M^{(2)} = \begin{pmatrix} 4 & 16 & 0 \\ 9 & 28 & -2 \\ 5 & 12 & -2 \\ 6 & 16 & -4 \\ 6 & 16 & -4 \\ 6 & 16 & -4 \end{pmatrix}$$

which tells how to find the pullback of $A\,dp\,dq + B\,dp\,dr + C\,dq\,dr$, given the numbers A, B, C. The columns of $M^{(2)}$ correspond to the three basic 2-forms $dp\,dq$, $dp\,dr$, $dq\,dr$ in that order, and the rows correspond to the six basic 2-forms $dS\,dT$, $dS\,dU$, etc. The entries in each column simply give the components of the pullback of the corresponding 2-form.

3-forms

A 3-form on pqr-space can be written $A\,dp\,dq\,dr$. Its pullback is

$$0 \cdot dS\,dT\,dU + 0 \cdot dS\,dT\,dV + 0 \cdot dS\,dU\,dV + 0 \cdot dT\,dU\,dV$$

as was seen in §4.1; hence for this example

$$M^{(3)} = \begin{pmatrix} 0 \\ 0 \\ 0 \\ 0 \end{pmatrix}.$$

This matrix has one column for each of the basic 3-forms on the range, one row for each of the basic 3-forms on the domain.

In general, the kth exterior power $M^{(k)}$ of an $m \times n$ matrix M can be described as follows: Consider M as the matrix of coefficients of an affine map from n-space to m-space. Consider the coordinates on n-space and m-space as being ordered, and use the lexicographic ordering in representing k-forms so that k-forms on m-space are represented by their $\binom{m}{k}$ coefficients (in order) and k-forms on n-space by their $\binom{n}{k}$ coefficients (in order). Then the pullback map is represented as a map from $\binom{m}{k}$-space to $\binom{n}{k}$-space. $M^{(k)}$ is the $\binom{n}{k} \times \binom{m}{k}$ matrix of coefficients of this map. That is, $M^{(k)}$ has one column for each of the $\binom{m}{k}$ basic k-forms on m-space and one row for each of the $\binom{n}{k}$ basic k-forms on n-space (both in lexicographic order). The entries of $M^{(k)}$ give the coefficient of the corresponding k-form on n-space in the pullback of the corresponding k-form on m-space. [If $k > n$ or $k > m$ then $M^{(k)}$ is a 'matrix with no rows' or 'matrix with no columns'.]

It is useful to have a notation for the individual entries of $M^{(k)}$. Since

$$dp = \frac{\partial p}{\partial S}dS + \frac{\partial p}{\partial T}dT + \frac{\partial p}{\partial U}dU + \frac{\partial p}{\partial V}dV$$

$$dq = \frac{\partial q}{\partial S}dS + \frac{\partial q}{\partial T}dT + \frac{\partial q}{\partial U}dU + \frac{\partial q}{\partial V}dV$$

$$dr = \frac{\partial r}{\partial S}dS + \frac{\partial r}{\partial T}dT + \frac{\partial r}{\partial U}dU + \frac{\partial r}{\partial V}dV$$

the notation $\frac{\partial p}{\partial S}$ is already used for the coefficient of dS in the pullback of dp, which is the entry of $M^{(1)}$ in the row corresponding to dS and the column corresponding to dp. This notation is generalized to 2-forms by writing*

*The symbol $\frac{\partial(p, q)}{\partial(S, T)}$ is read 'the Jacobian of p, q with respect to S, T'.

$$dp\,dq = \frac{\partial(p, q)}{\partial(S, T)}dS\,dT + \frac{\partial(p, q)}{\partial(S, U)}dS\,dU + \cdots$$

$$dp\,dr = \frac{\partial(p, r)}{\partial(S, T)}dS\,dT + \frac{\partial(p, r)}{\partial(S, U)}dS\,dU + \cdots$$

$$dq\,dr = \frac{\partial(q, r)}{\partial(S, T)}dS\,dT + \frac{\partial(q, r)}{\partial(S, U)}dS\,dU + \cdots$$

so that, for example, $\frac{\partial(p, r)}{\partial(S, U)}$ denotes the coefficient of $dS\,dU$ in the pullback of $dp\,dr$, that is, the entry of $M^{(2)}$ in the row corresponding to $dS\,dU$ and the column corresponding to $dp\,dr$; in the example above $\frac{\partial(p, r)}{\partial(S, U)} = 28$. In general, the entries of $M^{(k)}$ are denoted by symbols of the form

$$\frac{\partial(k \text{ variables on the range})}{\partial(k \text{ variables on the domain})}$$

called the *Jacobian* (or Jacobian determinant) of the k variables in the 'numerator' with respect to the k variables in the 'denominator'.

*So called because this number determines whether the n equations in n unknowns $y_i = \sum a_{ij}x_j$ can be solved for the x's as functions of the y's. This fact, which is a special case of the Implicit Function Theorem, is proved in the next section. If the determinant of the matrix of coefficients is $\neq 0$ then solution is possible, otherwise it is impossible.

If M is an $n \times n$ matrix then $M^{(n)}$ is a 1×1 matrix, i.e. a *number*. This number is called the *determinant** of the square matrix M. [Note that only square matrices have determinants.] More generally, if M is an $m \times n$ matrix then the entries of $M^{(k)}$ are the determinants of $k \times k$ matrices, namely the $k \times k$ matrices obtained by striking out all but k rows and all but k columns of M. (A matrix obtained in this way is called a '$k \times k$ minor' of the matrix M.) For example, the coefficient $\dfrac{\partial(p, r)}{\partial(S, U)} = 28$ in $M^{(2)}$ above is found by ignoring q, T, V and writing

$$\begin{aligned} p &= 4S + 3U + \text{other terms} \\ r &= 4S + 10U + \text{other terms} \\ dp\,dr &= (4\,dS + 3\,dU + \cdots)(4\,dS + 10\,dU + \cdots) = 28\,dS\,dU + \cdots \end{aligned}$$

that is, by finding the second exterior power of the 2×2 matrix

$$\begin{pmatrix} 4 & 3 \\ 4 & 10 \end{pmatrix}$$

which is a 2×2 minor of M.

The computation of the determinant of a $k \times k$ matrix—and hence the computation of the exterior power $M^{(k)}$ of a matrix—normally involves a prohibitive amount of arithmetic when k is at all large (say $k > 3$). Fortunately it is seldom necessary or even useful to carry out such a computation. What is essential is the idea of the pullback map and the fact that the exterior power $M^{(k)}$ of a matrix is the matrix of coefficients of the pullback map. (For the technique of computing determinants see Exercise 2.)

In summary, the exterior power $M^{(k)}$ is merely a new notation with which to describe the pullback map, just as the matrix product was a new notation with which to describe compositions of affine maps.

The Chain Rule, which gives the relationship between the operations of composition of maps and formation of pullbacks can therefore be stated as a relationship between the operations of matrix product and exterior powers, namely

$$(NM)^{(k)} = M^{(k)}N^{(k)}. \tag{5}$$

That is, the pullback under the composed† map is the

†Note the reversal of order. M maps n-space to m-space, N maps m-space to p-space, hence NM maps n-space to p-space. $N^{(k)}$ maps forms on p-space to forms on m-space and $M^{(k)}$ maps forms on m-space to forms on n-space, hence $M^{(k)}N^{(k)}$ maps forms on p-space to forms on n-space.

pullback of the pullback. N and M must, of course, be matrices such that the composition NM is defined; that is, the number of columns in N must be equal to the number of rows in M—say M is an $m \times n$ matrix and N a $p \times m$ matrix. Two cases of (5) are particularly important in that they are used frequently:

(i) $k = 1$. Then $M^{(k)}$, $N^{(k)}$, $(NM)^{(k)}$ are merely the transposes of M, N, NM and (5) states that the transpose of a product of two matrices is the product of the transposes in the reverse order. This is easily proved directly from the definition of the product of matrices.

(ii) $m = n = p = k$. In this case, M, N, NM are square matrices and $M^{(k)}$, $N^{(k)}$, $(NM)^{(k)}$ are 1×1 matrices, i.e. numbers. These numbers are by definition the determinants of M, N, NM respectively and (5) states that the determinant of the product of two square matrices is equal to the product of their determinants. This fact is not at all obvious from the definition of the product of matrices and the definition of the determinant of a square matrix, except insofar as the Chain Rule itself is obvious from the geometrical meaning of pullbacks.

The chain rule in Jacobian notation is $\frac{\partial z}{\partial x} = \sum \frac{\partial z}{\partial y} \cdot \frac{\partial y}{\partial x}$.

In terms of Jacobians these two instances of the rule (5) can be stated as follows: If $z_1, z_2, \ldots, z_p$ are affine functions of $y_1, y_2, \ldots, y_m$ and if $y_1, y_2, \ldots, y_m$ are affine functions of $x_1, x_2, \ldots, x_n$ then the partial derivatives $\frac{\partial z_i}{\partial x_j}$ of the composed function are given by

$$\frac{\partial z_i}{\partial x_j} = \sum_{\nu=1}^{m} \frac{\partial z_i}{\partial y_\nu} \frac{\partial y_\nu}{\partial x_j} \qquad \begin{array}{l}(i = 1, 2, \ldots, p)\\ (j = 1, 2, \ldots, n).\end{array} \tag{6}$$

If $n = m = p$, then

$$\frac{\partial(z_1, z_2, \ldots, z_n)}{\partial(y_1, y_2, \ldots, y_n)} \frac{\partial(y_1, y_2, \ldots, y_n)}{\partial(x_1, x_2, \ldots, x_n)} = \frac{\partial(z_1, z_2, \ldots, z_n)}{\partial(x_1, x_2, \ldots, x_n)}. \tag{7}$$

The Chain Rule derives its name from the resemblance of these formulas to the Chain Rule of Differentiation

$$\frac{dz}{dx} = \frac{dz}{dy} \frac{dy}{dx}.$$

In summary, the notion of an $m \times n$ matrix has been defined, the operations of product and exterior power of matrices have been defined, the notion of the Jacobian of k coordinates on the range with respect to k coordinates of the domain relative to a given affine map has been defined, and the interrelation of these concepts and operations has been expressed by the Chain Rule (5).

Exercises

1 (a) Write the matrix of coefficients of each of the 8 affine maps in Exercise 1, §4.1, labeling them $M_a, M_b, \ldots, M_h$.

(b) Make a complete list of all products which can be formed from pairs of these matrices (e.g. M_bM_c is meaningful, M_cM_b is not).

(c) Compute all the products in (b).

(d) Find all exterior powers of the matrices $M_a, M_b, \ldots, M_h$.

(e) Verify the chain rule (6) in all cases listed under (b).

2 *Computation of determinants.* The determinant of a square matrix is denoted by writing the matrix between straight lines rather than curved ones, e.g.

$$\begin{pmatrix} 3 & 4 & 7 \\ 2 & 1 & -2 \\ 6 & 2 & -3 \end{pmatrix} \quad \text{and} \quad \begin{vmatrix} 3 & 4 & 7 \\ 2 & 1 & -2 \\ 6 & 2 & -3 \end{vmatrix}$$

denote a 3×3 matrix and a number respectively. The problem at hand is to develop methods of computing numbers which are given as determinants. In Chapter 1 the explicit formulas

$$\begin{vmatrix} a & b \\ a' & b' \end{vmatrix} = ab' - a'b$$

and

$$\begin{vmatrix} a & b & c \\ a' & b' & c' \\ a'' & b'' & c'' \end{vmatrix} = ab'c'' + a'b''c + a''bc' - cb'a'' - c'b''a - c''ba'$$

were found. Thus for example the determinant above is

$$3 \cdot 1 \cdot (-3) + 2 \cdot 2 \cdot 7 + 6 \cdot 4 \cdot (-2) - 7 \cdot 1 \cdot 6 - (-2) \cdot 2 \cdot 3 - (-3) \cdot 4 \cdot 2$$
$$= -9 + 28 - 48 - 42 + 12 + 24 = -35.$$

The analogous formula for $n \times n$ matrices contains $n!$ terms and is therefore out of the question computationally for large n. A better method of computing large determinants is to prove the following rules:

(a) Interchanging any two rows or any two columns of a square matrix changes the sign of its determinant.
(b) Adding any multiple of one row to another row or adding any multiple of one column to another column leaves the determinant unchanged.
(c) Multiplying all entries of one row or all entries of one column by a number c multiplies the determinant* by c.
(d) If the first row of a matrix is 1, 0, 0, . . . , 0 then its determinant is the determinant of the $(n - 1) \times (n - 1)$ matrix obtained by striking out the first row and first column of the matrix. The same is true if the first column is 1, 0, 0, . . . , 0.

*Thus there is a very important distinction between multiplying a matrix by c and multiplying a determinant by c. To multiply a matrix by c means to multiply all entries by c (see §4.5), which multiplies its determinant by c^n.

Prove these rules. [The new matrices in (a)–(c) can be written as products of the given matrix with certain simple matrices whose determinants can be computed directly from the definition. The result then follows from the Chain Rule. The rule (d) follows directly from the definition if the first row *and* the first column are of the form 1, 0, 0, . . . , 0. Then use (b).] Using these rules the determinant above can be found by

$$\begin{vmatrix} 3 & 4 & 7 \\ 2 & 1 & -2 \\ 6 & 2 & -3 \end{vmatrix} = \begin{vmatrix} -5 & 0 & 15 \\ 2 & 1 & -2 \\ 6 & 2 & -3 \end{vmatrix} = \begin{vmatrix} -5 & 0 & 15 \\ 2 & 1 & -2 \\ 2 & 0 & 1 \end{vmatrix}$$

$$= -\begin{vmatrix} 2 & 1 & -2 \\ -5 & 0 & 15 \\ 2 & 0 & 1 \end{vmatrix} = \begin{vmatrix} 1 & 2 & -2 \\ 0 & -5 & 15 \\ 0 & 2 & 1 \end{vmatrix}$$

$$= \begin{vmatrix} -5 & 15 \\ 2 & 1 \end{vmatrix} = -5\begin{vmatrix} 1 & -3 \\ 2 & 1 \end{vmatrix} = -5\begin{vmatrix} 1 & -3 \\ 0 & 7 \end{vmatrix}$$

$$= (-5)(7) = -35.$$

Compute the following determinants:

$$\begin{vmatrix} 3 & 0 & 1 \\ 1 & 2 & 5 \\ -1 & 4 & 2 \end{vmatrix}, \quad \begin{vmatrix} 3 & -1 & 5 \\ -1 & 2 & 1 \\ -2 & 4 & 3 \end{vmatrix}$$

$$\begin{vmatrix} -1 & 1 & 2 & 0 \\ 0 & 3 & 2 & 1 \\ 0 & 4 & 1 & 2 \\ 3 & 1 & 5 & 7 \end{vmatrix}, \quad \begin{vmatrix} 0 & 1 & 0 & 0 \\ 1 & 0 & 1 & 0 \\ 0 & 1 & 0 & 1 \\ 0 & 0 & 1 & 0 \end{vmatrix}$$

$$\begin{vmatrix} 8 & 4 & 7 & 3 \\ 2 & 4 & -5 & -7 \\ 1 & 3 & 2 & 9 \\ -6 & -1 & 2 & 5 \end{vmatrix}$$

[The last one is −2242.]

3 *Decomposition of a matrix.* In Exercise 8, §1.3, it was shown that every affine map from a plane to a plane can be written as a composition of rotations $(x, y) \to (y, -x)$ of 90°, reflections $x \to -x$, shears, scale factors (including 0) in coordinate directions, and translations. Prove this again and generalize to n dimensions as follows: Describe the matrices of coefficients of affine maps of n-space to n-space of the above simple types (rotations, reflections, shears, etc.). Show that it suffices to show that every $n \times n$ matrix can be written as a product of such matrices. Prove that by multiplying an $n \times n$ matrix on the left and/or right by such matrices it can be reduced to a matrix whose first row and column are $1, 0, 0, \ldots, 0$ (unless the matrix is identically zero). Then use induction on n.

4 Show that the determinant of the transpose of a square matrix is equal to the determinant of the matrix itself. [Use Exercise 3.]

4.4 *The Implicit Function Theorem for Affine Maps*

The conclusions of §4.1 (pp. 83–84) concerning the results of step-by-step elimination can be summarized as follows:

Implicit Function Theorem*

**The theorem asserts that the relations (1) imply relationships of the form (2) without giving (2) explicitly, hence the name.*

A system of equations

$$(1) \qquad y_i = \sum_{j=1}^{n} a_{ij}x_j + \text{const.} \qquad (i = 1, 2, \ldots, m)$$

is equivalent to a system of the form

$$(2) \quad \begin{cases} \text{(2a)} \quad x_i = \sum_{j=1}^{r} A_{ij}y_j + \sum_{j=r+1}^{n} B_{ij}x_j + \text{const.} \qquad (i = 1, 2, \ldots, r) \\ \text{(2b)} \quad y_i = \sum_{j=1}^{r} C_{ij}y_j + \text{const.} \qquad (i = r + 1, \ldots, m) \end{cases}$$

if and only if the affine mapping defined by (1) has the properties that

(i) $\dfrac{\partial(y_1, \ldots, y_r)}{\partial(x_1, \ldots, x_r)} \neq 0$ and

(ii) the pullback of every k-form in the y's is zero for $k > r$.

As was pointed out in §4.1, equations (2) constitute a

solution of the problem "given $y_1, y_2, \ldots, y_m$ find all $x_1, x_2, \ldots, x_n$ satisfying (1)." If $y_1, y_2, \ldots, y_m$ do not satisfy (2b), then there are no such $x_1, x_2, \ldots, x_n$; if $y_1, y_2, \ldots, y_n$ do satisfy (2b), then there are such $x_1, x_2, \ldots, x_n$ and all of these can be found by choosing $x_{r+1}, x_{r+2}, \ldots, x_n$ arbitrarily and using (2a) to determine $x_1, x_2, \ldots, x_r$.

Proof

Assume first that relations of the form (2) are given which are equivalent to (1). It is to be shown that then (i) and (ii) must be satisfied. Fixing values of $x_{r+1}, x_{r+2}, \ldots, x_n$ in the first r equations of (1) and in (2a) gives equivalent relations of the form

$$y_i = \sum_{j=1}^{r} a_{ij}x_j + \text{const.}, \quad x_i = \sum_{j=1}^{r} A_{ij}y_j + \text{const.}$$

The equivalence of these relations implies that if the first set of equations is used to define $(y_1, y_2, \ldots, y_r)$ given a set of values $(x_1, x_2, \ldots, x_r)$, and if these values of $(y_1, y_2, \ldots, y_r)$ are substituted into the second set of equations, they will yield the original values of $(x_1, x_2, \ldots, x_r)$. In short, the composition of these two maps $(x) \to (y) \to (x)$ is the identity map. Thus by the Chain Rule

$$\frac{\partial(x_1, x_2, \ldots, x_r)}{\partial(y_1, y_2, \ldots, y_r)} \frac{\partial(y_1, y_2, \ldots, y_r)}{\partial(x_1, x_2, \ldots, x_r)} = \frac{\partial(x_1, x_2, \ldots, x_r)}{\partial(x_1, x_2, \ldots, x_r)} = 1$$

where the Jacobians are the Jacobians of the two affine maps above and of the identity map. Thus

$$\frac{\partial(y_1, y_2, \ldots, y_r)}{\partial(x_1, x_2, \ldots, x_r)}$$

cannot be zero* and (i) is proved. To prove (ii), note that by (2b) the map (1) can be written as a composition $(x_1, x_2, \ldots, x_n) \to (y_1, y_2, \ldots, y_r) \to (y_1, y_2, \ldots, y_m)$ where the first map is the first r equations of (1) and where the second map is (2b) together with the identity map $(y_1, \ldots, y_r) \to (y_1, \ldots, y_r)$. By the Chain Rule the pullback under the composed map—i.e. the pullback under (1)—is the pullback of the pullback. Since any $(r + 1)$-form in the r variables $(y_1, y_2, \ldots, y_r)$ is zero, (ii) follows.

*$ab = 1$ implies $a \neq 0$.

Now assume that (i) and (ii) are satisfied. It is to be

shown that then there exist relations of the form (2) equivalent to (1). This will be done by showing that such relations (2) can be derived by step-by-step elimination.

In order to avoid difficulties arising from the need to rearrange x's or y's during the elimination process, it is useful to rearrange the equations at the outset as follows: Since the pullback of $dy_1\, dy_2 \ldots dy_r$ has a non-zero term which involves none of the factors $dx_{r+1}, dx_{r+2}, \ldots,$ dx_n $\left(\text{namely the term } \dfrac{\partial(y_1, \ldots, y_r)}{\partial(x_1, \ldots, x_r)}\, dx_1\, dx_2 \ldots dx_r\right)$, the pullback of $dy_1\, dy_2 \ldots dy_{r-1}$ must also have a non-zero term which involves none of the factors dx_{r+1}, $dx_{r+2}, \ldots, dx_n$; that is, it must have a non-zero term in $dx_2\, dx_3 \ldots dx_r$ or $dx_1\, dx_3 \ldots dx_r$ or $\ldots$ or $dx_1\, dx_2 \ldots dx_{r-1}$. Therefore by rearranging the first r of the x's it can be assumed that

$$\frac{\partial(y_1, y_2, \ldots, y_{r-1})}{\partial(x_1, x_2, \ldots, x_{r-1})} \neq 0.$$

In the same way, by rearranging the first $r - 1$ of the x's it can be assumed that

$$\frac{\partial(y_1, y_2, \ldots, y_{r-2})}{\partial(x_1, x_2, \ldots, x_{r-2})} \neq 0.$$

In the same way it follows that the first $r - 2$ of the x's can be rearranged (if necessary) so that the Jacobian of $y_1, y_2, \ldots, y_{r-3}$ with respect to $x_1, x_2, \ldots, x_{r-3}$ is non-zero, then the Jacobian of $y_1, y_2, \ldots, y_{r-4}$ with respect to $x_1, x_2, \ldots, x_{r-4}$, and so forth. That is, by rearranging the first r of the x's (if necessary) it can be assumed that the given system (1) satisfies the stronger assumption

$$\text{(i}'\text{)} \qquad \frac{\partial(y_1, y_2, \ldots, y_k)}{\partial(x_1, x_2, \ldots, x_k)} \neq 0 \quad \text{for} \quad k = 1, 2, \ldots, r.$$

It will be shown that if (i′) and (ii) are satisfied then relations of the form (2) can be found by step-by-step elimination (without rearrangement).

Since $a_{11} = \dfrac{\partial y_1}{\partial x_1}$ is not zero (by (i′)) the first equation of (1) can be solved for x_1 in terms of $y_1, x_2, x_3, \ldots, x_n$ and substituted into the remaining equations to give $y_2, y_3, \ldots, y_m$ in terms of $y_1, x_2, x_3, \ldots, x_n$. This is the first step of the elimination process. Suppose that k steps of the process have been carried out to put the equations in the form

$$
(3) \qquad
\begin{aligned}
x_i &= \sum_{j=1}^{k} \mathcal{A}_{ij} y_j + \sum_{j=k+1}^{n} \mathcal{B}_{ij} x_j + \text{const.} \\
&\qquad\qquad (i = 1, 2, \ldots, k) \\
y_i &= \sum_{j=1}^{k} \mathcal{C}_{ij} y_j + \sum_{j=k+1}^{n} \mathcal{D}_{ij} x_j + \text{const.} \\
&\qquad\qquad (i = k+1, k+2, \ldots, m).
\end{aligned}
$$

The next step of the process requires that the equation for y_{k+1} be solved for x_{k+1}, which is possible if and only if $\mathcal{D}_{k+1,k+1} \neq 0$.

Fixing values of $x_{k+2}, x_{k+3}, \ldots, x_n$ and considering the first $k+1$ equations of (1) as functions of $x_1, x_2, \ldots, x_{k+1}$, this map $(x) \to (y)$ can be written as a composition $(x_1, x_2, \ldots, x_{k+1}) \to (y_1, y_2, \ldots, y_k, x_{k+1}) \to (y_1, y_2, \ldots, y_{k+1})$ where the first map is the first k equations of (1) together with $x_{k+1} = x_{k+1}$ and where the second map is $y_i = y_i$ $(i = 1, 2, \ldots, k)$ together with the $(k+1)$st equation of (3). Using the Chain Rule,

$$
\begin{aligned}
\frac{\partial(y_1, y_2, \ldots, y_{k+1})}{\partial(x_1, x_2, \ldots, x_{k+1})} &= \frac{\partial(y_1, y_2, \ldots, y_k, y_{k+1})}{\partial(y_1, y_2, \ldots, y_k, x_{k+1})} \cdot \frac{\partial(y_1, y_2, \ldots, y_k, x_{k+1})}{\partial(x_1, x_2, \ldots, x_k, x_{k+1})} \\
&= \mathcal{D}_{k+1,k+1} \frac{\partial(y_1, y_2, \ldots, y_k)}{\partial(x_1, x_2, \ldots, x_k)}.
\end{aligned}
$$

Thus (i′) implies $\mathcal{D}_{k+1,k+1} \neq 0$ for $k = 1, 2, 3, \ldots, r-1$, and the process continues until $k = r$. At this point $\mathcal{D}_{r+1,r+1} = 0$ by the above argument (using (ii)). In fact, this conclusion holds even if $x_{r+1}, x_{r+2}, \ldots, x_n$ or $y_{r+1}, y_{r+2}, \ldots, y_m$ are rearranged. Hence all of the $\mathcal{D}$'s must be zero—that is, the equations (3) must have the desired form (2)—and the Implicit Function Theorem follows.

This existence proof (if (i) and (ii) then *there exists* (2)) is constructive in the sense that it tells exactly how to construct the relations (2) in a finite number of steps—first arrange the equations (1) as indicated and then perform step-by-step elimination. However, this finite number is immense, even for relatively small systems of equations (1), and the process prescribed by the proof is actually wholly impractical in most cases. Even with the aid of a computing machine this method of solution is usually inadvisable because the amount of arithmetic is so great as to make the error due to roundoff intolerably large. Therefore the proof by elimination is constructive only in a very theoretical sense, and no really practical method of constructing a solution has in fact been given here. More practical methods are discussed in Chapter 7.

In theoretical work it is often useful to have a formula which expresses the solution in *closed form*. Specifically, the Implicit Function Theorem implies that the coefficients a_{ij} determine the coefficients A_{ij}, B_{ij}, C_{ij}; a solution in closed form is a formula expressing the A's, B's, and C's as functions of the a's. Although the result, known as *Cramer's Rule* or the *formula for the inverse of a matrix*, is of no practical significance and will not be used in the remainder of this book, its proof is included here to demonstrate the usefulness of forms in linear algebra.

Solution in Closed Form

Suppose that the systems of equations

$$(1) \qquad y_i = \sum_{j=1}^{n} a_{ij}x_j + \text{const.} \qquad (i = 1, 2, \ldots, m)$$

and

$$(2) \begin{cases} \text{(2a)} \quad x_i = \sum_{j=1}^{r} A_{ij}y_j + \sum_{j=r+1}^{n} B_{ij}x_j + \text{const.} \\ \qquad\qquad (i = 1, 2, \ldots, r) \\ \text{(2b)} \quad y_i = \sum_{j=1}^{r} C_{ij}y_j + \text{const.} \\ \qquad\qquad (i = r+1, r+2, \ldots, m) \end{cases}$$

are equivalent; that is, suppose that a set of $n + m$ numbers $(x_1, x_2, \ldots, x_n, y_1, y_2, \ldots, y_m)$ satisfies (1) if and only if it satisfies (2). The problem is to express the numbers A, B, C in terms of the numbers a. The method will be to apply the Chain Rule to a suitably chosen composite map. Consider the map defined by

$$(4) \quad \begin{aligned} x_i &= \sum_{j=1}^{r} A_{ij}u_j + \sum_{j=r+1}^{n} B_{ij}u_j + \text{const.} && (i = 1, 2, \ldots, r) \\ x_i &= u_i && (i = r+1, r+2, \ldots, n) \\ y_i &= u_i && (i = 1, 2, \ldots, r) \\ y_i &= \sum_{j=1}^{r} C_{ij}u_j + \text{const.} && (i = r+1, r+2, \ldots, m). \end{aligned}$$

where the constants are as in (2). The points of the image of this map satisfy the relations (2), so the assumption that (1) and (2) are equivalent implies that all points in the image satisfy the relations (1). Hence the y-coordinates of such a point can be obtained from its x-coordinates by using (1). In other words (4) can be written as the composition of the maps

$$
(5)\qquad \begin{aligned}
x_i &= \sum_{j=1}^{r} A_{ij}u_j + \sum_{j=r+1}^{n} B_{ij}u_j + \text{const.} \\
&\qquad\qquad\qquad\qquad (i = 1, 2, \ldots, r) \\
x_i &= u_i \qquad\qquad\qquad (i = r+1, r+2, \ldots, n)
\end{aligned}
$$

and

$$
(6)\qquad \begin{aligned}
x_i &= x_i & (i = 1, 2, \ldots, n) \\
y_i &= \sum_{j=1}^{n} a_{ij}x_j + \text{const.} & (i = 1, 2, \ldots, m)
\end{aligned}
$$

where the constants are as in (1). The pullback of the n-form $dy_1\,dy_2 \ldots dy_r\,dx_{r+1}\,dx_{r+2} \ldots dx_n$ under (4) is $du_1\,du_2 \ldots du_n$, whereas if just one of the factors in this n-form is changed then the pullback is $du_1\,du_2 \ldots du_n$ multiplied by one of the numbers A, B, C to be found. The first two columns of the table list the pullbacks of such n-forms under (4). On the other hand, the pullbacks of these n-forms under (6) are found by replacing the dy's by their expressions $dy_i = \sum a_{ij}\,dx_j$ and multiplying out to obtain a function of the a's times $dx_1\,dx_2 \ldots dx_n$. The resulting function of the a's can be expressed as a Jacobian of the y's with respect to the x's. For example, the pullback of the form in the second row of the table is found by moving the new factor dx_i to the front, which involves $(j - 1)$ interchanges and therefore a factor $(-1)^{j-1}$, by finding the pullback of the remaining $(r - 1)$-form in the dy's, by ignoring all terms except the term which contains none of the factors $dx_i\,dx_{r+1}\,dx_{r+2} \ldots dx_n$, and finally by moving dx_i to its natural position, which involves $i - 1$ interchanges and therefore a factor of $(-1)^{i-1}$. The result is that the pullback under (6) of the n-form in the second row is

$$
\begin{aligned}
(-1)^{j-1}\,dx_i\left[\frac{\partial(y_1, \ldots, \cancel{y_j}, \ldots\ldots, y_r)}{\partial(x_1, \ldots\ldots, \cancel{x_i}, \ldots, x_r)}\,dx_1 \ldots \cancel{dx_i} \ldots dx_r\right] dx_{r+1} \ldots dx_n& \\
= (-1)^{i+j}\,\frac{\partial(y_1, \ldots, \cancel{y_j}, \ldots\ldots, y_r)}{\partial(x_1, \ldots\ldots, \cancel{x_i}, \ldots, x_r)}\,dx_1\,dx_2 \ldots dx_n.&
\end{aligned}
$$

The pullback of	under (4) is $du_1\, du_2 \ldots du_n$ times	and under (6) is $dx_1\, dx_2 \ldots dx_n$ times
$dy_1\, dy_2 \ldots dy_r\, dx_{r+1}\, dx_{r+2} \ldots dx_n$	1	$\dfrac{\partial(y_1, y_2, \ldots, y_r)}{\partial(x_1, x_2, \ldots, x_r)}$
$dy_1 \ldots \overset{dx_i}{\cancel{dy_j}} \ldots dy_r\, dx_{r+1} \ldots dx_n$ $i \leqq r,\ j \leqq r$	A_{ij}	$(-1)^{i+j} \dfrac{\partial(y_1, \ldots, \cancel{y_j}, \ldots, y_r)}{\partial(x_1, \ldots, \cancel{x_i}, \ldots, x_r)}$
$dy_1 \ldots dy_r\, dx_{r+1} \ldots \overset{dx_i}{\cancel{dx_j}} \ldots dx_n$ $i \leqq r,\ j > r$	B_{ij}	$-\dfrac{\partial(y_1, y_2, \ldots, y_r)}{\partial(x_1, \ldots, \underset{x_j}{\cancel{x_i}}, \ldots, x_r)}$
$dy_1 \ldots \overset{dy_i}{\cancel{dy_j}} \ldots dy_r\, dx_{r+1} \ldots dx_n$ $i > r,\ j \leqq r$	C_{ij}	$\dfrac{\partial(y_1, \ldots, \overset{y_i}{\cancel{y_j}}, \ldots, y_r)}{\partial(x_1, x_2, \ldots, x_r)}$

This result is given in the third column of the table. The corresponding results for the other 3 rows, also given in the table, are found in the same way.

Now by the Chain Rule the pullback of the first column under (4) can also be written as the pullback under (5) of the pullback under (6). Since the pullback of $dx_1\, dx_2 \ldots dx_n$ under (5) is a constant $\dfrac{\partial(x_1, \ldots, x_n)}{\partial(u_1, \ldots, u_n)}$ times $du_1\, du_2 \ldots du_n$ it follows that

$$\text{second column} = \text{const.} \cdot \text{third column};$$

that is, the two columns are proportional, which means that the ratios of corresponding entries are equal. Thus A_{ij} is equal to the ratio of the last entry of the second row of the table to the last entry of the first row, which is a function of the a's. In the same way B_{ij}, C_{ij} can be expressed as ratios of entries in the third column and hence as functions of the a's as desired.

For example, if $m = n = r$ so that $M = (a_{ij})$ is an $n \times n$ matrix with non-zero determinant, then the coefficient $A_{ij} = \dfrac{\partial x_i}{\partial y_j}$ of the inverse function (2a) is given by

$$A_{ij} = \frac{A_{ij}}{1} = \frac{(-1)^{i+j}\det(M_{ij})}{\det(M)}$$

where M_{ij} is the $(n - 1) \times (n - 1)$ minor of M obtained by striking out the column corresponding to x_i and the row corresponding to y_j. This is the formula for the inverse of a matrix.

Exercises **1** Invert the equations

$$\begin{aligned} u &= x + 2y + 2z \\ v &= 5x + y + 3z \\ w &= -2x + 2y + z \end{aligned}$$

by (a) step-by-step elimination and by (b) using the formula for the inverse of a matrix.

2 Suppose that the system (1) is given and that the A's, B's, and C's of (2) are known; how can the constants in (2) be determined? For example, extending Exercise 1, invert the equations

$$\begin{aligned} u &= x + 2y + 2z + 7 \\ v &= 5x + y + 3z - 2 \\ w &= -2x + 2y + z + 1. \end{aligned}$$

3 (a) A mapping of the form

$$\begin{aligned} u &= a_{11}x + a_{12}y + a_{13}z + b_1 \\ v &= a_{21}x + a_{22}y + a_{23}z + b_2 \end{aligned}$$

cannot possibly be one-to-one. How could you find two points which have the same image point?

(b) A mapping

$$\begin{aligned} u &= a_{11}x + a_{12}y + b_1 \\ v &= a_{21}x + a_{22}y + b_2 \\ w &= a_{31}x + a_{32}y + b_3 \end{aligned}$$

cannot be onto. How would you find a point (u, v, w) which is not the image of any point (x, y)?

4 Show that an affine map of n-space to m-space cannot be one-to-one if $n > m$ and cannot be onto if $n < m$. This implies the 'geometrically obvious' statement that such a map can be one-to-one and onto only if $n = m$.

5 *Proposition.* A matrix whose entries are all integers has an inverse whose entries are all integers if and only if its determinant is ± 1.

(a) Which half of this proposition is an immediate consequence of the Chain Rule?

(b) Prove the other half using the formula for the inverse of a matrix.

4.5

Abstract Vector Spaces

The definitions and theorems of this section will not be used to any appreciable extent in the remainder of the book, and readers who are primarily interested in calculus may prefer to skip this section entirely. The theory of vector spaces provides essentially a new *vocabulary* for formulating the basic facts about the solution of linear equations. Once one becomes accustomed to the new terms, this vocabulary is natural, simple, and useful. It is used in virtually all branches of mathematics.

A vector* space is a set in which any two elements can be added and any element can be multiplied by a number. The set of all k-forms on n-space is an excellent example of a vector space and is the principal example studied in this book. To write a 2-form on 3-space as $A\,dy\,dz + B\,dz\,dx + C\,dx\,dy$ means that A (a number) times $dy\,dz$ (a 2-form on 3-space—namely, oriented area of the projection on the yz-plane) is another 2-form $A\,dy\,dz$; that similarly $B\,dz\,dx$, $C\,dx\,dy$ are 2-forms; and that the sum of these 2-forms is again a 2-form. More generally, the sum of two k-forms and a number times a k-form are defined as in §4.2.

**'Vector' is the Latin word for 'carrier'. Originally it referred to a flow (convection) represented by an arrow. It then came to mean any quantity (velocity, force) represented by an arrow, i.e., any quantity with magnitude and direction. Finally it came to refer to quantities which could be added and multiplied by numbers.*

Another natural example of a vector space is the set of all functions from any set S to the real numbers $\mathbf{R}$,† with the operations of addition and multiplication by numbers defined in the obvious way. For example, let S be the interval $\{0 \le x \le 1\}$ and let V be the set of all real-valued functions defined on this interval. If f and g are elements of V, that is, if f and g are real-valued functions defined on $\{0 \le x \le 1\}$, then $f + g$ is again an element of V; namely, $f + g$ is the function which assigns to the point x of $\{0 \le x \le 1\}$ the value $f(x) + g(x)$. This operation is so natural that one writes $F(x) = 3x^2 + 2x + 1$ without stopping to point out that one is adding the *functions* $3x^2$, $2x$, 1. Similarly $3x^2$ is the number 3 times the function x^2; and, in general, if a is a number and f an element of V, then $a \cdot f$ denotes the element which assigns to each x in the interval $\{0 \le x \le 1\}$ the value $a \cdot f(x)$. These definitions of $f + g$ and $a \cdot f$ have nothing to do with the fact that S is $\{0 \le x \le 1\}$ and serve to make $V = \{\text{functions } S \to \mathbf{R}\}$ into a vector space for *any* set S.

†The letter **R** *denotes the set of real numbers.*

An important special case is the case in which S is the finite set consisting of the first n integers $S = \{1, 2, \ldots, n\}$. For the sake of definiteness, let $S = \{1, 2, 3\}$. A real-valued function $f\colon S \to \mathbf{R}$ is a rule which assigns a real number $f(1), f(2), f(3)$ to each of the three elements of S. It is customary to write f_1 instead of $f(1)$, to write f_2, f_3 instead of $f(2), f(3)$, and to describe the function f by listing its three values $f = (f_1, f_2, f_3)$. Thus, the list $(7, -4, 2)$ represents the function which assigns the value 7 to 1, the value -4 to 2, and the value 2 to 3. In this way functions from the set $S = \{1, 2, 3\}$ to $\mathbf{R}$ are represented simply by triples of real numbers. This set will be denoted by $V_3 = \{$all functions from the set $\{1, 2, 3\}$ to $\mathbf{R}\}$. The sum of the function $f = (7, -4, 2)$ and the function $g = (3, 2, 1)$ is the function which assigns to 1 the value $7 + 3$ to 2 the value $-4 + 2$ and to 3 the value $2 + 1$, i.e.

$$(7, -4, 2) + (3, 2, 1) = (10, -2, 3).$$

In short, elements of V_3 are added componentwise. Similarly, 2 times $f = (7, -4, 2)$ is $(14, -8, 4)$; that is, multiplication of elements of V_3 by numbers is carried out in the obvious way.

A *subspace* of a vector space is a subset with the property that sums of elements in the subset are again in the subset and multiples of elements of the subset are again in the subset. For example, the vector space of continuous functions on $\{0 \leq x \leq 1\}$ is a subspace of the space of all functions on $\{0 \leq x \leq 1\}$ considered above; that is, the sum of two continuous functions is a continuous function and any multiple of a continuous function is a continuous function. The vector space of all polynomial functions on $\{0 \leq x \leq 1\}$ is a subspace of the vector space of continuous functions, and *a fortiori* of the vector space of all functions on $\{0 \leq x \leq 1\}$, because polynomials are continuous functions and sums and multiples of polynomials are polynomials. The space of all polynomials of degree $\leqq 5$, for example, is in turn a subspace of the vector space of all polynomials because sums and multiples of polynomials of degree at most five have degree at most five.* (Note that the same is not true of the set of polynomials of degree exactly five because $(3x^5 + x^2 - 1) + (-3x^5 + x^4 - 2x + 1) = x^4 + x^2 - 2x$.) An example of a subspace of the space V_5 (= real-valued functions on the set $\{1, 2, 3, 4, 5\}$, represented as quintuples of numbers) is the space of functions the sum of whose values is 0, i.e. all $(f_1, f_2, f_3, f_4, f_5)$ such that $f_1 + f_2 + f_3 + f_4 + f_5 = 0$. Another example is the space of all symmetric functions, i.e. all $(f_1, f_2, f_3, f_4, f_5)$ such that $f_1 = f_5, f_2 = f_4$. The inter-

*Another example of a subspace is the following: Intuitively, k-forms on n-space are functions assigning numbers to oriented k-dimensional surfaces in n-space (e.g. $n = 3$, $k = 2$). Sums and multiplies of k-forms are k-forms; hence the vector space of k-forms on n-space is a subspace of the vector space of all functions assigning numbers to oriented k-dimensional surfaces in n-space.

section of these two subspaces of V_5 is again a subspace, namely the subspace of all $(f_1, f_2, f_3, f_4, f_5)$ satisfying the three relations

$$
\begin{aligned}
f_1 + f_2 + f_3 + f_4 + f_5 &= 0 \\
f_1 &= f_5 \\
f_2 &= f_4.
\end{aligned} \tag{1}
$$

One immediately verifies that if f, g are elements of V_5 satisfying any one of these relations, then $f + g$ and $a \cdot f$ satisfy the same relation.

The most important concepts relating to vector spaces are those of *basis* and *dimension*. It was shown in §4.2 that every 2-form on $xyzt$-space can be written in exactly one way in the form

$$A\,dx\,dy + B\,dx\,dz + C\,dx\,dt + D\,dy\,dz + E\,dy\,dt + F\,dz\,dt.$$

This is what it means to say that the 2-forms $dx\,dy$, $dx\,dz$, $dx\,dt$, $dy\,dz$, $dy\,dt$, $dz\,dt$ are a basis of the space of 2-forms on $xyzt$-space. The space of 2-forms on 4-space is said to be *six-dimensional*, because 2-forms are uniquely described by the 6 numbers (A, B, C, D, E, F). The space V_3 is three-dimensional since elements of V_3 are uniquely described by their three values f_1, f_2, f_3; if one takes $\delta_1, \delta_2, \delta_3$ to be the elements $(1, 0, 0)$, $(0, 1, 0)$, $(0, 0, 1)$ of V_3—that is, the three functions on $S = \{1, 2, 3\}$ which are 1 on one element of S and 0 on the others—then every element of V_3 can be written in exactly one way in the form

$$
\begin{aligned}
f_1\,\delta_1 + f_2\,\delta_2 + f_3\,\delta_3 &= f_1(1, 0, 0) + f_2(0, 1, 0) + f_3(0, 0, 1) \\
&= (f_1, f_2, f_3)
\end{aligned}
$$

and $\delta_1, \delta_2, \delta_3$ are a basis of V_3.

In general, a *basis* of a vector space is a set of elements $v_1, v_2, \ldots, v_n$ of the space, with the property that every element of the space can be written in exactly one way as a combination

$$x_1v_1 + x_2v_2 + \cdots + x_nv_n \tag{2}$$

where $x_1, x_2, \ldots, x_n$ are numbers. When a vector space has a basis consisting of n elements it is said to be *n-dimensional*. The vector space of k-forms on n-space is $\binom{n}{k}$-dimensional and a basis is given by the basic k-forms in lexicographic order. The vector space V_n of real-valued functions on $S = \{1, 2, \ldots, n\}$ is n-dimensional and a basis is given by the 'δ-functions' $\delta_1, \delta_2, \ldots, \delta_n$ where δ_k is the function whose value is 1 on k and 0 on all other integers $1, 2, \ldots, n$. The space of all functions on $\{0 \leq x \leq 1\}$ is *infinite-dimensional* (i.e. not n-dimensional for any integer n) because no finite number of functions can possibly constitute a basis. In fact, the space of all continuous functions and even the smaller space of all polynomial functions on $\{0 \leq x \leq 1\}$ are both infinite-dimensional; for the latter space one has an infinite 'basis' consisting of the functions $1, x, x^2, x^3, \ldots$. The space of all polynomial functions of degree $\leqq 5$ on $\{0 \leq x \leq 1\}$ is, however, six-dimensional and a basis is given by the functions $1, x, x^2, x^3, x^4, x^5$. The subspace of V_5 defined by the relations (1) is 2-dimensional and a basis is given by the vectors $(1, 1, -4, 1, 1)$ and $(1, -1, 0, -1, 1)$. The subspace consisting of all symmetric functions on $\{1, 2, 3, 4, 5\}$ is three-dimensional and a basis is given by the two vectors above together with $(0, 0, 1, 0, 0)$. Another basis for the space of symmetric functions is given by the three vectors $(1, 0, 0, 0, 1)$, $(0, 1, 0, 1, 0)$, $(0, 0, 1, 0, 0)$.

The subspace of V_5 which consists of all functions the sum of whose values is zero is 4-dimensional; there are many natural choices of a basis for this space, one of them being the set $(1, -1, 0, 0, 0)$, $(0, 1, -1, 0, 0)$, $(0, 0, 1, -1, 0)$, $(0, 0, 0, 1, -1)$. To prove that this is a basis it must be shown that every vector $(f_1, f_2, f_3, f_4, f_5)$ satisfying $f_1 + f_2 + f_3 + f_4 + f_5 = 0$ can be written in just one way as a sum

$$\begin{aligned}(f_1, f_2, f_3, f_4, f_5) &= x_1(1, -1, 0, 0, 0) + x_2(0, 1, -1, 0, 0) \\ &\quad + x_3(0, 0, 1, -1, 0) + x_4(0, 0, 0, 1, -1) \\ &= (x_1, x_2 - x_1, x_3 - x_2, x_4 - x_3, -x_4).\end{aligned}$$

That is to say, the equations

$$\begin{aligned} f_1 &= x_1 \\ f_2 &= x_2 - x_1 \\ f_3 &= x_3 - x_2 \\ f_4 &= x_4 - x_3 \\ f_5 &= -x_4 \end{aligned} \tag{3}$$

have one and only one solution (x_1, x_2, x_3, x_4) for each

$(f_1, f_2, f_3, f_4, f_5)$ provided $\sum f_i = 0$. This follows from the explicit solution

$$
\begin{aligned}
x_1 &= f_1 \\
x_2 &= f_1 + f_2 \\
x_3 &= f_1 + f_2 + f_3 \\
x_4 &= f_1 + f_2 + f_3 + f_4 \\
f_5 &= -f_1 - f_2 - f_3 - f_4.
\end{aligned} \tag{4}
$$

A *linear** *map*† $f\colon V \to W$ from a vector space V to a vector space W is a mapping which preserves the operations of addition and multiplication by numbers. That is, f carries $v_1 + v_2$ (the sum of two elements of V) to $f(v_1) + f(v_2)$ (the sum of their images in W) and av (multiple of an element of V) to $af(v)$ (the same multiple of its image in W). In other words, a linear map $f\colon V \to W$ is a function which assigns an element $f(v)$ of W to each element v of V in such a way that

$$
\begin{aligned}
f(v_1 + v_2) &= f(v_1) + f(v_2) \qquad &&(\text{any } v_1, v_2 \text{ in } V) \\
f(av) &= af(v) \qquad &&(\text{any } v \text{ in } V, a \text{ in } \mathbf{R}).
\end{aligned} \tag{5}
$$

*Because a line is described by an equation $Ax + By = \text{const.}$, a polynomial of the form $Ax + By$ is called a linear form. More generally, a polynomial of any number of variables $A_1x_1 + A_2x_2 + \cdots + A_nx_n$ in which all terms are of degree one is called a linear form. For this reason, any mapping which can be expressed by linear forms is called a linear mapping. It will be seen in Exercise 9 that the abstract definition given here amounts to saying that the mapping can be expressed by linear forms.

†The term 'linear operator' or 'operator' is also used in certain contexts, principally when the range and domain of the map are the same vector space.

Pullback maps are linear; that is, given an affine map, the pullback map carrying the vector space of k-forms on the range to the vector space of k-forms on the domain is a linear map. This means that in finding the pullback of, say, $A\,dy\,dz + B\,dz\,dx + C\,dx\,dy$ one can first find the pullbacks of the basic forms $dy\,dz$, $dz\,dx$, $dx\,dy$ and then multiply and add. This fact was used in §4.3 when the pullback map was described by the matrix $M^{(k)}$ giving the pullbacks of the basic k-forms.

If V is the vector space of functions $\{S \to \mathbf{R}\}$ and W the vector space of functions $\{T \to \mathbf{R}\}$, and if F is a function from S to T, then the composition of elements of W with F defines a linear map from W to V called the *pullback* of elements of W under the map F. For example, if W is the vector space of real-valued functions on the xy-plane, if V is the vector space of functions on the interval $\{0 \le t \le 1\}$, and if $F(t) = (x(t), y(t))$ is a function from the interval $\{0 \le t \le 1\}$ to the xy-plane then the composed function $f(x(t), y(t))$ assigns an element of V to each element f of W. The fact that this map of W to V is linear is immediate from the definitions. Intuitively the pullback of k-forms is such a pullback—the set T being oriented k-dimensional surfaces in the range of an affine map, the set S being oriented k-dimensional surfaces in the domain, and the map F being the rule which assigns to each element of S its image under the given affine map. (The same holds for pull-backs of non-constant forms under non-affine maps.)

The map 'derivative' assigning functions to functions is a linear map, as is expressed by the familiar identities

$$(f + g)' = f' + g', \qquad (af)' = af'.$$

Similarly the map 'integral from a to b' assigning numbers to 1-forms on the interval $\{a \leq x \leq b\}$ is linear because

$$\int_a^b [f(x) + g(x)]\,dx = \int_a^b f(x)\,dx + \int_a^b g(x)\,dx$$

$$\int_a^b \text{const.}\, f(x)\,dx = \text{const.} \int_a^b f(x)\,dx.$$

The same is true of the generalizations

$$d\colon \{k\text{-forms}\} \to \{(k+1)\text{-forms}\}$$

and

$$\int_D\colon \{k\text{-forms}\} \to \{\text{numbers}\}$$

discussed in Chapters 2, 3 and 6. Generally speaking, any natural mapping whose range and domain are vector spaces will preserve addition and multiplication by numbers, that is, it will be linear.

It is useful to rephrase the definition of 'basis' in terms of linear mappings. Given a basis $\{v_1, v_2, \ldots, v_n\}$ of a vector space V, the mapping

$$(6) \quad f(x_1, x_2, \ldots, x_n) = x_1v_1 + x_2v_2 + \cdots + x_nv_n$$

carries the set of n-tuples of numbers to V. Regarding the n-tuple $(x_1, x_2, \ldots, x_n)$ as an element of V_n this map is a *linear* map $V_n \to V$ because

$$\begin{aligned}(x_1 + x_1')v_1 + \cdots + (x_n + x_n')v_n &= (x_1v_1 + \cdots + x_nv_n) + (x_1'v_1 + \cdots + x_n'v_n),\\ (ax_1)v_1 + (ax_2)v_2 + \cdots + (ax_n)v_n &= a(x_1v_1 + \cdots + x_nv_n),\end{aligned}$$

by the usual rules of distributivity, associativity, and commutativity. By the definition of 'basis' this linear map $f\colon V_n \to V$ is one-to-one and onto—that is, every element

v of V is the image of exactly one n-tuple $(x_1, x_2, \ldots, x_n)$. In other words, the map $f\colon V_n \to V$ has an *inverse* $f^{-1}\colon V \to V_n$ and, since the statement '$f(x_1) = y_1$, $f(x_2) = y_2$ implies $f(x_1 + x_2) = y_1 + y_2$ and $f(ax_1) = ay_1$' is identical to the statement '$x_1 = f^{-1}(y_1)$, $x_2 = f^{-1}(y_2)$ implies $x_1 + x_2 = f^{-1}(y_1 + y_2)$ and $ax_1 = f^{-1}(ay_1)$', the inverse is linear. In short, if $\{v_1, v_2, \ldots, v_n\}$ is a basis of the vector space V then the equation (6) defines an invertible linear map $V_n \leftrightarrow V$. Conversely, if f is any invertible linear map $V_n \leftrightarrow V$ then the set

$$\begin{aligned} v_1 &= f(1, 0, 0, \ldots, 0) \\ v_2 &= f(0, 1, 0, \ldots, 0) \\ &\vdots \\ v_n &= f(0, 0, 0, \ldots, 1) \end{aligned}$$

is a basis of V because every element of v can, by the invertibility of f, be written in exactly one way in the form

$$\begin{aligned} v &= f(x_1, x_2, \ldots, x_n) \\ &= f(x_1, 0, 0, \ldots, 0) + f(0, x_2, 0, \ldots, 0) + \cdots + f(0, 0, 0, \ldots, x_n) \\ &= x_1v_1 + x_2v_2 + \cdots + x_nv_n. \end{aligned}$$

Therefore the two notions are equivalent, that is, a basis determines an invertible linear map $V_n \leftrightarrow V$ and such a map determines a basis.

The basic facts about the solution of linear equations can be summarized by the following theorem which is the real *raison d'être* of the theory of vector spaces.

Theorem

Canonical form for linear maps. Given a linear mapping $f\colon V \to W$ of an n-dimensional vector space V to an m-dimensional vector space W, it is possible to choose bases $v_1, v_2, \ldots, v_n$ of V and $w_1, w_2, \ldots, w_m$ of W such that

$$f(x_1v_1 + x_2v_2 + \cdots + x_nv_n) = x_1w_1 + x_2w_2 + \cdots + x_rw_r$$

for some integer r. This integer r depends only on the map $f\colon V \to W$ and not on the choice of the bases $v_1, v_2, \ldots, v_n; w_1, w_2, \ldots, w_m$. It is called the *rank* of the map f. By its definition, $r \leq \min(n, m)$.

Corollary

If $n > m$ then f cannot be one-to-one ($r \leq m < n$ and $f(v_{r+1}) = f(2v_{r+1})$, whereas $v_{r+1} \neq 2v_{r+1}$), and if $n < m$ then f cannot be onto ($r \leq n < m$, and w_{r+1} is not in the image of f). Thus if f is one-to-one and onto, then $n = m$. Taking $f: V \to V$ to be the identity map, this implies that if V is n-dimensional (has a basis of n elements) and m-dimensional (has a basis of m elements) then $n = m$. In short, the *dimension* of a finite-dimensional vector space is well-defined.

Proof

To say that V is n-dimensional means that there is a basis of n elements $v_1, v_2, \ldots, v_n$, and to say that W is m-dimensional means that there is a basis of m elements $w_1, w_2, \ldots, w_m$. The method of proof is to start with such bases and to use a process of step-by-step elimination to replace them with new bases having the desired property.

First define numbers a_{ij} by

$$f(v_i) = \sum_{j=1}^{m} a_{ij}w_j \qquad (i = 1, 2, \ldots, n)$$

(every element of W can be expressed in just one way as a combination of the w's). Unless all of the a's are zero (in which case $r = 0$ and these bases have the desired form) the bases can be rearranged to make $a_{11} \neq 0$. Set $w_1' = f(v_1) = a_{11}w_1 + a_{12}w_2 + \cdots + a_{1m}w_m$. Then $w_1', w_2, w_3, \ldots, w_m$ can be shown to be a basis of W as follows: Since $w_1 = \dfrac{1}{a_{11}}[w_1' - a_{12}w_2 - \cdots - a_{1m}w_m]$ can be expressed in terms of $w_1', w_2, \ldots, w_n$, it follows that every element of W can be expressed in terms of $w_1', w_2, \ldots, w_n$. This expression is unique because

$$x_1w_1' + x_2w_2 + \cdots + x_mw_m = y_1w_1' + y_2w_2 + \cdots + y_mw_m$$

implies

$$x_1a_{11}w_1 + (x_2 + x_1a_{12})w_2 + \cdots = y_1a_{11}w_1 + (y_2 + y_1a_{12})w_2 + \cdots,$$

which implies $x_1a_{11} = y_1a_{11}$, $x_i + x_1a_{1i} = y_i + y_1a_{1i}$ $(i > 1)$, which in turn implies $x_1 = y_1$ $(a_{11} \neq 0)$, and, finally, $x_i = y_i$ $(i > 1)$. Therefore it can be assumed at the outset that $w_1', w_2, \ldots, w_m$ was the chosen basis of

W, i.e. that $f(v_1) = w_1$. If $f(v_i)$ for $i > 1$ contains a non-zero term in w_j for $j > 1$ then the process can be repeated: By rearrangement it can be assumed that $a_{22} \neq 0$. If w_2' is defined by $w_2' = f(v_2)$ then the above argument shows that $w_1, w_2', w_3, \ldots, w_m$ is a basis of W. Therefore it can be assumed at the outset that the chosen basis of W satisfies $f(v_1) = w_1, f(v_2) = w_2$. Continuing in this way gives a new basis $w_1, w_2, \ldots, w_m$ of W such that $f(v_1) = w_1, f(v_2) = w_2, \ldots, f(v_r) = w_r$, and such that $f(v_i)$ for $i > r$ contains no non-zero terms in $w_{r+1}, w_{r+2}, \ldots, w_m$. Assume this has been done and let $f(v_i) = \sum_{j=1}^{r} a_{ij}w_j$ for $i > r$. Set $v_i' = v_i - \sum_{j=1}^{r} a_{ij}v_j$ for $i > r$. Then $v_1, v_2, \ldots, v_r, v_{r+1}', \ldots, v_n'$ can be shown to be a basis of V as follows: Since $v_i = v_i' + \sum_{j=1}^{r} a_{ij}v_j$ can be expressed as a combination of $v_1, v_2, \ldots, v_r, v_{r+1}', \ldots, v_n'$ $(i > r)$ it follows that every element of V can be expressed as a combination of these elements. This expression is unique because the assumption

$$x_1v_1 + \cdots + x_rv_r + x_{r+1}v_{r+1}' + \cdots + x_nv_n' = y_1v_1 + \cdots + y_rv_r + y_{r+1}v_{r+1}' + \cdots + y_nv_n'$$

implies first of all, when v_i' is rewritten as $v_i - \sum_{j=1}^{r} a_{ij}v_j$ and the coefficients of v_i $(i > r)$ are compared, that $x_{r+1} = y_{r+1}, \ldots, x_n = y_n$. On the other hand,

$$f(v_i') = f(v_i) - \sum_{j=1}^{r} a_{ij}f(v_j) = \sum_{j=1}^{r} a_{ij}w_j - \sum_{j=1}^{r} a_{ij}w_j = \sum_{j=1}^{r} 0 \cdot w_j.$$

Hence when f is applied to the equation above it gives

$$x_1w_1 + \cdots + x_rw_r = y_1w_1 + \cdots + y_rw_r$$

and therefore $x_1 = y_1, \ldots, x_r = y_r$. Thus $v_1, \ldots, v_r, v_{r+1}', \ldots, v_n'$ is a basis and, since it was just shown that

$$f(x_1v_1 + \cdots + x_rv_r + x_{r+1}v_{r+1}' + \cdots + x_nv_n') = x_1w_1 + \cdots + x_rw_r,$$

this basis has the desired property. It remains only to show that the value of r is independent of the choices.

Suppose that $v_1', v_2', \ldots, v_n'$ and $w_1', w_2', \ldots, w_m'$ are bases of V and W such that

$$f(x_1v_1' + \cdots + x_nv_n') = x_1w_1' + \cdots + x_sw_s'.$$

It is to be shown that $s = r$. Define a linear map $V_s \to V$ by $(x_1, x_2, \ldots, x_s) \to x_1v_1' + \cdots + x_sv_s'$ and a linear map $W \to V_r$ by $y_1 w_1 + \cdots + y_m w_m \to (y_1, y_2, \ldots, y_r)$, and consider the composed map $V_s \to V \to W \to V_r$ with f. The first two maps carry V_s onto the image of f which the last map carries onto V_r. Hence this map carries V_s onto V_r. But the portion of the theorem already proved shows that if $s < r$ then $V_s \to V_r$ cannot be onto.* Therefore $s \geq r$. By the same token $r \geq s$ and the theorem is proved.

*Using the obvious fact that a composition of linear maps is linear.

Because most vector spaces which occur naturally are vector spaces of functions $\{f: S \to \mathbf{R}\}$ in one guise or another, the informal definition of vector space given in the beginning of this section is adequate for most applications. However, it is always advisable to have precise definitions.

Definition

A *vector space* is a set V together with two operations, *addition* and *multiplication by numbers*, satisfying certain axioms. The addition operation assigns to each (ordered) pair of elements v_1, v_2 of V a third element $v_1 + v_2$ of V, and multiplication by numbers assigns to each element v of V and each number a an element av of V. The axioms are:

I. *Commutative law.* The addition operation is commutative, i.e.

$$v_1 + v_2 = v_2 + v_1.$$

II. *Associative law.* Both addition and multiplication by numbers are associative, i.e.

$$(v_1 + v_2) + v_3 = v_1 + (v_2 + v_3) \quad \text{and} \quad a_1(a_2 v) = (a_1 a_2)v.$$

III. *Distributive law.* Multiplication by numbers is distributive over both addition of elements of V and addition of numbers, i.e.

$$a(v_1 + v_2) = av_1 + av_2 \quad \text{and} \quad (a_1 + a_2)v = a_1 v + a_2 v.$$

IV. *Solution of equations**. An equation of the form

$$v = a_1 v_1 + a_2 v_2 + \cdots + a_n v_n$$

in which $a_1 \neq 0$ has a unique solution v_1 in V given elements $v, v_2, v_3, \ldots, v_n$ of V and numbers $a_1, a_2, \ldots, a_n$.

*This axiom can also be stated "$v_1 = av + v_2$ has a unique solution v given v_1, v_2 in V and given a number $a \neq 0$." The statement given in the text emphasizes the relation of the axiom to the elimination process.

It is an interesting exercise to show that these axioms imply all the expected facts about addition and multiplication by numbers, for example, that they imply:

V. $1 \cdot v = v$ (any v in V).

VI. $v + 0 \cdot w = v$ (any v, w in V).

VII. The solution in IV is given by

$$v_1 = (1/a_1)[v - a_2v_2 - \cdots - a_nv_n].$$

Glossary

The following terms are used in connection with vector spaces.

Vector space. A *vector space* is a set with operations of addition and multiplication by numbers subject to the axioms I–IV above.

Linear map. A *linear map* is a function $f\colon V \to W$ whose range and domain are vector spaces and which preserves the vector space operations, i.e. $f(v_1 + v_2) = f(v_1) + f(v_2)$ and $f(av) = af(v)$ for v_1, v_1, v in V, a in $\mathbf{R}$.

Subspace of a vector space. A *subspace* of a vector space is a subset with the property that sums and multiples of elements of the subset are again in the subset. By statement VII above it follows that a subspace of a vector space is itself a vector space.

V_n. This symbol, for $n = 1, 2, \ldots,$ denotes the standard vector space consisting of all functions from the finite set $\{1, 2, \ldots, n\}$ to $\mathbf{R}$, added and multiplied by numbers in the natural way. Elements of V_n are conveniently described by listing their n values in order.

Basis. A *basis* of a vector space V is a set of n elements $v_1, v_2, \ldots, v_n$ (n a positive integer) such that the linear map $V_n \to V$ defined by $(x_1, x_2, \ldots, x_n) \to x_1v_1 + x_2v_2 + \cdots + x_nv_n$ is one-to-one and onto.

Dimension. A vector space is said to be *n-dimensional* if it has a basis containing n elements. (If so then the integer n is the same for all bases.) It is said to be 0-dimensional if it consists of a single element. It is said to be *finite-dimensional* if it is n-dimensional for some integer $n = 0, 1, 2, \ldots$. Otherwise it is said to be *infinite-dimensional.*

Canonical form. A linear map $f\colon V \to W$ whose range and domain are finite-dimensional is said to be in *canonical form* relative to bases $v_1, v_2, \ldots, v_n$ of V and $w_1, w_2, \ldots, w_m$ of W if there is an integer r such that $f(x_1v_1 + \cdots + x_nv_n) = x_1w_1 + \cdots + x_rw_r$.

Rank. Every linear map $f\colon V \to W$ whose range and domain are finite-dimensional can be put in canonical form. The resulting integer r, which depends only on the map, is called its *rank*.

Linearly independent. A set of vectors $v_1, v_2, \ldots, v_n$ in a vector space V is said to be *linearly independent* if the map $(x_1, \ldots, x_n) \to x_1v_1 + \cdots + x_nv_n$ of $V_n \to V$ is one-to-one. Otherwise the set is said to be *linearly dependent*.

Span. A set of vectors $v_1, v_2, \ldots, v_n$ in a vector space V is said to *span* V if the map $(x_1, \ldots, x_n) \to x_1v_1 + \cdots + x_nv_n$ of $V_n \to V$ is onto. Thus a basis of V is a linearly independent set of vectors which spans V.

Zero vector. The axioms imply that a vector space V contains a unique element 0 with the property that $v + a \cdot 0 = v$ for any v in V and any number a. This element is called the *zero vector* of V. Every vector space has a unique 0-dimensional subspace, namely the subspace consisting of the zero vector alone. The zero vector of any vector space of functions $\{f\colon S \to \mathbf{R}\}$ is the function which is identically zero; in particular, the zero vector of V_n is $(0, 0, \ldots, 0)$.

Kernel of a linear map. If $f\colon V \to W$ is a linear map, then the set of all elements v of V whose images under f are the zero vector of W is a subspace of V. It is called the *kernel* of f.

Exercises

1 The following sets of linear relations define subspaces of V_3. In each case determine the dimension of the subspace and find a basis.

(a) $f_1 + f_2 + f_3 = 0$

(b) $f_1 - 2f_2 + f_3 = 0$

(c) $f_1 - f_2 = 0$
$f_2 - f_3 = 0$
$f_3 - f_1 = 0$

(d) $2f_1 + f_2 - f_3 = 0$
$f_1 - 2f_2 + f_3 = 0$

(e) $4f_1 - 3f_2 + f_3 = 0$
$f_1 + f_2 - 3f_3 = 0$
$2f_1 + f_2 + f_3 = 0$

2 The following sets of linear relations define subspaces of V_6. In each case determine the dimension of the subspace and find a basis.

(a) $f_1 + f_2 + f_3 + f_4 + f_5 + f_6 = 0$
(b) $f_1 - 2f_2 + f_3 = 0$
$f_2 - 2f_3 + f_4 = 0$
$f_3 - 2f_4 + f_5 = 0$
$f_4 - 2f_5 + f_6 = 0$

3 Consider the set of all 'mobile planar arrows', that is, arrows in the plane which can be translated from point to point.

(a) How can two such arrows be added?
(b) How can such an arrow be multiplied by a number?
(c) What is the dimension of the resulting vector space? (Give a basis.)
(d) What physical quantities can be represented by such arrows? Give a physical application of the addition operation of (a).

4 Mobile arrows in space are a vector space in the same way as in Exercise 3. Give a geometrical interpretation of the statement that three arrows in space are linearly dependent (see Glossary).

5 Prove that V, VI, VII are consequences of the axioms I–IV. [Use the uniqueness statement of IV.]

6 Prove that the existence of a zero vector (see Glossary) is a consequence of the axioms I–IV and prove that the zero vector of any vector space of functions is the function which is identically zero.

7 Prove that the kernel (see Glossary) of a linear map is a subspace. Restate Exercise 1 in terms of kernels.

8 Why does VIII prove that a subspace of a vector space is a vector space?

9 If V, W are vector spaces then Hom(V, W) denotes the space of all linear maps from V to W.

(a) Show that Hom(V, W) is a vector space when addition and multiplication by numbers is defined in the obvious way.
(b) Is the set of all functions $\{f: S \to W\}$ from an arbitrary set S to a vector space W a vector space?
(c) Show that a linear map from V_n to V_m is a function of the form $y_i = \sum_{j=1}^{n} a_{ij}x_j$ where $(x_1, x_2, \ldots, x_n)$ and

$(y_1, y_2, \ldots, y_m)$ are coordinates on V_n and V_m respectively. In other words, show that a linear map is an affine map in which the constants are zero. Elements of $\mathrm{Hom}(V_n, V_m)$ are therefore represented by $m \times n$ matrices. What are the operations of addition and multiplication by numbers of elements of $\mathrm{Hom}(V_n, V_m)$ in terms of matrices?

(d) What is the dimension of $\mathrm{Hom}(V_n, V_m)$? (Give a basis.)

10 If V is any vector space, then the set of all linear maps from V to the one-dimensional space V_1, i.e. $\mathrm{Hom}(V, V_1)$, is a vector space by Exercise 9. It is called the *dual* of V and is denoted $V^* = \mathrm{Hom}(V, V_1)$.

(a) Show how a linear map $f: V \to W$ gives rise to a pullback mapping $f^*: W^* \to V^*$.

(b) How is the dimension of V^* related to that of V?

(c) Given a basis of V show how to obtain a basis of V^*. This is called the *dual basis* of V.

11 It is often convenient to represent elements of V_n as *column matrices*, i.e. as $n \times 1$ matrices. Then an $m \times n$ matrix gives a linear map $V_n \to V_m$ simply by multiplication of an $n \times 1$ matrix in the usual way to obtain an $m \times 1$ matrix.

(a) How then are elements of $(V_n)^*$ represented?

(b) Given a map $V_n \to V_m$ how is the pullback map $(V_m)^* \to (V_n)^*$ represented?

12 The 'Fredholm Alternative' states that a system of linear equations $Mx = y$ in which range and domain have the same dimension, i.e.

$$\begin{aligned} a_{11}x_1 + a_{12}x_2 + \cdots + a_{1n}x_n &= y_1 \\ a_{21}x_1 + a_{22}x_2 + \cdots + a_{2n}x_n &= y_2 \\ \vdots \qquad \vdots \qquad\qquad \vdots \quad & \\ a_{n1}x_1 + a_{n2}x_2 + \cdots + a_{nn}x_n &= y_n, \end{aligned}$$

has one of the two following properties: *Either* the associated homogeneous equations

$$\begin{aligned} a_{11}x_1 + a_{12}x_2 + \cdots + a_{1n}x_n &= 0 \\ a_{21}x_1 + a_{22}x_2 + \cdots + a_{2n}x_n &= 0 \\ \vdots \qquad \vdots \qquad\qquad \vdots \quad & \\ a_{n1}x_1 + a_{n2}x_2 + \cdots + a_{nn}x_n &= 0 \end{aligned}$$

have only the solution $x = 0$, i.e. $(x_1, x_2, \ldots, x_n) = (0, 0, \ldots, 0)$, in which case the given system has a unique solution $(x_1, x_2, \ldots, x_n)$ for every $(y_1, y_2, \ldots, y_n)$, *or* the homogeneous equations have some solution other than $x = 0$, in which case the given system never has a unique solution (either there is no solution or there are many). Deduce this statement from the canonical form for linear maps.

13 The canonical form for linear maps $V_n \to V_m$ can be stated in terms of matrices as follows: Let M be an $m \times n$ matrix. Then an $m \times m$ matrix P representing an invertible (one-to-one, onto) linear map $V_m \leftrightarrow V_m$ and an $n \times n$ matrix Q representing an invertible linear map $V_n \leftrightarrow V_n$ can be found such that

$$PMQ = C_r$$

where the canonical matrix C_r is the $m \times n$ matrix (c_{ij}) which is zero in all places except $c_{11}, c_{22}, \ldots, c_{rr}$ where it is one. The process of finding such P, Q given M can be carried out as follows: Starting with M, multiply on the left (by P_1) or on the right (by Q_1) by a simple matrix (shear, interchange of coordinates, scale factor) to obtain a new matrix M_1 which is more like C_r. Apply the same process to M_1 to obtain M_2 more like C_r and continue until C_r is obtained. Going back and collecting all the P's and Q's then gives an equation of the form

$$P_i P_{i-1} \ldots P_2 P_1 M Q_1 Q_2 \ldots Q_j = C_r$$

and hence P, Q are obtained by multiplying. Apply this method to find P, Q for each of the matrices

$$(1 \quad 1 \quad 1) \qquad \begin{pmatrix} 4 & -3 & 1 \\ 1 & 1 & -3 \\ 2 & 1 & 1 \end{pmatrix}$$

$$(1 \quad -2 \quad 1)$$

$$\begin{pmatrix} 1 & -1 & 0 \\ 0 & 1 & -1 \\ -1 & 0 & 1 \end{pmatrix} \qquad (1 \quad 1 \quad 1 \quad 1 \quad 1 \quad 1)$$

$$\begin{pmatrix} 2 & 1 & -1 \\ 1 & -2 & 1 \end{pmatrix} \qquad \begin{pmatrix} 1 & -2 & 1 & 0 & 0 & 0 \\ 0 & 1 & -2 & 1 & 0 & 0 \\ 0 & 0 & 1 & -2 & 1 & 0 \\ 0 & 0 & 0 & 1 & -2 & 1 \end{pmatrix}$$

of Exercises 1 and 2. Show the relationship between this 'canonical form' for matrices and the canonical form for linear maps.

4.6

Summary. Affine Manifolds

The set of n-tuples of real numbers is denoted by $\mathbf{R}^n$. A function $f: \mathbf{R}^n \to \mathbf{R}^m$, which assigns to each n-tuple of real numbers an m-tuple of real numbers, is described by naming the coordinates on range and domain—say $(y_1, y_2, \ldots, y_m)$ on the range and $(x_1, x_2, \ldots, x_n)$ on the domain—and by giving the m component functions

$$\begin{aligned} y_1 &= f_1(x_1, x_2, \ldots, x_n) \\ y_2 &= f_2(x_1, x_2, \ldots, x_n) \\ &\vdots \\ y_m &= f_m(x_1, x_2, \ldots, x_n). \end{aligned} \tag{1}$$

Such a function $f: \mathbf{R}^n \to \mathbf{R}^m$ is said to be *affine* if each of the component functions f_i is a polynomial of the first degree

$$f_i(x_1, x_2, \ldots, x_n) = \sum_{j=1}^{n} a_{ij}x_j + b_i \qquad (i = 1, 2, \ldots, m).$$

The Implicit Function Theorem deals with the question "given an affine function (1) and given values for $y_1, y_2, \ldots, y_m$, find all possible values for $x_1, x_2, \ldots, x_n$." It can be stated as follows: The x's and y's can be rearranged, an integer $r \geq 0$ can be chosen, and affine functions $g: \mathbf{R}^n \to \mathbf{R}^r$, $h: \mathbf{R}^r \to \mathbf{R}^{m-r}$ can be found such that the equations (1) are equivalent to the equations

$$(2)\quad \begin{cases} \text{(2a)} & (x_1, x_2, \ldots, x_r) = g(y_1, \ldots, y_r, x_{r+1}, \ldots, x_n), \\ \text{(2b)} & (y_{r+1}, \ldots, y_m) = h(y_1, y_2, \ldots, y_r); \end{cases}$$

that is, a set of $n + m$ numbers $x_1, x_2, \ldots, x_n, y_1, y_2, \ldots, y_m$ satisfies the m conditions (1) if and only if it satisfies the m conditions (2). This solves the problem 'given y find x' by stating that (2b) is a necessary and sufficient condition for the existence of a solution and that all solutions are then given by (2a).

This form of the Implicit Function Theorem was proved in §4.1 by simple step-by-step elimination. The more detailed version of the theorem given in §4.4 describes the integer r and the possible rearrangements of x's and y's for which (2) is possible in terms of *pullbacks of forms* under the given map (1). The integer r is the largest integer such that the pullback of some r-form under (1) is not zero. It is called the *rank* of the affine map (1). The solution (2) is possible (without rearrangement) if and only if the rank is r and the $dx_1\, dx_2 \ldots dx_r$-component of the pullback of $dy_1\, dy_2 \ldots dy_r$ is not zero.

The algebra of forms described in §4.2 is summarized by the formula $dx_i\, dx_j = -dx_j\, dx_i$. (Geometrically, this formula says that the rotation of 90° which carries (x_i, x_j) to $(-x_j, x_i)$ preserves oriented areas.) Setting $i = j$ gives $dx_i\, dx_i = 0$. The pullback operation is summarized by the formula $d(\sum a_i x_i + b) = \sum a_i\, dx_i$.

In §4.3, new notation was introduced. Among the new symbols defined there, the most important is the Jacobian notation

$$\frac{\partial(y_1, y_2, \ldots, y_k)}{\partial(x_1, x_2, \ldots, x_k)}$$

for 'the coefficient of $dx_1\,dx_2\ldots dx_k$ in the pullback of $dy_1\,dy_2\ldots dy_k$' under a given affine map $(x)\to(y)$, with an analogous notation for any set of k of the y's and k of the x's. The $\binom{n}{k}\times\binom{m}{k}$ possible Jacobians form the coefficients of a $\binom{n}{k}\times\binom{m}{k}$ matrix $M^{(k)}$ called the kth exterior power of the $m\times n$ matrix of coefficients M of the given map $(y)\to(x)$.

The most important fact about computation with forms and pullbacks is the Chain Rule: "The pullback under a composed map is the pullback of the pullback." In terms of Jacobians this is summarized by the formula

$$\frac{\partial z}{\partial x} = \sum \frac{\partial z}{\partial y}\frac{\partial y}{\partial x}$$

where $\frac{\partial z}{\partial x}$ denotes a $k\times k$ Jacobian of the composed map $(x)\to(y)\to(z)$ and where $\sum$ denotes a sum over all $\binom{m}{k}$ selections of k of the coordinates $(y_1, y_2, \ldots, y_m)$. In terms of exterior powers, the Chain Rule is

$$(NM)^{(k)} = M^{(k)}N^{(k)}.$$

The proofs of §4.4 used only the Chain Rule.

A reformulation of the Implicit Function Theorem which sheds considerable light on its geometrical meaning is the following formulation in terms of *affine manifolds:* A k-dimensional affine manifold in $\mathbf{R}^n$ *defined by parameters* is a subset of $\mathbf{R}^n$ which is the image of a one-to-one affine map $f\colon \mathbf{R}^k\to\mathbf{R}^n$. For example, the line $x = 3t + 1$, $y = 7t$ in the xy-plane defined by the parameter t. A k-dimensional affine manifold in $\mathbf{R}^n$ *defined by equations* is a subset of $\mathbf{R}^n$ which is a level surface of an onto affine map $f\colon \mathbf{R}^n\to\mathbf{R}^{n-k}$. For example the line $7x - 3y = 7$, which is the same as the parameterized line above. The dimensions can be remembered by the rule that $n-k$ independent (onto) conditions on n variables leave k degrees of freedom. The Implicit Function Theorem implies (Exercise 1) that every k-dimensional affine manifold in $\mathbf{R}^n$ defined by parameters can also be defined by equations and vice versa; hence the mode of definition is irrelevant and the notion of a *k-dimensional affine manifold in* $\mathbf{R}^n$ is well-defined. Geometrically these are lines* in the plane, lines in space, planes* in space, etc.

*Note that the usual terms imply a dimension, e.g. line ≡ 1-dimensional manifold, plane ≡ 2-dimensional manifold. The word 'manifold' is useful precisely because it leaves the dimension unspecified (many).

If A is a k-dimensional affine manifold in $\mathbf{R}^n$, then the projection of A on at least one of the k-dimensional coordinate planes* of $\mathbf{R}^n$ is one-to-one and A is parameterized by such a coordinate plane. That is, if $(x_1, x_2, \ldots, x_n)$ are the coordinates on $\mathbf{R}^n$ and if the projection of A on the $x_1x_2 \ldots x_k$-plane is one-to-one, then there is an affine map $F: \mathbf{R}^k \to \mathbf{R}^{n-k}$ such that A is the graph of F

*Strictly speaking it would be better to use the word 'manifold' since 'plane' implies $k = 2$. Similarly, 'level manifold' would be better than 'level surface'. Nonetheless, 'coordinate plane' and 'level surface' will be used in this book, with the understanding that these words do not imply 'two-dimensional'.

$$\{(x, y): x = (x_1, \ldots, x_k), y = F(x)\}$$

in $\mathbf{R}^n$. For example, the line $7x - 3y = 7$ is the line $y = \frac{7}{3}(x - 1)$ parameterized by x or the line $x = \frac{3}{7}y + 1$ parameterized by y. Two affine manifolds in $\mathbf{R}^n$ are said to be *parallel* if there is a translation of $\mathbf{R}^n$ $(x_1, x_2, \ldots, x_n) \to (x_1 + c_1, x_2 + c_2, \ldots, x_n + c_n)$ which carries one to the other.

In terms of these definitions the Implicit Function Theorem has the following geometrical meaning: The image of an affine map $f: \mathbf{R}^n \to \mathbf{R}^m$ is an affine manifold. Its dimension r is equal to the rank of the map. The level surfaces of an affine map $f: \mathbf{R}^n \to \mathbf{R}^m$ are a family of parallel affine manifolds in $\mathbf{R}^n$. Their dimension is $n - r$ where r is the rank of the map. The image of f is parameterized by $(y_1, y_2, \ldots, y_r)$ if and only if the pullback of $dy_1\, dy_2 \ldots dy_r$ is not zero. The level surfaces of f are parameterized by $(x_{r+1}, x_{r+2}, \ldots, x_n)$ if and only if the pullback of some r-form in $(y_1, y_2, \ldots, y_n)$ contains a non-zero term in which none of the factors dx_{r+1}, $dx_{r+2}, \ldots, dx_n$ appear, i.e. a non-zero term in $dx_1\, dx_2 \ldots dx_r$.

Exercises

1 Use the Implicit Function Theorem to prove that every k-dimensional affine manifold defined by parameters can be defined by equations and vice versa. ['Onto' means r = dimension of range, 'one-to-one' means r = dimension of domain. A graph $\{(x, y): y = F(x)\}$ is defined by the parameters x or by the equations $y - F(x) = 0$.]

2 Show that the rank of $M^{(k)}$ is $\binom{r}{k}$ where r is the rank of M. [Use the canonical form $PMQ = C_r$ of Exercise 13, §4.5. Use the definition of 'rank' to show that if $P_1M_1Q_1 = M_2$ where P_1, Q_1 are invertible then the rank of M_1 is equal to the rank of M_2. It suffices then to find the rank of $(C_r)^{(k)}$.] This includes as a special case the theorem that M has the same rank as its transpose.

3 Prove the following statements relating to orientations of $\mathbf{R}^n$ (see §1.4): A set of $n+1$ points $P_0P_1P_2\ldots P_n$ in $\mathbf{R}^n$ is said to be in general position (non-coplanar) if it can be written as the image of the set of points $(0, 0, \ldots, 0)$, $(1, 0, 0, \ldots, 0)$, $(0, 1, 0, \ldots, 0)$, $\ldots$, $(0, 0, \ldots, 0, 1)$ (in that order) under a one-to-one (hence onto) map $f_P\colon \mathbf{R}^n \to \mathbf{R}^n$. If $P_0P_1\ldots P_n$ and $Q_0Q_1\ldots Q_n$ are two sets of points in $\mathbf{R}^n$ in general position, then it is said that their 'orientations agree' if the affine map $f_P \circ f_Q^{-1}\colon \mathbf{R}^n \to \mathbf{R}^n$ carrying one set to the other has *positive* Jacobian.

(a) Show that the orientations of $P_0P_1P_2\ldots P_n$ and $P_1P_0P_2\ldots P_n$ do not agree.
(b) Show that if the orientations of $P_0P_1\ldots P_n$ and $Q_0Q_1\ldots Q_n$ do not agree then the orientations of $P_1P_0P_2\ldots P_n$ and $Q_0Q_1\ldots Q_n$ do agree.
(c) Show that if the orientations of $P_0P_1\ldots P_n$ and $Q_0Q_1\ldots Q_n$ agree then $P_0P_1\ldots P_n$ can be moved continuously to $Q_0Q_1\ldots Q_n$ in such a way that they remain in general position all the while. [Applying f_P^{-1}, one assumes $P_0P_1\ldots P_n$ is the standard set of $n+1$ points. Write f_Q as a composition of shears, rotations, translations and scale factors, as in Exercise 3, §4.3. The number of negative scale factors must be even, hence making all of them + does not change f_Q. This reduces (c) to four simple cases.]
(d) Show that if the orientations disagree then the motion of (c) is impossible. [A non-zero continuous function cannot change sign.]
(e) Show that two quadruples $(x_0, y_0, z_0), \ldots, (x_3, y_3, z_3)$, and $(x_0', y_0', z_0'), \ldots, (x_3', y_3', z_3')$ describe the same orientation of xyz-space if and only if the determinants

$$\begin{vmatrix} 1 & x_0 & y_0 & z_0 \\ 1 & x_1 & y_1 & z_1 \\ 1 & x_2 & y_2 & z_2 \\ 1 & x_3 & y_3 & z_3 \end{vmatrix}, \quad \begin{vmatrix} 1 & x_0' & y_0' & z_0' \\ 1 & x_1' & y_1' & z_1' \\ 1 & x_2' & y_2' & z_2' \\ 1 & x_3' & y_3' & z_3' \end{vmatrix}$$

have the same sign.

4 Suppose that a sequence of affine maps is given

$$\{\text{pt.}\} \xrightarrow{f_1} \mathbf{R}^{n_1} \xrightarrow{f_2} \mathbf{R}^{n_2} \xrightarrow{f_3} \mathbf{R}^{n_3} \to \cdots \xrightarrow{f_\nu} \mathbf{R}^{n_\nu} \xrightarrow{f_{\nu+1}} \{\text{pt.}\}$$

with the property that the image of f_i is a level surface of f_{i+1} (so in particular the level surfaces of f_2 are points—f_2 is one-to-one—and the image of f_ν is all of $\mathbf{R}^{n_\nu}$—f_ν is onto). Such a sequence of maps is said to be *exact*. Show that the alternating sum of the dimensions is zero, i.e. $n_1 - n_2 + n_3 - \cdots \pm n_\nu = 0$. [Write n_i as a sum of two terms and cancel.]

chapter 5

differential calculus

5.1

The Implicit Function Theorem for Differentiable Maps

Consider the problem of solving a system of m equations in n unknowns

(1) $$y_i = f_i(x_1, x_2, \ldots, x_n) \qquad (i = 1, 2, \ldots, m)$$

for all possible values of $(x_1, x_2, \ldots, x_n)$ given $(y_1, y_2, \ldots, y_m)$. Chapter 4 deals with the solution of this problem in the special case where the functions f_i are affine functions, that is, functions of the form

$$f_i(x_1, x_2, \ldots, x_n) = a_{i1}x_1 + a_{i2}x_2 + \cdots + a_{in}x_n + b_i.$$

The solution is given by reducing the equations to the form

$$(2) \quad \begin{cases} \text{(2a)} \quad x_i = g_i(y_1, \ldots y_r, x_{r+1}, \ldots, x_n) & (i = 1, 2, \ldots, r) \\ \text{(2b)} \quad y_i = h_i(y_1, \ldots, y_r) & (i = r + 1, \ldots, m) \end{cases}$$

by step-by-step elimination (where the functions g_i, h_i are affine functions). The equations (2b) are then necessary and sufficient conditions for the given $(y_1, y_2, \ldots, y_m)$ to be of the form $y = f(x)$ for some x; and, when they are satisfied, all possible values of $(x_1, x_2, \ldots, x_n)$

can be found by choosing $x_{r+1}, x_{r+2}, \ldots, x_n$ arbitrarily and substituting in (2a). A solution of this form is possible if and only if

(i) $$\frac{\partial(y_1, y_2, \ldots, y_r)}{\partial(x_1, x_2, \ldots, x_r)} \neq 0$$

and

(ii) the pullback of every k-form is zero for $k > r$.

It is shown in this chapter that the same is true locally for systems (1) in which the functions f_i are differentiable* functions. This is the Implicit Function Theorem for differentiable maps, which states that *locally* near a given solution there is a solution of the form (2a), (2b) in which g_i, h_i, are differentiable functions if and only if the conditions (i), (ii) are satisfied at all points near the given point.

**A function of n variables is said to be differentiable if its first partial derivatives exist and are continuous. Jacobians and pullbacks of forms under such functions are defined in §5.2.*

Implicit Function Theorem

Let

(1) $$y_i = f_i(x_1, x_2, \ldots, x_n) \qquad (i = 1, 2, \ldots, m)$$

be a system of equations in which the functions f_i are defined† and (continuously) differentiable near the point $(\bar{x}_1, \bar{x}_2, \ldots, \bar{x}_n)$ in $\mathbf{R}^n$, and let $(\bar{y}_1, \bar{y}_2, \ldots, \bar{y}_m)$, where

$$\bar{y}_i = f_i(\bar{x}_1, \bar{x}_2, \ldots, \bar{x}_n) \qquad (i = 1, 2, \ldots, m)$$

be the corresponding point in $\mathbf{R}^m$. If the conditions

(i) $\dfrac{\partial(y_1, y_2, \ldots, y_r)}{\partial(x_1, x_2, \ldots, x_r)} \neq 0$ at $(\bar{x}_1, \bar{x}_2, \ldots, \bar{x}_n)$.

and

(ii) The pullback of every k-form for $k > r$ is identically zero near $(\bar{x}_1, \bar{x}_2, \ldots, \bar{x}_n)$.

are satisfied, then there are differentiable functions g_i, h_i and a number $\epsilon > 0$ such that the relations (1) and the relations

(2) $$\begin{cases} \text{(2a)} & x_i = g_i(y_1, \ldots, y_r, x_{r+1}, \ldots, x_n) \qquad (i = 1, 2, \ldots, r) \\ \text{(2b)} & y_i = h_i(y_1, \ldots, y_r) \qquad (i = r+1, \ldots, m) \end{cases}$$

are defined and equivalent at all points $(x_1, x_2, \ldots, x_n,$

†Here, and in the remainder of this chapter, only local properties are considered and functions are assumed to be defined locally near the points in question. In these contexts the notation (1) and the notation $f\colon \mathbf{R}^n \to \mathbf{R}^m$ mean that the domain of f is contained in $\mathbf{R}^n$, not that it is necessarily all of $\mathbf{R}^n$. For example the function $f(x) = \frac{1}{x}$ is permitted.

$y_1, y_2, \ldots, y_m$) of $\mathbf{R}^{n+m}$ within ϵ of $(\bar{x}_1, \bar{x}_2, \ldots, \bar{x}_n, \bar{y}_1, \bar{y}_2, \ldots, \bar{y}_m)$

$$|x_i - \bar{x}_i| < \epsilon \qquad (i = 1, 2, \ldots, n)$$
$$|y_i - \bar{y}_i| < \epsilon \qquad (i = 1, 2, \ldots, m).$$

That is, the functions f_i, g_i, h_i are all defined at such points and the relations (1) are satisfied if and only if the relations (2) are satisfied. Conversely, if (1) can be reduced in this way to the form (2) near $(\bar{x}_1, \bar{x}_2, \ldots, \bar{x}_n, \bar{y}_1, \bar{y}_2, \ldots, \bar{y}_m)$ then the conditions (i) and (ii) must be satisfied.

An important difference between the Implicit Function Theorem for affine maps and the present theorem is the *possibility of singularities* in the differentiable case. Given an affine system (1) it is always possible to rearrange the variables x_i and the equations y_i in such a way that conditions (i) and (ii) are satisfied, but such a rearrangement is not always possible for differentiable systems (1). (If (i) and (ii) are to be satisfied, then r must be the largest integer such that the pullback of some r-form is not identically zero; it follows that the x's and y's can be rearranged so that $\dfrac{\partial(y_1, y_2, \ldots, y_r)}{\partial(x_1, x_2, \ldots, x_r)}$ is not identically zero, but this Jacobian is a function of $(x_1, x_2, \ldots, x_n)$ and the statement that it is not identically zero does not imply (i). If the system (1) is affine, then the Jacobians are constant and the statement that $\dfrac{\partial(y_1, y_2, \ldots, y_r)}{\partial(x_1, x_2, \ldots, x_r)}$ is not identically zero does imply (i).) A point $(\bar{x}_1, \bar{x}_2, \ldots, \bar{x}_n)$ is called a *singularity* of (1) if there is no solution of the form (2) no matter how the x's and y's are rearranged. The following examples illustrate the theorem and show that singularities occur even for very simple systems (1).

As a first example, consider the mapping f: $\mathbf{R}^2 \to \mathbf{R}$ defined by the equation

$$y = u^2 + v^2. \tag{3}$$

Since

$$dy = 2u\,du + 2v\,dv$$

the relationship (3) can, by the Implicit Function Theorem, be written in the form

$$u = g(y, v)$$

locally near any point $(\bar{u}, \bar{v}, \bar{u}^2 + \bar{v}^2)$ where $\dfrac{\partial y}{\partial u}$ is not zero, i.e. where $\bar{u} \neq 0$. Similarly, (3) can be written in the form

$$v = g(y, u)$$

locally near any point where $\bar{v} \neq 0$. All points $(\bar{u}, \bar{v}, \bar{u}^2 + \bar{v}^2)$ are covered by at least one of these two cases except for the point $\bar{u} = 0, \bar{v} = 0$. This point is, by the above definition, a singularity.

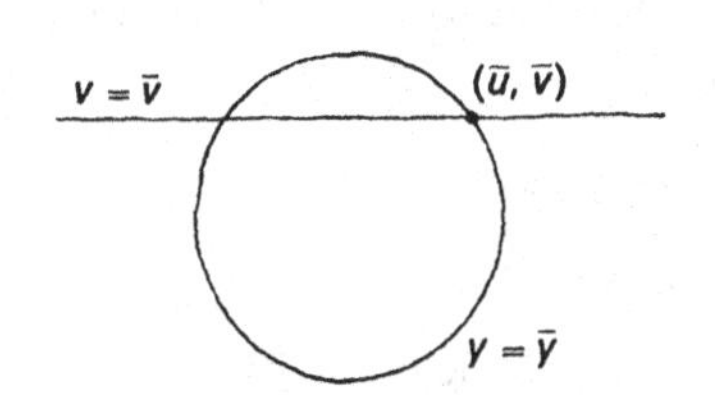

The geometrical significance of these solutions can easily be seen. Let $(\bar{u}, \bar{v})$ be a given point other than the singularity (0, 0). The set where $y = \bar{y}$ is the circle whose center is (0, 0) and which passes through $(\bar{u}, \bar{v})$. Consider this circle and the horizontal line $v = \bar{v}$ through $(\bar{u}, \bar{v})$. Circle and line intersect in *two* points unless $\bar{u} = 0$, in which case the line is tangent to the circle and there is only one point of intersection (or, as it is sometimes stated, the two points of intersection *coincide*). Assume now that $\bar{u} \neq 0$. It is clear geometrically that if $\bar{y}$ is changed slightly and $\bar{v}$ is changed slightly, then the corresponding circle and line will still intersect in two points, one of them near $(\bar{u}, \bar{v})$ and the other (relatively) far away. In this way, every (y, v) near $(\bar{y}, \bar{v})$ determines a point of the uv-plane near $(\bar{u}, \bar{v})$. The u-coordinate of this point is therefore a function of (y, v) near $(\bar{y}, \bar{v})$ and this is the solution $u = g(y, v)$. It is clear that there is no such solution $u = g(y, v)$ if $\bar{u} = 0$ because then a small change in circle and line can result in either no point of intersection or in two points of intersection equally near to $(\bar{u}, \bar{v})$.

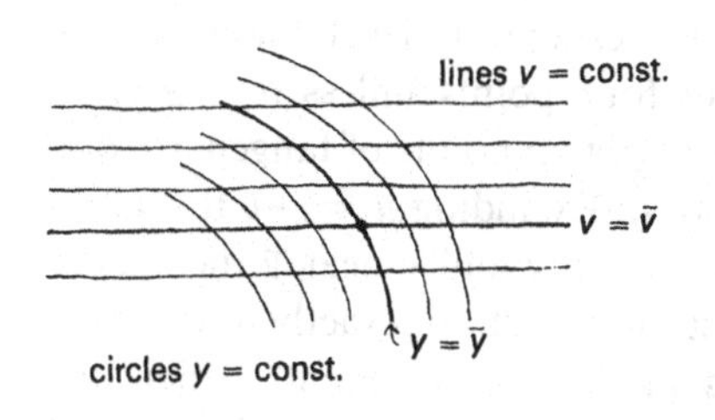

In this simple case the function $g(y, v)$ can be given explicitly, namely

$$g(y, v) = \begin{cases} \sqrt{y - v^2} & \text{if } \bar{u} > 0 \\ -\sqrt{y - v^2} & \text{if } \bar{u} < 0. \end{cases} \tag{4}$$

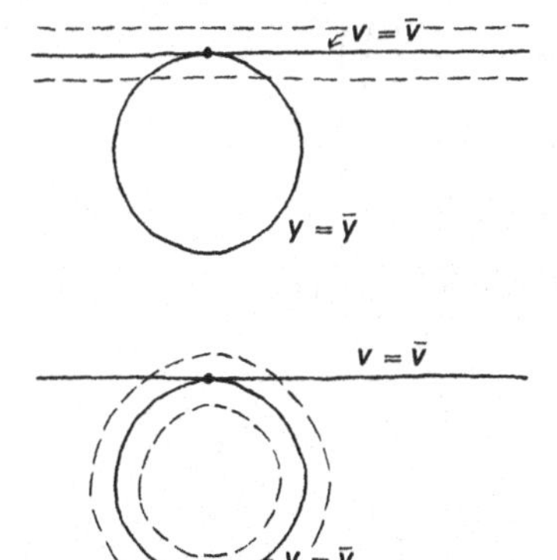

These functions are defined and differentiable provided $y - v^2 > 0$. Thus $g(y, v)$ is defined at all points (y, v) near $(\bar{y}, \bar{v})$ and '$u = g(y, v)$' is equivalent to '$y = u^2 + v^2$ and u has the same sign as $\bar{u}$'. Therefore $u = g(y, v)$ is defined and equivalent to $y = u^2 + v^2$ for all (u, v, y) sufficiently near $(\bar{u}, \bar{v}, \bar{y})$.

As a second example, consider the mapping f: $\mathbf{R}^2 \to \mathbf{R}^2$ defined by

$$\begin{aligned} x &= uv \\ y &= u^2 + v^2. \end{aligned} \tag{5}$$

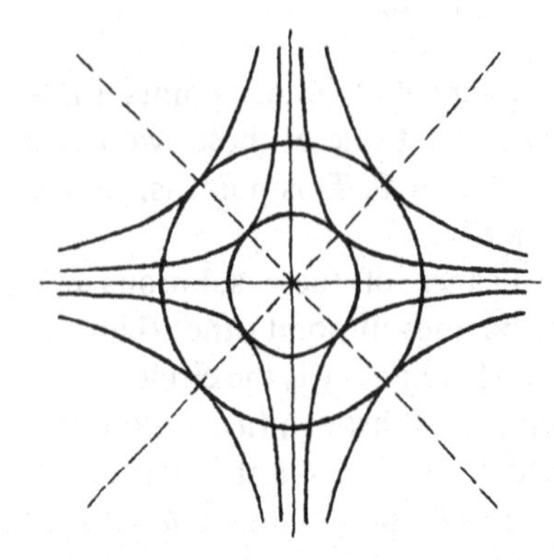

Here

$$dx\,dy = (u\,dv + v\,du)(2u\,du + 2v\,dv) = 2(v^2 - u^2)\,du\,dv.$$

Thus $r = 2$ because this 2-form is not identically zero near any point (u, v). However, it is equal to zero *at* all points along the lines $u = \pm v$ and these points are singularities; that is, (i) and (ii) cannot be satisfied by rearranging variables. At all other points they are satisfied and the theorem states that locally the equations (5) can be solved to give

$$(6) \qquad \begin{aligned} u &= g_1(x, y) \\ v &= g_2(x, y) \end{aligned}$$

for (x, y) near $(\bar{x}, \bar{y})$ and (u, v) near $(\bar{u}, \bar{v})$.

By sketching the level curves $x =$ const. (hyperbolae with axes $u = \pm v$, asymptotes $u = 0$, $v = 0$) and $y =$ const. (circles with center the origin) it can be seen that, given $(\bar{u}, \bar{v})$, the corresponding level curves $x = \bar{x}$ and $y = \bar{y}$ intersect in four points unless $\bar{u} = \pm\bar{v}$, in which case they intersect in two points of tangency (four points coincident in pairs). Excluding $\bar{u} = \pm\bar{v}$ it is clear geometrically that for x near $\bar{x}$ and y near $\bar{y}$ the curves $x =$ const., $y =$ const. intersect in exactly one point near $(\bar{u}, \bar{v})$ as well as in three other points which are (relatively) far away. The uv-coordinates of this point of intersection are therefore functions of x, y for x near $\bar{x}$ and y near $\bar{y}$; hence (6).

x = const.

y = const.

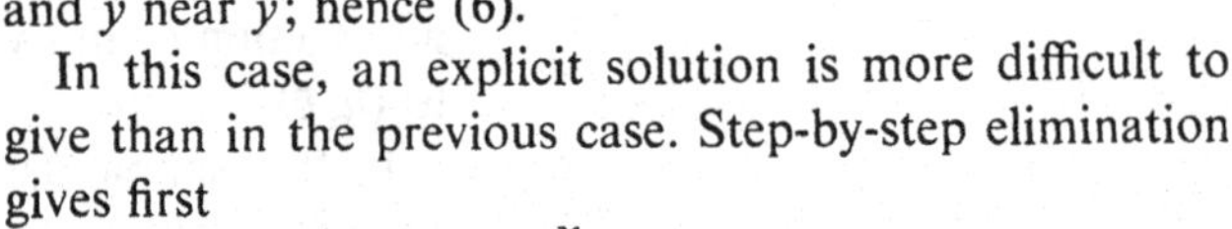

In this case, an explicit solution is more difficult to give than in the previous case. Step-by-step elimination gives first

$$v = \frac{x}{u}$$

$$y = u^2 + \left(\frac{x}{u}\right)^2$$

and then, solving for u,

$$u^2 y = u^4 + x^2$$

$$u^2 = \frac{y \pm \sqrt{y^2 - 4x^2}}{2}$$

$$u = \pm\sqrt{\frac{y \pm \sqrt{y^2 - 4x^2}}{2}}$$

and for v,

$$v = \frac{x}{u} = \pm\sqrt{\frac{2x^2}{y \pm \sqrt{y^2 - 4x^2}}} = \pm\sqrt{\frac{2x^2(y \mp \sqrt{y^2 - 4x^2})}{y^2 - (y^2 - 4x^2)}} = \pm\sqrt{\frac{y \mp \sqrt{y^2 - 4x^2}}{2}}$$

giving the final result

$$(6') \qquad \begin{aligned} u &= \pm\sqrt{\frac{y \pm \sqrt{y^2 - 4x^2}}{2}} \\ v &= \pm\sqrt{\frac{y \mp \sqrt{y^2 - 4x^2}}{2}} \end{aligned}$$

where the formula for u involves two choices of sign and where the signs in v are then determined using $v = x/u$. For each (x, y) the formula (6′) gives four points (u, v) as expected; the signs must then be determined in such a way as to select the one near $(\bar{u}, \bar{v})$.

This explicit solution, cumbersome as it is, is possible only because the mapping (5) is particularly simple. If the fourth-degree equation for u had involved u^3 and u as well as u^4, u^2, then it could not have been solved by means of the quadratic formula, and the final solution (6′) would have been very much more complicated. From this it is clear that an explicit algebraic solution of equations of the form

$$(7) \qquad \begin{aligned} x &= p_1(u, v) \\ y &= p_2(u, v), \end{aligned}$$

where p_1 and p_2 are polynomials, will not be feasible in general. It can be shown that in fact the solution of (7) for (u, v) as functions of (x, y), even when p_1 and p_2 are polynomials of relatively low degree, is in general *impossible* in the sense that the values of (u, v) cannot be obtained from those of (x, y) by arithmetic operations and the extraction of roots. Nonetheless, the Implicit Function Theorem proves that locally the solution exists in the sense that if a point $(\bar{u}, \bar{v}, \bar{x}, \bar{y})$ is given with $\bar{x} = p_1(\bar{u}, \bar{v})$, $\bar{y} = p_2(\bar{u}, \bar{v})$, then every point (x, y) near $(\bar{x}, \bar{y})$ is the image of one and only one point (u, v) near $(\bar{u}, \bar{v})$ under (7) (unless $(\bar{u}, \bar{v})$ is a singularity of (7)), and that the mapping $(x, y) \to (u, v)$ so defined is differentiable. This statement about the existence of a solution is all the more important in cases where the solution cannot be found explicitly, since it gives information about the mapping (7) which could not be derived by algebra alone.

For the case of a single function of a single variable

$$(8) \qquad y = f(x)$$

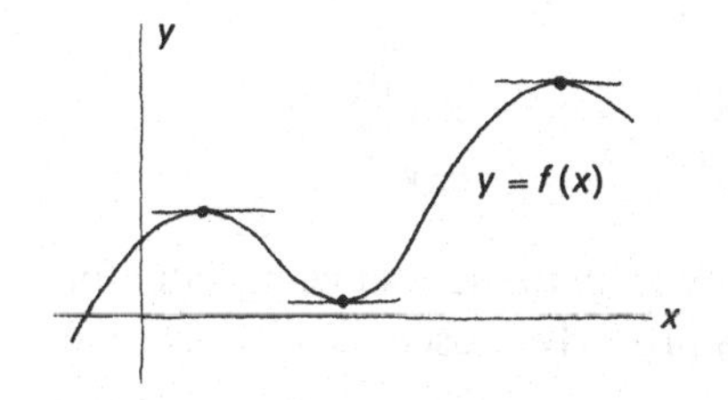

the conclusions of the Implicit Function Theorem are familiar facts of elementary calculus. If $n = m = r = 1$, then the theorem states that a (continuously) differentiable function (8) admits a differentiable inverse function

$$(9) \qquad x = g(y)$$

near a point $\bar{y} = f(\bar{x})$ if and only* if the pullback $dy = f'(x)\,dx$ is not zero at $\bar{x}$, that is, if and only if the derivative of f is not zero at $\bar{x}$. On the other hand, if the derivative is identically zero near $\bar{x}$ then $r = 0$, equation (2a) is not present, and equation (2b) gives y as a 'function of no variables'. That is, $y =$ const. and the statement reduces to the familiar fact that a function of one variable is constant if and only if its derivative is identically zero. The function (8) has a singularity at $\bar{x}$ when $f'(\bar{x}) = 0$ but f' is not identically zero near $\bar{x}$. Such a point is also called a 'critical point' of f, for reasons discussed in §5.4. The fact that a simple function $y = f(x)$ can have a very complicated inverse function is also familiar. For example, the function

The statement that the inverse function is to be differentiable is crucial here. The function $y = x^3$ has an inverse function $x = \sqrt[3]{y}$ valid for all y, but this function is not, of course, differentiable at $y = 0$.

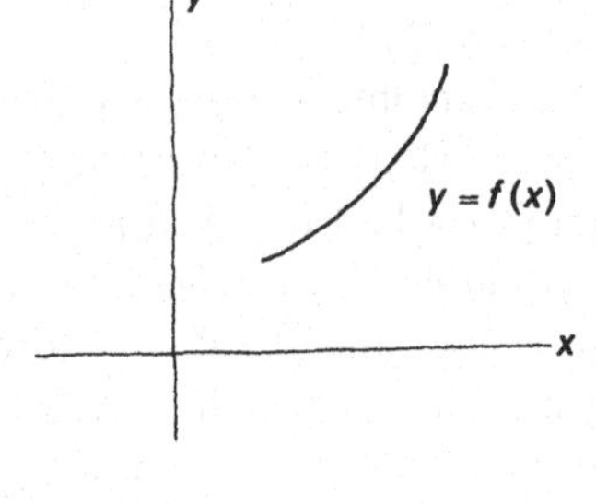

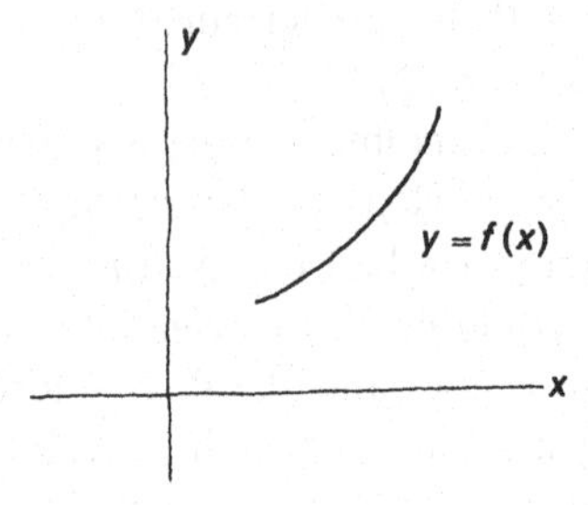

$$y = 2x^3 + x^2 + 2x - 1$$

has a positive derivative for all x, which implies that the function always increases, and hence that each y is the image of one and only one x, even though the explicit solution for x as a function of y is not at all simple.

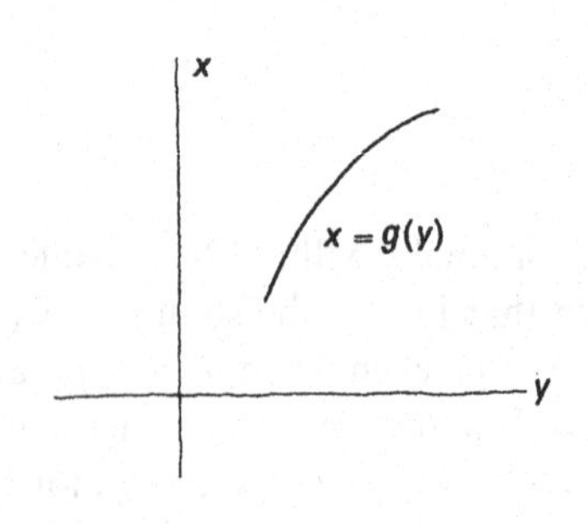

A different type of singularity is presented by the map $f\colon \mathbf{R} \to \mathbf{R}^2$ defined by

$$(10) \qquad \begin{aligned} x &= t^3 \\ y &= t^2. \end{aligned}$$

Here the solutions would be of the form

$$(11) \qquad \begin{aligned} t &= g(x) \\ y &= h(x) \end{aligned} \quad \text{or} \quad \begin{aligned} t &= g(y) \\ x &= h(y). \end{aligned}$$

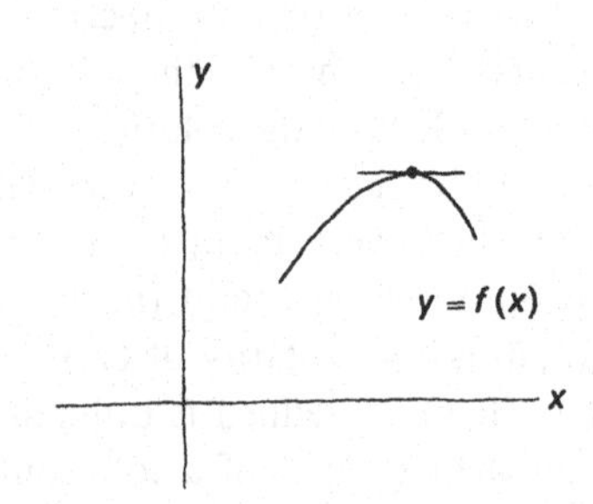

There is a solution of the first type provided $dx = 3t^2\,dt \neq 0$ and a solution of the second type provided $dy = 2t\,dt \neq 0$. Thus there is a solution of either type near a point $t \neq 0$ but the point $t = 0$ is a singularity. The explicit solutions are

$$(11') \qquad \begin{aligned} t &= \sqrt[3]{x} \\ y &= x^{2/3} \end{aligned} \quad \text{or} \quad \begin{aligned} t &= \pm y^{1/2} \\ x &= \pm y^{3/2} \end{aligned}$$

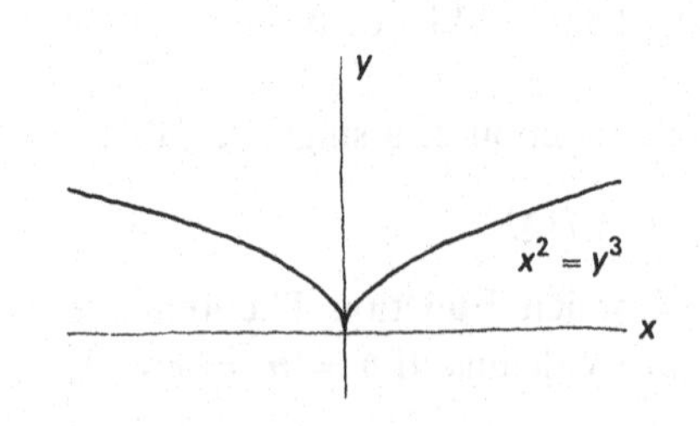

the first being valid for $x \neq 0$, the second being valid for $y > 0$ when the sign is properly chosen. A sketch of the curve $x^2 = y^3$ shows the meaning of these solutions and shows that there is indeed a 'singularity' at (0, 0).

As a final application of the Implicit Function Theorem consider the map f: $\mathbf{R}^2 \to \mathbf{R}^2$ given by

$$\begin{aligned} x &= \cos(u + v) \\ y &= \sin(u + v). \end{aligned} \tag{12}$$

Since $dx\,dy \equiv 0$ the integer r in any reduction must be ≤ 1. On the other hand,

$$dx = -\sin(u + v)\,du - \sin(u + v)\,dv$$

has neither component equal to zero except on the lines where $u + v$ is a multiple of π; hence there are solutions

$$\begin{aligned} u &= g(x, v) \\ y &= h(x) \end{aligned} \quad \text{or} \quad \begin{aligned} v &= g(x, u) \\ y &= h(x) \end{aligned}$$

near points $(\bar{u}, \bar{v})$ with $\bar{u} + \bar{v} \neq n\pi$. Similarly there are solutions

$$\begin{aligned} u &= g(y, v) \\ x &= h(y) \end{aligned} \quad \text{or} \quad \begin{aligned} v &= g(y, u) \\ x &= h(y) \end{aligned}$$

near points $(\bar{u}, \bar{v})$ with $\bar{u} + \bar{v} \neq n\pi + \frac{\pi}{2}$. Note that at least one of the two is valid at every $(\bar{u}, \bar{v})$. Hence there are *no singularities*. Explicit solutions can be given in terms of the inverse trigonometric functions Arccos* and Arcsin, e.g.

$$\begin{aligned} u &= \operatorname{Arccos} x - v + n\pi \\ y &= \pm\sqrt{1 - x^2} \end{aligned}$$

where the integer n and the sign $\pm$ depend on the point $(\bar{u}, \bar{v})$.

If the point $(\bar{x}_1, \bar{x}_2, \ldots, \bar{x}_n)$ is not a singularity of (1) then, by definition, the x's and y's can be rearranged (if necessary) so that there is a solution of the form (2). If this is the case then the integer r is determined by the map (1) (as the largest integer such that the pullback of some r-form is not zero at $(\bar{x}_1, \bar{x}_2, \ldots, \bar{x}_n)$) and is independent of the particular solution (2). It is called the *rank* of (1) at $(\bar{x}_1, \bar{x}_2, \ldots, \bar{x}_n)$ and (1) is said to be *non-singular of rank r at* $(\bar{x}_1, \bar{x}_2, \ldots, \bar{x}_n)$. If (1) is singular at $(\bar{x}_1, \bar{x}_2, \ldots, \bar{x}_n)$ then the rank is not defined. By

*The symbol Arccos x, defined for x a real number satisfying $|x| \leq 1$, denotes the unique real number y in the interval $\{0 \leq y \leq \pi\}$ such that $x = \cos y$. The symbol Arcsin x is defined similarly.

the Implicit Function Theorem, the map (1) is non-singular of rank r at $(\bar{x}_1, \bar{x}_2, \ldots, \bar{x}_n)$ if and only if the pullback of every $(r+1)$-form in the y's is identically zero near $(\bar{x}_1, \bar{x}_2, \ldots, \bar{x}_n)$, but the pullback of some r-form in the y's is not zero at $(\bar{x}_1, \bar{x}_2, \ldots, \bar{x}_n)$.

Exercises

1 Near what points $(\bar{u}, \bar{v})$ can the equation $y = u^2 - v^2$ be solved to give $u = g(y, v)$? To give $v = g(u, y)$? Give explicit solutions. Sketch the level curves $y =$ const. and relate their geometry to the existence of solutions $u = g(y, v)$, $v = g(u, y)$. For what values of y can $y = u^2 - v^2$ be solved? What is 'singular' about the solution if $y = 0$?

2 Simplify the solution of (5) for u, v as functions of x, y by considering $y \pm 2x$. For what values of (x, y) is there a (u, v), i.e. what is the image of the map (5)? For each of the eight possible choices of sign in (6′) find the points $(\bar{u}, \bar{v})$, if any, at which (6′) then gives a solution of (5).

3 Describe geometrically the solution of $y = u^2 + v^2 + w^2$ for $u = g(y, v, w)$. Near what points $(\bar{u}, \bar{v}, \bar{w})$ is this possible?

4 Discuss the solution of the equations

$$x = u^2 - v^2$$
$$y = 2uv$$

for (u, v) given (x, y). Sketch the level curves $x =$ const., $y =$ const.

5 Discuss the solution of the equations

$$x = e^u \cos v$$
$$y = e^v \sin v$$

for (u, v) given (x, y). [Compare to polar coordinates.]

6 The Implicit Function Theorem $(n = m = r)$ says that a differentiable map $f\colon \mathbf{R}^n \to \mathbf{R}^n$ which is non-singular of rank n is locally both one-to-one and onto. Example 5 shows that neither is true globally. Elaborate on this statement. Which of the two (one-to-one or onto) is true globally of non-singular maps $f\colon \mathbf{R} \to \mathbf{R}$?

7 Describe the example (12) geometrically as a map $\mathbf{R}^2 \to \mathbf{R} \to \mathbf{R}^2$.

8 Consulting a book on elementary calculus, prove the case $m = n = r = 1$ of the Implicit Function Theorem. Give a complete definition of the function Arcsin x (assuming the function sin x has been defined).

9 The *folium of Descartes* is defined by the equation $x^3 + y^3 = xy$. A good picture of the curve can be obtained from the following observations: The equation is unchanged when x and y are interchanged, therefore the curve is symmetric about the line $x = y$. The curve intersects each line $x =$ const. in at most three points and at least one point. Similarly, the curve intersects each line $ax + by =$ const. in at most three points and at least one point unless $a = b$, in which case it intersects in at most two points. The intersections of the line $x + y = C$ with the curve are most easily found by using the substitution $x = (C/2) + t$, $y = (C/2) - t$. Prove these statements and sketch the folium of Descartes. Find the singularities of the function $F = x^3 + y^3 - xy$. Where is $F > 0$? Where < 0? Sketch the curves $F =$ const. Near what points $(\bar{x}, \bar{y})$ on the folium $F = 0$ can the equation $F = 0$ be taken as defining y implicitly as a function of x? Indicate in a sketch the various functions $y = f(x)$ which satisfy $F(x, f(x)) \equiv 0$.

10 Using the Implicit Function Theorem, give a sufficient condition for an equation $F(x, y, z) \equiv 0$ to determine z locally (near a given solution) as a function of x and y.

11 Let $F(x, y, z)$, $G(x, y, z)$ be two (differentiable) functions of three variables, and let $(\bar{x}, \bar{y}, \bar{z})$ be a point where $dF \neq 0$, $dG \neq 0$. F and G are said to be *functionally related* (or dependent) near $(\bar{x}, \bar{y}, \bar{z})$ if there is a (differentiable) function f of two variables, defined and not identically zero near $(F(\bar{x}, \bar{y}, \bar{z}), G(\bar{x}, \bar{y}, \bar{z}))$ such that $f(F, G) \equiv 0$ for (x, y, z) near $(\bar{x}, \bar{y}, \bar{z})$. Use the Implicit Function Theorem to show that this is true if $dF\,dG \equiv 0$ near $(\bar{x}, \bar{y}, \bar{z})$. Conversely, show that if F and G are functionally related then $dF\,dG = 0$ at $(\bar{x}, \bar{y}, \bar{z})$, and hence that if F, G are functionally related then $dF\,dG \equiv 0$.

12 Sketch the graph of the function

$$f(x) = \begin{cases} x^2 \sin\left(\dfrac{1}{x^2}\right), & x \neq 0 \\ 0, & x = 0. \end{cases}$$

Show that the derivative $f'(x) = \lim_{h \to 0} [f(x + h) - f(x)]/h$ exists for all x but that $f'(x)$ is not a continuous function. (Therefore, this function is not 'differentiable' in the sense defined in §5.2.)

13 Define the 'local rank' of a system of equations (1) at the point $(\bar{x}_1, \bar{x}_2, \ldots, \bar{x}_n)$ to be the largest integer r such that the pullback of some r-form is not identically zero *near* $(\bar{x}_1,$

$\bar{x}_2, \ldots, \bar{x}_n)$ and define the 'infinitesimal rank' to be the largest r such that the pullback of some r-form is not zero *at* the point $(\bar{x}_1, \bar{x}_2, \ldots, \bar{x}_n)$. Define 'singular' and 'non-singular of rank r' in terms of 'local rank' and 'infinitesimal rank.'

5.2

k-Forms on n-Space. Differentiable Maps

Let $(x_1, x_2, \ldots, x_n)$ denote the coordinates on $\mathbf{R}^n$. A (variable) k-form on n-space $\mathbf{R}^n$ is a function assigning k-forms in $x_1, x_2, \ldots, x_n$ to points $(x_1, x_2, \ldots, x_n)$ of $\mathbf{R}^n$. Since constant k-forms are represented by sums of terms of the form $A\, dx_{i_1}\, dx_{i_2} \ldots dx_{i_k}$ in which A is a number, (variable) k-forms can be represented as sums of such terms in which A is a function $A(x_1, x_2, \ldots, x_n)$. In short, the definitions of §2.1 are to be extended in the obvious way to k-forms in n variables.

As before, it is computations with k-forms which are of primary importance. In addition to the rules $dx_i\, dx_i = 0$, $dx_i\, dx_j = -dx_j\, dx_i$, these computations are governed by the *rule for computing pullbacks* which is summarized by the formula

$$dy_i = \frac{\partial y_i}{\partial x_1} dx_1 + \frac{\partial y_i}{\partial x_2} dx_2 + \cdots + \frac{\partial y_i}{\partial x_n} dx_n.$$

More formally, the *pullback* of a given k-form

$$(1) \qquad A(y_1, y_2, \ldots, y_n)\, dy_1\, dy_2 \ldots dy_k + \cdots$$

in $y_1, y_2, \ldots, y_m$ under a given map

$$(2) \qquad y_i = f_i(x_1, x_2, \ldots, x_n) \qquad (i = 1, 2, \ldots, m)$$

is defined to be the k-form in $x_1, x_2, \ldots, x_n$ obtained by carrying out the substitution (2) expressing the y's in terms of the x's and the substitution

$$(3) \qquad dy_i = \sum_{j=1}^{n} \frac{\partial y_i}{\partial x_j} dx_j$$

expressing the dy's in terms of the x's and dx's. Here $\frac{\partial y_i}{\partial x_j}$ denotes the partial derivative

$$\frac{\partial y_i}{\partial x_j}(x_1, x_2, \ldots, x_n) = \lim_{h \to 0} \frac{f_i(x_1, \ldots, x_j + h, \ldots, x_n) - f_i(x_1, \ldots, x_n)}{h}$$

which is a function of $(x_1, x_2, \ldots, x_n)$. The resulting expression is a sum of multiples of products of 1-forms in the x's. Multiplying out according to the rules of §4.2 then gives a k-form in the x's which is defined to be the pullback. As in §4.2, this definition is valid only after one observes that if the given expression (1) is replaced by another expression representing the same k-form, then the resulting expression for the pullback represents the same k-form as before; this follows from the observation that the rules $\omega\omega = 0$, $\omega\sigma = -\sigma\omega$ hold for arbitrary (variable) 1-forms in the x's as well as for the 1-forms dx_i.

As in Chapter 2, all k-forms considered will be assumed to be *continuous*, that is, the coefficient functions $A(x_1, x_2, \ldots, x_n)$ will be assumed to be continuous*. In particular, all mappings (2) considered will be assumed to have the property that the 1-forms (3) are continuous. In other words, the functions (2) will be assumed to have the property that all mn first partial derivatives $\dfrac{\partial y_i}{\partial x_j}$ exist and are continuous. Such a function (2) is said to be *differentiable*. (To be more precise, a function $f\colon \mathbf{R}^n \to \mathbf{R}^m$ whose mn first partial derivatives all exist and are continuous is said to be 'C^1-differentiable' or 'continuously differentiable'. In this book the word 'differentiable' is used *only* in this sense unless otherwise stated. This will be emphasized by occasionally writing '(continuously) differentiable' instead of 'differentiable'.) The pullback of a continuous k-form under a differentiable map is a continuous k-form (because compositions, sums, and products of continuous functions are continuous).

*That is, given $(\bar{x}_1, \bar{x}_2, \ldots, \bar{x}_n)$ and given $\epsilon > 0$ there is a $\delta > 0$ such that $|A(x_1, x_2, \ldots, x_n) - A(\bar{x}_1, \bar{x}_2, \ldots, \bar{x}_n)| < \epsilon$ whenever $|x_1 - \bar{x}_1| < \delta$, $|x_2 - \bar{x}_2| < \delta, \ldots, |x_n - \bar{x}_n| < \delta$. See §2.3, p. 31.

The central theorem concerning computations with k-forms is the Chain Rule.

Chain Rule

Let

$$(4) \qquad y_i = f_i(x_1, x_2, \ldots, x_n) \qquad (i = 1, 2, \ldots, m)$$

and

$$(5) \qquad z_i = g_i(y_1, y_2, \ldots, y_m) \qquad (i = 1, 2, \ldots, p)$$

be differentiable maps $\mathbf{R}^n \to \mathbf{R}^m$ and $\mathbf{R}^m \to \mathbf{R}^p$ respectively. Then the composed map $\mathbf{R}^n \to \mathbf{R}^p$ is differentiable and the pullback of any k-form under the composed map is equal to the pullback of the pullback.

The Chain Rule will be proved in the following section. In Jacobian notation the statement of the Chain Rule is

$$\frac{\partial z}{\partial x} = \sum \frac{\partial z}{\partial y}\frac{\partial y}{\partial x}$$

where the left side is a Jacobian of k of the z's with respect to k of the x's and where the right side is a sum of $\binom{m}{k}$ terms as in Chapter 4. When $k = 1$ this is the formula

$$\frac{\partial z_i}{\partial x_j} = \sum_{\nu=1}^{m} \frac{\partial z_i}{\partial y_\nu}\frac{\partial y_\nu}{\partial x_j}$$

which is the usual chain rule of differentiation. Note that the differentiability of $z(x)$ is asserted to be a consequence of the differentiability of $z(y)$ and $y(x)$.

Many formulas of differential calculus are simple consequences of the Chain Rule. For example, applying the Chain Rule to the composition of the maps

$$p = xyz, \quad \begin{aligned} x &= t \\ y &= t \\ z &= t \end{aligned}$$

gives $dp = yz\,dx + xz\,dy + xy\,dz = tt\,dt + tt\,dt + tt\,dt = 3t^2\,dt$, i.e. $d(t^3) = 3t^2\,dt$. In the same way $d(t^n) = nt^{n-1}\,dt$ for any positive integer n. Less obvious formulas can be obtained from the Chain Rule by *implicit differentiation.* As will be shown in the next section, this process results from the application of the Chain Rule to implicitly defined functions.

Implicit Differentiation

If the conditions of the Implicit Function Theorem are satisfied and the functions f_i determine functions g_i, h_i, then the partial derivatives of the implicit functions g_i, h_i at a given point can be found in terms of the partial derivatives of the given functions f_i at that point by writing

$$dy_i = \sum_{j=1}^{n} \frac{\partial f_i}{\partial x_j}\,dx_j, \tag{6}$$

solving formally for $dx_1, \ldots, dx_r, dy_{r+1}, \ldots, dy_m$ in terms of $dy_1, \ldots, dy_r, dx_{r+1}, \ldots, dx_n$, and writing the result as

$$(7)\qquad \begin{aligned} dx_i &= \sum_{j=1}^{r} \frac{\partial g_i}{\partial y_j} dy_j + \sum_{j=r+1}^{n} \frac{\partial g_i}{\partial x_j} dx_j \qquad (i = 1, 2, \ldots, r) \\ dy_i &= \sum_{j=1}^{r} \frac{\partial h_i}{\partial y_j} dy_j \qquad (i = r+1, \ldots, m) \end{aligned}$$

thereby determining the partial derivatives in question.

Note that the result is expressed in terms of the variables $x_1, x_2, \ldots, x_n$ of the functions f. If $\frac{\partial g}{\partial y}, \frac{\partial g}{\partial x}, \frac{\partial h}{\partial y}$ are to be found as functions of the variables $(y_1, \ldots, y_r, x_{r+1}, \ldots, x_n)$, then the explicit solutions g, h for the x's as functions of $(y_1, \ldots, y_r, x_{r+1}, \ldots, x_n)$ must be used. Note also that in this context the Implicit Function Theorem can be stated simply "The functions g, h exist if and only if they can be differentiated implicitly and the partials of h with respect to $x_{r+1}, x_{r+2}, \ldots, x_n$ are identically zero," because (i) is true if and only if the first r equations of (6) can be solved for $dx_1, dx_2, \ldots, dx_r$ and (ii) is true if and only if the remaining equations are then independent of $dx_{r+1}, \ldots, dx_n$.

Examples

If $y = u^2 + v^2$ is solved for $u = g(y, v)$ then the derivatives of g are found by

$$\begin{aligned} dy &= 2u\,du + 2v\,dv \\ du &= \frac{1}{2u} dy - \frac{v}{u} dv \end{aligned}$$

giving

$$\frac{\partial g}{\partial y} = \frac{1}{2u}, \quad \frac{\partial g}{\partial v} = -\frac{v}{u}.$$

The Implicit Function Theorem says that g is defined *locally* near any point where this solution is possible, i.e. where $u \neq 0$. If the derivatives of $g(y, v)$ are to be expressed as functions of (y, v) then the solution $g(y, v) = \pm\sqrt{y - v^2}$ must be used. This gives

$$\begin{aligned} \frac{\partial}{\partial y}[\pm\sqrt{y - v^2}] &= \pm \frac{1}{2\sqrt{y - v^2}} \\ \frac{\partial}{\partial v}[\pm\sqrt{y - v^2}] &= -\frac{v}{\pm\sqrt{y - v^2}}. \end{aligned}$$

If $x = t^3$, $y = t^2$ is solved for $t = g(x)$, $y = h(x)$ then the derivatives of g, h are found by setting

$$dx = 3t^2\,dt$$
$$dy = 2t\,dt$$

and solving

$$dt = \frac{1}{3t^2}\,dx \qquad g'(x) = \frac{1}{3t^2}$$
$$dy = \frac{2}{3t}\,dx \qquad h'(x) = \frac{2}{3t}.$$

This is possible if and only if $t \neq 0$, which is the condition required by the Implicit Function Theorem to guarantee (locally) the existence of g, h. To express g', h' as functions of x the explicit solution $t = x^{1/3}$ must be used, giving

$$\frac{d}{dx}[\sqrt[3]{x}] = \frac{1}{3(\sqrt[3]{x})^2} = \frac{1}{3}x^{-2/3}$$
$$\frac{d}{dx}[x^{2/3}] = \frac{2}{3\sqrt[3]{x}} = \frac{2}{3}x^{-1/3}.$$

If polar coordinates $x = r\cos\theta$, $y = r\sin\theta$ are solved for $r = g_1(x, y)$, $\theta = g_2(x, y)$, then the derivatives can be found by solving

$$dx = \cos\theta\,dr - r\sin\theta\,d\theta$$
$$dy = \sin\theta\,dr + r\cos\theta\,d\theta$$

for dr, $d\theta$. Since

$$\cos\theta\,dx + \sin\theta\,dy = dr$$
$$-\sin\theta\,dx + \cos\theta\,dy = r\,d\theta$$

the solution is

$$dr = \cos\theta\,dx + \sin\theta\,dy = \frac{\partial g_1}{\partial x}dx + \frac{\partial g_1}{\partial y}dy$$
$$d\theta = -\frac{\sin\theta}{r}dx + \frac{\cos\theta}{r}dy = \frac{\partial g_2}{\partial x}dx + \frac{\partial g_2}{\partial y}dy.$$

The functions g_1, g_2 are defined locally near any point where $r \neq 0$. If the derivatives of g_1, g_2 are to be expressed as functions of (x, y) the solution

$$r = \sqrt{x^2 + y^2} = g_1(x, y)$$
$$\theta = \text{Arctan}\left(\frac{y}{x}\right) = g_2(x, y) \qquad (\text{if } x \neq 0)$$

must be used, giving

$$\frac{\partial}{\partial x}\sqrt{x^2+y^2} = \cos\theta = \frac{x}{r} = \frac{x}{\sqrt{x^2+y^2}}$$

$$\frac{\partial}{\partial y}\sqrt{x^2+y^2} = \sin\theta = \frac{y}{r} = \frac{y}{\sqrt{x^2+y^2}}$$

$$\frac{\partial}{\partial x}\left[\operatorname{Arctan}\left(\frac{y}{x}\right)\right] = -\frac{\sin\theta}{r} = -\frac{y}{r^2} = -\frac{y}{x^2+y^2}$$

$$\frac{\partial}{\partial y}\left[\operatorname{Arctan}\left(\frac{y}{x}\right)\right] = \frac{\cos\theta}{r} = \frac{x}{r^2} = \frac{x}{x^2+y^2}.$$

Exercises **1** If a point moves in the xy-plane according to parametric equations $x = f(t)$, $y = g(t)$ and if $F(x, y)$ is a quantity (temperature, pressure, altitude) depending on (x, y), then the rate of change of F with respect to t is expressed by the Chain Rule in terms of

$$\begin{aligned}(f(\bar{t}), g(\bar{t})) &= \text{location at time } \bar{t}\\ f'(\bar{t}) &= \text{velocity in } x\text{-direction at time } \bar{t}\\ g'(\bar{t}) &= \text{velocity in } y\text{-direction at time } \bar{t}\end{aligned}$$

and the partial derivatives of F.

(a) Let $F(x, y) = xy$, let $(\bar{x}, \bar{y}) = (f(\bar{t}), g(\bar{t}))$, and let $a = f'(\bar{t})$, $b = g'(\bar{t})$. Find the rate of change of F in terms of x, y, a, b.

(b) Let an arrow starting at $(\bar{x}, \bar{y})$ and ending at $(\bar{x} + a, \bar{y} + b)$ represent position $(\bar{x}, \bar{y})$ and velocity (a, b); such an arrow is called a 'velocity vector'. Draw the velocity vectors corresponding to

position	velocity
$(1, 0)$	$(0, 1)$
$(2, 2)$	$(0, 1)$
$(-3, 1)$	$(0, 3)$
$(-4, -2)$	$(1, 1)$
$(\frac{1}{2}, -1\frac{1}{2})$	$(1, -2)$

(c) For each of these velocity vectors find the rate of change of $F(x, y) = xy$, paying particular attention to the sign.

(d) If the position is $(2, -3)$ and if the point is moving along the hyperbola $xy = -6$ what condition must be satisfied by the velocity (a, b) according to part (a)? Draw several velocity vectors satisfying this condition.

(e) Using (d) find the equation of the line tangent to $xy = -6$ at $(2, -3)$. [A line through $(2, -3)$ can be written in the form $A(x - 2) + B(y + 3) = 0$.]

(f) Find the equation of the line tangent to $xy = \bar{x}\bar{y}$ at the point $(\bar{x}, \bar{y})$.
(g) Why is this not valid at $(0, 0)$?
(h) Find the *slope* of the tangent line to $xy = 1$ at the point $(\bar{x}, 1/\bar{x})$, hence the derivative of the function $f(x) = 1/x$.
(i) Find the derivative of $1/x$ by implicit differentiation of $xy = 1$.

2 (a) If $F(x, y) = 2x^2 + y^2$ and if a point at $(\bar{x}, \bar{y})$ moves with the velocity (a, b), what is the rate of change of F?
(b) Sketch the ellipse $2x^2 + y^2 = 12$ near the point $(2, 2)$. On which side of the ellipse is $F > 12$ and on which is it < 12?
(c) Draw six velocity vectors at $(2, 2)$, two for which $2x^2 + y^2$ has positive derivative, two for which it has negative derivative, and two for which its derivative is zero.
(d) Find the equation of the tangent line to the ellipse at $(2, 2)$. [Write it in the form $A(x - 2) + B(y - 2) = 0$.]
(e) Find the equation of the tangent line to the ellipse $2x^2 + y^2 = \text{const.}$ at the point $(\bar{x}, \bar{y})$. [Write it in the form $A(x - \bar{x}) + B(y - \bar{y}) = 0$.]
(f) Use (e) to find the slope of the tangent to the graph of $y = -\sqrt{1 - 2x^2}$ $(|x| < 1\sqrt{2})$ as a function of x.
(g) Find the derivative of $f(x) = -\sqrt{1 - 2x^2}$ by implicit differentiation of $z = 2x^2 + y^2$.

3 (a) Find the line which is tangent to the curve $x^3 + y^3 - xy = \text{const.}$ at the point $(\bar{x}, \bar{y})$. [Write the answer in the form $A(x - \bar{x}) + B(y - \bar{y}) = 0$.]
(b) Find all points at which the tangent line to the folium of Descartes $\{x^3 + y^3 = xy\}$ is vertical.
(c) Is horizontal.
(d) Has slope -1.

4 (a) What is the general formula for the line tangent to $F(x, y) = \text{const.}$ at $(\bar{x}, \bar{y})$? [Mnemonically the answer can be written $dF = 0$.]
(b) If $f(x)$ is a function such that $F(x, f(x)) = \text{const.}$, what formula does this give for $f'(x)$ in terms of the derivatives of F? When is it valid?
(c) Deduce the same formula by implicit differentiation.

5 Give a rigorous statement of the formula

$$\frac{dy}{dx} = \frac{1}{\dfrac{dx}{dy}}$$

[for the derivative of an inverse function $x = g(y)$ of $y =$

$f(x)]$ as an implicit differentiation. Given that $\frac{d}{dx}[e^x] = e^x$, use this formula to find the derivative of $\log x$ (the inverse function to e^x).

6 Let p and q be positive integers. The Implicit Function Theorem implies that the curve $y = t^p$, $x = t^q$ can be solved locally (except near the origin) to give $y = h(x)$.

(a) Show that for $x \geq 0$ there is a unique $y \geq 0$ such that (x, y) is on the curve. The function so defined is denoted $y = x^{p/q}$. Note that it is defined only for $x \geq 0$.
(b) For which values of p, q can $x^{p/q}$ be defined in this way for all x?
(c) By the Implicit Function Theorem $x^{p/q}$ is differentiable for $x > 0$; find its derivative by implicit differentiation and express the result as a function of x.
(d) The function $|x|^r$ is defined for all x provided r is a positive rational number $r = p/q$ and, by the Chain Rule, it is differentiable except possibly at $x = 0$. Sketch the graph of $|x|^r$ for $r = \frac{1}{2}$, 1, 3/2, 2, 2/3, 1/1,000, 1,000.
(e) Show that $|x|^r$ is differentiable if and only if $r > 1$.
(f) Give a formula for the derivative of $|x|^r$ which is valid for all x.

7 *Directional derivatives.* If a point in xyz-space moves according to parametric equations $x = f(t)$, $y = g(t)$, $z = h(t)$ then the rate of change of $F(x, y, z)$ with respect to t at time $\bar{t}$ is expressed by the Chain Rule in terms of

$(f(\bar{t}), g(\bar{t}), h(\bar{t}))$ = location at time $\bar{t}$
$f'(\bar{t}) = a$, $g'(\bar{t}) = b$, $h'(\bar{t}) = c$, the three components of the velocity at time $\bar{t}$

and the partial derivatives of F. This is called the 'directional derivative of F in the direction of the vector (a, b, c) at the point $(f(\bar{t}), g(\bar{t}), h(\bar{t}))$'.

(a) Find the directional derivative of $F(x, y, z) = 3x^2 + 2y^2 + z$ in the direction of the vector (a, b, c) at the point $(2, -3, 1)$.
(b) Which vectors (a, b, c) point into the ellipsoid $F =$ const., which point out of, and which are tangent to $F =$ const. at $(2, -3, 1)$?
(c) What is the general formula for the directional derivative of $F(x, y, z)$ at $(\bar{x}, \bar{y}, \bar{z})$ in the direction of the vector (a, b, c)?

8 *Tangent planes.* Assuming that $dF \neq 0$ at $(\bar{x}, \bar{y}, \bar{z})$, by the Implicit Function Theorem, $F =$ const. can be solved for one of the variables as a function of the other two, hence $F =$ const. is a two-dimensional surface near $(\bar{x}, \bar{y}, \bar{z})$.

(a) Justify the statement that the plane defined by the shorthand equation $dF = 0$ (see Exercise 4) is tangent to the surface $F =$ const. at $(\bar{x}, \bar{y}, \bar{z})$ by showing that the velocity vector $f'(\bar{t}) = a$, $g'(\bar{t}) = b$, $h'(\bar{t}) = c$, of any motion $(f(t), g(t), h(t))$ through $(\bar{x}, \bar{y}, \bar{z}) = (f(\bar{t}), g(\bar{t}), h(\bar{t}))$ lying in the surface $F =$ const. lies in the plane $dF = 0$.

(b) Find the equation of the plane which is tangent to the ellipsoid of Exercise 7 at $(2, -3, 1)$.

9 *Implicit differentiation.* Let $F(x, y, z)$ be a function of three variables with $\frac{\partial F}{\partial z}(\bar{x}, \bar{y}, \bar{z}) \neq 0$. Then locally the equation $F =$ const. can be solved to give $z = f(x, y)$, i.e. f is determined by the condition

$$F(x, y, f(x, y)) \equiv F(\bar{x}, \bar{y}, \bar{z}).$$

(a) Describe the tangent plane to the graph of f in terms of its partial derivatives.

(b) Since the graph of f is a surface $F =$ const. the equation of its tangent plane can also be written $dF = 0$. Use the fact that two planes

$$A_1x + B_1y + C_1z + D_1 = 0$$
$$A_2x + B_2y + C_2z + D_2 = 0$$

are identical if and only if $A_1:B_1:C_1:D_1 = A_2:B_2:C_2:D_2$, i.e. if and only if their equations differ by a non-zero multiple, to express the partial derivatives of f in terms of those of F.

(c) Find the same formula directly by implicit differentiation.

10 *Spherical coordinates.* The map

$$x = r\cos\theta\cos\phi$$
$$y = r\sin\theta\cos\phi$$
$$z = r\sin\phi$$

defines spherical coordinates on xyz-space. [r = distance from origin, θ = longitude, ϕ = latitude.]

(a) Find $\dfrac{\partial(x, y, z)}{\partial(r, \theta, \phi)}$.

(b) At what points is the map non-singular of rank three? Are the remaining points singularities, or are they non-singular of a different rank?

(c) Give ranges for (r, θ, ϕ) such that all points of xyz-

space except those on the half plane $y = 0, x \geq 0$ are covered exactly once.

(d) Using implicit differentiation, show that the derivatives of the inverse function are

$$\frac{\partial r}{\partial x} = \frac{x}{r}, \frac{\partial r}{\partial y} = \frac{y}{r}, \frac{\partial r}{\partial z} = \frac{z}{r}$$

$$\frac{\partial \theta}{\partial x} = -\frac{y}{x^2 + y^2}, \frac{\partial \theta}{\partial y} = \frac{x}{x^2 + y^2}, \frac{\partial \theta}{\partial z} = 0$$

$$\frac{\partial \phi}{\partial x} = \frac{-xz}{(x^2 + y^2 + z^2)\sqrt{x^2 + y^2}}, \frac{\partial \phi}{\partial y} = \frac{-yz}{(x^2 + y^2 + z^2)\sqrt{x^2 + y^2}}, \frac{\partial \phi}{\partial z} = \frac{\sqrt{x^2 + y^2}}{x^2 + y^2 + z^2}.$$

[Use the combinations $\cos\theta\, dx + \sin\theta\, dy$ and $-\sin\theta\, dx + \cos\theta\, dy$ as in the text. The above formulas assume that $r > 0$ and $\cos\phi > 0$.]

5.3 Proofs

A good intuitive understanding of the meaning of differentiability of maps $f: \mathbf{R}^n \to \mathbf{R}^m$ can be achieved by imagining that the map is being examined under a microscope.

The idea of a 'microscope of power $1:s$ directed at the point P of $\mathbf{R}^n$' can be described mathematically as a new coordinate system on $\mathbf{R}^n$ in which P is the origin and in which the scale is changed in the ratio $1:s$. Thus the new coordinates of a point P' are the amounts by which P' differs from P multiplied by the (large) factor $1/s$ (where s is small). In short,

$$Q = \frac{P' - P}{s}$$

gives the location of P' as seen through a microscope of power $1:s$ directed at P. Conversely, the point which 'appears' at Q under the microscope is the point

$$P' = P + sQ.$$

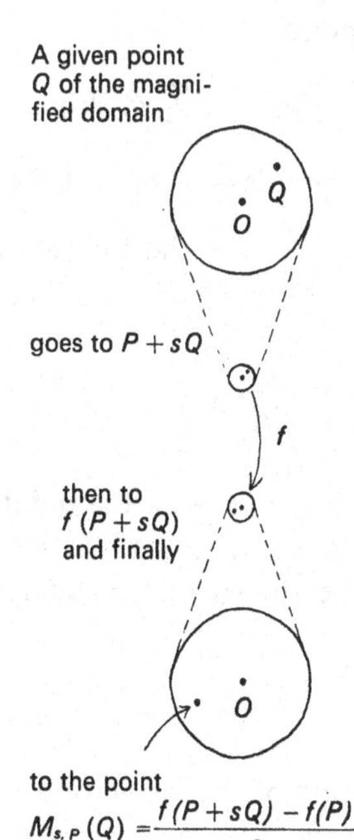

A given point Q of the magnified domain goes to $P + sQ$ then to $f(P + sQ)$ and finally to the point $M_{s,P}(Q) = \frac{f(P+sQ) - f(P)}{s}$ of the magnified range.

Now to examine a map f: $\mathbf{R}^n \to \mathbf{R}^m$ under a microscope naturally means to place both the range and the domain under microscopes and to examine the map of the *magnified* domain to the *magnified* range. Specifically, microscopes of power $1:s$ are directed at P and $f(P)$. Then a point Q of the magnified domain corresponds to a point $P + sQ$ of the domain; its image is $f(P + sQ)$, which appears at

$$(1) \qquad M_{s,P}(Q) = \frac{f(P + sQ) - f(P)}{s}$$

under the microscope directed at $f(P)$. In short, the map $M_{s,P}$ defined by (1) can be thought of as 'f at P under microscopes of power $1{:}s$'.

Intuitively *a differentiable map is one which is nearly affine under sufficiently high-powered microscopes.* If the original map f is in fact affine, then its appearance under a microscope is extremely simple: If $P = (\bar{x}_1, \bar{x}_2, \ldots, \bar{x}_n)$, if $Q = (h_1, h_2, \ldots, h_n)$, and if $f\colon \mathbf{R}^n \to \mathbf{R}^m$ is the affine map

$$y_1 = \sum_{j=1}^{n} a_{ij}x_j + b_i$$

then the ith coordinate of $M_{s,P}(Q)$ is

$$\frac{f_i(P + sQ) - f_i(P)}{s} = \frac{1}{s}\left(\left(\sum_{j=1}^{n} a_{ij}(\bar{x}_j + sh_j) + b_i\right) - \left(\sum_{j=1}^{n} a_{ij}\bar{x}_j + b_i\right)\right)$$

$$= \frac{1}{s}\sum_{j=1}^{n} a_{ij}(sh_j) = \sum_{j=1}^{n} a_{ij}h_j.$$

Hence the point $(h_1, h_2, \ldots, h_n)$ of the magnified domain is carried to the point

$$(\textstyle\sum a_{1j}h_j, \sum a_{2j}h_j, \ldots, \sum a_{mj}h_j)$$

of the magnified range, regardless of the ratio of magnification $1{:}s$. Here $a_{ij} = \dfrac{\partial y_i}{\partial x_j}$ and the formula can be abbreviated by

$$dy_i = \sum_{j=1}^{n} \frac{\partial y_i}{dx_j}\, dx_j$$

where the $dx_j = h_j$ give the location of the point in the magnified domain, and the dy_i gives its location in the magnified range. It will be shown that a differentiable map

$$(2) \qquad y_i = f_i(x_1, x_2, \ldots, x_n) \qquad (i = 1, 2, \ldots, m)$$

is one for which the location of the image of a point of the magnified range (dy_i) can be expressed by a similar formula in terms of its location in the magnified domain

(dx_i) in the limit as $s \to 0$. Specifically, if the map (2) is differentiable then

$$\text{(3)} \quad \lim_{s\to 0} M_{s,P}(Q) = \left(\sum_{j=1}^{n} \frac{\partial y_1}{\partial x_j} h_j, \ldots, \sum_{j=1}^{n} \frac{\partial y_m}{\partial x_j} h_j\right)$$

where $Q = (h_1, h_2, \ldots, h_n)$ and where the partial derivatives $\frac{\partial y_i}{\partial x_j}$ are evaluated at $P = (\bar{x}_1, \bar{x}_2, \ldots, \bar{x}_n)$. This formula can be abbreviated:

$$\text{(4)} \qquad dy_i = \sum_{j=1}^{n} \frac{\partial y_i}{\partial x_j} dx_j.$$

The formula (3), which is proved below, is the real meaning of the formula (4) by which pullbacks are defined.

Proof of the Chain Rule

The essence of the proof is to show that the equation (3) holds for differentiable maps. From this it is not difficult to prove the Chain Rule for 1-forms, from which the Chain Rule for k-forms follows algebraically as in §4.2.

The proof of (3) is based on the Fundamental Theorem of Calculus

$$\int_a^b F'(t)\,dt = F(b) - F(a).$$

Going from P to $P + sQ$ in n steps along n line segments parallel to coordinate axes—

from	$(\bar{x}_1, \bar{x}_2, \ldots, \bar{x}_n)$
to	$(\bar{x}_1 + sh_1, \bar{x}_2, \ldots, \bar{x}_n)$
to	$(\bar{x}_1 + sh_1, \bar{x}_2 + sh_2, \bar{x}_3, \ldots, \bar{x}_n)$
to ... to	$(\bar{x}_1 + sh_1, \bar{x}_2 + sh_2, \ldots, \bar{x}_n + sh_n)$

—and writing the change in the value of f_i along each such line segment as the integral of the corresponding partial derivative gives the formula

$$\text{(5)} \quad \begin{aligned} f_i(P + sQ) - f_i(P) &= f_i(\bar{x}_1 + sh_1, \ldots, \bar{x}_n + sh_1) - f_i(\bar{x}_1, \bar{x}_2, \ldots, \bar{x}_n) \\ &= \sum_{j=1}^{n} \int_{\bar{x}_j}^{\bar{x}_j + sh_j} \left\{\frac{\partial y_i}{\partial x_j}(\bar{x}_1 + sh_1, \ldots, \bar{x}_{j-1} + sh_{j-1}, t, \bar{x}_{j+1}, \ldots, \bar{x}_n)\right\} dt. \end{aligned}$$

To say that the partial derivatives are continuous means that for very small s the integrands in these integrals are practically constant, and, specifically, practically equal to $\frac{\partial y_i}{\partial x_j}(\bar{x}_1, \bar{x}_2, \ldots, \bar{x}_n)$. If the integrands actually did have this constant value then the integrals would be $sh_j \frac{\partial y_i}{\partial x_j}(\bar{x}_1, \bar{x}_2, \ldots, \bar{x}_n)$ and, dividing by s,

$$(6) \quad M_{s,P}(Q) \sim \left(\sum_{j=1}^{n} \frac{\partial y_1}{\partial x_j}(\bar{x})h_j, \ldots, \sum_{j=1}^{n} \frac{\partial y_m}{\partial x_j}(\bar{x})h_j \right)$$

where the approximation consists in replacing the integrands in (5) by the constants $\frac{\partial y_i}{\partial x_j}(\bar{x})$. The equation (3) to be proved is the assertion that the error in the approximation (6) goes to zero as $s \to 0$. Very briefly, this is true because the difference between two integrals is at most the maximum difference between the two integrands times the length of the path of integration; this is an arbitrarily small number times

$$s(|h_1| + |h_2| + \cdots + |h_n|);$$

hence the error divided by s is arbitrarily small and (3) follows.

In detail: Given a map (2), a point $P = (\bar{x}_1, \bar{x}_2, \ldots, \bar{x}_n)$, a bound B (to be thought of as the size of the eyepiece of the microscope magnifying the domain), and a margin for error ϵ, there is a degree of magnification δ such that the error in the approximation (6) is less than ϵ in each of the m coordinates whenever $|s| < \delta$ and $|h_i| < B$ $(i = 1, 2, \ldots, n)$. That is

$$\left| \frac{f_i(\bar{x} + sh) - f_i(\bar{x})}{s} - \sum_{j=1}^{n} \frac{\partial y_i}{\partial x_j}(\bar{x})h_j \right| < \epsilon$$

for $i = 1, 2, \ldots, m$. The proof is as follows: Since $\frac{\partial y_i}{\partial x_j}$ is a continuous function there is a δ_{ij} such that its value at $(x_1, x_2, \ldots, x_n)$ differs by less than ϵ/nB from its value at $(\bar{x}_1, \bar{x}_2, \ldots, \bar{x}_n)$ whenever $|x_1 - \bar{x}_1| < \delta_{ij}, \ldots, |x_n - \bar{x}_n| < \delta_{ij}$. Let such δ_{ij} be chosen for each i and j, and let δ_0 be the smallest of these mn numbers. Finally, let $\delta = \delta_0/B$ so that the path of integration in (5) lies entirely inside the region where $\frac{\partial y_i}{\partial x_j}$ is known to differ by less than ϵ/nB from its value at P, provided $|s| < \delta$.

Changing the integrands in (5) to $\frac{\partial y_i}{\partial x_j}(\bar{x})$ then introduces an error of at most $(\epsilon/nB) \times (nsB) = \epsilon s$, so the error in (6) is at most ϵ, as was to be shown.

Now let differentiable maps $f: \mathbf{R}^n \to \mathbf{R}^m, g: \mathbf{R}^m \to \mathbf{R}^p$ be given as in the statement of the Chain Rule in §5.2. The notation of the proof of the Chain Rule is inevitably somewhat cumbersome, but the idea of the proof is essentially contained in (6). The composed map $g \circ f$: $\mathbf{R}^n \to \mathbf{R}^p$ under a microscope is

$$\frac{g[f(P + sQ)] - g[f(P)]}{s}, \tag{7}$$

which can be written

$$\frac{g[f(P) + sQ'] - g[f(P)]}{s}$$

where

$$Q' = \frac{f(P + sQ) - f(P)}{s}.$$

When P is fixed and s is very small, Q' is very nearly $\left(\sum_{j=1}^{n} a_{1j}h_j, \ldots, \sum_{j=1}^{m} a_{mj}h_j\right)$ where $Q = (h_1, h_2, \ldots, h_n)$ and $a_{ij} = \frac{\partial y_i}{\partial x_j}(P)$, so that (7) is very nearly $\left(\sum_{\nu=1}^{m} \sum_{j=1}^{n} b_{1\nu}a_{\nu j}h_j, \ldots, \sum_{\nu=1}^{m} \sum_{j=1}^{n} b_{p\nu}a_{\nu j}h_j\right)$ where b_{ij} is the value of $\frac{\partial z_i}{\partial y_j}$ at $f(P)$. As $s \to 0$ all approximations improve and the formula becomes

$$dz_i = \sum_{j=1}^{n} \left(\sum_{\nu=1}^{m} b_{i\nu}a_{\nu j}\right) dx_j$$

which is the Chain Rule.

To fill in the details so that this becomes a proof of the Chain Rule, note first that it will suffice to prove that the partial derivatives $\frac{\partial z_i}{\partial x_j}$ of the composed function exist and are given by the formula

$$\frac{\partial z_i}{\partial x_j} = \sum_{\nu=1}^{m} \frac{\partial z_i}{\partial y_\nu} \frac{\partial y_\nu}{\partial x_j} \tag{8}$$

since this implies that the composed function is differentiable (sums and products of continuous functions are continuous) and that the pullback of dz_i under the composed map is the pullback of the pullback. The same is then true of the pullback of any 1-form $\sum A_i\, dz_i$ under the composed map; hence, by the argument of §4.2, the same is also true of the pullback of any k-form. Thus the formula to be proved is (8), which is the assertion that if Q is 0 except for a 1 in the jth place then the limit of the ith component of (7) as $s \to 0$ exists and is equal to $\sum_{\nu=1}^{m} b_{i\nu}a_{\nu j}$ where the a's and b's are defined as above.

To simplify notation, the case $i = j = 1$ will be considered, the other cases being exactly the same. Then $Q = (1, 0, 0, \ldots, 0)$ and it was proved above that $Q' = [f(P + sQ) - f(P)]/s$ differs from $(a_{11}, a_{21}, \ldots, a_{m1})$ by less than any prescribed ϵ once s is sufficiently small. Let h'_ν $(\nu = 1, 2, \ldots, m)$ denote the actual coordinates of Q', and let B be a number larger than $\max(|a_{11}|, |a_{21}|, \ldots, |a_{m1}|)$ so that $|h'_\nu| < B$ whenever s is sufficiently large. It was shown above that then the first coordinate of (7) is within any prescribed ϵ of $\sum_{\nu=1}^{m} b_{1\nu}h'_\nu$ once s is sufficiently small. But h'_ν differs from $a_{\nu 1}$ by less than any prescribed ϵ once s is small, hence $\sum_{\nu=1}^{m} b_{1\nu}h'_\nu$ differs from $\sum_{\nu=1}^{m} b_{1\nu}a_{\nu 1}$ by less than $\sum_{\nu=1}^{m} |b_{1\nu}\epsilon|$. Thus the first coordinate of (7) differs from $\sum_{\nu=1}^{m} b_{1\nu}a_{\nu 1}$ by less than $\epsilon \sum_{\nu=1}^{m} |b_{1\nu}| + \epsilon$ for arbitrarily small ϵ once s is sufficiently small. Therefore $\lim_{s\to 0}$ of the first coordinate of (7) exists and is $\sum_{\nu=1}^{m} b_{1\nu}a_{\nu 1}$ as was to be shown. This completes the proof of the Chain Rule.

Implicit Differentiation

Let f_i, g_i, h_i be differentiable functions satisfying the relations of the Implicit Function Theorem and let

$$a_{ij} = \frac{\partial f_i}{\partial x_j}, \quad A_{ij} = \frac{\partial g_i}{\partial y_j}, \quad B_{ij} = \frac{\partial g_i}{\partial x_j}, \quad C_{ij} = \frac{\partial h_i}{\partial y_j},$$

all derivatives being evaluated at $(\bar{x}_1, \bar{x}_2, \ldots, \bar{x}_n, \bar{y}_1, \bar{y}_2, \ldots, \bar{y}_m)$. The relations satisfied by f_i, g_i, h_i can be stated:

$$f_i(g_1(u), \ldots, g_r(u), u_{r+1}, \ldots, u_n) \equiv \begin{cases} u_i & (i = 1, 2, \ldots, r) \\ h_i(u) & (i = r + 1, \ldots, m) \end{cases}$$

for all $(u_1, u_2, \ldots, u_n)$ near $(\bar{y}_1, \ldots, \bar{y}_r, \bar{x}_{r+1}, \ldots, \bar{x}_n)$. Differentiating both sides of this identity with respect to each of the variables u_i using the chain rule on the left gives the matrix equation

$$(7)\quad m\left\{\overbrace{\begin{bmatrix} & & \\ & a & \\ & & \end{bmatrix}}^{n}\right. \begin{bmatrix} \overbrace{A}^{r} & \overbrace{B}^{n-r} \\ 0 & I \end{bmatrix}\begin{matrix} \}r \\ \}n-r \end{matrix} = \begin{bmatrix} \overbrace{I}^{r} & \overbrace{0}^{n-r} \\ C & 0 \end{bmatrix}\begin{matrix} \}r \\ \}m-r \end{matrix}$$

relating the a's, A's, B's, and C's. This means that the A's, B's, C's give the solution of the equations

$$dy_i = \sum_{j=1}^{n} a_{ij}\, dx_j \qquad (i = 1, 2, \ldots, m)$$

as was to be shown.

Implicit Function Theorem

The proof that (2) implies (i) and (ii) is deduced from the Chain Rule exactly as in the affine case. The functions

$$y_i = f_i(x_1, \ldots, x_r, \bar{x}_{r+1}, \ldots, \bar{x}_n) \qquad (i = 1, 2, \ldots, r)$$

and

$$x_i = g_i(y_1, \ldots, y_r, \bar{x}_{r+1}, \ldots, \bar{x}_n) \qquad (i = 1, 2, \ldots, r)$$

are by assumption inverse to each other, consequently

$$\frac{\partial(y_1, \ldots, y_r)}{\partial(x_1, \ldots, x_r)} \frac{\partial(x_1, \ldots, x_r)}{\partial(y_1, \ldots, y_r)} = 1$$

and (i) follows. On the other hand, the functions h_i can be used to write (1) as a composed map $(x_1, x_2, \ldots, x_n) \to (y_1, y_2, \ldots, y_r) \to (y_1, y_2, \ldots, y_m)$ and (ii) follows from the Chain Rule and the fact that an $(r + 1)$-form in r variables is necessarily zero.

To show that (i) and (ii) imply (2) one uses a process of step-by-step elimination exactly as in the affine case, first rearranging the first r of the x's so that the stronger condition

$$\text{(i}'\text{)}\quad \frac{\partial(y_1, y_2, \ldots, y_k)}{\partial(x_1, x_2, \ldots, x_k)} \neq 0 \qquad (k = 1, 2, \ldots, r)$$

is satisfied. However, the elimination of one variable from one equation cannot be accomplished by subtracting and dividing as it was in the affine case and one must appeal to the following theorem.

Elimination Theorem

Given an equation of the form

(9) $$y = f(x_1, x_2, \ldots, x_n)$$

in which f is a differentiable function, and given a point

$$\bar{y} = f(\bar{x}_1, \bar{x}_2, \ldots, \bar{x}_n)$$

at which $\dfrac{\partial y}{\partial x_1} \neq 0$, there exist a number $\epsilon > 0$ and a differentiable function $g(y, x_2, x_3, \ldots, x_n)$ such that the equation

(10) $$x_1 = g(y, x_2, x_3, \ldots, x_n)$$

is equivalent to (9) at all points within ϵ of $(\bar{y}, \bar{x}_1, \bar{x}_2, \ldots, \bar{x}_n)$; that is, f and g are both defined at all points $(y, x_1, x_2, \ldots, x_n)$ satisfying $|y - \bar{y}| < \epsilon$, $|x_1 - \bar{x}_1| < \epsilon, \ldots, |x_n - \bar{x}_n| < \epsilon$ and the equation (9) holds at such a point if and only if (10) does.

Using the Elimination Theorem to eliminate variables one-by-one (reducing ϵ whenever necessary) the condition (i′) guarantees, exactly as in the affine case, that the next equation can always be solved for the next unknown until the equations have the form

$$x_i = g_i(y_1, y_2, \ldots, y_r, x_{r+1}, \ldots, x_n) \qquad (i = 1, 2, \ldots, r)$$

$$y_i = h_i(y_1, y_2, \ldots, y_r, x_{r+1}, \ldots, x_n) \qquad (i = r + 1, \ldots, m).$$

The condition (ii) is now to be used to show that the functions h_i are independent of the x's, i.e. that $\dfrac{\partial h_i}{\partial x_j} \equiv 0$ $(i = r + 1, \ldots, m;\ j = r + 1, \ldots, n)$. But $\dfrac{\partial h_i}{\partial x_j} = \dfrac{\partial(y_1, y_2, \ldots, y_r, y_i)}{\partial(y_1, y_2, \ldots, y_r, x_j)}$ and, by the Chain Rule,

$$0 \equiv \frac{\partial(y_1, y_2, \ldots, y_r, y_i)}{\partial(x_1, x_2, \ldots, x_r, x_j)}$$

$$\equiv \frac{\partial h_i}{\partial x_j} \cdot \frac{\partial(y_1, y_2, \ldots, y_r)}{\partial(x_1, x_2, \ldots, x_r)}$$

so $\dfrac{\partial h_i}{\partial x_j} \equiv 0$ by (i) and (ii) exactly as before.

Thus the proof of the Implicit Function Theorem is reduced to the proof of the Elimination Theorem (which is one half of the special case $m = r = 1$ of the Implicit Function Theorem). The proof of the Elimination Theorem by the method of successive approximations is given in §7.1. (For an alternative proof see the exercise below.)

Exercise **1** Fill in the details of the following proof of the Elimination Theorem: Let $a = \dfrac{\partial y}{\partial x_1}(\bar{x}_1, \bar{x}_2, \ldots, \bar{x}_n)$ and let $F(x_1, x_2, \ldots, x_n) = \dfrac{1}{a}[f(x_1, x_2, \ldots, x_n) - f(0, x_2, x_3, \ldots, x_n)]$. If the equation $z = F(x_1, x_2, \ldots, x_n)$ can be solved locally to give $x_1 = G(z, x_2, x_3, \ldots, x_n)$ then the equation $y = f(x_1, x_2, \ldots, x_n)$ can be solved locally by solving $F(x_1, x_2, \ldots, x_n) = \dfrac{1}{a}[y - f(0, x_2, \ldots, x_n)]$, to obtain

$$x_1 = G\left(\frac{1}{a}[y - f(0, x_2, \ldots, x_n)], x_2, \ldots, x_n\right) = g(y, x_2, x_3, \ldots, x_n)$$

where g is defined by this equation. Conclude that one can assume at the outset that the function $f(x_1, x_2, \ldots, x_n)$ of the Elimination Theorem satisfies the additional conditions

$$\frac{\partial y}{\partial x_1}(\bar{x}_1, \bar{x}_2, \ldots, \bar{x}_n) = 1$$

$$f(0, x_2, x_3, \ldots, x_n) \equiv 0.$$

Let $\sigma > 0$ be a small number (any number less than 1 will do) and let $\delta > 0$ be such that $\dfrac{\partial y}{\partial x_1}$ is within σ of 1 whenever $(x_1, x_2, \ldots, x_n)$ is within δ of $(\bar{x}_1, \bar{x}_2, \ldots, \bar{x}_n)$, i.e. $|x_i - \bar{x}_i| < \delta$ $(i = 1, 2, \ldots, n)$. For all $(x_2, x_3, \ldots, x_n)$ within δ of $(\bar{x}_2, \bar{x}_3, \ldots, \bar{x}_n)$ the function $f(x_1, x_2, \ldots, x_n)$ is increasing for x_1 in the interval $\{\bar{x}_1 - \delta \le x_1 \le \bar{x}_1 + \delta\}$. At the left end of this interval its value is at most $-(1 - \sigma)\delta$ and at the right end the value is at least $(1 - \sigma)\delta$. [Use the Fundamental Theorem.] By the *Intermediate Value Theorem** conclude that for all $(y, x_2, x_3, \ldots, x_n)$ such that $|y| < (1 - \sigma)\delta$ and $|x_i - \bar{x}_i| < \delta$ there is a unique x_1 in the interval $|x_1 - \bar{x}_1| < \delta$ such that $y = f(x_1, x_2, \ldots, x_n)$. Denote this uniquely determined value by $g(y, x_2, x_3, \ldots, x_n)$.

A continuous function $f(x)$ on an interval $\{a \le x \le b\}$ assumes all values between $f(a)$ and $f(b)$.

Set $\epsilon = (1 - \sigma)\delta$ and show that the conclusions of the Elimination Theorem are satisfied, except that it remains to show that $g(y, x_2, x_3, \ldots, x_n)$ is differentiable. Given $(y, x_2, x_3, \ldots, x_n)$ within ϵ of $(0, \bar{x}_2, \bar{x}_3, \ldots, \bar{x}_n)$ and given $(h_1, h_2, \ldots, h_n)$ the quantity

$$\frac{\Delta x_1}{s} = \frac{g(y + sh_1, x_2 + sh_2, \ldots, x_n + sh_n) - g(y, x_2, \ldots, x_n)}{s}$$

can be estimated by setting

$$x_1' = g(y + sh_1, x_2 + sh_2, \ldots, x_n + sh_n)$$
$$x_1'' = g(y, x_2, x_3, \ldots, x_n)$$

so that

$$y + sh_1 = f(x_1', x_2 + sh_2, \ldots, x_n + sh_n)$$
$$y = f(x_1'', x_2, x_3, \ldots, x_n)$$

and using the estimate

$$sh_1 \sim \frac{\partial y}{\partial x_1}(x_1' - x_1'') + \frac{\partial y}{\partial x_2} sh_2 + \cdots + \frac{\partial y}{\partial x_n} sh_n$$

$$\frac{\Delta x_1}{s} \sim \left(\frac{\partial y}{\partial x_1}\right)^{-1}\left(h_1 - \frac{\partial y}{\partial x_2} h_2 - \cdots - \frac{\partial y}{\partial x_n} h_m\right)$$

where $\dfrac{\partial y}{\partial x_i}$ is evaluated at $(g(y, x_2, \ldots, x_n), x_2, x_3, \ldots, x_n)$. Show that the error in this estimate goes to zero as $s \to 0$. Conclude that g is continuous, that its first partial derivatives exist, and that they are continuous functions of $(y, x_2, x_3, \ldots, x_n)$. [The composition of continuous functions is continuous.] Thus g is differentiable and the theorem is proved.

5.4

Application: Lagrange Multipliers

One of the most direct and useful applications of differential calculus is to the problem of finding maxima and minima. It is based on this observation: *If a differentiable function of one variable $y = f(x)$ assumes a maximum or a minimum value at a point $\bar{x}$ inside its domain then $dy = f'(x)\,dx$ must be zero at that point, i.e. $f'(\bar{x}) = 0$.* This theorem can be deduced from the Implicit Function Theorem as follows: If $dy \neq 0$ at $\bar{x}$ then the equation $y = f(x)$ can be solved for $x = g(y)$ at all points y sufficiently near $\bar{y} = f(\bar{x})$. Setting $x = g(\bar{y} \pm \epsilon)$ then gives points where $f(x) = \bar{y} \pm \epsilon$, which

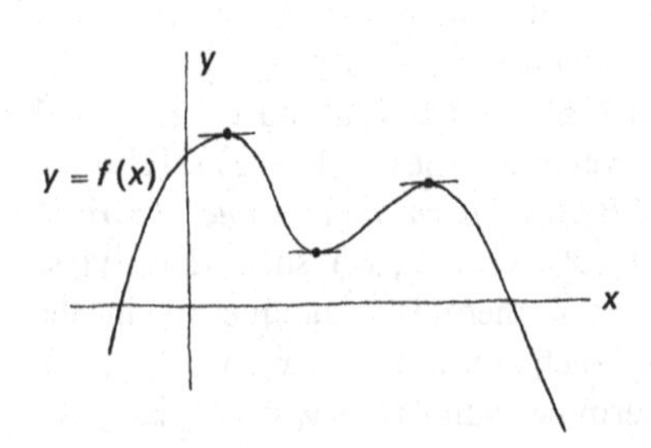

shows that $\bar{y}$ is neither a maximum nor a minimum value of $y = f(x)$. Thus, if $\bar{y}$ is a maximum or a minimum, then $dy = 0$.

The usefulness of this theorem lies in the fact that the equation $f'(x) = 0$ will normally have only a few solutions x and, therefore, that the set where $y = f(x)$ could possibly assume a maximum or a minimum value is reduced to a few points. One can then evaluate the function at each of these points and know that the largest of them is the maximum, and the smallest the minimum, among all values of $y = f(x)$. The steps in the solution are:

I. Find the equation $f'(x) = 0$ which must be satisfied at maxima and minima.

II. Find all solutions of the equation $f'(x) = 0$.

III. Test each solution to determine whether or not it is a maximum (or minimum).

Step I of this program is the application of differential calculus to the problem, and it is Step I which is generalized to functions of several variables by the method of Lagrange multipliers.

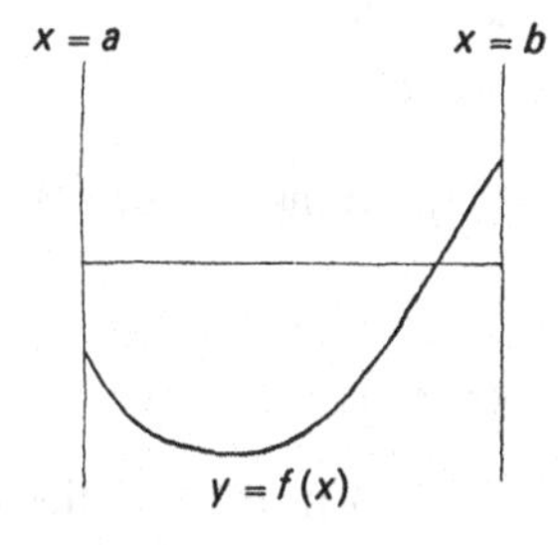

Note that if one is looking for the maximum of $y = f(x)$ on an interval $\{a \le x \le b\}$ then the condition $f'(x) = 0$ need not be satisfied at the maximum if the maximum occurs at an end point $x = a$ or $x = b$. This is not a serious problem since it means only that these two points cannot be excluded by the condition $f'(x) = 0$, and that the number of points which require further investigation at Step III is increased by two. However, in maximizing a function of more than one variable over a domain with boundary, the fact that the condition $dy = 0$ need not be satisfied if the maximum occurs on the boundary means that there remains an infinite set of points which are not excluded by the condition $dy = 0$. This means that further conditions must be found which will exclude as many of these boundary points as possible.

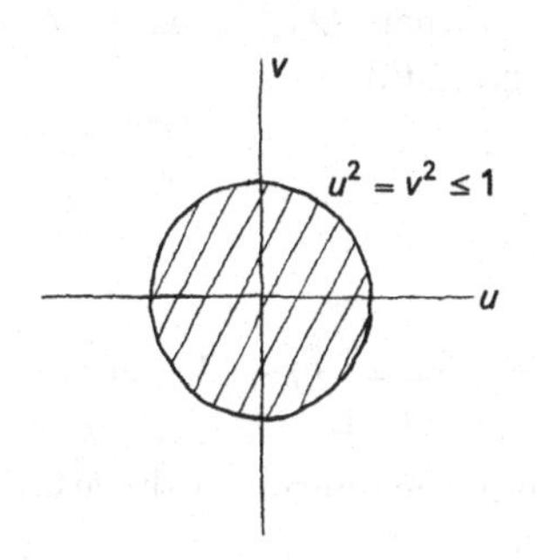

Consider for example the problem of finding a* point where a differentiable function $y = f(u, v)$ of two variables assumes a maximum value on the disk $\{u^2 + v^2 \le 1\}$. If a maximum occurs at an interior point $(\bar{u}, \bar{v})$ of the disk, then $dy = 0$ at $(\bar{u}, \bar{v})$ by the same argument as before: If $dy \neq 0$ then the equation $y = f(u, v)$ can be solved $u = g(y, v)$ or $v = g(u, y)$ by

**It is not difficult to show that there exists a point where y is a maximum (see Chapter 9) but this is not the question. The question is: Assuming there is a maximum, how can it be found?*

the Implicit Function Theorem; then $u = g(\bar{y} + \epsilon, v)$ or $v = g(u, \bar{y} + \epsilon)$ gives points (u, v) where $f(u, v) > f(\bar{u}, \bar{v})$; hence if $f(\bar{u}, \bar{v})$ is a maximum the assumption $dy \neq 0$ must be false. Thus the equations

$$\text{(1)} \qquad \frac{\partial y}{\partial u} = 0, \; \frac{\partial y}{\partial v} = 0$$

must be satisfied at any maximum which occurs inside the disk. The equations (1) give two equations in two unknowns (u, v) and normally will exclude most points of the disk as possible maxima. However, as before, the equations (1) need not be satisfied at a maximum if the maximum occurs on the bounding circle $\{u^2 + v^2 = 1\}$ of the disk. (The argument that $dy = 0$ at a maximum fails because the solutions (u, v) of $\bar{y} + \epsilon = f(u, v)$ may all lie outside the disk and $\bar{y}$ may still be the largest value assumed by $f(u, v)$ for $\{u^2 + v^2 \leq 1\}$.)

In order to find conditions satisfied by a point on the bounding circle $\{u^2 + v^2 = 1\}$ at which a maximum occurs (Step I), consider the map $\mathbf{R}^2 \to \mathbf{R}^2$ defined by

$$X = u^2 + v^2$$
$$y = f(u, v).$$

If $dX\, dy \neq 0$ at $(\bar{u}, \bar{v})$ these equations can be solved

$$u = g_1(X, y)$$
$$v = g_2(X, y)$$

for all (X, y) near $(\bar{u}^2 + \bar{v}^2, f(\bar{u}, \bar{v}))$. If $(\bar{u}, \bar{v})$ lies on the circle $X = 1$ then $\bar{y} = f(\bar{u}, \bar{v})$ is not a maximum on the disk because

$$u = g_1(1, \bar{y} + \epsilon)$$
$$v = g_2(1, \bar{y} + \epsilon)$$

gives a point of the disk (in fact a point of the circle $X = 1$) at which $f(u, v) = \bar{y} + \epsilon > f(\bar{u}, \bar{v})$. Therefore if $\bar{y} = f(\bar{u}, \bar{v})$ is a maximum then $dX\, dy$ must be zero at $(\bar{u}, \bar{v})$. This is the desired condition.

Example

Find the maximum value of $y = 8u + 3v$ on the disk $\{u^2 + v^2 \leq 1\}$. Here $dy = 8\, du + 3\, dv$ is never zero so there can be no maximum in the interior of the disk. The maximum must therefore occur at a point on the circle $\{u^2 + v^2 = 1\}$ at which

$$dX\,dy = 0$$
$$(2u\,du + 2v\,dv)(8\,du + 3\,dv) = 0$$
$$(6u - 16v)\,du\,dv = 0$$
$$v = \tfrac{3}{8}u$$
$$u^2 + (\tfrac{3}{8}u)^2 = 1$$
$$u = \pm\frac{8}{\sqrt{73}},\ v = \pm\frac{3}{\sqrt{73}}$$
$$y = 8u + 3v = \pm\sqrt{73}.$$

Thus there are only two points at which a maximum could occur. Clearly the maximum is $\sqrt{73}$ at $(8/\sqrt{73}, 3/\sqrt{73})$ and the minimum is $-\sqrt{73}$ at $(-8/\sqrt{73}, -3/\sqrt{73})$. (Strictly speaking, it has been proved that *if* there is a maximum *then* it must be $\sqrt{73}$. The fact that there must be a maximum is proved in §9.4.)

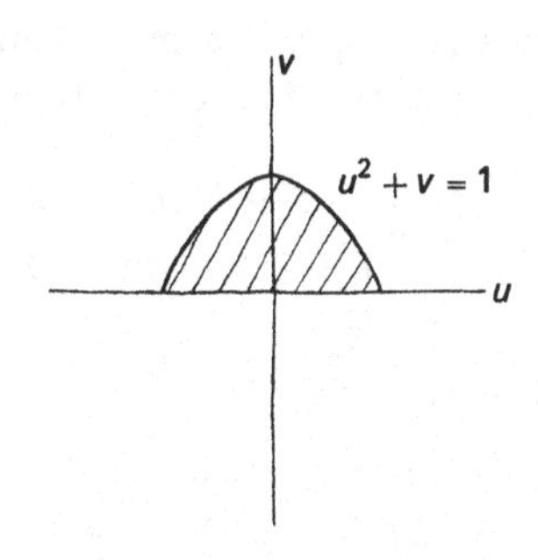

A similar technique applies to the problem of finding maxima and minima of functions $y = f(u, v)$ on other domains of the uv-plane. Consider, for example, the domain consisting of all points which lie under the parabola $u^2 + v = 1$ and above the u-axis. If a maximum occurs at a point $(\bar{u}, \bar{v})$ inside the domain then $dy = 0$ at $(\bar{u}, \bar{v})$. However, this condition need not be satisfied at a maximum which occurs on the boundary. If a maximum occurs at a point inside the line segment $\{-1 < u < 1, v = 0\}$ then the Implicit Function Theorem applied to the equations

$$X = v$$
$$y = f(u, v)$$

shows as before that $dX\,dy = 0$ at the point where the maximum occurs, i.e. $\left(\frac{\partial y}{\partial u}du + \frac{\partial y}{\partial v}dv\right)dv = 0$ or simply $\frac{\partial y}{\partial u} = 0$ at the maximum. If a maximum occurs at a point inside the parabolic arc $\{u^2 + v = 1, -1 < u < 1\}$ then the Implicit Function Theorem applied to

$$X = u^2 + v$$
$$y = f(u, v)$$

implies that $dX\,dy = 0$ at the maximum. However, a

maximum value can also occur at either of the points $(\pm 1, 0)$ without any further condition being satisfied.

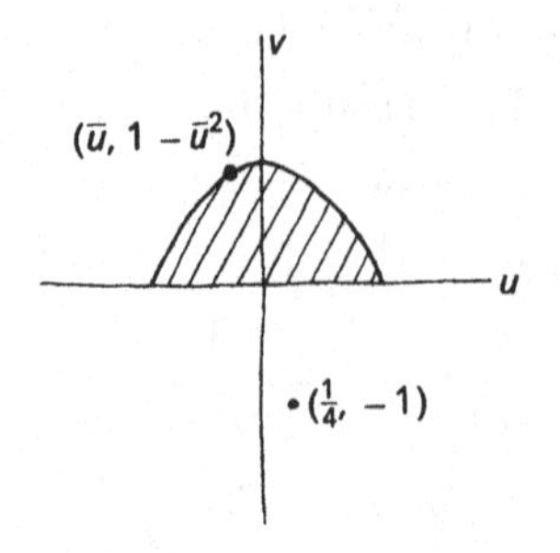

Example

Find the point of the above domain which is farthest from the point $(\frac{1}{4}, -1)$. To maximize the distance $\sqrt{(u-\frac{1}{4})^2 + (v+1)^2}$ is the same as to maximize the square of the distance

$$y = (u - \tfrac{1}{4})^2 + (v+1)^2.$$

Since

$$dy = 2(u - \tfrac{1}{4})\, du + 2(v+1)\, dv$$

is zero only at $(u, v) = (\frac{1}{4}, -1)$, there is no maximum inside the domain. A maximum inside the line segment $\{-1 < u < 1, v = 0\}$ could occur only where $\dfrac{\partial y}{\partial u} = 0$, i.e. only at $(\frac{1}{4}, 0)$, but this is clearly the minimum. A maximum inside the parabolic arc could occur only where

$$\begin{aligned}(2u\, du + dv)(2(u - \tfrac{1}{4})\, du + 2(v+1)\, dv) &= 0\\ (4u(v+1) - 2(u - \tfrac{1}{4}))\, du\, dv &= 0.\end{aligned}$$

Using $u^2 + v = 1$ this gives

$$\begin{aligned}4u(2-u^2) - 2(u - \tfrac{1}{4}) &= 0\\ 8u^3 - 12u - 1 &= 0.\end{aligned}$$

Evaluating this polynomial at $u = -2, -1, 0, 1$, and 2 shows that it has a root between -2 and -1, another between -1 and 0, and a third between 1 and 2. Thus only one root lies in the range being considered, and if the farthest point lies inside the parabolic arc then it must be the point $(\bar{u}, 1 - \bar{u}^2)$ where $\bar{u}$ is the root of $8u^3 - 12u - 1 = 0$ between -1 and 0. Finally, the farthest point could be one of the points $(\pm 1, 0)$. However, it is easily verified that $(-\frac{1}{2}, \frac{3}{4})$ is farther from $(\frac{1}{4}, -1)$ than either of these points. Therefore the farthest point is the point $(\bar{u}, 1 - \bar{u}^2)$ above. (Strictly speaking, it has been shown that *if* there is a farthest point *then* it is $(\bar{u}, 1 - \bar{u}^2)$. The geometrically evident fact that there must be a farthest point is proved in §9.4.)

Many physics problems involve finding maxima and minima subject to constraints. For example, if a particle

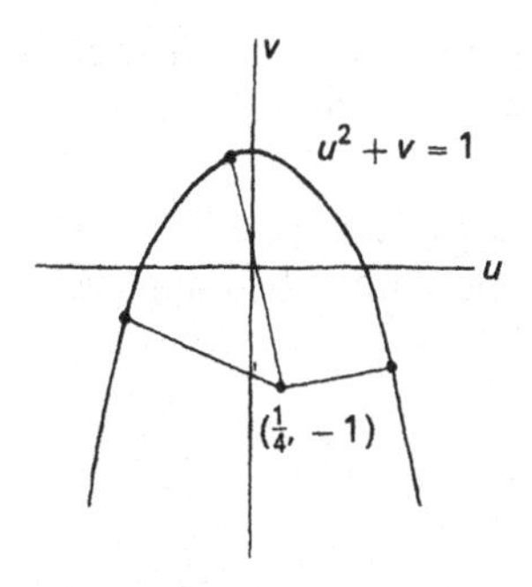

which is constrained to move along the parabola $u^2 + v = 1$ is being attracted toward the point $(\frac{1}{4}, -1)$, it will eventually come to rest at one of the two points of the parabola which are (locally) nearest to $(\frac{1}{4}, -1)$, that is, at one of the points $(\bar{u}, 1 - \bar{u}^2)$ where $\bar{u}$ is the root of $8u^3 - 12u - 1 = 0$ between -2 and -1 or the root between 1 and 2. (The third root corresponds to a point of unstable equilibrium and the particle could conceivably stop there as well.) In such problems there are often many variables and many constraints. As a simple example consider the following: A particle constrained to lie on the ellipse $4u^2 + v^2 = 4$ attracts a particle constrained to lie on the line $3u + 2v = 25$. At what position do they come to rest? Let (x, y) denote the coordinates of the particle on the line and let (u, v) denote the coordinates of the particle on the ellipse. Then the position of the 'system' is described by the four numbers (u, v, x, y) subject to the two constraints

$$(2) \qquad \begin{aligned} 4u^2 + v^2 &= 4 \\ 3x + 2y &= 25 \end{aligned}$$

which leave two degrees of freedom. At a point of equilibrium the function

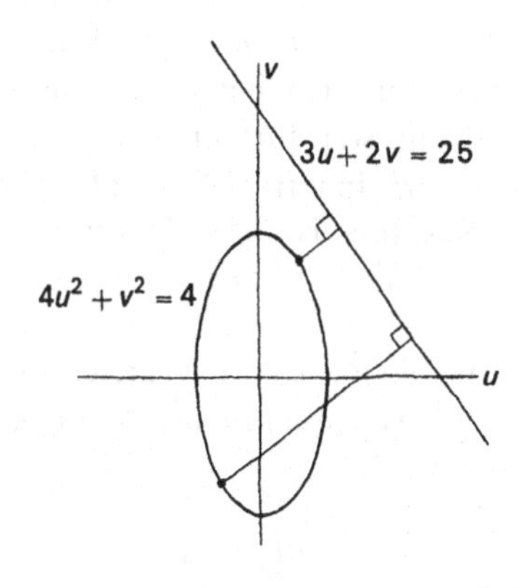

$$y = (u - x)^2 + (v - y)^2$$

must have a local minimum subject to the constraints (2). Applying the Implicit Function Theorem to the map $\mathbf{R}^4 \to \mathbf{R}^3$ defined by

$$\begin{aligned} U &= 4u^2 + v^2 \\ V &= 3x + 2y \\ Y &= (u - x)^2 + (v - y)^2 \end{aligned}$$

implies, by the same argument as before, that equilibrium can occur only at points where $dU\,dV\,dY = 0$. Therefore the problem is reduced to the *algebraic* problem of finding all points at which $dU\,dV\,dY = 0$ which also satisfy the constraints (2).

This algebraic problem is, however, considerably more difficult than the algebraic problems which had to be solved in the simpler examples above. The straightforward method of solution would be to write $dU\,dV\,dY$ in terms of (u, v, x, y), to set the 4 coëfficients of this 3-form equal to zero, and to try to find solutions (u, v, x, y) of these four equations which also satisfy the two equations (2). Although this can be done and a solution can be found by this method, there is another method, called *the method of Lagrange multipliers*, which is much simpler.

*λ = *lambda*.

The method of Lagrange multipliers is to observe that at any equilibrium point there must be numbers* λ_1, λ_2 (the multipliers) such that

$$dY = \lambda_1\, dU + \lambda_2\, dV. \tag{3}$$

To prove this, write

$$\begin{aligned} dU &= 8u\, du + 2v\, dv \\ dV &= \qquad\qquad\qquad\qquad 3\, dx + 2\, dy \\ dY &= \frac{\partial Y}{\partial u} du + \frac{\partial Y}{\partial v} dv + \frac{\partial Y}{\partial x} dx + \frac{\partial Y}{\partial y} dy. \end{aligned}$$

At any point (u, v, x, y) satisfying the constraints (2) the first two of these equations can be solved for two of the 1-forms du, dv, dx, dy in terms of dU, dV and the remaining two 1-forms. (If $u \neq 0$ then du, dx can be expressed in terms of dU, dV, dv, dy. If $u = 0$ then by (2) it follows that $v \neq 0$, so dv, dx can be expressed in terms of dU, dV, du, dy.) Substituting these solutions into the third equation expresses dY in terms of dU, dV, and the remaining two of the 1-forms du, dv, dx, dy. If this expression for dY is *not* of the form (3) then the elimination process can be carried one step further to express three of the 1-forms du, dv, dx, dy in terms of dU, dV, dY and the remaining one. But this is possible only if $dU\, dV\, dY \neq 0$, so it is not possible at a maximum, and at a maximum (3) must hold.

Using the equation (3), the possible solutions are easily found. Writing out the du, dv, dx, dy components of (3), it becomes four equations

$$\begin{matrix} 2(u - x) = \lambda_1 8u & 2(x - u) = \lambda_2 3 \\ 2(v - y) = \lambda_1 2v & 2(y - v) = \lambda_2 2. \end{matrix} \tag{3'}$$

These equations, together with (2), give 6 equations in the 6 unknowns $(u, v, x, y, \lambda_1, \lambda_2)$. Equating the two expressions for $2(u - x)$ and the two for $2(v - y)$ gives

$$8u\lambda_1 = -3\lambda_2, \qquad 2v\lambda_1 = -2\lambda_2.$$

Equating two expressions for $-6\lambda_2$ then gives

$$16u\lambda_1 = 6v\lambda_1.$$

Now $\lambda_1 \neq 0$ because if λ_1 were zero then (3′) would give

$u = x$, $v = y$ which is incompatible with (2). Therefore division by $6\lambda_1$ is possible and gives

$$\tfrac{8}{3}u = v.$$

Using this in (2) gives

$$4u^2 + \tfrac{64}{9}u^2 = 4$$
$$u^2 = 4 \cdot \frac{9}{100} = \frac{3^2}{5^2}$$
$$u = \pm\tfrac{3}{5}, \quad v = \pm\tfrac{8}{5}.$$

Thus equilibrium can occur only when the particle on the ellipse is at one of the two positions $(u, v) = (3/5, 8/5)$ or $(u, v) = (-3/5, -8/5)$. If it is at $(3/5, 8/5)$ then the position (x, y) of the particle on the line must satisfy

$$4(x - u) = 6\lambda_2 = 6(y - v),$$
$$4x - 6y = 4u - 6v = -\tfrac{36}{5},$$
$$2x - 3y = -\tfrac{18}{5}.$$

Together with $3x + 2y = 25$ this gives $(x, y) = (339/65, 304/65)$. In the same way the position $(u, v) = (-3/5, -8/5)$ determines a unique position (x, y). It is clear geometrically that this second equilibrium position is unstable, and that the unique position of stable equilibrium is $(3/5, 8/5, 339/65, 304/65)$ which solves the given problem. (Strictly speaking, it has been *assumed* that there is a position of stable equilibrium and it has been *proved* that no position other than the two found above can be a position of equilibrium.)

The method of Lagrange multipliers can also be used in the first two examples, even though they were simple enough to be done without it. For example, in the first problem

$$u^2 + v^2 = 1$$
$$8u + 3v = \max.$$

the condition $dX\,dy = 0$ can hold at a point only if $dy = \lambda\,dX$ (for some multiplier λ) at this point; this follows from the observation that

$$dX = 2u\,du + 2v\,dv$$
$$dy = 8\,du + 3\,dv$$

can always be solved for one of the 1-forms du, dv in terms of dX and the other (u and v cannot both be zero on $u^2 + v^2 = 1$) but that the elimination can be carried no further at a maximum (because $dX\,dy = 0$ at a maximum). Hence $dy = \lambda\,dX$ at a maximum. Equating the coefficients of du and dv gives two equations

$$8 = \lambda \cdot 2u$$
$$3 = \lambda \cdot 2v$$

in addition to

$$u^2 + v^2 = 1.$$

Eliminating λ gives $v = \frac{3}{8}u$ and the remainder of the solution is as before.

These techniques can be formulated in general terms as follows: Suppose that a differentiable function

$$y = f(x_1, x_2, \ldots, x_n)$$

is to be maximized subject to k constraints $X_i = \text{const.}$ $(i = 1, 2, \ldots, k)$ where

$$X_i = g_i(x_1, x_2, \ldots, x_n) \qquad (i = 1, 2, \ldots, k)$$

are differentiable functions.* This problem will be abbreviated

*Many more general maximization problems decompose into several problems of this type—for example the problem of maximizing $f(u, v)$ on $\{(u, v): v \geq 0,\ u^2 + v \leq 1\}$ considered above.

$$\text{(4)} \qquad \begin{aligned} X_i &= \text{const.} \qquad (i = 1, 2, \ldots, k) \\ y &= \max. \end{aligned}$$

At any point $(x_1, x_2, \ldots, x_n)$ where a maximum value is achieved, the $(k + 1)$-form $dX_1\,dX_2 \ldots dX_k\,dy$ must be zero. (Otherwise, by the Implicit Function Theorem, the equations $X_i = \text{const.}$, $y = \bar{y} + \epsilon$ can be solved to give an $(x_1, x_2, \ldots, x_n)$ where y is larger.) The constraints $X_i = \text{const.}$ are said to be *non-singular of rank k at a point* $P = (x_1, x_2, \ldots, x_n)$ which satisfies them if $dX_1\,dX_2 \ldots dX_k \neq 0$ at P, that is, if $X_i = g_i(x_1, x_2, \ldots, x_n)$ defines a map $\mathbf{R}^n \to \mathbf{R}^k$ which is non-singular of rank k at P. The constraints $X_i = \text{const.}$ are said to be *non-singular of rank k* if this holds at all points P which satisfy them. Thus the constraint $u^2 + v^2 = 1$ is non-singular of rank 1, and the constraints (2) are non-singular of rank 2. When the constraints are non-singular of rank k the problem of finding points at which $dX_1\,dX_2 \ldots dX_k\,dy = 0$ is simplified by the method of Lagrange multipliers.

Method of Lagrange Multipliers

If $P = (x_1, x_2, \ldots, x_n)$ is a solution of the problem

$$(4) \qquad \begin{aligned} X_i &= \text{const.} \qquad (i = 1, 2, \ldots, k) \\ y &= \max. \end{aligned}$$

and if the constraints $X_i = \text{const.}$ are non-singular of rank k at P then there exist numbers $\lambda_1, \lambda_2, \ldots, \lambda_k$ (the multipliers) such that

$$(5) \qquad dy = \lambda_1\, dX_1 + \lambda_2\, dX_2 + \cdots + \lambda_k\, dX_k$$

at P. The same is true if P is a solution of the problem

$$\begin{aligned} X_i &= \text{const.} \qquad (i = 1, 2, \ldots, k) \\ y &= \min. \end{aligned}$$

at which the constraints are non-singular of rank k.

Proof

Since $dX_1\, dX_2 \ldots dX_k \neq 0$ at P by assumption, the equations

$$dX_i = \sum_{j=1}^{n} \frac{\partial X_i}{\partial x_j}\, dx_j \qquad (i = 1, 2, \ldots, k)$$

can be solved for k of the 1-forms $dx_1, dx_2, \ldots, dx_n$. By rearranging the x's if necessary, it can be assumed that $dx_1, dx_2, \ldots, dx_k$ can be expressed in terms of $dX_1, dX_2, \ldots, dX_k, dx_{k+1}, \ldots, dx_n$. Substituting these expressions in $dy = \sum_{j=1}^{n} \frac{\partial y_i}{\partial x_j}\, dx_j$ gives

$$dy = \lambda_1\, dX_1 + \cdots + \lambda_k\, dX_k + \lambda_{k+1}\, dx_{k+1} + \cdots + \lambda_n\, dx_n.$$

If this expression is not of the form (5) then another of the dx's can be eliminated and the $(k + 1)$-form $dX_1\, dX_2 \ldots dX_k\, dy$ is not zero at P. But then, by the previous argument, P is not a maximum or a minimum. Hence if P is a maximum or a minimum then (5) holds, as was to be shown.

When the n components of the 1-forms in (5) are written out, (5) gives n equations in addition to the k equations $X_i = \text{const.}$, hence there are $n + k$ equations in the $n + k$ unknowns $(x_1, x_2, \ldots, x_n, \lambda_1, \lambda_2, \ldots, \lambda_k)$. Explicitly,

$$\frac{\partial y}{\partial x_i} = \sum_{j=1}^{k} \lambda_j \frac{\partial X_j}{\partial x_i} \qquad (i = 1, 2, \ldots, n)$$

$$X_i = \text{const.} \qquad (i = 1, 2, \ldots, k).$$

This generalizes Step I of the process above to problems of the form (4) in which the constraints are non-singular of rank k.* Step II is to *solve* these $n + k$ equations in $n + k$ unknowns, and Step III is to test each solution to determine whether or not it is a maximum.

*If $k = 0$, that is, if there are no constraints, then (5) becomes $dy = 0$. If $n = 1$ this is the case considered at the beginning of the section.

A point $(x_1, x_2, \ldots, x_n)$ is said to be a *critical point* of the problem (4) if it satisfies the constraints and if there exist numbers $\lambda_1, \lambda_2, \ldots, \lambda_k$ such that (5) holds† at $(x_1, x_2, \ldots, x_n)$. The steps are then:

†Picturesquely speaking, the equation $dy = \lambda_1\, dX_1 + \cdots + \lambda_k\, dX_k$ implies 'if $dX_1 = 0, \ldots, dX_k = 0$ then $dy = 0$', that is, 'no change in the X_i implies no change (infinitesimally) in y'. For this reason, critical points are also called 'stationary points' of y relative to the constraints $X_i = \text{const.}$

I. Set up the equations satisfied by the critical points.
II. Find all critical points.
III. Look for a maximum (or minimum) among the critical points.

Step II involves solving $n + k$ equations in $n + k$ unknowns and is therefore not feasible in most cases. However, there are many cases in which these equations have a particularly simple form and can either be solved explicitly or can be used to derive useful information about the critical points. The remainder of this section is devoted to a few such cases which are particularly useful in applications.

Example

Find the minimum and maximum values of $Au + Bv$ on the disk $u^2 + v^2 \leq 1$. (The case $A = 8$, $B = 3$ was solved above.) There are no critical points in the interior of the disk unless $A = B = 0$, that is, unless the function $Au + Bv$ is identically zero and 0 is both maximum and minimum. Otherwise the only critical points are on the bounding circle $u^2 + v^2 = 1$. The constraint $u^2 + v^2 = 1$ is non-singular of rank 1 and the method of Lagrange multipliers applies, giving

$$A\, du + B\, dv = \lambda(2u\, du + 2v\, dv).$$

Hence

$$A = 2\lambda u$$
$$B = 2\lambda v$$
$$u^2 + v^2 = 1.$$

If these equations are satisfied then

$$Au + Bv = 2\lambda u^2 + 2\lambda v^2 = 2\lambda.$$

so the maximum and minimum values are 2λ. But

$$A^2 + B^2 = (2\lambda)^2$$

hence

$$2\lambda = \pm\sqrt{A^2 + B^2}.$$

If 2λ has either of these two values then $u = A/2\lambda$, $v = B/2\lambda$ are determined; hence there are exactly two critical points and they occur when $u:v = A:B$. The desired maximum and minimum values are therefore $\pm\sqrt{A^2 + B^2}$. Note that this formula holds when $A = B = 0$ as well.

Example

Find the minimum and maximum values of $A_1x_1 + A_2x_2 + \cdots + A_nx_n$ on the n-dimensional ball $x_1^2 + x_2^2 + \cdots + x_n^2 \leq K$ where $A_1, A_2, \ldots, A_n, K$ are fixed numbers with $K > 0$. As above, if not all A_i are zero then the only critical points are on the $(n-1)$-dimensional bounding sphere $\{x_1^2 + x_2^2 + \cdots + x_n^2 = K\}$. The constraint $x_1^2 + \cdots + x_n^2 = K$ is nonsingular of rank 1 because $d(x_1^2 + \cdots + x_n^2)$ is zero only at $(0, 0, \ldots, 0)$, which does not satisfy the constraint. By the method of Lagrange multipliers the maximum and minimum must therefore occur at points which satisfy

$$A_i = 2\lambda x_i \qquad (i = 1, 2, \ldots, n)$$
$$x_1^2 + x_2^2 + \cdots + x_n^2 = K$$

for some λ. Since

$$A_1x_1 + A_2x_2 + \cdots + A_nx_n = 2\lambda\sum x_i^2 = 2\lambda K$$

and

$$A_1^2 + A_2^2 + \cdots + A_n^2 = (2\lambda)^2 \cdot K$$

there are only two possible values,

$$\lambda = \pm\frac{\sqrt{A_1^2 + \cdots + A_n^2}}{2\sqrt{K}}$$

of λ, each of which corresponds to a unique critical point

$$x_i = \frac{A_i}{2\lambda} \qquad (i = 1, 2, \ldots, n)$$

at which the value of the function is

$$\begin{aligned} 2\lambda K &= \pm\sqrt{A_1^2 + \cdots + A_n^2}\sqrt{K} \\ &= \pm\sqrt{A_1^2 + \cdots + A_n^2}\sqrt{x_1^2 + \cdots + x_n^2}. \end{aligned}$$

It follows that these are the maximum and minimum values and that each is assumed at exactly one point. These conclusions can also be stated: The inequality

$$(6)\quad |A_1x_1 + \cdots + A_nx_n| \le \sqrt{A_1^2 + \cdots + A_n^2}\sqrt{x_1^2 + \cdots + x_n^2}$$

holds for all pairs of n-tuples of numbers $(A_1, \ldots, A_n)$ and $(x_1, \ldots, x_n)$. Moreover, equality holds only if

$$x_1^2 + \cdots + x_n^2 = 0, \qquad \text{i.e. } x_i = 0 \text{ for } i = 1, 2, \ldots, n$$

or

$$A_1^2 + \cdots + A_n^2 = 0, \qquad \text{i.e. } A_i = 0, \text{ for } i = 1, 2, \ldots, n$$

*μ = *mu*.

or there is a non-zero number μ* $(=2\lambda)$ such that

$$A_i = \mu x_i \qquad (i = 1, 2, \ldots, n).$$

The inequality (6) is called the *Schwarz inequality*. The conditions under which equality can hold can be stated more symmetrically: There exist numbers μ_1, μ_2 not both zero such that

$$\mu_1 A_i = \mu_2 x_i \qquad (i = 1, 2, \ldots, n).$$

This includes all three of the cases $A_i \equiv 0$, $x_i \equiv 0$, $A_i = \mu x_i$.

Geometrically $\sqrt{x_1^2 + \cdots + x_n^2}$ can be regarded as the length of the line segment from the origin to $(x_1, x_2, \ldots, x_n)$ and $\sqrt{A_1^2 + \cdots + A_n^2}$ can be regarded as the length of the line segment from the origin to $(A_1, A_2, \ldots, A_n)$. The Schwarz inequality then says that the absolute value of the so-called 'dot product' $A_1x_1 + \cdots + A_nx_n$ is at most the product of the lengths and that equality holds if and only if the line segments are collinear.†

†*A 'line segment' which degenerates to a single point is considered to be collinear with any line segment.*

Example

Find the minimum and maximum values of $Ax + By$ on the curve $|x|^3 + |y|^3 = K$ where A, B, K are fixed and $K > 0$. It was shown in Exercise 6, §5.2, that $|x|^3$ is a

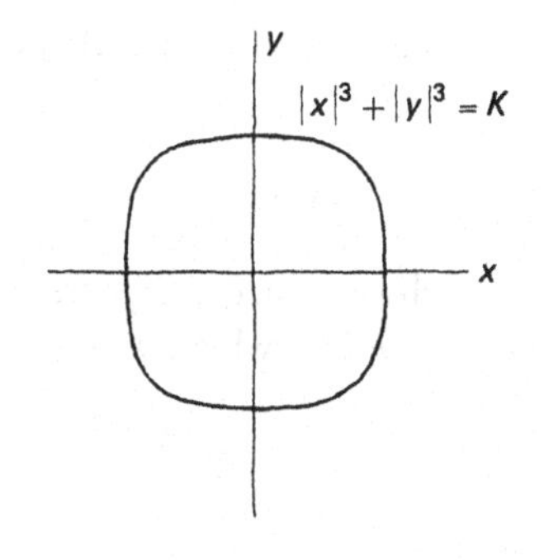

differentiable function with derivative $3x|x|$. Since the 1-form $d(|x|^3 + |y|^3) = 3x|x|\,dx + 3y|y|\,dy$ is not zero at any point satisfying the constraint $|x|^3 + |y|^3 = K > 0$, the method of Lagrange multipliers applies and says that if a maximum or minimum value is assumed at (x, y) then there must be a λ such that

$$A\,dx + B\,dy = \lambda(3x|x|\,dx + 3y|y|\,dy).$$

The critical points are therefore defined by the equations

$$\begin{aligned} A &= 3\lambda x|x| \\ B &= 3\lambda y|y| \\ |x|^3 + |y|^3 &= K. \end{aligned}$$

Since

$$Ax + By = 3\lambda|x|^3 + 3\lambda|y|^3 = 3\lambda K$$

the problem is to find $3\lambda K$. If the equations $|A| = |3\lambda|\,|x|^2$, $|B| = |3\lambda|\,|y|^2$ are raised to the power 3/2 and added, the result is

$$\begin{aligned} |A|^{3/2} + |B|^{3/2} &= |3\lambda|^{3/2}K \\ (|A|^{3/2} + |B|^{3/2})K^{1/2} &= |3\lambda K|^{3/2}. \end{aligned}$$

Hence, raising to the power 2/3 gives

$$(7)\qquad |3\lambda K| = (|A|^{3/2} + |B|^{3/2})^{2/3}(|x|^3 + |y|^3)^{1/3}.$$

If $A = B = 0$ then $\lambda = 0$, and the function is identically zero on $|x|^3 + |y|^3 = K$. Otherwise (7) gives two non-zero values for λ, and hence two critical points (x, y)—one at which the value of $Ax + By$ is

$$(|A|^{3/2} + |B|^{3/2})^{2/3}(|x|^3 + |y|^3)^{1/3}$$

and one at which the value is minus this number. These are clearly the maximum and minimum values respectively. If (x, y) is a critical point, then

$$\begin{aligned} |A| &= |3\lambda|\,|x|^2 & |B| &= |3\lambda|\,|y|^2 \\ |A|^{1/2} &= |3\lambda|^{1/2}|x| & |B|^{1/2} &= |3\lambda|^{1/2}|y| \\ A|A|^{1/2} &= \mu x|x|^2 & B|B|^{1/2} &= \mu y|y|^2 \end{aligned}$$

where $\mu = 3\lambda|3\lambda|^{1/2}$. These conclusions can all be summarized: For any (A, B) and (x, y) the inequality

$$(8)\qquad |Ax + By| \le (|A|^{3/2} + |B|^{3/2})^{2/3}(|x|^3 + |y|^3)^{1/3}$$

holds. Equality holds only when there are numbers μ_1, μ_2 not both zero such that

$$\mu_1(A|A|^{1/2}, B|B|^{1/2}) = \mu_2(x|x|^2, y|y|^2).$$

The inequality (8) and the Schwarz inequality are both special cases of the *Hölder inequality*, which will now be derived.

Example

The function $|x|^p$ has been defined so far only for rational values of p. If $|x|^p$ is defined to be $e^{p \log|x|}$ for $x \neq 0$ and 0 for $x = 0$ then it is differentiable for $p > 1$ and has derivative $px|x|^{p-2}$.

Let p be any* number greater than 1. Find the maximum values of $A_1x_1 + A_2x_2 + \cdots + A_nx_n$ subject to the constraint $|x_1|^p + |x_2|^p + \cdots + |x_n|^p = K$ where $A_1, A_2, \ldots, A_n$, K are fixed numbers with $K > 0$. The method of Lagrange multipliers applies and gives the equations

$$A_i = \lambda p \cdot x_i \cdot |x_i|^{p-2} \qquad (i = 1, 2, \ldots, n)$$
$$|x_1|^p + \cdots + |x_n|^p = K$$

for the critical points. Then

$$A_1x_1 + \cdots + A_nx_n = \lambda pK$$

is the value at the critical point and

$$|A_1|^{p/p-1} + \cdots + |A_n|^{p/p-1} = |\lambda p|^{p/p-1} \cdot K$$
$$(|A_1|^{p/p-1} + \cdots + |A_n|^{p/p-1})K^{1/p-1} = |\lambda pK|^{p/p-1}.$$

Hence the value at the critical point is

$$\pm(|A_1|^{p/p-1} + \cdots + |A_n|^{p/p-1})^{p-1/p}(|x_1|^p + \cdots + |x_n|^p)^{1/p}.$$

If the A's are not all zero then there are only two possible values of λ, each of which corresponds to a single critical point. Thus there are just two critical points, one the maximum and one the minimum. Set $q = p/(p-1)$. Then the critical points satisfy

$$|A_i| = |\lambda p|\, |x_i|^{p-1}$$
$$|A_i|^{q-1} = |\lambda p|^{q-1}|x_i| \quad \text{(because } (p-1)(q-1) = 1\text{)}$$
$$A_i|A_i|^{q-1} = \mu \cdot x_i|x_i|^{p-1}$$

where $\mu = \lambda p \cdot |\lambda p|^{q-1}$. These conclusions are summarized by:

The Hölder Inequality

Let $(x_1, x_2, \ldots, x_n)$ and $(A_1, A_2, \ldots, A_n)$ be given n-tuples and let $p > 1, q > 1$ be two numbers such that $(p-1)(q-1) = 1$. Then

$$(9) \qquad |A_1x_1 + A_2x_2 + \cdots + A_nx_n| \le |A|_q|x|_p$$

where

$$|A|_q = (|A_1|^q + \cdots + |A_n|^q)^{1/q}$$
$$|x|_p = (|x_1|^p + \cdots + |x_n|^p)^{1/p}.$$

Equality holds in (9) if and only if there exist numbers μ_1, μ_2 not both zero such that $\mu_1 A|A|^{q-1} \equiv \mu_2 x|x|^{p-1}$, i.e.

$$\mu_1 \cdot A_i \cdot |A_i|^{q-1} = \mu_2 \cdot x_i \cdot |x_i|^{p-1} \qquad (i = 1, 2, \ldots, n).$$

Example

Find the minimum and maximum values of the quadratic form* $Q(x, y) = 3x^2 - 2xy + 4y^2$ on the disk $x^2 + y^2 \le 1$. If a maximum or a minimum occurs in the interior of the disk then it must be at a point where $dQ = 0$, i.e. $6x\,dx - 2x\,dy - 2y\,dx + 8y\,dy = 0$, i.e.

*So called because it is a form (=homogeneous polynomial = polynomial whose terms all have the same degree) of degree two (quadrat = square = second power). Not to be confused with 'k-forms' in any way.

$$6x - 2y = 0$$
$$-2x + 8y = 0.$$

This occurs only at the point $(x, y) = (0, 0)$. If a maximum or a minimum occurs on the boundary $x^2 + y^2 = 1$ it must be at a point where $dQ = \lambda d(x^2 + y^2)$ for some λ, that is

$$(10) \qquad \begin{aligned} 6x - 2y &= 2\lambda x \\ -2x + 8y &= 2\lambda y. \end{aligned}$$

Rewriting these equations as

$$(11) \qquad \begin{aligned} (3 - \lambda)x - y &= 0 \\ -x + (4 - \lambda)y &= 0 \end{aligned}$$

and noting that $(x, y) \ne (0, 0)$ it follows that the map

is not one-to-one, hence that the rank is not two and the matrix of coefficients

$$\begin{pmatrix} 3-\lambda & -1 \\ -1 & 4-\lambda \end{pmatrix}$$

has determinant zero; hence $(3-\lambda)(4-\lambda)-1=0$, $\lambda^2-7\lambda+11=0$,

$$\lambda = \frac{7 \pm \sqrt{49-44}}{2} = \frac{1}{2}(7 \pm \sqrt{5}).$$

For each of these two values of λ the linear equations (11) are satisfied by all points (x, y) on the line $(3-\lambda)x - y = 0$ (which is identical to the line $-x + (4-\lambda)y = 0$). Hence there are *two* critical points for each of these two values of λ, namely the two points on the circle $x^2 + y^2 = 1$ which lie on the line (11), which gives *four* critical points in all. Multiplying the first equation of (10) by $x/2$, the second by $y/2$ and adding gives $Q(x, y) = \lambda(x^2 + y^2) = \lambda$. Hence λ is the value of Q at the corresponding critical points. Therefore $\frac{1}{2}(7+\sqrt{5})$ is the greatest value of Q on the disk and $\frac{1}{2}(7-\sqrt{5})$ the least value on the circle, but 0 the least value on the disk. This solves the given problem.

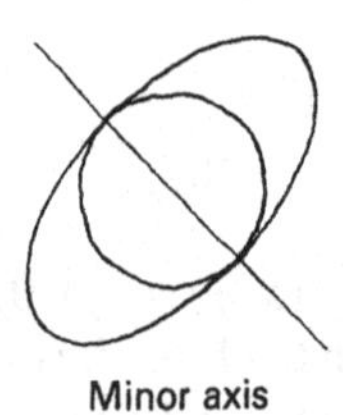

Minor axis

Major axis

The distribution of values of Q can be seen quite clearly by sketching the level curves $Q = \text{const}$. The curve $Q = 1$ is defined by a second-degree equation, and hence must be one of the conic sections. It never crosses the circle (because $\frac{1}{2}(7-\sqrt{5})$ is the least value on the circle) but touches points inside the circle (e.g. $(\pm 1/\sqrt{3}, 0)$, $(0, \pm\frac{1}{2})$) which means that it can only be an ellipse. Since $Q(ax, ay) = a^2 Q(x, y)$, the other level curves of Q are concentric ellipses with center at (0, 0). The largest of these ellipses which touches the circle $x^2 + y^2 = 1$ is $Q = \frac{1}{2}(7+\sqrt{5})$, which touches it at the points where the line $(3 - \frac{1}{2}(7+\sqrt{5}))x - y = 0$ intersects the circle. This line $y = -\frac{1}{2}(1+\sqrt{5})x$ is therefore the minor axis of the concentric ellipses. The major axis must be the line through (0, 0) perpendicular to this one, which is the line $x = \frac{1}{2}(1+\sqrt{5})y$. This agrees with the observation that it must also be the line

$$(3 - \tfrac{1}{2}(7-\sqrt{5}))x - y = 0.$$

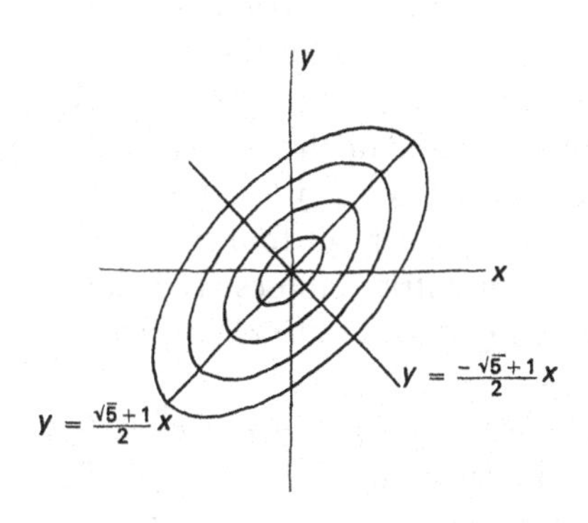

Example

Find the minimum and maximum values of the quadratic form $Q(x, y) = Ax^2 + By^2 + 2cxy$ on the disk $x^2 + y^2 \leq 1$. As before, a critical point in the interior must satisfy

$$(12) \qquad \begin{aligned} Ax + cy &= 0, \\ cx + By &= 0. \end{aligned}$$

Thus $(0, 0)$ is a critical point, and there are no other critical points in the interior unless $AB - c^2 = 0$. In this case either $Q \equiv 0$ or there is a line through the origin where (12) is satisfied. Critical points on the boundary must satisfy the equations

$$(13) \qquad \begin{aligned} Ax + cy &= \lambda x \\ cx + By &= \lambda y \\ x^2 + y^2 &= 1 \end{aligned}$$

which imply that the determinant of

$$\begin{pmatrix} A - \lambda & c \\ c & B - \lambda \end{pmatrix}$$

is zero; that is, $\lambda^2 - (A + B)\lambda + (AB - c^2) = 0$, which means that λ must be one of the values

$$\lambda_1 = \frac{A + B + \sqrt{(A - B)^2 + 4c^2}}{2}, \quad \lambda_2 = \frac{A + B - \sqrt{(A - B)^2 + 4c^2}}{2}.$$

These values are real numbers and they are distinct unless $A = B$ and $c = 0$. If $\lambda_1 \neq \lambda_2$, then each corresponds, as before, to two solutions of (13), giving four critical points on the boundary. If $\lambda_1 = \lambda_2$ then all points of $x^2 + y^2 = 1$ are critical points. Multiplying the first equation of (13) by x, the second by y, and adding, gives $Q(x, y) = \lambda(x^2 + y^2) = \lambda$; hence λ_1, λ_2 are the values of Q at the critical points. The possibilities can now be enumerated as follows:

$AB - c^2 > 0$. In this case $\lambda_1\lambda_2 = AB - c^2 > 0$, hence λ_1, λ_2 have the same sign. Since $\lambda_2 \leq \lambda_1$ the possibilities are that $0 < \lambda_2 < \lambda_1$, that $\lambda_2 < \lambda_1 < 0$, that $0 < \lambda_2 = \lambda_1$, or that $\lambda_2 = \lambda_1 < 0$. In the first case there is a maximum value λ_1 at the points where the line $(A - \lambda_1)x + cy = 0$ intersects the circle $x^2 + y^2 = 1$ and a minimum value 0 at $(0, 0)$. In the second case there is a minimum value λ_2 at the points where $(A - \lambda_2)x + cy = 0$ intersects the circle and a maximum value 0 at $(0, 0)$. In the third case there is a maximum value $\lambda_1 = \lambda_2 = A = B$ assumed at all points $x^2 + y^2 = 1$ and a minimum value 0 at $(0, 0)$. In the last case the minimum is assumed at all points $x^2 + y^2 = 1$ and the maximum 0 is assumed at $(0, 0)$. In all four cases the curves $Q =$ const. are concentric ellipses (possibly circular) and for this reason Q is said to be *elliptic* when $AB - c^2 > 0$.

$AB - c^2 < 0$. In this case $\lambda_1\lambda_2 = AB - c^2 < 0$, hence λ_1, λ_2 have opposite signs, i.e. $\lambda_1 > 0$, $\lambda_2 < 0$. In this case the maximum value occurs at the points where $(A - \lambda_1)x + cy = 0$ intersects the circle $x^2 + y^2 = 1$ and the minimum value where $(A - \lambda_2)x + cy = 0$ intersects it. The critical point at $(0, 0)$ is neither a maximum nor a minimum. The curves $Q =$ const. are hyperbolas whose axes are the lines $(A - \lambda_i)x + cy$ $(i = 1, 2)$. Q is said to be *hyperbolic* if $AB - c^2 < 0$.

$AB - c^2 = 0$. In this case λ_1 or λ_2 is zero; hence either $0 = \lambda_2 < \lambda_1$ or $\lambda_2 < \lambda_1 = 0$ or $\lambda_1 = \lambda_2 = 0$. In the first case $\lambda_1 = A + B$ is the maximum value and it is assumed at two points on the boundary while the minimum 0 is assumed all along the line $Ax + cy = 0$ (which is the same as $cx + By = 0$). In the second case the minimum is assumed at two points on the boundary and the maximum on $Ax + cy = 0$. The third case can occur only when $Q \equiv 0$ and will be ignored. The level curves $Q =$ const. are parallel straight lines, as is seen by writing $Q(x, y) = \frac{1}{A}(Ax + cy)^2 = \frac{1}{B}(cx + By)^2$. If $AB - c^2 = 0$ then Q is said to be *parabolic* because its graph is parabolic.

Example

Show that the axes of the conic sections $Q =$ const. (where $Q(x, y)$ is a quadratic form in two variables) are perpendicular to each other. From the fact that Q is symmetric about each of its axes this is clear geometrically and the problem is to prove it analytically. Let (x_1, y_1) be a critical point of Q and let $\lambda_1 = Q(x_1, y_1)$, i.e.

$$\begin{aligned} Ax_1 + cy_1 &= \lambda_1 x_1 \\ cx_1 + By_1 &= \lambda_1 y_1 \\ x_1^2 + y_1^2 &= 1. \end{aligned} \tag{13'}$$

Let (x_2, y_2) be one of the two points on the circle $x^2 + y^2 = 1$ which lie at right angles to (x_1, y_1), i.e.

$$x_2^2 + y_2^2 = 1$$
$$x_1x_2 + y_1y_2 = 0.$$

($x_1x + y_1y$ on the circle goes from a maximum value of 1 at (x_1, y_1) to -1 at $(-x_1, -y_1)$ and is 0 at the two points halfway in-between.) Since

$$\begin{pmatrix} x_1 & y_1 \\ x_2 & y_2 \end{pmatrix}\begin{pmatrix} x_1 & x_2 \\ y_1 & y_2 \end{pmatrix} = \begin{pmatrix} 1 & 0 \\ 0 & 1 \end{pmatrix}$$

it follows that the equations

$$x = ux_1 + vx_2$$
$$y = uy_1 + vy_2$$

can be solved for (u, v) as functions of (x, y), namely

$$u = x_1x + y_1y$$
$$v = x_2x + y_2y.$$

Therefore any (x, y) can be written in this form and, in particular,

$$Ax_2 + cy_2 = ux_1 + vx_2$$
$$cx_2 + By_2 = uy_1 + vy_2$$

for some pair of numbers (u, v). In fact u is given by

$$u = x_1(Ax_2 + cy_2) + y_1(cx_2 + By_2) = (Ax_1 + cy_1)x_2 + (cx_1 + By_1)y_2$$
$$= \lambda_1x_1x_2 + \lambda_1y_1y_2 = 0.$$

Since $u = 0$ the equations (13) are satisfied by (x_2, y_2, v); hence (x_2, y_2) is a critical point as was to be shown.

Example

Find the minimum and maximum values of the quadratic form $Q(x, y, z) = Ax^2 + By^2 + Cz^2 + 2ayz + 2bxz + 2cxy$ on the ball $x^2 + y^2 + z^2 \leq 1$. As in the case of two variables, the critical points in the interior are the solutions of

$$\begin{aligned} Ax + cy + bz &= 0 \\ cx + By + az &= 0 \\ bx + ay + Cz &= 0 \end{aligned} \tag{14}$$

and the critical points on the boundary are the solutions of

$$(15)\qquad \begin{aligned} Ax + cy + bz &= \lambda x \\ cx + By + az &= \lambda y \\ bx + ay + Cz &= \lambda z \\ x^2 + y^2 + z^2 &= 1. \end{aligned}$$

(The 1-form $d(x^2 + y^2 + z^2)$ is never zero on the sphere $x^2 + y^2 + z^2 = 1$.) These equations imply as before that

$$(16)\qquad \begin{vmatrix} A - \lambda & c & b \\ c & B - \lambda & a \\ b & a & C - \lambda \end{vmatrix} = 0$$

which gives a polynomial of degree 3 which must be satisfied by λ in (15). As before, $\lambda = Q(x, y, z)$ at any point which satisfies (15). Thus if Q assumes a maximum or a minimum value on $x^2 + y^2 + z^2 \leq 1$, and it is intuitively plausible that it must, then this value must be either 0 or a root of the polynomial (16). The solution of given problem is therefore reduced to finding the largest and smallest roots of the cubic polynomial (16) and comparing them to zero.

Example

Given $Q(x, y, z)$ as above, show that there exist three mutually perpendicular lines through the origin whose six points of intersection with $x^2 + y^2 + z^2 = 1$ are critical points of Q on $x^2 + y^2 + z^2 \leq 1$. Taking it for granted that Q must assume maximum and minimum values on $x^2 + y^2 + z^2 = 1$, it follows that the equations (15) must have at least two solutions. Let (x_1, y_1, z_1) be a solution. Consider now the problem of finding the maximum and minimum values of Q on the circle

$$\begin{aligned} x^2 + y^2 + z^2 &= 1 \\ x_1x + y_1y + z_1z &= 0 \end{aligned}$$

which is the 'equator' when the 'poles' are $\pm(x_1, y_1, z_1)$. The method of Lagrange multipliers applies (see Exercise 13) and implies that a maximum or a minimum of Q on the circle can occur only at a point (x, y, z) for which there are numbers λ_1, λ_2 such that

$$\begin{aligned}
Ax + cy + bz &= \lambda_1 x + \lambda_2 x_1/2 \\
cx + By + az &= \lambda_1 y + \lambda_2 y_1/2 \\
bx + ay + Cz &= \lambda_1 z + \lambda_2 z_1/2 \\
x^2 + y^2 + z^2 &= 1 \\
x_1 x + y_1 y + z_1 z &= 0.
\end{aligned}$$

Since the function Q must have a maximum and a minimum on the circle these equations must have at least two solutions. Let $(x_2, y_2, z_2, \lambda_1, \lambda_2)$ be a solution. Multiplying the first equation by x_1, the second by y_1, the third by z_1, and adding gives $\lambda_2/2 = x_1(Ax_2 + cy_2 + bz_2) + y_1(cx_2 + By_2 + az_2) + z_1(bx_2 + ay_2 + Cz_2) = (Ax_1 + cy_1 + bz_1)x_2 + (cx_1 + By_1 + az_1)y_2 + (bx_1 + ay_1 + Cz_1)z_2 = \lambda x_1 x_2 + \lambda y_1 y_2 + \lambda z_1 z_2 = \lambda(x_1 x_2 + y_1 y_2 + z_1 z_2) = 0$ where $\lambda = Q(x_1, y_1, z_1)$. Therefore $\lambda_2 = 0$ and $(x_2, y_2, z_2, \lambda_1)$ is a solution of (15); that is, (x_2, y_2, z_2) is a critical point of the original problem. If (x_3, y_3, z_3) is one of the two solutions of

$$\begin{aligned}
x_1 x + y_1 y + z_1 z &= 0 \\
x_2 x + y_2 y + z_2 z &= 0 \\
x^2 + y^2 + z^2 &= 1
\end{aligned}$$

then

$$(17)\qquad \begin{pmatrix} x_1 & y_1 & z_1 \\ x_2 & y_2 & z_2 \\ x_3 & y_3 & z_3 \end{pmatrix}\begin{pmatrix} x_1 & x_2 & x_3 \\ y_1 & y_2 & y_3 \\ z_1 & z_2 & z_3 \end{pmatrix} = \begin{pmatrix} 1 & 0 & 0 \\ 0 & 1 & 0 \\ 0 & 0 & 1 \end{pmatrix}.$$

Hence there exist numbers u, v, w such that

$$\begin{aligned}
Ax_3 + cy_3 + bz_3 &= ux_1 + vx_2 + wx_3 \\
cx_3 + By_3 + az_3 &= uy_1 + vy_2 + wy_3 \\
bx_3 + ay_3 + Cz_3 &= uz_1 + vz_2 + wz_3.
\end{aligned}$$

Multiplying by x_1, y_1, z_1, respectively, and adding gives $u = 0$ as before. Similarly $v = 0$ so (x_3, y_3, z_3, w) is a solution of (15); that is, (x_3, y_3, z_3) is a critical point of the original problem. The lines through the three points (x_i, y_i, z_i) $(i = 1, 2, 3)$ therefore have the desired properties.

Example

Find all critical points of $Q(x, y, z)$ on $x^2 + y^2 + z^2 \leq 1$. Let (x_i, y_i, z_i) $(i = 1, 2, 3)$ be mutually perpendicular critical points of Q as above, and let $\lambda_i = Q(x_i, y_i, z_i)$. Then Q has at least the seven critical points $(0, 0, 0)$, $\pm(x_i, y_i, z_i)$ $(i = 1, 2, 3)$, at which the values of Q are $0, \lambda_1, \lambda_2, \lambda_3$.

By (17) the equations

$$\begin{aligned} x &= ux_1 + vx_2 + wx_3 \\ y &= uy_1 + vy_2 + wy_3 \\ z &= uz_1 + vz_2 + wz_3 \end{aligned} \tag{18}$$

have the unique solution

$$\begin{aligned} u &= x_1x + y_1y + z_1z \\ v &= x_2x + y_2y + z_2z \\ w &= x_3x + y_3y + z_3z. \end{aligned} \tag{19}$$

Now if (x, y, z) is a critical point of $Q(x, y, z)$, that is, if

$$\begin{aligned} Ax + cy + bz &= \lambda x \\ cx + By + az &= \lambda y \\ bx + ay + Cz &= \lambda z \end{aligned}$$

for some λ, then the substitution (18) in these equations gives

$$\begin{aligned} u\lambda_1x_1 + v\lambda_2x_2 + w\lambda_3x_3 &= \lambda ux_1 + \lambda vx_2 + \lambda wx_3 \\ u\lambda_1y_1 + v\lambda_2y_2 + w\lambda_3y_3 &= \lambda uy_1 + \lambda vy_2 + \lambda wy_3 \\ u\lambda_1z_1 + v\lambda_2z_2 + w\lambda_3z_3 &= \lambda uz_1 + \lambda vz_2 + \lambda wz_3 \end{aligned}$$

using the fact that $(x_i, y_i, z_i, \lambda_i)$ satisfies (15). By the uniqueness of the solution of (18) this gives

$$\begin{aligned} u\lambda_1 &= u\lambda \\ v\lambda_2 &= v\lambda \\ w\lambda_3 &= w\lambda. \end{aligned}$$

If the numbers $0, \lambda_1, \lambda_2, \lambda_3$ are distinct, this implies that at least two of the numbers u, v, w are zero. Thus in this case the critical point (x, y, z) must be a multiple of (x_i, y_i, z_i) for $i = 1, 2$, or 3. Since the multiple must be 0 or ± 1, this shows that *if* $0, \lambda_1, \lambda_2, \lambda_3$ *are distinct then there are exactly seven critical points.* If two or more of these numbers $0, \lambda_1, \lambda_2, \lambda_3$, coincide, then the set of critical points also includes a line segment through the interior ($\lambda_i = 0$), a great circle on the sphere ($\lambda_i = \lambda_j$), a disk through the interior ($0 = \lambda_i = \lambda_j$), the entire sphere ($\lambda_1 = \lambda_2 = \lambda_3$), or the entire ball ($Q \equiv 0$).

Example

Determine whether a quadratic form $Q(x, y, z)$ has a unique minimum value at the origin $(0, 0, 0)$; i.e. determine whether $Q(x, y, z) > 0$ for $(x, y, z) \neq (0, 0, 0)$, *without* solving the cubic polynomial (16). (If $Q(x, y, z) > 0$ for $(x, y, z) \neq (0, 0, 0)$ then Q is said to be *positive definite*. The importance of this notion in physics is that it is the necessary and sufficient condition for $(0, 0, 0)$ to be a point of *stable equilibrium*.) This problem is solved in the exercises. The solution is that $Q(x, y, z)$ is positive definite if and only if

$$\begin{vmatrix} A & c & b \\ c & B & a \\ b & a & C \end{vmatrix} > 0, \quad \begin{vmatrix} A & c \\ c & B \end{vmatrix} > 0, \quad \text{and} \quad A > 0.$$

(The obvious generalization to quadratic forms in n variables is also true. This is particularly important for $n > 3$ because the generalization to n variables of the cubic (16) is a polynomial of degree n, and it can be very difficult to determine whether or not such a polynomial has any negative roots.)

Exercises

1 Find the point of the line

$$\begin{aligned} 3x - y + 2z &= 13 \\ x + 4y - 7z &= 36 \end{aligned}$$

which lies nearest the origin. What is the general procedure for finding the point of the line

$$(*) \qquad \begin{aligned} A_1x + B_1y + C_1z &= D_1 \\ A_2x + B_2y + C_2z &= D_2 \end{aligned}$$

nearest the origin? Under what circumstances would the method of Lagrange multipliers not apply to the constraints (*)?

2 *Snell's law.* In the diagram on the following page let l represent water level, let A be a point above water and B be a point below the water level. Show that if v_1 is the velocity of light in air and v_2 the velocity of light in water, and if the time

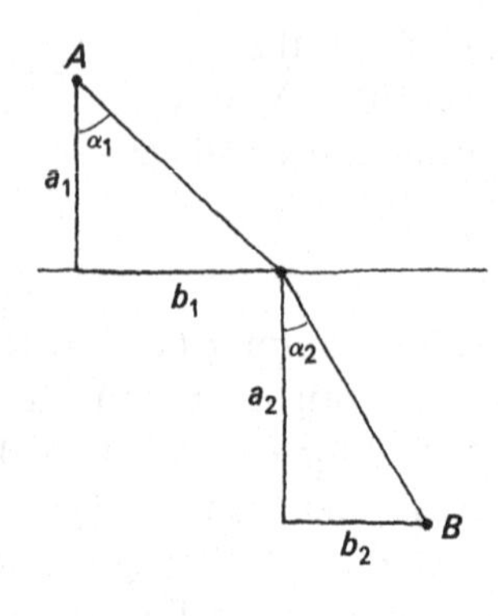

required for light to travel from the point A to the point B along the path indicated is minimal then

$$\frac{b_1}{v_1\sqrt{a_1^2 + b_1^2}} = \frac{b_2}{v_2\sqrt{a_2^2 + b_2^2}}.$$

In other words,

$$\frac{\sin \alpha_1}{\sin \alpha_2} = \frac{v_1}{v_2}.$$

3 A rectangular box with an open top is to have total surface area 24 (four sides and the bottom) and to have the largest possible volume. Find the dimensions of the box.

4 Find the maximum and minimum values of xyz on the sphere $x^2 + y^2 + z^2 = 1$ and all points at which they are assumed.

5 Let two attracting particles be constrained to lie on curves, the first on a curve $f(x, y) = 0$ and the other on the circle $u^2 + v^2 = 1$. Assuming that $f(x, y) = 0$ lies entirely outside the circle, show that the equilibrium positions of the particle on $f(x, y) = 0$ are the same as they would be if the other particle were fixed at $(0, 0)$, and that the corresponding equilibrium position of the particle on the circle is the point of the circle which lies on the line segment from $(0, 0)$ to the particle on the curve.

6 (a) Find the maximum value of xy on the line $x + y = 1$.
(b) Does xyz have a maximum value on $x + y + z = 1$?
(c) Find all critical points of xyz on $x + y + z = 1$.
(d) Let A, B, C be given positive numbers. If $x^A y^B z^C$ assumes a maximum value on the set where $x > 0$, $y > 0$, $z > 0$, $Ax + By + Cz = 1$, what must this maximum value be?
(e) Describe this set $\{x > 0, y > 0, z > 0, Ax + By + Cz = 1\}$ geometrically and evaluate $x^A y^B z^C$ on its boundary. Assuming there is a minimum, does it occur on the interior or does it occur on the boundary?
(f) Find the maximum value of $x_1^{A_1} x_2^{A_2} \ldots x_n^{A_n}$ on $A_1 x_1 + \cdots + A_n x_n = K$ $(A_i > 0, x_i > 0, K > 0)$ and state the result as an inequality. [First assume $\sum A_i = 1$ and then reduce the general case to this case.]
(g) Compare the *geometric mean* $\sqrt[n]{x_1 x_2 \ldots x_n}$ and the *arithmetic mean* $\frac{1}{n}(x_1 + x_2 + \cdots + x_n)$ of a set of n positive numbers.
(h) When can equality hold in the inequality of (f)?

7 (a) Sketch very roughly the curve $|x|^p + |y|^p = 1$ for $p = 2, 4, 999, \frac{3}{2}, \frac{1000}{999}, 1$.
(b) Let $|(x, y)|_\infty = \lim_{p\to\infty} |(x, y)|_p = \lim_{p\to\infty} [|x|^p + |y|^p]^{1/p}$. How else can $|(x, y)|_\infty$ be described?

(c) For very large values of p, approximately where on the curve $|x|^p + |y|^p = 1$ does $3x + 10y$ assume its largest value and approximately what is that value?

(d) Approximately where does $2x - 7y$ assume its largest value on $|x|^p + |y|^p = 1$ (p large), and approximately what is that value?

(e) What is $\lim_{p \to \infty}$ of the largest value of $Ax + By$ on $|x|^p + |y|^p = 1$ and near what point or points does $Ax + By$ have this value? Draw a diagram. Consider in particular the case $A = 0$, $B \neq 0$.

8 (a) Find the maximum and minimum values of a linear form $A_1x_1 + A_2x_2 + \cdots + A_nx_n$ on the 'cube' $|x_i| \leq 1$ $(i = 1, 2, \ldots, n)$.

(b) At what points is the maximum assumed?

(c) Comparing to Exercise 7, state the Hölder inequality in the case $p = 1, q = \infty$.

(d) What is the maximum value of $A_1x_1 + A_2x_2 + \cdots + A_nx_n$ on the 'octahedron' $|x_1| + |x_2| + \cdots + |x_n| = 1$?

9 (a) Prove that for any two points (x_1, x_2) and (y_1, y_2) in the plane the inequality

$$[(x_1 + y_1)^4 + (x_2 + y_2)^4]^{1/4} \leq [x_1^4 + x_2^4]^{1/4} + [y_1^4 + y_2^4]^{1/4}$$

holds.

(b) When does equality hold?

(c) Show that for any two n-tuples $x = (x_1, x_2, \ldots, x_n)$, $y = (y_1, y_2, \ldots, y_n)$ the inequality $|x + y|_p = |x|_p + |y|_p$ holds; that is

$$\left[\sum_{i=1}^{n} |x_i + y_i|^p\right]^{1/p} \leq \left[\sum_{i=1}^{n} |x_i|^p\right]^{1/p} + \left[\sum_{i=1}^{n} |y_i|^p\right]^{1/p}.$$

This is known as *Minkowski's Inequality*.

(d) When does equality hold?

10 Is the quadratic form

$$Q(x, y) = 5x^2 + 6xy + 2y^2$$

elliptic, hyperbolic, or parabolic? Sketch the curves $Q =$ const. showing axes and maximum and minimum values of Q on the circle $x^2 + y^2 = 1$.

11 Sketch the curves $Q =$ const. for

$$Q(x, y) = 3x^2 + 4xy + y^2$$

as in Exercise 10.

12 If $Q(x, y, z)$ is a quadratic form in three variables, what are the possible configurations of the surfaces $Q =$ const.? Give examples and verbal descriptions. [The terms 'ellipsoid', 'hyperboloid of one sheet' and 'hyperboloid of two sheets' will be useful.]

13 Show that the constraints

$$x^2 + y^2 + z^2 = 1$$
$$x_1x + y_1y + z_1z = 0$$

(x_1, y_1, z_1 = numbers not all zero) are non-singular of rank 2.

14 Let $Q(x, y, z)$ be a quadratic form in three variables and let (x_i, y_i, z_i), $(i = 1, 2, 3)$ be mutually orthogonal critical points of Q. The numbers (u, v, w) in (18) can be regarded as *new coordinates* on $\mathbf{R}^3$.

(a) Find the equation of the sphere $x^2 + y^2 + z^2 = 1$ in uvw-coordinates.
(b) Find the expression of $Q(x, y, z)$ in uvw-coordinates.
(c) Find all critical points of Q on the ball $u^2 + v^2 + w^2 \leq 1$ in uvw-coordinates.

15 Prove the test for positive definiteness stated in the text.

(a) Let Q be a quadratic form in n variables ($n = 1, 2, 3$). Prove first that the polynomial analogous to (16) has a root ≤ 0 if its constant term is ≤ 0. Conclude that *if* Q is positive definite, *then* all the determinants in the test must be positive.
(b) The proof of the converse is by induction. Since $A > 0$ is necessary and sufficient for Ax^2 to be positive definite, the test works for $n = 1$. For $n = 2$ the formula $AB - c^2 = \lambda_1\lambda_2$ shows that if the determinant $AB - c^2$ is positive then either λ_1, λ_2 are both positive or both are negative. If both are negative, then the maximum value on $x^2 + y^2 = 1$ is negative, hence A is negative and the desired conclusion follows. For $n = 3$, if the numbers $\lambda_1, \lambda_2, \lambda_3$ are distinct then the polynomial $(\lambda_1 - \lambda)(\lambda_2 - \lambda)(\lambda_3 - \lambda)$ has the same roots and the same leading coefficient as the polynomial (16); hence they must be identical and, equating their constant terms.

$$\lambda_1\lambda_2\lambda_3 = \begin{vmatrix} A & c & b \\ c & B & a \\ b & a & C \end{vmatrix}.$$

This argument is no longer valid if two of the λ_i coincide, but the formula still holds and can be proved by proving

$$\begin{pmatrix} A & c & b \\ c & B & a \\ b & a & C \end{pmatrix}\begin{pmatrix} x_1 & x_2 & x_3 \\ y_1 & y_2 & y_3 \\ z_1 & z_2 & z_3 \end{pmatrix} = \begin{pmatrix} x_1 & x_2 & x_3 \\ y_1 & y_2 & y_3 \\ z_1 & z_2 & z_3 \end{pmatrix}\begin{pmatrix} \lambda_1 & 0 & 0 \\ 0 & \lambda_2 & 0 \\ 0 & 0 & \lambda_3 \end{pmatrix}.$$

Thus if the determinant is positive then either Q is positive at all critical points on the sphere (hence has a positive minimum) or *two* of the numbers $\lambda_1, \lambda_2, \lambda_3$ are negative. But if two of them are negative—say λ_1, λ_2—then Q is easily shown to be negative on the entire circle

$$x = ux_1 + vx_2$$
$$y = uy_1 + vy_2$$
$$z = uz_1 + vz_2$$
$$x^2 + y^2 + z^2 = 1$$

and hence to be negative somewhere on the plane $z = 0$; that is, $Q(x, y, 0)$ is not positive definite. Therefore if $Q(x, y, 0)$ is positive definite and if the determinant is positive then $Q(x, y, z)$ is positive definite.

16 Give a complete proof *ab initio* that if $Q(x_1, x_2, \ldots, x_n)$ is a quadratic form in n variables, say

$$Q(x_1, \ldots, x_n) = \sum_{i,j=1}^{n} A_{ij}x_ix_j$$

where $A_{ij} = A_{ji}$, if $Q(x_1, \ldots, x_{n-1}, 0)$ is a positive definite quadratic form in $n - 1$ variables, and if $\det(A_{ij}) > 0$, then Q is positive definite.

17 A quadratic form in n variables is said to be *positive semi-definite* if it assumes no negative values. State and prove a necessary and sufficient condition for a quadratic form to be positive semi-definite.

18 *Second derivative test*. Let $F(x, y, z)$ be a twice differentiable function—that is, a function whose first partial derivatives have (continuous) first partial derivatives—and let $(\bar{x}, \bar{y}, \bar{z})$ be a critical point of F. Let

$$A = \frac{1}{2}\frac{\partial^2 F}{\partial x^2}(\bar{x}, \bar{y}, \bar{z}) \qquad B = \frac{1}{2}\frac{\partial^2 F}{\partial y^2}(\bar{x}, \bar{y}, \bar{z}) \qquad C = \frac{1}{2}\frac{\partial^2 F}{\partial z^2}(\bar{x}, \bar{y}, \bar{z})$$

$$a = \frac{1}{2}\frac{\partial^2 F}{\partial y \partial z}(\bar{x}, \bar{y}, \bar{z}) = \frac{1}{2}\frac{\partial^2 F}{\partial z \partial y}(\bar{x}, \bar{y}, \bar{z})$$

$$b = \frac{1}{2}\frac{\partial^2 F}{\partial x \partial z}(\bar{x}, \bar{y}, \bar{z}) = \frac{1}{2}\frac{\partial^2 F}{\partial z \partial x}(\bar{x}, \bar{y}, \bar{z})$$

$$c = \frac{1}{2}\frac{\partial^2 F}{\partial x \partial y}(\bar{x}, \bar{y}, \bar{z}) = \frac{1}{2}\frac{\partial^2 F}{\partial y \partial x}(\bar{x}, \bar{y}, \bar{z})$$

(using the equality of the mixed partials—see Exercise 6,

§3.2), and let $Q(x, y, z)$ be the corresponding quadratic form.

(a) Show that

$$F(\bar{x} + su, \bar{y} + sv, \bar{z} + sw) = F(\bar{x}, \bar{y}, \bar{z}) + s^2 Q(u, v, w) + 0(s^2)$$

that is, show that given a tolerance $\epsilon > 0$ and a bound $B > 0$ there is an $S > 0$ such that

$$\left| \frac{F(\bar{x} + su, \bar{y} + sv, \bar{z} + sw) - F(\bar{x}, \bar{y}, \bar{z})}{s^2} - Q(u, v, w) \right| < \epsilon$$

whenever $|u| < B$, $|v| < B$, $|w| < B$, $0 < s < S$. [Estimate the first partials of F with an error less than ϵs, then the value of F with an error less than ϵs^2.]

(b) Conclude that if Q is positive definite then there is a ball $(x - \bar{x})^2 + (y - \bar{y})^2 + (z - \bar{z})^2 \leq \delta^2$ on which $F(\bar{x}, \bar{y}, \bar{z})$ is the unique minimum value of F.

(c) Conclude that if there is a ball $(x - \bar{x})^2 + (y - \bar{y})^2 + (z - \bar{z})^2 \leq \delta^2$ on which $F(\bar{x}, \bar{y}, \bar{z})$ is the unique minimum value of F then Q is positive semi-definite.

(d) Suppose the numbers

$$\Delta_3 = \begin{vmatrix} A & c & b \\ c & B & a \\ b & a & C \end{vmatrix}, \quad \Delta_2 = \begin{vmatrix} A & c \\ c & B \end{vmatrix}, \quad \Delta_1 = A$$

are known. For which values $(\Delta_3, \Delta_2, \Delta_1)$ can one conclude that there is a ball $(x - \bar{x})^2 + (y - \bar{y})^2 + (z - \bar{z})^2 \leq \delta^2$ on which $F(\bar{x}, \bar{y}, \bar{z})$ is a unique minimum of F? For which values $(\Delta_3, \Delta_2, \Delta_1)$ can one conclude that this is *not* the case? For which values $(\Delta_3, \Delta_2, \Delta_1)$ can neither conclusion be drawn?

19 *Isoperimetric inequality.* Given n points $P_1 = (x_1, y_1)$, $P_2 = (x_2, y_2), \ldots, P_n = (x_n, y_n)$ in the xy-plane, set $P_0 = P_n$ and imagine $P_1P_2 \ldots P_n$ as describing a closed, oriented, n-sided polygonal curve in the plane. The length of this curve is

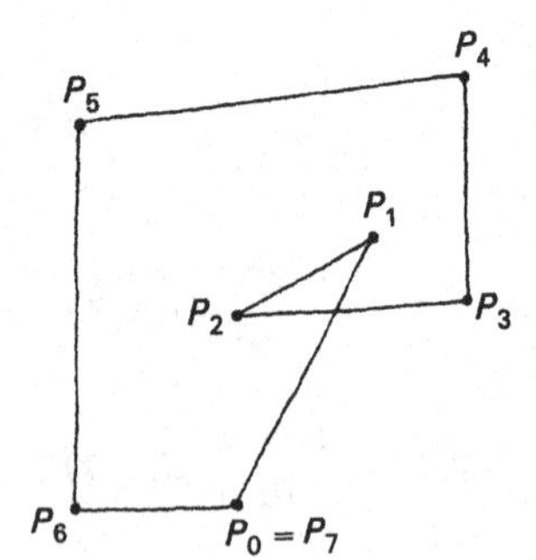

$$L = \sum_{i=1}^{n} |P_iP_{i-1}| = \sum_{i=1}^{n} \sqrt{(x_i - x_{i-1})^2 + (y_i - y_{i-1})^2}$$

and the oriented area it encloses is

$$A = \sum_{i=1}^{n} [\text{oriented area of } OP_{i-1}P_i]$$

$$= \frac{1}{2} \sum_{i=1}^{n} \begin{vmatrix} x_{i-1} & y_{i-1} \\ x_i & y_i \end{vmatrix}.$$

Show that if a given n-gon has a maximum (or minimum) value of A among all polygons with the same L then it must be a regular n-gon. Hence, assuming that the problem $A = \text{max.}, L = \text{const.}$ has a solution, the value of A for any n-gon is at most the area of the regular n-gon with the same L, that is,

$$A \leq \frac{1}{4n}\left(\cot\frac{\pi}{n}\right)L^2.$$

As $n \to \infty$ this gives

$$A \leq \frac{1}{4\pi}L^2$$

as an upper bound on the area A of a polygon (of any number of sides) whose perimeter is L. This is called the *isoperimetric inequality*. [If $n = 2$ the inequality $A \leq \frac{1}{4n}\left(\cot\frac{\pi}{n}\right)L^2$ holds trivially. Suppose it has been proved for all values less than a given value of n, and suppose a polygon $(x_1, y_1, x_2, y_2, \ldots, x_n, y_n)$ is given such that $A = \text{max.}$ for $L = \text{const.}$ It is to be shown that this must be a regular n-gon. First use the inductive hypothesis and the fact that $\frac{1}{4n}\left(\cot\frac{\pi}{n}\right)L^2$ increases as n increases to show that the given n-gon (assumed to be a maximum) cannot have the property that two consecutive vertices (x_{i-1}, y_{i-1}), (x_i, y_i) coincide. Conclude that L is differentiable at the point $(x_1, y_1, x_2, y_2, \ldots, x_n, y_n)$ and that the method of Lagrange multipliers applies to the problem $A = \text{max.}$, $L = \text{const.}$ Express the resulting equations $dA = \lambda\, dL$ in terms of $u_i = x_i - x_{i-1}$, $v_i = y_i - y_{i-1}$, and $l_i = \sqrt{u_i^2 + v_i^2}$. Derive two different expressions for $(u_{i+1} + u_i)(u_{i+1} - u_i) + (v_{i+1} + v_i)(v_{i+1} - v_i)$ and conclude that $l_i = l_{i+1}$, i.e. all sides of the given polygon must have the same length $l = L/n$. Then simplify and solve for u_{i+1}, v_{i+1} as functions of u_i, v_i, λ (and n, L) of the form

$$\begin{pmatrix} u_{i+1} \\ v_{i+1} \end{pmatrix} = M(\lambda)\begin{pmatrix} u_i \\ v_i \end{pmatrix}$$

where $M(\lambda)$ is a 2×2 matrix depending on λ. Use the formulas for $M(\lambda)$ to conclude that

$$M(\lambda) = \begin{pmatrix} \cos\theta & -\sin\theta \\ \sin\theta & \cos\theta \end{pmatrix}$$

for some number θ in the interval $\{0 < \theta < 2\pi\}$. From

$$[M(\lambda)]^n\begin{pmatrix} u_1 \\ v_1 \end{pmatrix} = \begin{pmatrix} u_1 \\ v_1 \end{pmatrix}$$

conclude that $\theta = j \cdot 2\pi/n$ for some integer j, $0 < j < n$. Given θ, which must be one of these $n - 1$ values, the first side (x_0, y_0), (x_1, y_1) of the polygon determines the rest of the polygon. Find A as a function of j and of

$$L = n\sqrt{(x_1 - x_0)^2 + (y_1 - y_0)^2}.$$

Conclude that $j = 1$ and that therefore the polygon is a regular n-gon.]

5.5

Summary. Differentiable Manifolds

A function $f: \mathbf{R}^n \to \mathbf{R}^m$ is said to be *differentiable** if it has the property that its mn first partial derivatives exist and are continuous. That is, a function

$$(1) \qquad y_i = f_i(x_1, x_2, \ldots, x_n) \qquad (i = 1, 2, \ldots, m)$$

is differentiable if

$$\lim_{h \to 0} \frac{f_i(x_1, \ldots, x_{j-1}, x_j + h, x_{j+1}, \ldots, x_n) - f_i(x_1, x_2, \ldots, x_n)}{h}$$

exists and depends continuously on $(x_1, x_2, \ldots, x_n)$ for all i, j. When this is the case, these limits (which are functions of $x_1, x_2, \ldots, x_n$) are denoted $\frac{\partial y_i}{\partial x_j}$. The *pullback* of the 1-form dy_i on the range $\mathbf{R}^m$ is defined to be the 1-form

$$(2) \quad dy_i = \frac{\partial y_i}{\partial x_1} dx_1 + \frac{\partial y_i}{\partial x_2} dx_2 + \cdots + \frac{\partial y_i}{\partial x_n} dx_n$$

on the domain $\mathbf{R}^n$. The pullback of a (variable) k-form

$$(3) \qquad A(y_1, y_2, \ldots, y_m)\, dy_1\, dy_2 \ldots dy_k + \cdots$$

on the range $\mathbf{R}^m$ under the differentiable map (1) is defined to be the k-form in $x_1, x_2, \ldots, x_n$ obtained by performing the substitutions (1) and (2) in (3) and using the usual rules ($dx_i\, dx_j = -dx_j\, dx_i$, $dx_i\, dx_i = 0$) for computing sums and products of 1-forms in $x_1, x_2, \ldots, x_n$. The result is a (variable) k-form on the domain $\mathbf{R}^n$ of the map (1).

The principal theorem of differential calculus is the *Chain Rule*, which states that the composition $\mathbf{R}^n \to \mathbf{R}^m \to \mathbf{R}^p$ of two differentiable maps is a differentiable map and that the pullback of a k-form under the composed map is equal to the pullback of the pullback.

In solving a system of equations (1) for $(x_1, x_2, \ldots, x_n)$

*The concept defined here is also called 'continuously differentiable' or 'C^1-differentiable'. In this book the word 'differentiable' is used only in this sense unless otherwise stated.

given $(y_1, y_2, \ldots, y_m)$ by a process of step-by-step elimination one uses the following theorem, which is proved in §7.1.

Elimination Theorem

Given an equation of the form

(4) $$y = f(x_1, x_2, \ldots, x_n)$$

in which f is a differentiable function and given a point

$$\bar{y} = f(\bar{x}_1, \bar{x}_2, \ldots, \bar{x}_n)$$

at which $\dfrac{\partial y}{\partial x_1} \neq 0$, there exist a number $\epsilon > 0$ and a differentiable function $g(y, x_2, x_3, \ldots, x_n)$ such that the equation

(5) $$x_1 = g(y, x_2, x_3, \ldots, x_n)$$

is equivalent to (4) at all points within ϵ of $(\bar{y}, \bar{x}_1, \bar{x}_2, \ldots, \bar{x}_n)$, that is, f and g are both defined at all points $(y, x_1, x_2, \ldots, x_n)$ where $|y - \bar{y}| < \epsilon$, $|x_1 - \bar{x}_1| < \epsilon$, $\ldots, |x_n - \bar{x}_n| < \epsilon$ and such a point satisfies (4) if and only if it satisfies (5).

Using the Elimination Theorem and the Chain Rule, one can prove the *Implicit Function Theorem:* A system of equations (1) can be reduced to the form

(2a) $$x_i = g_i(y_1, y_2, \ldots, y_r, x_{r+1}, \ldots, x_n) \qquad (i = 1, 2, \ldots, r)$$

(2b) $$y_i = h_i(y_1, y_2, \ldots, y_r) \qquad (i = r + 1, \ldots, k)$$

locally near a given point $(\bar{x}_1, \bar{x}_2, \ldots, \bar{x}_n)$ by differentiable functions g_i, h_i if and only if (i) the coefficient of $dx_1\, dx_2 \ldots dx_r$ in the pullback of $dy_1\, dy_2 \ldots dy_r$ under (1) is not zero at $(\bar{x}_1, \bar{x}_2, \ldots, \bar{x}_n)$ and (ii) the pullback of any k-form under (1) is identically zero near $(\bar{x}_1, \bar{x}_2, \ldots, \bar{x}_n)$ when $k > r$.

The map (1) is said to be *non-singular of rank r* at the point $(\bar{x}_1, \bar{x}_2, \ldots, \bar{x}_n)$ if (i) and (ii) of the Implicit Function Theorem are satisfied by some rearrangement of the x's and y's. This is true if and only if the pullback of some r-form is not zero at $(\bar{x}_1, \bar{x}_2, \ldots, \bar{x}_n)$, but the pullback of every $(r + 1)$-form is identically zero near $(\bar{x}_1, \bar{x}_2, \ldots, \bar{x}_n)$. If this is not the case for any r then $(\bar{x}_1, \bar{x}_2, \ldots, \bar{x}_n)$ is said to be a *singularity* of the map (1).

Just as the geometrical meaning of the Implicit Function Theorem for affine maps was clarified by the concept of an 'affine manifold', the geometrical meaning of the Implicit Function Theorem for differentiable maps is clarified by the concept of a 'differentiable manifold': A *k-dimensional differentiable manifold in* $\mathbf{R}^n$ *defined by local parameters* is a subset M of $\mathbf{R}^n$ with the property that for each point P of M there is a number $\epsilon > 0$ and a differentiable map f: $\mathbf{R}^k \to \mathbf{R}^n$ which is non-singular of rank k at all points where it is defined and which has the property that a point of $\mathbf{R}^n$ which lies within ϵ of P is in M if and only if it is in the image of f. In short, M is parameterized near P by k independent real variables. A *k-dimensional differentiable manifold in* $\mathbf{R}^n$ *defined by local equations* is a subset M of $\mathbf{R}^n$ with the property that for each point P of M there is an $\epsilon > 0$ and a differentiable map f: $\mathbf{R}^n \to \mathbf{R}^{n-k}$ which is defined and non-singular of rank $n - k$ at all points Q within ϵ of P, and which has the property that such a point Q is in M if and only if $f(Q) = f(P)$. In short, M is defined near P by $n - k$ independent equations.

The Implicit Function Theorem shows that every manifold defined by local parameters can be defined by local equations and vice versa (Exercise 1); hence the concept of a *k-dimensional differentiable manifold in* $\mathbf{R}^n$ is well defined.

In terms of this concept, the equations (2b) of the Implicit Function Theorem can be interpreted as stating that the *image* of a non-singular differentiable map is locally a differentiable manifold. The equations (2b) define the image both by parameters $(y_1, y_2, \ldots, y_r)$ and by equations $(y_i - h_i(y_1, y_2, \ldots, y_r) = 0)$. The dimension of the image is equal to the rank of the map. Similarly, the equations (2a) state that the *level surfaces* of a non-singular differentiable map are differentiable manifolds. The equations (2a) define the level surfaces both by parameters $(x_{r+1}, x_{r+2}, \ldots, x_n)$ and by equations $(x_i - g_i(y_1, y_2, \ldots, y_r, x_{r+1}, \ldots, x_n) = 0$ where $(y_1, y_2, \ldots, y_r)$ is fixed). The dimension of the level surfaces is $n - r$ where r is the rank of the map.

The *method of Lagrange Multipliers* can be stated in terms of the concept of 'differentiable manifold' as follows: Let M be a differentiable manifold in $\mathbf{R}^n$, let

$y = f(x_1, x_2, \ldots, x_n)$ be a differentiable function defined at all points of M, and consider the problem

$$P \text{ in } M$$
$$y = \text{max. at } P.$$

If P is a solution of this problem and if $X_1, X_2, \ldots, X_k$ are differentiable functions such that M near P is defined by the equations

$$X_i = \text{const.} \qquad (i = 1, 2, \ldots, k) \tag{6}$$

(hence M is $(n - k)$-dimensional) then there exist numbers $\lambda_1, \lambda_2, \ldots, \lambda_k$ such that

$$dy = \lambda_1\, dX_1 + \lambda_2\, dX_2 + \cdots + \lambda_k\, dX_k \tag{7}$$

at P. The equations (6) and (7) give $n + k$ equations in the $n + k$ unknowns $(x_1, x_2, \ldots, x_n, \lambda_1, \lambda_2, \ldots, \lambda_k)$ where $(x_1, x_2, \ldots, x_n)$ are the coordinates of P. These equations can be solved explicitly to find the coordinates of P in certain simple cases, and can be used to deduce useful statements about the possible solutions P in many others. The equation (7) was deduced from the Implicit Function Theorem in §5.4.

Exercises

1 Prove that a k-dimensional differentiable manifold defined by local parameters can be defined by local equations and vice versa.

2 Show that the equation $x_1^2 + x_2^2 + \cdots + x_n^2 = K$ defines a differentiable manifold in $\mathbf{R}^n$ whose dimension is $n - 1$ when $K > 0$ and whose dimension is 0 when $K = 0$.

3 Is the folium of Descartes [Exercise 9, §5.1] a 1-dimensional differentiable manifold in the xy-plane?

4 Prove that the set of all invertible 2×2 matrices

$$\begin{pmatrix} x_1 & x_2 \\ x_3 & x_4 \end{pmatrix}$$

is a differentiable manifold in $\mathbf{R}^4$. What is its dimension?

5 A 2×2 matrix is said to be an *orthogonal* matrix if it satisfies

$$\begin{pmatrix} x_1 & x_2 \\ x_3 & x_4 \end{pmatrix}\begin{pmatrix} x_1 & x_3 \\ x_2 & x_4 \end{pmatrix} = \begin{pmatrix} 1 & 0 \\ 0 & 1 \end{pmatrix},$$

that is, if its transpose is equal to its inverse. Prove that the set of all 2×2 orthogonal matrices is a differentiable manifold in $\mathbf{R}^4$. What is its dimension? Give an explicit definition of this manifold by parameters near $\begin{pmatrix} 1 & 0 \\ 0 & 1 \end{pmatrix}$.

6 An $n \times n$ matrix is said to be an *orthogonal* matrix if its transpose is equal to its inverse. [For instance, the equation (17) of §5.4 says that a certain 3×3 matrix is an orthogonal matrix.] Show that the set of orthogonal matrices is a differentiable manifold in $\mathbf{R}^{n^2}$ and find its dimension.

7 *Envelopes.* An equation of the form $f(x, y, \alpha) = \text{const.}$ can be imagined as defining a curve in the xy-plane for each fixed value of α; as α varies the equation then defines a *family of curves* in the xy-plane. For example,

$$(*) \qquad (x - \alpha)^2 + y^2 = 1$$

describes the family of circles in the xy-plane whose radii are 1 and whose centers lie on the x-axis. An *envelope* of a family of curves in the xy-plane is a curve C with the property that for each point P of C there is a curve of the family through P tangent to C. [Thus, in particular, any curve of the family is an envelope of the family.] The standard method of finding envelopes C of $f(x, y, \alpha) = \text{const.}$ which are not themselves curves of the family is to eliminate α from the equations

$$f(x, y, \alpha) = \text{const.}$$
$$f_\alpha(x, y, \alpha) = 0$$

where f_α denotes the partial derivative of f with respect to α. This gives an equation of the form $g(x, y) = \text{const.}$ which defines an envelope of the family.

(a) Apply this process to find envelopes of the family (*) above.

(b) Prove that this process indeed gives an envelope under the following conditions: If $(\bar{x}, \bar{y}, \bar{\alpha})$ is a point where the equations $f = \text{const.}$, $f_\alpha = 0$ are satisfied, and if f_α is a differentiable function with $f_{\alpha\alpha} \neq 0$ near $(\bar{x}, \bar{y}, \bar{\alpha})$ then the equation $f_\alpha(x, y, \alpha) = 0$ can be solved locally near $(\bar{x}, \bar{y}, \bar{\alpha})$ to give α as a function of x and y. Substituting in the equation $f(x, y, \alpha) = \text{const.}$ gives an equation of the form $g(x, y) = \text{const.}$ If the determinant of partial derivatives

$$\begin{vmatrix} f_x & f_y \\ f_{x\alpha} & f_{y\alpha} \end{vmatrix}$$

is not zero near $(\bar{x}, \bar{y}, \bar{\alpha})$ then the curve $g(x, y) = \text{const.}$ is an envelope.

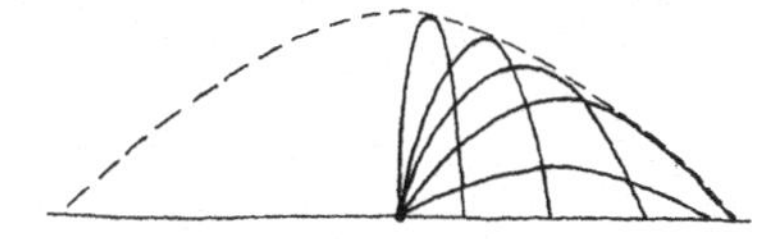

(c) The trajectory of a shell fired from (0, 0) is the curve

$$(\dagger) \qquad \begin{aligned} x &= v_x t \\ y &= v_y t - \tfrac{1}{2} g t^2 \end{aligned}$$

where (v_x, v_y) is the velocity with which the shell is fired and g is the acceleration of gravity. Show that the family of all trajectories (†) for a fixed value of $v = \sqrt{v_x^2 + v_y^2}$ has an envelope which is a parabola. This is the 'parabola of safety' beyond which the shell cannot be fired.*

*For further examples of envelopes see W. F. Osgood, Advanced Calculus, The Macmillan Co., 1925, Chapter VIII.

integral calculus

chapter 6

6.1

Summary

The integral $\int_S \omega$ was defined in Chapter 2 for ω a continuous k-form on n-space and for S a k-dimensional domain in n-space, parameterized on a bounded domain D of k-space. (In Chapter 2 only the cases $n \leq 3$ were considered, but the same definitions apply, with only minor modifications, to cases where $n > 3$.) The definition had very serious defects: it was necessary to parameterize a domain in order to define integrals over it, and it was never proved that such a parameterization can be given nor that the resulting number $\int_S \omega$ is independent of the choice of the parameterization. This chapter is devoted to the rigorous definition of $\int_S \omega$ for a continuous k-form ω on n-space and for a suitably general class of k-dimensional domains S in n-space—namely, the class of compact, oriented, differentiable, k-dimensional manifolds-with-boundary.

The subject of this chapter is merely the *definition* of $\int_S \omega$, an intuitive definition of which is very easily given in simple cases, and the *proof* that this definition has all the properties which would have been expected on the basis of the intuitive definition. Such definitions and proofs are of great theoretical importance, but have little practical significance. In practice, integrals $\int_S \omega$ are rarely evaluated. The *integrand* (the field) and the *concept* of integration (which gives the field its meaning) are the important ideas, and the actual number $\int_S \omega$ is usually of no interest. When it is actually necessary to perform an integration, the domain of integration S (as well as the integrand ω) must be relatively simple for there to be any hope of success—simple enough that the parameterization of S, and hence a definition of $\int_S \omega$, would present no problem.

6.2 *k-Dimensional Volume*

The intuitive meaning of the 2-form $dx\,dy$ on xyz-space is 'oriented area of the projection on the xy-plane'—a function assigning numbers to surfaces in xyz-space. It was on the basis of this intuitive idea that the algebraic rules governing 2-forms and their pullbacks under affine maps were derived in Chapter 1. Similarly, the algebra of 3-forms was based on 'oriented volume'. In Chapter 4, the algebra of k-forms was defined as a natural extension of the algebra of 2-forms and 3-forms, and this algebra was found to be very useful in stating and proving such basic theorems as the Chain Rule, the Implicit Function Theorem, and the method of Lagrange multipliers in Chapter 5. However, it has not yet been proved that 2-forms actually do describe areas or that 3-forms describe volumes, when areas and volumes are defined—as they must be—by integrals. This section is devoted to proving that the pullback operation on k-forms, as defined algebraically in Chapter 4, does indeed have a meaning in terms of 'k-dimensional volume'*, as defined by an integral in the obvious way:

**Of course 'volume' normally means 'three-dimensional'. Note that 'one-dimensional volume' is length and 'two-dimensional volume' is area. The term 'k-dimensional content' is often used instead of 'k-dimensional volume'.*

Definition

Let D be a bounded subset of $x_1x_2 \ldots x_k$-space. The *k-dimensional volume* of D, denoted $\int_D dx_1\,dx_2 \ldots dx_k$, is defined as follows: Let B be a number such that all coordinates of all points of D are less than B in absolute value. In other words, let B be a number such that D is contained in the k-dimensional cube $\{(x_1, x_2, \ldots, x_k)\colon |x_i| \le B, i = 1, 2, \ldots, k\}$. An *approximating sum* $\sum(\alpha)$ to $\int_D dx_1\,dx_2 \ldots dx_k$ is formed by choosing

(α)
- (i) a subdivision of each of the k intervals $\{-B \le x_i \le B\}$ into small subintervals, thereby subdividing the cube $\{|x_i| \le B\}$ into k-dimensional 'rectangles' which will be denoted generically by R_α, and
- (ii) a point P_α in each of the 'rectangles' R_α,

and by setting

$$\sum(\alpha) = \sum_{P_\alpha \text{ is in } D} k\text{-volume of } (R_\alpha)$$

where the k-volume of R_α is defined to be the product of the k dimensions of R_α. The *mesh size* of the choices α, denoted $|\alpha|$, is defined to be the largest number which occurs as one of the dimensions of one of the rectangles R_α. The integral $\int_D dx_1\, dx_2 \ldots dx_k$ is said to *converge* if the Cauchy Convergence Criterion is satisfied:

For every $\epsilon > 0$ there is a $\delta > 0$ such that $|\sum(\alpha) - \sum(\alpha')| < \epsilon$ whenever $|\alpha| < \delta$, $|\alpha'| < \delta$.

When this is the case the approximating sums determine a real number, called the k-dimensional volume of D and denoted by $\int_D dx_1\, dx_2 \ldots dx_k$.* When this is not the case it is said that $\int_D dx_1\, dx_2 \ldots dx_k$ does not exist or that the k-dimensional volume of D is not defined.

*To find $\int_D dx_1\, dx_2 \ldots dx_k$ to an accuracy of n decimal places set $\epsilon = 10^{-n}$, choose δ as in the Cauchy Criterion, choose α with $|\alpha| < \delta$, and compute $\sum(\alpha)$.

Theorem

Let

$$(1) \qquad y_i = \sum_{j=1}^{k} a_{ij}x_j + b_i \qquad (i = 1, 2, \ldots, k)$$

be an affine map $\mathbf{R}^k \to \mathbf{R}^k$, let D be a bounded domain in $x_1x_2 \ldots x_k$-space such that $\int_D dx_1\, dx_2 \ldots dx_k$ converges, and let $f(D)$ denote the image of D under the map (1). Then $\int_{f(D)} dy_1\, dy_2 \ldots dy_k$ converges and

$$\int_{f(D)} dy_1\, dy_2 \ldots dy_k = \pm \frac{\partial(y_1, y_2, \ldots, y_k)}{\partial(x_1, x_2, \ldots, x_k)} \int_D dx_1\, dx_2 \ldots dx_k$$

where the sign is determined by the condition that both integrals are by definition not negative.

This theorem gives a precise formulation of the intuitive meaning of the pullback operation, namely, that the pullback is a composed function assigning numbers to k-dimensional domains D in $x_1x_2 \ldots x_k$-space by 'evaluating' the given form $dy_1\, dy_2 \ldots dy_k$ on the image of D under the given map (1). The ambiguity of sign arises from the fact that $dy_1\, dy_2 \ldots dy_k$ and $dx_1\, dx_2 \ldots dx_k$ represent *oriented* volumes, whereas $\int_D dx_1\, dx_2 \ldots dx_k$

and $\int_{f(D)} dy_1\, dy_2 \ldots dy_k$ were defined without reference to any orientation of D or $f(D)$.

Proof

If $f\colon \mathbf{R}^k \to \mathbf{R}^k$ and $g\colon \mathbf{R}^k \to \mathbf{R}^k$ are affine maps for which the theorem is true, then the theorem is obviously true for the affine map $f \circ g$:

$$\int_{g(D)} dy \text{ converges and is } \left|\frac{\partial y}{\partial x}\right| \int_D dx$$

$$\int_{f(g(D))} dz \text{ converges and is } \left|\frac{\partial z}{\partial y}\right| \int_{g(D)} dy$$

$$= \left|\frac{\partial z}{\partial y} \cdot \frac{\partial y}{\partial x}\right| \int_D dx = \left|\frac{\partial z}{\partial x}\right| \int_D dx$$

as desired. Therefore it suffices to show that every affine map $\mathbf{R}^k \to \mathbf{R}^k$ can be written as a composition of simple maps for which the theorem is true. Now every affine map $\mathbf{R}^k \to \mathbf{R}^k$ can be written (see Exercise 4) as a composition of:

(i) Translations $y_i = x_i + \text{const.}$
(ii) Multiplication of coordinates by scale factors $y_i = \text{const.}\ x_i$ (possibly negative or zero).
(iii) Interchanges of coordinates. (Each y_i is a different x_i.)
(iv) The shear $y_1 = x_1 + x_2, y_2 = x_2, y_3 = x_3, \ldots, y_k = x_k$.

The types (i)–(iii) carry rectangles to rectangles and approximating sums to approximating sums, which fact makes the theorem for these maps an immediate consequence of the definition of $\int_D dy_1\, dy_2 \ldots dy_k$. Thus it suffices to prove the theorem for the map (iv). This can be done by the following steps which are left to the reader to prove:

(a) If D is a rectangle then $\int_{f(D)} dy_1\, dy_2 \ldots dy_k$ converges (where f is the map (iv)).
(b) The integral of (a) is equal to $\int_D dx_1\, dx_2 \ldots dx_k$.
(c) The formula

$$\int_{f(D)} dy_1\, dy_2 \ldots dy_k = \int_D dx_1\, dx_2 \ldots dx_k$$

holds whenever D is a finite union of rectangles.

(d) If $\int_D dx_1\, dx_2 \ldots dx_k$ converges to the number V and if $\epsilon > 0$ then there exist domains $\underline{D}, \overline{D}$ which are finite unions of rectangles such that $\underline{D}$ is contained in D, such that D is contained in $\overline{D}$, and such that

$$V - \epsilon < \int_{\underline{D}} dx_1 \ldots dx_k, \int_{\overline{D}} dx_1 \ldots dx_k < V + \epsilon.$$

(e) All sufficiently fine approximating sums to $\int_{f(D)} dy_1\, dy_2 \ldots dy_k$ are at least $V - 2\epsilon$ and at most $V + 2\epsilon$ where ϵ is arbitrarily small; hence $\int_{f(D)} dy_1\, dy_2 \ldots dy_k$ converges to V, as was to be shown.

Exercises

1 The number π is defined to be the area of the unit circle $\{x^2 + y^2 \leq 1\}$. Find the area of the ellipse

$$\left(\frac{x}{a}\right)^2 + \left(\frac{y}{b}\right)^2 = 1 \qquad a \neq 0, b \neq 0.$$

2 Find the formula for the volume of the tetrahedron with vertices (x_0, y_0, z_0), (x_1, y_1, z_1), (x_2, y_2, z_2), (x_3, y_3, z_3).

3 Prove that if D is the 'unit cube' $\{0 \leq x_i \leq 1;\ i = 1, 2, \ldots k\}$ then $\int_D dx_1\, dx_2 \ldots dx_n$ converges to 1. [Give an explicit estimate of the error $|\Sigma(\alpha) - 1|$ in terms of $|\alpha|$.]

4 Prove that every affine map $\mathbf{R}^k \to \mathbf{R}^k$ can be written as a composition of the types (i)–(iv).

5 Prove the statements (a)–(e) of the text.

6.3

Independence of Parameter and the Definition of $\int_S \omega$

The technical difficulties involved in giving a precise definition of the integral of a 2-form over a surface in 3-space are just as great as those involved in giving a precise definition of the integral of a k-form over a k-dimensional manifold in n-space. Therefore, in order to simplify the notation and in order to make the geometrical ideas involved in the definition as clear as possible, the definitions and proofs will be given for the case $k = 2$, $n = 3$. The definitions and proofs for general k and n are identical to these.

Let $\omega = A\, dy\, dz + B\, dz\, dx + C\, dx\, dy$ be a continuous 2-form on xyz-space and let S be an oriented surface in xyz-space. The greatest difficulty in defining the number $\int_S \omega$ actually lies in defining precisely what is meant by 'an oriented surface in xyz-space'. This difficulty is essentially overcome by the Implicit Function Theorem, which makes it possible to define a 2-dimensional differentiable manifold in xyz-space as a set which locally can be described by two independent parameters or by one non-singular equation. Since the integral of a 2-form over a parameterized surface can be defined (§2.4) to be the integral of the pullback over the parameter space, this leaves just the following problems:

> How can several local parameterizations of a 2-dimensional differentiable manifold S be put together to give a definition of $\int_S \omega$?
> How are orientations to be described?
> How can the number $\int_S \omega$ be proved to be independent of the choices of local parameterizations used to define it?

In §2.5 it was shown that when S is the sphere $\{x^2 + y^2 + z^2 = 1\}$ the integral $\int_S \omega$ can be defined in various ways by cutting S into several pieces (say into two hemispheres), by parameterizing each piece (say by stereographic projection of the hemispheres onto disks), and by defining $\int_S \omega$ to be the sum of the integrals over the pieces. Although this is a natural and simple procedure for a *particular* surface such as the sphere, it is extremely difficult to prove that an arbitrary 2-dimensional differentiable manifold can be cut into simple pieces, each of which can be parameterized. Therefore, for the case of an arbitrary surface, some other solution to the first problem above would be desirable.

Instead of *decomposing* S (writing S as a union of pieces S_i so that $\left(\int_S \omega = \sum_i \int_{S_i} \omega\right)$, one can *decompose* ω $\left(\text{write } \omega = \sum \omega_i \text{ so that } \int_S \omega = \sum_i \int_S \omega_i\right)$. If this can be done in such a way that $\omega = \omega_1 + \omega_2 + \cdots + \omega_n$ is written as a sum of 2-forms ω_i, each of which is identically zero except on a small portion of S which can be parameterized, then $\int_S \omega$ can be defined to be $\sum_i \int_S \omega_i$, and each integral $\int_S \omega_i$ can be defined by parameterizing that portion of S where ω_i is not zero. This process is very unwieldy from a practical standpoint because the only 2-forms which are identically zero except in a small region are very artificial and the integrals $\int_S \omega_i$, although they are easy to define, are impossible to compute. Nonetheless, from a theoretical standpoint the method of decomposing ω is much simpler than that of decomposing S.

The main reason that decomposing ω is simpler than decomposing S is that the parameterized portions may now overlap and in fact should overlap in such a way that every point of S is inside one parameterized portion. Hence there is no need to cut S, and locally S is simply a surface which can be defined by two local parameters. Globally, the definition of $\int_S \omega$ makes two further assumptions about S, namely, that all of S can be covered by a *finite number* of simple parameterized portions, and that these portions can be *oriented* in a consistent way. Specifically, it will be assumed that S is a subset of xyz-space which is described in the following way:

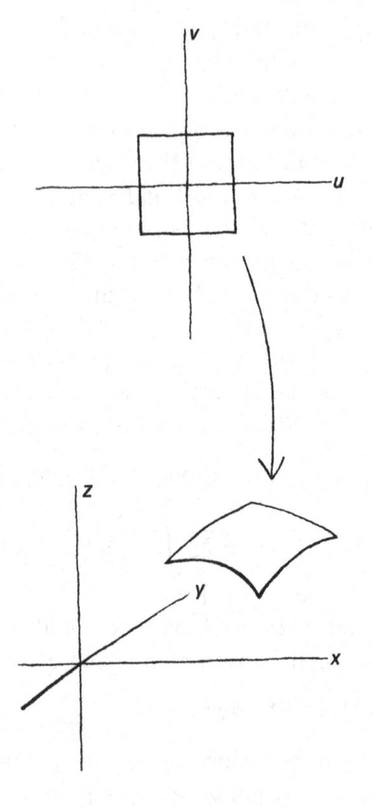

(a) There is given a finite number of differentiable maps $F_1, F_2, \ldots, F_N$ of the uv-plane to xyz-space. Each map F_i is non-singular of rank 2 at all points of the square $\{|u| \leq 1, |v| \leq 1\}$ and carries points of this square one-to-one to points of S. The maps F_i are called 'charts'.

(b) For each point P of S there exists at least one chart F_i such that P lies *inside* the image of $\{|\bar{u}| \leq 1, |\bar{v}| \leq 1\}$ under F_i and such that the surface S near P coincides with the image of F_i. That is, $P = F_i(\bar{u}, \bar{v})$ where $|\bar{u}| < 1$, $|\bar{v}| < 1$ and there is an $\epsilon > 0$ such that a point $Q = (x, y, z)$ within ϵ of $P = (\bar{x}, \bar{y}, \bar{z})$ is a point of S if and only if it is of the form $Q = F_i(u, v)$ for some (u, v) with $|u| \leq 1$, $|v| \leq 1$.

(c) The charts $F_1, F_2, \ldots, F_N$ are *consistently oriented* in the following sense: If P is a point of S which lies inside the image of two different charts F_i, F_j (that is, $P = F_i(\bar{u}, \bar{v})$, $P = F_j(\bar{\bar{u}}, \bar{\bar{v}})$ where $|\bar{u}| < 1$, $|\bar{v}| < 1$, $|\bar{\bar{u}}| < 1$, $|\bar{\bar{v}}| < 1$) then by the Implicit Function Theorem (see Exercise 2) the map $F_j^{-1} \circ F_i: \mathbf{R}^2 \to \mathbf{R}^2$ is defined near $F_i^{-1}(P) = (\bar{u}, \bar{v})$. The charts $F_1, F_2, \ldots, F_N$ are said to be consistently oriented if all such maps $F_j^{-1} \circ F_i$ of the uv-plane to itself preserve the sign of $du\,dv$, in other words, if the pullback of $du\,dv$ under $F_j^{-1} \circ F_i$ near $(\bar{u}, \bar{v})$ is a positive multiple of $du\,dv$. The same condition can be stated more simply (but less geometrically) as follows: If P is a point of S which lies in the image of two different charts F_i, F_j and if ω is a 2-form on xyz-space defined at P, then the pullback of ω under F_i at $F_i^{-1}(P)$ is a positive multiple of the pullback of ω under F_j at $F_j^{-1}(P)$. (For the proof of the equivalence of these two definitions see Exercise 2.)

Example

The sphere $\{x^2 + y^2 + z^2 = 1\}$ oriented by $x\,dy\,dz + y\,dz\,dx + z\,dx\,dy$ can be described by charts in this way. To do this it suffices to use the stereographic projection of the sphere minus the 'north pole' $(0, 0, 1)$ onto the uv-plane and the stereographic projection of the sphere minus the 'south pole' $(0, 0, -1)$ onto the uv-plane. By introducing a scale factor in the uv-plane the square $\{|u| \leq 1, |v| \leq 1\}$ can be made to parameterize all but a small area around the omitted pole and, in particular, each of these charts can be made to parameterize an entire hemisphere. By orienting each of these two charts (e.g. changing u to $-u$ if the orientation is wrong) to agree with $x\,dy\,dz + y\,dz\,dx + z\,dx\,dy$ they agree with each other and the requirements (a)–(c) are fulfilled. More picturesquely, it is useful to think of a geographical atlas, which consists of a finite number of maps, oriented in a consistent way (the earth as seen from above), such that each point of the earth's surface is pictured in at least one map (and not at the edge of the map).

A surface S in xyz-space which can be described by charts $F_1, F_2, \ldots, F_N$ in this way is called a *compact, oriented, differentiable surface* in xyz-space. The word 'compact' is being used here in a technical sense which will not be defined (see §9.4 for this definition*); only the entire phrase 'compact, oriented, differentiable surface' as defined by (a)–(c) will be used here. The word 'oriented' means geometrically that given three nearby non-collinear points $P_0P_1P_2$ of S the rotational direction on S which they describe can be classed as 'positive' or 'negative' according to whether the corresponding points of the uv-plane under some (and hence under any) chart F_i describe a counterclockwise or a clockwise direction. However, 'oriented', like 'compact', will be used here only in the entire phrase above and will not be defined separately. Two sets of charts $F_1, F_2, \ldots, F_N$ and $G_1, G_2, \ldots, G_M$ satisfying (a)–(c) will be said to

*Loosely speaking, 'compact' means exactly what it means in everyday English—firmly united, arranged within a relatively small space.

describe *the same* compact, oriented, differentiable surface if they describe the same set S, and if at any point P of S contained in charts F_i, G_j the orientations of the charts F_i, G_j agree in the sense of (c).

Theorem

Let S be a compact, oriented, differentiable surface in xyz-space, and let $\omega = A\,dy\,dz + B\,dz\,dx + C\,dx\,dy$ be a continuous 2-form on xyz-space which is defined at all points of S. Then a number $\int_S \omega$ depending only on ω and S can be defined as follows: Let $F_1, F_2, \ldots, F_N$ be a specific description of S by charts satisfying (a)–(c) above. Then there exist continuous 2-forms $\omega_1, \omega_2, \ldots, \omega_N$ defined at all points of S such that $\omega = \omega_1 + \omega_2 + \cdots + \omega_N$ and such that ω_i is zero at all points of S other than those which lie in the image under F_i of a square $\{|u| \le 1 - \delta_i, |v| \le 1 - \delta_i\}$ contained in $\{|u| \le 1, |v| \le 1\}$ (δ_i a small positive number). Let $F_i^*(\omega_i)$ denote the pullback of ω_i under the map F_i. Then the integrals

$$\int_{\{|u| \le 1, |v| \le 1\}} F_i^*(\omega_i)$$

in which the domain of integration $\{|u| \le 1, |v| \le 1\}$ is oriented by $du\,dv$, are defined as in §2.3. These integrals all converge, hence the number $\sum_{i=1}^{N} \int F_i^*(\omega_i)$ is defined (all integrals being over $\{|u| \le 1, |v| \le 1\}$ oriented $du\,dv$). This number can be defined to be $\int_S \omega$ because it is in fact independent of the choices; that is, if $G_1, G_2, \ldots, G_M$ is another set of charts describing the same compact, oriented, differentiable surface S and if $\omega = \sigma_1 + \sigma_2 + \cdots + \sigma_M$ where σ_j is a continuous 2-form on S which is zero at all points of S other than the image of $\{|u| \le 1 - d_j, |v| \le 1 - d_j\}$ under G_j (d_j a small positive number) then

$$\sum_{i=1}^{N} \int F_i^*(\omega_i) = \sum_{j=1}^{M} \int G_j^*(\sigma_j) \tag{1}$$

where all integrals are over $\{|u| \le 1, |v| \le 1\}$ oriented $du\,dv$.

Proof

The remainder of this section is devoted to the proof of this theorem. The first statement to be proved is that if $F_1, F_2, \ldots, F_N$ and ω are given, then there is a decomposition $\omega = \omega_1 + \omega_2 + \cdots + \omega_N$ as stated in the theorem.

For each point P of S let c_P be a continuous function on xyz-space such that: (a) all values of c_P are ≥ 0, (b) the value of c_P at P is > 0, and (c) all points of S where $c_P > 0$ are inside any chart F_i which contains P in its inside. More precisely this condition (c) can be formulated: If P is the image of a point inside $\{|u| < 1, |v| < 1\}$ under the chart F_i then there is an $\epsilon > 0$ such that all points of S where $c_P > 0$ are contained in the image of $\{|u| \leq 1 - \epsilon, |v| \leq 1 - \epsilon\}$ under F_i. It is not difficult to prove that for every point P of S there is such a function c_P [Exercise 1].

Lemma

If such a function c_P is chosen for each point P of S then it is possible to select a *finite number* of the functions c_P, say $c_1, c_2, \ldots, c_K$, such that $c_1 + c_2 + \cdots + c_K > 0$ at all points of S.

This lemma is a standard application of the Heine-Borel Theorem (§9.4). It can be proved as follows: If the lemma is false then at least one of the charts $F_1, F_2, \ldots, F_N$ must have the property that no finite number of the c's can be chosen such that their sum is positive on the image of $\{|u| \leq 1, |v| \leq 1\}$ under F_i (because otherwise there would be a finite number for each F_i and the sum of all of these for all i would give a finite number whose sum was positive on all of S). Choose such an F_i. Dividing $\{|u| \leq 1, |v| \leq 1\}$ into four quarters by the lines $u = 0, v = 0$, it follows in the same way that the image of at least one of the four quarters under F_i has the property that no finite number of c's can be chosen whose sum is positive on this subset of S. Choose such a quarter, divide it into quarters, and repeat the argument. One obtains in this way a nested sequence of squares $R_0 \supset R_1 \supset R_2 \supset R_3 \supset \cdots$ such that:

(i) R_0 is $\{|u| \leq 1, |v| \leq 1\}$.
(ii) For each n, the square R_{n+1} is one of the four quarters of the square R_n.
(iii) No finite number of the c's can be chosen such that their sum is positive on the image of R_n under the map F_i.

The squares R_n close down to a single point P_0 in the uv-plane. Let $P = F_i(P_0)$. Then the function c_P is positive near P and the condition (iii) is contradicted for large n because a *single* function c_P is positive on $F_i(R_n)$. This contradicts the assumption that the lemma was false and thereby proves the lemma.

Using the lemma, let $c_1, c_2, \ldots, c_K$ be a finite collection of the c's such that $c_1 + c_2 + \cdots + c_K$ is positive at all points of S, and set

$$a_\mu = \frac{c_\mu}{c_1 + c_2 + \cdots + c_K} \qquad (\mu = 1, 2, \ldots, K).$$

Then a_μ is a continuous function on xyz-space defined at all points of S. The a's have the property that the function $\sum_{\mu=1}^{K} a_\mu$ is identically 1 on S, for which reason they are called a *continuous partition of unity on S*. They can be used to effect the desired decomposition $\omega = \omega_1 + \omega_2 + \cdots + \omega_N$ as follows: For each $\mu = 1, 2, \ldots, K$ it is possible to choose an $i(\mu)$, $1 \leq i(\mu) \leq N$ such that the points of S where $a_\mu > 0$ are all contained in the image of some square $\{|u| \leq 1 - \epsilon, |v| \leq 1 - \epsilon\}$ under the chart $F_{i(\mu)}$. This follows from the fact that a_μ is a multiple of c_μ, from the conditions which were imposed on the c's, and from the fact that every P in S is inside the image of at least one chart. Now set $\omega_i = \sum_{i(\mu)=i} a_\mu \omega$. Then ω_i is zero at all points of S except those which are contained in the image of some square $\{|u| \leq 1 - \epsilon, |v| \leq 1 - \epsilon\}$ under F_i $(i = 1, 2, \ldots, N)$ and $\omega_1 + \omega_2 + \cdots + \omega_N = \sum_{\mu=1}^{N} a_\mu \omega = \omega$ as desired.

The second statement of the theorem to be proved is that the integrals $\int F_i^*(\omega_i)$ converge. Here the integrands are continuous 2-forms and the domain of integration is a square $\{|u| \leq 1, |v| \leq 1\}$. The proof of §2.3 therefore proves that these integrals converge except that the *uniform* continuity of the integrands must be proved. This is easily done using a subdivision argument like the one used to prove the lemma above; this argument is given in §9.4 (Theorem 2).

The hard part of the theorem is of course the final statement (1) that $\int_S \omega$ is independent of the choices. This will be proved by first reducing it to the following simple case and by then proving this case.

Independence of Parameter

Let S be a compact, oriented, differentiable surface in xyz-space described by charts $F_1, F_2, \ldots, F_N$ as above, and let ω be a continuous 2-form defined at all points of S. If two different charts F_i, F_j both have the property that they parameterize the portion of S where ω is not zero—that is, if ω is zero at all points of S which are not in the image of $\{|u| \le 1 - \epsilon, |v| \le 1 - \epsilon\}$ under F_i *and* if ω is zero at all points of S which are not in the image of $\{|u| \le 1 - \epsilon, |v| \le 1 - \epsilon\}$ under F_j (ϵ a small positive number)—then

$$(2) \qquad \int F_i^*(\omega) = \int F_j^*(\omega)$$

where the integrals are over $\{|u| \le 1, |v| \le 1\}$ oriented $du\, dv$.

To reduce the general case (1) to the special case (2) one can argue as follows: If $F_1, F_2, \ldots, F_N$ and $G_1, G_2, \ldots, G_M$ describe the same S, then the set of all $N + M$ charts $F_1, F_2, \ldots, F_N, G_1, G_2, \ldots, G_M$ also describes S, in that it satisfies the conditions (a)–(c). The proof above therefore gives a continuous partition of unity $a_1, a_2, \ldots, a_K$ on S built up out of functions c_P chosen to have the property that if P is inside the image of $\{|u| < 1, |v| < 1\}$ under any chart F_i or G_j, then all points of S where $c_P > 0$ are contained in the image of $\{|u| \le 1 - \epsilon, |v| \le 1 - \epsilon\}$ under this chart. This implies that for every $\mu = 1, 2, \ldots, K$ there are integers $i(\mu)$, $1 \le i(\mu) \le N$, and $j(\mu)$, $1 \le j(\mu) \le M$, such that all points of S where $a_\mu > 0$ are contained in the image of $\{|u| \le 1 - \epsilon, |v| \le 1 - \epsilon\}$ under $F_{i(\mu)}$ as well as in the image of this set under $G_{j(\mu)}$. Now if $\omega = \omega_1 + \omega_2 + \cdots + \omega_N$ is any decomposition of ω as in the theorem, then for all i and μ

$$\int F_i^*(a_\mu \omega_i) = \int F_{i(\mu)}^*(a_\mu \omega_i)$$

by (2) because a_μ is zero outside $F_{i(\mu)}$ and ω_i is zero outside F_i. Summing over all μ gives

$$\int F_i^*(\omega_i) = \sum_{\mu=1}^{K} \int F_{i(\mu)}^*(a_\mu \omega_i)$$

because $\sum_{\mu=1}^{K} a_\mu = 1$. Summing this over i gives

$$\sum_{i=1}^{N} \int F_i^*(\omega_i) = \sum_{\mu=1}^{K} \int F_{i(\mu)}^*(a_\mu \omega)$$

because $\sum_{i=1}^{N} \omega_i = \omega$. Similarly

$$\sum_{j=1}^{M} \int G_j^*(\sigma_j) = \sum_{\mu=1}^{K} \int G_{j(\mu)}^*(a_\mu \omega).$$

Finally, (2) gives

$$\int F_{i(\mu)}^*(a_\mu \omega) = \int G_{j(\mu)}^*(a_\mu \omega)$$

and summing over μ gives (1). This completes the reduction of the formula (1) to the formula (2).

The formula (2) will be proved by the same line of argument used to prove the lemma above; namely, it will be assumed that (2) is false, and this assumption will be contradicted by successively dividing the square $\{|u| \leq 1, |v| \leq 1\}$ into quarters.

To simplify notation, set $F = F_i$ and $G = F_j$. The statement to be proved is that if F and G both parameterize the part of S where $\omega \neq 0$, then $\int F^*(\omega) = \int G^*(\omega)$. Assume this is false and let E be a positive number such that

$$\left| \int F^*(\omega) - \int G^*(\omega) \right| > E.$$

Let R_0 denote the square $\{|u| \leq 1, |v| \leq 1\}$, let R_0 be divided into quarters by the lines $u = 0, v = 0$, and let $\sum(\alpha)$ be an approximating sum to $\int G^*(\omega)$. (Thus α represents a subdivision of $\{|u| \leq 1, |v| \leq 1\}$ into small rectangles R_{ij} and a choice of a point P_{ij} in each R_{ij}. The sum $\sum(\alpha)$ contains one term for each rectangle R_{ij}, namely, $A(P_{ij})$ times the area of R_{ij} where $A(u, v)\, du\, dv = G^*(\omega)$.) Any term of the sum $\sum(\alpha)$ which is not zero corresponds to a chosen point P_{ij} at which $G^*(\omega)$ is not zero; hence, by the assumption on ω, it corresponds to a chosen point P_{ij} such that $G(P_{ij})$ lies in the portion of S parameterized by F. Therefore $F^{-1}[G(P_{ij})]$ is defined and must lie in at least one of the four quarters of R_0. By moving P_{ij} slightly if necessary—which causes an arbitrarily small change in $\sum(\alpha)$—it can be assumed that $F^{-1}[G(P_{ij})]$ lies in just one of the four quarters of R_0. In this way, every approximating sum $\sum(\alpha)$ to $\int G^*(\omega)$ differs arbitrarily little from one which falls into four parts $\sum(\alpha) = \sum_1(\alpha) + \sum_2(\alpha) + \sum_3(\alpha) + \sum_4(\alpha)$, each part consisting of those terms of $\sum(\alpha)$ for which $F^{-1}[G(P_{ij})]$ lies in the corresponding quarter of R_0. Since $\int_{R_0} F^*(\omega)$ can be written as a sum of four integrals, one for each quarter of R_0, and since $\sum(\alpha) \to \int G^*(\omega)$ as $|\alpha| \to 0$, the assumption $|\int F^*(\omega) - \int G^*(\omega)| > E$ implies that there must be at least one quarter of R_0, say R_1, with the following property: For every $\delta > 0$ there is an approximating sum $\sum(\alpha)$ to $\int G^*(\omega)$ such that $|\alpha| < \delta$ and such that

$$\left| \int_{R_1} F^*(\omega) - \sum_1(\alpha) \right| > \frac{E}{4}$$

where $\sum_1(\alpha)$ consists of those (non-zero) terms of $\sum(\alpha)$ for which $F^{-1}[G(P_{ij})]$ is in R_1. (Otherwise, for each of the four quarters there would be a δ beyond which this was impossible. Let δ be the smallest of these four values. Choose $\sum(\alpha)$ such that $|\alpha| < \delta$, such that each of the non-zero terms of $\sum(\alpha)$ corresponds unambiguously to just one quarter of R_0, and such that $\sum(\alpha)$ is close enough to $\int G^*(\omega)$ that $|\int F^*(\omega) - \sum(\alpha)| > E$. Then the assumption on δ would also give $\left| \int F^*(\omega) - \sum(\alpha) \right| \leq 4\left(\frac{E}{4}\right) = E$ so there is no such δ.)

Now let R_1 be divided into quarters. The contention is that at least one of these four quarters, say R_2, must have the property that for every $\delta > 0$ there is an approximating sum $\sum(\alpha)$ to $\int G^*(\omega)$ such that $|\alpha| < \delta$ and such that

$$\left| \int_{R_2} F^*(\omega) - \sum_2(\alpha) \right| > \frac{E}{16}$$

where $\sum_2(\alpha)$ consists of those (non-zero) terms of $\sum(\alpha)$ for which $F^{-1}[G(P_{ij})]$ is in R_2.

This is proved by noting that if δ is given then there is a $\sum(\alpha)$ such that $\left| \int_{R_1} F^*(\omega) - \sum_1(\alpha) \right| > \frac{E}{4}$. By changing the chosen points P_{ij} slightly it can be assumed that for each term of $\sum_1(\alpha)$ the corresponding point $F^{-1}[G(P_{ij})]$ lies in only one quarter of R_1. Splitting $\int_{R_1} F^*(\omega)$ into four parts it follows that the desired inequality holds for this $\sum(\alpha)$ on at least one of the four quarters of R_1. If the inequality were impossible for sufficiently small δ on all four quarters the assumption on R_1 would then be contradicted.

Repeating this process, it follows that there is a nested sequence of squares $R_0 \supset R_1 \supset \cdots \supset R_n \supset \cdots$, each square being a quarter of the preceding square, such that each R_n has the following property: Given $\delta > 0$ there is an approximating sum $\sum(\alpha)$ to $\int G^*(\omega)$ such that $|\alpha| < \delta$ and such that

$$\left| \int_{R_n} F^*(\omega) - \sum_n (\alpha) \right| > \frac{E}{4^n} \tag{3}$$

where $\sum_n(\alpha)$ is the sum of those (non-zero) terms of $\sum(\alpha)$ for which $F^{-1}[G(P_{ij})]$ is in R_n. It is to be shown that this is impossible.

The sums $\sum_n(\alpha)$ are approximating sums to the integral* $\int_{G^{-1}[F(R_n)]} G^*(\omega)$ where $G^{-1}[F(R_n)]$ denotes the set of points in $\{|u| \leq 1, |v| \leq 1\}$ whose images under G lie in $F(R_n)$. The next step of the proof is to simplify the description of $G^{-1}[F(R_n)]$.

**It has not been shown, however, that this integral converges.*

Lemma

Let P_0 be the point of the uv-plane common to all the rectangles R_n above. Then P_0 lies inside $\{|u| < 1, |v| < 1\}$, there is a point P_1 inside $\{|u| < 1, |v| < 1\}$ such that $F(P_0) = G(P_1)$, and $G^{-1} \circ F$ is a well-defined differentiable function near P_0.

Proof

Once the first two statements are proved, the third statement is an immediate consequence of the Implicit Function Theorem because $(x, y, z) = G(u, v)$ can then be solved near $F(P_0)$ for $(u, v, z) = g(x, y)$ [or $(u, y, v) = g(x, z)$ or $(x, u, v) = g(y, z)$] in such a way that '$(x, y, z) = G(u, v)$' is equivalent to '(x, y, z) is in S and $(u, v, z) = g(x, y)$'. Since $F(u, v)$ is in S, the equation $F(u_1, v_1) = G(u_2, v_2)$ is therefore solved for (u_2, v_2) as differentiable functions of (u_1, v_1).

To prove the first two statements note first that each of the squares R_n must contain a point Q_n such that ω is not zero at $F(Q_n)$, since otherwise $F^*(\omega)$ would be identically zero on R_n, $\sum_n(\alpha)$ could contain no non-zero terms, and (3) would be false. By the assumption on ω it follows that there is an ϵ such that Q_n is inside $\{|u| \leq 1 - \epsilon, |v| \leq 1 - \epsilon\}$ for all n. Since $P_0 = \lim_{n\to\infty} Q_n$

it follows that P_0 is inside $\{|u| \leq 1 - \epsilon, |v| \leq 1 - \epsilon\}$. Moreover, for each Q_n there is a unique Q_n' in $\{|u| \leq 1 - \epsilon, |v| \leq 1 - \epsilon\}$ such that $F(Q_n) = G(Q_n')$. By repeated subdivision of $\{|u| \leq 1, |v| \leq 1\}$ (or by the Bolzano-Weierstrass Theorem of §9.4) it can be shown that there is a point P_1 of $\{|u| \leq 1 - \epsilon, |v| \leq 1 - \epsilon\}$ with the property that for every $\delta > 0$ there are infinitely many of the points Q_n' which lie within δ of P_1. It follows, then, that $G(P_1) = F(P_0)$ because any point P of xyz-space other than $F(P_0)$ has the property that there is an $\epsilon > 0$ such that only a finite number of the points $F(Q_n) = G(Q_n')$ lie within ϵ of P (because the points $F(Q_n)$ all lie near $F(P_0)$) and hence $G(P_1) \neq P$ unless $P = F(P_0)$. This completes the proof of the lemma.

Using the lemma it follows that $G^{-1} \circ F$ is a differentiable function of (u, v) defined on all of R_n when n is sufficiently large. Let $f = G^{-1} \circ F$ and $A\,du\,dv = G^*(\omega)$ so that $F^*(\omega) = (G \circ f)^*(\omega) = f^*[G^*(\omega)] = f^*[A\,du\,dv]$. The $\sum_n(\alpha)$ are approximating sums to $\int_{f(R_n)} A\,du\,dv$ and it is to be shown that

$$\left| \int_{R_n} f^*(A\,du\,dv) - \sum_n(\alpha) \right| > \frac{E}{4^n}$$

cannot hold for arbitrarily small $|\alpha|$. (If it were known that the integrals $\int_{f(R_n)} A\,du\,dv$ converged this would be a matter of proving that $\int_{f(R_n)} A\,du\,dv = \int_{R_n} f^*(A\,du\,dv)$ holds—at least with an error less than const./4^n as $n \to \infty$—but it has not been proved that these integrals converge and the approximating sums $\sum_n(\alpha)$ must be estimated instead.)

The final step of the proof is to examine $\int_{R_n} f^*(A\,du\,dv)$ and $\sum_n(\alpha)$ 'under a microscope of power 2^n:1'. To this end, let $P_0 = (u_0, v_0)$, $P_1 = (u_1, v_1)$ and set

$$u = u_0 + \frac{h}{2^n} \qquad h = 2^n(u - u_0)$$

$$v = v_0 + \frac{k}{2^n} \qquad k = 2^n(v - v_0)$$

on the domain of f, and

$$u = u_1 + \frac{p}{2^n} \qquad p = 2^n(u - u_1)$$

$$v = v_1 + \frac{q}{2^n} \qquad q = 2^n(v - v_1)$$

on the range of f. The square R_n in the uv-plane corresponds to a square in the hk-plane, say $\hat{R}_n$. The square $\hat{R}_n$ is 2×2 and contains the point $(0, 0)$; therefore $\hat{R}_n$ is contained in $\{|h| \leq 2, |k| \leq 2\}$. Let $A_n\, dp\, dq$ be $\frac{1}{4^n}$ times the expression of $A\, du\, dv$ in the coordinates (p, q). Then $\sum_n(\alpha)$ is $\frac{1}{4^n}$ times an approximating sum to $\int_{f(\hat{R}_n)} A_n\, dp\, dq$ and (3) becomes

$$(4) \qquad \left| \int_{\hat{R}_n} f^*(A_n\, dp\, dq) - \widehat{\sum}_n(\alpha) \right| > E$$

where $\widehat{\sum}_n(\alpha)$ is an approximating sum to $\int_{f(\hat{R}_n)} A_n\, dp\, dq$. It is to be shown that this cannot be true for arbitrarily small $|\alpha|$ for all n.

The idea is that 'under a microscope' f is nearly affine and $A_n\, dp\, dq$ is nearly constant. If f were actually affine and $A_n\, dp\, dq$ were actually constant, then, since $f = G^{-1} \circ F$ has positive Jacobian by assumption, the theorem of §6.2 would assert that as $|\alpha| \to 0$ the approximating sums $\widehat{\sum}_n(\alpha)$ actually converge to $\int_{f(\hat{R}_n)} A\, dp\, dq = \int_{\hat{R}_n} f^*(A\, dp\, dq)$, and (4) would be contradicted. The objective is to show that for n large enough, the error in this approximation can be made less than E for any $E > 0$.

Let L denote the map

$$p = a_1 h + a_2 k$$
$$q = a_3 h + a_4 k$$

(a_1, a_2, a_3, a_4 constants) which has the same partial derivatives as f at $P_0 = (0, 0)$ (relative to h, k, p, q). It was shown in §5.3 that $L(h, k)$ can be made to differ from $f(h, k)$ by less than any preassigned ϵ for all (h, k) in $\{|h| \leq 2, k \leq 2\}$ by making n sufficiently large (that is, by making the scale factors $s = 2^{-n}$ sufficiently small). In particular, if $\hat{R}_n{}^+$ is a square slightly larger than $\hat{R}_n$, say the square with the same center but with the side $2 + 2\epsilon$ instead of the side 2, then $f(\hat{R}_n)$ is contained in $L(\hat{R}_n{}^+)$ for all sufficiently large n. Similarly, if $\hat{R}_n{}^-$ is slightly smaller than $\hat{R}_n$, then $f(\hat{R}_n)$ contains $L(\hat{R}_n{}^-)$.

The number $\int_{\hat{R}_n} f^*(A_n\, dp\, dq)$ will now be compared to $\widehat{\sum}_n(\alpha)$ in several steps and it will be shown that each of these steps can be made small by making n large and $|\alpha|$

small, hence contradicting (4). In the first place, $\int_{\hat{R}_n} f^*(A_n\, dp\, dq)$ differs arbitrarily little from

$$\int_{\hat{R}_n} L^*(A(P_1)\, dp\, dq)$$

because these integrands differ arbitrarily little from each other throughout $\hat{R}_n$ when n is large. Next,

$$\int_{\hat{R}_n} L^*(A(P_1)\, dp\, dq) = \int_{L(\hat{R}_n)} A(P_1)\, dp\, dq$$

by the theorem of §6.2. Next, $\widehat{\sum}_n(\alpha)$ differs arbitrarily little from the corresponding approximating sum $\widehat{\sum}'_n(\alpha)$ to $\int_{f(\hat{R}_n)} A(P_1)\, dp\, dq$ because the integrands differ by arbitrarily little on $f(\hat{R}_n)$ and because the total area of the squares specified by α for which P_{ij} is in $f(\hat{R}_n)$ is bounded when $|\alpha|$ is bounded. It remains to show that the approximating sums $\widehat{\sum}'_n(\alpha)$ to $\int_{f(\hat{R}_n)} A(P_1)\, dp\, dq$ differ arbitrarily little from $\int_{L(\hat{R}_n)} A(P_1)\, dp\, dq$. If $A(P_1) = 0$, then both are zero and this is trivially true; otherwise one can divide by $A(P_1)$ and it suffices to show that the approximating sums to $\int_{f(R_n)} dp\, dq$ differ arbitrarily little from $\int_{L(\hat{R}_n)} dp\, dq$. But if $f(\hat{R}_n)$ is contained in $L(\hat{R}_n^+)$ and contains $L(\hat{R}_n^-)$ it follows that as $|\alpha| \to 0$ the approximating sums to $\int_{f(\hat{R}_n)} dp\, dq$ all lie between $\int_{L(\hat{R}_n^-)} dp\, dq = \int_{\hat{R}_n^-} L^*(dp\, dq)$ and $\int_{L(\hat{R}_n^+)} dp\, dq = \int_{\hat{R}_n^+} L^*(dp\, dq)$. Since these two numbers lie arbitrarily close to $\int_{\hat{R}_n} L^*(dp\, dq) = \int_{L(\hat{R}_n)} dp\, dq$ the desired conclusion follows. This completes the proof of the theorem.

Exercises

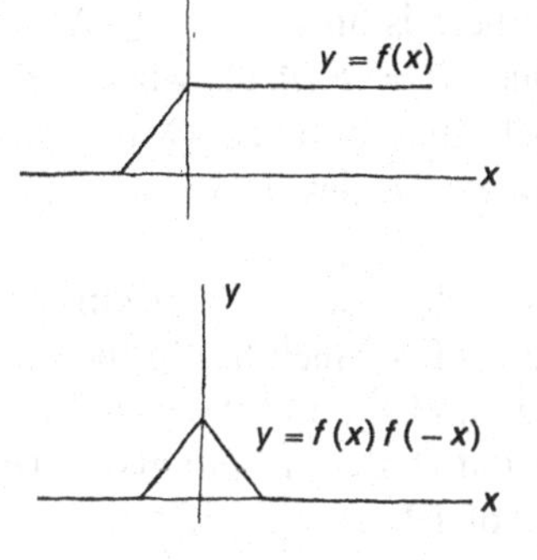

1 Prove that there exists a function c_P for each P as stated in the proof of the theorem. [Remember that c_P can be—in fact must be—a very artificial function. Use the function $f(x)$ which is 0 for $x \leq -1$, which is 1 for $x \geq 0$, and which is $1 + x$ for x between -1 and 0.] Prove that in fact c_P can be chosen to be differentiable, and that this gives a differentiable partition of unity $a_1, a_2, \ldots, a_K$. This fact is used in the proof of Stokes' Theorem.

2 Show that if P is inside the image of two different charts F_i, F_j then $F_j^{-1} \circ F_i$ is a well-defined differentiable function of the uv-plane to itself defined near $F_i^{-1}(P)$. [This is actually proved in the text.] Prove that this map $F_j^{-1} \circ F_i$ has positive

Jacobian if and only if $F_j^*(\omega)$ is a positive multiple of $F_i^*(\omega)$ for all 2-forms ω. [Use the Chain Rule.]

3 Show that the torus of Exercise 7, §2.5, is a compact, oriented, differentiable surface in xyz-space, that is, that it can be described by charts $F_1, F_2, \ldots, F_N$ satisfying (a)–(c). [This is not as simple as one might expect. Three charts can be used.]

6.4

Manifolds-with-Boundary and Stokes' Theorem

To state and prove Stokes' Theorem $\int_{\partial S} \omega = \int_S d\omega$ (see §3.4) one must first define the integrals it involves. This is an extremely difficult definition to make, and even the very elaborate definition of §6.3 is not yet adequate because it applies only to integrals $\int_S \omega$ in which the domain of integration S has no boundary ∂S. However, the needed definitions require only a slight modification of the definition of §6.3.

Definition

A compact, oriented, differentiable surface-with-boundary in xyz-space is a set S which can be described by a finite number of oriented charts as follows:

(a) A finite number of differentiable maps $F_1, F_2, \ldots, F_N$ of the uv-plane to xyz-space are given. Each F_i is one-to-one and non-singular of rank 2 on the square $\{|u| \le 1, |v| \le 1\}$. The maps F_i are called 'charts'.

(b) For each chart F_i there is specified a closed* rectangle R_i in $\{|u| \le 1, |v| \le 1\}$ such that the image of a point of $\{|u| \le 1, |v| \le 1\}$ under F_i is a point of S if and only if the point lies in R_i.

(c) For each P in S there is at least one chart F_i such that P lies *inside* the image of $\{|u| \le 1, |v| \le 1\}$ under F_i and such that S near P is $F_i(R_i)$. That is, for each P in S there is an i, $1 \le i \le N$, and an $\epsilon > 0$, such that $P = F_i(\bar{u}, \bar{v})$ where $|\bar{u}| < 1$, $|\bar{v}| < 1$ and such that a point Q in xyz-space which lies within ϵ of P lies in S if and only if it lies in $F_i(R_i)$.

(d) The orientations of the charts F_i agree in the sense that if P is a point of S which lies in the image of two charts, say F_i and F_j, and if ω is a 2-form on xyz-space defined at P, then $F_i^*(\omega)$ at $F_i^{-1}(P)$ is a positive multiple of $F_j^*(\omega)$ at $F_j^{-1}(P)$.

*This means that the rectangle includes its boundary. This is analogous to the definition of a closed interval $\{a \le x \le b\}$ to be an interval which includes its endpoints $x = a$, $x = b$. Specifically, each R_i is to be a set of the form $\{a \le u \le b, c \le v \le d\}$ where $-1 \le a < b \le 1$ and $-1 \le c < d \le 1$.

If S is a compact, oriented, differentiable surface-with-boundary, then the *boundary* of S, denoted ∂S, consists of those points of S which are the image under some F_i of a point inside $\{|u| \leq 1, |v| \leq 1\}$ which lies on a boundary of R_i.

Note that a 'surface' in the sense of §6.3 is also a 'surface-with-boundary' in the sense just defined. [Take R_i to be all of $\{|u| \leq 1, |v| \leq 1\}$ for all $i = 1, 2, \ldots, N$.] Thus the boundary is optional and a surface-with-boundary need not have a boundary. Unfortunately there is no accepted terminology which avoids this linguistic absurdity.

Theorem

Let S be a compact, oriented, differentiable surface-with-boundary in xyz-space, and let $\omega = A\,dy\,dz + B\,dz\,dx + C\,dx\,dy$ be a continuous 2-form on xyz-space which is defined at all points of S. Then a number $\int_S \omega$ depending only on ω and S can be defined as follows: Let $F_1, F_2, \ldots, F_N$ and $R_1, R_2, \ldots, R_N$ be a specific description of S by charts F_i and rectangles R_i satisfying (a)–(d) above. Then there exist continuous 2-forms $\omega_1, \omega_2, \ldots, \omega_N$ defined at all points of S such that $\omega = \omega_1 + \omega_2 + \cdots + \omega_N$ and such that ω_i is zero at all points of S other than those which lie in the image under F_i of a square $\{|u| \leq 1 - \delta_i, |v| \leq 1 - \delta_i\}$ contained in $\{|u| \leq 1, |v| \leq 1\}$. The integrals

$$\int_{R_i} F_i^*(\omega_i),$$

where $F_i^*(\omega_i)$ is the pullback of ω_i under F_i and where R_i is oriented by $du\,dv$, all converge; their sum depends only on ω and S, so that the definition

$$\int_S \omega = \sum_{i=1}^{N} \int_{R_i} F_i^*(\omega_i)$$

is valid. Similarly, if $\omega = A\,dx + B\,dy + C\,dz$ is a continuous 1-form on xyz-space which is defined at all points of ∂S, then a number $\int_{\partial S} \omega$ depending only on ω and S can be defined as follows: Let $F_1, F_2, \ldots, F_N$ and $R_1, R_2, \ldots, R_N$ be a specific description of S. Then there exist continuous 1-forms $\omega_1, \omega_2, \ldots, \omega_N$ defined at all points of ∂S such that $\omega = \omega_1 + \omega_2 + \cdots + \omega_N$ and such that ω_i is zero at all points of ∂S other than those which lie in the image under F_i of a square $\{|u| \leq 1 - \delta_i,$

$|v| \leq 1 - \delta_i\}$ contained in $\{|u| \leq 1, |v| \leq 1\}$. Each of the integrals

$$\int_{\partial R_i} F_i^*(\omega_i)$$

can be defined as a sum of 4 simple integrals of the form $\int_a^b A(t)\, dt$ by orienting each of the 4 sides of ∂R_i by the counterclockwise convention. The sum

$$\int_{\partial S} \omega = \sum_{i=1}^{N} \int_{\partial R_i} F_i^*(\omega_i)$$

is independent of the choices and is therefore a valid definition of $\int_{\partial S} \omega$.

Proof

The proof of this theorem is virtually identical to the proof of the theorem of §6.3. The decompositions $\omega = \omega_1 + \omega_2 + \cdots + \omega_N$ can be accomplished using a partition of unity as before, the convergence of the integrals $\int_{R_i} F_i^*(\omega_i)$ and $\int_{\partial R_i} F_i^*(\omega_i)$ follows as before, and the proof that the definition is independent of the choices is reduced as before to the case where the 2-form (resp. 1-form) is zero except at points of S (resp. ∂S) which are contained inside two different charts. The only change in the proof that $\int_S \omega$ is independent of the choices is that the map $f = G^{-1} \circ F$ is not necessarily defined at all points near P_0 (if $F(P_0)$ is on ∂S); this causes no difficulty because there is still a differentiable map f defined near P_0 which agrees with $G^{-1} \circ F$ at all points of $F^{-1}(S)$. In the proof that $\int_{\partial S} \omega$ is independent of the choices it must be shown that if F, G are charts in which S corresponds to R_F, R_G respectively then locally the map $G^{-1} \circ F$, when it is defined, carries the sides of R_F inside $\{|u| < 1, |v| < 1\}$ (if there are any) to sides of R_G in a non-singular orientation-preserving way. This is not difficult to prove (see Exercise 1).

Thus integrals over S and ∂S are defined in terms of integrals over rectangles and their boundaries. Since Stokes' Theorem for rectangles was proved in Chapter 3, it is only a short step to the proof of the general Stokes theorem. For the sake of simplicity it will be assumed that the surface S is *twice* differentiable, that is, that S can be described by charts $F_1, F_2, \ldots, F_N$ which have the property that $F^*(\omega)$ is differentiable whenever ω is differentiable.*

**This is equivalent to saying that the 3 functions of 2 variables which describe F have continuous second partial derivatives—see Exercise 2.*

Stokes' Theorem

Let S be a compact, oriented, twice differentiable surface-with-boundary in xyz-space, and let $\omega = A\,dx + B\,dy + C\,dz$ be a differentiable 1-form defined at all points of S. Then

$$\int_{\partial S} \omega = \int_S d\omega.$$

Proof

Let charts $F_1, F_2, \ldots, F_N$ and rectangles $R_1, R_2, \ldots, R_N$ be given describing S as above. Then it is possible to write $\omega = \omega_1 + \omega_2 + \cdots + \omega_N$ where ω_i is a *differentiable* 1-form which is zero on S except inside $F_i(R_i)$. This is done by constructing a differentiable partition of unity by the method of §6.3; to do this one need only begin with differentiable functions c_P (see Exercise 1 of §6.3).

Thus it suffices to prove

$$\int_{\partial R_i} F_i^*(\omega_i) = \int_{R_i} F_i^*(d\omega_i)$$

since the sums over i of these numbers are $\int_{\partial S} \omega$ and $\int_S d\omega$ by the definition of these integrals ($d\omega = d\omega_1 + d\omega_2 + \cdots + d\omega_N$ and $d\omega_i$ is zero on S except on $F_i(R_i)$). But it was proved in §3.2 that

$$\int_{\partial R_i} F_i^*(\omega_i) = \int_{R_i} d[F_i^*(\omega_i)]$$

provided the 1-form $F_i^*(\omega_i)$ on the uv-plane is differentiable. Thus it suffices to show that $d[F_i^*(\omega_i)] = F_i^*(d\omega_i)$. Let

$$\omega_i = A\,dx + B\,dy + C\,dz,$$
$$d\omega_i = dA\,dx + dB\,dy + dC\,dz.$$

$\mathbf{R}^2 \xrightarrow{(x,y,z)} \mathbf{R}^3 \xrightarrow{A} \mathbf{R}$

The Chain Rule implies that the pullback of dA is d of the pullback of A. (Consider A as a map from xyz-space to a line on which the coordinate is A so that dA is the pullback of oriented length on the A-line.) Now if D is any differentiable function (0-form) on the uv-plane and if σ is any differentiable 1-form then $d(D\sigma) = dD \cdot \sigma + D \cdot d\sigma$ as is easily seen from the definition of the operation d. Thus, considering A, dx, B, dy, C, dz as forms on the uv-plane, d of the pullback of $A\,dx + B\,dy + C\,dz$ is $dA\,dx + dB\,dy + dC\,dz + Ad[dx] + Bd[dy] +$

$Cd[dz]$ and the desired formula is reduced to the formula $d[dx] = 0$ (and similarly $d[dy] = 0$, $d[dz] = 0$) where dx is considered as a 1-form on the uv-plane. The assumption that F_i is twice differentiable implies that dx is differentiable; hence, by Stokes' Formula,

$$\int_R d[dx] = \int_{\partial R} dx$$

for all rectangles R. But the Fundamental Theorem applied to the 4 integrals of $\int_{\partial R} dx$ gives cancellations at the 4 vertices; hence $\int_{\partial R} dx = 0$, hence $\int_R d[dx] = 0$ for all rectangles R. Dividing by the area of R and letting R shrink to a point P gives $d[dx] = 0$ at P for all P; hence $d[dx] \equiv 0$. This completes the proof of Stokes' Theorem.

Exercises

1 Show that if $R = \{a \leq x \leq b, c \leq y \leq d\}$ is a rectangle in the xy-plane and if $f: \mathbf{R}^2 \to \mathbf{R}^2$ is a differentiable map, say

$$x = f_1(u, v)$$
$$y = f_2(u, v),$$

defined and orientation preserving $\left(\dfrac{\partial(x, y)}{\partial(u, v)} > 0\right)$ near $(u, v) = (0, 0)$, which carries points (u, v) with $u \leq 0$ to points of R and points (u, v) with $u > 0$ to points not in R, then f carries the line $u = 0$ near $(0, 0)$ to one of the sides of R and carries the orientation dv of $u = 0$ to the counterclockwise orientation of ∂R. [First consider the case where f is affine. Then note that if the conditions hold for f then they also hold for f under a 'microscope'.]

2 Show that a differentiable map

$$y_i = f_i(x_1, x_2, \ldots, x_n) \qquad (i = 1, 2, \ldots, m)$$

for which the functions f_i have continuous second partial derivatives has the property that the pullback $f^*(\omega)$ of a differentiable k-form ω is differentiable. Show that, conversely, if the pullback of every differentiable 1-form is differentiable then the f_i have continuous second partial derivatives.

3 Show that if P_0, P_1, P_2 are three non-collinear points in xyz-space then the oriented triangle they describe is a compact, oriented, differentiable surface-with-boundary in xyz-space, that is, can be described by charts $F_1, F_2, \ldots, F_N$ satisfying (a)–(d). [The description of this simple surface-with-boundary by charts is not very simple.]

6.5 General Properties of Integrals

The case $k = 2$, $n = 3$ was considered in §6.3 and §6.4 merely to simplify the notation. The generalization to arbitrary k and n presents no additional difficulties.

Definition

A compact, oriented, differentiable, k-dimensional manifold-with-boundary in $x_1x_2 \ldots x_n$-space is a subset of $x_1x_2 \ldots x_n$-space which can be described by a finite number of oriented charts as follows:

(a) A finite number of differentiable maps $F_1, F_2, \ldots, F_N$ of $u_1u_2 \ldots u_k$-space to $x_1x_2 \ldots x_n$-space are given. Each F_i is one-to-one and non-singular of rank k on the k-dimensional cube $\{|u_i| \leq 1;\ i = 1, 2, \ldots, k\}$. The maps F_i are called 'charts'.

(b) For each chart F_i there is specified a closed* k-dimensional rectangle R_i in $\{|\bar{u}_j| \leq 1;\ j = 1, 2, \ldots, k\}$ such that the image of a point of $\{|u_j| \leq 1; j = 1, 2, \ldots, k\}$ under F_i is a point of S if and only if the point lies in R_i.

(c) For each P in S there is an i, $1 \leq i \leq N$, such that $P = F_i(\bar{u}_1, \bar{u}_2, \ldots, \bar{u}_k)$ where $|\bar{u}_j| < 1$ ($j = 1, 2, \ldots, k$) and there is an $\epsilon > 0$ such that any point $Q = (x_1, x_2, \ldots, x_n)$ which lies within ϵ of P (each of the n coordinates x_j of Q differs by less than ϵ from the corresponding coordinate of P) lies in S if and only if it lies in $F_i(R_i)$.

(d) The orientations of the charts F_i agree in the sense that if P is a point of S which lies in the image of two charts, say F_i and F_j, and if ω is a k-form on $x_1x_2 \ldots x_n$-space defined at P then $F_i^*(\omega)$ at $F_i^{-1}(P)$ is a positive multiple of $F_j^*(\omega)$ at $F_j^{-1}(P)$.

*Specifically, each R_i is to be a set of the form $\{a_1 \leq u_1 \leq b_1, a_2 \leq u_2 \leq b_2, \ldots, a_k \leq u_k \leq b_k\}$ where $-1 \leq a_j < b_j \leq 1$ for $j = 1, 2, \ldots, k$.

Theorem

If S is a compact, oriented, differentiable, k-dimensional manifold-with-boundary in $x_1x_2 \ldots x_n$-space and if ω is a continuous k-form on $x_1x_2 \ldots x_n$-space defined at all points of S, then a number $\int_S \omega$ depending only on ω and S can be defined in the same way as before. Similarly, if ω is a continuous $(k - 1)$-form defined at all points of ∂S then a number $\int_{\partial S} \omega$ depending only on ω and S can be defined. This second definition depends on a convention for orienting the boundary ∂R of a k-dimensional rectangle R in $u_1u_2 \ldots u_k$-space. This convention is that the side $u_1 = b_1$ of the rectangle

$\{a_1 \leq u_1 \leq b_1, a_2 \leq u_2 \leq b_2, \ldots, a_k \leq u_k \leq b_k\}$ is oriented by $du_2\, du_3 \ldots du_k$ and the remaining sides are oriented accordingly. (The orientation of the remaining sides is determined by the convention that the map $u_i \to u_j$, $u_j \to -u_i$ preserves orientations. Note that if $k = 1$ then S is a curve, and $\int_{\partial S} \omega$ for ω a 0-form (function) is a sum over the endpoints ∂S of S, each endpoint being counted with a sign determined by the orientation of S.)

The proof of this theorem is exactly the same as in the case $n = 3$, $k = 2$ of §6.4.

A manifold-with-boundary is said to be *twice* differentiable if it can be described by charts F_i which have the property that the pullback of any differentiable form under F_i is itself differentiable. This is equivalent to saying that the first partial derivatives of F_i are functions which have continuous first partial derivatives.

Stokes' Theorem

Let S be a compact, oriented, twice differentiable, k-dimensional manifold-with-boundary in $\mathbf{R}^n$ and let ω be a differentiable $(k-1)$-form on $\mathbf{R}^n$ defined at all points of S. Then

$$\int_{\partial S} \omega = \int_S d\omega$$

where $d\omega$ is the k-form on $\mathbf{R}^n$ defined by the formula

$$d(A\, dx_1\, dx_2 \ldots dx_{k-1} + \cdots) = dA\, dx_1\, dx_2 \ldots dx_{k-1} + \cdots$$

of Chapter 3.

(The assumption that S is twice differentiable is not actually necessary and is made for the sake of convenience.)

Proof

Using a differentiable partition of unity the theorem is reduced immediately to the formula

$$\int_{\partial R_i} F_i^*(\omega_i) = \int_{R_i} F_i^*(d\omega_i).$$

By examining the definition of the operation d and by using the fact that F_i is twice differentiable it can be shown that $d[F_i^*(\omega_i)] = F_i^*(d\omega_i)$. Setting $\sigma = F_i^*(\omega_i)$ this reduces the theorem to

$$\int_{\partial R} \sigma = \int_R d\sigma$$

i.e. to Stokes' Formula for a 'rectangle'. This special case was proved in Chapter 3. (It suffices to consider the case $\sigma = A\,du_2\,du_3 \ldots du_k$. Then, letting R' denote the rectangle which is the projection of R on the $u_2u_3 \ldots u_k$-plane, the orientation of ∂R is defined in such a way that

$$\int_{\partial R} \sigma = \int_{R'} [A(b_1, u_2, \ldots, u_k) - A(a_1, u_2, \ldots, u_k)]\,du_2\,du_3 \ldots du_k.$$

By the Fundamental Theorem this is

$$\int_{R'} \left(\int_{a_1}^{b_1} \frac{\partial A}{\partial u_1}(u_1, u_2, \ldots, u_n)\,du_1\right) du_2 \ldots du_k$$

which, by the formula for a double integral as an iterated integral, is

$$\int_R \frac{\partial A}{\partial u_1}\,du_1\,du_2 \ldots du_k = \int_R d\sigma$$

as was to be shown.)

In addition to Stokes' Theorem, the following properties of integrals can now be stated and proved:

I. *Linearity in* ω. If $\omega = a_1\omega_1 + a_2\omega_2$ where a_1, a_2 are numbers and ω_1, ω_2 are continuous k-forms on S then

$$\int_S \omega = a_1 \int_S \omega_1 + a_2 \int_S \omega_2.$$

II. *Decomposition of S.* If the (compact, oriented, differentiable) k-dimensional manifold (-with-boundary) S is divided into two k-dimensional manifolds (compact, etc., -with-boundary) S_1, S_2 by a $(k - 1)$-dimensional manifold, then

$$\int_S \omega = \int_{S_1} \omega + \int_{S_2} \omega.$$

III. *Orientation.* If the orientation of S is reversed, then the integral $\int_S \omega$ changes sign. (The orientation of integrals is a constant source of aggravation. However, without orientations there can be no cancellation on the interior boundaries and hence no Stokes' Theorem or Fundamental Theorem of Calculus.)

IV. *Independence of parameter.* The formula

$$\int_{f(S)} \omega = \int_S f^*(\omega)$$

holds whenever S is a compact, oriented, differentiable, k-dimensional manifold-with-boundary, f: $\mathbf{R}^n \to \mathbf{R}^m$ is a differentiable map which is one-to-one and non-singular of rank k on S, ω is a continuous k-form defined at all points of $f(S)$, and $f(S)$ is oriented in accord with S.*

*The case $n = m = k$ of this theorem is known as the formula for change of variable in a multiple integral. Note that the conditions guarantee that $f(S)$ is a compact, oriented, differentiable, k-dimensional manifold-with-boundary, so that $\int_{f(S)} \omega$ is defined.

V. *Microscope.* If M_s: $\mathbf{R}^n \to \mathbf{R}^n$ is the map of $(h_1, h_2, \ldots, h_n)$ to $(x_1, x_2, \ldots, x_n)$ defined by

$$x_i = \bar{x}_i + sh_i \qquad (i = 1, 2, \ldots, n)$$

if S is a compact, oriented, differentiable, k-dimensional manifold-with-boundary in $h_1h_2 \ldots h_n$-space, and if ω is a continuous k-form on $x_1x_2 \ldots x_n$-space defined near $(\bar{x}_1, \bar{x}_2, \ldots, \bar{x}_n)$, then

$$\lim_{s \to 0} \frac{1}{s^k} \int_{M_s(S)} \omega = \int_S \bar{\omega}$$

where $\bar{\omega}$ is the constant k-form in $h_1, h_2, \ldots, h_n$ obtained by evaluating ω at $(\bar{x}_1, \bar{x}_2, \ldots, \bar{x}_n)$ and changing dx_i to dh_i. (This gives a precise meaning to the statement that the integral of a k-form over a small k-dimensional manifold is nearly the value of a constant k-form on the manifold.) In particular, if S is the boundary of a $(k + 1)$-dimensional manifold, then

$$\lim_{s \to 0} \frac{1}{s^k} \int_{M_s(S)} \omega = 0.$$

VI. *Integration by parts.* If ω_1 is a differentiable k_1-form, if ω_2 is a differentiable k_2-form, and if S is a compact, oriented, twice-differentiable $(k_1 + k_2 + 1)$-dimensional manifold-with-boundary such that ω_1, ω_2 are defined at all points of S then

$$\int_S d\omega_1 \cdot \omega_2 = \int_{\partial S} \omega_1 \cdot \omega_2 - (-1)^{k_1} \int_S \omega_1 \cdot d\omega_2.\dagger$$

†Taking $\omega_2 \equiv 1$, this is Stokes' Theorem. On the other hand, setting $\omega = \omega_1 \cdot \omega_2$ in Stokes' Theorem and using the formula $d(\omega_1 \cdot \omega_2) = d\omega_1 \cdot \omega_2 + (-1)^{k_1}\omega_1 \cdot d\omega_2$ gives the formula for integration by parts.

VII. *Approximating sums.* If S is a compact, oriented, differentiable, 2-dimensional manifold-with-boundary in the xy-plane, if the orientation of S agrees with $dx\,dy$ at all points, and if $\omega = A\,dx\,dy$ is a continuous 2-form defined at all points of S, then the definition of $\int_S \omega$ given above coincides with the definition given in §2.3. That is, if S is enclosed in a rectangle (which is possible by compactness—see §9.4) and if approximating sums $\sum(\alpha)$ are formed as in §2.3, then the approximating sums converge to the number $\int_S \omega$ defined in this chapter. Analogous theorems apply to integrals $\int_S \omega$ in which S is a k-dimensional manifold-with-boundary in a k-dimensional space.

Proofs

I, III, IV, V and VI follow immediately from the preceding theorems. II requires a more rigorous formulation of the manner in which S is divided into $S_1 \cup S_2$ and will be omitted. To prove VII one can first write $\omega = \omega_1 + \omega_2 + \cdots + \omega_N$ where ω_i is zero outside of one chart; it is then essentially the statement that $\int \omega_i$ can be computed using either of two charts, which was proved in §6.3.

Exercises **1** Show that if S is a compact, oriented, differentiable, k-dimensional manifold-with-boundary with the additional property that for each chart F_i the rectangle R_i has at most one side inside $\{|u_j| \leq 1;\ h = 1, 2, \ldots, k\}$, then ∂S is a compact, oriented, differentiable, $(k-1)$-dimensional manifold-with-boundary (although ∂S has no boundary) and can be oriented so that the two definitions of $\int_{\partial S} \omega$ agree.

2 The integral $\int_{\partial R} (A\,dx + B\,dy)$ of a 1-form over the boundary of a rectangle $\{a \leq x \leq b, c \leq y \leq d\}$ can be defined explicitly by the formula

$$\int_{\partial R} (A\,dx + B\,dy) = \int_c^d [B(b, y) - B(a, y)]\,dy + \int_a^b [A(x, c) - A(x, d)]\,dx.$$

Give the analogous formula for the integral of a $(k-1)$-form over the boundary of a k-dimensional rectangle.

6.6

Integrals as Functions of S

A useful insight into the meaning of $\int_S \omega$ can be obtained by considering it as a function of S for fixed ω.

Definition

A *k-integral* on $\mathbf{R}^n$ is a function assigning numbers to compact, oriented, differentiable, k-dimensional manifolds-with-boundary in $\mathbf{R}^n$, which is of the form $S \to \int_S \omega$ for some continuous k-form ω on $\mathbf{R}^n$.

By property V of §6.5 the integrand ω is determined by the values of its integrals $\int_S \omega$. One can think of the k-form $\omega = A\, dx_1\, dx_2 \ldots dx_k + \cdots$ as giving the value of the k-integral $\int_S \omega$ on 'infinitesimal' oriented rectangles—for instance, as $A(P)$ times the oriented area of S when S is an infinitesimal rectangle in the $x_1x_2 \ldots x_k$-direction at the point P. These values are then sufficient to determine all values of $\int_S \omega$ by a process of integration.

From this point of view, Stokes' Theorem says essentially that if ω is a differentiable k-form, then the function $S \to \int_{\partial S} \omega$ is a $(k + 1)$-integral. The integrand $d\omega$ which gives this integral is found by examining the value of $S \to \int_{\partial S} \omega$ on 'infinitesimal' rectangles S—essentially a process of differentiation.

Similarly, the independence of parameter of $\int_S \omega$ can be regarded as the statement that if f: $\mathbf{R}^n \to \mathbf{R}^n$ is nonsingular of rank n then the function $S \to \int_{f(S)} \omega$ is a k-integral. The integrand $f^*(\omega)$ which gives this integral is found by examining the value of $S \to \int_{f(S)} \omega$ on infinitesimal rectangles. For infinitesimal rectangles, f is an affine map and ω is a constant form, so $f^*(\omega)$ can be found by the algebraic methods of Chapter 4.

The elaborate definitions of this chapter are necessary in order to define the *domain* of a k-integral $S \to \int_S \omega$—that is, to define the category of k-dimensional domains S for which integrals $\int_S \omega$ are defined. It is intuitively rather clear what sorts of domains S are to be considered as domains of integration, so that the essential aspects of the theory of k-integrals can be developed, as they were in the preceding chapters, without first defining precisely the category of compact, oriented, differentiable, k-dimensional manifolds-with-boundary.

In most applications it is as a k-integral that $\int_S \omega$ arises; that is, ω is fixed and $\int_S \omega$ is considered as a function of S. For example, 'work' is a 1-integral assigning numbers to oriented curves in xyz-space. The integrand $\omega = A\,dx + B\,dy + C\,dz$ which describes this 1-integral (by giving its values on infinitesimal line segments in the coordinate directions) is essentially the *force field*. More generally, a k-form ω is what physicists call a *field** (of alternating covariant tensors); the meaning of the 'field' ω is that it is an integrand whose integrals $\int_S \omega$ are physically meaningful quantities. The field gives the values of these quantities for infinitesimal k-dimensional rectangles in coordinate directions, and the values for more general domains S are found by integration.

However, physicists also consider other sorts of 'fields' which are not k-forms.

Exercises

1 What is the natural definition of a '0-integral'? Of a 'compact, oriented 0-dimensional manifold'?

2 Describe 'mass' as a 3-integral in which the integrand is $\rho\,dx\,dy\,dz$ where ρ = density.

practical methods of solution

chapter 7

7.1

Successive Approximation

Unlike Chapters 4 and 5, which were devoted to statements about the *nature* of the solution of an equation $y = f(x)$ for x given y, this chapter is devoted to methods of *actually solving* equations $y = f(x)$ for x given y. The phrase 'actually solving' will be interpreted to mean 'describing a numerical process for determining the solution to any prescribed degree of accuracy'. Thus the equation $x^2 = 2$ is not 'solved' by writing $x = \pm\sqrt{2}$, but only by giving some method of extracting the root $x = \pm 1.4142\ldots$ to any prescribed number of decimal places. Defining 'solution' in this way is tantamount to saying that it must take the form of a process of *successive approximation;* that is, it must take the form of a procedure which enables one to compute a succession of approximate solutions $x^{(1)}, x^{(2)}, x^{(3)}, \ldots$* which give the actual solution to a greater and greater degree of accuracy. This first section is devoted to proving the Elimination Theorem by such a method of successive approximation. Recall the statement of the theorem:

The letter x will denote a point $x = (x_1, x_2, \ldots, x_n)$ in R^n, which prevents the use of subscripts to denote 'more than one x'. The superscript notation $x^{(1)} = (x_1^{(1)}, x_2^{(1)}, \ldots, x_n^{(1)})$ will be used instead.

Elimination Theorem

If $y = f(x_1, x_2, \ldots, x_n)$ is a differentiable function and if $\bar{y} = f(\bar{x}_1, \bar{x}_2, \ldots, \bar{x}_n)$ is a point where $\dfrac{\partial y}{\partial x_1} \neq 0$ then there is a differentiable function $g(y, x_2, x_3, \ldots, x_n)$ defined near $(\bar{y}, \bar{x}_2, \bar{x}_3, \ldots, \bar{x}_n)$ such that the relations

$$y = f(x_1, x_2, \ldots, x_n) \qquad x_1 = g(y, x_2, \ldots, x_n)$$

are equivalent near $(\bar{y}, \bar{x}_1, \bar{x}_2, \ldots, \bar{x}_n)$.

Proof

Let a be the value of $\dfrac{\partial y}{\partial x_1}$ at $(\bar{x}_1, \bar{x}_2, \ldots, \bar{x}_n)$. The successive approximations will be based on the estimate

$$(1) \qquad \Delta y \sim a\,\Delta x_1$$

of the change in y which results from changing x_1 and keeping $x_2, x_3, \ldots, x_n$ fixed. Since $a \neq 0$ by assumption, this gives

$$(2) \qquad \Delta x_1 \sim \frac{\Delta y}{a}$$

as an approximation to the Δx_1 which will produce a given Δy. Thus if $(y, x_2, x_3, \ldots, x_n)$ are given and if $x_1^{(N)}$ is an approximate solution of $y = f(x_1, x_2, \ldots, x_n)$, that is, if $y \sim f(x_1^{(N)}, x_2, x_3, \ldots, x_n)$, then the formula (2) says that to produce

$$\text{desired } \Delta y = y - f(x_1^{(N)}, x_2, x_3, \ldots, x_n)$$

one should use approximately

$$\Delta x_1 \sim \frac{y - f(x_1^{(N)}, x_2, x_3, \ldots, x_n)}{a}.$$

That is, the 'correction' of the approximate solution $x_1^{(N)}$ should be

$$(3) \qquad x_1^{(N+1)} = x_1^{(N)} + \frac{y - f(x_1^{(N)}, x_2, \ldots, x_n)}{a}.$$

The given $(y, x_2, x_3, \ldots, x_n)$ will be assumed to be near $(\bar{y}, \bar{x}_2, \bar{x}_3, \ldots, \bar{x}_n)$ so a 'zeroth approximation' to a solution x_1 of $y = f(x_1, x_2, \ldots, x_n)$ would be

$$(4) \qquad x_1^{(0)} = \bar{x}_1.$$

The main step in the proof of the Elimination Theorem is the proof of the statement that *if the given* $(y, x_2, x_3, \ldots, x_n)$ *lie sufficiently near to* $(\bar{y}, \bar{x}_2, \bar{x}_3, \ldots, \bar{x}_n)$ *then the sequence of successive approximations* (3) *with the initial approximation* (4) *converges*, i.e. $\lim_{N\to\infty} x_1^{(N)}$ exists, *and the limit of the sequence is a solution x_1 of the equation* $y = f(x_1, x_2, \ldots, x_n)$. It will then be shown that the function $g(y, x_2, x_3, \ldots, x_n)$ defined by $\lim_{N\to\infty} x_1^{(N)}$ is differentiable.

The first step in proving that (3) converges to a solution is of course to estimate the error in the approximation (1) on which it is based. This is done, as usual, by using the Fundamental Theorem to write

$$\Delta y = \int_{x_1^{(N)}}^{x_1^{(N+1)}} \frac{\partial y}{\partial x_1}\, dx_1.$$

If $\dfrac{\partial y}{\partial x_1}$ differs from a by less than ϵ at all points of the interval of integration, then this integral shows that Δy lies between $(a+\epsilon)(x_1^{(N+1)} - x_1^{(N)})$ and $(a-\epsilon)(x_1^{(N+1)} - x_1^{(N)})$. Hence

$$(5) \qquad |\Delta y - a\,\Delta x| \leq \epsilon|\Delta x|$$

where Δx is the difference between the two given values $x_1^{(N)}, x_1^{(N+1)}$ and where Δy is the difference between the corresponding values $f(x_1^{(N)}, x_2, x_3, \ldots, x_n)$, $f(x_1^{(N+1)}, x_2, x_3, \ldots, x_n)$. This estimate, together with (3), gives

$$\begin{aligned}
|x_1^{(N+1)} - x_1^{(N)}| &= \left| x_1^{(N)} + \frac{y - f(x_1^{(N)}, x_2, \ldots, x_n)}{a} - x_1^{(N-1)} - \frac{y - f(x_1^{(N-1)}, x_2, \ldots, x_n)}{a} \right| \\
&= \left| x_1^{(N)} - x_1^{(N-1)} - \frac{f(x_1^{(N)}, x_2, \ldots, x_n) - f(x_1^{(N-1)}, x_2, \ldots, x_n)}{a} \right| \\
&= \left| \Delta x - \frac{\Delta y}{a} \right| = \frac{1}{|a|}|\Delta y - a\,\Delta x| \leq \frac{\epsilon}{|a|}|x_1^{(N)} - x_1^{(N-1)}|
\end{aligned}$$

which shows that *the Nth step is at most $\epsilon/|a|$ times as long as the $(N-1)$st step*, provided $x_1^{(N)}, x_1^{(N-1)}$ lie in the region where the estimate (5) is valid.

Choose a positive number* $\rho < 1$ and set $\epsilon = \rho|a|$. Then there is a $\delta > 0$ such that $\dfrac{\partial y}{\partial x_1}$ differs from a by less than ϵ at all points $(x_1, x_2, \ldots, x_n)$ within δ of $(\bar{x}_1, \bar{x}_2, \ldots, \bar{x}_n)$; hence (5) applies in this region. If $(x_2, x_3, \ldots, x_n)$ lies within δ of $(\bar{x}_2, \bar{x}_3, \ldots, \bar{x}_n)$ and if $x_1^{(N)}, x_1^{(N-1)}$ both lie within δ of $\bar{x}_1$ then the estimate

$$(6) \qquad |x_1^{(N+1)} - x_1^{(N)}| \leq \rho|x^{(N)} - x^{(N-1)}|$$

*ρ = rho. In this context it stands for 'ratio', namely, the ratio of $|x_1^{(N+1)} - x_1^{(N)}|$ to $|x_1^{(N)} - x_1^{(N-1)}|$.

holds; that is, the size of the step is decreased by a factor of ρ.

Now let $(y, x_2, x_3, \ldots, x_n)$ be given. Using the initial approximation (4) gives

$$x_1^{(1)} = \bar{x}_1 + \frac{y - f(\bar{x}_1, x_2, x_3, \ldots, x_n)}{a}.$$

Thus $x_1^{(1)}$ lies within δ of $\bar{x}_1$ provided the number

$$d = \left| \frac{y - f(\bar{x}_1, x_2, x_3, \ldots, x_n)}{a} \right|$$

is less than δ. When this is the case the estimate (6) applies to $|x_1^{(2)} - x_1^{(1)}|$ and gives

$$\begin{aligned} |x_1^{(2)} - \bar{x}_1| &\le |x_1^{(2)} - x_1^{(1)}| + |x_1^{(1)} - \bar{x}_1| \\ &\le (\rho + 1)|x_1^{(1)} - \bar{x}_1| \\ &= (\rho + 1)d. \end{aligned}$$

Therefore $x_1^{(2)}$ lies within δ of $\bar{x}_1$ provided

$$(\rho + 1)d < \delta.$$

When this is the case the estimate (6) applies to $|x_1^{(3)} - x_1^{(2)}|$ and gives

$$\begin{aligned} |x_1^{(3)} - \bar{x}_1| &\le |x_1^{(3)} - x_1^{(2)}| + |x_1^{(2)} - \bar{x}_1| \\ &\le \rho|x_1^{(2)} - x_1^{(1)}| + (\rho + 1)d \\ &\le \rho^2|x_1^{(1)} - \bar{x}_1| + (\rho + 1)d \\ &= (\rho^2 + \rho + 1)d. \end{aligned}$$

Therefore $x_1^{(3)}$ lies within δ of $\bar{x}_1$ provided

$$(\rho^2 + \rho + 1)d < \delta.$$

Repeating this argument N times, one finds that $x_1^{(N)}$ lies within δ of $\bar{x}_1$ provided

$$(\rho^{N-1} + \rho^{N-2} + \cdots + \rho^2 + \rho + 1)d < \delta.$$

Multiplying by the positive number $1 - \rho$ this inequality can be restated

$$(1 - \rho^N)d < (1 - \rho)\delta.$$

This holds for *all* N if (and only if)

$$d \le (1 - \rho)\delta.$$

This proves: If $(y, x_2, x_3, \ldots, x_n)$ are given such that $|x_i - \bar{x}_i| < \delta$ $(i = 2, 3, \ldots, n)$ and such that

$$\left|\frac{y - f(\bar{x}_1, x_2, x_3, \ldots, x_n)}{a}\right| < (1 - \rho)\delta$$

(where $0 < \rho < 1$ and where δ is chosen as above when ρ is given) then the sequence $x_1^{(N)}$ defined by (3) with the initial approximation (4) lies entirely in the interval $\{|x_1 - \bar{x}_1| < \delta\}$ and the estimate (6) always applies. But the estimate (6) shows that the Cauchy Convergence Criterion is satisfied, because for $M > N$ it gives

$$\begin{aligned}
|x_1^{(M)} - x_1^{(N)}| &\le |x_1^{(M)} - x_1^{(M-1)}| + |x_1^{(M-1)} - x_1^{(M-2)}| + \cdots + |x_1^{(N+1)} - x_1^{(N)}| \\
&\le \rho^{M-1}|x_1^{(1)} - x_1^{(0)}| + \rho^{M-2}|x_1^{(1)} - x_1^{(0)}| + \cdots + \rho^{N}|x_1^{(1)} - x_1^{(0)}| \\
&= (\rho^{M-1} + \rho^{M-2} + \cdots + \rho^{N})\left|\frac{y - f(\bar{x}_1, x_2, \ldots, x_n)}{a}\right| \\
&\le (\rho^{M-1} + \rho^{M-2} + \cdots + \rho^{N})(1 - \rho)\delta \\
&= (\rho^{N} - \rho^{M})\delta < \rho^{N}\delta.
\end{aligned}$$

This not only proves that the sequence $x_1^{(N)}$ is convergent but also gives an explicit estimate of the *rate* of convergence: All terms past $x_1^{(N)}$ differ from $x_1^{(N)}$ by at most $\rho^N\delta$.

Let $x_1^{(\infty)} = \lim_{N\to\infty} x_1^{(N)}$. Then passing to the limit in (3) gives

$$x_1^{(\infty)} = x_1^{(\infty)} + \frac{y - f(x_1^{(\infty)}, x_2, \ldots, x_n)}{a}$$

which implies

$$y = f(x_1^{(\infty)}, x_2, x_3, \ldots, x_n).$$

That is, $\lim_{N\to\infty} x_1^{(N)}$ is indeed a solution of the given equation. Moreover, if $\tilde{x}_1$ is any solution of the given equation which satisfies $|\tilde{x}_1 - \bar{x}_1| < \delta$, then

$$\begin{aligned}
|x_1^{(N+1)} - \tilde{x}_1| &= \left|x_1^{(N)} + \frac{y - f(x_1^{(N)}, x_2, \ldots, x_n)}{a} - \tilde{x}_1\right| \\
&= \left|x_1^{(N)} - \tilde{x}_1 + \frac{f(\tilde{x}_1, x_2, \ldots, x_n) - f(x_1^{(N)}, x_2, \ldots, x_n)}{a}\right| \\
&= \left|\Delta x - \frac{\Delta y}{a}\right| \le \frac{\epsilon}{|a|}|\Delta x| \\
&= \rho|x_1^{(N)} - \tilde{x}_1|.
\end{aligned}$$

Thus $x_1^{(N+1)}$ is nearer to $\tilde{x}_1$ than $x_1^{(N)}$ is, which implies that $\lim_{N\to\infty} x_1^{(N)} = \tilde{x}_1$. [More formally, passing to the limit in the inequality above gives $|x_1^{(\infty)} - \tilde{x}_1| \le \rho|x_1^{(\infty)} - \tilde{x}_1|$, $(1-\rho)|x_1^{(\infty)} - \tilde{x}_1| \le 0$, $|x_1^{(\infty)} - \tilde{x}_1| \le 0$, $x_1^{(\infty)} = \tilde{x}_1$.]

In summary, if $(y, x_2, x_3, \ldots, x_n)$ satisfies

$$|x_2 - \bar{x}_2| < \delta, |x_3 - \bar{x}_3| < \delta, \ldots, |x_n - \bar{x}_n| < \delta$$
$$|y - f(\bar{x}_1, x_2, x_3, \ldots, x_n)| < |a|(1-\rho)\delta$$

(which is true for all $(y, x_2, x_3, \ldots, x_n)$ near $(\bar{y}, \bar{x}_2, \bar{x}_3, \ldots, \bar{x}_n)$) then there is a unique solution x_1 in the interval $\{|x_1 - \bar{x}_1| < \delta\}$ of the equation $y = f(x_1, x_2, \ldots, x_n)$, and this solution is the limit of the sequence defined by (3) and (4). Here a is $\dfrac{\partial y}{\partial x_1}$ at $(\bar{x}_1, \bar{x}_2, \ldots, \bar{x}_n)$, ρ is an arbitrary number in the range $0 < \rho < 1$, and δ is chosen so that $\dfrac{\partial y}{\partial x_1}$ differs from a by less than $\rho|a|$ throughout the 'cube' $\{|x_i - \bar{x}_i| < \delta;\ i = 1, 2, \ldots, n\}$.

*In the strict sense that the sequence (3), (4) can be used to compute the value of g to any prescribed degree of accuracy.

This defines* a function $g(y, x_2, x_3, \ldots, x_n)$ near $(\bar{y}, \bar{x}_2, \bar{x}_3, \ldots, \bar{x}_n)$ with the desired property. It remains to show that this function is differentiable. Now

$$\frac{g(\bar{y} + sh_1, \bar{x}_2 + sh_2, \ldots, \bar{x}_n + sh_n) - g(\bar{y}, \bar{x}_2, \ldots, \bar{x}_n)}{s}$$
$$= \frac{x_1^{(\infty)} - \bar{x}_1}{s}$$
$$= \frac{1}{s}[(x_1^{(1)} - \bar{x}_1) + (x_1^{(2)} - x_1^{(1)}) + (x_1^{(3)} - x_1^{(2)}) + \cdots].$$

The first term is the principal term, and the remaining terms satisfy

$$\left|\frac{1}{s}[(x_1^{(2)} - x_1^{(1)}) + (x_1^{(3)} - x_1^{(2)}) + \cdots\right|$$
$$\le \frac{1}{|s|}(\rho + \rho^2 + \rho^3 + \cdots)|x_1^{(1)} - \bar{x}_1|$$
$$= \frac{\rho}{1-\rho}|\text{first term}|.$$

The first term is given explicitly by the formula

$$\frac{x_1^{(1)} - \bar{x}_1}{s} = \frac{\bar{y} + sh_1 - f(\bar{x}_1, \bar{x}_2 + sh_2, \ldots, \bar{x}_n + sh_n)}{sa}$$
$$= \frac{1}{a}\left[h_1 - \frac{f(\bar{x}_1, \bar{x}_2 + sh_2, \ldots, \bar{x}_n + sh_n) - f(\bar{x}_1, \bar{x}_2, \ldots, \bar{x}_n)}{s}\right].$$

By making s small, this can be made to differ arbitrarily little from

$$\left(\frac{\partial y}{\partial x_1}\right)^{-1}\left(h_1 - \frac{\partial y}{\partial x_2}h_2 - \frac{\partial y}{\partial x_3}h_3 - \cdots - \frac{\partial y}{\partial x_n}h_n\right)$$

where the partial derivatives are evaluated at $(\bar{x}_1, \bar{x}_2, \ldots, \bar{x}_n)$. The remaining terms are at most $\rho/(1 - \rho)$ times this term in absolute value; but ρ can be made arbitrarily small (when δ is sufficiently small, which means s must also be small) so this proves that

$$\lim_{s\to 0} \frac{g(\bar{y} + sh_1, \bar{x}_2 + sh_2, \ldots, \bar{x}_n + sh_n) - g(\bar{y}, \bar{x}_2, \ldots, \bar{x}_n)}{s}$$

exists* and is equal to

*Moreover, if f is uniformly differentiable in the sense of §9.3 these estimates prove that this limit is approached uniformly; that is, g is uniformly differentiable.

$$\left(\frac{\partial y}{\partial x_1}\right)^{-1}\left(h_1 - \frac{\partial y}{\partial x_2}h_2 - \cdots - \frac{\partial y}{\partial x_n}h_n\right).$$

That is, the partial derivatives of g at $(\bar{y}, \bar{x}_2, \bar{x}_3, \ldots, \bar{x}_n)$ exist and can be found by implicit differentiation. But if $(\tilde{y}, \tilde{x}_2, \tilde{x}_3, \ldots, \tilde{x}_n)$ is any point at which g is defined, then g is the solution of $y = f(x_1, x_2, \ldots, x_n)$ near $\tilde{y} = f(\tilde{x}_1, \tilde{x}_2, \ldots, \tilde{x}_n)$ where $\tilde{x}_1 = g(\tilde{y}, \tilde{x}_2, \ldots, \tilde{x}_n)$ and, by the above argument, the partial derivatives of g at $(\tilde{y}, \tilde{x}_2, \ldots, \tilde{x}_n)$ therefore exist and are equal to

$$\left(\frac{\partial y}{\partial x_1}\right)^{-1}, -\left(\frac{\partial y}{\partial x_1}\right)^{-1}\left(\frac{\partial y}{\partial x_2}\right), \ldots, -\left(\frac{\partial y}{\partial x_1}\right)^{-1}\left(\frac{\partial y}{\partial x_n}\right)$$

where these partial derivatives are evaluated at $(g(\tilde{y}, \tilde{x}_2, \ldots, \tilde{x}_n), \tilde{x}_2, \ldots, \tilde{x}_n)$. It follows that g is continuous (otherwise its partial derivatives could not exist) and therefore that its partial derivatives are continuous (because they are compositions of continuous functions). Therefore g is differentiable and the proof is complete.

Using the Elimination Theorem, the Implicit Function Theorem can now be proved by the method of step-by-step elimination as in §5.3. The Implicit Function Theorem can also be proved directly by the method of successive approximations as follows:

It suffices to show that r equations in n unknowns

$$(7) \qquad y_i = f_i(x_1, x_2, \ldots, x_n) \qquad (i = 1, 2, \ldots, r)$$

can be solved locally for

$$x_i = g_i(y_1, \ldots, y_r, x_{r+1}, \ldots, x_n) \qquad (i = 1, 2, \ldots, r)$$

provided

$$\frac{\partial(y_1, y_2, \ldots, y_r)}{\partial(x_1, x_2, \ldots, x_r)} \neq 0 \tag{8}$$

because the remaining statements of the Implicit Function Theorem follow from the Chain Rule. To solve the equations (7) near a given point

$$\bar{y}_i = f_i(\bar{x}_1, \bar{x}_2, \ldots, \bar{x}_n) \qquad (i = 1, 2, \ldots, r)$$

at which (8) is satisfied, one can use the method of successive approximations as follows: Given $(y_1, y_2, \ldots, y_r, x_{r+1}, \ldots, x_n)$ near $(\bar{y}_1, \bar{y}_2, \ldots, \bar{y}_r, \bar{x}_{r+1}, \ldots, \bar{x}_n)$ and given an approximate solution $(x_1^{(N)}, x_2^{(N)}, \ldots, x_r^{(N)})$ of (7), the approximation can be 'corrected' by setting

$$\Delta y_i = \sum_{j=1}^{r} \frac{\partial y_i}{\partial x_j} \Delta x_j \qquad (i = 1, 2, \ldots, r) \tag{9}$$

where the partial derivatives are evaluated at* $(\bar{x}_1, \bar{x}_2, \ldots, \bar{x}_n)$, by setting Δy_i equal to

It would seem more reasonable to evaluate the partial derivatives at $(x_1^{(N)}, x_2^{(N)}, \ldots, x_r^{(N)}, x_{r+1}, \ldots, x_n)$. This is Newton's method (see §7.3); it gives a more efficient process for solving the equations, but the proof of convergence is more difficult.

$$\text{desired } \Delta y_i = y_i - f_i(x_1^{(N)}, \ldots, x_r^{(N)}, x_{r+1}, \ldots, x_n) \tag{10}$$

by solving (9) [which is possible by the assumption (8)] to obtain values for Δx_j $(j = 1, 2, \ldots, r)$ and by setting

$$x_j^{(N+1)} = x_j^{(N)} + \Delta x_j \qquad (j = 1, 2, \ldots, r). \tag{11}$$

Together with the initial approximation

$$x_j^{(0)} = \bar{x}_j \qquad (j = 1, 2, \ldots, r) \tag{12}$$

the equations (9), (10), (11) define a sequence of successive approximations $(x_1^{(N)}, x_2^{(N)}, \ldots, x_r^{(N)})$ to the solution of the given equations. When $r = 1$ this is precisely the sequence (3), (4) used in the proof above. By the same method of proof as that which was used above, it can be shown that the sequence $(x_1^{(N)}, x_2^{(N)}, \ldots, x_r^{(N)})$ converges to a solution $(x_1^{(\infty)}, x_2^{(\infty)}, \ldots, x_r^{(\infty)})$ of the given equations for $(y_1, y_2, \ldots, y_r, x_{r+1}, \ldots, x_n)$ sufficiently near $(\bar{y}_1, \bar{y}_2, \ldots, \bar{y}_r, \bar{x}_{r+1}, \ldots, \bar{x}_n)$ and that this solution depends differentiably on $(y_1, y_2, \ldots, y_r, x_{r+1}, \ldots, x_n)$. The only added complication is that when $r > 1$, the solution of (9) requires a matrix inversion and the relation

$$|\Delta y - L(\Delta x)| \leq \epsilon |\Delta x|$$

[where L is the matrix of partial derivatives in (9)] must be shown to imply an inequality of the form

$$|L^{-1}(\Delta y) - \Delta x| \leq \rho|\Delta x|$$

($0 < \rho < 1$) when ϵ is sufficiently small.

Exercises

1 Solve $y = u^2 + v^2$ for u given $y = 1, v = .01$ by setting $(\bar{y}, \bar{u}, \bar{v}) = (1, 1, 0)$ and using the successive approximations (3), (4) to find $u^{(3)}$. What accuracy is guaranteed for this answer by the formula $|u^{(N)} - u^{(3)}| \leq \rho^3\delta$ of the text? Do the same calculations for $v = \epsilon$ instead of $v = .01$ and compare the result to the Taylor series expansion of the function $\sqrt{1 - x^2}$. [The terms in ϵ^6 agree, but the term of ϵ^8 in $u^{(3)}$ is incorrect.]

2 The sequence (3), (4) can be seen graphically as follows: Let $y = f(x)$ be a real-valued function of one variable, let $\bar{y} = f(\bar{x})$ be a given value and let $\bar{\bar{y}}$ be a value near $\bar{y}$. Draw the graph of $y = f(x)$, draw the tangent line to the graph at $(\bar{x}, \bar{y})$, and draw the horizontal line $y = \bar{\bar{y}}$. The point $(x^{(1)}, \bar{\bar{y}})$ is the point of intersection of these two lines. To construct the point $(x^{(2)}, \bar{\bar{y}})$, follow the vertical line $x = x^{(1)}$ to the point $(x^{(1)}, f(x^{(1)}))$, then follow the line parallel to the tangent line back to $y = \bar{\bar{y}}$. In the same way, one goes from $(x^{(N)}, \bar{\bar{y}})$ to $(x^{(N+1)}, \bar{\bar{y}})$ by first going to the point $(x^{(N)}, f(x^{(N)}))$, and by then constructing the line parallel to the original tangent line [at $(\bar{x}, \bar{y})$] and following it back to $y = \bar{\bar{y}}$. Show that this process indeed corresponds to the formula (2). Draw this picture for a few specific functions and show that it converges in all reasonable cases. Looked at under a microscope directed at the limit point $(x^{(\infty)}, \bar{\bar{y}})$, the graph of f appears as a (nearly) straight line. Let $y = b(x - x^{(\infty)}) + \bar{\bar{y}}$ be the equation of this line; that is, let $b = f'(x^{(\infty)})$, and let a be the slope of the original tangent at $\bar{x}$, i.e. $a = f'(\bar{x})$. Draw a picture showing the process 'under a microscope' as a polygonal path consisting of vertical lines and lines of slope a bouncing back and forth between a horizontal line and a line of slope b. Write exact formulas for this process (it is essentially the geometric series) and give an exact criterion for its convergence or divergence for various values of a, b. Give an estimate then of the asymptotic rate of convergence of the iteration (2)—that is, the rate at which the error decreases with each step after a large number of steps have been taken—in terms of $f'(\bar{x})$ and $f'(x^{(\infty)})$.

3 Show that if f is k times continuously differentiable then so is g. [The partial derivatives of f are $k - 1$ times differentiable and g is continuously differentiable which shows, by the explicit expression for the partials of g, that g is twice continuously differentiable when $k > 1$. If $k > 2$, then the partial derivatives of g are twice continuously differentiable because they are a composition of twice differentiable functions; hence g is thrice differentiable. Using the fact that a composition of two functions which are j times continuously differentiable is itself j times continuously differentiable (prove this using the Chain Rule) the process continues inductively to show that g is k times continuously differentiable.]

7.2 Solution of Linear Equations

This section deals with the problem of finding n numbers $(x_1, x_2, \ldots, x_n)$ satisfying

$$\begin{aligned} a_{11}x_1 + a_{12}x_2 + \cdots + a_{1n}x_n &= y_1 \\ a_{21}x_1 + a_{22}x_2 + \cdots + a_{2n}x_n &= y_2 \\ \vdots \qquad \vdots \qquad\qquad \vdots \quad &\quad \vdots \\ a_{n1}x_1 + a_{n2}x_2 + \cdots + a_{nn}x_n &= y_n \end{aligned}$$

when the n numbers $(y_1, y_2, \ldots, y_n)$ and the $n \times n$ matrix of coefficients (a_{ij}) are given. The letter x will denote the n-tuple $x = (x_1, x_2, \ldots, x_n)$, the letter y the n-tuple $y = (y_1, y_2, \ldots, y_n)$ and the letter L the $n \times n$ matrix of coefficients (a_{ij}), so that the given equations can be written simply $Lx = y$. Given L and y, the problem is to find x.

Virtually all procedures for solving $Lx = y$ assume that an *approximate* inverse to L can be found*; that is, they assume that an $n \times n$ matrix M can be found such that for all x the approximation $MLx \sim x$ is true in some sense. If the exact inverse $M = L^{-1}$ can be found, then of course the problem is completely solved by setting $x = L^{-1}y$. Normally it is not feasible to find an exact inverse, however, and in fact the necessity of rounding makes it virtually impossible to find an exact inverse in most cases. On the other hand, it is often not difficult to find an approximate inverse. In many cases which occur in practice the mapping L can be well approximated by a mapping whose inverse is known explicitly—that is, L can be regarded as a perturbation of an explicitly invertible mapping—and in any case a process of step-by-step elimination can be carried out with some degree of approximation (often a very crude approximation will suffice) to give an approximate inverse M.

*Or that L itself is 'nearly the identity', i.e. $Lx \sim x$.

Assume therefore that a matrix M has been found such that $MLx \sim x$. There are two basic techniques for constructing successive approximations to a solution of $Lx = y$, namely:

(a) *Correction of* $x^{(N)}$. If $x^{(N)}$ is an approximate solution $Lx^{(N)} \sim y$ then the method of §7.1 gives

$$\begin{aligned}\text{desired } \Delta y &= L(\text{desired } \Delta x)\\ M(\text{desired } \Delta y) &= ML(\text{desired } \Delta x)\\ &\sim \text{desired } \Delta x\\ \text{desired } \Delta x &\sim M(y - Lx^{(N)}).\end{aligned}$$

Hence one defines

$$x^{(N+1)} = x^{(N)} + M(y - Lx^{(N)}) \tag{1}$$

as the next approximation to a solution of $Lx = y$.

(b) *Correction of* M_N. If M_N is an approximate inverse $M_N L \sim I$ (where I is the identity map) then

$$\begin{aligned}M_N L - I &\sim 0\\ (M_N L - I)^2 &\sim 0\\ M_N L M_N L - 2M_N L + I &\sim 0\\ [2M_N - M_N L M_N]L &\sim I\end{aligned}$$

hence one defines

$$M_{N+1} = 2M_N - M_N L M_N \tag{2}$$

as the next approximation to a solution of $ML = I$. (If $ML - I$ is 'small' in some sense, then its square is 'smaller'.)

Clearly the correction (2) of M, which requires two matrix multiplications, involves much more computation than the correction (1) of x. On the other hand, if (1) is used repeatedly with the same approximation M, then one is repeatedly committing the same error, which means that the rate of convergence may be poor in relation to the amount of arithmetic which is required. Thus, as is always the case in large-scale computation, judgment must be exercised in choosing a method of computation, taking into account the accuracy of the data, the degree of accuracy required of the solution, and the type of computing machinery being used.

The procedure (1) is improved considerably by the

following simple observation: The numbers $x_i^{(N+1)}$ are defined by the equations

$$x_i^{(N+1)} = x_i^{(N)} + i\text{th component of } My - i\text{th component of } MLx^{(N)}.$$

The normal procedure for carrying out the computation indicated by this formula would be to find $x_1^{(N+1)}$, $x_2^{(N+1)}, \ldots, x_n^{(N+1)}$ in turn. However, this means that *at the time* $x_i^{(N+1)}$ *is being computed, the values* $x_1^{(N+1)}$, $x_2^{(N+1)}, \ldots, x_{i-1}^{(N+1)}$ *are already known.* Since these new values are presumably more accurate than the old values $x_1^{(N)}, x_2^{(N)}, \ldots, x_{i-1}^{(N)}$, it is only reasonable to use them in place of the old values in (1), i.e. to change (1) to

$$(1') \quad x_i^{(N+1)} = x_i^{(N)} + i\text{th component of } My - i\text{th component of } ML \text{ applied to } (x_1^{(N+1)}, \ldots, x_{i-1}^{(N+1)}, x_i^{(N)}, \ldots, x_n^{(N)}).$$

Although the formula (1′) is very clumsy, the computation it prescribes is in fact much more simple—as well as more accurate—than that prescribed by (1). The method it prescribes can be stated: Correct each of the approximations $x_1, x_2, \ldots, x_n$ in turn and then begin again correcting $x_1, x_2, \ldots$, etc., *ad infinitum.* More briefly: Correct the components of x in cyclic order.

This algorithm is eminently practical. Given L, y, and an approximate inverse M, one computes the n numbers My and the n^2 numbers ML. One then chooses a first approximation $(x_1, x_2, \ldots, x_n)$ to a solution of $Lx = y$ on the basis of whatever information is available ($(0, 0, \ldots, 0)$ is the simplest choice) and performs the correction

$$(1'') \quad \text{new } x_i = x_i + i\text{th component of } (My - MLx)$$

in cyclic order. After performing each correction (1″) the old value of x_i is discarded, so that at any time only $n^2 + 2n$ numbers need to be 'remembered' ($n^2 + n$ for the statement of the problem and n for the latest approximation to the answer). If the approximate inverse M is sufficiently good, then the process in fact converges, as will be proved below, to a solution of the equation $Lx = y$. In other words, beyond a certain point one will find that further corrections do not significantly change $(x_1, x_2, \ldots, x_n)$. This set of numbers is then, except for roundoff errors, a solution of the given equation $Lx = y$.

This algorithm is rendered even more practical by the observation that after the correct values of the x_i have been obtained to a few significant digits they can be set aside and a new process of successive approximation can be set up for the remaining digits. This is done merely by setting $x_i = \bar{x}_i + z_i$ where $\bar{x}_i$ is the approximation already obtained and where z_i is the remainder to be found. Then (1″) becomes

$$\bar{x}_i + \text{new } z_i = \bar{x}_i + z_i + i\text{th component of } (My - ML(\bar{x} + z)).$$

Hence

$$\text{new } z_i = z_i + i\text{th component of } (w - MLz)$$

where $w = My - ML\bar{x}$. Using this method for computing the less significant digits z_i of x_i eliminates the redundant effort of computing $w = My - ML\bar{x}$ to several significant digits with each application of (1″). A further modification of the method (1″) which may be useful is that the order in which the components x_i are corrected can be altered from a strict cyclical order if it appears in the course of the computation that some components are more in need of correction than others. Such a modification of the order of correction is called a *relaxation*.

In proving the convergence of the algorithm (1″) it is useful to notice that it is in fact the algorithm (1) with M changed to $M' = (I + T)^{-1}M$ where T is the 'lower triangular' matrix which is equal to ML below the main diagonal (see Exercise 6). Thus to find criteria for the convergence of (1″) it suffices to find criteria for the convergence of (1) and to apply them to the case where $M' = (I + T)^{-1}M$.

The simplest condition on the approximate inverse M which guarantees the convergence of (1) is the condition* that the approximation $MLx \sim x$ on which it is based satisfy

*This condition is exactly analogous to the condition $|\frac{\Delta y}{a} - \Delta x| \le \rho|\Delta x|$ of §7.1 where the 'approximate inverse was $\frac{1}{a}$.

(3) $$|MLx - x| \le \rho|x|$$

for some fixed $\rho < 1$ where $|x|$ denotes the maximum of $|x_i|$ for $i = 1, 2, \ldots, n$ (that is, $B = |x|$ describes the smallest cube $\{|x_i| \le B;\ i = 1, 2, \ldots, n\}$ which contains x) and where $|MLx - x|$ is defined accordingly. If the condition (3) is satisfied, if $x^{(0)}$ is an arbitrary n-tuple of numbers, if the sequence $x^{(N)}$ is defined by (1), and if $\bar{x}$ is any solution of $L\bar{x} = y$, then

$$\begin{aligned}|x^{(N+1)} - \bar{x}| &= |x^{(N)} - \bar{x} + M(L\bar{x} - Lx^{(N)})| \\ &\le \rho|x^{(N)} - \bar{x}|.\end{aligned}$$

Hence each step reduces the 'distance' from $x^{(N)}$ to $\bar{x}$ by a factor of ρ and

$$\lim_{N\to\infty} x^{(N)} = \bar{x}. \tag{4}$$

Thus *if* there is a solution of $Lx = y$ it is the limit of $x^{(N)}$ regardless of the choice of $x^{(0)}$. This implies in particular that there is *at most one* solution, i.e. L is one-to-one. But, by dimensionality, L must therefore be onto. That is to say, $Lx = y$ always does have a solution $\bar{x}$ which is therefore given by (4). This proves that *if condition* (3) *is satisfied and if y is given, then the sequence* (1) *is convergent and its limit is the unique solution x of the equation $Lx = y$ regardless of the choice of the initial approximation $x^{(0)}$.*

The same condition

$$|(ML - I)x| \le \rho|x| \qquad (\text{fixed } \rho < 1, \text{ all } x) \tag{3}$$

implies that the sequence M_N defined by (2) with $M_0 = M$ converges to L^{-1}. This is proved by noting that

$$\begin{aligned} M_N L - I = (M_{N-1}L - I)^2 &= (M_{N-2}L - I)^4 \\ = \cdots &= (ML - I)^{2^N}. \end{aligned}$$

Hence for any y the solution x of $Lx = y$ guaranteed by (3) satisfies

$$|M_N y - x| = |(M_N L - I)x| \le \rho^{2^N}|x|,$$

and hence $\lim_{N\to\infty} M_N y = x = L^{-1}y$. This holds for all y, which is precisely the meaning of the statement that $\lim_{N\to\infty} M_N = L^{-1}$.

It is easily seen that the condition (3) is satisfied if (and only if) the matrix M is such that the matrix $(b_{ij}) = ML - I$ satisfies

$$\max_{i=1,2,\ldots,n} \left\{ \sum_{j=1}^{n} |b_{ij}| \right\} < 1. \tag{5}$$

For a given matrix M it is a simple matter to check whether or not this condition is satisfied. Thus (5) is an easily verifiable condition which is *sufficient* for the convergence of the processes (1) and (2). Other *sufficient* conditions for the convergence of (1) and (2) can be obtained by observing that if $|x|$ denotes any norm (see §9.8) then condition (3) guarantees the convergence of

(1) and (2) by exactly the same proof (see Exercises 7, 8). A *necessary* and sufficient condition for convergence is that

$$\text{(6)} \qquad \lim_{N\to\infty} (ML - I)^N \to 0$$

where by convergence of (1) one means regardless of the choice of $x^{(0)}$ (since obviously (1) converges if $x^{(0)}$ is a solution to begin with, no matter what M might be). However, the condition (6) is usually difficult to affirm or deny and the condition (3) is more useful in practice, even though convergence can still occur when it is not fulfilled.

Exercises

1 *Simple iteration.* A simple iterative procedure for solving $\sum a_{ij}x_j = y_i$ is given by moving all but the diagonal term to the right and setting

$$a_{ii}x_i^{(N+1)} = y_i - \sum_{j\neq i} a_{ij}x_j^{(N)}.$$

This gives n equations, one for each i, defining $x^{(N+1)}$ in terms of $x^{(N)}$. Find an 'approximate inverse' M of $L = (a_{ij})$ such that this process is the process (1). For which matrices L is the condition (5) satisfied?

2 *Gauss-Seidel iteration.* A more elementary method than that of simple iteration is the corresponding iteration (1′) i.e.

$$a_{ii}x_i^{(N+1)} = y_i - \sum_{j<i} a_{ij}x_j^{(N+1)} - \sum_{j>i} a_{ij}x_j^{(N)}.$$

What general characteristics should the matrix $L = (a_{ij})$ have in order for this iteration to converge rapidly? Note that in this case the condition (5) is more difficult to state explicitly in terms of L. Which method, this one or that of Exercise 1, would you expect to converge more rapidly and why?

3 Solve the system

$$\begin{aligned} 10x - y + 2z &= 11 \\ x - 9y + z &= -5 \\ -3x + 4y + 34z &= 15 \end{aligned}$$

by explicit elimination (expressing the answer first in terms of rational numbers, then as a decimal fraction) by simple iteration, and by Gauss-Seidel iteration. Use a desk calculator if possible.

4 Set

$$M_0 = \begin{pmatrix} \frac{1}{10} & 0 & 0 \\ 0 & -\frac{1}{9} & 0 \\ 0 & 0 & \frac{1}{34} \end{pmatrix}$$

and use the iteration (2) to find an approximation to L^{-1} where L is the matrix of the preceding exercise. Use a desk calculator. Using the answer, solve the system of Exercise 3.

5 Having found the solution of the system of Exercise 3 to three places by the Gauss-Seidel iteration, set aside these three places by the method suggested in the text and set up a new iteration for the higher-order decimals.

6 Show that the method (1′) of the text is the same as the method (1) with M changed to $M' = (I + T)^{-1}M$ where T is the matrix whose entry in the ith row and the jth column is the corresponding entry of ML if $j < i$ and is zero if $j \geq i$. [Add and subtract $Tx^{(N)}$ on the right and add $Tx^{(N+1)}$ on both sides.] Show that $(I + T)^{-1} = I - T + T^2 - T^3 + \cdots \pm T^{n-1}$.

7 If $|x|$ denotes $\sum_{i=1}^{n} |x_i|$ what condition analogous to (5) guarantees (3)?

8 If $|x|$ denotes $\left[\sum_{i=1}^{n} |x_i|^p\right]^{1/p}$ $(p > 1)$ what condition analogous to (5) guarantees (3)? [Use the Hölder inequality, let μ_i be the 'q-norm' of the ith row, $\mu_i = \left[\sum_j |b_{ij}|^q\right]^{1/q}$, where $(1 - q)(1 - p) = 1$, and let ρ be the 'p-norm' of the μ's, $\rho = [\sum \mu_i^p]^{1/p}$.]

9 Show that the condition (3) implies that $|x^{(N)} - \bar{x}| \leq \frac{\rho^N}{1 - \rho} |x^{(1)} - x^{(0)}|$ where $\bar{x}$ is the limit of (1). In what way is this estimate more useful than the simpler estimate $|x^{(N)} - \bar{x}| \leq \rho^N |x^{(0)} - \bar{x}|$?

10 Prove directly from the condition (3) that L is one-to-one and onto. [This is easy.] Prove directly from condition (3) that the sequence (1) converges and that its limit is a solution. [Follow the method of §7.1. It must be shown that M is one-to-one and onto if (3) holds.]

11 Let A_N be a sequence of $n \times n$ matrices and let B be an $n \times n$ matrix. Show that $\lim_{N\to\infty} A_N x = Bx$ for all x if and only if all entries of A_N converge to the corresponding entries of B as $N \to \infty$.

12 Set up and solve several systems of linear equations choosing systems of a size and complexity suitable to the computing machinery at your disposal.

13 Prove that the Gauss-Seidel method (Exercise 2) converges to a solution of $Lx = y$ regardless of the choice of $x^{(0)}$ whenever L is a *positive definite symmetric matrix*, i.e. $L = (a_{ij})$ where $a_{ij} = a_{ji}$ and where the quadratic form $Q(x_1, x_2, \ldots, x_n) = \sum a_{ij}x_ix_j$ is positive except when $x_1 = x_2 = \cdots = x_n = 0$. [Let $\bar{x}$ be the solution of $L\bar{x} = y$. Each 'correction' of the Gauss-Seidel procedure moves x along a line parallel to a coordinate axis to the point on that line where $Q(x - \bar{x})$ has its minimum. Therefore the value of $Q(x - \bar{x})$ is non-increasing as the process continues. If $Q(x - \bar{x})$ does not decrease during a complete cycle of n steps then x must be $\bar{x}$. By Exercise 6 there is a matrix N such that a complete cycle of n steps carries x to $\bar{x} + N(x - \bar{x})$. The function $Q[N(x - \bar{x})]$ must have a maximum on the 'ellipsoid' $Q(x - \bar{x}) = 1$ (see §9.4) and, by the above, this maximum is less than 1. Call it ρ. Then a complete cycle of n corrections decreases $Q(x - \bar{x})$ by a factor of ρ so that $Q(x - \bar{x}) \to 0$ which implies $x \to \bar{x}$ as desired.]

7.3 Newton's Method

In §7.1 the equation $y = f(x)$ was solved for x near $\bar{x}$ given y near $\bar{y} = f(\bar{x})$ by taking $x^{(0)} = \bar{x}$ as the initial approximation and by obtaining successive 'corrections' of the approximate solution $x^{(N)}$ using

$$(1) \quad f(x^{(N+1)}) - f(x^{(N)}) \sim f'(\bar{x})(x^{(N+1)} - x^{(N)})$$

and the desired formula

$$f(x^{(N+1)}) = y$$

to obtain

$$(2) \qquad x^{(N+1)} = x^{(N)} + \frac{y - f(x^{(N)})}{f'(\bar{x})}.$$

The approximation (1) is improved by using $f'(x^{(N)})$ instead of $f'(\bar{x})$, which would lead one to believe that the method (2) would be improved by changing it to

$$(3) \qquad x^{(N+1)} = x^{(N)} + \frac{y - f(x^{(N)})}{f'(x^{(N)})}.$$

This is Newton's method.

Similarly, to solve r equations in r unknowns

$$y_i = f_i(x_1, x_2, \ldots, x_r) \qquad (i = 1, 2, \ldots, r)$$

Newton's method is to 'correct' an approximate solution $(x_1, x_2, \ldots, x_r)$ in three steps:

$$\Delta y_i = y_i - f_i(x_1, x_2, \ldots, x_r)$$
$$\Delta y_i = \sum_{j=1}^{r} \frac{\partial y_i}{\partial x_j} \Delta x_j$$
$$\text{new } x_j = x_j + \Delta x_j$$

where the first equation defines the quantities Δy_i, the second implicitly defines the Δx_j (the partial derivatives are evaluated at $(x_1, x_2, \ldots, x_r)$) and the third defines the new x_j's.

Newton's method (3) is more accurate and more natural than the method (2), but it does not lend itself to the purposes of §7.1 because it is not easily proved to be convergent.* However, it is easily shown that *if* $y = f(x)$ *has a solution, if the initial approximation* $x^{(0)}$ *is sufficiently close to this solution, and if* f *is 'reasonable', then the procedure (3) converges very rapidly to the solution*: Let $\bar{x}$ denote the solution $y = f(\bar{x})$ whose existence is assumed. Then the distance of the $(N + 1)$st approximation from $\bar{x}$ is estimated by

*This is entirely analogous to the situation in §7.2, where the method (1') was more accurate and more natural than the method (1) but could not as easily be proved to be convergent.

$$\begin{aligned} x^{(N+1)} - \bar{x} &= x^{(N)} + \frac{y - f(x^{(N)})}{f'(x^{(N)})} - \bar{x} \\ &= \int_{\bar{x}}^{x^{(N)}} dx + \frac{1}{f'(x^{(N)})} \int_{x^{(N)}}^{\bar{x}} f'(x)\, dx \\ &= \int_{\bar{x}}^{x^{(N)}} \frac{f'(x^{(N)}) - f'(x)}{f'(x^{(N)})}\, dx. \end{aligned}$$

Thus, if $x^{(N)} - \bar{x}$ is small, it follows that $x^{(N+1)} - \bar{x}$ is doubly small because it is the integral of a small function over a small interval. Specifically, if 'reasonableness' of f near $\bar{x}$ is defined to mean that f'' exists and is continuous and that $f'(\bar{x}) \neq 0$ (otherwise one could set $g(x) = f'(x)$ and find $\bar{x}$ by solving $g(x) = 0$ using Newton's method, unless $g'(\bar{x}) = f''(\bar{x}) = 0$, in which case one could set $g(x) = f''(x)$, etc.) and if '$x^{(0)}$ sufficiently near $\bar{x}$' is defined to mean that $|x^{(0)} - \bar{x}| < \delta$ where δ is a number satisfying

$$|f''(x)| \leq B \quad \text{for} \quad |x - \bar{x}| < \delta$$
$$|f'(x)| \geq A \quad \text{for} \quad |x - \bar{x}| < \delta$$
$$\delta < \frac{A}{B}$$

then

$$|f'(x^{(N)}) - f'(x)| = \left|\int_{\bar{x}}^{x^{(N)}} f''(x)\,dx\right| \leq B|x - x^{(N)}|.$$

Hence the integrand in the integral above is bounded by $\frac{B}{A}|\bar{x} - x^{(N)}|$ and the integral itself is bounded by $\frac{B}{A}|\bar{x} - x^{(N)}|^2$. Hence

$$(4) \qquad |x^{(N+1)} - \bar{x}| \leq \frac{B}{A}|x^{(N)} - \bar{x}|^2$$

provided $|x^{(N)} - \bar{x}| < \delta$. Therefore $x^{(N+1)}$ is nearer to $\bar{x}$ by a factor of $\frac{B}{A}|x^{(N)} - \bar{x}| \leq \frac{B}{A}\delta < 1$. This proves not only that (3) converges provided $|x^{(0)} - \bar{x}| < \delta$ but also that the convergence is very rapid: The error after $N + 1$ steps is a constant times the square of the error after N steps, which means roughly that the number of decimal places of accuracy in $x^{(N)}$ doubles with each step, whereas in the cruder method (2) the number of decimal places of accuracy increases by roughly the same amount with each step (for instance, if $\rho = .01$ it increases by 2). Although these estimates assume that the solution $\bar{x}$ is known, the method (3) itself does not. Thus in practice one can merely apply (3) and observe whether the successive $x^{(N)}$ are converging; if they are, then (4) can be used to estimate the rate of convergence.

In the case of r equations in r unknowns the same argument shows that if $x^{(N)}$ is sufficiently near to a solution $\bar{x}$ where the Jacobian is not zero and if the functions are twice differentiable then

$$(4') \qquad |x^{(N+1)} - \bar{x}| \leq \text{const.}|x^{(N)} - \bar{x}|^2$$

and convergence is again very rapid once it begins.

Exercises

1 Show that Newton's method applied to the equation $y = x^2$ gives

$$x^{(N+1)} = \tfrac{1}{2}\left(x^{(N)} + \frac{y}{x^{(N)}}\right).$$

This method of finding square roots was known and used in ancient times. Use it to find $\sqrt{2}$ to 11 decimal places. [Use a desk calculator if possible.]

2 Give Newton's method for finding $\sqrt[3]{y}$. Find $\sqrt[3]{2}$ to 7 decimal places.

3 How many decimal places of π must be retained in finding $\sqrt{\pi}$ to 6 decimal places? Find $\sqrt{\pi}$ to 6 places.

4 Find five decimal places of the root of the cubic polynomial $8x^3 - 12x - 1$ of §5.4, p. 164, which lies between -1 and 0.

5 Draw a diagram analogous to that of Exercise 2, §7.1, showing the convergence of Newton's method to a solution of $y = f(x)$ where f is a real-valued function of one variable.

6 Show that if Newton's method is used to solve $y = \dfrac{1}{x}$ the result is essentially the iteration (2) of §7.2. Find an explicit formula for the Nth approximation to $(1 - \epsilon)^{-1}$, beginning with $x^{(0)} = 1$. [It can be written as a product of N simple factors.]

7.4 Solution of Ordinary Differential Equations

The method of successive approximations which was used to construct a solution of an equation of the form $y = f(x)$ can also be used to construct a solution of a differential* equation

$$\frac{dx}{dt} = f(x) \tag{1}$$

in which the unknown is a function $x(t)$.

*Actually the equation $dx/dt = f(x)$ should be called a 'derivative equation', the differential equation being $dx - f(x)\,dt = 0$ (see §8.5). However, the term 'differential equation' is universally used in the sense of 'equation involving derivatives'. An ordinary differential equation is one which involves ordinary derivatives, as opposed to partial differential equations, which involve partial derivatives.

Theorem

Let f: $\mathbf{R}^n \to \mathbf{R}^n$ be a differentiable function defined near a point $\bar{x}$ of $\mathbf{R}^n$. Then the differential equation (1) together with the initial condition $x(0) = \bar{x}$ defines a unique function $x(t)$ for t near zero and this function depends differentiably on $\bar{x}$. More specifically, there is a number $\delta > 0$ and a differentiable function x: $\mathbf{R} \to \mathbf{R}^n$ defined for $\{|t| \leq \delta\}$ such that $(x_1(0), x_2(0), \ldots, x_n(0)) = (\bar{x}_1, \bar{x}_2, \ldots, \bar{x}_n)$ and such that

$$\begin{aligned} \frac{dx_1}{dt} &= f_1(x_1(t), x_2(t), \ldots, x_n(t)) \\ \frac{dx_2}{dt} &= f_2(x_1(t), x_2(t), \ldots, x_n(t)) \\ &\vdots \\ \frac{dx_n}{dt} &= f_n(x_1(t), x_2(t), \ldots, x_n(t)). \end{aligned} \tag{1'}$$

Any other function defined on $\{|t| \leq \delta\}$ and satisfying these conditions is identical with $x(t)$ for $\{|t| \leq \delta\}$. Finally, if this function is written $(x_1(\bar{x}_1, \bar{x}_2, \ldots, \bar{x}_n, t), \ldots, x_n(\bar{x}_1, \bar{x}_2, \ldots, \bar{x}_n, t))$, making explicit its dependence on $(\bar{x}_1, \bar{x}_2, \ldots, \bar{x}_n)$, then each $x_i(\bar{x}_1, \bar{x}_2, \ldots, \bar{x}_n, t)$ is a differentiable function of its $n + 1$ variables.

Proof

The unknown in this theorem is a *function* $\mathbf{R} \to \mathbf{R}^n$ rather than a point of $\mathbf{R}^n$ as it was in the preceding sections. Let F denote the operation $F(x(t)) = \dfrac{dx}{dt} - f[x(t)]$ assigning to each function x: $\mathbf{R} \to \mathbf{R}^n$ a new function $F(x)$: $\mathbf{R} \to \mathbf{R}^n$. The problem is then to find a function x such that $F(x) = 0$. Thus in the formula

$$\text{desired } \Delta y = L(\text{chosen } \Delta x)$$

the desired Δy is

$$0 - F(x^{(N)}) = -\frac{dx^{(N)}}{dt} + f[x^{(N)}(t)].$$

The operation L assigning to each function x: $\mathbf{R} \to \mathbf{R}^n$ a new function $L(x)$: $\mathbf{R} \to \mathbf{R}^n$ should be an *invertible* operation which approximates the operation F. If L is taken to be ordinary differentiation $L(x) = \dfrac{dx}{dt}$, then the inverse of L is given by the Fundamental Theorem of Calculus. Taking all approximate solutions $x^{(N)}$ to satisfy $x^{(N)}(0) = \bar{x}$ this leads to

$$\text{desired } \Delta y = L(\text{chosen } \Delta x)$$

$$-\frac{dx^{(N)}}{dt} + f[x^{(N)}(t)] = \frac{dx^{(N+1)}}{dt} - \frac{dx^{(N)}}{dt}$$

$$\int_0^t \left[-\frac{dx^{(N)}}{dt}(u) + f[x^{(N)}(u)]\right] du = x^{(N+1)}(t) - x^{(N)}(t)$$

which gives $x^{(N+1)}$ in terms of $x^{(N)}$ as

(2) $$x^{(N+1)}(t) = \bar{x} + \int_0^t f[x^{(N)}(u)]\, du.$$

Together with the initial approximation

(3) $$x^{(0)}(t) \equiv \bar{x}$$

the formula* (2) defines an infinite sequence of functions

*The method of successive approximations was used to prove the existence of solutions of differential equations by Emile Picard. The iterative formula (2) is known as Picard's iteration.

$x^{(N)}(t)$, each of which is defined on some interval containing $t = 0$. (From the fact that $x^{(N)}(u)$ is defined for u sufficiently near zero it follows that $x^{(N+1)}(u)$ is defined for u sufficiently near zero, hence that $x^{(N+2)}(u)$ is defined for u sufficiently near zero, etc.) The main step in the proof of the theorem is to show that *there is a* $\delta > 0$ *such that the functions* $x^{(N)}(t)$ *are all defined for* $|t| \leq \delta$, *the limit* $\lim_{N\to\infty} x^{(N)}(t) = x^{(\infty)}(t)$ *exists for* $|t| \leq \delta$, *the limit function* $x^{(\infty)}(t)$ *is differentiable and satisfies the differential equation* (1), *and any function* $\tilde{x}(t)$ *which satisfies* (1) *and* $\tilde{x}(0) = \bar{x}$ *must be identical with* $x^{(\infty)}$ *for* $|t| \leq \delta$. It will then be shown that $x^{(\infty)}(t)$ depends differentiably on $\bar{x}$.

Let B be a number such that $f: \mathbf{R}^n \to \mathbf{R}^n$ is defined at all points x within B of $\bar{x}$, i.e. at all points x satisfying $\{|x - \bar{x}| \leq B\}$ where $|x|$ denotes $\max\{|x_1|, |x_2|, \ldots, |x_n|\}$, and let* A be a number which is larger than all values of $|f_i(x_1, x_2, \ldots, x_n)|$ at points x in $\{|x - \bar{x}| \leq B\}$ for $i = 1, 2, \ldots, n$. (Intuitively, A is a bound on the velocity specified by (1).) Setting $\delta = B/A$ and assuming $x^{(N)}(t)$ is defined and lies in $\{|x - \bar{x}| \leq B\}$ for $|t| \leq \delta$ gives

*Such a number A exists because a continuous function on a cube $\{|x - \bar{x}| \leq B\}$ is bounded (see §9.4).

$$|x_i^{(N+1)}(t) - \bar{x}_i| = \left| \int_0^t f_i[x^{(N)}(u)]\, du \right| \leq A|t| \leq B.$$

That is, $x^{(N+1)}(t)$ is defined and lies in $\{|x - \bar{x}| \leq B\}$. Since $x^{(0)} \equiv \bar{x}$ is defined and lies in $\{|x - \bar{x}| \leq B\}$ for $|t| \leq \delta$, this proves that the entire sequence $x^{(0)}(t)$, $x^{(1)}(t), x^{(2)}(t), \ldots$ is defined and lies inside $\{|x - \bar{x}| \leq B\}$ for all t in the interval $|t| \leq \delta$. (Intuitively, a particle whose velocity never exceeds A in any direction cannot move a distance of B in any direction in less than time B/A.)

The fact that the sequence $\bar{x}, x^{(1)}(t), x^{(2)}(t), \ldots$ converges for $|t| \leq \delta$ is proved, as before, by estimating the size of each step $|x^{(N+1)}(t) - x^{(N)}(t)|$. First of all

$$|x^{(1)}(t) - x^{(0)}(t)| = \left| \int_0^t f[x^{(0)}(u)]\, du \right| \leq A \cdot |t|$$

as above. Then

$$|x^{(2)}(t) - x^{(1)}(t)| = \left| \int_0^t \{f[x^{(1)}(u)] - f[x^{(0)}(u)]\}\, du \right| \cdot$$

The integrand is estimated, as in §5.3, by writing

$f(x^{(1)}) - f(x^{(0)})$ as an integral of the partial derivatives of f over a broken line from $x^{(0)}$ to $x^{(1)}$, which gives

$$(4) \qquad |f(x^{(1)}) - f(x^{(0)})| \leq K|x^{(1)} - x^{(0)}|$$

*See preceding note.

where* K is a number such that K/n is larger than all values of $|\partial f_i/\partial x_j|$ at points of $\{|x - \bar{x}| \leq B\}$ for all i, j. This gives

$$\begin{aligned} |x^{(2)}(t) - x^{(1)}(t)| &\leq \left| K \int_0^t |x^{(1)}(u) - x^{(0)}(u)|\, du \right| \\ &\leq \left| K \int_0^t A \cdot |u|\, du \right| \\ &= KA \frac{|t|^2}{2}. \end{aligned}$$

Similarly

$$\begin{aligned} |x^{(3)}(t) - x^{(2)}(t)| &= \left| \int_0^t \{f[x^{(2)}(u)] - f[x^{(1)}(u)]\}\, du \right| \\ &\leq \left| K \int_0^t |x^{(2)}(u) - x^{(1)}(u)|\, du \right| \\ &\leq \left| K \int_0^t KA \frac{|u|^2}{2}\, du \right| = K^2 A \frac{|t|^3}{3 \cdot 2} \end{aligned}$$

and in general

$$(5) \qquad |x^{(N+1)}(t) - x^{(N)}(t)| \leq K^N A \frac{|t|^{N+1}}{(N+1)!}.$$

Choose an integer J such that $J > K \cdot |t|$ and let $\rho = \dfrac{K \cdot |t|}{J} < 1$. Then for $M > N \geq J$ the estimate

$$\begin{aligned} &|x^{(M)}(t) - x^{(N)}(t)| \\ &\quad \leq |x^{(N+1)}(t) - x^{(N)}(t)| + \cdots + |x^{(M)}(t) - x^{(M-1)}(t)| \\ &\quad \leq K^N A \frac{|t|^{N+1}}{(N+1)!} + \cdots + K^{M-1} A \frac{|t|^M}{M!} \\ &\quad = \frac{K^J A |t|^{J+1}}{(J+1)!} \left[\frac{K|t|}{J+2} \cdot \frac{K|t|}{J+3} \cdot \cdots \cdot \frac{K|t|}{N+1} + \cdots + \frac{K|t|}{J+2} \cdot \frac{K|t|}{J+3} \cdot \cdots \cdot \frac{K|t|}{M} \right] \\ &\quad \leq \frac{K^J A |t|^{J+1}}{(J+1)!} [\rho^{N-J} + \cdots + \rho^{M-J-1}] \\ &\quad = \text{const.}\, (\rho^{N-J} - \rho^{M-J}) \leq \text{const.}\, \rho^N \end{aligned}$$

proves that the Cauchy Criterion is satisfied; hence the sequence $x^{(N)}(t)$ converges.

Moreover, since this estimate is uniform for t in the interval $|t| \leq \delta$—that is, given $\epsilon > 0$ there is a N such that $|x^{(m)}(t) - x^{(n)}(t)| < \epsilon$ whenever $m, n \geq N$ and whenever $|t| \leq \delta$—it follows* not only that the limit function $x^{(\infty)}(t)$ is continuous but also that one can pass to the limit under the integral sign in the equation

*See §9.6 and §9.7. Also Exercise 7.

$$x^{(N+1)}(t) = \bar{x} + \int_0^t f[x^{(N)}(u)]\,du$$

to obtain the equation

$$x^{(\infty)}(t) = \bar{x} + \int_0^t f[x^{(\infty)}(u)]\,du.$$

This proves that

$$\frac{dx^{(\infty)}}{dt} = \lim_{h\to 0} \frac{x^{(\infty)}(t+h) - x^{(\infty)}(t)}{h} = f[x^{(\infty)}(t)].$$

Thus $x^{(\infty)}(t)$ is differentiable and satisfies the differential equation (1).

If $\tilde{x}(t)$ is any other function on $\{|t| \leq \delta\}$ which satisfies the differential equation (1) and satisfies the 'initial condition' $\tilde{x}(0) = \bar{x}$, then as long as $\tilde{x}(t)$ remains in $\{|x - \bar{x}| \leq B\}$ the estimates

$$|\tilde{x}(t) - x^{(0)}(t)| = \left|\int_0^t f[\tilde{x}(u)]\,du\right| \leq A|t|$$

$$\begin{aligned}|\tilde{x}(t) - x^{(1)}(t)| &= \left|\int_0^t \{f[\tilde{x}(u)] - f[x^{(0)}(u)]\}\,du\right| \\ &\leq \left|K\int_0^t |\tilde{x}(u) - x^{(0)}(u)|\,du\right| \\ &\leq \left|K\int_0^t A|u|\,du\right| = KA\,\frac{|t|^2}{2}\end{aligned}$$

and similarly

$$|\tilde{x}(t) - x^{(N)}(t)| \leq K^N A\,\frac{|t|^{N+1}}{(N+1)!}$$

all hold. Hence, $\tilde{x}(t) = \lim_{N\to\infty} x^{(N)}(t) = x^{(\infty)}(t)$ for as long as $\tilde{x}(t)$ remains in $\{|x - \bar{x}| \le B\}$. But since $x^{(\infty)}(t)$ lies inside $\{|x - \bar{x}| < B\}$ for $|t| < \delta$ this means $\tilde{x}$ cannot leave $\{|x - \bar{x}| \le B\}$ without becoming discontinuous. Since $\tilde{x}$ is differentiable by assumption, it cannot be discontinuous, hence $\tilde{x}$ stays in $\{|x - \bar{x}| \le B\}$ and $\tilde{x}(t) = x^{(\infty)}(t)$ for all t in $\{|t| \le \delta\}$. (More precisely, $\tilde{x}(t)$ lies inside $\{|x - \bar{x}| < B\}$ for some open interval of t containing $t = 0$. The largest such open interval must contain $\{|t| \le \delta\}$ because otherwise the argument above would show that it could be extended.)

It remains only to show that $x^{(\infty)}(t)$ depends differentiably on $\bar{x}$. Now $x^{(0)}(t) \equiv \bar{x}$ clearly depends differentiably on $\bar{x}$, and if $x^{(N)}(t)$ depends differentiably on $\bar{x}$ then so does

$$x^{(N+1)}(t) = \bar{x} + \int_0^t f[x^{(N)}(u)]\, du$$

by differentiation under the integral sign (see Exercise 5, §9.4). Thus the functions $x^{(N)}(t)$ all depend differentiably on $\bar{x}$ and the problem is to show that this is also true of their limit $x^{(\infty)}(t)$. Let $(h_1, h_2, \ldots, h_n)$ be a fixed n-tuple, and let $x_s^{(N)}(t)$ denote the Nth approximation to the solution of (1) for the initial condition $\bar{x} + sh$ where s is a (small) real number. Then

$$x_s^{(N)} - x_0^{(N)} = \int_0^s \frac{\partial}{\partial s}[x_s{}^{(N)}(t)]\, ds$$

for all N. The method of proof is to show that the functions $\dfrac{\partial}{\partial s}[x^{(N)}(t)]$ approach a limit $F(s, t)$ as $N \to \infty$, that $F(s, t)$ depends continuously on $\bar{x}$, t, and that integration can be interchanged with passage to the limit as $N \to \infty$ to give

$$x_s^{(\infty)}(t) - x_0^{(\infty)}(t) = \int_0^s F(s, t)\, ds.$$

Then dividing by s and letting $s \to 0$ will give

$$\frac{\partial}{\partial s} x_s^{(\infty)} = F(s, t)$$

at $s = 0$, and the theorem will follow. Thus the essence of the proof is to examine the dependence of the functions $\dfrac{\partial}{\partial s}[x_s^{(N)}(t)]$ on N for large N.

Now

$$\left|\frac{\partial}{\partial s}[x_s^{(1)} - x_s^{(0)}]\right| = \left|\frac{\partial}{\partial s}\int_0^t f[\bar{x} + sh]\,du\right|$$
$$\leq \left|K|h|\int_0^t du\right| = K|h|\,|t|.$$

$$\left|\frac{\partial}{\partial s}[x_s^{(2)} - x_s^{(1)}]\right| = \left|\frac{\partial}{\partial s}\int_0^t \{f[x_s^{(1)}] - f[x_s^{(0)}]\}\,du\right|$$
$$\leq \left|\int_0^t K\left|\frac{\partial}{\partial s}[x_s^{(1)} - x_s^{(0)}]\right|du\right|$$
$$\leq \left|\int_0^t K^2|h|\,|u|\,du\right| = K^2|h|\,\frac{|t|^2}{2}$$

and similarly

$$\left|\frac{\partial}{\partial s}[x_s^{(N+1)} - x_s^{(N)}]\right| \leq |h|K^N\frac{|t|^N}{N!}.$$

Therefore

$$\left|\frac{\partial}{\partial s}x_s^{(M)} - \frac{\partial}{\partial s}x_s^{(N)}\right| \leq |h|K^{M-1}\frac{|t|^{M-1}}{(M-1)!} + \cdots + |h|K^N\frac{|t|^N}{N!}.$$

Since this number can be made small for large N and for all $M > N$ (as was shown above), it follows that the functions $\frac{\partial}{\partial s}x_s^{(N)}$ approach a limit as $N \to \infty$ and that this limit is approached in such a way (namely, the error is uniformly small for all s) that the limit function is continuous and that the integral of the limit is the limit of the integrals. This completes the proof of the theorem.

Exercises **1** *The trigonometric functions.* Geometrically, a differential equation can be imagined as a rule specifying an arrow at each point of $\mathbf{R}^n$ and a solution can be imagined as a parameterized curve whose velocity at each point is equal to the specified arrow.

(a) Sketch the arrows in the plane specified by the equation

$$\frac{dx}{dt} = -y$$
$$\frac{dy}{dt} = x.$$

That is, the velocity in the x-direction is minus the y-coordinate, and in the y-direction it is plus the x-coordinate.

(b) Prove that if $(x(t), y(t))$ is any solution of the given equation then the value of $[x(t)]^2 + [y(t)]^2$ is constant. Interpret this geometrically.

(c) *Define* the functions $\cos t$, $\sin t$ by taking them to be the two coordinates of that solution of the above equation which begins at $(1, 0)$ at time 0. [Hence by (b) $\cos^2 t + \sin^2 t \equiv 1$.] Use Picard's iteration (2) to give a convergent power series representation of $(\cos t, \sin t)$ valid for all t. [Show that $x^{(N)}(t)$ converges as $N \to \infty$ for all t.]

(d) Express the solution which begins at a given point $(\bar{x}, \bar{y})$ in terms of $\bar{x}$, $\bar{y}$, $\cos t$, $\sin t$.

(e) Use (d) and the uniqueness of the solution of a differential equation to prove the addition formula of the trigonometric functions, i.e. the formulas for $\cos(a + b)$, $\sin(a + b)$.

(f) Express the addition formula as a statement about the product of two matrices of the form

$$\begin{pmatrix} \cos a & \sin a \\ -\sin a & \cos a \end{pmatrix}.$$

(g) Express the addition formulas as De Moivre's law

$$\cos(a + b) + i \sin(a + b) = [\cos a + i \sin a][\cos b + i \sin b].$$

(h) Prove that $(\cos t, \sin t)$ moves around the circle in the monotone way suggested by the differential equation. [$\cos t$ decreases as long as $\sin t > 0$ and $\sin t$ can change sign only when $\cos t = \pm 1$.]

(i) Use (h) to *define* the number 2π; then *prove* that π is the area of the disk $x^2 + y^2 \leq 1$. [Stokes' theorem.]

(j) *Scaling.* In tabulating the functions cos, sin defined above there is a great practical advantage in listing the values $\cos t$, $\sin t$ for t equal to evenly-spaced rational multiples of 2π rather than for t equal to evenly-spaced rational numbers. Describe this advantage.

(k) Assuming that the value of π is known to five decimal places $\pi \sim 3.14159$, estimate $\sin 1°$ to four decimal places. [Use the power series of part (c).]

(l) In the same way, estimate $\sin 6°$ to four decimal places.

(m) Use the power series to estimate $\cos 6°$.

(n) Use these results and the addition formula to estimate $\sin 12°$. Thus trigonometric tables are quite easily constructed on the basis of an accurate estimate of π, the power series for $\sin t$, $\cos t$ for small values of t, and the addition formulas.

2 *The exponential function.*

(a) Define the exponential function $\exp(t) = e^t$ as the solution of a differential equation.

(b) Prove the 'addition formula' $\exp(x + y) = \exp(x)\exp(y)$.

(c) Find the power series expansion as in Exercise 1.

(d) If the number e is *defined* to be $\exp(1)$ then the statement

$$e = \lim_{n\to\infty}\left(1 + \frac{1}{n}\right)^n$$

becomes a theorem to be proved. Do so. [By the Fundamental Theorem of Calculus $\exp\left(\frac{1}{n}\right) - 1 \geq \frac{1}{n}$, $1 - \exp\left(-\frac{1}{n}\right) \leq \frac{1}{n}$. Use this to show that $\left(1 + \frac{1}{n}\right)^n$ is less than e but that if it is multiplied by $1 + \frac{1}{n}$ it becomes greater than e.]

(e) *Scaling.* In tabulating the function $\exp(x)$ there is a great practical advantage in first finding the number α such that $\exp(\alpha) = 10$ and listing the values of $\exp(t)$ for t equal to evenly-spaced rational multiples of α. Describe this advantage. Such a table is called a 'table of anti-logarithms to the base 10'.

(f) The number α above is called the natural logarithm of 10. Given that its value is 2.302585 to six places, estimate $10^{1.001}$ with bounds on the error. [$10^{1.001}$ is defined to be the one-thousandth root of 10^{1001}. Using $\exp(x + y) = \exp(x)\exp(y)$ it is easily expressed in terms of exp and α.]

3 *The hyperbolic functions.* Consider the solution of the differential equation

$$\frac{dx}{dt} = y$$

$$\frac{dy}{dt} = x$$

satisfying $x(0) = 1$, $y(0) = 0$. This solution is $(\cosh t, \sinh t)$ by definition of these functions, called the hyperbolic functions. The formula $\cos^2 t + \sin^2 t \equiv 1$, the addition formulas of the trigonometric functions, and the power series expansions of $\cos t$, $\sin t$ all have analogs for $\cosh t$, $\sinh t$. Derive these analogs.

4 *The exponential of a 2×2 matrix.* Show that Picard's method applied to the differential equation

$$\frac{dx}{dt} = ax + by$$

$$\frac{dy}{dt} = cx + dy$$

with the initial value $(x(0), y(0)) = (\bar{x}, \bar{y})$ yields the solution

$$\exp(tM)\begin{pmatrix}\bar{x}\\ \bar{y}\end{pmatrix}$$

where

$$M = \begin{pmatrix} a & b \\ c & d \end{pmatrix}$$

and where $\exp(tM)$ is a 2×2 matrix defined by an infinite series. Let $a + bi$ represent the 2×2 matrix

$$\begin{pmatrix} a & -b \\ b & a \end{pmatrix}$$

and find $\exp(a + bi)$. (Such a 2×2 matrix is called a 'complex number'.) Show that $\exp(z_1 + z_2) = \exp(z_1)\exp(z_2)$ for z_1, z_2 complex numbers. Show, on the other hand, that if

$$M_1 = \begin{pmatrix} 0 & 1 \\ 0 & 0 \end{pmatrix} \qquad M_2 = \begin{pmatrix} 0 & 0 \\ 1 & 0 \end{pmatrix}$$

then $\exp(M_1 + M_2) \neq \exp(M_1)\exp(M_2)$. Prove that if M_1, M_2 are 2×2 matrices such that $M_1M_2 = M_2M_1$ then $\exp(M_1 + M_2) = \exp(M_1)\exp(M_2)$. If the binomial theorem dealt with the functions $x^n/n!$ rather than the functions x^n what would its statement be?

5 *Cauchy's polygon.* The geometrical meaning of a differential equation $dx/dt = f(x)$ can be seen by constructing polygonal approximations to the solutions as follows: Given a large integer N let $P_0^{(N)}$ be the given initial point $\bar{x}$, let $P_1^{(N)}$ be the point $P_0^{(N)} + \frac{1}{N} f(P_0^{(N)})$, let $P_2^{(N)}$ be the point $P_1^{(N)} + \frac{1}{N} f(P_1^{(N)})$, and, in general, let

$$P_i^{(N)} = P_{i-1}^{(N)} + \frac{1}{N} f(P_{i-1}^{(N)})$$

for integers $i > 0$. The definition for $i < 0$ is

$$P_i^{(N)} = P_{i+1}^{(N)} - \frac{1}{N} f(P_{i+1}^{(N)}).$$

(a) For the equation $dx/dt = x$, $x(0) = 1$, plot the points $(i/N, P_i^{(N)})$ in the tx-plane, for $N = 10$ and $i = 0, \pm 1, \pm 2, \ldots, \pm 10$. Give the exact coordinates of these 21 points.

(b) Plot the 21 points $P_i^{(N)}$ for $N = 10$, $|i| \leq 10$ of the

*It can be shown that this is in fact true for an arbitrary equation $dx/dt = f(x)$; it was by this method that Cauchy proved the existence of a solution.

Cauchy polygon of the equation of Exercise 1 giving their coordinates exactly.

(c) Prove in the case of (a) that as $N \to \infty$ the polygon approaches the actual solution.*

6 Note that the functions $x(t) = (t + c)^3$ and $x(t) \equiv 0$ all satisfy $dx/dt = 3x^{2/3}$. Plot these curves in the tx-plane. Show that the equation

$$\frac{dx}{dt} = 3x^{2/3}, \quad x(0) = -1,$$

has infinitely many solutions. [One can stop for an arbitrary period of time at $x = 0$.] Why doesn't this contradict the theorem?

7 Given that a sequence of continuous functions $x^{(N)}(t)$ defined for $|t| \leq \delta$ has the property that "for every $\epsilon > 0$ there is an N such that for $n, m > N$ and $|t| \leq \delta$ the inequality $|x^{(n)}(t) - x^{(m)}(t)| < \epsilon$ holds," show that there is a uniquely determined limit function $x^{(\infty)}(t)$ defined for $|t| \leq \delta$ and that this function is continuous. Show, moreover, that if the condition (4) is satisfied then $\lim_{N\to\infty} \int_0^t f[x^{(N)}(s)]\, ds$ exists, $\int_0^t f[x^{(\infty)}(s)]\, ds$ exists, and the two are equal.

8 *Equations of higher order.* A kth order ordinary differential equation is an equation of the form

$$(*) \qquad F\left(\frac{d^k x}{dt^k}, \frac{d^{k-1}x}{dt^{k-1}}, \ldots, \frac{dx}{dt}, x, t\right) \equiv 0.$$

where F is a differentiable function of $k + 2$ variables. Prove that if numbers

$$x(0), \frac{dx}{dt}(0), \frac{d^2x}{dt^2}(0), \ldots, \frac{d^kx}{dt^k}(0)$$

are given which satisfy (*) with $t = 0$, and if (*) can be solved locally near the given point for d^kx/dt^k as a function of the remaining variables, then there is a $\delta > 0$ and a curve $x(t)$ defined for $|t| \leq \delta$ such that the equation (*) is satisfied and the derivatives d^jx/dt^j $(j = 0, 1, \ldots, k)$ have the specified values for $t = 0$. Show, moreover, that any solution of (*) with the specified values of $d^jx/dt^j(0)$ coincides with this one. [Introduce new variables $y_1 = dx/dt$, $y_2 = d^2x/dt^2, \ldots, y_k = d^kx/dt^k$ and apply the theorem of the text.]

9 Combine Exercise 8 and Exercise 1 to give an explicit solution of the equation

$$\frac{d^2x}{dt^2} = -x$$

where $x(t)$ is a real-valued function. This equation governs simple oscillations in which there is a 'restoring force' equal and opposite to the distance x from equilibrium.

10 *The exponential of an $n \times n$ matrix.* Generalize Exercise 4 to $n \times n$ matrices M by defining $\exp(M)$ and by giving a differential equation whose solutions can be expressed in terms of $\exp(tM)$. Under what circumstances is the formula $\exp(M_1)\exp(M_2) = \exp(M_1 + M_2)$ valid?

11 Generalizing Exercise 9, express the solution of the equation

$$(*) \qquad \frac{d^k x}{dt^k} + a_{k-1}\frac{d^{k-1}x}{dt^{k-1}} + \cdots + a_1\frac{dx}{dt} + a_0 x = 0,$$

in which $a_{k-1}, a_{k-2}, \ldots, a_1, a_0$ are numbers and $x(t)$ is a real-valued function, in terms of the exponential of a $k \times k$ matrix. An equation of the form (*) is described in words by saying that it is a *linear*, *homogeneous*, ordinary differential equation of order k with *constant coefficients*—linear because the left side is a linear function of the derivatives, homogeneous because the right side is zero, with constant coefficients because the a's are constant.

7.5 Three Global Problems

This section deals with three specific problems which are not of a local nature and which therefore cannot be solved by successive approximations.

Problem 1

Find π to several decimal places. (By Exercise 1, §7.4, this is virtually the same as the problem 'construct a trigonometric table'. Because of the practical value of trigonometric tables—in astronomy and navigation, for example—the historical importance of this problem is very great.)

By definition,

$$\pi = \int_D dx\,dy$$

where D is the disk $\{x^2 + y^2 \leq 1\}$ oriented $dx\,dy$. By Stokes' Theorem this is

$$\pi = \tfrac{1}{2}\int_{\partial D} (x\,dy - y\,dx)$$

which can be written as a single integral by parameterizing ∂D. A simple way of parameterizing the circle ∂D is

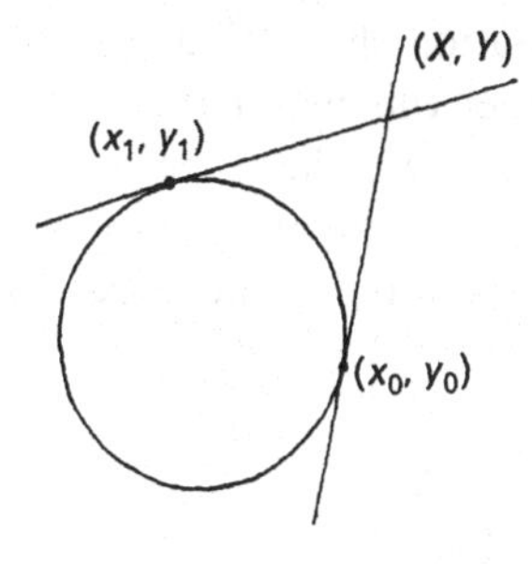

the following: Given two points (x_0, y_0), (x_1, y_1) on the circle, consider the point (X, Y) where the tangent lines to the circle at (x_0, y_0) and (x_1, y_1) intersect. The tangent line to the circle $x^2 + y^2 = \text{const.}$ at (x_0, y_0) is $x_0\,dx + y_0\,dy = 0$ (that is, $x_0(x - x_0) + y_0(y - y_0) = 0$) which gives the equation

$$x_0x + y_0y = 1$$

for this line. Parametrically it is given by

$$X = x_0 - ty_0$$
$$Y = y_0 + tx_0$$

(where the sign of t is chosen to agree with the counterclockwise orientation of the circle). If this point also lies on the line $x_1x + y_1y = 1$ tangent to the circle at (x_1, y_1) then

$$x_1(x_0 - ty_0) + y_1(y_0 + tx_0) = 1$$

from which

$$t = \frac{1 - x_0x_1 - y_0y_1}{x_0y_1 - x_1y_0}.$$

When (x_0, y_0) is fixed this establishes a correspondence between points (x_1, y_1) of the circle (other than $(x_1, y_1) = \pm(x_0, y_0)$) and real numbers t. To parameterize the circle, that is, to write (x_1, y_1) in terms of t, it suffices to note that by symmetry

$$X = x_1 + ty_1$$
$$Y = y_1 - tx_1$$

which gives

$$\begin{pmatrix} 1 & -t \\ t & 1 \end{pmatrix}\begin{pmatrix} x_0 \\ y_0 \end{pmatrix} = \begin{pmatrix} X \\ Y \end{pmatrix} = \begin{pmatrix} 1 & t \\ -t & 1 \end{pmatrix}\begin{pmatrix} x_1 \\ y_1 \end{pmatrix}$$

$$\begin{pmatrix} x_1 \\ y_1 \end{pmatrix} = \begin{pmatrix} 1 & t \\ -t & 1 \end{pmatrix}^{-1}\begin{pmatrix} 1 & -t \\ t & 1 \end{pmatrix}\begin{pmatrix} x_0 \\ y_0 \end{pmatrix}$$

$$= \begin{pmatrix} \dfrac{1}{1+t^2} & \dfrac{-t}{1+t^2} \\ \dfrac{t}{1+t^2} & \dfrac{1}{1+t^2} \end{pmatrix}\begin{pmatrix} 1 & -t \\ t & 1 \end{pmatrix}\begin{pmatrix} x_0 \\ y_0 \end{pmatrix}$$

$$x_1 = \frac{(1 - t^2)x_0 - 2ty_0}{1 + t^2}$$

$$y_1 = \frac{2tx_0 + (1 - t^2)y_0}{1 + t^2}.$$

A straightforward computation shows that the pullback of $\frac{1}{2}[x_1\,dy_1 - y_1\,dx_1]$ under this map (when x_0, y_0 are fixed) is simply $(1+t^2)^{-1}\,dt$. This proves that the integral of $\frac{1}{2}[x\,dy - y\,dx]$ over the arc from (x_0, y_0) to (x_1, y_1), oriented so that the arc does not pass through $(-x_0, -y_0)$, is

$$\int_0^T \frac{dt}{1+t^2}$$

where

$$T = \frac{1 - x_0x_1 - y_0y_1}{x_0y_1 - x_1y_0}.$$

If $|T| < 1$, then this integral can be computed using the geometric series

$$\frac{1}{1+t^2} = \frac{1}{1-(-t^2)} = 1 - t^2 + t^4 - t^6 + \cdots.$$

Specifically,

$$\begin{aligned}\int_0^T \frac{dt}{1+t^2} &= \int_0^T \frac{1-(-t^2)^n}{1-(-t^2)} + \int_0^T \frac{(-t^2)^n}{1-(-t^2)}\,dt \\ &= \int_0^T [1 - t^2 + t^4 - \cdots + (-t^2)^{n-1}]\,dt + (-1)^n \int_0^T \frac{t^{2n}}{1+t^2}\,dt \\ &= T - \frac{T^3}{3} + \frac{T^5}{5} - \cdots + (-1)^{n-1}\frac{T^{2n-1}}{2n-1} + R_n(T)\end{aligned}$$

where

$$|R_n(t)| = \left|\int_0^T \frac{t^{2n}}{1+t^2}\,dt\right| \le \left|\int_0^T t^{2n}\,dt\right| = \frac{|T|^{2n+1}}{2n+1}.$$

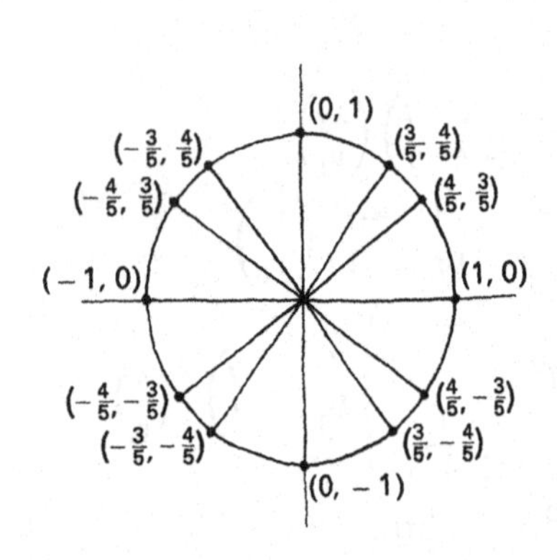

Thus the integral can be computed using the series

$$\int_0^T \frac{dt}{1+t^2} = T - \frac{T^3}{3} + \frac{T^5}{5} - \frac{T^7}{7} + \cdots$$

and the error resulting from taking only a finite number of terms is less than the size of the first term omitted.

If the entire circle is divided into arcs using the points $(\pm 1, 0)$, $(0, \pm 1)$, $(\pm\frac{4}{5}, \pm\frac{3}{5})$, $(\pm\frac{3}{5}, \pm\frac{4}{5})$ then there are 8 arcs like the arc from $(1, 0)$ to $(\frac{4}{5}, \frac{3}{5})$ for which

$$T = \frac{1 - 1\cdot(\frac{4}{5}) - 0\cdot(\frac{3}{5})}{1\cdot(\frac{3}{5}) - 0\cdot(\frac{4}{5})} = \frac{1}{3}$$

and 4 arcs like the arc from $(\frac{4}{5}, \frac{3}{5})$ to $(\frac{3}{5}, \frac{4}{5})$ for which

$$T = \frac{1 - (\frac{4}{5})(\frac{3}{5}) - (\frac{3}{5})(\frac{4}{5})}{(\frac{4}{5})(\frac{4}{5}) - (\frac{3}{5})(\frac{3}{5})} = \frac{1}{7}.$$

Hence

$$\pi = 8\int_0^{1/3} \frac{dt}{1+t^2} + 4\int_0^{1/7} \frac{dt}{1+t^2}$$
$$= 8[(\tfrac{1}{3}) - \tfrac{1}{3}(\tfrac{1}{3})^3 + \tfrac{1}{5}(\tfrac{1}{3})^5 - \cdots] + 4[(\tfrac{1}{7}) - \tfrac{1}{3}(\tfrac{1}{7})^3 + \tfrac{1}{5}(\tfrac{1}{7})^5 - \cdots].$$

Using these series it is easy to find π with 3-place accuracy:

$$\frac{8}{3} = 2.6667 \qquad \frac{8}{3\cdot 3^3} = \frac{8}{81} = .0988$$
$$\frac{8}{5\cdot 3^5} = \frac{8}{1215} = .0066 \qquad \frac{8}{7\cdot 3^7} = \frac{8}{15{,}309} = .0005$$
$$\frac{8}{9\cdot 3^9} = .0000 \qquad \cdots$$
$$\frac{4}{7} = .5714 \qquad \frac{4}{3\cdot 7^3} = \frac{4}{1029} = .0039$$
$$\frac{4}{5\cdot 7^5} = .0000 \qquad \cdots$$

2.6667	.0988	3.2447
.0066	.0005	− .1032
.5714	.0039	3.1415
3.2447	.1032	

Hence $\pi = 3.141 \pm 1$.

Many more decimal places of π can be found with relative ease by using the following decomposition of the circle into arcs: The arc from $(1, 0)$ to $(\frac{12}{13}, \frac{5}{13})$ corresponds to

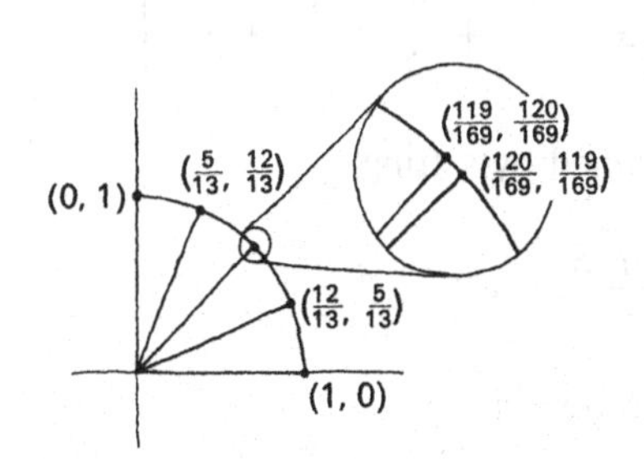

$$T = \frac{1 - 1\cdot(\frac{12}{13}) - 0\cdot(\frac{5}{13})}{1\cdot(\frac{5}{13}) - 0\cdot(\frac{12}{13})} = \frac{1}{5}.$$

Twice this arc is very nearly an eighth of the circle because $(x_0, y_0) = (\frac{12}{13}, \frac{5}{13})$ and $t = \frac{1}{5}$ gives

$$\begin{pmatrix} x_1 \\ y_1 \end{pmatrix} = \begin{pmatrix} \frac{1-t^2}{1+t^2} & \frac{-2t}{1+t^2} \\ \frac{2t}{1+t^2} & \frac{1-t^2}{1+t^2} \end{pmatrix} \begin{pmatrix} \frac{12}{13} \\ \frac{5}{13} \end{pmatrix} = \begin{pmatrix} \frac{12}{13} & -\frac{5}{13} \\ \frac{5}{13} & \frac{12}{13} \end{pmatrix} \begin{pmatrix} \frac{12}{13} \\ \frac{5}{13} \end{pmatrix} = \begin{pmatrix} \frac{119}{169} \\ \frac{120}{169} \end{pmatrix}$$

which is a point just past the point where $x = y$. Thus

16 of these arcs cover the entire circle with an excess of 4 times the arc from $(\frac{120}{169}, \frac{119}{169})$ to $(\frac{119}{169}, \frac{120}{169})$ for which

$$T = \frac{1 - 2(\frac{119}{169})(\frac{120}{169})}{(\frac{120}{169})^2 - (\frac{119}{169})^2} = \frac{(169)^2 - 2(119)(120)}{(120 - 119)(120 + 119)}$$

$$= \frac{28{,}561 - 28{,}560}{239} = \frac{1}{239}.$$

Thus

$$\pi = 16 \int_0^{1/5} \frac{dt}{1 + t^2} - 4 \int_0^{1/239} \frac{dt}{1 + t^2}.$$

Using this formula it is easy to compute 10 decimal places of π. (The fact that $\frac{1}{5} = 2 \cdot 10^{-1}$ simplifies the computation even further.)

Problem 2

Find log 10 to several decimal places. [As was seen in Exercise 2 of §7.4, this is the essential step in tabulating the function 10^x, that is, in constructing a table of antilogarithms to the base 10.]

The basic formulas here are

$$e^{\log x} = x \qquad \text{(definition of } \log x\text{)}$$

$$\log(xy) = \log x + \log y \qquad \text{(addition formula for } e^x\text{)}$$

and

$$\log\left(\frac{1+x}{1-x}\right) = 2\left[x + \frac{x^3}{3} + \frac{x^5}{5} + \frac{x^7}{7} + \cdots\right].$$

The last formula is obtained by setting

$$\frac{1+x}{1-x} = e^{y(x)}$$

and by differentiating to obtain

$$\frac{d}{dx}\left(\frac{1+x}{1-x}\right) = y'(x)e^{y(x)} = y'(x)\,\frac{1+x}{1-x}$$

$$y'(x) = \frac{1-x}{1+x} \cdot \frac{2}{(1-x)^2} = \frac{2}{1-x^2}$$

$$y(x) = y(x) - y(0) = \int_0^x \frac{2dt}{1-t^2}.$$

The error resulting from setting

$$\log\left(\frac{1+x}{1-x}\right) \sim 2\left[x + \frac{x^3}{3} + \cdots + \frac{x^{2n+1}}{2n+1}\right]$$

has absolute value at most

$$\left|2\int_0^x \frac{t^{2n}\,dt}{1-t^2}\right| \le \frac{2}{1-x^2}\int_0^x |t^{2n}|\,dt = \frac{2|x|^{2n+1}}{(1-x^2)(2n+1)}.$$

These formulas give an effective means of computing logarithms of numbers near 1. The computation of log 10 then becomes a matter of ingenuity in writing 10 as a product of several factors near 1. One way of doing this is the following:

$$2^{10} = 1024 = 10^3\left(\tfrac{1024}{1000}\right)$$

$$10^3 = 2^{10}\,\frac{1012-12}{1012+12} = 2^{10}\left(\frac{1-\frac{3}{253}}{1+\frac{3}{253}}\right).$$

To find the log of 10 it of course suffices to find the log of the number on the right in the preceding equation and divide by 3. The log of the second factor can be found to several places using the series above. For example, to five decimal places,

$$\begin{aligned}
\tfrac{3}{253} &= .011858\\
\left(\tfrac{3}{253}\right)^3 &= .000002\\
\log 10 &= \tfrac{10}{3}\log 2 - \tfrac{2}{3}\left[\left(\tfrac{3}{253}\right) + \tfrac{1}{3}\left(\tfrac{3}{253}\right)^3 + \cdots\right]\\
&= \tfrac{10}{3}\log 2 - \tfrac{2}{253} \pm 10^{-6}\\
&= \tfrac{10}{3}\log 2 - .007905 \pm 1.
\end{aligned}$$

Now to find log 2 one can write

$$\begin{aligned}
2 &= \left(\frac{6}{5}\right)^4 \cdot \frac{5^4}{3^4\cdot 2^3}\\
&= \left(\frac{12}{10}\right)^4 \frac{625}{648} = \left(\frac{12}{10}\right)^4\left(\frac{27}{28}\right)\left(\frac{4375}{4374}\right)\\
&= \left(\frac{11+1}{11-1}\right)^4\left(\frac{55-1}{55+1}\right)\left(\frac{8749+1}{8749-1}\right)
\end{aligned}$$

$$\log 2 = 4\log\left(\frac{1+\frac{1}{11}}{1-\frac{1}{11}}\right) - \log\left(\frac{1+\frac{1}{55}}{1-\frac{1}{55}}\right) + \log\left(\frac{1+\frac{1}{8749}}{1-\frac{1}{8749}}\right)$$

for example. Then $\frac{10}{3}\log 2$ can be computed to 6 decimal places relatively easily.

$$\frac{40}{3}\log\left(\frac{1+\frac{1}{11}}{1-\frac{1}{11}}\right) = \frac{80}{3}\left[\frac{1}{11} + \frac{1}{3\cdot(11)^3} + \frac{1}{5\cdot(11)^5} + \frac{1}{7\cdot(11)^7} + \cdots\right]$$

$$= \frac{80}{3\cdot 11} + \frac{80}{9\cdot(11)^3} + \frac{16}{3(11)^5} = \begin{array}{r} 2.424242 \\ .006678 \\ \underline{.000033} \\ 2.430953 \end{array}$$

$$\frac{10}{3}\log\left(\frac{1+\frac{1}{55}}{1-\frac{1}{55}}\right) = \frac{20}{3}\left[\frac{1}{55} + \frac{1}{3}\left(\frac{1}{55}\right)^3\right] = \begin{array}{r} .121212 \\ \underline{+\ .000013} \\ .121225 \end{array}$$

$$\frac{10}{3}\log\left(\frac{1+\frac{1}{8749}}{1-\frac{1}{8749}}\right) = \frac{20}{3\cdot 8749} = .000762.$$

Hence for log 10

$$\begin{array}{rr} 2.430953 & .007905 \\ \underline{.000762} & \underline{.121225} \\ 2.431715 & .129130 \end{array}$$

$$\begin{array}{r} 2.431715 \\ \underline{-\ .129130} \\ 2.302585 \end{array} \qquad \log 10 = 2.30259 \pm 1.$$

Problem 3

Prove that Newton's method converges to a solution of $y = x^n$ for any positive number y and any positive integer n regardless of the choice of the initial approximation $x_0 > 0$. (This proves in a very constructive way that every positive number has an nth root.)

Newton's method in this case can be written

$$\text{desired } \Delta y = y - x_N^n$$

$$\Delta y \sim n x_N^{n-1}\,\Delta x$$

$$\text{chosen } \Delta x = \frac{y - x_N^n}{n x_N^{n-1}} = \delta_N$$

$$x_{N+1} = x_N + \delta_N.$$

Then

$$x_{N+1}^n = (x_N + \delta_N)^n$$

$$= x_N^n + n x_N^{n-1}\delta_N + \frac{n(n-1)}{2} x_N^{n-2}\delta_N^2 + \cdots$$

$$= y + \frac{n(n-1)}{2} x_N^{n-2}\delta_N^2 + \binom{n}{3} x_N^{n-3}\delta_N^3 + \cdots$$

If x_N was too small ($x_N^n < y$) then $\delta_N > 0$ and x_{N+1} is too large. If x_N was too large then

$$\begin{aligned} 0 < y &\quad < x_N^n \\ -x_N^n < y - x_N^n &\quad < 0 \\ -x_N^n < n\delta_N x_N^{n-1} &\quad < 0 \\ -x_N < n\delta_N &\quad < 0. \end{aligned}$$

Thus δ_N is negative but $|\delta_N| < x_N/n$. In the expansion

$$x_{N+1}^n - y = \sum_{j=2}^{n} \binom{n}{j} x_N^{n-j}\delta_N^j$$

it is easily shown that the terms alternate in sign and decrease in size. Thus the sign of $x_{N+1}^n - y$ is the sign of the term $j = 2$ which is $+$; therefore x_{N+1} is again too large, and δ_{N+1} is negative. Regardless of the choice of x_0, then, the first approximation x_1 and all subsequent approximations are too large ($x_1 > x_2 > x_3 > \cdots$). Furthermore, the first term in the expansion of $x_{N+1}^n - y$ is larger than the actual value; hence

$$\begin{aligned} |\delta_{N+1}| &= \frac{x_{N+1}^n - y}{nx_{N+1}^{n-1}} < \frac{n-1}{2}\frac{x_N^{n-2}}{x_{N+1}^{n-1}}\delta_N^2 \\ &< \frac{n-1}{2}\frac{x_N^{n-2}x_{N+1}}{y}\delta_N^2 < K\delta_N^2 \end{aligned}$$

where

$$K = \frac{n-1}{2}\frac{x_1^{n-1}}{y}.$$

Now $x_N = x_1 + \delta_1 + \delta_2 + \cdots + \delta_N > 0$. If $|\delta_i| \geq \dfrac{1}{K}$ for $i = 1, 2, \ldots, N$, then

$$\begin{aligned} x_1 &\geq \frac{N}{K} \\ N &\leq Kx_1. \end{aligned}$$

Therefore, after more than Kx_1 steps at least one step must occur in which $|\delta_i| < \dfrac{1}{K}$. Then $|\delta_{i+1}| < K|\delta_i|\,|\delta_i| \leq \rho|\delta_i|$ where $\rho < 1$. Each succeeding step is smaller by a factor of at least ρ, the Cauchy Criterion is satisfied, and the sequence converges.

Exercises

1 Find π with 10-place accuracy.

2 Find log 2 with 8-place accuracy.

3 Find log 3 with 5-place accuracy.

4 Show that if $x = i = \sqrt{-1}$ is substituted in the series for $\log\left(\frac{1+x}{1-x}\right)$, and that if the series is considered as i times the integral of $\frac{1}{2}(x\,dy - y\,dx)$ over an arc, the result is $\log i = i\frac{\pi}{2}$. Show that this is a true formula if i is taken to be the 'complex number'

$$i = \begin{pmatrix} 0 & -1 \\ 1 & 0 \end{pmatrix},$$

if $\log x = y$ is defined to mean $x = e^y$, and if the exponential of a complex number is defined as in Exercise 4, §7.4.

applications

chapter 8

8.1

Vector Calculus

The so-called 'vector calculus' is not a separate calculus at all, but a particular notation for the calculus of three variables. Many physical quantities, notably force fields, flows, and gradients, are naturally imagined as vector fields, that is, as sets of arrows in xyz-space, one arrow to describe the quantity (of force, flow, or gradient) at each point of space. Analytically, a vector field is written $A\mathbf{i} + B\mathbf{j} + C\mathbf{k}$, where $\mathbf{i}$, $\mathbf{j}$, $\mathbf{k}$ are imagined as arrows of unit length in the x-, y-, z-directions respectively, and where the components A, B, C of the arrow

$$A\mathbf{i} + B\mathbf{j} + C\mathbf{k}$$

are functions of (x, y, z). In short, a vector field on xyz-space is described by three functions A, B, C on xyz-space, $A = A(x, y, z)$, etc., and is imagined geometrically as a field of arrows, of which A, B, C are the components. The letter $\mathbf{X} = A\mathbf{i} + B\mathbf{j} + C\mathbf{k}$ will denote such a vector field on xyz-space.

Line Integrals

The integral $\int_\Gamma (A\,dx + B\,dy + C\,dz)$ of a 1-form over an oriented curve Γ in xyz-space is also denoted by $\int_\Gamma \mathbf{X} \cdot d\mathbf{x}$, or some similar notation, where $\mathbf{X}$ is the vector field $A\mathbf{i} + B\mathbf{j} + C\mathbf{k}$ determined by the 1-form $A\,dx + B\,dy + C\,dz$, $d\mathbf{x}$ is an 'infinitesimal vector along the curve Γ', and $\mathbf{X} \cdot d\mathbf{x}$ is the 'dot product' of these two vectors. In whatever way the 'infinitesimal vector' $d\mathbf{x}$ is defined, the net effect is that the symbol $\int_\Gamma \mathbf{X}\cdot d\mathbf{x}$ *means* $\int_\Gamma (A\,dx + B\,dy + C\,dz)$. Thus if Γ is given parametrically by

$$(1)\qquad \begin{aligned} x &= f(t)\\ y &= g(t)\\ z &= h(t)\end{aligned} \qquad (a \le t \le b)$$

$\int_\Gamma \mathbf{X}\cdot d\mathbf{x}$ means the number $\int_a^b \left[A\frac{dx}{dt} + B\frac{dy}{dt} + C\frac{dz}{dt}\right] dt$ where A, B, C are functions of t by composition with (1).

Surface Integrals

The integral $\int_S (A\,dy\,dz + B\,dz\,dx + C\,dx\,dy)$ of a 2-form over an oriented surface S in xyz-space is also denoted by $\int_S \mathbf{X}\cdot\mathbf{n}\,d\sigma$, or some similar notation, where $\mathbf{X}$ is the vector field $A\mathbf{i} + B\mathbf{j} + C\mathbf{k}$ determined by the 2-form $A\,dy\,dz + B\,dz\,dx + C\,dx\,dy$, $\mathbf{n}$ the unit vector normal to the surface S (the sign of $\mathbf{n}$ is determined by the orientation of S according to the right hand rule), $d\sigma$ the area of an 'infinitesimal element' of the surface S, and $\mathbf{X}\cdot\mathbf{n}$ denotes the dot product. In whatever way the individual symbols $\mathbf{n}$ and $d\sigma$ are defined, the net effect is that the entire symbol $\int_S \mathbf{X}\cdot\mathbf{n}\,d\sigma$ *means* $\int_S (A\,dy\,dz + B\,dz\,dx + C\,dx\,dy)$ as defined in §6.3, or, less precisely, in §2.2.

Volume Integrals

The integral $\int_D F\,dx\,dy\,dz$ of a 3-form $F\,dx\,dy\,dz$ over a three-dimensional domain D in xyz-space is also denoted $\int_D F\,dV$ where dV is the 'element of volume' $dV = dx\,dy\,dz$. Normally an orientation of D is not given and it is understood that $\int_D F\,dx\,dy\,dz$ is defined by using the standard orientation ($dx\,dy\,dz$ positive) for D.

In this notation the three cases of Stokes' Theorem $\int_{\partial S} \omega = \int_S d\omega$ (ω a 0-, 1-, or 2-form) have very dissimilar forms:

ω a 0-form. If $\omega(x, y, z) = f(x, y, z)$ is a function (0-form) and if C is an oriented curve with end points C^+, C^-, then

$$f(C^+) - f(C^-) = \int_C \mathbf{X}\cdot d\mathbf{x},$$

where $\mathbf{X}$ is the vector field $\frac{\partial f}{\partial x}\mathbf{i} + \frac{\partial f}{\partial y}\mathbf{j} + \frac{\partial f}{\partial z}\mathbf{k}$ determined by the 1-form $df = \frac{\partial f}{\partial x} dx + \frac{\partial f}{\partial y} dy + \frac{\partial f}{\partial z} dz$. For this reason one defines the *gradient* of a function f to be the vector field

$$\text{grad } f = \frac{\partial f}{\partial x} \mathbf{i} + \frac{\partial f}{\partial y} \mathbf{j} + \frac{\partial f}{\partial z} \mathbf{k},$$

so that the formula becomes

$$f(C^+) - f(C^-) = \int_C \text{grad } f \cdot d\mathbf{x}. \tag{2}$$

ω a 1-form. If $\omega = A\,dx + B\,dy + C\,dz$ is the 1-form corresponding to the vector field $A\mathbf{i} + B\mathbf{j} + C\mathbf{k}$, and if S is an oriented two-dimensional surface with oriented boundary ∂S, then the formula is

$$\int_{\partial S} \mathbf{X} \cdot d\mathbf{x} = \int_S \mathbf{Y} \cdot \mathbf{n}\, d\sigma,$$

where $\mathbf{Y}$ is the vector field determined by the 2-form

$$\begin{aligned} d\omega &= \frac{\partial A}{\partial y} dy\,dx + \frac{\partial A}{\partial z} dz\,dx + \frac{\partial B}{\partial x} dx\,dy + \frac{\partial B}{\partial z} dz\,dy + \frac{\partial C}{\partial x} dx\,dz + \frac{\partial C}{\partial y} dy\,dz \\ &= \left(\frac{\partial C}{\partial y} - \frac{\partial B}{\partial z}\right) dy\,dz + \left(\frac{\partial A}{\partial z} - \frac{\partial C}{\partial x}\right) dz\,dx + \left(\frac{\partial B}{\partial x} - \frac{\partial A}{\partial y}\right) dx\,dy. \end{aligned}$$

For this reason one defines the *curl* of a vector field $\mathbf{X} = A\mathbf{i} + B\mathbf{j} + C\mathbf{k}$ to be the vector field

$$\text{curl } \mathbf{X} = \left(\frac{\partial C}{\partial y} - \frac{\partial B}{\partial z}\right)\mathbf{i} + \left(\frac{\partial A}{\partial z} - \frac{\partial C}{\partial x}\right)\mathbf{j} + \left(\frac{\partial B}{\partial x} - \frac{\partial A}{\partial y}\right)\mathbf{k}$$

so that the formula becomes

$$\int_{\partial S} \mathbf{X} \cdot d\mathbf{x} = \int_S (\text{curl } \mathbf{X}) \cdot \mathbf{n}\, d\sigma. \tag{3}$$

ω a 2-form. If $\omega = A\,dy\,dz + B\,dz\,dx + C\,dx\,dy$ is the 2-form corresponding to the vector field $A\mathbf{i} + B\mathbf{j} + C\mathbf{k}$, and if D is a three-dimensional region whose boundary ∂D is oriented by the usual rules, then

$$\int_{\partial D} \mathbf{X} \cdot \mathbf{n}\, d\sigma = \int_D F\, dV,$$

where F is the function $\frac{\partial A}{\partial x} + \frac{\partial B}{\partial y} + \frac{\partial C}{\partial z}$ determined by the 3-form $d\omega = \left(\frac{\partial A}{\partial x} + \frac{\partial B}{\partial y} + \frac{\partial C}{\partial z}\right) dx\, dy\, dz$. For this reason, one defines the *divergence* of a vector field $\mathbf{X} = A\mathbf{i} + B\mathbf{j} + C\mathbf{k}$ to be the function

$$\operatorname{div} \mathbf{X} = \frac{\partial A}{\partial x} + \frac{\partial B}{\partial y} + \frac{\partial C}{\partial z},$$

so that the formula becomes

$$(4) \qquad \int_{\partial D} \mathbf{X} \cdot \mathbf{n}\, d\sigma = \int_D (\operatorname{div} \mathbf{X})\, dV.$$

The formulas (2), (3), (4) are all special cases of the generalized Stokes' Theorem $\int_{\partial S} \omega = \int_S d\omega$. Formula (3) is called Stokes' Theorem (proper). Formula (4) is called the Divergence Theorem or Gauss' Theorem. Formula (2) has no special name unless perhaps it is called the Fundamental Theorem of Calculus.

The formula for integration by parts

$$(5) \quad \int_S d\omega_1 \cdot \omega_2 = \int_{\partial S} \omega_1 \cdot \omega_2 - (-1)^j \int_S \omega_1 \cdot d\omega_2$$

(where ω_1 is a j-form, ω_2 a k-form, S a $(j + k + 1)$-manifold) can also be translated into vector notation, and takes various forms for various values of j, k. For example, for $j = 2$, $k = 0$ it takes the form

$$\int_D f \operatorname{div} \mathbf{X}\, dV = \int_{\partial D} f\mathbf{X} \cdot \mathbf{n}\, d\sigma - \int_D \operatorname{grad} f \cdot \mathbf{X}\, dV.$$

There are many such formulas, frequently encountered in physics books, all of them expressing the single idea (5), which itself is just the Fundamental Theorem of Calculus

$$\int_S d(\omega_1 \cdot \omega_2) = \int_{\partial S} \omega_1 \cdot \omega_2.$$

Much of the literature of mathematical physics is, regrettably, written in this vector notation of grad, curl, and div, so students of physics must become acquainted with it. In summary, grad f, curl $\mathbf{X}$, div $\mathbf{X}$ are $d\omega$ where ω is a 0-, 1-, 2-form respectively. This notation applies *only* to three dimensions.

Exercises **1** Find the vector field grad f for (a)–(c).

(a) $f = x^2 + y^2 + z^2$

(b) $f = \frac{1}{r} = (x^2 + y^2 + z^2)^{-1/2}$

(c) $f = \log(x^2 + y^2)$ (independent of z)

Describe these vector fields geometrically.

2 Show that the curl of each of the vector fields found in Exercise 1 is zero. Find the divergence of each of these vector fields.

3 Find div $\mathbf{X}$ and curl $\mathbf{X}$ for

$$\mathbf{X} = x\mathbf{i} + y\mathbf{j} + z\mathbf{k}$$
$$\mathbf{X} = (x^2 + 1)\mathbf{i} + xyz\mathbf{j} + \sin(x + y)\mathbf{k}$$

4 Show that the value of a constant 2-form $A\,dy\,dz + B\,dz\,dx + C\,dx\,dy$ on a parallelogram in space is $\mathbf{X} \cdot \mathbf{n}\,d\sigma$, where $\mathbf{X} = A\mathbf{i} + B\mathbf{j} + C\mathbf{k}$, $\mathbf{n}$ is the unit vector normal to the parallelogram (suitably oriented), and $d\sigma$ is the area of the parallelogram. [Since n, $d\sigma$ are vaguely defined in terms of geometrical notions, the proof is necessarily imprecise. Since both functions are linear in (A, B, C), it suffices to consider the cases '$\mathbf{X}$ parallel to $\mathbf{n}$' and '$\mathbf{X}$ perpendicular to $\mathbf{n}$', for which the desired formula is easily established.]

5 The 'cross product' $\mathbf{a} \times \mathbf{b}$ of two vectors $\mathbf{a} = a_1\mathbf{i} + a_2\mathbf{j} + a_3\mathbf{k}$, $\mathbf{b} = b_1\mathbf{i} + b_2\mathbf{j} + b_3\mathbf{k}$ is defined to be the vector

$$\begin{vmatrix} a_1 & a_2 & a_3 \\ b_1 & b_2 & b_3 \\ \mathbf{i} & \mathbf{j} & \mathbf{k} \end{vmatrix} = (a_2b_3 - a_3b_2)\mathbf{i} + (a_3b_1 - a_1b_3)\mathbf{j} + (a_1b_2 - a_2b_1)\mathbf{k}$$

where the determinant on the left is merely a mnemonic. The dot product of two vectors is defined as the sum of the products of corresponding components [i.e., $\mathbf{a} \cdot \mathbf{b} = a_1b_1 + a_2b_2 + a_3b_3$]. Show that the value of $A\,dy\,dz + B\,dz\,dx + C\,dx\,dy$ on a parallelogram with sides $\mathbf{a}$, $\mathbf{b}$ is $\mathbf{X} \cdot (\mathbf{a} \times \mathbf{b})$ where $\mathbf{X} = A\mathbf{i} + B\mathbf{j} + C\mathbf{k}$. This number $\mathbf{X} \cdot (\mathbf{a} \times \mathbf{b})$ is called the 'scalar triple product' of $\mathbf{X}$, $\mathbf{a}$, $\mathbf{b}$.

6 Combining 4 and 5 show that $\mathbf{a} \times \mathbf{b}$ is perpendicular to $\mathbf{a}$ and $\mathbf{b}$ and that its length is equal to the area (which has not been mathematically defined) of the parallelogram enclosed by $\mathbf{a}$, $\mathbf{b}$.

7 Instead of $\int_S \mathbf{X} \cdot \mathbf{n}\,d\sigma$, the integral of a vector field over a surface is often denoted

$$\int_S \mathbf{X} \cdot (\mathbf{du} \times \mathbf{dv})$$

where u and v are parameters describing the surface $S = \{x(u, v), y(u, v), z(u, v)\}$. Justify this notation in terms of 'infinitesimal parallelograms'.

8.2 Elementary Differential Equations

Although a differential equation and an initial condition

$$\frac{dy}{dx} = f(x, y) \qquad (y(\bar{x}) = \bar{y}) \tag{1}$$

determine* y locally as a function of x (see §7.4), it is impossible to give a *formula* for the solution $y(x)$ except in certain very simple cases.

*Apply the theorem of §7.4 to the pair of equations $\frac{dx}{dt} \equiv 1$, $\frac{dy}{dt} = f(x, y)$.

The difficulty of the problem of finding an explicit solution $y(x)$ of (1) can be gauged by observing that even in the very simple cases where f does not depend on y—$\frac{dy}{dx} = f(x)$—the problem is to 'find a function with a given derivative'. As was mentioned in §3.1 in connection with the Fundamental Theorem of Calculus, the fact that there exists a solution of this problem in no way implies that a *formula* for the solution can be given in terms of known functions. The same remark applies *a fortiori* to the solution of the more general equation $\frac{dy}{dx} = f(x, y)$. This section deals with a few of the basic techniques of finding formulas for solutions of differential equations of the type (1). Since it is not necessarily true that the solution can be expressed by a formula, these techniques obviously cannot be expected to apply to all cases, but only to particularly simple equations (1).

The problem is greatly simplified if one allows the solution $y(x)$ of (1) to be given *implicitly* rather than explicitly, that is, if one accepts as a 'solution' a relation of the form $F(x, y) =$ const. which, when solved for y as a function of x, gives a solution of (1). The rule for the implicit differentiation of $F(x, y) =$ const. is (see §5.2)

$$F(x, y) = C$$

$$\frac{\partial F}{\partial x} dx + \frac{\partial F}{\partial y} dy = dC$$

$$dy = \frac{1}{\frac{\partial F}{\partial y}} dC - \frac{\frac{\partial F}{\partial x}}{\frac{\partial F}{\partial y}} dx.$$

Hence

$$\frac{dy}{dx} = -\frac{\dfrac{\partial F}{\partial x}}{\dfrac{\partial F}{\partial y}}$$

where it is assumed that $\dfrac{\partial F}{\partial y} \neq 0$ in order that $F(x, y) =$ const. can be solved for y as a function of x. Thus $F(x, y) = C$ *gives a solution of* (1) when solved for y *if and only if at all points of the curve* $F(x, y) = C$ *the 1-form* dF *is a non-zero multiple of the 1-form* $dy - f(x, y)\,dx$. This can be seen geometrically as follows:

The tangent line to the curve $F =$ const. through $(\bar{x}, \bar{y})$ is the line

$$\frac{\partial F}{\partial x}(\bar{x}, \bar{y})(x - \bar{x}) + \frac{\partial F}{\partial y}(\bar{x}, \bar{y})(y - \bar{y}) = 0$$

which can be remembered simply as $dF = 0$. If $y(x)$ is a function satisfying (1), then the tangent line to its graph is the line

$$\frac{y - \bar{y}}{x - \bar{x}} = f(\bar{x}, \bar{y}),$$

that is, the line

$$(y - \bar{y}) - f(\bar{x}, \bar{y})(x - \bar{x}) = 0$$

or, mnemonically, the line $dy - f\,dx = 0$. Since two lines

$$A_1(x - \bar{x}) + B_1(y - \bar{y}) = 0$$
$$A_2(x - \bar{x}) + B_2(y - \bar{y}) = 0$$

through a point $(\bar{x}, \bar{y})$ coincide if and only if the equation of one line is a non-zero multiple of the equation of the other line, it follows that the function $y(x)$ defined implicitly by $F(x, y) =$ const. satisfies $\dfrac{dy}{dx} = f(x, y)$ if and only if the lines $dF = 0$, $dy - f\,dx = 0$ coincide, that is, if and only if the 1-form dF is a non-zero multiple of $dy - f\,dx$ at all points of the curve.

More generally, a curve $F(x, y) =$ const. is said to be a *solution of the differential equation* $A\,dx + B\,dy = 0$, where A, B are functions of x, y, if and only if the 1-form dF is a non-zero multiple of the 1-form $A\,dx + B\,dy$ at all points of the curve. Geometrically this means that at each point $(\bar{x}, \bar{y})$ of the curve the tangent line $dF = 0$ is the line $A\,dx + B\,dy = 0$; that is, for each $(\bar{x}, \bar{y})$ on the curve the equations

$$\frac{\partial F}{\partial x}(\bar{x}, \bar{y})(x - \bar{x}) + \frac{\partial F}{\partial y}(\bar{x}, \bar{y})(y - \bar{y}) = 0$$

and

$$A(\bar{x}, \bar{y})(x - \bar{x}) + \quad B(\bar{x}, \bar{y})(y - \bar{y}) = 0$$

define the same line through $(\bar{x}, \bar{y})$.

According to this definition the solution of $\frac{dy}{dx} = f(x, y)$ is equivalent to the solution of $dy - f(x, y)\,dx = 0$. In the remainder of the section the original equation (1) will be dropped and in its place the differential equation

$$A\,dx + B\,dy = 0 \tag{2}$$

will be considered. The problem is to find explicit formulas for functions $F(x, y)$ such that $F =$ const. solves (2) in the sense defined above. That is, the problem is to find $F(x, y)$ such that dF is a non-zero multiple of the given 1-form $A\,dx + B\,dy$ at each point.

Example

Instead of the 'derivative equation' $\frac{dy}{dx} = \frac{x}{y}$ it is easier to consider the 'differential equation' $dy - \frac{x}{y}\,dx = 0$. Multiplying by y this becomes $y\,dy - x\,dx = 0$, or $d[y^2 - x^2] = 0$; hence $y^2 - x^2 =$ const. gives solutions of the equation. The factor y is non-zero at all points where the given equation is meaningful. By implicit differentiation the curves $y = \pm\sqrt{C + x^2}$ satisfy the original equation $\frac{dy}{dx} = \frac{x}{y}$.

Example

To solve $\frac{dy}{dx} = -\sqrt{\frac{1 - y^2}{1 - x^2}}$ rewrite it as

$$dy + \sqrt{\frac{1 - y^2}{1 - x^2}}\,dx = 0$$

$$\frac{dy}{\sqrt{1 - y^2}} + \frac{dx}{\sqrt{1 - x^2}} = 0$$

$$d(\text{Arcsin } y) + d(\text{Arcsin } x) = 0$$

$$\text{Arcsin } y + \text{Arcsin } x = \text{const.}$$

Given (x, y) inside the square $\{|x| < 1, |y| < 1\}$ where the equation is defined, the equation

$$\operatorname{Arcsin} x + \operatorname{Arcsin} y = \operatorname{Arcsin} \bar{x} + \operatorname{Arcsin} \bar{y}$$

can be solved for y as a function of x for x near $\bar{x}$, giving a function $y(x)$ which satisfies the given equation and the initial condition $y(\bar{x}) = \bar{y}$.

If a differential equation can be written in the form

$$A(x)\,dx + B(y)\,dy = 0 \tag{3}$$

then it can be solved by applying the techniques of elementary calculus to find (if possible) functions $g(x)$, $h(y)$ such that $g'(x) = A(x)$, $h'(y) = B(y)$, after which the solution is given by $g(x) + h(y) = \text{const.}$ An equation of the form (3) is said to have *variables separated*, and a differential equation which can be put in the form (3) by multiplying by a non-zero function $\rho(x, y)$ is said to have *variables separable*.

Example

The homogeneous first order linear equation $\dfrac{dy}{dx} + f(x)y = 0$ has variables separable. Writing

$$\begin{aligned} dy + f(x)y\,dx &= 0 \\ \frac{dy}{y} + f(x)\,dx &= 0 \\ \log|y| + F(x) &= \text{const.} \\ y &= \text{const. } e^{-F(x)} \end{aligned}$$

gives the solution for $y \neq 0$ if a function $F(x)$ can be found such that $F'(x) = f(x)$. The solution for $y = 0$ is $y \equiv 0$.

A 1-form is said to be *exact* if it is equal to dF for some function $F(x, y)$. Thus the differential equation $A\,dx + B\,dy = 0$ is solved by writing the 1-form $\omega = A\,dx + B\,dy$ as a non-zero multiple of an exact 1-form dF, after which $F = \text{const.}$ gives the solutions.

A 1-form ω is said to be *closed* if $d\omega = 0$. Since $d[dF] = 0$, every exact 1-form is closed. Locally the converse is true—every closed form is exact—and, moreover, if $d\omega = 0$ then a function F such that $\omega = dF$ can be found by 'integration', i.e. by antidifferentiation.

Example

The 1-form $y\,dx + (x + y)\,dy$ is (locally) exact because $d[y\,dx + (x + y)\,dy] = dy\,dx + dx\,dy = 0$. To find $F(x, y)$ such that $dF = y\,dx + (x + y)\,dy$ one can use the equation $\frac{\partial F}{\partial x} = y$ to obtain $F(x, y) = xy + C(y)$, where the constant of integration may depend on y. Then $\frac{\partial F}{\partial y} = x + y$ gives $C'(y) = y$; hence $C(y) = \frac{1}{2}y^2 + \text{const}$. Thus $F(x, y) = xy + \frac{1}{2}y^2$ satisfies $dF = y\,dx + (x + y)\,dy$ and the differential equation

$$y\,dx + (x + y)\,dy = 0$$

is solved by

$$xy + \tfrac{1}{2}y^2 = \text{const.}$$

The same method applies to the solution of any equation $A\,dx + B\,dy = 0$ for which $d(A\,dx + B\,dy) \equiv 0$, i.e. for which $\frac{\partial A}{\partial y} = \frac{\partial B}{\partial x}$. One finds (if possible) a function $F_0(x, y)$ such that $\frac{\partial F_0}{\partial x} = A$. Setting $F(x, y) = F_0(x, y) + C(y)$ the equation $\frac{\partial F}{\partial y} = B(x, y)$ becomes $\frac{\partial F_0}{\partial y} + C'(y) = B$, $C'(y) = B - \frac{\partial F_0}{\partial y}$. The right side does not involve x because $\frac{\partial}{\partial x}\left[B - \frac{\partial F_0}{\partial y}\right] = \frac{\partial B}{\partial x} - \frac{\partial A}{\partial y} = 0$; hence $C'(y) = B - \frac{\partial F_0}{\partial y}$ can be solved (theoretically) for $C(y)$ so that $F(x, y)$ satisfies $dF = A\,dx + B\,dy$.

A differential equation $A\,dx + B\,dy = 0$ in which $d(A\,dx + B\,dy) \equiv 0$ is said to be *exact* (although it would be more consistent to call it 'closed'). The above shows that the solution of an exact differential equation can be achieved by solving

$$\frac{\partial F_0}{\partial x} = A, \; C'(y) = B - \frac{\partial F_0}{\partial y}$$

and setting $F(x, y) = F_0(x, y) + C(y)$. Alternatively, one can solve

$$\frac{\partial F_0}{\partial y} = B, \; C'(x) = A - \frac{\partial F_0}{\partial x}$$

and set $F(x, y) = F_0(x, y) + C(x)$. Thus the solution of an exact equation requires two antidifferentiations.

Example

An equation $A(x)\,dx + B(y)\,dy = 0$ with variables separated is exact and the function $F(x, y)$ is found merely by antidifferentiating $A(x)$ and $B(y)$ separately as before.

Example

The 1-form $d\theta = \dfrac{-y}{x^2 + y^2}\,dx + \dfrac{x}{x^2 + y^2}\,dy$ is closed. It is dF for $F(x, y) = \text{Arctan}(y/x)$ or $\text{Arccot}(x/y)$. These functions are not defined for $x = 0$ or $y = 0$ respectively, and there is *no* function $F(x, y)$ such that dF is the given 1-form at *all* points $(x, y) \neq (0, 0)$. (In short, the 1-form $d\theta$ is closed and therefore is exact *locally*, but θ is not a global function of (x, y).) The differential equation $d\theta = 0$ is solved by setting $\text{Arctan}(y/x) = \text{const.}$, $y = Cx$, which gives rays through $(0, 0)$ as expected. The equation $d\theta = 0$ can also be solved by multiplying by $(x^2 + y^2)/xy$, which separates the variables and gives $-\dfrac{dx}{x} + \dfrac{dy}{y} = 0$, $-d\log|x| + d\log|y| = 0$, $\log|y/x| = \text{const.}$, $y = \text{const.}\ x$.

Example

The 1-form $ye^{xy}\,dx + (xe^{xy} + \cos y)\,dy$ is closed. To find $F(x, y)$ write $\dfrac{\partial F}{\partial x} = ye^{xy}$, $F = e^{xy} + C(y)$, $\dfrac{\partial F}{\partial y} = xe^{xy} + \cos y = xe^{xy} + C'(y)$, $C(y) = \sin y + \text{const.}$, $F = e^{xy} + \sin y$. Thus

$$ye^{xy}\,dx + (xe^{xy} + \cos y)\,dy = 0$$

is solved by

$$e^{xy} + \sin y = \text{const.}$$

To solve an arbitrary differential equation $\omega = 0$ it suffices to find a non-zero function $\rho(x, y)$ such that $\rho\omega$ is closed and then to find F satisfying $dF = \rho\omega$ by the method above. Such a function ρ is called an 'integrating factor' for the 1-form ω because multiplication by the factor ρ makes it possible to 'integrate', that is, to find an F.

Example

The inhomogeneous first order linear equation $\frac{dy}{dx} + f(x)y = g(x)$ always has an integrating factor of the form $\rho(x)$. Writing

$$\rho\omega = \rho(x)\,dy + \rho(x)[f(x)y - g(x)]\,dx$$

and setting $d(\rho\omega) = 0$ gives $\rho'(x) - \rho(x)f(x) \equiv 0$. This is a homogeneous first order linear equation for ρ; the solution of this equation was found above to be $\rho(x) =$ const. $e^{F(x)}$, where $F'(x) = f(x)$. Thus $e^{F(x)}\,dy + e^{F(x)}[f(x)y - g(x)]\,dx$ is closed and can be integrated, that is, can be written as dF for some $F(x, y)$. Then $F(x, y) =$ const. solves the given differential equation.

It is shown in §8.5 that locally there exists an integrating factor for any equation $A\,dx + B\,dy = 0$. Specifically, it is shown that if A, B are differentiable functions defined near $(\bar{x}, \bar{y})$ and if A, B are not both zero at $(\bar{x}, \bar{y})$, then there is a differentiable function ρ defined near $(\bar{x}, \bar{y})$ such that $d(\rho A\,dx + \rho B\,dy) \equiv 0$ near $(\bar{x}, \bar{y})$. This fact is obviously of no avail, however, in the problem of finding a formula for F. The integrating factor ρ may itself be a function for which there is no simple formula; moreover, even if there is a simple integrating factor ρ it may be very difficult to find.

In summary, a differential equation $A\,dx + B\,dy = 0$ is solved as follows: First find an integrating factor ρ, that is, find a function ρ such that the equation $\rho A\,dx + \rho B\,dy = 0$ is exact. If $A\,dx + B\,dy = 0$ has variables separable or if it is already exact, then such a function ρ can be found immediately. Otherwise it may not be possible to find a ρ even though one always 'exists'. Once ρ is found, the problem is reduced to solving an exact equation $A\,dx + B\,dy = 0$. By two antidifferentiations, which in simple cases can be done using the techniques of elementary calculus, one obtains a function $F(x, y)$ such that $dF = A\,dx + B\,dy$. The curves $F =$ const. are then the solutions of the given equation.

Exercises

1 Find functions $F(x, y)$ such that the curves $F =$ const. solve the following differential equations.

(a) $\dfrac{x\,dx + y\,dy}{x^2 + y^2} = 0$

(b) $x\,dx + \dfrac{1 + x^2}{1 + y^2}\,dy = 0$

(c) $x^2\,dy + y^2\,dx = xy\,dy$

(d) $(2x + y)\,dx + (x + 2y)\,dy = 0$

(e) $\dfrac{dy}{dx} + y\cos x = 0$

(f) $e^x(x^2 + y^2 + 2x)\,dx + 2ye^x\,dy = 0$

(g) $\dfrac{dy}{dx} + 2xy = 2x^3$

2 A differential equation gives a concise description of a *family of curves*. For example, $dx = 0$ describes lines parallel to the y-axis, $x\,dy - y\,dx = 0$ describes lines through the origin, $x\,dx + y\,dy = 0$ describes concentric circles about the origin, etc. Find differential equations describing the following families of curves.

(a) hyperbolas whose asymptotes are the lines $x = \pm y$
(b) hyperbolas whose asymptotes are the coordinate axes
(c) circles with center $(1, 0)$
(d) ellipses with center at $(0, 0)$, with major axis along x-axis, and with major axis twice as long as the minor axis
(e) all circles passing through both of the points $(1, 0)$ $(-1, 0)$

3 *Orthogonal trajectories.* A curve is an 'orthogonal trajectory' of a family of curves if at each point $(\bar{x}, \bar{y})$ of the curve its tangent line is perpendicular to the tangent line to the curve of the family through that point. Thus if the family is described by a differential equation $A\,dx + B\,dy = 0$, a curve is an orthogonal trajectory of the family if and only if it is a solution of the differential equation $A\,dy - B\,dx = 0$. Sketch the orthogonal trajectories of the families $dx = 0$, $x\,dy - y\,dx = 0$, $x\,dx + y\,dy = 0$, and (a)–(e) above. Then give formulas for the orthogonal trajectories. [The 'integration' of $A\,dy - B\,dx = 0$ for the family (e) is not easy. Show that the curves $(x - 1)^2 + y^2 = C[(x + 1)^2 + y^2]$ $(C > 0)$ are orthogonal trajectories of (e) and show that these curves are circles.]

4 Sketch the curves $y = \text{const.}|x|^p$ for $p > 0$ and find their orthogonal trajectories.

8.3 Harmonic Functions and Conformal Coordinates

A function of two variables $u(x, y)$ is said to be *harmonic* if it is continuous and if its average value over any disk $D = \{(x - \bar{x})^2 + (y - \bar{y})^2 \leq r^2\}$ where it is defined is equal to its value at the center of the disk:

$$(1) \qquad u(\bar{x}, \bar{y}) = \frac{1}{\pi r^2} \int_D u(x, y)\, dx\, dy.$$

Similarly, a function of n variables $u(x_1, x_2, \ldots, x_n)$ is said to be harmonic if it is continuous and if its average value over any n-dimensional ball $D = \{(x_1 - \bar{x}_1)^2 + (x_2 - \bar{x}_2)^2 + \cdots + (x_n - \bar{x}_n)^2 \leq r^2\}$ where it is defined is equal to its value at the center of the ball:

$$(1') \qquad u(\bar{x}_1, \bar{x}_2, \ldots, \bar{x}_n) = \frac{\int_D u(x_1, x_2, \ldots, x_n)\, dx_1\, dx_2 \ldots dx_n}{\int_D dx_1\, dx_2 \ldots dx_n}.$$

For the sake of simplicity, only harmonic functions of two variables $u(x, y)$ will be considered in this section.

Theorem

A harmonic function $u(x, y)$ is necessarily (continuously) differentiable and its first partial derivatives $\frac{\partial u}{\partial x}, \frac{\partial u}{\partial y}$ are themselves harmonic functions. Therefore a harmonic function is necessarily infinitely differentiable* and all its partial derivatives, e.g. $\frac{\partial^2 u}{\partial x^2}, \frac{\partial^3 u}{\partial x\, \partial y^2}$, etc., are harmonic functions. Moreover, the second partial derivatives of a harmonic function $u(x, y)$ necessarily satisfy the relation

$$\frac{\partial^2 u}{\partial x^2} + \frac{\partial^2 u}{\partial y^2} \equiv 0$$

known as *Laplace's equation*. Conversely, any twice differentiable function $u(x, y)$ which satisfies Laplace's equation is harmonic.

*In §8.4 it will be proved that a harmonic function is in fact analytic, that is, it can be written locally as the sum of a power series. This conclusion is even stronger than the conclusion that it is infinitely differentiable.

This theorem makes it easy to find *examples* of harmonic functions, which is not at all easy to do on the basis of the defining property (1). For example, the theorem says that a quadratic form $ax^2 + 2bxy + cy^2$ is a harmonic function if and only if $2a + 2c = 0$. In the same way a cubic form $ax^3 + 3bx^2y + 3cxy^2 + ey^3$ is harmonic if and only if $6ax + 6by + 6cx + 6ey \equiv 0$, that is, if and only if $a + c = 0$, $b + e = 0$. Thus the theorem implies immediately that a cubic polynomial in x and y is a harmonic function if and only if it is a polynomial of the form

$$A + Bx + Cy + D(x^2 - y^2) + E(2xy) + F(x^3 - 3xy^2) + G(y^3 - 3yx^2).$$

This is not at all obvious from the definition (1).

Proof

Let $(\bar{x}, \bar{y})$ be a point of the domain of u and let R be a number such that the disk $D = \{(x - \bar{x})^2 + (y - \bar{y})^2 \leq R^2\}$ lies entirely inside the domain of u. The first step of the proof is to show that

$$\lim_{h \to 0} \frac{u(\bar{x} + h, \bar{y}) - u(\bar{x}, \bar{y})}{h}$$

exists and is equal to $\frac{1}{\pi R^2} \int_{\partial D} u(x, y)\, dy$. To this end, consider the map

$$x = a + c$$
$$y = b$$

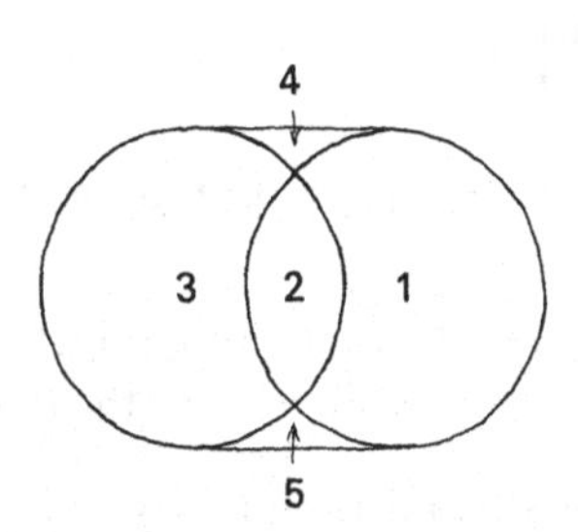

of abc-space to the xy-plane, and consider the cylinder $C = \{(a - \bar{x})^2 + (b - \bar{y})^2 \leq R^2, 0 \leq c \leq h\}$ in abc-space. The integral over ∂C of the pullback of any 2-form on the xy-plane is zero. This can be seen geometrically from the fact that the map carries ∂C to the xy-plane in such a way that every point in the image is covered twice with opposite orientations so that the whole integral cancels.* More specifically, the image of ∂C is divided in a natural way into five regions and ∂C is then divided into ten pieces such that each region of the image is the image of two pieces of ∂C with opposite orientations. These pieces can easily be described in detail and the above statement can thereby be proved rigorously using the principle of independence of parameter. (A slight technical difficulty arises from the fact that some of the pieces are not differentiable surfaces-with-boundary because their corners are 'spikes' where the boundary curves are tangent, but these details will be omitted.)

*A k-form is said to be 'weakly closed' if its integral over the boundary of any $(k + 1)$-dimensional domain is zero. The statement here is a special case of a general theorem which states that the pullback of any m-form under a differentiable map $\mathbf{R}^n \to \mathbf{R}^m$ is weakly closed. If the function u is assumed to be differentiable, rather than merely continuous, this follows from $f^*[d(u\, dx\, dy)] = d[f^*(u\, dx\, dy)]$.

Applying this observation to the 2-form $u(x, y)\, dx\, dy$ gives

$$\int_{\partial C} [u(a + c, b)\, da\, db + u(a + c, b)\, dc\, db] = 0.$$

Orienting C by $da\, db\, dc$ and orienting ∂C accordingly, the integrals over the two disks at the ends of C are equal to $\pi R^2 u(\bar{x} + h, \bar{y})$ and $-\pi R^2 u(\bar{x}, \bar{y})$ by (1). Thus

$$\pi R^2 \frac{u(\bar{x} + h, \bar{y}) - u(\bar{x}, \bar{y})}{h} = \frac{1}{h} \int_{\text{sleeve}} u(a + c, b)\, db\, dc,$$

where the 'sleeve' is the portion $\{(a - \bar{x})^2 + (b - \bar{y})^2 = R^2,\ 0 \le c \le h\}$ of ∂C. Parameterizing the sleeve by $a = \bar{x} + R\cos\theta$, $b = \bar{y} + R\sin\theta$, $c = t$, this integral becomes

$$\frac{1}{h} \int u(\bar{x} + R\cos\theta + t, \bar{y} + R\sin\theta) R\cos\theta\, d\theta\, dt,$$

where the integral is over the rectangle $\{0 \le \theta \le 2\pi,\ 0 \le t \le h\}$ oriented $d\theta\, dt$. (Checking the orientation of the rectangle is a good exercise in orientations.) Writing this as an iterated integral and passing to the limit as $h \to 0$ gives

$$\lim_{h \to 0} \frac{1}{h} \int_0^h \left[\int_0^{2\pi} u(\bar{x} + R\cos\theta + t, \bar{y} + R\sin\theta) R\cos\theta\, d\theta \right] dt$$
$$= \int_0^{2\pi} u(\bar{x} + R\cos\theta, \bar{y} + R\sin\theta) R\cos\theta\, d\theta.$$

But this is just $\int_{\partial D} u\, dy$ when ∂D is parameterized by $x = \bar{x} + R\cos\theta$, $y = \bar{y} + R\sin\theta$. Thus the limit which defines $\frac{\partial u}{\partial x}$ at $(\bar{x}, \bar{y})$ exists and is equal to $\frac{1}{\pi R^2} \int_{\partial D} u\, dy$, where D is any disk with center $(\bar{x}, \bar{y})$ lying in the domain of u, and where R is the radius of D. But for fixed R the function $\int_{\partial D} u\, dy$ is clearly a continuous function of $(\bar{x}, \bar{y})$ (pass to the limit under the integral sign using the theorem of §9.7); hence $\frac{\partial u}{\partial x}$ exists and is a continuous function. Similarly $\frac{\partial u}{\partial y}$ exists and is continuous. Therefore u is differentiable and by Stokes' Theorem

$$\frac{\partial u}{\partial x}(\bar{x}, \bar{y}) = \frac{1}{\pi R^2} \int_{\partial D} u\, dy = \frac{1}{\pi R^2} \int_D \frac{\partial u}{\partial x}\, dx\, dy.$$

That is, $\frac{\partial u}{\partial x}$ is harmonic. By symmetry between x and y it follows that $\frac{\partial u}{\partial y}$ also exists and is a harmonic function.

This completes the proof of the first statement of the theorem. The remaining statements of the theorem are summarized by the statement: *A twice differentiable function u satisfies the mean value property* (1) *if and only if* $\dfrac{\partial^2 u}{\partial x^2} + \dfrac{\partial^2 u}{\partial y^2} \equiv 0$. To prove this let D be a disk with center $(\bar{x}, \bar{y})$ and let $u(r, \theta)$ denote $u(\bar{x} + r\cos\theta, \bar{y} + r\sin\theta)$. Then (1) gives

$$\pi R^2 u(\bar{x}, \bar{y}) = \int_D u(r, \theta) r\, dr\, d\theta = \int_0^R \left[\int_0^{2\pi} u(r, \theta) r\, d\theta \right] dr$$

where R is the radius of D. Differentiating with respect to R gives

$$2\pi R u(\bar{x}, \bar{y}) = \int_0^{2\pi} u(R, \theta) R\, d\theta$$

$$u(\bar{x}, \bar{y}) = \frac{1}{2\pi} \int_0^{2\pi} u(R, \theta)\, d\theta.$$

Thus the average value of u over any *circle* with center $(\bar{x}, \bar{y})$ is also equal to $u(\bar{x}, \bar{y})$. Differentiating this relation with respect to R

$$0 = \frac{1}{2\pi} \int_0^{2\pi} \left(\frac{\partial u}{\partial x} \cdot \cos\theta + \frac{\partial u}{\partial y} \cdot \sin\theta \right) d\theta$$

$$= \frac{1}{2\pi R} \int_0^{2\pi} \left(\frac{\partial u}{\partial x} \cdot \frac{\partial y}{\partial \theta} - \frac{\partial u}{\partial y} \cdot \frac{\partial x}{\partial \theta} \right) d\theta$$

$$= \frac{1}{2\pi R} \int_{\partial D} \left(\frac{\partial u}{\partial x}\, dy - \frac{\partial u}{\partial y}\, dx \right)$$

$$= \frac{1}{2\pi R} \int_D \left(\frac{\partial^2 u}{\partial x^2} + \frac{\partial^2 u}{\partial y^2} \right) dx\, dy$$

$$\int_D \left(\frac{\partial^2 u}{\partial x^2} + \frac{\partial^2 u}{\partial y^2} \right) dx\, dy = 0.$$

Since this is true for any disk, it follows that

$$\frac{\partial^2 u}{\partial x^2} + \frac{\partial^2 u}{\partial y^2} \equiv 0.$$

But the steps in this argument are reversible: If $\dfrac{\partial^2 u}{\partial x^2} + \dfrac{\partial^2 u}{\partial y^2} \equiv 0$ then the derivative with respect to R of

$\frac{1}{2\pi}\int_0^{2\pi} u(R, \theta)\, d\theta$ is identically zero. Hence

$$\frac{1}{2\pi}\int_0^{2\pi} u(R, \theta)\, d\theta$$

is constant. For small values of R it is an average of numbers all of which are nearly $u(\bar{x}, \bar{y})$; hence

$$u(\bar{x}, \bar{y}) \equiv \frac{1}{2\pi}\int_0^{2\pi} u(R, \theta)\, d\theta.$$

Multiplying both sides by $2\pi R\, dR$ and integrating from 0 to r then gives (1). This completes the proof of the theorem.

If u is harmonic then the 1-form $\omega = \frac{\partial u}{\partial x} dy - \frac{\partial u}{\partial y} dx$ is *closed*, i.e., satisfies $d\omega = 0$. It follows that locally ω is *exact*, that is, $\omega = dv$ for some function $v(x, y)$ which can be found from ω by integration. For example, if $u(x, y) = x^2 - y^2$, then $dv = \frac{\partial u}{\partial x} dy - \frac{\partial u}{\partial y} dx = 2x\, dy + 2y\, dx$ gives $v = 2xy + \text{const.}$ The relationship between these two functions $u = x^2 - y^2$, $v = 2xy$ can be seen very clearly by plotting the hyperbolas $u = \text{const.}$, $v = \text{const.}$ First of all, these hyperbolas are *orthogonal;* that is, the curves $u = \text{const.}$ and $v = \text{const.}$ intersect at right angles (see Exercise 3, §8.2). Moreover, if the constants are evenly spaced, say $u = 1, 1.01, 1.02, 1.03, \ldots$ and $v = 1, 1.01, 1.02, 1.03, \ldots$, then the network of curves very nearly forms *squares*. The same phenomenon will occur if one starts with any harmonic function u and defines v by $dv = \frac{\partial u}{\partial x} dy - \frac{\partial u}{\partial y} dx$. To verify this, note that if $\bar{u} = u(\bar{x}, \bar{y})$, $\bar{v} = v(\bar{x}, \bar{y})$ and if δ is a small number, then the curves $u = \bar{u} + \delta$, $v = \bar{v}$ intersect approximately at the point $(\bar{x} + \Delta x, \bar{y} + \Delta y)$ where Δx, Δy are defined by

$$\delta = \Delta u = \frac{\partial u}{\partial x}(\bar{x}, \bar{y})\, \Delta x + \frac{\partial u}{\partial y}(\bar{x}, \bar{y})\, \Delta y$$

$$0 = \Delta v = \frac{\partial v}{\partial x}(\bar{x}, \bar{y})\, \Delta x + \frac{\partial v}{\partial y}(\bar{x}, \bar{y})\, \Delta y$$

from which

$$\Delta x = \delta \frac{\dfrac{\partial u}{\partial x}}{\left(\dfrac{\partial u}{\partial x}\right)^2 + \left(\dfrac{\partial u}{\partial y}\right)^2} = \delta \frac{A}{\Delta}$$

$$\Delta y = \delta \frac{\dfrac{\partial u}{\partial y}}{\left(\dfrac{\partial u}{\partial x}\right)^2 + \left(\dfrac{\partial u}{\partial y}\right)^2} = \delta \frac{B}{\Delta}$$

where

$$A = \frac{\partial u}{\partial x}(\bar{x}, \bar{y}) = \frac{\partial v}{\partial y}(\bar{x}, \bar{y}),\ B = \frac{\partial u}{\partial y}(\bar{x}, \bar{y}) = -\frac{\partial v}{\partial x}(\bar{x}, \bar{y})$$

$$\Delta = \begin{vmatrix} A & B \\ -B & A \end{vmatrix} = A^2 + B^2.$$

Similarly the curves $u = \bar{u}$, $v = \bar{v} + \delta$ intersect approximately at $(\bar{x} + \Delta x, \bar{y} + \Delta y)$, where

$$0 = A \cdot \Delta x + B \cdot \Delta y$$
$$\delta = -B \cdot \Delta x + A \cdot \Delta y$$
$$\Delta x = -\delta \frac{B}{\Delta},\ \Delta y = \delta \frac{A}{\Delta}.$$

Thus the four points of intersection of the four curves $u = \bar{u}$, $u = \bar{u} + \delta$, $v = \bar{v}$, $v = \bar{v} + \delta$ lie approximately at the corners of the square

$$(\bar{x}, \bar{y}), \left(\bar{x} + \delta \frac{A}{\Delta}, y + \delta \frac{B}{\Delta}\right),$$
$$\left(\bar{x} - \delta \frac{B}{\Delta}, \bar{y} + \delta \frac{A}{\Delta}\right), \left(\bar{x} + \delta \frac{A - B}{\Delta}, \bar{y} + \delta \frac{A + B}{\Delta}\right).$$

Moreover, since $du\,dv = (A\,dx + B\,dy)(-B\,dx + A\,dy) = (A^2 + B^2)\,dx\,dy$, the orientations $du\,dv$ and $dx\,dy$ agree and the area of the 'square' is approximately $(A^2 + B^2)\delta^2$.

A pair of differentiable functions $u(x, y)$, $v(x, y)$ is said to define *conformal coordinates* near a point $(\bar{x}, \bar{y})$ if*

$$\frac{\partial u}{\partial x} = \frac{\partial v}{\partial y},\ \frac{\partial u}{\partial y} = -\frac{\partial v}{\partial x} \tag{2}$$

and if $du\,dv \neq 0$, i.e., $\left(\dfrac{\partial u}{\partial x}\right)^2 + \left(\dfrac{\partial u}{\partial y}\right)^2 \neq 0$. Geometri-

*These equations (2) are called the 'Cauchy-Riemann equations'.

cally this means that the curves u = const., v = const. are orthogonal and divide the xy-plane into squares (approximately). If $u(x, y)$ is a harmonic function, then $dv = -\frac{\partial u}{\partial y}\,dx + \frac{\partial u}{\partial x}\,dy$ can be integrated (locally) to give a function v such that (u, v) are conformal coordinates provided $du = \frac{\partial u}{\partial x}\,dx + \frac{\partial u}{\partial y}\,dy \neq 0$ at the point under consideration. (If $du = 0$ at the point, then v still exists but (u, v) are not conformal coordinates because $du\,dv = 0$.) Conversely, if (u, v) are conformal coordinates, then the integral of the 1-form $dv = -\frac{\partial u}{\partial y}\,dx + \frac{\partial u}{\partial x}\,dy$ over any circle is zero and the proof above shows that u is a harmonic function. If (u, v) are conformal coordinates, then so are $(v, -u)$. Hence v must also be harmonic. Thus *if (u, v) are conformal coordinates, then both u and v are harmonic functions and either of them suffices to determine the other up to an additive constant.*

By the Implicit Function Theorem the equations $u = u(x, y)$, $v = v(x, y)$ can be solved locally for x, y as functions of u, v and the rule for implicit differentiation gives

$$du = A\,dx + B\,dy$$

$$dv = -B\,dx + A\,dy$$

$$dx = \frac{A}{A^2 + B^2}\,du + \frac{-B}{A^2 + B^2}\,dv$$

$$dy = \frac{B}{A^2 + B^2}\,du + \frac{A}{A^2 + B^2}\,dv$$

$$\frac{\partial x}{\partial u} = \frac{A}{A^2 + B^2} = \frac{\partial y}{\partial v}, \quad \frac{\partial x}{\partial v} = \frac{-B}{A^2 + B^2} = -\frac{\partial y}{\partial u}.$$

Hence $x(u, v)$, $y(u, v)$ are conformal coordinates on the uv-plane. By the chain rule, if (r, s) are conformal coordinates on the uv-plane and if (u, v) are conformal coordinates on the xy-plane,

$$dr = C\,du + D\,dv \qquad du = A\,dx + B\,dy$$

$$ds = -D\,du + C\,dv \qquad dv = -B\,dx + A\,dy$$

then

$$dr = (CA - DB)\,dx + (CB + DA)\,dy$$

$$ds = (-DA - CB)\,dx + (-DB + CA)\,dy$$

which shows that the composite functions (r, s) define conformal coordinates on the xy-plane.

It follows in particular that if $r(u, v)$ is any harmonic function of u and v, then the composite function $r(u(x, y), v(x, y))$ is a harmonic function of x and y whenever (u, v) are conformal coordinates. This makes it possible to find many harmonic functions very easily. For example, setting $r = u^2 - v^2$, $u = x^2 - y^2$, $v = 2xy$ shows that $(x^2 - y^2)^2 - (2xy)^2 = x^4 - 6x^2y^2 + y^4$ is harmonic.

An extremely important set of conformal coordinates on the xy-plane can be derived from the observation that the radial lines $ax + by = 0$ (a, b const.) are orthogonal to the circles $r = \sqrt{x^2 + y^2} =$ const. If a large number of evenly spaced radial lines are drawn, and if circles are then to be drawn in such a way as to form squares, the circles must be drawn more densely near the origin. Specifically, the density must be inversely proportional to the circumference, which is proportional to r. This leads to the conjecture that setting $du = \dfrac{dr}{r}$ will lead to a set of conformal coordinates on the plane.

Now $du = dr/r$ gives $u = \log r$, a function defined at all points of the plane other than $(0, 0)$. This function is indeed harmonic as is seen by writing

$$du = \frac{dr}{r} = \frac{1}{2}\frac{d(r^2)}{r^2} = \frac{x\,dx + y\,dy}{x^2 + y^2}$$

$$-\frac{\partial u}{\partial y}dx + \frac{\partial u}{\partial x}dy = \frac{x\,dy - y\,dx}{x^2 + y^2}.$$

Since this 1-form is closed (it is $d(\text{Arctan}(y/x))$ or $d(\text{Arccot}\,(x/y))$), it follows not only that $u = \log r$ is harmonic but also that $(\log r, \theta)$ are conformal coordinates, where θ denotes the 'multiple-valued function' $\text{Arctan}(y/x)$. Near any point $(\bar{x}, \bar{y}) \neq (0, 0)$ one can define a function $\theta +$ const., for example, $\text{Arctan}(y/x)$ or $\text{Arccot}(x/y)$, such that $(\log r, \theta)$ are conformal coordinates; but no one function with this property can be defined for all $(x, y) \neq (0, 0)$.

Inverting the equations $u = \log r$, $v = \theta$ gives $r = e^u$, $x = r\cos\theta = e^u \cos v$, $y = r \sin\theta = e^u \sin v$, which leads to the conclusion that the functions $(e^u \cos v, e^u \sin v)$ are conformal coordinates on the uv-plane (because the inverse of conformal coordinates gives conformal coordinates). This is immediately verified. Thus

$e^u \cos v$ and $e^u \sin v$ are both harmonic functions of u and v.

Since the difference of two harmonic functions is obviously a harmonic function, the function

$$(3)\quad u = \tfrac{1}{2}(\log[(x - x_0)^2 + (y - y_0)^2] - \log[(x - x_1)^2 + (y - y_1)^2])$$

is a harmonic function defined for all (x, y) other than (x_0, y_0), (x_1, y_1). The curves $u =$ const. are the curves

$$\frac{(x - x_0)^2 + (y - y_0)^2}{(x - x_1)^2 + (y - y_1)^2} = \text{positive const.}$$

It is easily seen that such a curve is a circle except when the positive constant is 1 (i.e., $u = 0$), in which case it is the perpendicular bisector of the line segment from (x_0, y_0) to (x_1, y_1). This harmonic function, which can be regarded as describing a 'source' of some kind at (x_0, y_0) and a 'sink' of equal magnitude at (x_1, y_1), has many physical applications.

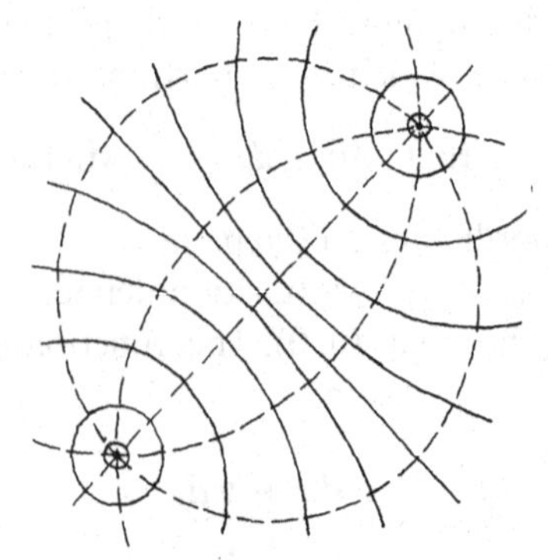
The curves u = const. are solid. The orthogonal curves are dotted.

An important mathematical application of the harmonic function (3) is the following: The curves $u =$ const. near (x_0, y_0) are circles around (x_0, y_0) but (x_0, y_0) is not at the center of these circles. The curves orthogonal to $u =$ const. can be shown (see Exercise 3, §8.2) to be circles which pass through (x_0, y_0) and (x_1, y_1). Near (x_0, y_0) these are like radial lines through (x_0, y_0), and it is easily imagined that there is a conformal change of coordinates near (x_0, y_0) in which these curves actually are radial lines, and hence in which the circles $u =$ const. have their centers at (x_0, y_0). If $U(x, y)$ is any harmonic function, then with respect to this imagined new coordinate system $U(x_0, y_0)$ is equal to the average value of U on the circle $u =$ const. This average value is found by drawing a large number, say N, of equally spaced radial lines (circles in xy-coordinates) from (x_0, y_0), evaluating U at the N points where they intersect the circle $u =$ const., adding, and dividing by N. As $N \to \infty$ the average so obtained would be expected to approach $U(x_0, y_0)$. With respect to xy-coordinates these 'radial lines' are actually circles through (x_0, y_0) and the N points of intersection with $u =$ const. are clustered more densely on the part of the circle nearest (x_0, y_0). As $N \to \infty$ one then obtains $U(x_0, y_0)$ as a *weighted* average of the values of U on the eccentric circle around (x_0, y_0), weighting the values of the points nearer (x_0, y_0) more heavily. This leads to

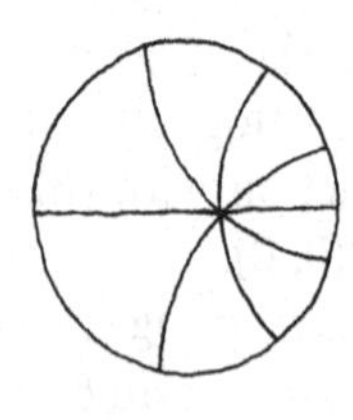

the expectation that *if $U(x, y)$ is harmonic on a disk $\{(x-a)^2 + (y-b)^2 \le r^2\}$, then the value of U at any point (x_0, y_0) inside the disk can be written as a weighted average of the values of U on the circle $\{(x-a)^2 + (y-b)^2 = r^2\}$.* The explicit formula, called *Poisson's Integral Formula,* is

$$U(x_0, y_0) = \frac{1}{2\pi}\int_0^{2\pi} U(a + r\cos\theta, b + r\sin\theta)\frac{r^2 - (x_0-a)^2 - (y_0-b)^2}{(x_0 - a - r\cos\theta)^2 + (y_0 - b - r\sin\theta)^2}\,d\theta.$$

It is derived in §8.4. Note that if $x_0 = a$, $y_0 = b$, the formula reduces to the statement that $U(a, b)$ is the (unweighted) average value of U on the circle $\{(x-a)^2 + (y-b)^2 = r^2\}$.

More generally, if D is any compact differentiable two-dimensional manifold-with-boundary in the xy-plane, and if $U(x, y)$ is harmonic throughout D, then the values of U on ∂D determine the values of U throughout D. This is proved as follows: If U_1 is another harmonic function on D which agrees with U on ∂D, then $U - U_1$ is harmonic throughout D and is identically zero on ∂D. If $U - U_1$ is not identically zero on D, then $|U - U_1|$ assumes a non-zero maximum at some point (x_0, y_0) inside D. Let r be such that the circle $\{(x-x_0)^2 + (y-y_0)^2 = r^2\}$ lies inside D and touches ∂D. Then, since $U - U_1$ is harmonic, its value at (x_0, y_0) is the average of its values on the circle $\{(x-x_0)^2 + (y-y_0)^2 = r^2\}$. On the other hand, the absolute value of $U - U_1$ on the circle is at most $|U(x_0, y_0) - U_1(x_0, y_0)|$ (by assumption) and strictly less than this at some points (near ∂D). Since averaging decreases the absolute value, this is a contradiction unless the assumption $U \not\equiv U_1$ is false. Thus $U = U_1$ as was to be shown.

Thus the values of a harmonic function on the boundary ∂D of a compact domain D determine the values throughout D and theoretically one should be able to find a formula for the values inside D in terms of the values on the boundary. The Poisson Integral Formula accomplishes this when D is a disk, but for other domains, even for rectangles, an explicit formula is very difficult to give, although such a formula does 'exist' for all domains D.

Another question of considerable interest is: Given a compact 2-dimensional domain D, and given a function on ∂D, is there a harmonic function U on D whose values on ∂D are the given function? On physical grounds the answer is "yes" for the following reason: Imagine D to be a sheet of tin and imagine the edge ∂D of the sheet to be maintained at a temperature equal to the value of the given function on ∂D. After a sufficient period of time one would expect to arrive at a state of thermal equilibrium, at which time the temperature of the tin sheet would be the desired harmonic function $U(x, y)$ on D. The problem of proving that this statement is true mathematically, that is, the problem of proving that *there exists a harmonic function with given boundary values* is called the 'Dirichlet Problem'. Its successful solution (for reasonable domains D there exists a harmonic function on D with arbitrarily specified boundary values) has played a major role in the history of mathematics.

Exercises

1 Discuss harmonic functions of one variable. What is the analogue of Laplace's equation? What is the analogue of the Poisson Integral Formula?

2 Prove that a twice differentiable function $u(x, y, z)$ of three variables is harmonic [i.e. has the mean value property (1')] if and only if it satisfies Laplace's equation

$$\frac{\partial^2 u}{\partial x^2} + \frac{\partial^2 u}{\partial y^2} + \frac{\partial^2 u}{\partial z^2} \equiv 0.$$

[The proof of the two variable case given in the text uses polar coordinates (r, θ) which do not generalize nicely to three dimensions. What generalizes nicely is the 1-form

$$d\theta = \frac{x\,dy - y\,dx}{x^2 + y^2}$$

which becomes

$$\omega = \frac{x\,dy\,dz + y\,dz\,dx + z\,dx\,dy}{(x^2 + y^2 + z^2)^{3/2}}.$$

Show first that $dx\,dy\,dz = r^2\omega\,dr$ and that $d\omega = 0$. The identity

$$u(\bar{x}, \bar{y}, \bar{z}) \int_D r^2\omega\,dr \equiv \int_D ur^2\omega\,dr$$

is shown to be equivalent to

$$u(\bar{x}, \bar{y}, \bar{z}) \int_{\partial D} r^2\omega \equiv \int_{\partial D} ur^2\omega$$

(D a ball with center $(\bar{x}, \bar{y}, \bar{z})$), by considering both sides as integrals with respect to r. Since r is constant on ∂D, this is

$$u(\bar{x}, \bar{y}, \bar{z}) \int_{\partial D} \omega \equiv \int_{\partial D} u\omega.$$

The left side is constant and the right side is nearly equal to the left when the radius of D is small; hence the original identity is equivalent to $\int_{\partial D} u\omega = \text{const.}$ as a function of the radius of D. From

$$\int_{\partial D} u\omega \equiv \int_{a^2+b^2+c^2=1} u(\bar{x} + ra, \bar{y} + rb, \bar{z} + rc)(a\, db\, dc + b\, dc\, da + c\, da\, db)$$

this can be shown to be equivalent to

$$\int_{\partial D} \left(\frac{\partial u}{\partial x} dy\, dz + \frac{\partial u}{\partial y} dz\, dx + \frac{\partial u}{\partial z} dx\, dy \right) \equiv 0$$

from which the desired conclusion follows.]

3 Prove that a function $u(x, y, z)$ which is harmonic on a reasonable domain D in $\mathbf{R}^3$ is determined by its values on ∂D.

4 Using the fact that $d\omega = 0$ (where ω is as in Exercise 2), conclude that r^{-1} is a harmonic function on $\mathbf{R}^3$ (not defined at the origin). Prove in the same way that r^{2-n} is a harmonic function on $\mathbf{R}^n$ for $n > 3$.

5 Define 'average over a sphere' in such a way that $u(x, y, z)$ is harmonic if and only if its average over any sphere is equal to its value at the center of the sphere.

6 The conformal coordinates $(u, v) = (x^2 - y^2, 2xy)$ arise from squaring the matrix

$$\begin{pmatrix} x & -y \\ y & x \end{pmatrix} \begin{pmatrix} x & -y \\ y & x \end{pmatrix} = \begin{pmatrix} x^2 - y^2 & -2xy \\ 2xy & x^2 - y^2 \end{pmatrix}.$$

Show that the nth power of this matrix gives in the same way polynomials $P_n(x, y)$, $Q_n(x, y)$ of degree n which are conformal coordinates except at $(0, 0)$.

7 Show that if (u, v), (r, s) are conformal coordinates on the xy-plane, then $(ur - vs, us + vr)$ are conformal coordinates provided $d(ur - vs) \neq 0$. Express this in terms of 2×2 matrices.

8.4 Functions of a Complex Variable

A *complex number* is a 2×2 matrix of the form

$$\begin{pmatrix} x & -y \\ y & x \end{pmatrix}$$

where x, y are real numbers. Such matrices are added to

each other, multiplied by each other, and multiplied by real numbers, according to the rules of matrix algebra. Letting 1, i denote the complex numbers

$$\begin{pmatrix} 1 & 0 \\ 0 & 1 \end{pmatrix}, \begin{pmatrix} 0 & -1 \\ 1 & 0 \end{pmatrix}$$

respectively, every complex number z can be written in just one way as $z = x \cdot 1 + y \cdot i$, where x and y are real numbers. All the usual rules of arithmetic apply to the addition and multiplication of complex numbers, namely:

The sum of two complex numbers is a complex number. Addition is associative and commutative $(z_1 + z_2) + z_3 = z_1 + (z_2 + z_3)$, $z_1 + z_2 = z_2 + z_1$. Given any two complex numbers z_1, z_2 there is a unique complex number z such that $z_1 + z = z_2$ (subtraction axiom). The solution z of $z_1 + z = z_1$ is the same for all z_1, namely the complex number

$$\begin{pmatrix} 0 & 0 \\ 0 & 0 \end{pmatrix}.$$

This complex number is denoted by 0. The product of two complex numbers is a complex number and the operation of multiplication of complex numbers is associative, distributive over addition, and *commutative*. The commutative law does not apply to the multiplication of arbitrary 2×2 matrices, but it does apply to complex numbers:

$$\begin{pmatrix} x_1 & -y_1 \\ y_1 & x_1 \end{pmatrix}\begin{pmatrix} x_2 & -y_2 \\ y_2 & x_2 \end{pmatrix} = \begin{pmatrix} x_1x_2 - y_1y_2 & -x_1y_2 - y_1x_2 \\ y_1x_2 + x_1y_2 & -y_1y_2 + x_1x_2 \end{pmatrix} = \begin{pmatrix} x_2 & -y_2 \\ y_2 & x_2 \end{pmatrix}\begin{pmatrix} x_1 & -y_1 \\ y_1 & x_1 \end{pmatrix}.$$

This computation can also be written in the form $(x_1 \cdot 1 + y_1 \cdot i)(x_2 \cdot 1 + y_2 \cdot i) = x_1x_2 \cdot 1 \cdot 1 + x_1y_2 \cdot 1 \cdot i + y_1x_2 \cdot i \cdot 1 + y_1y_2 \cdot i \cdot i = (x_1x_2 - y_1y_2) \cdot 1 + (x_1y_2 + y_1x_2) \cdot i = (x_2 \cdot 1 + y_2 \cdot i)(x_1 \cdot 1 + y_1 \cdot i)$. Given two complex numbers z_1, z_2 with $z_1 \neq 0$, there is a unique complex number z such that $z_1z = z_2$ (division axiom). This is proved by setting $z_1 = x_1 \cdot 1 + y_1 \cdot i$ and multiplying both sides of $z_1z = z_2$ by $x_1 \cdot 1 - y_1 \cdot i$

to find $(x_1^2 + y_1^2)z = (x_1 \cdot 1 - y_1 \cdot i)z_2$. By assumption, $x_1^2 + y_1^2 \neq 0$; hence if $z_1 z = z_2$, then z can only be the complex number $(x_1^2 + y_1^2)^{-1}(x_1 \cdot 1 - y_1 \cdot i)z_2$. But this complex number z does satisfy $z_1 z = z_2$, which proves that the division axiom holds. If z is any complex number, then $1 \cdot z = z$, $0 \cdot z = 0$.

A complex number z is said to be 'real' if it has the form $z = x \cdot 1$, where x is a real number. No substantial ambiguity results if the distinction between the real number x and the real complex number $x \cdot 1$ is dropped. Then every complex number z can be written in only one way as $z = x + iy$, where x, y are real numbers (that is, real complex numbers).

The size or *modulus* of a complex number is defined to be the square root of its determinant

$$|x + iy| = \sqrt{\begin{vmatrix} x & -y \\ y & x \end{vmatrix}} = \sqrt{x^2 + y^2}.$$

For real numbers x the modulus is the ordinary absolute value $\sqrt{x^2} = |x|$. The modulus has the familiar properties of absolute values:

(i) $|z| = 0$ if and only if $z = 0$.
(ii) $|z_1 z_2| = |z_1| \cdot |z_2|$.
(iii) $|z_1 + z_2| \leq |z_1| + |z_2|$.

The first two of these statements are immediate from the definitions, and the third is easily verified by squaring and applying the Schwarz inequality $x_1x_2 + y_1y_2 \leq \sqrt{x_1^2 + y_1^2}\sqrt{x_2^2 + y_2^2}$.

Thus the complex numbers form an arithmetic (a 'field' in the terminology of algebra) whose *arithmetic* operations are exactly like the arithmetic operations for real numbers. There are two major differences between these 'arithmetics': The first is that the real numbers have an *order relation*, and the complex numbers do not. For example, i is neither 'greater than' nor 'less than' 1. The second difference is that many algebraic equations which do not have solutions in the arithmetic of real numbers do have complex solutions. For example, the equation $x^2 = -1$ has no real solution but has the complex solutions $\pm i$.

The important similarity between the two arithmetics from the point of view of calculus is that the modulus $|z|$ can be used to define *limits* of complex numbers. In fact, the concepts of limit of a sequence, derivative, integral, etc., can be defined for complex numbers using exactly the same words as for real numbers, the only change being that $|z|$ denotes the modulus of a complex number rather than the absolute value of a real number. For example: A sequence of complex numbers $z_1, z_2, z_3, \ldots$ is said to converge to the limit z_∞ if for every (real) $\epsilon > 0$ there is an N such that $|z_n - z_\infty| < \epsilon$ whenever $n \geq N$. Thus the sequence

$$1, 1 + z, 1 + z + z^2, 1 + z + z^2 + z^3, \ldots$$

converges to the limit $1/(1 - z)$ whenever $|z| < 1$ because

$$\begin{aligned}\left|\frac{1}{1-z} - (1 + z + z^2 + \cdots + z^n)\right| \\ &= \left|\frac{1}{1-z} - \frac{1 - z^{n+1}}{1-z}\right| \\ &= \frac{|z|^{n+1}}{|1-z|} \to 0.\end{aligned}$$

The *Cauchy Convergence Criterion* states that a sequence $z_1, z_2, z_3, \ldots$ converges (to some limit z_∞) if and only if for every $\epsilon > 0$, there is an N such that $|z_n - z_m| < \epsilon$ whenever $n, m \geq N$. Thus the sequence

$$z, z + \frac{z^2}{2}, z + \frac{z^2}{2} + \frac{z^3}{3}, z + \frac{z^2}{2} + \frac{z^3}{3} + \frac{z^4}{4}, \ldots$$

converges for $|z| < 1$ because

$$\begin{aligned}\left|\frac{z^{m+1}}{m+1} + \frac{z^{m+2}}{m+2} + \cdots + \frac{z^{n-1}}{n-1} + \frac{z^n}{n}\right| \\ &\leq \frac{|z|^{m+1}}{m+1} + \frac{|z|^{m+2}}{m+2} + \cdots + \frac{|z|^n}{n} \\ &\leq |z|^{m+1} + |z|^{m+2} + \cdots + |z|^n \\ &\leq |z|^{m+1}(1 + |z| + |z|^2 + \cdots) \\ &= \frac{|z|^{m+1}}{1 - |z|} \to 0.\end{aligned}$$

As in the real case, this means that z_n is determined to within an arbitrarily small margin for error ϵ by taking n large, that is, increasing n does not significantly change z_n once n is large. The truth of the Cauchy Criterion for sequences of complex numbers is easily deduced from the corresponding statement for sequences of real numbers (Exercise 2).

A function f assigning complex numbers $f(z)$ to complex numbers z is said to be continuous at a point* z_0 if $\lim_{z\to z_0} f(z) = f(z_0)$, that is, if for every (real) $\epsilon < 0$, there is a $\delta > 0$ such that $|f(z) - f(z_0)| < \epsilon$ whenever $|z - z_0| < \delta$. A function $f(z)$ is said to be *continuous* if it is continuous at every point where it is defined. A function $f(z)$ is said to be *differentiable at a point* z_0 if $\lim_{h\to 0} \frac{f(z_0 + h) - f(z_0)}{h}$ exists; that is, if there is a complex number $f'(z_0)$ with the property that for every $\epsilon > 0$, there is a $\delta > 0$ such that $f(z_0 + h)$ is defined and

*Complex numbers are imagined as points in a plane in the same way that real numbers are imagined as points on a line. This is discussed below.

To be consistent with the notation of the rest of the book, the notation $\bar{z}$ should be used here to denote 'a particular complex number'. However, the notation $\bar{z}$ has another meaning (complex conjugate—see below), so the notation z_0 is used instead.

$$\left|\frac{f(z_0 + h) - f(z_0)}{h} - f'(z_0)\right| < \epsilon$$

whenever h is a non-zero complex number satisfying $|h| < \delta$. A function $f(z)$ is said to be (continuously) *differentiable* if it is differentiable at every point and if its derivative $f'(z)$ is a continuous function.

As in the real case, the sum $f(z) + g(z)$ of two differentiable functions $f(z)$, $g(z)$ is differentiable with derivative $f'(z) + g'(z)$, and their product $f(z)g(z)$ is differentiable with derivative $f'(z)g(z) + f(z)g'(z)$, because

$$\frac{[f(z + h) + g(z + h)] - [f(z) + g(z)]}{h}$$

$$= \frac{f(z + h) - f(z)}{h} + \frac{g(z + h) - g(z)}{h}$$

$$\to f'(z) + g'(z)$$

and

$$\frac{f(z + h)g(z + h) - f(z)g(z)}{h}$$

$$= \frac{f(z + h) - f(z)}{h} g(z + h) + f(z) \frac{g(z + h) - g(z)}{h}$$

$$\to f'(z)g(z) + f(z)g'(z).$$

(The precise proofs are exactly as in the real case.) Since a constant function $f(z) \equiv a$ is differentiable with derivative zero and the identity function $f(z) = z$ is differentiable with derivative identically 1, it follows that a polynomial function $f(z) = a_n z^n + a_{n-1} z^{n-1} + \cdots + a_1 z + a_0$ is differentiable with derivative $n a_n z^{n-1} + (n - 1) a_{n-1} z^{n-2} + \cdots + a_1$.

The function $f(z) = 1/z$, defined for $z \neq 0$, is differentiable with derivative

$$\lim_{h\to 0} \frac{1}{h}\left[\frac{1}{z+h} - \frac{1}{z}\right] = \lim_{h\to 0} \frac{z-(z+h)}{h(z+h)z}$$

$$= \lim_{h\to 0} \frac{-1}{(z+h)z} = -\frac{1}{z^2}.$$

The composition $f[g(z)]$ of differentiable functions is differentiable with derivative $f'[g(z)] \cdot g'(z)$. (As in the real case, the proof of this fact is a bit tricky. See Exercise 10.) Thus if $f(z)$ is differentiable and $f(z) \neq 0$, then $1/f(z)$ is differentiable with derivative $-f'(z)/[f(z)]^2$.

Geometrically, it is natural to represent complex numbers as points in a plane, letting the complex number $z = x + iy$ correspond to the point of the xy-plane (now called the z-plane) with coordinates (x, y). Thus 0 corresponds to the origin, 1 to the point $(1, 0)$, i to the point $(0, 1)$. Real numbers correspond to points on the x-axis. Multiplication by i carries $(1, 0)$ to $(0, 1)$, $(0, 1)$ to $(-1, 0)$, and (x, y) to $(-y, x)$; hence geometrically the operation of multiplication by i is a rotation of 90° in the positive sense. Multiplication by i^2 is a rotation of 180° carrying (x, y) to $(-x, -y)$. More generally, multiplication by the complex number $z = a + ib$ carries $x + iy$ to $(ax - by) + i(ay + bx)$; hence multiplication by $z = a + ib$ corresponds geometrically to the linear transformation of the xy-plane whose matrix of coefficients is

$$\begin{pmatrix} a & -b \\ b & a \end{pmatrix}.$$

It is easy to give a geometrical description of this map (Exercise 4). The operation of addition of complex numbers is performed by adding corresponding components, as $(x_1 + iy_1) + (x_2 + iy_2) = (x_1 + x_2) + i(y_1 + y_2)$, which can be expressed geometrically by saying that the points $0, z_1, z_2, z_1 + z_2$ form the vertices of a parallelogram.

A complex 1-form is an expression of the form $f(z)\,dz$ where $f(z)$ is a complex-valued function of a complex variable. The integral $\int_C f(z)\,dz$ of a 1-form $f(z)\,dz$ over a curve C in the z-plane is the limit of the approximating sums

$$\sum_j f(z_j)\,\Delta z_j$$

formed by subdividing C into a large number of small segments, choosing a point z_j on the jth segment and letting $\Delta z_j = \hat{z}_j - \hat{z}_{j-1}$ be the difference between the end points $\hat{z}_j$, $\hat{z}_{j-1}$ of the jth segment. If $f(z)\,dz$ is a continuous 1-form, that is, if $f(z)$ is a continuous function, and if C is a compact, oriented, differentiable curve in the z-plane, then it can be shown that the approximating sums have a limiting value $\int_C f(z)dz$. However, rather than prove that $\lim \sum_j f(z_j)\,\Delta z_j$ exists, one can instead define the integral $\int_C f(z)\,dz$ in terms of *real* integrals

$$(1) \qquad \int_C (u\,dx - v\,dy) + i \int_C (u\,dy + v\,dx)$$

where $u(x, y)$, $v(x, y)$ are the real-valued functions defined by $f(x + iy) = u(x, y) + iv(x, y)$. If $f(z)$ is a continuous function, then the real 1-forms $u\,dx - v\,dy$, $u\,dy + v\,dx$ are continuous and the integrals above are defined in Chapter 6 for any compact, oriented, differentiable curve C (with or without boundary points) in the z-plane. It is true that the number defined by (1) is indeed the limit of the approximating sums $\sum f\,\Delta z = \sum (u + iv) \cdot (\Delta x + i\,\Delta y) = \sum (u\,\Delta x - v\,\Delta y) + i\sum (u\,\Delta y + v\,\Delta x)$ but the proof of this fact will be omitted.

The Fundamental Theorem of Calculus

$$(2) \qquad f(z_1) - f(z_0) = \int_{z_0}^{z_1} f'(z)\,dz$$

(where the integral is taken over a curve from z_0 to z_1 along which f is continuously differentiable) is simply Stokes' Theorem

$$(2') \qquad \int_{\partial C} f = \int_C df = \int_C f'(z)\,dz$$

for this case. The 0-dimensional 'integral' on the left is a sum over the boundary points of the curve C, each being 'oriented' plus or minus depending on whether the oriented curve C leads into or out of the boundary point. To prove the formula (2) on the basis of the definition (1) of the right-hand side, one must note that the definition of differentiability implies that $f'(z)$ can be written in either the form

$$\lim_{h \to 0} \frac{f(z + h) - f(z)}{h} = \frac{\partial u}{\partial x} + i \frac{\partial v}{\partial x}$$

or the form

$$\lim_{h\to 0} \frac{f(z+ih)-f(z)}{ih} = \frac{1}{i}\left[\frac{\partial u}{\partial y} + i\frac{\partial v}{\partial y}\right]$$

where the limit is through real values of h. Thus the differentiability of f implies not only that the functions $\frac{\partial u}{\partial x}, \frac{\partial u}{\partial y}, \frac{\partial v}{\partial x}, \frac{\partial v}{\partial y}$ are continuous, but also that they satisfy

$$\frac{\partial u}{\partial x} + i\frac{\partial v}{\partial x} = \frac{1}{i}\left[\frac{\partial u}{\partial y} + i\frac{\partial v}{\partial y}\right]$$

which implies

$$\frac{\partial u}{\partial x} \equiv \frac{\partial v}{\partial y}, \ \frac{\partial v}{\partial x} \equiv -\frac{\partial u}{\partial y}.$$

Therefore

$$\begin{aligned}\int_C f'(z)\,dz &= \int_C \left(\frac{\partial u}{\partial x} + i\frac{\partial v}{\partial x}\right)(dx + i\,dy)\\ &= \int_C \left(\frac{\partial u}{\partial x}dx - \frac{\partial v}{\partial x}dy\right) + i\int_C \left(\frac{\partial u}{\partial x}dy + \frac{\partial v}{\partial x}dx\right)\\ &= \int_C du + i\int_C dv = \int_{\partial C} f\end{aligned}$$

as expected.

Now if $f(z)$ is differentiable, then formally $d[f(z)\,dz] = f'(z)\,dz\,dz = 0$. Hence one would expect that *if $f(z)$ is a differentiable function, then the 1-form $f(z)\,dz$ is closed.* This is immediately verified:

$$\begin{aligned}&d[(u\,dx - v\,dy) + i(u\,dy + v\,dx)]\\ &\quad = \left(-\frac{\partial u}{\partial y} - \frac{\partial v}{\partial x}\right)dx\,dy + i\left(\frac{\partial u}{\partial x} - \frac{\partial v}{\partial y}\right)dx\,dy \equiv 0.\end{aligned}$$

It follows that if f is differentiable on a two-dimensional domain* D, then

$$\int_{\partial D} f(z)\,dz = 0.$$

*In the remainder of this section a 'domain' D will be a compact, differentiable two-dimensional manifold-with-boundary in the z-plane oriented by $dx\,dy$, and ∂D will denote the boundary curve of D oriented by the usual convention.

This is known as *Cauchy's Theorem*.

Of course if $f(x)$ is a differentiable function of one *real* variable, then $d[f(x)\,dx] \equiv 0$. In this case, however, there are no two-dimensional domains, hence no boundary curves ∂D over which to integrate, and hence no Cauchy's Theorem. This difference—between the presence and the absence of Cauchy's Theorem—is the source of the vast difference between the calculus of complex functions and the calculus of real functions, which will be seen in the following paragraphs.

The complex 1-form dz/z plays a central role in the theory of functions of a complex variable. It is closely related to the local conformal coordinates $(\log r, \theta)$ of §8.3 as is seen from the equations

$$\begin{aligned}\frac{dz}{z} &= \frac{(x - iy)(dx + i\,dy)}{(x - iy)(x + iy)} \\ &= \frac{x\,dx + y\,dy}{x^2 + y^2} + i\frac{x\,dy - y\,dx}{x^2 + y^2} \\ &= d(\log r) + i\,d\theta.\end{aligned}$$

In particular, if $f(z)$ is any function of z, then the integral of $f(z)\,dz/z$ around the boundary of the disk $|z| \le r$ can be found by parameterizing $z = r\cos\theta + ir\sin\theta$, $0 \le \theta \le 2\pi$:

$$\begin{aligned}&\int_{|z|=r} f(z)\frac{dz}{z} \\ &\quad= \int_0^{2\pi} f(r\cos\theta + ir\sin\theta)\frac{(-r\sin\theta + ir\cos\theta)\,d\theta}{r\cos\theta + ir\sin\theta} \\ &\quad= \int_0^{2\pi} f(r\cos\theta + ir\sin\theta)i\,d\theta \\ &\quad= 2\pi i \times (\text{average value of } f \text{ on the circle } |z| = r).\end{aligned}$$

Consider now the integral of $f(z)\,dz/z$ around the boundary ∂D of an arbitrary domain D on which $f(z)$ is differentiable. If D does not contain the point 0, then $1/z$ is defined and differentiable throughout D; hence so is $f(z) \cdot 1/z$ and

$$\int_{\partial D} \frac{f(z)\,dz}{z} = 0$$

by Cauchy's Theorem. If D contains the point zero in its interior, then D can be decomposed into two domains D_1, D_2 where D_1 is the disk of radius ϵ with center at 0 and D_2 is the remainder of D; hence

$$\begin{aligned}\int_{\partial D} \frac{f(z)\,dz}{z} &= \int_{\partial D_1} \frac{f(z)\,dz}{z} + \int_{\partial D_2} \frac{f(z)\,dz}{z} \\ &= \int_{|z|=\epsilon} \frac{f(z)\,dz}{z} + 0 \\ &= 2\pi i \cdot (\text{average value of } f(z) \text{ on } |z| = \epsilon).\end{aligned}$$

Thus the integral over ∂D is the same for *all* domains D which contain 0, and this number is equal to $2\pi i$ times the average value of f on any circle with center at 0. By the continuity of f this average value must be arbitrarily near $f(0)$ for ϵ sufficiently small; hence it must be exactly $f(0)$ for all ϵ. The same argument applies to the evaluation of the integral of $f(z)\,dz/(z - a)$ over ∂D to give the *Cauchy Integral Formula*

$$(3)\quad \frac{1}{2\pi i}\int_{\partial D}\frac{f(z)\,dz}{z-a} = \begin{cases} f(a) \text{ if } a \text{ is inside } D \\ 0 \quad \text{if } a \text{ is outside } D \\ \text{not defined if } a \text{ is on } \partial D \end{cases}$$

for any function $f(z)$ which is differentiable throughout the domain D.

The Cauchy Integral Formula shows in particular that the value $f(a)$ of a differentiable function f at any point a inside D is determined by its values $f(z)$ on ∂D. Using the algebraic identity

$$1 - r^{n+1} = (1 - r)(1 + r + r^2 + \cdots + r^n)$$

with

$$r = \frac{z_0 - a}{z - a},\quad 1 - r = \frac{z - z_0}{z - a}$$

gives

$$1 - \left(\frac{z_0 - a}{z - a}\right)^{n+1} = \left(\frac{z - z_0}{z - a}\right)\left[1 + \frac{z_0 - a}{z - a} + \cdots + \left(\frac{z_0 - a}{z - a}\right)^n\right]$$

$$\frac{1}{z - z_0} = \frac{1}{z - a} + \frac{z_0 - a}{(z - a)^2} + \cdots + \frac{(z_0 - a)^n}{(z - a)^{n+1}} + \frac{(z_0 - a)^{n+1}}{(z - z_0)(z - a)^{n+1}}$$

which combines with the Cauchy Integral Formula to give

$$\begin{aligned} f(z_0) &= \frac{1}{2\pi i}\int_{\partial D}\frac{f(z)\,dz}{z - z_0} \\ &= f(a) + \frac{(z_0 - a)}{2\pi i}\int_{\partial D}\frac{f(z)\,dz}{(z - a)^2} + \cdots \\ &\quad + \frac{(z_0 - a)^n}{2\pi i}\int_{\partial D}\frac{f(z)\,dz}{(z - a)^{n+1}} + \frac{(z_0 - a)^{n+1}}{2\pi i}\int_{\partial D}\frac{f(z)\,dz}{(z - z_0)(z - a)^{n+1}} \\ &= c_0 + c_1(z_0 - a) + \cdots + c_n(z_0 - a)^n + R_n \end{aligned}$$

where D is any domain containing z_0, a and where

$$c_j = \frac{1}{2\pi i} \int_{\partial D} \frac{f(z)\,dz}{(z - a)^{j+1}} \qquad (j = 0, 1, 2, \ldots, n)$$

$$R_n = \frac{(z_0 - a)^{n+1}}{2\pi i} \int_{\partial D} \frac{f(z)\,dz}{(z - z_0)(z - a)^{n+1}}.$$

It will be shown that for all complex numbers z_0 sufficiently near a, the remainder R_n approaches 0 as $n \to \infty$. Hence

$$f(z_0) = \sum_{n=0}^{\infty} c_n(z_0 - a)^n$$

which shows that *every differentiable function is in fact analytic*. (A function is said to be *analytic* if it can be written locally as the sum of a power series.) More specifically:

Theorem

Let $f(z)$ be a complex function of a complex variable which is differentiable at all points inside the disk $\{|z - a| < r\}$ of radius r with center a. Then the value $f(z)$ of f at any point z of the disk is equal to the sum of the power series

$$f(z) = \sum_{n=0}^{\infty} c_n(z - a)^n = \lim_{N\to\infty} \sum_{n=0}^{N} c_n(z - a)^n$$

with coefficients

$$c_n = \frac{1}{2\pi i} \int_{\partial D} \frac{f(z)\,dz}{(z - a)^{n+1}}$$

where D is any domain containing a on which $f(z)$ is differentiable. (This number c_n is independent of the choice of D, by the argument above.) Conversely, if $c_0, c_1, c_2, c_3, \ldots$ are complex numbers such that the power series $c_0 + c_1(z - a) + c_2(z - a)^2 + \cdots + c_n(z - a)^n + \cdots$ is convergent for some $z_0 \neq a$, then it is convergent for all values of z inside the disk $\{|z - a| < |z_0 - a|\}$ and defines a differentiable function

$$f(z) = c_0 + c_1(z - a) + c_2(z - a)^2 + \cdots + c_n(z - a)^n + \cdots$$

whose derivative $f'(z)$ is equal to the sum of the power series

$$f'(z) = c_1 + 2c_2(z-a) + 3c_3(z-a)^2 + \cdots + nc_n(z-a)^{n-1} + \cdots$$

for all z in the disk $\{|z-a| < |z_0 - a|\}$.

Corollary

If $f(z)$ is (continuously) differentiable on a disk $\{|z-a| < r\}$, then it is in fact infinitely differentiable and the Taylor series converges to the function

$$f(z) = f(a) + f'(a)(z-a) + \frac{f''(a)}{2}(z-a)^2 + \cdots + \frac{f^{(n)}(a)}{n!}(z-a)^n + \cdots$$

for all z in $\{|z-a| < r\}$.

Proof of Corollary

By the first part of the theorem $f(z) = c_0 + c_1(z-a) + c_2(z-a)^2 + \cdots$; hence by the second part of the theorem $f'(z) = c_1 + 2c_2(z-a) + 3c_3(z-a)^2 + \cdots$. Hence, again by the second part of the theorem, $f'(z)$ is differentiable and $f''(z) = 2c_2 + 6c_3(z-a) + \cdots + n(n-1)c_n(z-a)^{n-2} + \cdots$. Repeating this argument it follows that all derivatives exist and that $f^{(n)}(z) = n!c_n + \cdots + \dfrac{m!c_m}{(m-n)!}(z-a)^{m-n} + \cdots$. In particular $f^{(n)}(a) = n!c_n$ and $f(z) = c_0 + c_1(z-a) + \cdots + c_n(z-a)^n + \cdots = f(a) + f'(a)(z-a) + \cdots + \dfrac{f^{(n)}(a)}{n!}(z-a)^n + \cdots$ as was to be shown.

As a further corollary, the two formulas for c_n give

$$f^{(n)}(a) = \frac{n!}{2\pi i}\int_{\partial D} \frac{f(z)\,dz}{(z-a)^{n+1}}$$

where D is any domain containing a on which f is differentiable.

Proof of Theorem

Let $f(z)$ be differentiable on the disk $\{|z-a| < r\}$, and let z_0 be a given point of this disk. The first statement of the theorem is that the remainder term

$$R_n = \frac{1}{2\pi i}\int_{\partial D} \frac{f(z)}{z-z_0}\left(\frac{z_0-a}{z-a}\right)^{n+1} dz$$

goes to zero as $n \to \infty$. Here D is any domain containing

a and z_0 on which $f(z)$ is differentiable. Let $\rho < r$ be a number such that the disk $\{|z - a| \leq \rho\}$ contains z_0 in its interior, parameterize the boundary of this disk $z = a + \rho \cos \theta + i\rho \sin \theta$, $0 \leq \theta \leq 2\pi$, and hence write R_n as the average value of

$$f(z) \cdot \frac{z_0 - a}{z - z_0} \left(\frac{z_0 - a}{z - a}\right)^n$$

on the circle $|z - a| = \rho$. The first two terms of this product are independent of n; the third has modulus

$$\frac{|z_0 - a|^n}{|z - a|^n} = \left(\frac{|z_0 - a|}{\rho}\right)^n \to 0.$$

Hence if $F_n(\theta)$ denotes the above function on $|z - a| = \rho$ and if $\epsilon > 0$ is given, there is an N such that $|F_n(\theta)| < \epsilon$ for all θ, $0 \leq \theta \leq 2\pi$, and for all $n \geq N$. It follows that $\left|\frac{1}{2\pi} \int_0^{2\pi} F_n(\theta)\, d\theta\right| < \epsilon$ (consider the integral as a limit of sums and use the triangle inequality $|z_1 + z_2| \leq |z_1| + |z_2|$); hence $|R_n| \to 0$ as $n \to \infty$ as was to be shown.

To prove the second half of the theorem, let $c_0, c_1, c_2, \ldots$ and $z_0 \neq a$ be given such that $\sum_{n=0}^{\infty} c_n(z_0 - a)^n$ is convergent, and let $r = |z_0 - a| > 0$. It is to be shown first of all that $\sum_{n=0}^{\infty} c_n(z - a)^n$ is convergent whenever $|z - a| < r$. This can be done by noting that the numbers $|c_n(z_0 - a)^n| = |c_n|r^n$ must be bounded, say $|c_n|r^n < K$ (otherwise $\sum c_n(z_0 - a)^n$ could not converge —see Exercise 3). Hence for $|z - a| = r_1 < r$ and for $N > M$,

$$\begin{aligned}
&\left[\sum_{n=0}^{N} c_n(z - a)^n - \sum_{n=0}^{M} c_n(z - a)^n\right] \\
&\qquad = |c_{M+1}(z - a)^{M+1} + \cdots + c_N(z - a)^N| \\
&\qquad \leq |c_{M+1}|r_1^{M+1} + \cdots + |c_N|r_1^N \\
&\qquad < K\left(\frac{r_1}{r}\right)^{M+1} + \cdots + K\left(\frac{r_1}{r}\right)^N \\
&\qquad < K\left(\frac{r_1}{r}\right)^{M+1}\left(1 + \left(\frac{r_1}{r}\right) + \left(\frac{r_1}{r}\right)^2 + \cdots\right) \\
&\qquad = K\left(\frac{r_1}{r}\right)^{M+1} \cdot \frac{1}{1 - \left(\frac{r_1}{r}\right)}.
\end{aligned}$$

As $M \to \infty$ this goes to zero, which shows that the sequence of partial sums $f_N(z) = \sum_{n=0}^{N} c_n(z-a)^n$ satisfies the Cauchy Criterion for $|z-a| < r$. Hence $\lim_{N\to\infty} f_N(z)$ exists for all z in the disk $\{|z-a| < r\}$. Let $f(z)$ denote this limit. It is to be shown that the function $f(z)$ so defined is differentiable and that its derivative is equal to the sum of the series

$$c_1 + 2c_2(z-a) + 3c_3(z-a)^2 + \cdots$$

for all z in the disk $\{|z-a| < r\}$.

It will be shown first that this series $\sum_{n=1}^{\infty} nc_n(z-a)^{n-1}$ is convergent in the disk $\{|z-a| < r\}$. To show this it suffices to show that the numbers $n|c_n|r_1^{n-1}$ are bounded for $r_1 < r$. Then the argument above shows that $\lim_{N\to\infty} \sum_{n=1}^{N} nc_n(z-a)^{n-1}$ exists for $|z-a| < r_1$ and, since r_1 is arbitrary, it follows that the limit exists for all z in the disk $\{|z-a| < r\}$. Now since $n|c_n|r_1^{n-1} < nKr^{-n}r_1^{n-1} = n(r_1/r)^{n-1} \cdot r^{-1} \cdot K$, it suffices to show that if $0 < \rho < 1$, then the numbers $n\rho^{n-1}$ are bounded. This follows from

$$\begin{aligned}
n\rho^{n-1} &= \rho^{n-1} + \rho^{n-1} + \cdots + \rho^{n-1} \\
&< 1 + \rho + \rho^2 + \rho^3 + \cdots + \rho^{n-1} + \rho^n + \cdots \\
&\qquad + \rho + \rho^2 + \rho^3 + \cdots + \rho^{n-1} + \rho^n + \cdots \\
&\qquad\quad + \rho^2 + \rho^3 + \cdots + \rho^{n-1} + \rho^n + \cdots \\
&\qquad\qquad + \rho^3 + \cdots + \rho^{n-1} + \rho^n + \cdots \\
&\qquad\qquad\qquad \vdots \\
&\qquad\qquad\qquad + \rho^{n-1} + \rho^n + \cdots \\
&= \frac{1}{1-\rho} + \frac{\rho}{1-\rho} + \frac{\rho^2}{1-\rho} + \cdots + \frac{\rho^{n-1}}{1-\rho} \\
&< \frac{1}{1-\rho}(1 + \rho + \rho^2 + \cdots) = (1-\rho)^{-2}.
\end{aligned}$$

It remains to show that $g(z) = \sum_{n=1}^{\infty} nc_n(z-a)^{n-1}$ is the derivative of $f(z)$ for z in the disk $\{|z-a| < r\}$. The idea is that

$$\frac{f_N(z+h) - f_N(z)}{h} \sim g_N(z) \tag{4}$$

where $g_N(z) = \sum_{n=1}^{N} nc_n(z-a)^{n-1}$. Hence passing to the limit as $N \to \infty$

$$f(z+h) - f(z) \over h \sim g(z). \tag{5}$$

One must estimate the error in the approximation (4) and use the estimate to conclude that the error in the approximation (5) goes to zero as $|h| \to 0$. The error in (4) is estimated, as usual, by writing it as an integral:

$$\begin{aligned} \frac{f_N(z+h) - f_N(z)}{h} - g_N(z) \\ &= \frac{1}{h}\int_z^{z+h} f_N'(w)\,dw - \frac{1}{h}\int_z^{z+h} g_N(z)\,dw \\ &= \frac{1}{h}\int_z^{z+h} [g_N(w) - g_N(z)]\,dw \end{aligned}$$

where the integral is over a path from z to $z+h$, say over the line segment joining them. Parameterizing this segment $w = z + th$, $0 \le t \le 1$, so that $dw = h\,dt$, gives

$$\begin{aligned} \frac{f_N(z+h) - f_N(z)}{h} - g_N(z) \\ &= \int_0^1 [g_N(z+th) - g_N(z)]\,dt \\ &= \sum_{n=1}^{N} nc_n \int_0^1 [(z+th-a)^{n-1} - (z-a)^{n-1}]\,dt. \end{aligned}$$

Now

$$\begin{aligned} \int_0^1 [(z+th-a)^{n-1} - (z-a)^{n-1}]\,dt \\ &= \int_0^1 [(z+th-a) - (z-a)][(z+th-a)^{n-2} + (z+th-a)^{n-3}(z-a) \\ &\quad + \cdots + (z-a)^{n-2}]\,dt \\ &= h\int_0^1 t[(z+th-a)^{n-2} + \cdots + (z-a)^{n-2}]\,dt. \end{aligned}$$

The integrand is a sum of $n-1$ terms, each of which

has modulus at most $(|z-a|+|h|)^{n-2}$. Setting $r_1 = |z-a| + |h|$ this gives

$$\left|\frac{f_N(z+h) - f_N(z)}{h} - g_N(z)\right| \le |h| \sum_{n=1}^{N} n(n-1)|c_n| r_1^{n-2}.$$

The convergence of the series $\sum_{n=1}^{\infty} n(n-1)|c_n| r_1^{n-2}$ for $r_1 < r$ follows from the convergence of $\sum_{n=1}^{\infty} n|c_n| r_1^{n-1}$ in the same way that the convergence of the latter series followed from the convergence of $\sum_{n=0}^{\infty} |c_n| r_1^n$. Letting $K' = \sum_{n=1}^{\infty} n(n-1)|c_n| r_1^{n-2}$ for some r_1 satisfying $|z-a| < r_1 < r$, it follows that

$$\left|\frac{f(z+h) - f(z)}{h} - g(z)\right| \le |h| \cdot K'$$

for all non-zero h such that $|z-a| + |h| < r_1$, and the theorem follows.

The theorem just proved is the central fact in the theory of functions of a complex variable. It has many applications to the theory of functions of *real* variables, of which the following are a few of the most important examples.

The Binomial Series

Let $\alpha = \pm\frac{p}{q}$ be any rational number (p, q positive integers), and let x^α be the function $\sqrt[q]{x^{\pm p}}$ defined for $x > 0$. Then the power series representation

$$(1+x)^\alpha = 1 + \alpha x + \frac{\alpha(\alpha-1)}{2}x^2 + \cdots + \binom{\alpha}{n} x^n + \cdots$$

where

$$\binom{\alpha}{n} = \frac{\alpha(\alpha-1)\cdots(\alpha-n+1)}{n!}$$

is valid for $|x| < 1$. This is proved by showing that $(1+x)^\alpha$ can be extended to a differentiable complex function of a complex variable as follows: Note first that the power series $\sum_{n=0}^{\infty} z^n/n!$ converges for all real values of z. Hence the function defined by

$$e^z = 1 + z + \frac{z^2}{2} + \cdots + \frac{z^n}{n!} + \cdots$$

is defined and differentiable for all complex z and its derivative is e^z. The function e^{z_0+z} is defined for all z and is its own derivative. Therefore its Taylor series for $a = 0$ is

$$\begin{aligned} e^{z_0+z} &= e^{z_0} + e^{z_0} \cdot z + \frac{e^{z_0}}{2!} z^2 + \cdots \\ &= e^{z_0}(e^z). \end{aligned}$$

Hence $e^{z_0+z_1} = e^{z_0}e^{z_1}$ for all z_0, z_1. In particular e^z is never zero and $1/e^z$ is e^{-z}. Since the series $1 - (\frac{1}{2}) + (\frac{1}{3}) - (\frac{1}{4}) + \cdots$ converges (by the alternating series test), it follows that the function defined by

$$\log(1 + z) = z - \frac{z^2}{2} + \frac{z^3}{3} - \frac{z^4}{4} + \cdots$$

is defined and differentiable for $|z| < 1$, and that its derivative is

$$1 - z + z^2 - z^3 + z^4 - \cdots = \frac{1}{1 + z}.$$

The composed function

$$e^{\log(1+z)}$$

is therefore defined and differentiable for $|z| < 1$. Its derivative is

$$e^{\log(1+z)} \cdot \frac{1}{1 + z}$$

and its second derivative is

$$e^{\log(1+z)} \cdot \frac{1}{(1 + z)^2} + e^{\log(1+z)} \cdot \frac{-1}{(1 + z)^2} \equiv 0.$$

Hence since $e^{\log(1+0)} = e^0 = 1$, the Taylor series of the composed function for $a = 0$ is

$$e^{\log(1+z)} = 1 + z.$$

The function $e^{\alpha \log(1+z)}$ for any (real or complex) number α is defined and differentiable for $|z| < 1$, and its derivative is

$$e^{\alpha \log(1+z)} \cdot \alpha \cdot \frac{1}{1 + z} = \alpha \frac{e^{\alpha \log(1+z)}}{e^{\log(1+z)}} = \alpha e^{(\alpha-1)\log(1+z)}.$$

Its second derivative is therefore

$$\alpha(\alpha - 1)e^{(\alpha-2)\log(1+z)},$$

etc., which gives the Taylor series

$$e^{\alpha \log(1+z)} = 1 + \alpha z + \frac{\alpha(\alpha-1)}{2} z^2 + \cdots + \binom{\alpha}{n} z^n + \cdots$$

valid for $|z| < 1$. On the other hand, if $\alpha = \pm p/q$ and if x is a real number with $|x| < 1$, then $e^{\alpha\log(1+x)}$ is a real number $\left(\text{the series has real coefficients } \binom{\alpha}{n}\right)$, which is positive (because it is the square of the non-zero real number $e^{\frac{1}{2}\alpha\log(1+x)}$), and satisfies

$$\begin{aligned}(e^{\alpha\log(1+x)})^q &= e^{q\alpha\log(1+x)} = e^{\pm p\log(1+x)}\\ &= (e^{\log(1+x)})^{\pm p}\\ &= (1+x)^{\pm p}.\end{aligned}$$

Hence $e^{\alpha\log(1+x)} = \sqrt[q]{(1+x)^{\pm p}} = (1+x)^\alpha$ as was to be shown.

The Fundamental Theorem of Algebra

Every polynomial $f(x) = x^n + a_{n-1}x^{n-1} + \cdots + a_1x + a_0$ with real coefficients a_i can be written as a product of factors of the form $(x - A)$ and $(x^2 + 2Bx + C)$ where A, B, C are real numbers and $C > B^2$ (so that $x^2 + 2Bx + C = (x + B)^2 + (C - B^2)$ is positive for all x). Every polynomial $f(z) = z^n + a_{n-1}z^{n-1} + \cdots + a_1z + a_0$ with complex coefficients can be written as a product of factors of the form $(z - A)$, where A is a complex number. To prove these statements, let $f(z) = \sum_{j=0}^{n} a_jz^j$ be considered as a function of a complex variable. If z_0 is any complex number, then the polynomial $f(z)$ can be written in the form $f(z) = (z - z_0)g(z) + r$, where $g(z)$ is a polynomial of degree $n - 1$, and where r is a complex number. (This is simple algebraic division of polynomials.) If z_0 can be chosen so that $f(z_0) = 0$, then $r = 0$ and $f(z) = (z - z_0)g(z)$. If z_0 is real or if the given polynomial has complex coefficients, then the factor $z - z_0$ is of the desired type and the factorization of the given polynomial is reduced to the factorization of a polynomial of lower degree. If $z_0 = x_0 + iy_0$ with $y_0 \neq 0$ and if the given polynomial has real coefficients, set $B = -x_0$ and $C = x_0^2 + y_0^2$. Use division to write $f(x) = (x^2 + 2Bx + C)g(x) + r_1x + r_0$, where $g(x)$ is a polynomial with real coefficients whose degree is 2 less than that of f, and where r_1, r_0 are real numbers. Since $f(z_0) = 0$ and $z_0^2 + 2Bz_0 + C = 0$, the complex substitution $x = z_0$ gives first $r_1 = 0$ and then $r_0 = 0$. Hence $f(x) = (x^2 + 2Bx + C)g(x)$ and the factorization of f is reduced to the factorization of the polynomial g of lower degree. Thus the factorization of any given f can always be reduced to the factorization of a polynomial of lower degree provided one can find a complex number z_0 such that $f(z_0) = 0$. The theorem will therefore be proved if it is shown that *every polynomial* $f(z) = z^n + a_{n-1}z^{n-1} + \cdots + a_0$ *with complex coefficients and with degree* $n > 0$ *has at least one root* z_0, that is, that there is at least one complex number z_0 such that $f(z_0) = 0$. This is proved as follows:

If $f(z)$ is never zero, then $1/f(z)$ is differentiable for all z; hence by the theorem

$$\frac{1}{f(z)} = c_0 + c_1z + c_2z^2 + \cdots,$$

where

$$c_m = \frac{m!}{2\pi i}\int_{|z|=\rho}\frac{dz}{f(z)z^{m+1}}$$

for arbitrarily large ρ. Thus c_m is $m!$ times the average value on the circle $|z| = \rho$ of the function

$$\frac{1}{f(z)z^m} = \frac{1}{z^{n+m}}\cdot\frac{1}{a_n + a_{n-1}\left(\frac{1}{z}\right) + \cdots + a_0\left(\frac{1}{z}\right)^n}.$$

For large ρ the modulus of the second factor is at most $(|a_n| - \epsilon)^{-1}$ and the modulus of the first factor is constantly $\rho^{-(n+m)}$. Thus $|c_m| \le m!(|a_n| - \epsilon)^{-1}\rho^{-(n+m)}$ for all sufficiently large ρ, and hence $c_m = 0$ unless $m = n = 0$. Thus if $f(z)$ is never zero, $f(z) \equiv c_0^{-1}$ and the degree of f must be zero.

The Implicit Function Theorem for Analytic Functions

If $f(x)$ is a differentiable function of a real variable, if $f'(x) \neq 0$ so that the Implicit Function Theorem can be applied to solve $y = f(x)$ for $x = g(y)$, and if the function $f(x)$ is analytic, then the inverse function $g(y)$ is also analytic. This is proved by using the power series

$$f(x) = c_0 + c_1(x - \bar{x}) + c_2(x - \bar{x})^2 + \cdots$$

to extend f to a differentiable function of a complex variable $f(z)$ defined for z near $\bar{x}$. The Implicit Function Theorem for complex functions $w = f(z)$ is proved by successive approximations exactly as for real functions and yields a differentiable complex function $g(w) = z$. But since g is differentiable in the complex sense, it is in fact analytic; hence its restriction to real values $g(y) = x$ is analytic, as was to be shown. The generalization to functions of several variables can be proved by showing in this way that at each stage of a step-by-step elimination the new functions are analytic if the original ones were.

Harmonic Functions are Analytic

If $u(x, y)$ is a harmonic function of two variables defined near x_0, y_0, then the Taylor series

$$\begin{aligned} u(x_0, y_0) &+ \frac{\partial u}{\partial x}(x_0, y_0)(x - x_0) \\ &+ \frac{\partial u}{\partial y}(x_0, y_0)(y - y_0) \\ &+ \frac{1}{2!}\frac{\partial^2 u}{\partial x^2}(x_0, y_0)(x - x_0)^2 + \cdots \\ &+ \frac{1}{j!k!}\frac{\partial^{j+k} u}{\partial x^j\,\partial y^k}(x_0, y_0)(x - x_0)^j(y - y_0)^k + \cdots \end{aligned}$$

converges to the function $u(x, y)$ for all (x, y) near (x_0, y_0). This is proved by defining $v(x, y)$ by $dv = -\frac{\partial u}{\partial y}dx + \frac{\partial u}{\partial x}dy$ (which determines v locally up to an additive constant) and noting that $f(x + iy) = u(x, y) + iv(x, y)$ is then a differentiable function of a complex variable. Hence f is analytic by the theorem and the power series for f gives power series for u and v.

Poisson Integral Formula

Let $U(x, y)$ be a harmonic function on the disk $\{x^2 + y^2 \leq 1\}$ and let (x_0, y_0) be a point inside this disk. The discussion of §8.3 suggests a change of coordinates of the form

$$w = \frac{z - z_0}{z - z_1}$$

where $z_0 = x_0 + iy_0$, and where z_1 is a point outside the disk chosen in such a way that the circle $|z| = 1$ is a circle of the form $|w| = \text{const}$. It can be shown (Exercise 13) that $z_1 = x_1 + iy_1$ has this property if and only if

$$\begin{aligned} x_0x_1 + y_0y_1 &= 1 \\ x_0y_1 - y_0x_1 &= 0 \end{aligned}$$

which can be expressed simply as $\bar{z}_0 z_1 = 1$, where $\bar{z}_0$ denotes the 'complex conjugate' $x_0 - iy_0$ of z_0. Expressing z as a function of w and hence expressing (x, y) as functions of (u, v), where $w = u + iy$, the functions (x, y) are conformal coordinates on the uv-plane. Hence the harmonic function $U(x, y)$ becomes a harmonic function $U(u, v)$ whose value at $u = 0$, $v = 0$ ($x = x_0$, $y = y_0$) is

$$\frac{1}{2\pi i}\int_{|w|=\text{const.}} U(u, v)\,\frac{dw}{w}.$$

To express this as an integral over $|z| = 1$ it suffices to express the 1-form dw/w in terms of z. Now

$$\begin{aligned} \frac{dw}{w} &= \left(\frac{z - z_1}{z - z_0}\right)\left(\frac{(z - z_1)\,dz - (z - z_0)\,dz}{(z - z_1)^2}\right) = \frac{dz}{z - z_0} - \frac{dz}{z - z_1} \\ &= dz\left(\frac{\bar{z} - \bar{z}_0}{z\bar{z} - z_0\bar{z} - \bar{z}_0 z + z_0\bar{z}_0} - \frac{\bar{z}z_0\bar{z}_0 - \bar{z}_0}{z\bar{z}z_0\bar{z}_0 - \bar{z}z_0 - z\bar{z}_0 + 1}\right) \end{aligned}$$

where $\bar{z} = z - iy$ and where the identity $\bar{z}_0 z_1 = 1$ is used. On the circle $|z| = 1$ one has $z\bar{z} = |z|^2 = 1$. Hence the denominators are equal, the z_0's in the numerator cancel, and

$$\frac{dw}{w} = dz\left(\frac{\bar{z}(1 - z_0\bar{z}_0)}{(z - z_0)(\bar{z} - \bar{z}_0)}\right) = \frac{dz}{z}\left(\frac{1 - z_0\bar{z}_0}{(z - z_0)(\bar{z} - \bar{z}_0)}\right)$$

on $z\bar{z} = 1$. Thus

$$\begin{aligned} U(x_0, y_0) &= \frac{1}{2\pi i}\int_{|w|=\text{const.}} U(u, v)\,\frac{dw}{w} \\ &= \frac{1}{2\pi i}\int_{|z|=1} U(x, y)\,\frac{1 - x_0^2 - y_0^2}{(x - x_0)^2 + (y - y_0)^2}\,\frac{dz}{z} \\ &= \frac{1}{2\pi}\int_0^{2\pi} U(\cos\theta, \sin\theta)\,\frac{1 - x_0^2 - y_0^2}{(\cos\theta - x_0)^2 + (\sin\theta - y_0)^2}\,d\theta \end{aligned}$$

which is the Poisson Integral Formula for $r = 1$, $a = b = 0$. The general formula is obtained by a simple change of coordinates. (Note that the proof of the Poisson Integral Formula uses only the arithmetic of complex numbers to simplify the expression of dw/w in terms of z, and does not depend on the theorem that differentiable complex functions are analytic.)

Exercises

1 Use the Schwarz inequality (for real numbers) to prove the triangle inequality $|z_1 + z_2| \le |z_1| + |z_2|$ for complex numbers.

2 Prove the Cauchy Convergence Criterion for sequences $z_1, z_2, z_3, \ldots$ of complex numbers using the Cauchy Convergence Criterion for sequences of real numbers. [Use

$$\max(|x|, |y|) \le \sqrt{x^2 + y^2} \le 2\max(|x|, |y|);$$

to prove that a complex number $z = x + iy$ is 'small' if and only if x, y are both 'small'.]

3 Show that if an infinite series $\lim\limits_{N\to\infty} (z_1 + z_2 + z_3 + \ldots + z_N)$ converges, then the terms are bounded, that is, $|z_n| < K$ (all n) for some K. [Use the Cauchy Criterion plus the fact that a finite set is bounded.]

4 In what way does the operation of multiplication by 2 transform the z-plane? Multiplication by -1? By 0? Show that a linear transformation

$$\begin{pmatrix} a & -b \\ b & a \end{pmatrix}$$

in which $a^2 + b^2 = 1$ is a *rotation* of the z-plane (defining 'rotation' in some suitable way). Show that any linear transformation $\begin{pmatrix} a & -b \\ b & a \end{pmatrix}$ is a rotation and a change of scale by a (positive) scale factor except when $a = b = 0$.

5 Let $a = \cos\dfrac{2\pi}{5}$, $b = \sin\dfrac{2\pi}{5}$. Express in terms of a and b the coordinates of the vertices of the regular pentagon inscribed in the circle $|z| = 1$ with one point at $z = 1$. Deduce algebraic relations satisfied by a and b and show that

$$a = \frac{\sqrt{5} - 1}{4} \qquad b = \frac{\sqrt{5 + \sqrt{5}}}{2\sqrt{2}}.$$

6 Find all complex numbers z which satisfy $z^2 + 1 = z$ and plot them in the z-plane. [$z^6 = 1$]

7 Show that if e^z is defined as in the text, then $e^{i\theta} = \cos\theta + i\sin\theta$ by showing that the components of $e^{i\theta}$ satisfy the differential equations which define the trigonometric functions.

8 Show that $e^z = \lim_{n\to\infty}\left(1 + \frac{z}{n}\right)^n$ for any complex number z.

9 Prove *Liouville's Theorem:* If $f(z)$ is differentiable for all z and bounded (there is a K such that $|f(z)| < K$ for all z), then $f(z)$ must be constant. [Use the method of the proof of the fundamental theorem of algebra.]

10 Prove the *Chain Rule:* If $f(z)$, $g(z)$ are (continuously) differentiable functions such that the composed function $f[g(z)]$ is defined (i.e., the domain of f includes the range of g), then the composed function is differentiable with derivative $f'[g(z)]g'(z)$. [Prove that $\lim_{s\to 0}\frac{f(x+sh) - f(x)}{s} = f'(x)h$ and that this holds uniformly in h; that is, show that for every $\epsilon > 0$, $K > 0$ there is a $\delta > 0$ such that

$$\left|\frac{f(x+sh) - f(x)}{s} - f'(x)\cdot h\right| < \epsilon$$

whenever $|h| \leq K$, $|s| \leq \delta$, $s \neq 0$. Then use

$$\frac{f[g(x+sh)] - f[g(x)]}{s} = \frac{f\left[g(x) + s\,\dfrac{g(x+sh) - g(x)}{s}\right] - f[g(x)]}{s}.$$

To prove that the derivative is continuous, it suffices to show that compositions and products of continuous functions are continuous.]

11 Give an example of a real function of a real variable which is (continuously) differentiable but not analytic.

12 Show that the integral formula for $f^{(n)}(a)$, obtained as a corollary of the theorem, can also be obtained by differentiating under the integral sign in the Cauchy Integral Formula.

13 Let z_0, z_1 be given complex numbers such that $z_0 \neq z_1$. Show that if c is any positive number, then the complex numbers z such that

$$\left|\frac{z - z_0}{z - z_1}\right| = c$$

form a circle in the z-plane except when $c = 1$. Show that this circle coincides with the circle $|z| = 1$ if and only if $\bar{z}_0 z_1 = 1$. [Express the equation in terms of x and y and use elementary Cartesian geometry.]

14 *Alternative proof of the Fundamental Theorem of Algebra.* As was shown in the text, the Fundamental Theorem of Algebra is essentially the statement that *every real polynomial has at least one complex root*, since a complex root gives a factor of the polynomial. Now a complex root of a real polynomial is evidenced by the fact that the power series for its reciprocal is not convergent—e.g., the complex roots of $x^2 + 1$ 'explain' the fact that the power series

$$\frac{1}{x^2+1} = 1 - x^2 + x^4 - x^6 + \cdots$$

does not converge for $|x| > 1$. More specifically, if $f(x)$ is a real polynomial and if f has no complex roots $f(z) = 0$ satisfying $|z| < K$, then the theorem of the text shows that the Taylor series of $1/f(x)$ converges to $1/f(x)$ for all x satisfying $|x| < K$. Hence, using the theorem of the text, the Fundamental Theorem of Algebra is reduced to the following statement about *real* numbers:

Let $f(x) = a_n x^n + a_{n-1}x^{n-1} + \cdots + a_0$ *be a polynomial of degree* ≥ 1 *with real coefficients. Then its reciprocal* $1/f(x)$, *considered as a function of* x, *cannot be expanded as a power series*

$$\frac{1}{f(x)} = c_0 + c_1 x + c_2 x^2 + c_3 x^3 + \cdots$$

which converges for all x.

To prove this, one first shows that the c's must satisfy the relations

$$\begin{aligned}
&a_0c_0 && && &&= 1\\
&a_0c_1 &&+ a_1c_0 && &&= 0\\
&a_0c_2 &&+ a_1c_1 &&+ a_2c_0 &&= 0\\
& &&\ \ \vdots && && \\
&a_0c_n &&+ a_1c_{n-1} &&+ \cdots + a_nc_0 &&= 0\\
&a_0c_{n+1} &&+ a_1c_n &&+ \cdots + a_nc_1 &&= 0\\
& &&\ \ \vdots && && \\
&a_0c_{n+k} &&+ a_1c_{n+k-1} &&+ \cdots + a_nc_k &&= 0.
\end{aligned}$$

That is, one shows that formal multiplication of the power series $c_0 + c_1x + c_2x^2 + \cdots$ by the polynomial $a_0 + a_1x + \cdots + a_nx^n$ gives the power series

$$1 + 0 \cdot x + 0 \cdot x^2 + 0 \cdot x^3 + \cdots \equiv 1.$$

This can be proved without reference to complex numbers (Exercise 8, §9.6) and will be assumed here. [Note in particular that a_0 must be non-zero.] The problem is to show that if these relations are used to *define* the c's, then $c_0 + c_1x +$

$c_2x^2 + \cdots$ cannot converge for all x. This is proved by noting that the c's so defined satisfy

$$\begin{pmatrix} c_{k+1} \\ c_{k+2} \\ \vdots \\ c_{k+n} \end{pmatrix} = \begin{pmatrix} 0 & 1 & 0 & \cdots & 0 \\ 0 & 0 & 1 & \cdots & 0 \\ \vdots & \vdots & \vdots & & \vdots \\ -\frac{a_n}{a_0} & -\frac{a_{n-1}}{a_0} & -\frac{a_{n-2}}{a_0} & \cdots & -\frac{a_1}{a_0} \end{pmatrix} \begin{pmatrix} c_k \\ c_{k+1} \\ \vdots \\ c_{n+k-1} \end{pmatrix}.$$

Let A denote the $n \times n$ matrix in this relation. Show that there is a number α such that $|Av| \geq \alpha|v|$, where $|v| = \max(|v_1|, (v_2|, \ldots, |v_n|)$ for any v in $\mathbf{R}^n$. Then show that there is a constant K such that any string $c_{k+1}, c_{k+2}, \ldots, c_{k+n}$ of n consecutive c's must contain at least one c satisfying $|c_{k+j}| \geq K\alpha^k$, and conclude that $c_0 + c_1x + c_2x^2 + \cdots$ does not converge for $x = \alpha^{-1}$.*

*This proof is closely related to a method, known as Bernoulli's method, of finding complex roots of polynomials. See F. B. Hildebrand, Introduction to Numerical Analysis, McGraw-Hill, 1956.

15 Show that if $a_0 \neq 0$, then the power series $c_0 + c_1x + c_2x^2 + \cdots$ of Exercise 14 does converge for all sufficiently small x. Use the theorem of the text to conclude that the limit is indeed $1/f(x)$ for all sufficiently small x.

8.5 Integrability Conditions

A curve $f(x, y) =$ const. in the xy-plane is said to be a solution of the differential equation $A\,dx + B\,dy = 0$ (where A, B are functions of x, y) if at each point $(\bar{x}, \bar{y})$ of the curve the 1-form df is a multiple of the 1-form $A\,dx + B\,dy$. As was explained in §8.2, the geometrical meaning of this definition is that the tangent line

$$\frac{\partial f}{\partial x}(\bar{x}, \bar{y})(x - \bar{x}) + \frac{\partial f}{\partial y}(\bar{x}, \bar{y})(\bar{y} - y) = 0$$

coincides with the line

$$A(\bar{x}, \bar{y})(x - \bar{x}) + B(\bar{x}, \bar{y})(y - \bar{y}) = 0$$

specified by the differential equation $A\,dx + B\,dy = 0$. The solution is said to be *non-singular* at $(\bar{x}, \bar{y})$ if df and $A\,dx + B\,dy$ are both non-zero at $(\bar{x}, \bar{y})$ (so that the above equations actually determine lines). Only non-singular solutions will be considered in what follows.

Similarly, a surface $f(x, y, z) =$ const. in xyz-space is said to be a non-singular solution of the differential equation $A\,dx + B\,dy + C\,dz = 0$ (where A, B, C are functions of x, y, z) if at each point $(\bar{x}, \bar{y}, \bar{z})$ of the surface the 1-forms df and $A\,dx + B\,dy + C\,dz$ are non-zero and are multiples of each other. Geometrically this means that the tangent plane to the surface ($df = 0$) coincides with the plane specified by the differential equation ($A\,dx + B\,dy + C\,dz = 0$).

A pair of differential equations

$$A_1\,dx + B_1\,dy + C_1\,dz = 0$$
$$A_2\,dx + B_2\,dy + C_2\,dz = 0$$

in 3 variables can be regarded as specifying a line

$$A_1(\bar{x}, \bar{y}, \bar{z})(x - \bar{x}) + B_1(\bar{x}, \bar{y}, \bar{z})(y - \bar{y}) + C_1(\bar{x}, \bar{y}, \bar{z})(z - \bar{z}) = 0$$
$$A_2(\bar{x}, \bar{y}, \bar{z})(x - \bar{x}) + B_2(\bar{x}, \bar{y}, \bar{z})(y - \bar{y}) + C_2(\bar{x}, \bar{y}, \bar{z})(z - \bar{z}) = 0$$

through each point $(\bar{x}, \bar{y}, \bar{z})$ of xyz-space. These 2 equations in 3 unknowns describe a line if and only if they have rank 2, which is true if and only if $\omega_1\omega_2 \neq 0$, where $\omega_1 = A_1\,dx + B_1\,dy + C_1\,dz$, $\omega_2 = A_2\,dx + B_2\,dy + C_2\,dz$. Accordingly, a pair of differential equations $\omega_1 = 0$, $\omega_2 = 0$ is said to be *non-singular* at a point $(\bar{x}, \bar{y}, \bar{z})$ if the 2-form $\omega_1\omega_2$ is not zero at that point. A curve $f =$ const., $g =$ const. in xyz-space is said to be a *non-singular solution* of the differential equations $\omega_1 = 0, \omega_2 = 0$ if at each point $(\bar{x}, \bar{y}, \bar{z})$ of the curve the 2-forms $df\,dg$ and $\omega_1\omega_2$ are non-zero and are multiples of each other. Geometrically this means that the tangent line to the curve ($df = 0$, $dg = 0$) is indeed a line and coincides with the line specified by the equation ($\omega_1 = 0$, $\omega_2 = 0$) (see Exercise 1).

In general, a set of k differential equations $\omega_1 = 0$, $\omega_2 = 0, \ldots, \omega_k = 0$ in n variables ($\omega_i = A_{i1}\,dx_1 + A_{i2}\,dx_2 + \cdots + A_{in}\,dx_n$) is said to be *non-singular* at a point $(\bar{x}_1, \bar{x}_2, \ldots, \bar{x}_n)$ if the k-form $\omega_1\omega_2 \ldots \omega_k$ is not zero at that point. A *non-singular solution* of such a set of equations is an $(n - k)$-dimensional manifold $f_1 =$ const., $f_2 =$ const., $\ldots$, $f_k =$ const. such that the k-form $df_1\,df_2 \ldots df_k$ is a non-zero multiple of $\omega_1\omega_2 \ldots \omega_k$ at each point of the manifold.*

*Solutions will be considered locally; that is, it will be assumed that in the neighborhood of any point of the solution manifold the points of the manifold can be described by k equations $f_i =$ const. satisfying the given condition.

A set of k differential equations $\omega_1 = 0, \omega_2 = 0, \ldots, \omega_k = 0$ in n variables $(x_1, x_2, \ldots, x_n)$ is said to be *integrable* near a point $(\bar{x}_1, \bar{x}_2, \ldots, \bar{x}_n)$ if they are non-singular at that point, and if it is true that through every point near $(\bar{x}_1, \bar{x}_2, \ldots, \bar{x}_n)$ there is a non-singular solution of the equations. In order that a set of differential equations be integrable it is necessary and sufficient (assuming that the 1-forms ω_i are differentiable) that certain conditions be satisfied. Specifically, the theorem is the following:

Theorem

Let $\omega_1, \omega_2, \ldots, \omega_k$ be differentiable 1-forms in n variables $(x_1, x_2, \ldots, x_n)$, and let $(\bar{x}_1, \bar{x}_2, \ldots, \bar{x}_n)$ be a point at which the k-form $\omega_1\omega_2 \ldots \omega_k$ is not zero. If there is a non-singular solution of the differential equations $\omega_1 = 0, \omega_2 = 0, \ldots, \omega_k = 0$ through the point $(\bar{x}_1, \bar{x}_2, \ldots, \bar{x}_n)$, then the $(k+2)$-forms $\omega_1\omega_2 \ldots \omega_k\, d\omega_i$ $(i = 1, 2, \ldots, k)$ must all be zero at $(\bar{x}_1, \bar{x}_2, \ldots, \bar{x}_n)$. Therefore if the equations $\omega_1 = 0, \omega_2 = 0, \ldots, \omega_k = 0$ are integrable near $(\bar{x}_1, \bar{x}_2, \ldots, \bar{x}_n)$, these $(k+2)$-forms must be identically zero near $(\bar{x}_1, \bar{x}_2, \ldots, \bar{x}_n)$. Conversely, if these $(k+2)$-forms are identically zero near $(\bar{x}_1, \bar{x}_2, \ldots, \bar{x}_n)$,

$$(1) \qquad \omega_1\omega_2 \ldots \omega_k\, d\omega_i \equiv 0, \qquad (i = 1, 2, \ldots, k)$$

then there exist $\epsilon > 0$ and functions $f_1, f_2, \ldots, f_k$ defined within ϵ of $(\bar{x}_1, \bar{x}_2, \ldots, \bar{x}_n)$ such that the manifolds $f_1 = \text{const.}, f_2 = \text{const.}, \ldots, f_k = \text{const.}$ are non-singular solutions of the differential equations $\omega_1 = 0, \omega_2 = 0, \ldots, \omega_k = 0$.

The conditions (1) are called the integrability conditions for the differential equations $\omega_1 = 0, \omega_2 = 0, \ldots, \omega_k = 0$. Note that for $k = n - 1$ they are automatically fulfilled (an $(n+1)$-form in n variables is necessarily zero). For one equation $A\,dx + B\,dy + C\,dz = 0$ in three variables there is a single condition

$$(A\,dx + B\,dy + C\,dz)\left[\left(\frac{\partial C}{\partial y} - \frac{\partial B}{\partial z}\right) dy\,dz + \left(\frac{\partial A}{\partial z} - \frac{\partial C}{\partial x}\right) dz\,dx + \left(\frac{\partial B}{\partial x} - \frac{\partial A}{\partial y}\right) dx\,dy\right] \equiv 0,$$

that is,

$$A\left(\frac{\partial C}{\partial y} - \frac{\partial B}{\partial z}\right) + B\left(\frac{\partial A}{\partial z} - \frac{\partial C}{\partial x}\right) + C\left(\frac{\partial B}{\partial x} - \frac{\partial A}{\partial y}\right) \equiv 0.$$

The theorem states that the differential equation $A\,dx + B\,dy + C\,dz = 0$ describes a family of surfaces in space (by specifying their tangent planes) if and only if this condition is satisfied. In general, if the integrability conditions are written out in terms of the coefficient functions, they give $k\binom{n}{k+2}$ conditions on the coefficient functions; by a method similar to the method of Lagrange multipliers these can be reduced to a set of just $k\binom{n-k}{2}$ conditions (Exercise 6). The remainder of this section is devoted to proving the above theorem.

Proof

Let $\omega_1 = 0, \ldots, \omega_k = 0$ be as in the theorem, and let $(\bar{x}_1, \bar{x}_2, \ldots, \bar{x}_n)$ be a point through which there is a solution manifold of these equations. It is to be shown that $\omega_1\omega_2 \ldots \omega_k \, d\omega_i = 0$ at $(\bar{x}_1, \bar{x}_2, \ldots, \bar{x}_n)$ for $i = 1, 2, \ldots, k$. By assumption the solution manifold is of the form $f_i(x_1, x_2, \ldots, x_n) = \text{const.}$ $(i = 1, 2, \ldots, k)$, where $f_1, f_2, \ldots, f_k$ are differentiable functions defined near $(\bar{x}_1, \bar{x}_2, \ldots, \bar{x}_n)$ such that $df_1\, df_2 \ldots df_k \neq 0$. Choose new coordinates $y_1, y_2, \ldots, y_n$ on R^n near $(\bar{x}_1, \bar{x}_2, \ldots, \bar{x}_n)$ in which the solution manifold is $\{y_1 = 0, y_2 = 0, \ldots, y_k = 0\}$. (For example, take $y_i = f(x_1, x_2, \ldots, x_n) - f(\bar{x}_1, \bar{x}_2, \ldots, \bar{x}_n)$ for $i = 1, 2, \ldots, k$ and select $y_{k+1}, \ldots, y_n$ from among the coordinates $x_1, x_2, \ldots, x_n$ so that $dy_1\, dy_2 \ldots dy_n \neq 0$.) Expressing $\omega_1, \omega_2, \ldots, \omega_k$ in terms of the coordinates y the k-form $\omega_1\omega_2 \ldots \omega_k$ is a multiple of $dy_1\, dy_2 \ldots dy_k$ on the solution manifold $\{y_1 = 0, y_2 = 0, \ldots, y_k = 0\}$; hence setting $\omega_i = A_{i1}\, dy_1 + \cdots + A_{in}\, dy_n$ it follows (see Exercise 2) that ω_i is a combination of $dy_1, dy_2, \ldots, dy_k$ on the solution manifold. Thus on the manifold $\{y_1 = 0, y_2 = 0, \ldots, y_k = 0\}$ the functions A_{ij} are identically zero for $j > 0$; hence their partial derivatives with respect to $y_{k+1}, y_{k+2}, \ldots, y_n$ are also zero on this plane. Thus when $d\omega_i$ is expressed in terms of the dy's, $d\omega_i = \sum B_{\mu\nu}\, dy_\mu\, dy_\nu$, the terms in which both μ and ν are greater than k are all zero at points of $\{y_1 = 0, \ldots, y_k = 0\}$. Since $\omega_1\omega_2 \ldots \omega_k$ is a multiple of $dy_1 dy_2 \ldots dy_k$ at such points it follows that $\omega_1\omega_2 \ldots \omega_k \, d\omega_i$ is zero at such points, as was to be shown.

The proof of the remaining half of the theorem is facilitated by the observation that the given equations $\omega_1 = 0, \omega_2 = 0, \ldots, \omega_k = 0$ can be replaced by any other system $\omega_1' = 0, \omega_2' = 0, \ldots, \omega_k' = 0$ in which

$$\text{(2)} \qquad \begin{aligned} \omega_i' &= \sum_{j=1}^{k} b_{ij}\omega_j \\ \omega_i &= \sum_{j=1}^{k} c_{ij}\omega_j' \end{aligned} \qquad (i = 1, 2, \ldots, k)$$

where the coefficients b_{ij}, c_{ij} are differentiable functions of $(x_1, x_2, \ldots, x_n)$. Since $\omega_1\omega_2 \ldots \omega_k$ is a multiple of $\omega_1'\omega_2' \ldots \omega_k'$ and vice versa, a solution manifold of $\{\omega_1 = 0, \omega_2 = 0, \ldots, \omega_k = 0\}$ is a solution manifold of $\{\omega_1' = 0, \omega_2' = 0, \ldots, \omega_k' = 0\}$ and conversely. On the other hand the equations $\{\omega_i = 0\}$ satisfy the integrability conditions if and only if the equations

$\{\omega_i' = 0\}$ do because $\omega_1'\omega_2' \ldots \omega_k'\, d\omega_i' = \sum_{j=1}^{k} A\omega_1\omega_2 \ldots \omega_k$ $[c_{ij}\, d\omega_j + dc_{ij}\omega_j] \equiv 0$ if $\omega_1\omega_2 \ldots \omega_k\, d\omega_j \equiv 0$ (all j), and similarly $\omega_1\omega_2 \ldots \omega_k\, d\omega_i, \equiv 0$ if $\omega_1'\omega_2' \ldots \omega_k'\, d\omega_j' \equiv 0$. Thus instead of showing that the given equations $\{\omega_i = 0\}$ are integrable it suffices to show that some combination of them is integrable.

The case $k = n - 1$ is the most important one. In this case the integrability conditions are automatically satisfied and the theorem is essentially the existence theorem for ordinary differential equations (§7.4). Specifically the case $k = n - 1$ can be deduced from the theorem of Chapter 7 as follows:

Given differentiable 1-forms $\omega_1, \omega_2, \ldots, \omega_{n-1}$ defined near $(\bar{x}_1, \bar{x}_2, \ldots, \bar{x}_n)$ with $\omega_1\omega_2 \ldots \omega_{n-1} \neq 0$ at $(\bar{x}_1, \bar{x}_2, \ldots, \bar{x}_n)$ one can assume, by reordering the coordinates if necessary, that the $dx_1\, dx_2 \ldots dx_{n-1}$ component of $\omega_1\omega_2 \ldots \omega_{n-1}$ is not zero. Then the equations

$$\begin{aligned} \omega_1 &= A_{11}\, dx_1 + A_{12}\, dx_2 + \cdots + A_{1n}\, dx_n \\ &\vdots \\ \omega_{n-1} &= A_{n-1,1}\, dx_1 + A_{n-1,2}\, dx_2 + \cdots + A_{n-1,n}\, dx_n \end{aligned}$$

can be solved near $(\bar{x}_1, \bar{x}_2, \ldots, \bar{x}_n)$ to give

$$\begin{aligned} dx_1 &= B_{11}\omega_1 + \cdots + B_{1,n-1}\omega_{n-1} + B_{1n}\, dx_n \\ &\vdots \\ dx_{n-1} &= B_{n-1,1}\omega_1 + \cdots + B_{n-1,n-1}\omega_{n-1} + B_{n-1,n}\, dx_n. \end{aligned}$$

Setting $\omega_i' = \sum_{j=1}^{n-1} B_{ij}\omega_j$ and $C_i = B_{in}$, this gives $\omega_i' = dx_i - C_i\, dx_n$. Since $\omega_1'\omega_2' \ldots \omega_{n-1}' \neq 0$ it suffices to prove that the equations $\{dx_i - C_i\, dx_n = 0; i = 1, 2, \ldots, n-1\}$ are integrable in order to prove that the given equations $\{\omega_i = 0\}$ are integrable.

Now the existence theorem for ordinary differential equations implies that there is a solution $(x_1(t), x_2(t), \ldots, x_n(t))$ of the differential equation

$$\begin{aligned} \frac{dx_1}{dt} &\equiv C_1(x_1(t), \ldots, x_n(t)) \\ &\vdots \\ \frac{dx_{n-1}}{dt} &\equiv C_{n-1}(x_1(t), \ldots, x_n(t)) \\ \frac{dx_n}{dt} &\equiv 1 \end{aligned}$$

satisfying $(x_1(0), \ldots, x_n(0)) = (x_1, x_2, \ldots, x_n)$ for any

$(x_1, x_2, \ldots, x_n)$ near $(\bar{x}_1, \bar{x}_2, \ldots, \bar{x}_n)$. Moreover, the solution depends differentiably on the initial condition $(x_1, x_2, \ldots, x_n)$. Taking the initial condition to be $(\bar{x}_1 + y_1, \bar{x}_2 + y_2, \ldots, \bar{x}_{n-1} + y_{n-1}, \bar{x}_n)$, it follows that there is a differentiable map

$$x_i = g_i(y_1, y_2, \ldots, y_{n-1}, t) \qquad (i = 1, 2, \ldots, n)$$

defined for $(y_1, y_2, \ldots, y_{n-1}, t)$ within ϵ of $(0, 0, \ldots, 0)$ such that

$$g_i(y_1, y_2, \ldots, y_{n-1}, 0) \equiv \bar{x}_i + y_i \qquad (i = 1, 2, \ldots, n - 1)$$

$$g_n(y_1, y_2, \ldots, y_{n-1}, 0) \equiv \bar{x}_n$$

and such that

$$\frac{\partial g_i}{\partial t}(y_1, y_2, \ldots, y_{n-1}, t) \equiv C_i[g(y_1, \ldots, y_{n-1}, t)]$$

$$\frac{\partial g_n}{\partial t}(y_1, y_2, \ldots, y_{n-1}, t) \equiv 1.$$

On the plane $t = 0$ the pullback of dx_i is identically dy_i $(i = 1, 2, \ldots, n - 1)$, whereas the coefficient of dt in the pullback of dx_n is identically 1 at all points. Hence the pullback of $dx_1\, dx_2 \ldots dx_n$ at $(0, 0, \ldots, 0)$ is $dy_1\, dy_2 \ldots dy_{n-1}\, dt$. Therefore the map g is invertible near $(0, 0, \ldots, 0)$, so $(y_1, y_2, \ldots, y_{n-1}, t)$ can be expressed as functions of $(x_1, x_2, \ldots, x_n)$ near $(\bar{x}_1, \bar{x}_2, \ldots, \bar{x}_n)$. Now the map g was chosen in such a way that the pullback of $dx_i - C_i\, dx_n$ has no term in dt. Thus

$$dx_i - C_i\, dx_n = \sum_{j=1}^{n-1} a_{ij}\, dy_j.$$

To prove that the given equations are integrable it suffices, therefore, to prove that the equations $\{dy_i = 0;\ i = 1, 2, \ldots, n - 1\}$ are integrable. But the lines $\{y_i = \text{const.};\ i = 1, 2, \ldots, n - 1\}$ are solutions of these equations and the case $k = n - 1$ follows.

The case $n - k > 1$ can be deduced from the case $n - k = 1$ as follows: Given $\omega_1, \omega_2, \ldots, \omega_k$ and given $(\bar{x}_1, \bar{x}_2, \ldots, \bar{x}_n)$, one can reorder the coordinates and put the equations in the form

$$\text{(3)} \qquad dx_i - \sum_{j=k+1}^{n} B_{ij}\, dx_j = 0 \qquad (i = 1, 2, \ldots, k)$$

as was done above in the case $n - k = 1$. Adding to these k equations the equations $dx_{k+1} = 0, \ldots,$

$dx_{n-1} = 0$ gives $n - 1$ equations which, by the above, can be solved. The solution can be described by giving a new system of coordinates $(y_1, y_2, \ldots, y_n)$ near $(\bar{x}_1, \bar{x}_2, \ldots, \bar{x}_n)$ in which the lines $y_i =$ const. $(i = 1, 2, \ldots, n - 1)$ are solution curves. This implies that the expression of ω_i in terms of $y_1, y_2, \ldots, y_n$ contains no term in dy_n. Since the reduction of $\omega_1 = 0, \omega_2 = 0, \ldots, \omega_k = 0$ to the form (3) relative to the coordinates y introduces no terms in dy_n, one can assume at the outset that the given system is in the form (3) with $B_{in} \equiv 0$ $(i = 1, 2, \ldots, k)$. Then for $k < j \leq n - 1$ the $(k + 2)$-form $\omega_1\omega_2 \ldots \omega_k\, d\omega_i$ contains just one term in $dx_1\, dx_2 \ldots dx_k\, dx_n\, dx_j$, and the coefficient of this term is $\dfrac{\partial B_{ij}}{\partial x_n}$. Hence the integrability conditions imply

$$\frac{\partial B_{ij}}{\partial x_n} \equiv 0 \qquad \begin{array}{c}(i = 1, 2, \ldots, k)\\ (j = k + 1, \ldots, n - 1).\end{array}$$

Thus the equations (3) do not involve the variable x_n and the solution of the given system is reduced to the solution of k equations in $n - 1$ variables. Repeating this process j times reduces the given equations to k equations in $n - j$ variables. When $j = n - k$ the equations (3) become simply $dx_i = 0$ $(i = 1, 2, \ldots, k)$, with the solutions $x_i =$ const. $(i = 1, 2, \ldots, k)$. This completes the proof of the theorem.

Exercises

1 (a) Show that two equations in three unknowns

$$A(x - \bar{x}) + B(y - \bar{y}) + C(z - \bar{z}) = 0$$
$$D(x - \bar{x}) + E(y - \bar{y}) + F(z - \bar{z}) = 0$$

(where $A, B, \ldots, F$ are numbers) define a line in xyz-space if and only if $(A\,dx + B\,dy + C\,dz)(D\,dx + E\,dy + F\,dz) \neq 0$. [The Implicit Function Theorem for Affine Maps.]

(b) Let ω_1, ω_2 and ω_1', ω_2' be two pairs of constant 1-forms in xyz such that $\omega_1\omega_2 \neq 0$ and $\omega_1'\omega_2' \neq 0$. Show that $\omega_1\omega_2$ is a multiple of $\omega_1'\omega_2'$ if and only if there exist numbers b_{ij} such that

$$(*) \qquad \begin{aligned}\omega_1' &= b_{11}\omega_1 + b_{12}\omega_2\\ \omega_2' &= b_{21}\omega_1 + b_{22}\omega_2,\end{aligned}$$

and that when this is the case, the numbers b_{ij} are uniquely determined. [If (*) holds, then obviously $\omega_1'\omega_2'$ is a multiple of $\omega_1\omega_2$. To prove the converse, choose a constant 1-form ω_3 such that $\omega_1\omega_2\omega_3 \neq 0$. Every 1-form ω' can then be written $\omega' = a_1\omega_1 + a_2\omega_2 + a_3\omega_3$. The number a_1 is determined by the fact that it is the coefficient of $\omega_1\omega_2\omega_3$ in $\omega'\omega_2\omega_3$; similarly for a_2, a_3.]

(c) Show that two pairs of equations as in (a) describe the same line in xyz-space if and only if the corresponding 2-forms are multiples of each other.

2 Let $\omega_1, \omega_2, \ldots, \omega_k$ and $\omega_1', \omega_2', \ldots, \omega_k'$ be two sets of of 1-forms in n variables such that $\omega_1\omega_2\ldots\omega_k \neq 0$ and $\omega_1'\omega_2'\ldots\omega_k' \neq 0$ at $(\bar{x}_1, \bar{x}_2, \ldots, \bar{x}_n)$. Show that $\omega_1\omega_2\ldots\omega_k$ is a multiple of $\omega_1'\omega_2'\ldots\omega_k'$ at $(\bar{x}_1, \bar{x}_2, \ldots, \bar{x}_n)$ if and only if there exist numbers b_{ij}, c_{ij} such that (2) is satisfied at $(\bar{x}_1, \bar{x}_2, \ldots, \bar{x}_n)$. Conclude that the analog of Exercise 1 is true for $(n-k)$-dimensional affine manifolds in n-space. [Use the method of 1(b).]

3 Find the integrability conditions for two differential equations in four variables

$$A\,dx + B\,dy + C\,dz + D\,dt = 0$$
$$E\,dx + F\,dy + G\,dz + H\,dt = 0$$

as explicit relations on the coefficient functions $A, B, \ldots, H$ and their derivatives.

4 Use the theorem of this section to prove that there exists an integrating factor of any non-singular equation $A\,dx + B\,dy = 0$ in two variables, as was stated in §8.2.

5 Deduce the theorem of §7.4 from the theorem of this section.

6 Given $\omega_1, \omega_2, \ldots, \omega_k$ with $\omega_1\omega_2\ldots\omega_k \neq 0$ as in the theorem, choose additional 1-forms $\omega_{k+1}, \omega_{k+2}, \ldots, \omega_n$ such that $\omega_1\omega_2\ldots\omega_n \neq 0$. Show that every 2-form can be written in just one way as $\sum a_{\mu\nu}\omega_\mu\omega_\nu$, where the sum is over all μ, ν satisfying $1 \leq \mu < \nu \leq n$. Use this observation to give $\binom{n-k}{2}k$ conditions which are equivalent to the integrability conditions (1).

8.6 Introduction to Homology Theory

Roughly speaking, homology theory is devoted to the question, "When is a k-form exact?" That is, "Given a k-form ω, under what conditions is there a $(k-1)$-form σ such that $\omega = d\sigma$?"

For the sake of simplicity it will be assumed that the given k-form ω is differentiable. Then if ω is exact, say $\omega = d\sigma$, it follows that $d\omega = d[d\sigma] = 0$, i.e. ω is closed.

In other words, a *necessary* condition for a differentiable k-form to be exact is that it be closed. This condition is not *sufficient*, however, as is shown by the 1-form

$$\frac{x\,dy - y\,dx}{x^2 + y^2} \tag{1}$$

which is closed but not exact. To prove that this 1-form is not exact it suffices to observe that its integral over the circle $x^2 + y^2 = 1$ is not zero (this integral is $\pm 2\pi$, depending on the orientation), whereas if it were exact its integral over this closed curve would have to be zero. Intuitively it is clear that this fact is related to the fact that the 1-form (1) is not defined at the origin; in fact, the integral over the circle of a closed 1-form ω which is defined at *all* points of the xy-plane is $\int_{\partial D} \omega = \int_D d\omega = 0$, where $D = \{x^2 + y^2 \leq 1\}$. Thus the existence of a closed form (1) which is not exact is related to the configuration or 'topology' of the domain of the form, which in this case is the xy-plane with the point $(0, 0)$ removed. An explicit theorem to this effect is the following.

Theorem

Let ω be a differentiable 1-form $\omega = A\,dx + B\,dy$ defined at all points of the xy-plane except (possibly) the origin. Then ω is exact if and only if (i) ω is closed, and (ii) the integral of ω over the circle $x^2 + y^2 = 1$ is zero.

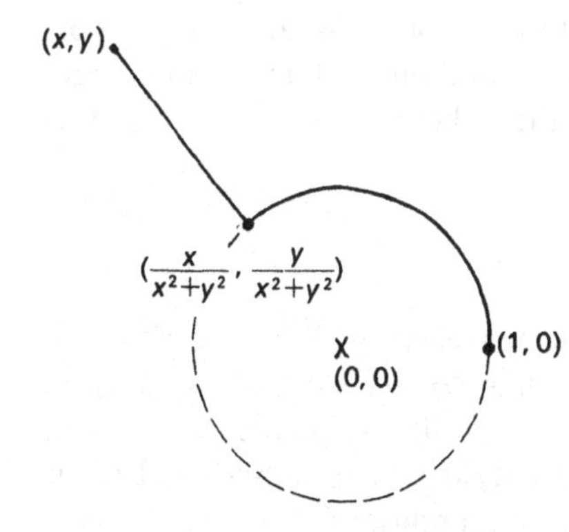

It is easy enough to see that this is the case: If ω is exact, i.e., if $\omega = df$ for some function f, then (i) and (ii) must hold. Conversely, given ω satisfying (i) and (ii) one can define a function $f(x, y)$ by

$$f(x, y) = \int_{(1,0)}^{\left(\frac{x}{x^2+y^2}, \frac{y}{x^2+y^2}\right)} \omega + \int_{\left(\frac{x}{x^2+y^2}, \frac{y}{x^2+y^2}\right)}^{(x,y)} \omega$$

where the first integral is over the circle and the second integral is over the radial line. The circular path from $(1, 0)$ to $\left(\dfrac{x}{x^2 + y^2}, \dfrac{y}{x^2 + y^2}\right)$ can be taken in either sense, clockwise or counterclockwise, and can consist of any number of circuits, because by (ii) the integral over a complete circuit in either sense is zero. Now if $(x + \Delta x, y + \Delta y)$ is near (x, y), then the integral of ω over the path consisting of the line segment from (x, y) to

$(x + \Delta x, y + \Delta y)$, the radial line from $(x + \Delta x, y + \Delta y)$ to the circle, the circular path to $\left(\frac{x}{x^2 + y^2}, \frac{y}{x^2 + y^2}\right)$, and the radial line back to (x, y) is zero, either because this path bounds a 2-dimensional region D and therefore $\int_{\partial D} \omega = \int_D d\omega = 0$ by (i), or because this path lies entirely on one radial line so that each point covered is covered twice with opposite orientations. Thus $\int_{(x,y)}^{(x+\Delta x, y+\Delta y)} \omega$ is $f(x + \Delta x, y + \Delta y) - f(x, y)$ from which $\omega = df$; hence (i) and (ii) imply that ω is exact. This geometrical argument does not of course constitute a rigorous proof of the theorem. The purpose here is merely to illustrate the sorts of ideas which are involved in homology theory.

In general, homology theory is concerned with the following type of problem: Given an n-dimensional domain D, what conditions (if any) must be added to the condition $d\omega = 0$ in order to obtain necessary and sufficient conditions for a differentiable k-form ω on D to be exact? The following theorems answer this question in a few specific cases.

Theorem

A differentiable 1-form $\omega = A\,dx + B\,dy + C\,dz$ which is defined at all points of xyz-space except (possibly) at the origin is exact if and only if it is closed. On the other hand, a differentiable 2-form $\omega = A\,dy\,dz + B\,dz\,dx + C\,dx\,dy$ defined at all points of xyz-space except (possibly) the origin is exact if and only if (i) it is closed, and (ii) its integral over the sphere $x^2 + y^2 + z^2 = 1$ is zero.

Theorem

A differentiable 1-form $\omega = A\,dx + B\,dy + C\,dz$ which is defined at all points of xyz-space except (possibly) on the circle $\{x^2 + y^2 = 4, z = 0\}$ and on the line $\{x = 0, y = 0\}$ is exact if and only if (i) it is closed, and (ii) it satisfies the two additional conditions

$$\int_{\gamma_1} \omega = 0 \qquad \int_{\gamma_2} \omega = 0$$

where γ_1 is the circle $\{x^2 + y^2 = 1, z = 0\}$ and γ_2 is the circle $\{(x - 2)^2 + z^2 = 1, y = 0\}$. A differentiable

2-form $\omega = A\,dy\,dz + B\,dz\,dx + C\,dx\,dy$ defined at all points of xyz-space except (possibly) on the circle $\{x^2 + y^2 = 4, z = 0\}$ and on the line $\{x = 0, y = 0\}$ is exact if and only if (i) it is closed, and (ii) it satisfies the additional condition $\int_\gamma \omega = 0$, where γ is the torus obtained by rotating the circle $\{(x - 2)^2 + z^2 = 1, y = 0\}$ around the z-axis.

Each of these theorems is of the form "a differentiable k-form ω defined at all points of the domain D is exact if and only if (i) it is closed, and (ii) it satisfies the additional conditions $\int_{\gamma_1} \omega = 0, \int_{\gamma_2} \omega = 0, \ldots, \int_{\gamma_\nu} \omega = 0$." (In the examples above, ν is 0, 1, or 2.) The k-dimensional domains of integration $\gamma_1, \gamma_2, \ldots, \gamma_\nu$ are called a 'homology basis' (in dimension k) of the n-dimensional domain D. In general a homology basis is defined as follows: Each of the elements $\gamma_1, \gamma_2, \ldots, \gamma_\nu$ of a (k-dimensional) homology basis is a k-dimensional 'domain of integration' contained in D. Specifically, each γ is either a compact, oriented, differentiable, k-dimensional manifold-with-boundary (for example, the circle $x^2 + y^2 = 1$ or the sphere $x^2 + y^2 + z^2 = 1$ in the examples above); or else γ is a collection of a finite number of such k-manifolds with $\int_\gamma \omega$ defined to be the sum of the integrals of ω over each k-manifold in the collection (for example, the boundary of the square $\{|x| \le 1, |y| \le 1\}$ or of the cube $\{|x| \le 1, |y| \le 1, |z| \le 1\}$). Such a collection of k-manifolds is called a *k-chain*. Thus a (k-dimensional) homology basis of D is a set $\gamma_1, \gamma_2, \ldots, \gamma_\nu$ of k-chains γ_i in D. To be a homology basis these k-chains must be such that:

(a) The conditions $\int_{\gamma_i} \omega = 0$ are *necessary* for ω to be exact, i.e., $\int_{\gamma_i} d\sigma = 0$ for all $(k - 1)$-forms σ. By Stokes' Theorem, this means that $\int_{\partial\gamma_i} \sigma = 0$ for all $(k - 1)$-forms σ; hence γ_i 'has no boundary'. If γ_i is a single k-manifold, this means literally that it is a manifold without boundary; whereas if γ_i consists of several k-manifolds, then it means that the boundaries of the pieces *cancel*. For example, the six pieces which form the boundary of a cube have boundaries which cancel. (Each of the 12 edges of the cube occurs twice, with opposite orientations, in the boundary of the 6 faces.) A chain with this property is called a *cycle*, and the requirement is that the chains γ_i

in a homology basis be cycles. The γ's in the theorems above are clearly cycles, because they are manifolds without boundary.

(b) The conditions $\int_{\gamma_i} \omega = 0$ $(i = 1, 2, \ldots \nu)$ and $d\omega = 0$ are *independent*. This means that for each $i = 1, 2, \ldots, \nu$ there is an ω such that $d\omega = 0$, $\int_{\gamma_j} \omega = 0$ $(j \neq i)$, but $\int_{\gamma_i} \omega \neq 0$. (Also, it means that there is an ω such that $\int_{\gamma_i} \omega = 0$ for all i and such that $d\omega \neq 0$; however, this imposes no conditions on the γ's at all.) In the only case where this applies in the theorems above—namely, the case of the two circles in the last theorem—this condition is easily shown to hold [Exercise 1].

(c) The conditions $d\omega = 0$ and $\int_{\gamma_i} \omega = 0$ $(i = 1, 2, \ldots, \nu)$ are *sufficient* for a differentiable k-form ω to be exact. That is, if ω is differentiable and satisfies these conditions, then there is a differentiable $(k - 1)$-form σ such that $d\sigma = \omega$.

This definition of 'homology basis' is only a definition and no assertion is made that there exists a homology basis for an arbitrary domain D. The purpose of the definition is to prescribe a specific form in which the answer to the question "When is a k-form exact?" can be given. Namely, the question is to be answered by giving a k-dimensional homology basis of the domain of the form; then the question "$\omega = d\sigma$?" is answered by testing the finite number of conditions $d\omega = 0$, $\int_{\gamma_1} \omega = 0, \ldots, \int_{\gamma_\nu} \omega = 0$ which are necessary and sufficient for $\omega = d\sigma$.

Note that unlike most of the subjects discussed in this book, the problem of finding a homology basis for a given domain is a *global* problem, depending on the whole domain and the way it fits together. Historically, the subject of algebraic topology (also called combinatorial topology) developed from the study of this global problem.

The theorems stated above are not especially difficult to prove, but the techniques of proof belong more to the subject of topology than to calculus, so it is not appropriate to enter into them here. A theorem of this type which does properly belong in a calculus book is the theorem that *every closed form is locally exact*. In terms of homology this is a consequence of the fact that an n-dimensional cube is homologically trivial ($\nu = 0$ for all k). This fact is known as Poincaré's Lemma.

Poincaré's Lemma

A differentiable k-form ω defined at all points of an n-dimensional cube $\{|x_1 - \bar{x}_1| \leq a, |x_2 - \bar{x}_2| \leq a, \ldots, |x_n - \bar{x}_n| \leq a\}$ in $\mathbf{R}^n$ is exact if and only if it is closed.

Proof

If $d\sigma = \omega$ then $d\omega = d[d\sigma] = 0$, so the condition $d\omega = 0$ is, as always, necessary for $\omega = d\sigma$. It is to be shown that when the domain is a cube it is also sufficient. To this end let f: $\mathbf{R}^{n+1} \to \mathbf{R}^n$ be the map defined by

$$x_1 = \bar{x}_1 + ty_1, x_2 = \bar{x}_2 + ty_2, \ldots, x_n = \bar{x}_n + ty_n,$$

where $(x_1, x_2, \ldots, x_n)$ are the coordinates of $\mathbf{R}^n$, $(\bar{x}_1, \bar{x}_2, \ldots, \bar{x}_n)$ is the center of the cube in question, and $(t, y_1, y_2, \ldots, y_n)$ are the coordinates on $\mathbf{R}^{n+1}$. This map is differentiable and carries points of the set $\{0 \leq t \leq 1, |y_1| \leq a, \ldots, |y_n| \leq a\}$ into the cube where ω is defined. Let $\int_0^1 f^*(\omega)$ denote the $(k-1)$-form in $y_1, y_2, \ldots, y_n$ found by dropping all terms in $f^*(\omega)$ which do not contain dt, writing terms which do contain dt with dt in the first position, integrating the coefficients of these terms dt from $t = 0$ to $t = 1$, and dropping dt. Finally, let σ be the pullback of $\int_0^1 f^*(\omega)$ under the map

$$y_1 = x_1 - \bar{x}_1, y_2 = x_2 - \bar{x}_2, \ldots, y_n = x_n - \bar{x}_n.$$

Then σ is a differentiable $(k-1)$-form in $x_1, x_2, \ldots, x_n$. It will be shown that $d\sigma = \omega$ as desired.

Let R be a small oriented k-dimensional rectangle lying in a coordinate direction in the cube $\{|x_i - \bar{x}_i| \leq a\}$. It will suffice to show that $\int_R \omega = \int_{\partial R} \sigma$ for every such rectangle R. Let $I \times R$ denote the oriented $(k+1)$-dimensional rectangle in $\mathbf{R}^{n+1}$ consisting of all points $(t, y_1, y_2, \ldots, y_n)$ such that $0 \leq t \leq 1$, and such that $\bar{x}_i + y_i$ is in R for $i = 1, 2, \ldots, n$. By Stokes' theorem $\int_{\partial(I \times R)} f^*(\omega) = \int_{I \times R} d[f^*(\omega)] = \int_{I \times R} f^*(d\omega) = 0$, because ω is closed. Now $\partial(I \times R)$ can be divided into three pieces, the two ends $\{0\} \times R$ and $\{1\} \times R$, and the sleeve $I \times \partial R$. The orientation of the sleeve as the boundary of $I \times R$ is opposite to the orientation of $I \times \partial R$ when ∂R is oriented as the boundary of R, hence

$$-\int_{\{0\} \times R} f^*(\omega) + \int_{\{1\} \times R} f^*(\omega) = \int_{I \times \partial R} f^*(\omega).$$

The first integral on the left is zero because the map f collapses this end to the single point $(\bar{x}_1, \bar{x}_2, \ldots, \bar{x}_n)$, whereas the second integral on the left is $\int_R \omega$ because the map f on this end is the translation $x_i = y_i + \bar{x}_i$ $(i = 1, 2, \ldots, n)$. Carrying out the integration with respect to t, the integral on the right becomes

$$\int_{\partial R} \left(\int_0^1 f^*(\omega) \right),$$

where $\int_0^1 f^*(\omega)$ is the $(k - 1)$-form defined above. Since this is $\int_{\partial R} \sigma$ the equation is $\int_R \omega = \int_{\partial R} \sigma$ as desired, and Poincaré's Lemma is proved.

Exercises

1 Show that the two circles γ_1, γ_2 of the third theorem satisfy the independence condition (b). [Use cylindrical coordinates (r, θ, z). Then $\int_{\gamma_1} d\theta \neq 0$, $\int_{\gamma_2} d\theta = 0$. Find a closed 1-form in r, z (defined for $(r, z) \neq (2, 0)$) whose integral over γ_2 is not zero, and show that its integral over γ_1 is zero. Express both 1-forms in terms of x, y, z.]

2 (a) Show that the boundary of the square $\{|x| \leq 1, |y| \leq 1\}$ is a cycle, that is, that the integral of $d\sigma$ over this set of oriented curves is zero for any differentiable function σ.

(b) By parameterizing each of the four sides of the square $\{|x| \leq 1, |y| \leq 1\}$, write the integral of

$$\frac{x\,dy - y\,dx}{x^2 + y^2}$$

over the boundary of the square as a sum of four definite integrals, combine into one term, and show that the integral is not zero. What is the value of this integral?

3 Give a geometrical 'proof', like the one in the text, that the circle $\{x^2 + y^2 = 1\}$ in the first theorem can be replaced by the boundary of the square $\{|x| \leq 1, |y| \leq 1\}$.

4 Give a geometrical 'proof' that the 1-dimensional homology of R^3 with the origin removed is trivial (i.e. $\nu = 0$) by arguing that every point can be joined to the point $(1, 0, 0)$ by a curve which does not pass through the origin, and that two such curves are the boundary of a 2-dimensional manifold. [This argument is essentially the one given in the text in the case of R^2, with the origin removed, but it leaves even more to the imagination.]

5 Give a geometrical 'proof' that the 2-dimensional homology of the cube $\{|x| \leq 1, |y| \leq 1, |z| \leq 1\}$ is trivial. [Given a closed 2-form ω, define a 1-form σ by defining the integral of σ over a curve to be the integral of ω over the 2-manifold swept out by a line segment from the origin to the curve as the far end moves along the given curve. Argue that $d\sigma = \omega$. This is the geometrical idea behind the proof in the text of Poincaré's Lemma.]

6 Give a geometrical 'proof' that a closed 2-form on the sphere $\{x^2 + y^2 + z^2 = 1\}$ is exact if its integral over the entire sphere is zero. [Repeat the argument of Exercise 5, replacing the line segment from the origin to the curve with the great circle arc from $(1, 0, 0)$ to the curve. The extra condition is needed to guarantee $\int_S d\sigma = \int_S \omega$ for small 2-dimensional manifolds which contain the antipodal point $(-1, 0, 0)$.]

7 'Prove' that the sphere $\{x^2 + y^2 + z^2 = 1\}$ is a 2-dimensional homology basis for R^3 with the origin removed. [This is a slight extension of Exercise 6.]

8 Find a basis of the 1-dimensional homology of the plane R^2 with the two points $(\pm 1, 0)$ removed.

9 Show that part II of the Fundamental Theorem of Calculus as stated in §3.1 is, except for a differentiability assumption, a special case of Poincaré's Lemma. Show the relationship between the proofs of these two facts.

10 *Betti numbers*. Show that if $\gamma_1, \gamma_2, \ldots, \gamma_\nu$ and $\gamma_1', \gamma_2', \ldots, \gamma_\mu'$ are two k-dimensional homology bases of the same domain D, then $\mu = \nu$. That is, any two homology bases contain the same number of cycles. This natural number (or zero) is called the kth Betti number of the domain D. [Let $\omega_1, \omega_2, \ldots, \omega_\nu$ be closed k-forms on D such that $\int_{\gamma_i}\omega_j$ is 0 if $i \neq j$, 1 if $i = j$. This is possible by condition (b) of the definition of homology basis. Define the 'period matrix' (a_{ij}) by $a_{ij} = \int_{\gamma_i'} \omega_j$, so that (a_{ij}) is a $\mu \times \nu$ matrix. Show that if ω is a closed k-form on D, then its 'γ' periods' can be obtained from its 'γ periods' $b_j = \int_{\gamma_j} \omega$ by the formula

$$\int_{\gamma_i'} \omega = \sum_{j=1}^{\nu} a_{ij} b_j.$$

(Show first that $\omega - \sum b_j \omega_j$ is exact.) In the same way the γ periods can be obtained from the γ' periods. Thus (a_{ij}) defines a map $R^\nu \to R^\mu$ which is invertible; hence $\nu = \mu$.]

8.7

Flows A 'flow' is an imaginary physical phenomenon in which space is filled with a moving fluid which consists of infinitely many particles. Such a phenomenon can be described mathematically in two quite different ways—by following the particles, and by standing still and counting the particles as they go by.

The first of these descriptions consists of three functions of four variables

$$\begin{aligned} x &= f(a, b, c, t) \\ y &= g(a, b, c, t) \\ z &= h(a, b, c, t). \end{aligned} \tag{1}$$

Here t is time and the three coordinates (a, b, c) can be considered as naming the particles; for fixed (a, b, c) the values of (1) for varying values of t describe a curve in xyz-space which is the trajectory of the particle named (a, b, c). One assumes that for each fixed value of t there is exactly one particle at each point of space and vice versa; that is, one assumes that the equations (1) can be put in the form

$$\begin{aligned} a &= F(x, y, z, t) \\ b &= G(x, y, z, t) \\ c &= H(x, y, z, t) \end{aligned} \tag{2}$$

giving the name of the particle which is at the point (x, y, z) at the time t. The trajectories are then defined implicitly by $a =$ const., $b =$ const., $c =$ const. Another flow

$$\begin{aligned} x &= f_1(\alpha, \beta, \gamma, t) \\ y &= g_1(\alpha, \beta, \gamma, t) \\ z &= h_1(\alpha, \beta, \gamma, t) \end{aligned} \tag{1'}$$

is considered to be the same as the given flow (1) if it describes the same trajectories. If this is true, then the curves $a =$ const., $b =$ const., $c =$ const. are identical to the curves $\alpha =$ const., $\beta =$ const., $\gamma =$ const., which implies that a, b, c can be written as functions of α, β, γ and vice versa (by the Implicit Function Theorem). Conversely, if a, b, c are functions of α, β, γ, then substitution of these functions into (1) gives a new flow (1′) which is the same as (1).

The second method of describing a flow is to give a 3-form in (x, y, z, t),

$$\omega = A\,dx\,dy\,dz + B\,dy\,dz\,dt + C\,dz\,dx\,dt + D\,dx\,dy\,dt \tag{3}$$

describing the 'number of trajectories which intersect a given 3-rectangle in $xyzt$-space'. The function $A(x, y, z, t)$ gives the 'number' of trajectories which intersect 3-rectangles in planes $t =$ const.; that is, $A(x, y, z, t)$ is the *density* of the particles (per volume $dx\,dy\,dz$) at the time t in the vicinity of the point (x, y, z). The function $B(x, y, z, t)$ gives the 'number' of trajectories which intersect 3-rectangles in planes $x =$ const., that is, the number of particles which cross 2-rectangles $x =$ const. during time intervals $\{\bar{t} \le t \le \bar{t} + \Delta t\}$. $B(x, y, z, t)$ is therefore the rate of flow (per time dt) of particles across rectangles $x =$ const. (per oriented area $dy\,dz$). Similarly, C and D are rates of flow across surfaces $y =$ const., $z =$ const. respectively. Since any trajectory that enters a 4-rectangle in $xyzt$-space must also leave it, it follows that the total value of ω on the boundary of any 4-rectangle is zero (when orientations are assigned consistently), i.e., $d\omega = 0$. It will be assumed therefore that a 3-form (3) describing a flow is *closed.* In more physical language, the assumption $d\omega = 0$ means that any change in the number of particles in a region D of xyz-space is accounted for by the flow of particles across ∂D.

Naturally one would suppose that any flow described by equations (1) could be described by a 3-form (3) and vice versa. The translation from one form of description to the other is accomplished as follows:

Using the map (1) the 3-form ω can be written in terms of (a, b, c, t). Since the curves $a =$ const., $b =$ const., $c =$ const. are trajectories, it follows that there is no flow across surfaces $a =$ const., $b =$ const., or $c =$ const., and hence that the $db\,dc\,dt$, $dc\,da\,dt$, and $da\,db\,dt$ components of ω are zero. Therefore ω is of the form $\omega = E\,da\,db\,dc$. Moreover, since $d\omega = 0$ implies $\dfrac{\partial E}{\partial t} \equiv 0$, the function E is independent of t and

$$\omega = E(a, b, c)\,da\,db\,dc.$$

The function $E(a, b, c)$ can be regarded as the "number of particles whose names (a, b, c) lie in a 3-rectangle of *abc*-space," i.e., the density (per $da\,db\,dc$) of the particles. This function is not determined by the description (1), which gives the trajectories but not the density of the particles; hence (1) must be supplemented by giving the 3-form $E(a, b, c)\,da\,db\,dc$. Then the closed 3-form ω in (x, y, z, t) is found by putting the equations (1) in the form (2) (which is possible by assumption) and using (2) to express $E(a, b, c)\,da\,db\,dc$ in terms of (x, y, z, t).

To find functions (1) and the 3-form $E(a, b, c)\,da\,db\,dc$ given the closed 3-form (3), one proceeds as follows: The flow does not determine the functions (1), but only the trajectories $a = \text{const.}$, $b = \text{const.}$, $c = \text{const.}$; that is, the flow determines the curves $da = 0$, $db = 0$, $dc = 0$. It is to be shown that the closed 3-form ω can be used to find these trajectories. In *xyzt*-coordinates the trajectories are naturally described by differential equations

$$\frac{dx}{dt} = u, \quad \frac{dy}{dt} = v, \quad \frac{dz}{dt} = w$$

where $u(x, y, z, t)$ is the x-component of the velocity of the particle at the point (x, y, z) at the time t, and similarly for v, w. These functions u, v, w must have the property that the curves defined by the differential equations

$$dx - u\,dt = 0, \; dy - v\,dt = 0, \; dz - w\,dt = 0$$

are identical to the curves defined by

$$da = 0, \; db = 0, \; dc = 0.$$

This is true if and only if the 1-forms da, db, dc can be expressed as combinations of the 1-forms $dx - u\,dt$, $dy - v\,dt$, $dz - w\,dt$ and vice versa, which is true if and only if the 3-form $da\,db\,dc$ is a multiple of

$$\begin{aligned}(dx - u\,dt)&(dy - v\,dt)(dz - w\,dt) \\ &= dx\,dy\,dz - u\,dy\,dz\,dt - v\,dz\,dx\,dt - w\,dx\,dy\,dt.\end{aligned}$$

Since ω is $E(a, b, c)\,da\,db\,dc$, this implies that ω is a multiple of $dx\,dy\,dz - u\,dy\,dz\,dt - v\,dz\,dx\,dt - w\,dx\,dy\,dt$; hence if ω is given, then the functions u, v, w can be determined immediately from

$$u = -\frac{B}{A}, \quad v = -\frac{C}{A}, \quad w = -\frac{D}{A},$$

provided $A \neq 0$. By the existence theorem for ordinary differential equations the equations

$$dx - u\,dt = 0,\ dy - v\,dt = 0,\ dz - w\,dt = 0$$

then have a solution of the form $a = \text{const.}$, $b = \text{const.}$, $c = \text{const.}$, where

$$\begin{aligned} a &= F(x, y, z, t) \\ b &= G(x, y, z, t) \\ c &= H(x, y, z, t). \end{aligned} \tag{2}$$

If ω is given with $A \neq 0$ and if u, v, w (hence a, b, c) are determined in this way, then it follows that $da\,db\,dc$ is a multiple of $(dx - u\,dt)(dy - b\,dt)(dz - w\,dt) = dx\,dy\,dz - u\,dy\,dz\,dt - v\,dz\,dx\,dt - w\,dx\,dy\,dt$, which is in turn a multiple of ω. In particular the $dx\,dy\,dz$ component of $da\,db\,dc$ is not zero and (locally) the equations (2) can be put in the form

$$\begin{aligned} x &= f(a, b, c, t) \\ y &= g(a, b, c, t) \\ z &= h(a, b, c, t). \end{aligned} \tag{1}$$

When these functions are used to express ω in terms of a, b, c, t, the result is of the form $E(a, b, c, t)\,da\,db\,dc$ (because ω is a multiple of $da\,db\,dc$ by the choice of a, b, c), hence of the form $E(a, b, c)\,da\,db\,dc$ (because $d\omega = 0$ by assumption).

These pseudo-physical statements can be summarized by the following mathematical theorem.

Theorem

Let $\omega = A\,dx\,dy\,dz + B\,dy\,dz\,dt + C\,dz\,dx\,dt + D\,dx\,dy\,dt$ be a closed 3-form on $xyzt$-space and let $(\bar{x}, \bar{y}, \bar{z}, \bar{t})$ be a point at which $A \neq 0$. Then there exist functions

$$\begin{aligned} a &= F(x, y, z, t) \\ b &= G(x, y, z, t) \\ c &= H(x, y, z, t) \end{aligned} \tag{2}$$

and a 3-form $E(a, b, c)\,da\,db\,dc$ such that ω is the pull-back of $E(a, b, c)\,da\,db\,dc$ under (2) at all points (x, y, z, t) near $(\bar{x}, \bar{y}, \bar{z}, \bar{t})$. Since the $dx\,dy\,dz$ component of $da\,db\,dc$ is not zero, these functions can be solved to give

$$(1)\qquad \begin{aligned} x &= f(a, b, c, t) \\ y &= g(a, b, c, t) \\ z &= h(a, b, c, t). \end{aligned}$$

The trajectories $a =$ const., $b =$ const., $c =$ const. determined by (1) depend only on the given 3-form ω. Hence the 3-form ω describes a flow.

It is customary to denote the density $A(x, y, z, t)$ by $\rho(x, y, z, t)$ so that the closed 3-form ω describing the flow is

$$\omega = \rho\,dx\,dy\,dz - \rho u\,dy\,dz\,dt - \rho v\,dz\,dx\,dt - \rho w\,dx\,dy\,dt.$$

The equation $d\omega = 0$ then takes the form

$$\frac{\partial \rho}{\partial t} + \frac{\partial}{\partial x}(\rho u) + \frac{\partial}{\partial y}(\rho v) + \frac{\partial}{\partial z}(\rho w) \equiv 0$$

which is known as the 'continuity equation'.

The density ρ is closely related to the integrating factor ρ of §8.2. Specifically, the argument above can be used to prove the following generalization to n dimensions.

Theorem

Let ω be an $(n - 1)$-form on $\mathbf{R}^n$ and let $(\bar{x}_1, \bar{x}_2, \ldots, \bar{x}_n)$ be a point of $\mathbf{R}^n$ at which $\omega \not\equiv 0$. Then there exist functions

$$\begin{aligned} y_1 &= f_1(x_1, x_2, \ldots, x_n) \\ &\vdots \\ y_{n-1} &= f_{n-1}(x_1, x_2, \ldots, x_n) \end{aligned}$$

and

$$\rho(x_1, x_2, \ldots, x_n)$$

defined near $(\bar{x}_1, \bar{x}_2, \ldots, \bar{x}_n)$ such that $\rho \not\equiv 0$ at $(\bar{x}_1, \bar{x}_2, \ldots, \bar{x}_n)$ and such that $\rho\omega \equiv dy_1\,dy_2 \ldots dy_{n-1}$. The $(n - 1)$-form ω is closed if and only if ρ is constant on these curves, that is, ρ can be expressed as a function of $(y_1, y_2, \ldots, y_{n-1})$. The curves $y_i =$ const. are determined by the $(n-1)$-form ω and if ω is closed then the number $\int_S \omega$ can be imagined as 'the number of curves $y_i =$ const. which intersect the surface S' when these curves are drawn with the correct density.

Exercises

1 A flow

$$\rho\, dx\, dy\, dz - \rho u\, dy\, dz\, dt - \rho v\, dz\, dx\, dt - \rho w\, dx\, dy\, dt$$

is said to be 'divergence-free' if the density ρ is independent of t. Show that this is true if and only if the 2-form $\rho u\, dy\, dz + \rho v\, dz\, dx + \rho w\, dx\, dy$ giving rate of flow (per time) across surfaces in xyz-space is closed.

2 The second theorem of the text says that locally a closed $(n-1)$-form on an n-dimensional domain can be described by curves. Give such a description of the following $(n-1)$-forms:

(a) dx on the xy-plane

(b) $\dfrac{x\, dy - y\, dx}{x^2 + y^2}$ on the xy-plane

(c) $\dfrac{x\, dx + y\, dy}{x^2 + y^2}$ on the xy-plane

(d) $dx\, dy$ on xyz-space

(e) $\dfrac{x\, dy\, dz + y\, dz\, dx + z\, dx\, dy}{(x^2 + y^2 + z^2)^{3/2}}$ on xyz-space

Be sure to include the *density* with which the curves must be drawn.

3 Prove the second theorem of the text. [Use $1/\rho$ in proving that ω is closed if and only if ρ is a function of $y_1, y_2, \ldots, y_{n-1}$.]

8.8 Applications to Mathematical Physics

The Heat Equation

Suppose that a solid body occupies a volume V of xyz-space, and let $T(x, y, z, t)$ be the temperature at the point (x, y, z) of the solid at the time t. The heat equation is a relationship which must be satisfied by the function $T(x, y, z, t)$ if the phenomenon of 'temperature' is adequately described by the following assumptions:

(a) Changes in temperature can be accounted for by the motion of a fictitious fluid called 'heat'.

(b) The density of heat at (x, y, z) at time t is proportional to the temperature. The constant of proportionality $c(x, y, z)$, which may depend on

(x, y, z) but not on t, is called the 'heat capacity' of the solid at (x, y, z):

amount of heat

$$= \text{(temp.)} \times \text{(heat capacity)} \times \text{(volume)}.$$

(c) The rate of flow of heat across any small rectangle in the solid is proportional to the area of the rectangle and to the normal derivative* of the temperature, the flow being in the direction of decreasing T. The constant of proportionality $k(x, y, z)$, which may depend on the point (x, y, z) but not on the time t or the direction of the small rectangle, is called the 'conductivity' of the solid at (x, y, z). (If the conductivity depends on the direction as well as the position of the rectangle, the medium is said to be *anisotropic* and the heat equation is slightly more complicated.)

(d) Heat is conserved.

*That is, the derivative in the direction perpendicular to the rectangle.

Expressed in equations, the above assumptions become (a) a flow $\omega = \rho\, dx\, dy\, dz - \rho u\, dy\, dz\, dt - \rho v\, dz\, dx\, dt - \rho w\, dx\, dy\, dt$, (b) $\rho = cT$, (c) $\rho u = -k \frac{\partial T}{\partial x}$, $\rho v = -k \frac{\partial T}{\partial y}$, $\rho w = -k \frac{\partial T}{\partial z}$, and (d) $d\omega = 0$. Putting them together gives

$$d\left[cT\, dx\, dy\, dz + k \frac{\partial T}{\partial x}\, dy\, dz\, dt + k \frac{\partial T}{\partial y}\, dz\, dx\, dt + k \frac{\partial T}{\partial z}\, dx\, dy\, dt\right] = 0$$

$$c \frac{\partial T}{\partial t} = \frac{\partial}{\partial x}\left(k \frac{\partial T}{\partial x}\right) + \frac{\partial}{\partial y}\left(k \frac{\partial T}{\partial y}\right) + \frac{\partial}{\partial z}\left(k \frac{\partial T}{\partial z}\right)$$

which is the heat equation. If k is constant (the solid is homogeneous), then the heat equation is simply $c \frac{\partial T}{\partial t} = k\nabla^2 T$, where $\nabla^2 T$ denotes $\frac{\partial^2 T}{\partial x^2} + \frac{\partial^2 T}{\partial y^2} + \frac{\partial^2 T}{\partial z^2}$ (the Laplacian of T). In particular, if a homogeneous solid is in thermal equilibrium $\left(\frac{\partial T}{\partial t} = 0\right)$, then the temperature $T(x, y, z)$ is a harmonic function.

Potential Theory

Newton's law of gravity states that two particles attract each other with a force which is proportional to the product of their masses and inversely proportional to the square of the distance between them. Coulomb's law states that two electrified particles repel each other with a force which is proportional to the product of their electrical charges (hence they attract when the charges have opposite signs) and inversely proportional to the square of the distance between them.

From a mathematical standpoint these two laws are the same and can be expressed as follows: The law 'work = force × displacement' says that the components of force are the components of a 1-form called work. In terms of work the inverse square law of attraction states that the work required to displace a particle of unit mass (or unit negative charge) in the presence of a particle of mass (charge) m at $(0, 0, 0)$ is given by the 1-form

$$\frac{\gamma m\, dr}{r^2} = -d\left(\frac{\gamma m}{r}\right)$$

where $r = \sqrt{x^2 + y^2 + z^2}$ and where the constant of proportionality γ depends on the units used to measure mass, work, and distance.

More generally, if there are N particles of mass (charge) $m_1, m_2, \ldots, m_N$ at the points (x_1, y_1, z_1), $(x_2, y_2, z_2), \ldots, (x_N, y_N, z_N)$ then, by the assumption that forces add vectorially, the amount of work required for displacements of a unit test particle is given by the 1-form

$$(1) \quad \sum_{i=1}^{N} \gamma m_i \frac{dr_i}{r_i^2} = -\sum_{i=1}^{N} d\left(\frac{\gamma m_i}{r_i}\right) = -d\left[\sum_{i=1}^{N} \frac{\gamma m_i}{r_i}\right]$$

where $r_i = \sqrt{(x - x_i)^2 + (y - y_i)^2 + (z - z_i)^2}$. If there is a *continuous* distribution of mass, then (1) is naturally replaced by

$$(1') \quad \text{work} = -d\left[\iiint \frac{\gamma\rho(\xi, \eta, \zeta)\, d\xi\, d\eta\, d\zeta}{\sqrt{(x - \xi)^2 + (y - \eta)^2 + (z - \zeta)^2}}\right]$$

where ρ is the density of the distribution of mass and where the 1-form work refers to displacements of a unit particle near a point (x, y, z) where there is no mass (so the integral is well-defined).

The fundamental property of the inverse square law, which distinguishes it from all other 'possible' laws of attraction that are radially symmetric, is the theorem of Newton that *the force exerted by a homogeneous ball of total mass M on a particle P outside the ball is the same as the force exerted on P by a particle of mass M located at the center of the ball.* To prove this it suffices to consider the case where P has unit mass, in which case the assertion is that

$$d\left[\frac{\gamma M}{\sqrt{(x-\bar{x})^2+(y-\bar{y})^2+(z-\bar{z})^2}}\right] = d\left[\int_B \frac{\gamma\rho\, d\xi\, d\eta\, d\zeta}{\sqrt{(x-\xi)^2+(y-\eta)^2+(z-\zeta)^2}}\right],$$

where B is the ball of radius r with center $(\bar{x}, \bar{y}, \bar{z})$ and where the density of mass in B is the constant ρ. Since $df = dg$ if and only if $f = g + \text{const.}$, and since both of the functions above approach zero as (x, y, z) moves to an infinite distance from B, it follows that the equation above holds if and only if

$$\frac{1}{\sqrt{(x-\bar{x})^2+(y-\bar{y})^2+(z-\bar{z})^2}} = \frac{1}{M}\int_B \frac{\rho\, d\xi\, d\eta\, d\zeta}{\sqrt{(x-\xi)^2+(y-\eta)^2+(z-\zeta)^2}}$$

or, since $M = \rho\int_B d\xi\, d\eta\, d\zeta$, if and only if the value of the function

$$u(\xi, \eta, \zeta) = [(x-\xi)^2+(y-\eta)^2+(z-\zeta)^2]^{-1/2}$$

at the center of B is equal to its average value over all of B. Thus Newton's theorem is essentially the statement that the function

$$\frac{1}{r} = \frac{1}{\sqrt{x^2+y^2+z^2}}$$

is a harmonic function on xyz-space, defined at all points except $(0, 0, 0)$. This can be proved (see Exercise 4 of §8.3) by showing that $u(x, y, z)$ is harmonic if and only if it satisfies Laplace's equation $\dfrac{\partial^2 u}{\partial x^2} + \dfrac{\partial^2 u}{\partial y^2} + \dfrac{\partial^2 u}{\partial z^2} = 0$, and by showing that $\dfrac{1}{r}$ satisfies this equation.

Another important theorem of Newton is that *the total force exerted by a homogeneous spherical shell on a particle in its interior is zero.* This is proved by using (1′) to show that the work required to displace a unit particle in the interior is $-du$, where

$$u(x, y, z) = \int_{r_1^2 \le \xi^2+\eta^2+\zeta^2 \le r_2^2} \frac{\rho \, d\xi \, d\eta \, d\zeta}{\sqrt{(x-\xi)^2 + (y-\eta)^2 + (z-\zeta)^2}}.$$

For each fixed (ξ, η, ζ) the function under the integral sign is a harmonic function of (x, y, z) for $x^2 + y^2 + z^2 < r_1^2$. Interchanging the order of integration it follows that the average of $u(x, y, z)$ over any ball in $x^2 + y^2 + z^2 < r_1^2$ is its value at the center of the ball, i.e., $u(x, y, z)$ is harmonic. By symmetry u is constant on spheres $x^2 + y^2 + z^2 = \text{const.} < r_1^2$. Since its average on any sphere is $u(0, 0, 0)$ it follows that $u \equiv u(0, 0, 0)$; hence $du \equiv 0$ as was to be shown.

The inverse square law determines the forces (work) given the masses. It is of obvious interest to be able to invert this relation, that is, to find the masses given the forces. The solution of this problem follows from the simple observation that the 2-form

$$\omega = \frac{x\,dy\,dz + y\,dz\,dx + z\,dx\,dy}{(x^2 + y^2 + z^2)^{3/2}},$$

defined at all points other than $(0, 0, 0)$, is closed and that its integral over the sphere $x^2 + y^2 + z^2 = 1$ is not zero. The actual value of its integral over the sphere is 4π, but the only important fact in what follows is that it is not zero. Thus for any volume V (a compact, differentiable, three-dimensional manifold-with-boundary, oriented by $dx\,dy\,dz$),

$$\int_{\partial V} \omega = \begin{cases} 0 \text{ if } (0, 0, 0) \text{ is not in } V \\ 4\pi \text{ if } (0, 0, 0) \text{ is inside } V \\ \text{not defined if } (0, 0, 0) \text{ is on } \partial V. \end{cases}$$

Now if the 1-form work is

$$\begin{aligned} A\,dx + B\,dy + C\,dz &= -d\left(\frac{\gamma m}{r}\right) \\ &= \frac{\gamma m\,dr}{r^2} = \gamma m \frac{1}{2} \frac{d(r^2)}{r^3} \\ &= \gamma m \frac{x\,dx + y\,dy + z\,dz}{(x^2 + y^2 + z^2)^{3/2}} \end{aligned}$$

so that

$$A = \frac{\gamma m x}{r^3}, \; B = \frac{\gamma m y}{r^3}, \; C = \frac{\gamma m z}{r^3}$$

it then follows that

$$A\,dy\,dz + B\,dz\,dx + C\,dx\,dy = \gamma m \omega$$

and hence that

$$\int_{\partial V} A\,dy\,dz + B\,dz\,dx + C\,dx\,dy = \begin{cases} 0 & \text{if the mass is not in } V \\ 4\pi\gamma m & \text{if the mass is inside } V \\ \text{not defined} & \text{if the mass is on } \partial V. \end{cases}$$

Thus the amount of mass and its location can be determined when A, B, C are known. More generally, if

$$A\,dx + B\,dy + C\,dz = -d\left(\sum_{i=1}^{N} \frac{\gamma m_i}{r_i}\right)$$

then

$$\frac{1}{4\pi\gamma}\int_{\partial V} A\,dy\,dz + B\,dz\,dx + C\,dx\,dy$$

is defined provided that none of the masses lie on ∂V and is equal to the total mass of the particles in V. This determines the distribution of masses when work $=$ $A\,dx + B\,dy + C\,dz$ is known (and when the constant γ is known).

This problem is simpler when the masses are continuously distributed, since then the 1-form

$$\text{work} = A\,dx + B\,dy + C\,dz$$

is defined even at points where mass is present. This fact, which will not be proved rigorously, can be seen as a result of the fact that a ball exerts no force on its center so that when mass is continuously distributed the (nearly constant) mass near the point cancels out and only masses away from the point, where $\frac{1}{r}$ is defined, need be counted. In short, the integral

$$u(x, y, z) = \int \frac{\gamma\rho(\xi, \eta, \zeta)\,d\xi\,d\eta\,d\zeta}{\sqrt{(x-\xi)^2 + (y-\eta)^2 + (z-\zeta)^2}}$$

converges in a generalized sense (see §9.7), even though the integrand is not defined at $(\xi, \eta, \zeta) = (x, y, z)$ and even though it is unbounded near this point. Then

$$A\,dx + B\,dy + C\,dz = -du$$

is a well-defined 1-form for all (x, y, z) which, by the inverse square law, is the work required for displacements of a unit test particle. By the above,

$$\frac{1}{4\pi\gamma}\int_{\partial V} A\,dy\,dz + B\,dz\,dx + C\,dx\,dy = \text{total mass inside } V = \int_V \rho(x, y, z)\,dx\,dy\,dz.$$

By Stokes' Theorem it follows that

$$\rho(x, y, z)\,dx\,dy\,dz = d\left[\frac{1}{4\pi\gamma}(A\,dy\,dz + B\,dz\,dx + C\,dx\,dy)\right]\cdot$$

Hence

$$\rho(x, y, z) = \frac{1}{4\pi\gamma}\left[\frac{\partial A}{\partial x} + \frac{\partial B}{\partial y} + \frac{\partial C}{\partial z}\right]$$

or, noting that $A\,dx + B\,dy + C\,dz = -du$,

$$\frac{\partial^2 u}{\partial x^2} + \frac{\partial^2 u}{\partial y^2} + \frac{\partial^2 u}{\partial z^2} + 4\pi\gamma\rho = 0. \tag{2}$$

The function u is called the *potential function* determined by the mass distribution ρ, and the equation (2) is called *Poisson's equation.*

The subject of *potential theory* is essentially the study of the inverse square law from the point of view of Poisson's equation (2). Given the potential function u, the 1-form $-du$ describes the work required to perform displacements of a unit test particle, or, as it is more commonly stated, the 1-form du describes the work *done by the masses* ρ during the displacement of the test particle. At points of space where there is no mass, Poisson's equation says that the potential function is harmonic. If the only mass is at (0, 0, 0), then u must be harmonic except for a singularity at (0, 0, 0); hence, by radial symmetry, u must be of the form $\dfrac{\text{const.}}{r} + \text{const.}$

(see Exercise 1). The condition that u 'vanish at ∞' gives $u = \text{const.}/r$. Rewriting (2) as

$$-d\left[\frac{\partial u}{\partial x}\,dy\,dz + \frac{\partial u}{\partial y}\,dz\,dx + \frac{\partial u}{\partial z}\,dx\,dy\right] = 4\pi\gamma\rho(x, y, z)\,dx\,dy\,dz$$

and integrating over the ball $x^2 + y^2 + z^2 = 1$, it follows that the constant is γm, where m is the mass. Therefore

$$u = \frac{\gamma m}{r},$$

which shows that the inverse square law follows from Poisson's equation and the assumption that $u(\infty) = 0$. In general this argument shows that if $u(\infty) = 0$ and if Poisson's equation is satisfied, then u can be determined from ρ by

$$(3)\quad u(x, y, z) = \int \frac{\gamma\rho(\xi, \eta, \zeta)\,d\xi\,d\eta\,d\zeta}{\sqrt{(x - \xi)^2 + (y - \eta)^2 + (z - \zeta)^2}},$$

which is the inverse square law. Thus the equations (2) and (3) are merely different ways of saying the same thing.

Historically the Newtonian formulation (1), (3) of course came first, and when one is dealing with a finite number of particles or balls (which behave like particles), this is the simpler formulation to apply. However, it involves the concept of 'action at a distance' between two bodies, which even to Newton was 'an absurdity'.* The formulation (2) reverses the roles and assigns the 'reality' to the potential u rather than to the masses ρ which are derived from u *locally* by (2). In his formulation of electrostatics Maxwell preferred the Poisson formulation both because it eliminated 'action at a distance' and because the exact nature of 'electrical charge' was even more obscure than that of 'mass', so that 'potential' or 'work' was a more satisfactory basis for the theory.

**"That gravity should be innate, inherent, and essential to matter, so that one body may act upon another at a distance through a vacuum, without the mediation of anything else, by and through which their action and force may be conveyed from one to another, is to me so great an absurdity, that I believe no man who has in philosophical matters a competent faculty of thinking can ever fall into it." (Newton's third letter to Bentley.)*

Maxwell's Equations

The statement of Coulomb's law assumes that the medium through which the forces act is homogeneous and isotropic; that is, the force between two bodies depends only on the distance which separates them and not on their particular locations. The constant $\epsilon = \frac{1}{4\pi\gamma}$ is called the *dielectric constant* of the medium. In terms of the dielectric constant Poisson's equation takes the form

$$\epsilon\nabla^2\phi + \rho = 0 \tag{4}$$

where ρ is the charge density and where the letter denoting 'potential' has been changed from u to ϕ to accord with standard usage. $\nabla^2\phi$ means, as before, $\frac{\partial^2\phi}{\partial x^2} + \frac{\partial^2\phi}{\partial y^2} + \frac{\partial^2\phi}{\partial z^2}$.

The 1-form $-d\phi$ is called the 'electric force field' and is denoted by E; it describes the work contributed by the force field toward displacements of a unit charged particle. The 2-form

$$-\epsilon\left(\frac{\partial\phi}{\partial x}\,dy\,dz + \frac{\partial\phi}{\partial y}\,dz\,dx + \frac{\partial\phi}{\partial z}\,dx\,dy\right)$$

is called the 'electric displacement' and is denoted by D. With these definitions Poisson's equation (4) is broken into three steps

$$E = -d\phi$$
$$D = \epsilon E$$
$$\rho\,dx\,dy\,dz = dD$$

where the second 'equation' between a 1-form and a 2-form means that the 2-form D is defined in terms of the 1-form

$$E = E_1\,dx + E_2\,dy + E_3\,dz$$

by the equation

$$D = \epsilon E_1\,dy\,dz + \epsilon E_2\,dz\,dx + \epsilon E_3\,dx\,dy.$$

These are the equations of electrostatics (no moving charge).

A *moving* charged particle is acted on by and exerts forces other than the forces described by Coulomb's law, namely *magnetic forces*. The basic facts concerning magnetic forces were discovered by Faraday, who found that *magnetic forces can be described by a closed 2-form*

$$B = B_1\, dy\, dz + B_2\, dz\, dx + B_3\, dx\, dy$$

and their relation to electrical forces can be described by the equation

$$d(E\, dt + B) = 0.$$

Needless to say, Faraday did not state his laws in this form; his description, which was deliberately physical and not mathematical, was as follows:

The 'magnetic field' can be visualized as consisting of oriented curves in space called the 'lines of force'.* The lines of force do not terminate; that is, the number of lines of force which enter any small three-dimensional region V of space is equal to the number which leave V. If S is a surface (compact, oriented, differentiable surface-with-boundary) in space and if the magnetic field changes over a time interval $\{\bar{t} \leq t \leq \bar{t} + \Delta t\}$, then an electromotive force acts around the curve ∂S. The total force around ∂S is proportional to the change during the time interval in the number of lines of force which cross S (Faraday's Law of Induction). (Like any force, this force around ∂S does not mean that anything necessarily moves; it means that if there were a circuit on ∂S, then one would observe a certain amount of work being done by the changing lines of force in moving charge around the circuit.) Since the lines of force do not end, the change in the number which cross S must be equal to the number which 'cut' across ∂S during the time interval, so an equivalent statement of Faraday's law of induction is that the total electromotive force around a closed curve is proportional to the number of lines of force which cut the curve during the time interval.

The physical reality of the lines of force can be seen in the way that iron filings distribute themselves in the presence of a magnetic field.

Mathematically, the quantity 'number of lines of force which cross the surface S at time t' is naturally represented as the integral of a closed 2-form $\int_S B$ depending on t, and Faraday's law is

$$\int_{\bar{t}}^{\bar{t}+\Delta t} \left(\int_{\partial S} E \right) dt = \text{const.} \times \text{change in} \int_S B.$$

When the unit of E is chosen ($E = -d\phi$ so the unit of E is potential and the units of E_1, E_2, E_3 are potential/length), the unit of B can be chosen to make the constant of proportionality equal to -1 so that the equation becomes

$$\int_{\bar{t}}^{\bar{t}+\Delta t}\left(\int_S dE\right)dt = -\int_{\bar{t}}^{\bar{t}+\Delta t}\left(\int_S \frac{\partial B}{\partial t}\right)dt$$

$$\int_{I\times S}\left(dE\,dt + \frac{\partial B}{\partial t}\,dt\right) = 0$$

$$\int_{I\times S} d[E\,dt + B] = 0.$$

Here $I \times S$ is the 3-manifold $\{(x, y, z, t) : (x, y, z) \text{ is in } S \text{ and } \bar{t} \le t \le \bar{t} + \Delta t\}$, and the fact that $dB = 0$ for each fixed t has been used. Thus the integral of $d(E\,dt + B)$ over any 3-rectangle in any of the 4 coordinate directions is 0 (in the xyz-direction it is zero because $dB = 0$), which shows that Faraday's law is indeed equivalent to

$$d(E\,dt + B) = 0$$

when the units of B are properly chosen. (In terms of the components of E and B this equation is

$$d[E_1\,dx\,dt + E_2\,dy\,dt + E_3\,dz\,dt + B_1\,dy\,dz + B_2\,dz\,dx + B_3\,dx\,dy] = 0$$

$$\left(\frac{\partial B_1}{\partial x} + \frac{\partial B_2}{\partial y} + \frac{\partial B_3}{\partial z}\right)dx\,dy\,dz + \left(\frac{\partial E_3}{\partial y} - \frac{\partial E_2}{\partial z} + \frac{\partial B_1}{\partial t}\right)dy\,dz\,dt$$

$$+ \left(\frac{\partial E_1}{\partial z} - \frac{\partial E_3}{\partial x} + \frac{\partial B_2}{\partial t}\right)dz\,dx\,dt + \left(\frac{\partial E_2}{\partial x} - \frac{\partial E_1}{\partial y} + \frac{\partial B_3}{\partial t}\right)dx\,dy\,dt = 0$$

or, as it is usually stated in physics books, $\operatorname{div} B = 0$, $\operatorname{curl} E + \dfrac{\partial B}{\partial t} = 0.$)

The closed 2-form $E\,dt + B$ on $xyzt$-space is called the *electromagnetic field.* The presence of E is related to the presence of *charges*, and the presence of B is related to the presence of *moving charges.* Altogether the charge and its motion are described by a 3-form

$$J = \rho\,dx\,dy\,dz - j_1\,dy\,dz\,dt - j_2\,dz\,dx\,dt - j_3\,dx\,dy\,dt$$

where ρ is the charge density and (j_1, j_2, j_3) are the components of the *current** giving the amount of charge crossing surfaces in xyz-space per unit area per unit time. The assumption that charge is conserved is the assumption that $dJ = 0$. The presence of the field $E\,dt + B$ is related to the presence of the flow of charge J;

*One cannot write $j_1 = \rho u$, $j_2 = \rho v$, $j_3 = \rho w$ because ρ (the net charge density) may be zero even when the current is not; this occurs, for example, when there is a stationary positive charge and a moving negative charge.

Maxwell's equations give this relationship in explicit form.

When the charge is stationary ($j_1 = j_2 = j_3 = 0$), there is no magnetic force ($B = 0$), and E is related to ρ by the equations $D = \epsilon E$, $dD = \rho\, dx\, dy\, dz$ as above. The desired relation (Maxwell's equations) between $E\, dt + B$ and J must be a generalization of this relation to the case in which J has non-zero terms in dt.

The explicit relation between moving charge and B is found by placing a wire on the z-axis and running current along the wire. This creates a magnetic field in the vicinity of the wire, as was discovered by Oersted. When this field is measured, its magnitude is found to be proportional to the current, and its lines of force are found to be circles in planes $z =$ const. whose centers are on the wire and whose density is inversely proportional to the distance from the wire. In short, the field is

$$B = \text{const.} \times \text{current} \times \frac{dr\, dz}{r}$$

where $r = \sqrt{x^2 + y^2}$. Or, denoting the constant of proportionality by α,

$$B = \alpha j_3 \frac{dr\, dz}{r} = \alpha j_3 \frac{x\, dx\, dz + y\, dy\, dz}{x^2 + y^2}.$$

If $B = B_1\, dx\, dz + B_2\, dy\, dz$ is known, then the rate of flow of charge across any small surface S can be found by the equation

$$\begin{aligned} \int_{\partial S} (B_2\, dx - B_1\, dy) \\ &= \alpha j_3 \int \frac{y\, dx - x\, dy}{x^2 + y^2} \\ &= \begin{cases} -2\pi\alpha j_3 & \text{if } \partial S \text{ goes around the } z \text{ axis} \\ 0 & \text{otherwise.} \end{cases} \end{aligned}$$

Therefore in all cases

$$-\frac{1}{2\pi\alpha} \int_{\partial S} (B_2\, dx - B_1\, dy) = \text{rate at which charge crosses } S.$$

The constant α is in fact negative (the orientation of B having been chosen above), so that the constant $\mu = -2\pi\alpha$ is positive. Since units of charge, length, and time, and hence of potential and of magnetic force have been chosen, μ is a well-defined quantity; it is called the 'magnetic permeability' of the medium.

If currents flow along any collection of wires parallel to the coordinate axes and if the resulting magnetic field $B = B_1\,dy\,dz + B_2\,dz\,dx + B_3\,dx\,dy$ is known, then the currents can be found:

$$\frac{1}{\mu}\int_{\partial S}(B_1\,dx + B_2\,dy + B_3\,dz)$$
$$= \text{rate of flow of charge across } S.$$

Assuming the charge flow to be continuously distributed this can be stated

$$\int_S dH = \int_{\partial S} H = \int_S (j_1\,dy\,dz + j_2\,dz\,dx + j_3\,dx\,dy)$$

where

$$H = \frac{1}{\mu}B_1\,dx + \frac{1}{\mu}B_2\,dy + \frac{1}{\mu}B_3\,dz.$$

The relation between B and H will be abbreviated $B = \mu H$ (analogous to the abbreviated equation $D = \epsilon E$). Thus

$$dH = j_1\,dy\,dz + j_2\,dz\,dx + j_3\,dx\,dy.$$

In cases in which this equation holds (currents on wires) the charge density ρ and the electrical field E are constant in time, so that this equation can be combined with $dD = \rho\,dx\,dy\,dz$ to give

$$d[D - H\,dt] = \rho\,dx\,dy\,dz - j_1\,dy\,dz\,dt - j_2\,dz\,dx\,dt - j_3\,dx\,dy\,dt.$$

Maxwell concluded that this relation must hold even when ρ is not constant in time, that is, that

$$d(D - H\,dt) = J$$

gives the desired relation between the motion of charge and the electromagnetic field.

In summary, the equations are

$$\text{field} = E\,dt + B$$
$$\text{moving charge} = J$$
$$d(E\,dt + B) = 0,\ dJ = 0$$
$$D = \epsilon E,\ B = \mu H$$
$$d(D - H\,dt) = J.$$

These are known collectively as Maxwell's equations, although only the last equation is actually Maxwell's discovery. (In terms of the components E_1, E_2, E_3, B_1, B_2, B_3 of the field this equation is

$$d\left[\epsilon E_1\,dy\,dz + \epsilon E_2\,dz\,dx + \epsilon E_3\,dx\,dy - \frac{1}{\mu}B_1\,dx\,dt - \frac{1}{\mu}B_2\,dy\,dt - \frac{1}{\mu}B_3\,dz\,dt\right]$$
$$= \rho\,dx\,dy\,dz - j_1\,dy\,dz\,dt - j_2\,dz\,dx\,dt - j_3\,dx\,dy\,dt,$$

which gives

$$\epsilon\left(\frac{\partial E_1}{\partial x} + \frac{\partial E_2}{\partial y} + \frac{\partial E_3}{\partial z}\right) = \rho$$

$$\frac{1}{\mu}\left(\frac{\partial B_3}{\partial y} - \frac{\partial B_2}{\partial z} - \epsilon\mu\frac{\partial E_1}{\partial t}\right) = j_1$$

$$\frac{1}{\mu}\left(\frac{\partial B_1}{\partial z} - \frac{\partial B_3}{\partial x} - \epsilon\mu\frac{\partial E_2}{\partial t}\right) = j_2$$

$$\frac{1}{\mu}\left(\frac{\partial B_2}{\partial x} - \frac{\partial B_1}{\partial y} - \epsilon\mu\frac{\partial E_3}{\partial t}\right) = j_3$$

which can be abbreviated ϵ div $E = \rho$, curl $B - \epsilon\mu\dfrac{\partial E}{\partial t} = \mu j$ or div $D = \rho$, curl $H - \dfrac{\partial D}{\partial t} = j$, where $j = (j_1, j_2, j_3)$.)

Maxwell* showed that the equations of electromagnetism can be formulated in a compact way in terms of a 'potential' so that they are analogous to Poisson's equation $\epsilon\nabla^2\phi + \rho = 0$ for electrostatics and reduce to this equation in cases in which there is no current. Since $d(E\,dt + B) = 0$ there is a 1-form $A = A_1\,dx + A_2\,dy + A_3\,dz + A_4\,dt$ such that $-dA = E\,dt + B$.† (There are many such 1-forms, for example, $A + df$, where f is any function.) Writing the components of $E\,dt + B$ in terms of the derivatives of the components of A and substituting into the equation $d(D - H\,dt) = J$ gives

*Actually Maxwell's formulation was somewhat different. The formulation given here is due to Lorentz.

†By Poincaré's Lemma (§8.6).

$$-\epsilon\left(\frac{\partial^2 A_4}{\partial x^2} + \frac{\partial^2 A_4}{\partial y^2} + \frac{\partial^2 A_4}{\partial z^2} - \frac{\partial}{\partial t}\left(\frac{\partial A_1}{\partial x} + \frac{\partial A_2}{\partial y} + \frac{\partial A_3}{\partial z}\right)\right) dx\,dy\,dz$$
$$- \frac{1}{\mu}\left(\frac{\partial^2 A_1}{\partial y^2} + \frac{\partial^2 A_1}{\partial z^2} - \epsilon\mu\frac{\partial^2 A_1}{\partial t^2} - \frac{\partial}{\partial x}\left(\frac{\partial A_2}{\partial y} + \frac{\partial A_3}{\partial z} - \epsilon\mu\frac{\partial A_4}{\partial t}\right)\right) dy\,dz\,dt$$
$$+ \text{etc.} = J.$$

If the components of A are assumed to satisfy the additional condition

$$(5) \qquad \frac{\partial A_1}{\partial x} + \frac{\partial A_2}{\partial y} + \frac{\partial A_3}{\partial z} - \epsilon\mu \frac{\partial A_4}{\partial t} = 0,$$

this equation is simply

$$\epsilon\Box^2 A_4\, dx\, dy\, dz + \frac{1}{\mu}\Box^2 A_1\, dy\, dz\, dt + \frac{1}{\mu}\Box^2 A_2\, dz\, dx\, dt + \frac{1}{\mu}\Box^2 A_3\, dx\, dy\, dt + J = 0,$$

where $\Box^2$ denotes the 'differential operator' assigning to a function $f(x, y, z, t)$ the function

$$\Box^2 f = \frac{\partial^2 f}{\partial x^2} + \frac{\partial^2 f}{\partial y^2} + \frac{\partial^2 f}{\partial z^2} - \epsilon\mu \frac{\partial^2 f}{\partial t^2}.$$

This differential operator is called the *D'Alembertian*. Defining Φ to be the 3-form

$$\Phi = A_4\, dx\, dy\, dz + \frac{1}{\epsilon\mu} A_1\, dy\, dz\, dt + \frac{1}{\epsilon\mu} A_2\, dz\, dx\, dt + \frac{1}{\epsilon\mu} A_3\, dx\, dy\, dt$$

the assumption (5) on the coefficients of A is simply the assumption

$$(5') \qquad d\Phi = 0$$

and the equations reduce to

$$(4') \qquad \epsilon\Box^2\Phi + J = 0$$

where the D'Alembertian $\Box^2$ is applied to a 3-form by applying it to each component separately.

A 3-form Φ is called a *vector potential* for the flow of charge J if the equations (4'), (5') are satisfied. When this is the case an electromagnetic field $E\, dt + B$ whose charge flow is J can be obtained by setting

$$\Phi = \phi\, dx\, dy\, dz - \psi_1\, dy\, dz\, dt - \psi_2\, dz\, dx\, dt - \psi_3\, dx\, dy\, dt$$

and

$$E\, dt + B = -d(\phi\, dt - \epsilon\mu\psi_1\, dx - \epsilon\mu\psi_2\, dy - \epsilon\mu\psi_3\, dz).$$

Conversely, it can be shown that any electromagnetic field $E\, dt + B$ can be derived from a vector potential Φ in this way.

The equation $d\Phi = 0$ means that Φ can be interpreted as a *flow* of a quantity called 'potential' which is preserved; ϕ is the density of this quantity in space and ψ_1, ψ_2, ψ_3 give the rate (per unit area per unit time) at which it crosses surfaces in space. When the potential is stationary ($\psi_1 = \psi_2 = \psi_3 = 0$), the equation $d\Phi = 0$ implies ϕ is independent of t and the equation $\epsilon\Box^2\Phi + J = 0$ reduces to $\epsilon\left(\frac{\partial^2\phi}{\partial x^2} + \frac{\partial^2\phi}{\partial y^2} + \frac{\partial^2\phi}{\partial z^2}\right) dx\, dy\, dz + \rho\, dx\, dy\, dz = 0$, that is, reduces to Poisson's equation (4).

When $J = 0$ (no net charge and no net motion of charge) the equation $\epsilon\Box^2\Phi + J = 0$ reduces to

$$\Box^2\phi = 0,\ \Box^2\psi_1 = 0,\ \Box^2\psi_2 = 0,\ \Box^2\psi_3 = 0.$$

A typical solution f of the equation

$$\frac{\partial^2 f}{\partial x^2} + \frac{\partial^2 f}{\partial y^2} + \frac{\partial^2 f}{\partial z^2} - \epsilon\mu\frac{\partial^2 f}{\partial t^2} = 0$$

(that is, $\Box^2 f = 0$) is a function of the form

$$f(x, y, z, t) = A\sin(B(a_1x + a_2y + a_3z - ct) + C)$$

where $a_1^2 + a_2^2 + a_3^2 = 1$, $c^2 = \frac{1}{\epsilon\mu}$, and A, B, C are arbitrary constants. This function can be described as a *wave* moving in the direction (a_1, a_2, a_3) with velocity c. For this reason the equation $\Box^2 f = 0$ is called the 'wave equation' and the number

$$c = \frac{1}{\sqrt{\epsilon\mu}}$$

is called the 'wave velocity'. A solution Φ of $\epsilon\Box^2\Phi = 0$ can be thought of as being made up of many such "waves moving with velocity $(\epsilon\mu)^{-1/2}$" which would lead one to expect that, when $J = 0$, electromagnetic disturbances propagate with the velocity $(\epsilon\mu)^{-1/2}$. The constants ϵ, μ can be measured by electromagnetic experiments and the wave velocity $(\epsilon\mu)^{-1/2}$ determined. Maxwell determined these velocities for various media and found that they agreed with the velocity of *light* in these media (to within the limits of experimental error). He concluded that *light is an electromagnetic phenomenon*, one of the most important discoveries in the history of physics. More generally, an 'electromagnetic wave' is an electromagnetic field for which $J = 0$. The possibility of generating electromagnetic waves by electrical means led to the invention of radio.

Lorentz Transformations

The formulation of Maxwell's equations as $d\Phi = 0$, $\epsilon\square^2\Phi + J = 0$ makes it easy to find the manner in which they transform under an affine change of coordinates

$$
\begin{aligned}
x' &= a_{11}x + a_{12}y + a_{13}z + a_{14}t + b_1 \\
y' &= a_{21}x + a_{22}y + a_{23}z + a_{24}t + b_2 \\
z' &= a_{31}x + a_{32}y + a_{33}z + a_{34}t + b_3 \\
t' &= a_{41}x + a_{42}y + a_{43}z + a_{44}t + b_4.
\end{aligned}
\tag{6}
$$

Expressing the 3-forms Φ, J in terms of (x', y', z', t') the equation $d\Phi = 0$ is unchanged (i.e., the statement that a 3-form is closed has intrinsic meaning independent of the coordinates), and the equation $\epsilon\square^2\Phi + J = 0$ becomes another equation of the same form in which $\square^2$ is the differential operator $\frac{\partial^2}{\partial x^2} + \frac{\partial^2}{\partial y^2} + \frac{\partial^2}{\partial z^2} - \epsilon\mu\frac{\partial^2}{\partial t^2}$ expressed in terms of (x', y', z', t'). This differential operator can be found explicitly by noting that

$$= \frac{\partial f}{\partial x'}\frac{\partial x'}{\partial x} + \frac{\partial f}{\partial y'}\frac{\partial y'}{\partial x} + \text{etc.} = a_{11}\frac{\partial f}{\partial x'} + a_{21}\frac{\partial f}{\partial y'} + \text{etc.}$$

Hence

$$= a_{11}^2\frac{\partial^2 f}{\partial x'^2} + a_{11}a_{21}\frac{\partial^2 f}{\partial x'\partial y'} + \cdots + a_{21}a_{11}\frac{\partial^2 f}{\partial y'\partial x'} + \cdots.$$

All together

$$\square^2 f = c_{11}\frac{\partial^2 f}{\partial x'^2} + c_{12}\frac{\partial^2 f}{\partial x'\,\partial y'} + \cdots + c_{44}\frac{\partial^2 f}{\partial t'^2},$$

where there are 16 terms and where the coefficients c_{ij} are the coefficients of the 4×4 symmetric matrix

$$C = AMA^t$$

where

$$M = \begin{pmatrix} 1 & 0 & 0 & 0 \\ 0 & 1 & 0 & 0 \\ 0 & 0 & 1 & 0 \\ 0 & 0 & 0 & -\epsilon\mu \end{pmatrix}$$

$$A = \begin{pmatrix} a_{11} & a_{12} & a_{13} & a_{14} \\ a_{21} & a_{22} & a_{23} & a_{24} \\ a_{31} & a_{32} & a_{33} & a_{34} \\ a_{41} & a_{42} & a_{43} & a_{44} \end{pmatrix}.$$

Thus the expression of Maxwell's equations $d\Phi = 0$, $\epsilon\Box^2\Phi + J = 0$ in terms of (x', y', z', t') is found essentially by computing the matrix AMA^t.

A transformation of coordinates (6) is called a *Lorentz transformation* if Maxwell's equations $d\Phi = 0$, $\epsilon\Box^2\Phi + J = 0$ have the same form with respect to (x', y', z', t') as they do with respect to (x, y, z, t); that is, a Lorentz transformation is an affine transformation of coordinates (6) such that $AMA^t = M$. Typical Lorentz transformations are translations ($x' = x +$ const., $y' = y +$ const., $z' = z +$ const., $t' = t +$ const.); rotations of xyz-space (x', y', z' a rotation of x, y, z while $t' = t$); and the transformation

$$(7) \qquad \begin{aligned} x' &= \gamma(x - vt) \\ y' &= y \\ z' &= z \\ t' &= \gamma(-\epsilon\mu v x + t) \end{aligned} \qquad \gamma = \frac{1}{\sqrt{1 - \epsilon\mu v^2}}$$

where v is any constant* less than the speed of light—i.e., $\epsilon\mu v^2 < 1$—and where γ is obtained from v as indicated. It is not difficult to show that every† Lorentz transformation is a composition of these three types: translations, rotations of xyz-space, and the transformation (7).

*Note that v is $\frac{dx}{dt}$ when $x' = 0$; that is, it is the velocity of the point $(x', y', z') = (0, 0, 0)$ in xyz coordinates. Note also that $\epsilon\mu$ is a very small number so that γ is nearly 1 and the transformation is nearly $x' = x - vt$, $y' = y$, $z' = z$, $t' = t$ unless v is very large.

†More precisely, the Lorentz transformations obtained in this way are proper Lorentz transformations. They do not include reflections in a plane of xyz-space or time reversal.

Not only are the equations $d\Phi = 0$, $\epsilon\Box^2\Phi + J = 0$ unchanged by a Lorentz transformation, but *all* of the equations

$$E\,dt + B = -d(\phi\,dt - \epsilon\mu\psi_1\,dx - \epsilon\mu\psi_2\,dy - \epsilon\mu\psi_3\,dz)$$
$$D = \epsilon E,\ B = \mu H,\ dJ = 0$$
$$d(D - H\,dt) = J$$

are unchanged‡ (see Exercise 3). Thus if the field $E\,dt + B$ is derived from Φ in $xyzt$-coordinates and then transformed to $x'y'z't'$-coordinates, the result is the same field (2-form in (x', y', z', t')) as is obtained by first converting Φ to $x'y'z't'$-coordinates and then deriving the field by the first equation above. Similarly, the 2-form $D - H\,dt$ can be obtained from the 2-form $E\,dt + B$ by the rules $D = \epsilon E$, $B = \mu H$ either before or after performing a (proper) Lorentz transformation of the coordinates and the result is the same.

‡Provided the change of coordinates (6) preserves the orientation, i.e. provided $\det A > 0$.

The fact that the laws of electromagnetism are unchanged by a Lorentz transformation of the coordinates is very useful in applications because it allows one to choose new coordinates in which the problem at hand is simpler. For example, to find the electromagnetic field generated by a moving magnet, one can first perform a Lorentz transformation of coordinates such that the magnet is stationary with respect to the new coordinates, find the field (now purely magnetic) which it generates in these coordinates, and return to the original coordinates to find the desired field (which is partly electrical).

Special Relativity

The fundamental postulate of Einstein's theory of special relativity is simply this: *All laws of physics should, like Maxwell's laws of electrodynamics, be unchanged by Lorentz transformations of the coordinates.* The motivation of this postulate is, briefly, as follows:

It is a fundamental postulate of Newtonian physics that the notion of velocity has no intrinsic physical meaning; that is, a body in uniform motion in a straight line cannot be distinguished from a body at rest. For example, it is not meaningful to say that 'the sun is stationary' but only that 'the motion of the sun is unaccelerated'. Since the velocity of light enters into the formulation of Maxwell's laws of electrodynamics in an essential way, Maxwell's laws are compatible with Newtonian physics only if there is a notion of 'rest' relative to which the velocity of light is defined. If there is such a physically meaningful notion of rest, then the application of Maxwell's laws to an actual physical system requires that one first determine its motion relative to absolute rest. This adaptation of the laws of electrodynamics to Newton's postulates is unsatisfactory both from a philosophical standpoint—because the notion of 'absolute rest' is contrary to the spirit of the Newtonian postulates which it is trying to salvage—and from a practical standpoint—because no 'absolute motion' of the earth has been detected experimentally (the Michelson-Morley experiment), even though the motion of the earth relative to the sun varies greatly during the course of the year. In this context, Einstein's postulate can be regarded as the postulate that the laws of electrodynamics take precedence over the Newtonian postulates, and that the latter, not the former, need to be revised.

Mass and Energy

The revision of Newtonian mechanics in accordance with the postulate of special relativity is not at all simple.

For example, in seeking a 'relativistic' version of the fundamental law

$$\text{force} = \text{mass} \times \text{acceleration}$$

one must make a fundamental change in one's conception of 'mass'. Einstein asserts, in fact, that "mass and energy are essentially alike", even though the original idea of mass was *inertia*, which is virtually the *opposite* of energy. The argument by which Einstein arrived at this amazing conclusion was roughly as follows:

Assuming that matter consists of charged particles and assuming that all forces are electromagnetic, the electricity of the system is described by a closed 3-form Φ (vector potential) from which the electromagnetic field and the distribution of moving charge $J = -\epsilon\Box^2\Phi$ can be derived.

The rule

$$\phi\, dx\, dy\, dz - \psi_1\, dy\, dz\, dt - \psi_2\, dz\, dx\, dt - \psi_3\, dx\, dy\, dt \leftrightarrow \phi\, dt - \epsilon\mu\psi_1\, dx - \epsilon\mu\psi_2\, dy - \epsilon\mu\psi_3\, dz$$

used in deriving the electromagnetic field from Φ is, as was stated above, 'Lorentz invariant'; that is, the 3-form determines the 1-form, and after a (proper) Lorentz transformation of coordinates the same 3-form determines the same 1-form. Similarly, the 1-form

$$\rho\, dt - \epsilon\mu j_1\, dx - \epsilon\mu j_2\, dy - \epsilon\mu j_3\, dz$$

determined by J is unchanged by a (proper) Lorentz transformation. Multiplying by the 2-form $D - H\,dt$ gives the 3-form

$$-\epsilon^2\mu(j_1E_1 + j_2E_2 + j_3E_3)\, dx\, dy\, dz + \epsilon(\rho E_1 + j_2B_3 - j_3B_2)\, dy\, dz\, dt + \epsilon(\rho E_2 + j_3B_1 - j_1B_3)\, dz\, dx\, dt + \epsilon(\rho E_3 + j_1B_2 - j_2B_1)\, dx\, dy\, dt. \tag{8}$$

This 3-form is determined by Φ (because $D - H\,dt$ and J are), and is unchanged by a (proper) Lorentz transformation of coordinates. When there is no net moving charge ($j = 0$), the first term in the 3-form (8) is zero and the coefficients of the remaining three terms are $\epsilon\rho E_i$ ($i = 1, 2, 3$). By Coulomb's law these are ϵ times the internal forces of the system, i.e., ϵ times the force exerted by the field E on the charge ρ. Now for any point $(\bar{x}, \bar{y}, \bar{z}, \bar{t})$ there is a Lorentz transformation of coordinates in which $j = 0$ at $(\bar{x}, \bar{y}, \bar{z}, \bar{t})$ (let the origin of the coordinates move with the charge), so that the 3-form

(8) divided by ϵ is 'force'. But, by the postulate of special relativity, if force is to have any physical meaning it must be unchanged by Lorentz transformations. Since ϵ^{-1} times the 3-form (8) is force relative to one coordinate system and is unchanged by Lorentz transformations, the only possible definition of force which is consistent with the theory of relativity is ϵ^{-1} times (8), that is

$$\begin{aligned} \text{force} = & -\epsilon\mu(j_1E_1 + j_2E_2 + j_3E_3)\,dx\,dy\,dz \\ & + (\rho E_1 + j_2B_3 - j_3B_2)\,dy\,dz\,dt \\ & + (\rho E_2 + j_3B_1 - j_1B_3)\,dz\,dx\,dt \\ & + (\rho E_3 + j_1B_2 + j_2B_1)\,dx\,dy\,dt. \end{aligned}$$

The last three components of this expression for the force were well-known before Einstein; they give, in addition to the electrical force ρE, the magnetic force exerted by the magnetic field B on the moving charge j. It was Einstein who showed that his 'principle of relativity' implied the existence of the remaining component

$$-\epsilon\mu(j_1E_1 + j_2E_2 + j_3E_3)\,dx\,dy\,dz.$$

The quantity $j_1E_1 + j_2E_2 + j_3E_3$ is the 'rate of work' being performed by the field $E_1\,dx + E_2\,dy + E_3\,dz$ on the net charge moving at the rate j. Hence the new component of force is $-\epsilon\mu$ times the rate of change (per time) of the internal energy of the system (per volume).

This gives the relativistic meaning of force. In finding the relativistic meaning of $F = ma$ Einstein first wrote $F = \frac{d}{dt}(mv)$ and integrated to obtain

$$\int \text{force} \cdot dt = \text{change in momentum}$$

where 'momentum' is mass times velocity. Now it is natural to describe the flow of mass by a 3-form

$$m\,dx\,dy\,dz - mu\,dy\,dz\,dt - mv\,dz\,dx\,dt - mw\,dx\,dy\,dt$$

where m is the density of mass and (u, v, w) are the components of its velocity. The last three components are the negative of momentum which leads to the conclusion that, just as force has a fourth component which was omitted from classical physics, so also momentum has a fourth component and the complete expression should be

$$\text{Momentum} = -m\,dx\,dy\,dz + mu\,dy\,dz\,dt + mv\,dz\,dx\,dt + mw\,dx\,dy\,dt.$$

Newton's law $F = ma$ is the equation

$$\int \text{force } dt = \text{change in momentum}$$

applied to the last three components of force and momentum. Applying the same equation to the newly found $dx\,dy\,dz$ components gives

$$-\epsilon\mu \int \frac{d}{dt}(\text{internal energy})\,dt = \text{change in } (-m)$$

or, letting E denote the internal energy and $\epsilon\mu = 1/c^2$, where c is the speed of light

$$\text{change in } E = \text{change in } mc^2.$$

This shows that energy and mass are 'essentially alike'. Of course the internal energy E of the system is determined only up to an additive constant, but if it is assumed that E can actually be reduced to zero and that the mass is then zero it follows that

$$E = mc^2.$$

Exercises **1** Show that if a function $u(x, y, z)$ defined for $\{x^2 + y^2 + z^2 \geq R^2\}$ is harmonic and radially symmetric, then it is of the form

$$u(x, y, z) = \frac{A}{r} + B$$

where A, B are constants and $r = \sqrt{x^2 + y^2 + z^2}$. [The 2-form $\frac{\partial u}{\partial x}\,dy\,dz + \frac{\partial u}{\partial y}\,dz\,dx + \frac{\partial u}{\partial z}\,dx\,dy$ is closed. By subtracting a suitable multiple of r^{-1} from u it can be assumed that its integral over any large sphere is zero. Thus it suffices to show that if $u(r) = u(x, y, z)$ is a radially symmetric harmonic function such that the integral of $\frac{\partial u}{\partial x}\,dy\,dz + \frac{\partial u}{\partial y}\,dz\,dx + \frac{\partial u}{\partial z}\,dx\,dy$ over large spheres is zero, then $u =$

const., i.e. $du = 0$. But $du = u'(r)\,dr$ and $u'(r)$ is constant on spheres. Hence the 2-form above is a multiple $f(r)$ of the 2-form $x\,dy\,dz + y\,dz\,dx + z\,dx\,dy$. Applying Stokes' Theorem gives $f(r) \equiv 0$. Hence $u'(r) \equiv 0$ as desired.]

2 Find the electromagnetic field generated by a single particle of charge e moving at uniform velocity v (assume $v < c$) along the x-axis, passing through $(0, 0, 0)$ at $t = 0$. [Take a Lorentz transformation of coordinates in which the particle is stationary, find the field in the new coordinates by Coulomb's law, and convert it to the original coordinates.]

3 Prove that all of Maxwell's laws are unchanged by a Lorentz transformation with positive determinant as follows:

(a) Show that it suffices to prove that the rules

$$\phi\,dx\,dy\,dz - \psi_1\,dy\,dz\,dt - \psi_2\,dz\,dx\,dt - \psi_3\,dx\,dy\,dt \leftrightarrow \phi\,dt - \epsilon\mu\psi_1\,dx - \epsilon\mu\psi_2\,dy - \epsilon\mu\psi_3\,dz$$

establishing a correspondence between 3-forms and 1-forms and

$$E_1\,dx\,dt + E_2\,dy\,dt + E_3\,dz\,dt + B_1\,dy\,dz + B_2\,dz\,dx + B_3\,dx\,dy \leftrightarrow \frac{1}{\mu}(\epsilon\mu E_1\,dy\,dz + \epsilon\mu E_2\,dz\,dx + \epsilon\mu E_3\,dx\,dy - B_1\,dx\,dt - B_2\,dy\,dt - B_3\,dz\,dt)$$

establishing a correspondence between 2-forms and 2-forms are unchanged by a Lorentz transformation with positive determinant, i.e. 'transform and correspond' is the same as 'correspond and transform.'

(b) Prove that this is true for the particular Lorentz transformation (7) of the text.

(c) Prove that this is true of the particular Lorentz transformation

$$x' = x\cos\theta + y\sin\theta$$
$$y' = -x\sin\theta + y\cos\theta$$
$$z' = z$$
$$t' = t.$$

(d) Prove that this is true of the Lorentz transformation

$$x' = -x,\ y' = y,\ z' = z,\ t' = t.$$

(e) It is true that every Lorentz transformation with positive determinant is a composition of rotations around coordinate axes in xyz-space, the transformation (7), translations, and the transformation of (d). Hence by appeal to this fact, (a), (b), (c), (d) suffice to prove the theorem. To prove it without appeal to this fact requires more linear algebra than is contained in this book. Readers who have some background in linear algebra can prove the needed facts stated in (a) as follows: The rule

$$(A_1\,dx + A_2\,dy + A_3\,dz + A_4\,dt)\cdot(B_1\,dx + B_2\,dy + B_3\,dz + B_4\,dt)$$
$$= A_1B_1 + A_2B_2 + A_3B_3 - \epsilon\mu A_4B_4$$

is a symmetric bilinear form from pairs of 1-forms to functions. It is unchanged by Lorentz transformations. Thus every 1-form determines a map {1-forms} → {functions} in a Lorentz invariant way. A 3-form also determines a map {1-forms} → {functions} by the rule

$$\omega_1 \to \frac{\omega_1\omega_3}{dx\,dy\,dz\,dt}$$

where ω_3 is the 3-form. This is unchanged by Lorentz transformations with positive determinant. The correspondence between 1-forms and 3-forms given in (a) is the correspondence between forms which give the same map {1-forms} → {functions} and is therefore invariant. There is also a Lorentz invariant bilinear form

$$\{\text{2-forms}\} \times \{\text{2-forms}\} \to \{\text{functions}\}$$

which can be used to prove that the correspondence of 2-forms is invariant.

further study of limits

chapter 9

9.1

The Real Number System

The fundamental operations of calculus—differentiation and integration—involve limits of real numbers. In order to give a rigorous definition of these operations one must therefore have a precise formulation of the concept of real number. This in turn requires a re-examination of the concept of number itself.

Natural Numbers

The most primitive notion of number occurs in the process of counting 'one, two, three, . . . ', which is the context in which one first learns these words for number. The numbers used in counting, that is, the positive whole numbers, are called *natural numbers*. These numbers are natural to us because we are in possession of such an efficient system of *representing* them and of *performing the operations of arithmetic* on them.

Numbers have been represented in a great variety of ways in various civilizations. Vestiges of ancient systems of numeration are encountered in the use of Roman numerals to denote years (used because MCMLXIX is, by its lack of clarity, so much more imposing than 1969), in the use of 23° 2′ 18″ to denote an angle (from the

Greek practice of using a prime to denote the 'place' of a number, necessitated by the lack of a symbol 0 to fill empty 'places'), and in the primitive but still useful method of tallying 𝍸 𝍸 𝍸 |||. These systems have now been replaced by the decimal notation based on the Arabic (originally Hindu) symbols 0, 1, 2, 3, 4, 5, 6, 7, 8, 9. The advantages of this system over all previous ones are so great that it has been incorporated into all the major languages of the world.

In decimal notation a natural number is represented by a symbol of the form $a_n a_{n-1} \ldots a_2 a_1$ consisting of a finite sequence of symbols a_i, each of which is one of the ten digits $0, 1, 2, \ldots, 9$. (Commas, or, in many languages periods, are often inserted dividing the digits into threes to increase legibility, e.g. 3,427,182. In theory there is no limit to the number of digits n although in practice n rarely exceeds fifteen.) Two such symbols $a_n a_{n-1} \ldots a_1$ and $b_m b_{m-1} \ldots b_1$ are considered to represent the same number if they differ only by the addition or deletion of 0's on the left end. For example, 10 means the same thing as 0010. For this reason all 0's on the left end are normally deleted so that $a_n \neq 0$. A symbol $a_n a_{n-1} \ldots a_1$ in which all digits are 0 is excluded.

214739
214379

$3 < 7$
$\therefore 214379 < 214739$

Given two such symbols $a = a_n a_{n-1} \ldots a_1$ and $b = b_m b_{m-1} \ldots b_1$, one writes $a < b$ if the following relation holds: When 0's are added to the left end in such a way that a and b have the same number of digits, a precedes b in the lexicographic ordering based on the ordering $0 < 1 < 2 < \cdots < 9$ of the digits; that is, reading from the left, the digit of a precedes the corresponding digit of b in the first position where they differ. Thus 1 is the least symbol, i.e. $1 < b$ for all $b \neq 1$; then 2 is the next least, etc., giving the usual order $1 < 2 < \cdots < 9 < 10 < 11 < \cdots < 99 < 100 < 101 < \cdots$ for the symbols. In this way the symbols serve as 'counters' and give a concise way of recording the result of any count, by doing a parallel count of the symbols.

1 2 3 4 5 6
1 2 3 4

$2 + 2 = 4$

1 2 3 4
5 6 7 8
9 10 11 12

$3 \times 4 = 12$

Given two symbols a and b, one finds a third symbol, denoted $a + b$ and called their *sum*, by the operation 'count first to a, then to b, and record the total count'. Given two symbols a and b, one finds a third symbol, denoted ab or $a \times b$ and called their product, by the operation 'count to a, repeat b times, and record the total count.'

The relation $a < b$ and the operations $a + b$, ab clearly have the following properties: If symbols a, b are

*$a > b$ means, of course, that $b < a$ in the sense defined above.

given, then exactly one of the relations $a < b$, $a = b$, $a > b$ holds.* The relation $a < b$ holds if and only if there is a symbol c such that $a + c = b$. The commutative, associative and distributive laws $a + b = b + a$, $ab = ba$, $(a + b) + c = a + (b + c)$, $(ab)c = a(bc)$, $a(b + c) = ab + ac$ all hold. Terms can be cancelled from sums and products; that is, if $a + c = b + c$ then $a = b$, and if $ac = bc$ than $a = b$. The symbol 1 has the property that $1 \cdot a = a$ for all a.

These properties are in fact properties of the counting process and have nothing to do with decimal representations. Rather, they tell *how + and × can be performed as operations on symbols.* The crucial observation is that $a_n a_{n-1} \dots a_2 a_1 = a_1 + 10 \cdot a_n a_{n-1} \dots a_2$. This implies that $a_n a_{n-1} \dots a_1 = a_1 + 10a_2 + 10 \cdot 10 a_n a_{n-1} \dots a_3 = a_1 + 10a_2 + 10^2 a_3 + \cdots + 10^{n-1} a_n$ where 10^{n-1} means ten multiplied by itself one less than n times (n itself is a natural number). Sums and products are then formed by using the commutative, associative and distributive laws together with an addition table and a multiplication table for one digit numbers, e.g. $31 + 94 = (3 \cdot 10 + 1) + (9 \cdot 10 + 4) = (3 + 9) \cdot 10 + 1 + 4 = 12 \cdot 10 + 5 = 125$ and $31 \times 94 = (3 \cdot 10 + 1) \times (9 \cdot 10 + 4) = 3 \cdot 9 \cdot 10^2 + 3 \cdot 4 \cdot 10 + 1 \cdot 9 \cdot 10 + 1 \cdot 4 = 27 \cdot 10^2 + 12 \cdot 10 + 9 \cdot 10 + 4 = 2 \cdot 10^3 + 8 \cdot 10^2 + 11 \cdot 10 + 4 = 2914$. In short, sums and products of *symbols* are found according to the familiar schemes. In fact, one could consider + and × as being *defined* by these schemes rather than by reference to the counting process. If such a procedure is followed, then the arithmetic of symbols can be defined very explicitly, but the arithmetic of natural numbers which it 'represents' remains undefined. This distinction between the arithmetic of symbols and the arithmetic of numbers may seem less artificial when the idea of 'number' is generalized beyond the natural numbers and hence can no longer be apprehended so concretely in terms of counting.

```
 31        31
 94        94
---       ---
125       124
         279
         ----
         2914
```

Rational Numbers

For example, fractions, negative numbers, and zero are normally considered to be numbers. Everyone learns how to record such numbers (e.g. -1, $\frac{2}{3}$, $-4\frac{7}{8}$, 0) and how to add and multiply them, but there is normally much less of a feeling of security about the meaning of the result. (The rule $(-1)(-1) = 1$ is particularly puzzling to most people.) For this reason, in describing

*That is, numbers which are ratios, although the term now includes zero and negatives as well.

the arithmetic of such numbers, called *rational numbers**, it is desirable to begin at the beginning and to make as few assumptions as possible.

The arithmetic of rational numbers can be founded on the arithmetic of natural numbers (which will now be assumed to be completely familiar) by observing that every rational number q satisfies an equation of the form $aq + b = c$, where a, b, c are natural numbers, and that the triad (a, b, c) of natural numbers uniquely determines q. Thus -1 is the only solution of $1 \cdot q + 2 = 1$, $\frac{2}{3}$ is the only solution of $3q + 1 = 3$, $-4\frac{7}{8}$ the only solution of $8q + 50 = 11$, 0 the only solution of $1 \cdot q + 3 = 3$, etc. In other words, *rational numbers can be described by triads of natural numbers*, the rational numbers -1, $\frac{2}{3}$, $-4\frac{7}{8}$, 0, being described, for example, by the triads $(1, 2, 1)$, $(3, 1, 3)$, $(8, 50, 11)$, $(1, 3, 3)$ respectively.

Of course this description is not unique, that is, the same rational number can be described by more than one triad. For example, $-4\frac{7}{8}$ can also be described by the triads $(8, 40, 1)$, $(16, 80, 2)$, $(16, 79, 1)$, $(32, 158, 2)$, etc., as well as by $(8, 50, 11)$. The relation 'describe the same rational number' between triads will be denoted $\equiv$, read 'is congruent to'. Thus $(8, 40, 1) \equiv (16, 80, 2) \equiv (8, 50, 11)$, etc. More generally $(a, b, c) \equiv (a, b + d, c + d)$ and $(a, b, c) \equiv (ad, bd, cd)$. Given (a, b, c), (a', b', c') one can determine whether $(a, b, c) \equiv (a', b', c')$ by rewriting

$$aq + b = c, \quad a'q' + b' = c'$$

as

$$(*) \qquad \begin{aligned} aa'q + ba' + ab' &= ca' + ab' \\ aa'q' + ba' + ab' &= ac' + ba'. \end{aligned}$$

If $q = q'$ then the right-hand sides are equal and conversely; hence

$$(1) \qquad \begin{gathered} (a, b, c) \equiv (a', b', c') \\ \text{if and only if } ca' + ab' = ac' + ba'. \end{gathered}$$

$8q + 40 = 1$
$16q + 79 = 1$

$16 \cdot 8q + 16 \cdot 40 = 16$
$16 \cdot 8q + 8 \cdot 79 = 8$

$16 + 8 \cdot 79 = 8 + 16 \cdot 40$

Checks.
$\therefore$ $8q + 40 = 1$ and $16q + 79 = 1$ describe the same q.

This gives a criterion, expressed solely in terms of the arithmetic of natural numbers, for determining whether two given triads of natural numbers describe the same rational number.

Given two triads (a, b, c), (a', b', c') describing rational numbers q, q', the relation $q < q'$ and the operations $q + q'$, qq' can be expressed in terms of the triads as follows: If $q < q'$ then the equations (*) give $ca' + ab' < ac' + ba'$. Moreover, the steps are reversible, and

$q = (3, 1, 2)$
$q' = (2, 2, 3)$
$ca' + ab' = 2 \cdot 2 + 3 \cdot 2 = 10$
$ac' + ba' = 3 \cdot 3 + 1 \cdot 2 = 11$
$\therefore q < q'$

hence

$$(2) \qquad (a, b, c) < (a', b', c') \text{ if and only if } ca' + ab' < ac' + ba'.$$

Adding the equations (*) gives an equation satisfied by $q + q'$; hence

$$(3) \quad (a, b, c) + (a', b', c') \equiv (aa', ba' + ab', ca' + ac').$$

$q + q' \equiv (6, 1 \cdot 2 + 3 \cdot 2, 2 \cdot 2 + 3 \cdot 3) \equiv (6, 8, 13) \equiv (6, 1, 6)$

That is, the rational number described by the triad on the right is the sum of the rational numbers described by the triads on the left. In the same way $(aq + b)(a'q' + b') = cc'$ gives $aa'qq' + ba'q' + b'aq + bb' = cc'$, $aa'qq' + b(a'q + b') + b'(aq + b) = cc' + bb'$, $aa'qq' + bc' + b'c = cc' + bb'$. Hence

$$(4) \quad (a, b, c)(a', b', c') \equiv (aa', bc' + cb', cc' + bb')$$

gives the rule for forming products.

$q \cdot q' \equiv (6, 1 \cdot 3 + 2 \cdot 2, 2 \cdot 3 + 1 \cdot 2) \equiv (6, 7, 8) \equiv (6, 1, 2)$

The rules (1), (2), (3), (4) completely describe the arithmetic of rational numbers in terms of the arithmetic of natural numbers. In the absence of a definition of 'the arithmetic of rational numbers' this statement has no meaning. However, it can be reversed and the 'arithmetic of rational numbers' can be defined to be 'that which the rules (1), (2), (3), (4) describe'. This arithmetic is as follows:

The arithmetic applies to triads of natural numbers. Two triads are considered to be 'the same' in the arithmetic if they are congruent in the sense defined by the rule (1). This use of the word 'same' is justified only if its basic meaning is valid, that is, only if it is true that '(a, b, c) is the same as (a, b, c)', that 'if (a, b, c) is the same as (a', b', c') then (a', b', c') is the same as (a, b, c)', and that 'if (a, b, c) is the same as (a', b', c') and (a', b', c') is the same as (a'', b'', c'') then (a, b, c) is the same as (a'', b'', c'')'. In other words the relation $\equiv$ defined by (1) must be shown to have the properties

$$(5) \quad \begin{array}{c} (a, b, c) \equiv (a, b, c) \text{ (reflexive)} \\ (a, b, c) \equiv (a', b', c') \\ \text{implies } (a', b', c') \equiv (a, b, c) \text{ (symmetric)} \\ (a, b, c) \equiv (a', b', c') \\ \text{and } (a', b', c') \equiv (a'', b'', c'') \\ \text{imply } (a, b, c) \equiv (a'', b'', c'') \text{ (transitive)} \end{array}$$

The first two of these statements are the tautologies

'$ca + ab = ac + ba$' and '$ca' + ab' = ac' + ba'$ implies $c'a + a'b = a'c + b'a$'. The third says '$ca' + ab' = ac' + ba'$ and $c'a'' + a'b'' = a'c'' + b'a''$ imply $ca'' + ab'' = ac'' + ba''$', which is proved by multiplying the first equation by a'', the second by a, adding, and performing cancellations.

Given two triads (a, b, c), (a', b', c') one writes $(a, b, c) < (a', b', c')$ if $ca' + ab' < ac' + ba'$. This relation $<$ is consistent with the convention that congruent triads are to be considered 'the same' because

$$(6)\quad \begin{aligned}&(a, b, c) \equiv (d, e, f),\ (a', b', c') \equiv (d', e', f')\\ &\text{and } (a, b, c) < (a', b', c')\\ &\text{imply } (d, e, f) < (d', e', f').\end{aligned}$$

Although this can be proved directly without difficulty, the following observation simplifies the proof somewhat: If $(a, b, c) \equiv (a', b', c')$ then it is possible to go from (a, b, c) to (a', b', c') by a finite number of applications (in fact four applications) of the rules $(a, b, c) \equiv (ga, gb, gc)$, $(a, b, c) \equiv (a, b + g, c + g)$. [Explicitly, $(a, b, c) \equiv (aa', ba', ca') \equiv (aa', ba' + ab', ca' + ab') \equiv (aa', ba' + ab', ac' + ba') \equiv (aa', ab', ac') \equiv (a', b', c')$.] Therefore in checking (5) it suffices to consider the cases $(d, e, f) = (ga, gb, gc)$, $(d, e, f) = (a, b + g, c + g)$ and similarly $(d', e', f') = (g'a', g'b', g'c')$, $(d', e', f') = (a', b' + g', c' + g')$. In these cases (6) is immediate, e.g. $(ga, gb, gc) < (a', b', c')$ means $gca' + gab' < gac' + gba'$ which, cancelling g, implies $(a, b, c) < (a', b', c')$ and conversely.

$(2, 7, 5) \equiv (6, 21, 15) \equiv (6, 12, 6) \equiv (1, 2, 1)$

Triads are added in the arithmetic by the rule (3). This definition is justified only if the sum is 'the same' whenever the summands are 'the same', that is, only if it is true that

$$(7)\quad \begin{aligned}&(a, b, c) \equiv (d, e, f) \text{ and } (a', b', c') \equiv (d', e', f')\\ &\text{imply } (a, b, c) + (a', b', c') \equiv (d, e, f) + (d', e', f')\end{aligned}$$

where $\equiv$ and $+$ are defined by (1) and (3). Again it suffices to consider the cases $(d, e, f) = (ag, bg, cg)$, $(d, e, f) = (a, b + g, c + g)$, for which (7) is immediately verified.

Triads are multiplied by the rule (4). Again this definition must be justified by showing that the product is 'the same' whenever the factors are 'the same', that is,

$$(8)\quad \begin{aligned}&(a, b, c) \equiv (d, e, f) \text{ and } (a', b', c') \equiv (d', e', f')\\ &\text{imply } (a, b, c) \cdot (a', b', c') \equiv (d, e, f) \cdot (d', e', f').\end{aligned}$$

This is again easily verified in the cases $(d, e, f) = (ag, bg, cg)$ and $(d, e, f) = (a, b + g, c + g)$ which suffice to prove (8) in general.

$q = (3, 1, 2)$
$q' = (2, 2, 3)$
$q'' = (1, 2, 1)$

$$q(q' + q'') \equiv (3, 1, 2)(2, 2 + 4, 3 + 2) \equiv (3, 1, 2)(2, 6, 5) \equiv (6, 5 + 12, 10 + 6) \equiv (6, 17, 16)$$

The arithmetic defined by (1), (2), (3), (4) and justified by (5), (6), (7), (8) is now seen to have the following properties:

Given two triads (a, b, c), (a', b', c'), exactly one of the relations $(a, b, c) < (a', b', c')$, $(a, b, c) \equiv (a', b', c')$, $(a, b, c) > (a', b', c')$ holds. This is called the *trichotomy law*. It is true because exactly one of the relations $<$, $=$, $>$ holds for the natural numbers $ca' + ab'$ and $ac' + ba'$.

$$qq' + qq'' \equiv (3, 1, 2)(2, 2, 3) + (3, 1, 2)(1, 2, 1) \equiv (6, 3 + 4, 6 + 2) + (3, 1 + 4, 2 + 2) \equiv (6, 7, 8) + (3, 5, 4) \equiv (18, 21 + 30, 24 + 24) \equiv (18, 51, 48) \equiv (6, 17, 16) \equiv q(q' + q'')$$

If three triads $q \equiv (a, b, c)$, $q' \equiv (a', b', c')$, $q'' \equiv (a'', b'', c'')$ are given, then the commutative, associative, and distributive laws $q + q' \equiv q' + q$, $qq' \equiv q'q$, $q + (q' + q'') \equiv (q + q') + q''$, $q(q'q'') \equiv (qq')q''$, $q(q' + q'') \equiv qq' + qq''$ all hold. These are verified by direct computation. For example, the distributive law is proved by

$$\begin{aligned}
&(a, b, c)[(a', b', c') + (a'', b'', c'')] \\
&\quad \equiv (a, b, c)(a'a'', b'a'' + a'b'', c'a'' + a'c'') \\
&\quad \equiv (aa'a'', bc'a'' + ba'c'' + cb'a'' + ca'b'', cc'a'' + ca'c'' + bb'a'' + ba'b'') \\
&(a, b, c)(a', b', c') + (a, b, c)(a'', b'', c'') \\
&\quad \equiv (aa', bc' + cb', cc' + bb') + (aa'', bc'' + cb'', cc'' + bb'') \\
&\quad \equiv (aa'aa'', bc'aa'' + cb'aa'' + aa'bc'' + aa'cb'', cc'aa'' + bb'aa'' + aa'cc'' + aa'bb'') \\
&\quad \equiv (a, b, c)[(a', b', c') + (a'', b'', c'')].
\end{aligned}$$

$$(1, 1, a + 1) + (1, 1, b + 1) \equiv (1, 1 + 1, a + 1 + b + 1) \equiv (1, 1, a + b + 1)$$

$$(1, 1, a + 1)(1, 1, b + 1) \equiv (1, b + 1 + a + 1, (a + 1)(b + 1) + 1) \equiv (1, 1, ab + 1)$$

Given a natural number a, let $\bar{a}$ denote the triad $(1, 1, a + 1)$. Such a triad (or, of course, any triad which is 'the same' as such a triad, i.e. congruent to such a triad) is called a *positive integer*. Now $a < b$ if and only if $\bar{a} < \bar{b}$, and $a = b$ if and only if $\bar{a} \equiv \bar{b}$. Moreover, $\bar{a} + \bar{b} \equiv \overline{a + b}$ and $\bar{a}\bar{b} \equiv \overline{ab}$.* Hence the arithmetic of positive integers has all the properties of the arithmetic of natural numbers.

*Note that there are two different kinds of addition here and two different kinds of multiplication, namely, addition and multiplication of natural numbers—$a + b$, ab—and addition and multiplication of triads $\bar{a} + \bar{b}$, $\overline{ab}$.

If it is given that $q \equiv (a, b, c)$, then it is to be expected that $\bar{a}q + \bar{b} \equiv \bar{c}$. This is immediately verified:

$$\begin{aligned}
\bar{a}q &\equiv (1, 1, a + 1)(a, b, c) \\
&\equiv (a, c + ab + b, ac + c + b) \\
&\equiv (a, ab, ac) \equiv (1, b, c) \\
\bar{a}q + \bar{b} &\equiv (1, b, c) + (1, 1, b + 1) \\
&\equiv (1, b + 1, c + b + 1) \equiv (1, 1, c + 1) \equiv \bar{c}.
\end{aligned}$$

Thus every q satisfies an equation of the form $\bar{a}q + \bar{b} \equiv \bar{c}$

where $\bar{a}$, $\bar{b}$, $\bar{c}$ are positive integers. Moreover, if $q' \equiv (a', b', c')$ is any solution of $\bar{a}q' + \bar{b} \equiv \bar{c}$ then this equation together with $\bar{a}'q' + \bar{b}' \equiv \bar{c}'$ gives $\overline{ca'} + \overline{ab'} \equiv \overline{ac'} + \overline{ba'}$ exactly as in the derivation of (1). Hence $ca' + ab' = ac' + ba'$, that is, $q \equiv q'$. In other words, within the arithmetic the solution q of $\bar{a}q + \bar{b} = \bar{c}$ is *unique*, meaning that any two solutions are congruent.

Consider now the problem of solving an equation of the form $qx + q' \equiv q''$ for x, given q, q', q''. It is to be expected that the arithmetic will contain a unique element 0 such that solution is not (in general) possible when $q \equiv 0$ but such that there is a unique solution x of $qx + q' \equiv q''$ whenever $q \not\equiv 0$. This can be verified as follows:

*Because $1 \cdot 0 + 1 = 1$.

One would expect* 0 to be represented by the triad $(1, 1, 1)$. It is easily verified that $(a, b, c) \equiv (1, 1, 1)$ if and only if $b = c$; moreover, $0 \cdot q \equiv 0$ and $0 + q \equiv q$ for all q where $0 \equiv (1, 1, 1)$. Thus $0 \cdot x + q' \equiv q''$ has no solution x unless $q' \equiv q''$, in which case any x is a solution.

$(a, b, c)\,(1, 1, 1)$
$\equiv (a, b + c, c + b)$
$\equiv (a, 1, 1) \equiv (a, a, a)$
$\equiv (1, 1, 1)$

Given three triads q, q', q'' one can first replace them with congruent triads of the form $q \equiv (a, b, c)$, $q' \equiv (a, b, c')$, $q'' \equiv (a, b, c'')$. (If $q \equiv (A, B, C)$, $q' \equiv (A', B', C')$, $q'' \equiv (A'', B'', C'')$ set $a = AA'A''$, $b = AA'B'' + AB'A'' + BA'A''$.) Moreover, if $q \not\equiv 0$ then $b \neq c$, and hence either $b + d = c$ or $b = d + c$ and q is either of the form $q \equiv (a, 1, d + 1)$ or $q \equiv (a, d + 1, 1)$. The solution of $qx + q' \equiv q''$ will be considered for these two cases separately.

$(a, b, c) + (1, 1, 1)$
$\equiv (a, a + b, a + c)$
$\equiv (a, b, c)$

If $q \equiv (a, 1, d + 1)$, $q' \equiv (a, b, c')$, $q'' \equiv (a, b, c'')$ then $\bar{a}q \equiv (1, 1, d + 1) \equiv \bar{d}$ and $\bar{a}q' + \bar{b} \equiv \bar{c}'$, $\bar{a}q'' + \bar{b} \equiv \bar{c}''$ as was seen above. Hence $qx + q' \equiv q''$ implies

$$\bar{a}qx + \bar{a}q' + \bar{b} \equiv \bar{a}q'' + \bar{b}$$
$$\bar{d}x + \bar{c}' \equiv \bar{c}''.$$

Since this equation has the unique solution $x \equiv (d, c', c'')$, the solution x of $qx + q' \equiv q''$, if it exists, can only be this triad (or, of course, a triad congruent to this one). Conversely, for this x one easily verifies $qx + q' \equiv (ad,\ c'd + c' + c'',\ c' + c'' + c''d) + (a,\ b,\ c') \equiv (a, c', c'') + (a, b, c') \equiv (a, b, c'') \equiv q''$.

If $q \equiv (a, d + 1, 1)$ then $\bar{a}q + \bar{d} \equiv 0$, and hence $qx + q' \equiv q''$ implies

$$\bar{a}qx + \bar{a}q' + \bar{b} + \bar{d}x \equiv \bar{a}q'' + \bar{b} + \bar{d}x$$
$$\bar{c}' \equiv 0 \cdot x + \bar{c}' \equiv \bar{c}'' + \bar{d}x.$$

Hence $qx + q' \equiv q$ implies $x \equiv (d, c'', c')$. It is easily shown as above that this x is indeed a solution, which completes the proof of the elimination theorem: *If q, q', q'' are given with $q \not\equiv 0$, then there is a unique solution x of the equation $qx + q' \equiv q''$.*

The elimination theorem can be used to prove all the familiar facts about subtraction and division: The unique solution x of $q + x \equiv 0$ is denoted by $x \equiv -q$. The identities $-(-q) \equiv q$, $-(q_1 + q_2) \equiv (-q_1) + (-q_2)$, $-(q_1q_2) \equiv (-q_1)q_2 \equiv q_1(-q_2)$, $(-q_1)(-q_2) \equiv -[q_1(-q_2)] \equiv -[-q_1q_2] \equiv q_1q_2$ are deduced immediately from this definition. The symbol $q_1 - q_2$ denotes $q_1 + (-q_2)$. Since $\bar{1} \cdot (\bar{1} \cdot q) \equiv (\bar{1} \cdot \bar{1})q \equiv \bar{1} \cdot q$ it follows from the uniqueness statement of the elimination theorem that $\bar{1} \cdot q \equiv q$ for all q. The unique solution x of the equation $qx \equiv \bar{1}$ $(q \not\equiv 0)$ is denoted by $x \equiv 1/q$ or $x \equiv q^{-1}$. The identities $(q^{-1})^{-1} \equiv q$, $(q_1q_2)^{-1} \equiv (q_1)^{-1}(q_2)^{-1}$, $(-q)^{-1} \equiv -q^{-1}$ are deduced immediately from this definition. The symbol q_2/q_1 denotes $q_2(q_1)^{-1}$. Now the unique solution x of $qx + q' \equiv q''$ $(q \not\equiv 0)$ can be written as $(q'' - q')/q$ meaning, of course, $(q'' + (-q'))q^{-1}$. In particular, $q \equiv (a, b, c)$ can be written as $(\bar{c} - \bar{b})/\bar{a}$.

A triad $q \equiv (a, b, c)$ satisfies $q > 0$ if and only if $a + b < a + c$, that is, if and only if there is a natural number d such that $c = b + d$, $\bar{a}q + \bar{b} \equiv \bar{b} + \bar{d}$, $\bar{a}q \equiv \bar{d}$. It follows that q is positive $(q > 0)$ if and only if q satisfies an equation of the form $\bar{a}q \equiv \bar{d}$ where $\bar{a}$, $\bar{d}$ are positive integers. This observation implies that if $q_1 > 0$, $q_2 > 0$ then $q_1 + q_2 > 0$, $q_1q_2 > 0$. It is to be expected that $q_1 < q_2$ if and only if $q_2 - q_1 > 0$; to verify this, set $q_3 = q_2 - q_1$ and observe that the statement to be proved is that $q_1 + q_3 > q_1$ if and only if $q_3 > 0$. By setting $q_1 \equiv (a, b, c)$, $q_3 \equiv (a, b, c')$, this becomes '$(a, b + b, c + c') > (a, b, c)$ if $c' > b$ and conversely' which is immediately verified. The familiar facts about inequalities can now be deduced: $q_1 + q_2 < q_1 + q_3$ if and only if $q_2 < q_3$; if $q_1 < q_2$ and $q_2 < q_3$ then $q_1 < q_3$; if $q_1 > 0$ and $q_2 < q_3$ then $q_1q_2 < q_1q_3$; if $q_1 > 0$ then $-q_1 < 0$ and conversely.

$(a, b, c) + (a, b, c')$
$= (aa, ab + ba, ca + ac')$
$= (a, b + b, c + c')$

The *absolute value* $|q|$ of q is defined to be q if $q > 0$, 0 if $q \equiv 0$, $-q$ if $q < 0$. Then in all cases $|q| \geq 0$, $|q| \geq q$, and $|-q| \equiv |q|$. The identity $|q_1q_2| \equiv |q_1|\,|q_2|$ is proved by choosing signs $\pm$ so that $\pm q_1$ and $\pm q_2$ are both positive, hence $|q_1q_2| \equiv |(\pm q_1)(\pm q_2)| \equiv |\pm q_1|\,|\pm q_2| \equiv |q_1|\,|q_2|$. To prove the *triangle inequality* $|q_1 + q_2| \leq$

$|q_1| + |q_2|$ one can assume, by changing both signs if necessary, that $q_1 + q_2 \geq 0$; then $|q_1 + q_2| \equiv q_1 + q_2 \leq |q_1| + |q_2|$.

In short, the arithmetic of rational numbers defined by the rules (1), (2), (3), (4) has all the expected properties. It is convenient now to drop the use of the bars and to write 1, 2, 3, ... meaning elements of the extended arithmetic. Also, the symbol $\equiv$ will be replaced by the more familiar symbol $=$.

An 'object' q in this extended arithmetic is called a 'rational number'. Thus 'rational number' is, like 'natural number', essentially an undefined concept. Rules have been given for representing rational numbers, for ordering them, and for forming sums, differences, products, and quotients, but no statements about *what* is being represented have been made. It is not the 'numbers' but the *operations* and *relations* among them which are paramount.

An *integer* is a rational number q which satisfies an equation of the form $q + b = c$ where b, c are positive integers; in other words the integers are the positive integers, zero, and minus positive integers, i.e. negative integers.

It can be shown that every rational number q can be written uniquely as

$$q = \pm((\text{a positive integer or zero}) + (\text{a proper fraction in lowest terms or zero}))$$

where a 'proper fraction in lowest terms' is defined to be an expression of the form a/b where $a < b$ and a, b are positive integers without common factors. The usual arithmetic operations applied to such expressions are then justified as operations in the arithmetic of rational numbers as defined by the rules (1), (2), (3), (4).

An important property of the arithmetic of rational numbers is the *Archimedean law:* If positive rational numbers q_1, q_2 are given, then there is a natural number (positive integer) n such that $nq_1 > q_2$. A useful variant of this law is: If q_1, q_2 are given with $|q_1| < 1$ and $q_2 > 0$, then there is a natural number n such that $|q_1^n| < q_2$ (where q_1^n means $q_1 \cdot q_1 \ldots q_1$ (n times)). These facts are easily proved (Exercise 11).

In actual practice the arithmetic of rational numbers is quite cumbersome and one often uses decimal fractions instead. A *decimal fraction* is a rational number which can be written with a denominator which is a power of ten, that is, a rational number q which satisfies an equation of the form

$$10^n q = \text{integer}$$

where n is a natural number. Using the familiar decimal point notation, the arithmetic of decimal fractions is reduced to the arithmetic of natural numbers with the small added complications of keeping track of the sign and of the position of the decimal point. Every rational number can be approximated arbitrarily closely by 'rounding' to a decimal fraction. (More specifically, if a rational number q and a natural number n are given, then there is a decimal fraction q' with n places (at most) to the right of the decimal point such that $|q - q'| \leq \frac{1}{2}10^{-n}$. This approximation process is called 'rounding to n places'.) The ease of computation with decimal fractions usually compensates for the loss of accuracy resulting from rounding.

Real Numbers

The arithmetic of rational numbers cannot be used as a basis for calculus because *a limit of rational numbers need not be a rational number*. This is illustrated by the following examples:

Newton's method (§7.3) gives the sequence $1, \frac{3}{2}, \frac{17}{12}, \frac{577}{408}, \ldots, q_{n+1} = \frac{1}{2}\left(q_n + \frac{2}{q_n}\right)$, of rational numbers converging to $\sqrt{2}$. However $\sqrt{2}$ is not a rational number because the square of a fraction in lowest terms is a fraction in lowest terms. Hence $p^2/q^2 = \frac{2}{1}$ could occur only if $q^2 = 1$, $p^2 = 2$, which is impossible.

$$\frac{1}{a+1} + \frac{1}{(a+1)(a+2)} + \cdots < \frac{1}{a+1} + \left(\frac{1}{a+1}\right)^2 + \left(\frac{1}{a+1}\right)^3 + \cdots = \frac{1}{a}$$

The number $e = 1 + 1 + \frac{1}{2!} + \frac{1}{3!} + \frac{1}{4!} + \cdots$ is the limit of the sequence of rational numbers $1, 2, 2\frac{1}{2}, 2\frac{2}{3}, 2\frac{17}{24}, 2\frac{43}{60}, \ldots, q_{n+1} = q_n + \frac{1}{n!}$. However, if e were rational then $a!e$ would be an integer for all sufficiently large integers a; but this would give

$$a!e = a! + a! + \frac{a!}{2!} + \cdots + \frac{a!}{a!} + \frac{a!}{(a+1)!} + \cdots$$

$$\text{integer} = \text{integer} + \frac{1}{a+1} + \frac{1}{(a+1)(a+2)} + \frac{1}{(a+1)(a+2)(a+3)} + \cdots$$

$$\text{integer} = \text{integer} + \left(\text{positive number smaller than } \frac{1}{a}\right)$$

which is impossible.

The number π, which is defined to be the ratio of the area of a circle to the square of its radius, is the limit of a sequence of rational numbers

$$q_n = \frac{\text{number of pairs } (\pm a, \pm b) \text{ of integers such that } a^2 + b^2 \leq n^2}{n^2}$$

as was seen in Exercises 2, 3, §2.3. However, as is well known, π is not a rational number. (There is no simple proof of this fact.)

In order for the operations of calculus to be meaningful it is necessary, therefore, to extend the concept of 'number' in such a way that these limits $\sqrt{2}$, e, π and others like them are 'numbers'. The extended concept is of course that of a *real number*. As in the case of natural numbers and rational numbers, what is needed is a system for *representing* real numbers and of *performing operations* on them.

Intuitively speaking, a real number is a number which can be approximated arbitrarily closely by rational numbers (usually decimal fractions are used). Therefore a real number can be 'represented' by giving a sequence of rational numbers converging to it. The Cauchy Convergence Criterion gives a definition of 'convergence' for sequences of rational numbers without reference to real numbers. Hence, the Cauchy Criterion tells which sequences of rational numbers 'represent real numbers': A sequence $q_1, q_2, \ldots$ of rational numbers is said to be *convergent* if it is true that for every rational number $\epsilon > 0$ there is a natural number N such that for any two natural numbers, $n, m \geq N$ the condition $|q_n - q_m| < \epsilon$ holds. Such a sequence will be said to 'represent a real number' and the 'arithmetic of real numbers' will be described as an arithmetic which applies to such sequences.

Since the remainder of the discussion is devoted to this arithmetic of convergent sequences of rational numbers it is important to have as clear a formulation of the idea of convergence as possible. An alternative statement of the Cauchy Criterion is the following: An interval $\{\bar{q} \leq x \leq \bar{\bar{q}}\}$ of rational numbers will be said to *contain* a given infinite sequence $\{q_1, q_2, q_3, \ldots\}$ of rational numbers if it contains all but a finite number of terms q_n of the sequence, that is, if there is an N such that $\bar{q} \leq q_n \leq \bar{\bar{q}}$ for all $n \geq N$. Using this terminology, a given sequence $\{q_n\}$ is convergent if and only if it is 'contained in arbitrarily small intervals'. That is, a sequence is convergent if and only if for every $\epsilon > 0$ there exist $\bar{q}, \bar{\bar{q}}$ such that $0 < \bar{\bar{q}} - \bar{q} \leq \epsilon$ and such that the interval $\{\bar{q} \leq x \leq \bar{\bar{q}}\}$ contains the sequence. This condition clearly implies the Cauchy Criterion ($\bar{q} \leq q_n \leq \bar{\bar{q}}$ and $\bar{q} \leq q_m \leq \bar{\bar{q}}$ imply $|q_n - q_m| \leq \bar{\bar{q}} - \bar{q}$) and, conversely, if the Cauchy Criterion is fulfilled, then an N can be found such that $|q_n - q_m| < \epsilon/2$ for $n, m \geq N$. Hence $q_N - \epsilon/2 < q_m < q_N + \epsilon/2$ for $m \geq N$, that is, the sequence is contained in the interval $\{q_N - (\epsilon/2) \leq x \leq q_N + (\epsilon/2)\}$ of length ϵ.

The above sequence representing $\sqrt{2}$ was proved to be convergent in §7.5. The above sequence for e can be shown to be convergent as follows: The sequence increases, and after the nth term the further increase is less than

$$\frac{1}{n!}\left[1 + \frac{1}{n+1} + \frac{1}{(n+1)(n+2)} + \cdots\right]$$
$$< \frac{1}{n!}\left[1 + \left(\frac{1}{n+1}\right) + \left(\frac{1}{n+1}\right)^2 + \left(\frac{1}{n+1}\right)^3 + \cdots\right]$$
$$= \frac{n+1}{n \cdot n!}.$$

Hence the interval $\left\{q_n \leq x \leq q_n + \dfrac{n+1}{n \cdot n!}\right\}$ contains the sequence and has arbitrarily small length. It is much more difficult to show that the above sequence for π is convergent; this is the substance of Exercise 2, §2.3. Note that for this example, unlike the preceding ones, the convergence is slow and each q_n difficult to compute; computationally this sequence is useless.

$$\begin{aligned} S &= 1 + \tfrac{1}{n+1} + (\tfrac{1}{n+1})^2 + \cdots \\ (n+1)\,S &= (n+1) + 1 + \tfrac{1}{n+1} + \cdots \\ &= n + 1 + S \\ nS &= n + 1 \\ S &= \tfrac{n+1}{n} \end{aligned}$$

Two convergent sequences can, of course, represent the same real number. For example, the decimal expansion $\sqrt{2} = 1.41421\ldots$ means that $\sqrt{2}$ can be represented by the convergent sequence 1, 1.4, 1.41, 1.414, 1.4142, ... as well as by the sequence given above. The intuitive notion of real number suggests that if this is the case then an arbitrarily small interval about this 'real number' would contain both sequences. This can be stated in terms of sequences of rational numbers:

(1′) $\{q_n\} \equiv \{q_n'\}$ means that there exist arbitrarily short intervals which contain both $\{q_n\}$ and $\{q_n'\}$.

Written out in full the condition (1′) is: Given any rational number $\epsilon > 0$ there exist rational numbers $\bar{q}, \bar{\bar{q}}$ and natural numbers N, M such that $0 < \bar{\bar{q}} - \bar{q} \leq \epsilon$

and such that $\bar{q} \leq q_n \leq \bar{\bar{q}}$ whenever $n \geq N$, $\bar{q} \leq q'_n \leq \bar{\bar{q}}$ whenever $n \geq M$.

Two convergent sequences of rational numbers will be considered to be 'the same' or congruent* in the arithmetic of real numbers if the condition (1′) is satisfied. As before, this terminology is justified only if it is shown that

*The term 'equivalent' is more commonly used; 'congruent' is used here because it suggests a notion from the domain of arithmetic and because the word 'equivalent' has other uses.

$$\{q_n\} \equiv \{q_n\} \qquad \text{(reflexive)}$$

(5′)
$$\{q_n\} \equiv \{q'_n\} \text{ implies } \{q'_n\} \equiv \{q_n\} \qquad \text{(symmetric)}$$

$$\{q_n\} \equiv \{q'_n\} \text{ and } \{q'_n\} \equiv \{q''_n\} \text{ imply } \{q_n\} \equiv \{q''_n\} \qquad \text{(transitive).}$$

The first two statements are tautologies. To prove the third it suffices to note that an interval of length $\epsilon/2$ can be found containing both $\{q_n\}$ and $\{q'_n\}$ and another of length $\epsilon/2$ can be found containing both $\{q'_n\}$ and $\{q''_n\}$. These intervals must overlap because they have points of $\{q'_n\}$ in common, and hence together they give an interval of length $\leq \epsilon$ which contains both $\{q_n\}$ and $\{q''_n\}$.

The formula $e = \lim_{n\to\infty} \left(1 + \frac{1}{n}\right)^n$ *means* that the sequence $q'_n = \left(1 + \frac{1}{n}\right)^n$ is congruent to the sequence for e given above; this can be proved directly (see Exercise 8). The problem of computing π can be described as the problem of finding other sequences congruent to the above sequence which are computationally more practical (see §7.5).

The relation $\{q_n\} < \{q'_n\}$ is very naturally defined by:

(2′) $\{q_n\} < \{q'_n\}$ means that there exist rational numbers $\bar{q} < \bar{\bar{q}} < \bar{q}' < \bar{\bar{q}}'$, such that $\{q_n\}$ is contained in the interval $\{\bar{q} \leq x \leq \bar{\bar{q}}\}$ and $\{q'_n\}$ in the interval $\{\bar{q}' \leq x \leq \bar{\bar{q}}'\}$.

Written out in full the condition (2′) states that, in addition to $\bar{q} < \bar{\bar{q}} < \bar{q}' < \bar{q}''$, there exist natural numbers N, M such that $\bar{q} \leq q_n \leq \bar{\bar{q}}'$ for $n \geq M$ and $\bar{q}' \leq q'_n \leq \bar{\bar{q}}'$ for $n \geq M$. The fact that

(6′)
$$\{q_n\} \equiv \{p_n\},\ \{q'_n\} \equiv \{p'_n\},\ \{q_n\} < \{q'_n\} \text{ imply } \{p_n\} < \{p'_n\}$$

is proved as follows: Let $\bar{q} < \bar{\bar{q}} < \bar{q}' < \bar{\bar{q}}'$ be given for $\{q_n\}$, $\{q'_n\}$ as above. Choose ϵ less than $(\bar{q}' - \bar{\bar{q}})/2$ and find an interval of length less than ϵ which contains

$\{q_n\}$, $\{p_n\}$ and another of length less than ϵ which contains $\{q'_n\}$, $\{p'_n\}$. These intervals cannot overlap (because then together they would form an interval of length less than $\bar{q}' - \bar{\bar{q}}$ which would have to contain rational numbers $\leq \bar{q}'$ and rational numbers $\geq \bar{\bar{q}}$, which is impossible) and serve to prove that $\{p_n\} < \{p'_n\}$.

If $\{q_n\}$ and $\{q'_n\}$ are convergent sequences of rational numbers then $\{q_n + q'_n\}$ is a convergent sequence of rational numbers. This follows immediately from the fact that $\bar{q} \leq q_n \leq \bar{\bar{q}}$ and $\bar{q}' \leq q'_n \leq \bar{\bar{q}}'$ imply that $\bar{q} + \bar{q}' \leq q_n + q'_n \leq \bar{\bar{q}} + \bar{\bar{q}}'$; since $(\bar{\bar{q}} + \bar{\bar{q}}') - (\bar{q} + \bar{q}') = (\bar{\bar{q}} - \bar{q}) + (\bar{\bar{q}}' - \bar{q}')$, this shows that if $\{q_n\}$, $\{q'_n\}$ can be contained in intervals of length ϵ then $\{q_n + q'_n\}$ can be contained in an interval of length 2ϵ. Thus

$$(3') \qquad \{q_n\} + \{q'_n\} \equiv \{q_n + q'_n\}$$

defines an operation on convergent sequences of rational numbers. The above argument also shows that

$$(7') \qquad \{q_n\} \equiv \{p_n\} \quad \text{and} \quad \{q'_n\} \equiv \{p'_n\} \quad \text{imply } \{q_n + q'_n\} \equiv \{p_n + p'_n\}$$

and hence that the definition (3′) is consistent with the definition (1′) of 'sameness' in the arithmetic of real numbers.

Similarly, if $\{q_n\}$, $\{q'_n\}$ are convergent sequences of rational numbers so is $\{q_n q'_n\}$, and hence

$$(4') \qquad \{q_n\} \cdot \{q'_n\} \equiv \{q_n q'_n\}$$

defines an operation on convergent sequences of rational numbers. The fact that

$$(8') \qquad \{q_n\} \equiv \{p_n\} \quad \text{and} \quad \{q'_n\} \equiv \{p'_n\} \quad \text{imply } \{q_n q'_n\} \equiv \{p_n p'_n\}$$

shows that this can be considered as a multiplication operation in the arithmetic of real numbers. To prove these statements let $\bar{q}$, $\bar{q}'$, ϵ be such that the interval $\{|x - \bar{q}| \leq \epsilon\}$ contains $\{q_n\}$, $\{p_n\}$ and the interval $\{|x - \bar{q}'| < \epsilon\}$ contains $\{q'_n\}$, $\{p'_n\}$ (always in the sense of 'contain' defined above, that is, all but a finite number of points of the sequences lie in the stated intervals); then $|q_n q'_n - \bar{q}\bar{q}'| = |q_n q'_n - q_n \bar{q}' + q_n \bar{q}' - \bar{q}\bar{q}'| \leq |q_n|\,|q'_n - \bar{q}'| + |\bar{q}'|\,|q_n - \bar{q}| \leq (|\bar{q}| + \epsilon)\epsilon + |\bar{q}'|\epsilon$ for all sufficiently large n, hence $\{q_n q'_n\}$ and $\{p_n p'_n\}$ are contained in the same interval of length $2\epsilon(|\bar{q}| + |\bar{q}'| + \epsilon)$. Since $|\bar{q}|$, $|\bar{q}'|$

are bounded and ϵ can be made arbitrarily small this completes the proof.

The rules (1′), (2′), (3′), (4′), justified by (5′), (6′), (7′), (8′), define the arithmetic of real numbers. This arithmetic is also called the 'real number field' or the 'real number system'. It has the following properties: The trichotomy law "given two elements $\{q_n\}$, $\{q'_n\}$ of the arithmetic, exactly one of the conditions $\{q_n\} < \{q'_n\}$, $\{q_n\} \equiv \{q'_n\}$, $\{q_n\} > \{q'_n\}$ holds" is proved as follows: If $\{q_n\} \not\equiv \{q'_n\}$ then there must exist a rational number $\epsilon > 0$ such that no interval of length $\leq \epsilon$ contains both $\{q_n\}$ and $\{q'_n\}$. Since $\{q_n\}$, $\{q'_n\}$ are convergent, there are intervals of length $\epsilon/2$ containing each of them. These intervals cannot overlap, since otherwise the assumption on ϵ would be contradicted. They therefore serve to prove either $\{q_n\} < \{q'_n\}$ or $\{q_n\} > \{q'_n\}$.

If r_1, r_2, r_3 are convergent sequences of rational numbers, then the commutative, associative, and distributive laws $r_1 + r_2 \equiv r_2 + r_1$, $r_1r_2 \equiv r_2r_1$, $r_1 + (r_2 + r_3) \equiv (r_1 + r_2) + r_3$, $r_1(r_2r_3) \equiv (r_1r_2)r_3$, $r_1(r_2 + r_3) \equiv r_1r_2 + r_1r_3$ follow trivially from the analogous laws for the arithmetic of rational numbers.

Given a rational number q, let $\bar{q}$ denote the constant (and therefore convergent) sequence $\{q, q, q, \ldots\}$. Then $q_1 < q_2$, $q_1 = q_2$, $q_1 > q_2$ are equivalent to $\bar{q}_1 < \bar{q}_2$, $\bar{q}_1 \equiv \bar{q}_2$, $\bar{q}_1 > \bar{q}_2$ respectively; moreover $\bar{q}_1 + \bar{q}_2 \equiv \overline{q_1 + q_2}$ and $\bar{q}_1\bar{q}_2 \equiv \overline{q_1q_2}$, and hence all rules of the arithmetic of rational numbers apply to such sequences.

The sequence $\bar{0}$ satisfies $r + \bar{0} \equiv r$ for all convergent sequences of rational numbers r. If $r \equiv \{q_n\}$ is a convergent sequence of rational numbers then $-r$ can be defined to be $\{-q_n\}$ because this is a convergent sequence of rational numbers which satisfies $r + (-r) \equiv \bar{0}$. The sequence $\bar{1}$ satisfies $\bar{1} \cdot r \equiv r$ for all r. If $r \equiv \{q_n\}$ is convergent and $r \not\equiv \bar{0}$ then r^{-1} can be defined to be $\{q_n^{-1}\}$ because this sequence is convergent and satisfies $r \cdot r^{-1} \equiv \bar{1}$. Hence the equation $rx + r' \equiv r''$ for $r \not\equiv \bar{0}$ has the unique* solution $x \equiv (r'' + (-r'))r^{-1}$. The only statement here which requires any proof is that $\{q_n^{-1}\}$ is convergent. In fact, individual terms q_n may be zero, hence q_n^{-1} may not be defined for some n. However, by the trichotomy law there is a δ such that $|q_n| \geq \delta$ for all sufficiently large n. Then $|q_n^{-1} - q_m^{-1}| = \left|\dfrac{q_m - q_n}{q_mq_n}\right| \leq \delta^{-2}|q_n - q_m|$, hence $\{q_n^{-1}\}$ satisfies the Cauchy Criterion if $\{q_n\}$ does.

*Meaning, of course, that if x, x' are both solutions then $x \equiv x'$.

Given r_1, r_2, the relation $r_1 < r_2$ holds if and only if $r_2 - r_1 > \bar{0}$. This is proved simply by setting $r_1 \equiv \{q_n\}$, $r_2 \equiv \{q_n'\}$, choosing disjoint intervals containing these sequences, letting δ be the distance which separates these intervals and noting that then $q_n' - q_n \geq \delta$ for all sufficiently large n.

This reduces the ordering relation $r_1 < r_2$ to the notion of positivity $r > \bar{0}$. Since $r > \bar{0}$ if and only if $r > \bar{\delta}$ for some rational $\delta > 0$ it follows that '$r_1, r_2 > \bar{0}$ implies $r_1 + r_2 > \bar{0}$ and $r_1 r_2 > \bar{0}$'. Hence '$r_1 < r_2$, $r_2 < r_3$ imply $r_1 < r_3$' and '$r_1 < r_2$, $\bar{0} < r_3$ imply $r_1 r_3 < r_2 r_3$'. Finally $r > \bar{0}$ if and only if $-r < \bar{0}$.

The absolute value $|r|$ of r is defined to be r if $r > \bar{0}$, $\bar{0}$ if $r = \bar{0}$, $-r$ if $r < \bar{0}$. The proofs of $r \leq |r|$, $|r_1 r_2| = |r_1|\,|r_2|$, $|r_1 + r_2| \leq |r_1| + |r_2|$ are as before. For every r there is a rational K such that $|r| < \bar{K}$ (take K to be the upper limit of an interval containing r). The Archimedean laws 'given r_1, r_2 with $r_1 > 0$ there is a natural number n such that $nr_1 > |r_2|$' and 'given r_1, r_2 with $r_1 > 0$ and $|r_2| < 1$ there is a natural number n such that $|r_2^n| < r_1$', are also proved as before.

If $r \equiv \{q_n\}$ then the sequence $\bar{q}_n$ in the arithmetic of real numbers converges to the limit r; that is, given any $\epsilon > 0$ there is an N such that $|r - \bar{q}_n| < \bar{\epsilon}$ for all $n \geq N$. In other words, every convergent sequence of rational numbers has a limit in the arithmetic of real numbers. This is, of course, a triviality since the limit is just the sequence itself, i.e. the 'real number' it represents. Slightly less trivial is the fact that every convergent sequence of *real* numbers has a limit in the arithmetic of real numbers. This basic property of the real number system is called *completeness*.

Theorem

Completeness of the real number system. If $r_1, r_2, r_3, \ldots$ is an infinite sequence in the arithmetic of real numbers with the property that for every* $\epsilon > 0$ there is an N such that $|r_n - r_m| < \epsilon$ whenever $n, m \geq N$ then there is an element r_∞ in the arithmetic of real numbers with the property that for every $\epsilon > 0$ there is an N such that $|r_n - r_\infty| < \epsilon$ whenever $n \geq N$. In short, every convergent sequence of real numbers (sequence satisfying the Cauchy Criterion) has a limit which is a real number. The limit is, of course, unique.

Here ϵ is a real number. Since $\epsilon > 0$ implies $\epsilon > \bar{\epsilon}_1$ for some rational number ϵ_1 this actually involves no greater generality than assuming that ϵ itself is rational.

Proof

The main idea of the proof is to 'weed out' sequences satisfying the Cauchy Criterion so that they have the property that the nth term of the sequence represents the limit to n decimal places. More specifically, the given sequence $r_1, r_2, r_3, \ldots$ can be assumed at the outset to have the property

$$(9) \qquad |r_n - r_m| < \overline{10^{-n}} \qquad (m > n)$$

(r_n represents all succeeding terms r_m with an accuracy of n decimal places) because if it does not then it can be replaced by one which does. This is done as follows: By the Cauchy Criterion there is an N such that $|r_N - r_m| < \overline{(1/10)}$ for all $m > N$. Throw out all terms $r_1, r_2, \ldots, r_{N-1}$ of the given sequence and renumber the remaining terms $r_i = r_{i+(N-1)}$ so that the term which was r_N is now r_1. Next select a term r_M such that $|r_M - r_n| < \overline{(1/100)}$ for $n > M$, throw out all of the terms $r_2, r_3, \ldots, r_{M-1}$, and renumber so that r_M becomes r_2. Continuing this 'weeding' process, the given sequence can be reduced to one which satisfies (9) for all n.

Thus it can be assumed not only that (9) holds for the given sequence $r_1, r_2, r_3, \ldots$, but also that each r_i of the sequence is represented by a sequence $r_i = \{q_n^{(i)}\}$ of rational numbers $q_n^{(i)}$ satisfying

$$|q_n^{(i)} - q_m^{(i)}| < 10^{-n} \qquad (m > n).$$

It will be shown that then the 'diagonal sequence' $q_1^{(1)}, q_2^{(2)}, q_3^{(3)}, \ldots$ represents a real number with the desired property.

Note that if $r \equiv \{q_n\}$ is a convergent sequence and if $\epsilon > 0$ is a rational number then $|r| < \bar{\epsilon}$ if and only if the sequence $\{q_n\}$ is contained in a subinterval of $\{|x| \le \epsilon\}$; this is simply a restatement of the conditions $r - \bar{\epsilon} < \bar{0}$, $r + \bar{\epsilon} > \bar{0}$. Thus the assumption $|r_i - r_j| < \overline{10^{-i}}$ $(j > i)$ implies $|q_n^{(i)} - q_n^{(j)}| < 10^{-i}$ for all sufficiently large n. Together with the assumption on the sequences $\{q_n^{(i)}\}$ this gives

$$\begin{aligned}
&|q_i^{(i)} - q_j^{(j)}| \\
&\quad \le |q_i^{(i)} - q_n^{(i)}| + |q_n^{(i)} - q_n^{(j)}| + |q_n^{(j)} - q_j^{(j)}| \\
&\quad \le 10^{-i} + 10^{-i} + 10^{-j} \le 3 \cdot 10^{-i}
\end{aligned}$$

for $j > i$ (and for all sufficiently large n but, since the conclusion is independent of n, this need not be stated).

This proves that the sequence $\{q_i^{(i)}\}$ is convergent. Let r_∞ denote the real number it represents. Then $|r_i - r_\infty|$ is estimated by estimating

$$|q_n^{(i)} - q_n^{(n)}| \le |q_n^{(i)} - q_i^{(i)}| + |q_i^{(i)} - q_n^{(n)}|$$
$$\le 10^{-i} + 3 \cdot 10^{-i} = 4 \cdot 10^{-i}$$

for all $n > i$. Thus $|r_i - r_\infty|$ can be made arbitrarily small and r_∞ has the desired property. Finally, the uniqueness of r_∞ follows from the triangle inequality $|r_\infty - r'_\infty| \le |r_\infty - r_n| + |r_n - r'_\infty|$ which shows that if $|r_\infty - r_n|$ and $|r'_\infty - r_n|$ can both be made arbitrarily small then $|r_\infty - r'_\infty|$ is less than every positive number, which cannot be true if $|r_\infty - r'_\infty| \not\equiv 0$, i.e. cannot be true if $r_\infty \not\equiv r'_\infty$.

This completes the discussion of the properties of the arithmetic of real numbers. It is convenient now to drop the use of the bars and to let a rational number q be considered as an element of the extended arithmetic. Moreover, the symbol $\equiv$ will again be replaced by the ordinary = sign.

An 'object' in the arithmetic of real numbers is called a real number. Rules for representing real numbers have been defined and relations and operations have been defined for the *representations*—which is what is important—but no attempt has been made here to define *what* the representations represent. The concept of 'real number', like those of 'natural number' and 'rational number' remains undefined.

Exercises **1** Let q_1, q_2, q_3 be the rational numbers which satisfy the equations

$$8q_1 + 2 = 5$$
$$5q_2 + 7 = 1$$
$$7q_3 + 4 = 9.$$

Find $q_1 + q_2 + q_3$ and $q_1q_2q_3$, first using the ordinary notation of the arithmetic of fractions and then using the triad notation of the text.

2 Given a triad (a, b, c) of natural numbers representing q, find a triad representing q^{-1}. [Begin by writing q as $(a, b, b + d)$ or minus such a number.]

3 Two natural numbers a, b are said to be 'congruent modulo 6', written '$a \equiv b \pmod 6$,' if there exist natural numbers p, q such that $a + 6p = b + 6q$. Show that this relation of congruence ('sameness') is reflexive, symmetric and transitive. Show that it is consistent with addition and multiplication of natural numbers, and that it therefore defines an arithmetic in which there are only six distinct 'objects'. Show that *subtraction* is possible in this arithmetic (an equation of the form $a + x = b$ has a unique solution x for all a, b) even though it is not possible in the arithmetic of natural numbers. What is zero in this arithmetic? Show that division by non-zero elements in this arithmetic is *not* possible (i.e. an equation $ax = b$ with $a \not\equiv 0$ need not have a unique solution x). For which elements a of this arithmetic does the equation $x^2 \equiv a$ have a solution x? (Give a list.)

4 'Arithmetic modulo n' (n a natural number) is the arithmetic defined by the congruence relation $a \equiv b \pmod n$, which means $a + pn = b + qn$ for some natural numbers p, q. All statements in Exercise 3 for the case $n = 6$ are true for arbitrary n, except for the impossibility of division. For which natural numbers n is division by non-zero elements possible in an arithmetic modulo n? For a given n, which natural numbers a are 'invertible mod n', i.e. have inverses b such that $ab \equiv 1 \pmod n$? [Find these answers experimentally.]

5 Picturesquely speaking, two natural numbers are congruent modulo n if and only if it is possible to go from one to the other in steps of size n. Show that a is invertible mod n if and only if it is possible to go from any natural number to any other natural number by a combination of steps of size n and steps of size a. Given a, n let S be the set of all natural numbers which can be reached from 1 by a combination of steps of size n and steps of size a. Show that there must be a natural number d such that S is the set of all natural numbers of the form $1 + pd$ ($p = 0, 1, 2, 3, \ldots$) and that d is the greatest common factor of a and n. Apply this to prove a theorem stating the conclusion of Exercise 4.

6 The proof of Exercise 5 is 'non-constructive' in that it makes appeal to 'the set of all numbers which can be reached in steps of size a and n'. It is made constructive by the following process known as the *Euclidean Algorithm*: Given a, n with $a < n$, then either n is a multiple of a or there are natural numbers p, b such that $n = pa + b$ where $b < a$. Then either a is a multiple of b or $a = qb + c$ where $c < b$. In this way one can construct a decreasing sequence $n > a > b > c > \cdots$ which, of course, eventually terminates. From $n = pa + b$ it follows that any number which can be reached in steps of size a, b can also be reached in steps of size n, a; moreover the answer is given by the constructive formula $b = n - pa$ telling how to take steps of size b if steps of size n, a are allowed. Similarly any number which can be reached in steps of size b, c can be reached in steps of size a, b (and hence in steps of size n, a) and an explicit formula can be given for the answer. Thus to take a step of size 1 when steps of size 49 and 32 are allowed, one first constructs the sequence $49 > 32 > 17 > 15 > 2 > 1$ and learns how to take steps of size 17, 15, 2, 1 successively. Use this method to solve the equation $49p = 1 + 32q$ (p, q natural numbers). In the same way solve $48p = 1 + 31q$ and $63p = 1 + 40q$.

$49 = 1 \cdot 32 + 17$
$32 = 1 \cdot 17 + 15$
$17 = 1 \cdot 15 + 2$
$15 = 7 \cdot 2 + 1$
$2 = 2 \cdot 1$

7 Is the fraction $\frac{1953}{5115}$ in lowest terms?

8 Prove that $\lim_{n\to\infty}\left(1 + \frac{1}{n}\right)^n = e$ where e is the real number represented by the sequence $q_n = 1 + 1 + \frac{1}{2} + \cdots + \frac{1}{n!}$. [Expand $p_n = (1 + 1/n)^n$ by the binomial theorem and show that $p_n < q_m$ for $m \geq n$. Then show that for every n, ϵ there is an N such that $p_m > q_n - \epsilon$ whenever $m \geq N$. Conclude that any interval which contains the sequence $\{q_n\}$ contains the sequence $\{p_n\}$ (in the sense of 'contains' defined in the text), hence that $\{p_n\}$ is a convergent sequence congruent to $\{q_n\}$.] Find $\lim_{n\to\infty}\left(1 + \frac{1}{n}\right)^n$ to three decimal places.

9 Prove the alternating series test: If $a_1 - a_2 + a_3 - a_4 + \cdots$ is an alternating series in which $a_n \geq a_{n+1} \geq 0$, and if $\lim_{n\to\infty} a_n = 0$, then the sequence of partial sums $s_n = \sum_{k=1}^{n} (-1)^{k-1}a_k$ is convergent.

10 Prove that an increasing sequence $q_{n+1} \geq q_n$ which is not convergent is not bounded (for every K there is an N such that $q_n > K$) hence that a bounded increasing sequence is convergent.

11 *Archimedean law.* Show that if a, b are natural numbers then there is a natural number n such that $na > b$. [This is easy.] Show that the same is true for positive rational numbers a, b. [Use a common denominator.] Show that if a, b are positive rational numbers then $(1 + a)^n > b$ for n sufficiently large. Conclude that if $|q_1| < 1$ and $q_2 > 0$ then $|q_1|^n < q_2$ for n sufficiently large.

12 *Continued fractions.* The answer to Exercise 7 can be formulated as follows: $\frac{1953}{5115}$ can be written as a fraction in which the numerator is 1 and the denominator is $2 + \frac{1209}{1953}$. Then $\frac{1209}{1953}$ can be written as a fraction in which the numerator is 1 and the denominator is $1 + \frac{744}{1209}$. Then $\frac{744}{1209}$ can be

$$\frac{1953}{5115} = \cfrac{1}{2+\cfrac{1209}{1953}} = \cfrac{1}{2+\cfrac{1}{1+\cfrac{1}{1+\cfrac{1}{1+\cfrac{1}{1+\cfrac{1}{1+\cfrac{1}{2}}}}}}}$$

*For an introduction to this theory see C. D. Olds, Continued Fractions, Random House, 1963.

treated similarly and the process repeated until the denominator is a whole number. This gives $\frac{1953}{5115}$ as a fraction whose denominator is a fraction whose denominator is a fraction whose denominator is a fraction . . . , the process eventually terminating. This is the representation of $\frac{1953}{5115}$ as a *continued fraction.* The theory of continued fractions has many profound and beautiful applications—for example Lambert's proof that π is not a rational number.* One of the principal difficulties in the theory of continued fractions, albeit a trivial one, is the notational awkwardness of writing fractions within fractions within fractions, *ad infinitum.* This is obviated by using 2×2 matrices instead of fractions. For example, the computations of the Euclidean algorithm can be written

$$\begin{pmatrix}1953\\5115\end{pmatrix} = \begin{pmatrix}0 & 1\\1 & 2\end{pmatrix}\begin{pmatrix}1209\\1953\end{pmatrix},$$

$$\begin{pmatrix}1209\\1953\end{pmatrix} = \begin{pmatrix}0 & 1\\1 & 1\end{pmatrix}\begin{pmatrix}744\\1209\end{pmatrix}, \ldots,$$

$$\begin{pmatrix}186\\279\end{pmatrix} = \begin{pmatrix}0 & 1\\1 & 1\end{pmatrix}\begin{pmatrix}93\\186\end{pmatrix},$$

$$\begin{pmatrix}93\\186\end{pmatrix} = \begin{pmatrix}0 & 1\\1 & 2\end{pmatrix}\begin{pmatrix}0\\93\end{pmatrix}$$

hence all together

$$\begin{pmatrix}1953\\5115\end{pmatrix} = \begin{pmatrix}0 & 1\\1 & 2\end{pmatrix}\begin{pmatrix}0 & 1\\1 & 1\end{pmatrix}\begin{pmatrix}0 & 1\\1 & 1\end{pmatrix}\begin{pmatrix}0 & 1\\1 & 1\end{pmatrix}\begin{pmatrix}0 & 1\\1 & 1\end{pmatrix}\begin{pmatrix}0 & 1\\1 & 1\end{pmatrix}\begin{pmatrix}0 & 1\\1 & 2\end{pmatrix}\begin{pmatrix}0\\93\end{pmatrix}.$$

When the matrix product is computed

$$\begin{pmatrix}1953\\5115\end{pmatrix} = \begin{pmatrix}8 & 21\\21 & 55\end{pmatrix}\begin{pmatrix}0\\93\end{pmatrix}$$

this gives the answer to Exercise 7. Note that the determinant of the product is $(-1)^7 = 8 \cdot 55 - 21^2$, which implies that 55 and 21 have no common factors, i.e. $\frac{21}{55}$ is in lowest terms. Use this method to reduce the fraction $\frac{1037}{2379}$ to lowest terms. Write it as a continued fraction.

13 *Continued fraction expansion of a ratio of real numbers.* Given two positive real numbers $r_0 > r_1 > 0$ one can define a sequence $r_0 > r_1 > r_2 > r_3 > \cdots$ of positive real numbers by the Euclidean algorithm

$$\begin{aligned} r_{n-1} &= a_n r_n + r_{n+1} \\ a_n &= \text{natural number} \\ 0 &\le r_{n+1} < r_n. \end{aligned}$$

If $r_{n+1} = 0$ for some n the sequence $r_0 > r_1 > r_2 > \cdots$ *terminates,* and otherwise continues indefinitely. Define integers p_n, q_n by the formula

$$\begin{pmatrix}0 & 1\\ 1 & a_1\end{pmatrix}\begin{pmatrix}0 & 1\\ 1 & a_2\end{pmatrix}\cdots\begin{pmatrix}0 & 1\\ 1 & a_n\end{pmatrix} = \begin{pmatrix}p_{n-1} & p_n\\ q_{n-1} & q_n\end{pmatrix}.$$

This gives two different definitions of p_n, q_n but there is no conflict because the equation

$$\begin{pmatrix}p_n & p_{n+1}\\ q_n & q_{n+1}\end{pmatrix} = \begin{pmatrix}p_{n-1} & p_n\\ q_{n-1} & q_n\end{pmatrix}\begin{pmatrix}0 & 1\\ 1 & a_{n+1}\end{pmatrix}$$

shows that the two definitions coincide. Show that p_n/q_n is a fraction in lowest terms for all n, that if $r_{n+1} = 0$ then $r_1/r_0 = p_n/q_n$, and that if $r_{n+1} > 0$ for all n then r_1/r_0 is an irrational number and the fractions p_n/q_n converge to r_1/r_0 as $n \to \infty$. The convergence in fact follows the pattern

$$0 = \frac{p_0}{q_0} < \frac{p_2}{q_2} < \frac{p_4}{q_4} < \cdots < \frac{r_1}{r_0} < \cdots$$

$$< \frac{p_5}{q_5} < \frac{p_3}{q_3} < \frac{p_1}{q_1} = \frac{1}{a_1}$$

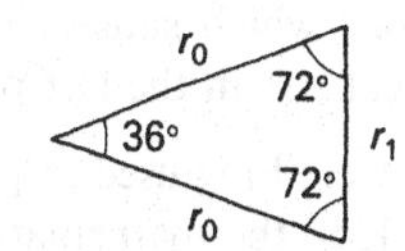

and the error is estimated by

$$\frac{p_n}{q_n} - \frac{p_{n-1}}{q_{n-1}} = (-1)^{n-1}\frac{1}{q_n q_{n-1}}.$$

[It is useful to show that if r, p, q, R, P, Q are positive numbers with $p/q \neq P/Q$ then $(rp + RP)/(rq + RQ)$ lies (strictly) between p/q and P/Q.]

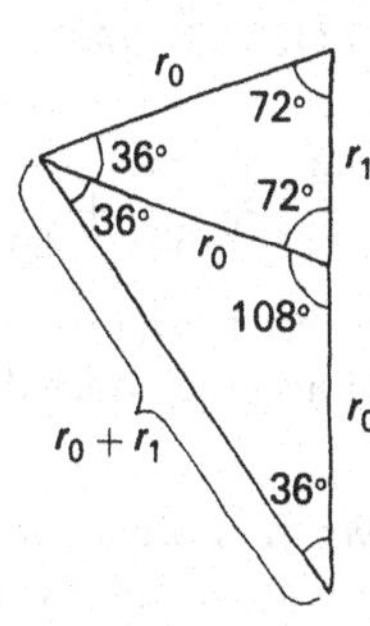

$r_1 : r_0 = r_0 : r_0 + r_1$

$$x = \frac{1}{1+x} = \frac{1}{1+\frac{1}{1+x}} = \frac{1}{1+\frac{1}{1+\frac{1}{1+}}}$$

14 The integer ratios p_n/q_n are called the *convergents* of the ratio of real numbers $0 < r_1/r_0 < 1$. Show that the convergents of r_1/r_0 are the 'best' rational approximations to r_1/r_0 in the sense that if p/q is any rational number for which

$$\left|\frac{r_1}{r_0} - \frac{p}{q}\right| < \left|\frac{r_1}{r_0} - \frac{p_n}{q_n}\right|,$$

where p_n/q_n is a convergent, then $q > q_n$. In short, a better approximation than p_n/q_n must have a larger denominator.

15 An important role in the theory of continued fractions is played by the ratio r_1/r_0 for which $a_1 = 1$, $a_2 = 1$, $a_3 = 1, \ldots$. Show that the convergents are $\frac{0}{1}, \frac{1}{1}, \frac{1}{2}, \frac{2}{3}, \frac{3}{5}, \frac{5}{8}, \frac{8}{13}, \frac{13}{21}, \ldots$ and that the ratio r_1/r_0 is a solution of the equation $x^2 + x = 1$ hence

$$\frac{r_1}{r_0} = \frac{\sqrt{5}-1}{2} = 2\sin 18^\circ$$

$$= \text{chord subtended by an angle of } \frac{\pi}{5}.^*$$

This is the 'golden section' of the Greeks; see H. S. M. Coxeter, Introduction to Geometry, John Wiley and Sons, 1961, Chapter 11, or L. Zippin, Uses of Infinity, Random House, 1963, Chapter 5.

Find this number as a decimal fraction accurate to three decimal places.

16 The ancient trigonometric tables of Ptolemy expressed fractions in sexagesimal form rather than decimal form. For example, Ptolemy gives the value 31′25″ for the length of the chord which subtends an angle of $\frac{1}{2}°$ in a circle of radius 60, i.e.

$$120 \sin\left(\frac{1}{4}\right)^\circ = \frac{31}{60} + \frac{25}{60^2}$$

or

$$2 \sin \frac{\pi}{720} = \frac{31}{60^2} + \frac{25}{60^3}.$$

Using an accurate value of π, verify that this value is correct (as Ptolemy proved that it was) to the nearest 60^{-3}.*

*For a lucid explanation of Ptolemy's techniques, see A. Aaboe, Episodes from the Early History of Mathematics, Random House, 1964, pp. 101–125.

17 A real number a (for approximate) is said to represent another real number t (for true) 'with an accuracy of n decimal places' if $|t - a| < 10^{-n}$. (See Exercise 2, §2.2.) Show that:

(a) Every real number can be represented with an accuracy of n decimal places by an n-place decimal fraction.
(b) Two representations which satisfy the condition of (a) differ by at most ± 1 in the last place.

18 Sexagesimal fractions were long used in place of decimal fractions. For example, in 1250 the mathematician Leonardo Pisano found the real root of $x^3 + 2x^2 + 10x = 20$ to be

$$x = 1^p 22'7''42'''33^{iv}4^v40^{vi}$$

i.e.

$$x = 1 + \frac{22}{60} + \frac{7}{60^2} + \frac{42}{60^3} + \frac{33}{60^4} + \frac{4}{60^5} + \frac{40}{60^6}.†$$

†See O. Toeplitz, The Calculus, Univ. of Chicago Press, 1963, p. 16. This book contains a wealth of information on the history of calculus.

Assuming this answer is correct to the nearest $(60)^{-6}$, with how many decimal places of accuracy does this represent the root of the equation?

19 *Infinite decimal fractions.* It is often convenient to think of real numbers as being represented by infinite decimal fractions, e.g. $\frac{1}{3} = .3333\ldots$, $\frac{2}{3} = .6666\ldots$, $\pi = 3.141592\ldots$, etc. Such an expression is said to represent the real number r if the sequence of rational numbers (decimal fractions) obtained by *truncating* the infinite decimal (taking the first n places without rounding) represents the real number r in the sense defined in the text.

(a) Show that every real number r can be represented by an infinite decimal fraction. [Assume $r > 0$ and consider the largest n-place decimal which is less than r.]
(b) Show that if two infinite decimals represent the same real number then either they are identical or the real number they represent is a (finite) decimal fraction and one terminates with all 0's and the other terminates with all 9's, e.g. $3.6000\ldots = 3.5999\ldots$.

9.2

Real Functions of Real Variables

The arithmetic operations of addition and multiplication can be regarded as functions $R^2 \to R$, that is, as rules by which an element of R is determined (the sum or product) when an element of R^2 (a pair of real numbers) is given.* This section is devoted to various types of functions $R^m \to R^n$ which arise in practice.

*Here R denotes the set of real numbers.

The simplest functions are those which can be described by formulas. Of these the simplest are those which can be described by formulas which involve only addition and multiplication, that is, by polynomial formulas. For example the formula $w = xy + yz + zx$ describes a function $R^3 \to R$ which assigns numbers (w) to triples of numbers (x, y, z). Similarly

$$x = uv$$
$$y = u^2 + v^2$$

is a function $R^2 \to R^2$ (see §5.1, for a discussion of this particular function, its singularities, etc.). More generally, a *polynomial function* f: $R^n \to R^m$ is one which can be described by m formulas

$$(1) \qquad y_i = f_i(x_1, x_2, \ldots, x_n) \qquad (i = 1, 2, \ldots, m)$$

in which the right-hand sides are polynomials in $(x_1, x_2, \ldots, x_n)$, that is, sums of multiples of products of the x_i's

$$(2) \qquad f_i(x_1, x_2, \ldots, x_n) = \sum A x_1^{p_1} x_2^{p_2} \ldots x_n^{p_n}$$

where $\sum$ indicates that there is a sum of a finite number of terms of the form $A x_1^{p_1} x_2^{p_2} \ldots x_n^{p_n}$. Here the p_i's are integers ≥ 0 so that $x_1^{p_1} x_2^{p_2} \ldots x_n^{p_n}$ is simply the result of several multiplications. The numbers A are called the *coefficients* of the polynomial, and a polynomial function f: $R^n \to R^m$ is said to have *integer coefficients* if the A's which occur in the formulas (1), (2) describing it are integers. A polynomial function f: $R^n \to R^m$ is said to have *rational coefficients* if this is true of the formulas which define it.

An *algebraic function* is one which is defined *implicitly* by polynomial functions, that is, a function $g_i(y_1, \ldots, y_r, x_{r+1}, \ldots, x_n)$ or $h_i(y_1, \ldots, y_r)$ obtained by applying the Implicit Function Theorem to a polynomial function $y_i = f_i(x_1, x_2, \ldots, x_n)$ at a point $\bar{y}_i = f_i(\bar{x}_1, \bar{x}_2, \ldots, \bar{x}_n)$ which is not a singularity. An algebraic function is by its very nature defined locally near a specified point.

The most familiar example of an algebraic function is $y = \sqrt{x}$, which is defined locally near $x = 1$ by saying that it is the inverse of $x = y^2$ near $(\bar{x}, \bar{y}) = (1, 1)$. Near $(\bar{x}, \bar{y}) = (1, -1)$ the inverse of $x = y^2$ is the algebraic function $y = -\sqrt{x}$. Similarly, near $(\bar{x}, \bar{y}) = (a^2, a)$ the inverse of $x = y^2$ is $y = \pm\sqrt{x}$ with the sign chosen to agree with the sign of a; if $a = 0$ the given function $x = y^2$ has a singularity at the given point and there is no inverse function. It was proved in §7.5 that the function $\sqrt{x}$ is in fact defined for all positive x; this is a special global property of this particular algebraic function.

A very important example of an algebraic function is the function obtained by inverting the relation

$$x = u^2 - v^2$$
$$y = 2uv.$$

(Writing $z = x + iy$ and $w = u + iv$ this relation is $z = w^2$, so the problem of inverting the relation is the problem of finding the 'complex square root.') This mapping $(u, v) \rightarrow (x, y)$ is non-singular of rank 2 except at $(u, v) = (0, 0)$. Hence near any point $\bar{x} = \bar{u}^2 - \bar{v}^2$, $\bar{y} = 2\bar{u}\bar{v}$ it defines (u, v) as functions of (x, y) provided $(\bar{u}, \bar{v}) \neq (0, 0)$. The important point which this example illustrates is that it is not possible to extend this algebraic function unambiguously to all values of (x, y) even if singularities are avoided. (This was possible for $y = \sqrt{x}$, which extended to all values of x which could be reached from $\bar{x} = 1$ without crossing the singularity at $x = 0$.) This can be seen from a geometrical examination of the given relation, which is left to the reader (Exercise 4); suffice it to say that each point $(x, y) \neq (0, 0)$ is the image of exactly two points (u, v), $(-u, -v)$ and that there is no way to choose one of these two for all $(x, y) \neq (0, 0)$ at once without having discontinuities. This shows that the 'localness' of algebraic functions is essential and cannot be avoided merely by avoiding singularities.

(More generally, if $f(w)$ is a polynomial function of one complex variable $w = u + iv$ then $x + iy = z = f(w) = f(u + iv)$ gives a map $(u, v) \rightarrow (x, y)$ of $R^2 \rightarrow R^2$. This map $(u, v) \rightarrow (x, y)$ is easily shown to be non-singular of rank 2 except at the finite number of points (u, v) where the derivative $f'(w) = f'(u + iv)$ is zero, e.g. where $2w$ is zero in the example $f(w) = w^2$ above. Thus, locally near any point (u, v), except a finite number, the relation $z = f(w)$ can be solved to give (u, v) as algebraic functions of (x, y). However, for each value of (x, y) there are n possible values for (u, v), where n is the degree of the polynomial f. In extending one of the local functions $(x, y) \to (u, v)$ one is led to the notions of *analytic continuation*, *Riemann surface*, and *n-sheeted covering of the xy-plane* which are fundamental to the more profound non-local study of algebraic functions.)

An important sub-class of the class of algebraic functions is the class of *rational functions*. A rational function is a function which can be expressed by a formula which is a quotient of polynomials, e.g.

$$f(x, y) = \frac{2xy}{x^2 - y^2}.$$

Such a function is defined at all points where the denominator is not zero. Rational functions arise in applying the Implicit Function Theorem to a polynomial in several variables which contains one variable to the first degree only. For example, the solution of

$$y^2z - x^2z + 2xy = u$$

for z as a function of (u, x, y) is

$$z = \frac{u - 2xy}{y^2 - x^2}$$

which for $u = 0$ is the function above. The zeros of the denominator are the singularities of the map, where such a solution $z = g(u, x, y)$ is impossible.

A function $f\colon R^n \to R^m$ which is not algebraic is said to be *transcendental*. The most important transcendental functions are the functions $R \to R$ given by the formulas $x = e^t$, $x = \cos t$, $x = \sin t$. These functions can be defined in a variety of ways, for example as the solutions of differential equations: Picard's method (§7.4) shows that the differential equation

$$\frac{dx}{dt} = x, \quad x(0) = 1$$

has a unique solution $x = x(t)$ defined for all values of t and that this function can be expanded as a power series

$$x(t) = 1 + t + \frac{t^2}{2} + \frac{t^3}{3!} + \cdots.$$

This function is by definition $x(t) = e^t$. Similarly, the differential equation

$$\frac{dx}{dt} = -y \qquad x(0) = 1$$
$$\frac{dy}{dt} = x \qquad y(0) = 0$$

has a unique solution $(x, y) = (x(t), y(t))$ defined for all t and these solutions can be expanded as power series

$$x(t) = 1 - \frac{t^2}{2} + \frac{t^4}{4!} - \frac{t^6}{6!} + \cdots$$
$$y(t) = t - \frac{t^3}{3!} + \frac{t^5}{5!} - \frac{t^7}{7!} + \cdots.$$

These functions are by definition $(x(t), y(t)) = (\cos t, \sin t)$.

A function which can be expressed explicitly or implicitly in terms of algebraic operations and the functions e^t, $\cos t$, $\sin t$ is called an *elementary function*. The elementary functions include

algebraic functions
$\log x$ (the inverse of $x = e^t$ defined for $x > 0$)
$\tan x$ $(=\sin x/\cos x)$
$\sinh x$ $(=\frac{1}{2}(e^x - e^{-x}))$
Arcsin x (the inverse of $x = \sin t$ at $(\bar{x}, \bar{t}) = (0, 0)$ defined for $-1 \leq x \leq 1$)
x^y $(=e^{y \log x}$ defined for $x > 0$ and all $y)$
etc.

and all functions $R^n \to R^m$ which can be expressed in terms of such functions.

Functions other than elementary functions which are frequently used include: *Bessel functions* and *elliptic functions* which are defined by differential equations which arise in mathematical physics; the *Γ-function*, which is discussed in Exercise 10 of §9.6; the *ζ-function*, which is important in the application of calculus to number theory (analytic number theory) and which is defined as the sum of the series

$$\zeta(x) = 1 + \frac{1}{2^x} + \frac{1}{3^x} + \cdots + \frac{1}{n^x} + \cdots$$

for $x > 1$; and *hypergeometric functions*, which can be defined by differential equations, and which are also used in physics.

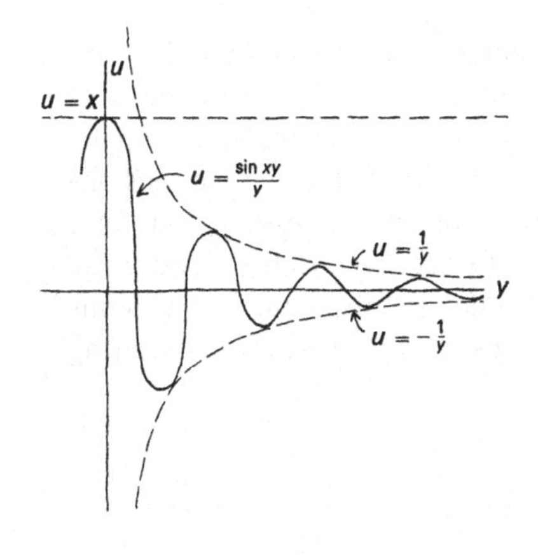

Of course the notion of a function $R^n \to R^m$ is a very general one and *any* rule assigning points in R^m to (some or all) points of R^n is included. An interesting example is the function

$$f(x) = \int_{-\infty}^{\infty} \frac{\sin xy}{y}\, dy. \tag{2}$$

For each fixed x the integrand on the right is a continuous function of y for all y except $y = 0$, where it is not defined; however, the fact that $\frac{d}{dt} \sin t = \cos t$ implies that $\lim_{y\to 0} \frac{\sin xy}{y} = x \lim_{h\to 0} \frac{\sin h}{h} = x \cos 0 = x$, hence if the integrand is given the value x at 0 it is a continuous function for all y. As $|y|$ becomes large, the value of the integrand oscillates between $\frac{1}{|y|}$ and $-\frac{1}{|y|}$, from which it is easily shown (an alternating series converges if its terms decrease to zero) that $\lim_{K\to\infty} \int_{-K}^{K} \frac{\sin xy}{y}\, dy$ exists. For each fixed x, then, the formula (2) defines a number $f(x)$. What is interesting about the function (2) is that it can be described in a very different way as

$$f(x) = \begin{cases} \pi & \text{if } x > 0 \\ 0 & \text{if } x = 0 \\ -\pi & \text{if } x < 0. \end{cases} \tag{3}$$

The fact that (2) and (3) define the same function is by no means obvious; it is proved in §9.6. Historically, this example and others like it were influential in the development of the modern concept of an *arbitrary* function which is not assumed to be expressible explicitly or implicitly by a single formula such as (2); generally speaking, even very 'artificial' functions such as (3) can often be expressed by formulas such as (2) but the description (3) is more manageable in practice than the formula.

The concept of an 'arbitrary' function is, however, far too broad. This was dramatically demonstrated by an example of Weierstrass* of a function $f: R \to R$ which is defined and continuous for all x but not differentiable for any x (i.e. $\lim_{h\to 0} f(x + h) = f(x)$ but

$$\lim_{h\to 0} \frac{f(x + h) - f(x)}{h}$$

does not exist for any x).

See Exercise 4, §9.6. Actually the first such example was discovered by Bolzano, but the Weierstrass example is more famous.

The theorems of calculus—for example the Implicit Function Theorem or Stokes' Theorem—do not apply to such 'arbitrary' functions, and in fact cannot even be *stated* without the assumption that the functions involved are at least differentiable. It is necessary, therefore, to develop a vocabulary of *conditions*, such as continuity, differentiability, etc., in order to formulate the theorems of calculus. Such conditions are discussed in the following section.

Exercises

1 What functions $R^n \to R^m$ can be expressed in terms of addition and multiplication *alone*?

2 Once the function e^x has been defined by its differential equation, the function $y = \log x$ can be defined as the inverse of $x = e^y$ $(x > 0)$. Show that the function so defined satisfies

$$\log x = \int_1^x \frac{1}{t}\,dt \qquad (x > 0)$$

and

$$\log x = \lim_{n\to\infty} n(\sqrt[n]{x} - 1) \qquad (x > 0).$$

Either of these relations can be used as an alternative definition of $\log x$ for $x > 0$.

3 Once the functions $(\cos x, \sin x)$ have been defined by their differential equation, the function $y = \text{Arcsin}\, x$ can be defined as the solution of $x = \sin y$, $-1 < x < 1$, $-\frac{\pi}{2} < y < \frac{\pi}{2}$. Write $\text{Arcsin}\, x$ as a definite integral. Similarly, give a domain of definition of $\text{Arctan}\, x$ and write it as a definite integral.

4 To prove that the function $x = u^2 - v^2$, $y = 2uv$ has the property stated in the text—i.e. that it cannot be inverted globally for $(x, y) \neq (0, 0)$—prove that the integral of $d\theta = (x\,dy - y\,dx)/(x^2 + y^2)$ over any differentiable closed curve is a multiple of 2π. Show that the pullback of $d\theta$ under $(u, v) \to (x, y)$ is $2\,d\theta$. Conclude that if there were an inverse $(x, y) \to (u, v)$ then the pullback of $d\theta$ would have to be $\frac{1}{2}\,d\theta$ and there would be a curve over which the integral of $d\theta$ was π.

5 Let a_2, a_1, a_0 be real numbers. Set $z = x + iy$, $w = u + iv$ and express the map $z = a_2w^2 + a_1w + a_0$ as an explicit function $R^2 \to R^2$ in terms of the coordinates (u, v), (x, y) and the constants a_2, a_1, a_0. Show that the singularities of this map occur at the points (u, v) such that $2a_2w + a_1 = 0$. Prove that the same is true if $a_n = b_n + ic_n$ $(n = 0, 1, 2)$. Using the arithmetic of complex numbers, prove that the singularities of $z = a_nw^n + a_{n-1}w^{n-1} + \cdots + a_0$ are at the points w where $na_nw^{n-1} + (n-1)a_{n-1}w^{n-2} + \cdots + a_1 = 0$.

6 Let $f: R \to R$ be defined by the formula

$$f(x) = \int_{-\infty}^{\infty} e^{-t^2} \cos xt\, dt = \lim_{K\to\infty} \int_{-K}^{K} e^{-t^2} \cos xt\, dt.$$

Sketch the integrand for various values of x and draw whatever conclusions you can about the function $f(x)$. An explicit formula for $f(x)$ is derived in §9.6. [It is not difficult to see for which values of x this function $f(x)$ is defined, for which x it is a maximum, a symmetry of the function, and its approximate value for very large values of x.]

9.3 *Uniform Continuity and Differentiability*

The condition of 'uniform continuity' of a function on a set occurs very naturally in many contexts, for example in the definition of the transcendental function 10^x: For natural numbers a, the product of 10 with itself a times is denoted 10^a. For *rational* numbers q, say for the solution q of $aq + b = c$ where a, b, c are natural numbers, the number 10^q is defined by the equation $(10^q)^a \cdot 10^b = 10^{aq+b} = 10^c$, that is, 10^q is the positive number y such that $y^a = 10^c/10^b$. As was shown in §7.5, this defines y uniquely. If (a', b', c') is another description of q then the solution of $10^by^a = 10^c$ also solves $10^{b'}y^{a'} = 10^{c'}$, hence 10^q is well-defined. The extension of 10^x to *irrational* values of x is 'by continuity' which means that for a sequence of rational numbers x_n converging to x, one defines 10^x to be $\lim 10^{x_n}$. It is this 'extension by continuity' which requires the notion of uniform continuity.

Intuitively, the definition $10^{\lim x_n} = \lim 10^{x_n}$ means that, for example, 10^π is approximated by 10^{x_n} where x_n is a rational number approximating π (e.g. by the seventh root of 10^{22}) and that any number of decimal places of 10^π can be found by taking x_n to be sufficiently near to π. As a practical matter, the actual computation of 10^{x_n}, where x_n is a rational number with a large denominator, involves sophisticated numerical techniques, logarithms, etc. What is at issue, however, is not the practicality of finding 10^π but the question of whether $10^\pi = \lim 10^{x_n}$ is a valid *definition* of 10^π. In order to prove that it is, one must show that if $\lim x_n = \pi$ then 10^{x_n} is a convergent sequence, and that if $\lim x_n' = \pi$ then $\lim 10^{x_n} = \lim 10^{x_n'}$. This essentially reduces to the question "If x, x' are rational numbers such that $|x - x'|$ is small, how large can $|10^x - 10^{x'}|$ be?" The estimates needed to define 10^π are provided by the following observations: Let x, x' be rational numbers in the interval $\{3 \leq x \leq 4\}$ and let $\delta = |x - x'|$. Suppose $x' > x$ so that $x' = x + \delta$. Then $0 \leq 10^{x'} - 10^x = 10^x(10^\delta - 1) \leq 10^4(10^\delta - 1)$. Given any margin for error ϵ, choose a natural number N so large that $10^{1/N} - 1 < \epsilon \cdot 10^{-4}$, i.e. $10 < (1 + \epsilon \cdot 10^{-4})^N$. For example, since $(1 + \epsilon \cdot 10^{-4})^N > 1 + N \cdot \epsilon \cdot 10^{-4}$, it suffices to take $N > 9 \cdot 10^4 \cdot \epsilon^{-1}$. Then $|x' - x| < 1/N$ implies $|10^{x'} - 10^x| \leq 10^4(10^\delta - 1) \leq 10^4 \cdot (10^{1/N} - 1) < \epsilon$, hence 10^x is determined to within ϵ when x is determined to within $1/N$. Thus to determine 10^π to within ϵ it suffices to find 10^x for x a rational number such that $|x - \pi| < 1/2N$. Any other $10^{x'}$ found by the same method will differ by less than ϵ from this one. Thus 10^π can be found to any prescribed degree of accuracy and 10^π is therefore a well-defined real number.

Letting X denote the set of rational numbers in the interval $\{3 \leq x \leq 4\}$, these observations are a special case of the following theorem:

Theorem

Let f: $R^n \to R^m$ be a function which is defined at all points of a subset X of R^n. The function f is said to be *uniformly continuous on* X if it is true that for every $\epsilon > 0$ there is a $\delta > 0$ such that $|f(x) - f(x')| < \epsilon$ whenever x, x' are points of X satisfying $|x - x'| < \delta$ (where, as usual, $|x - x'| < \delta$ means that each of the n coordinates of x differs by less than δ from the corresponding coordinate of x' and similarly for $|f(x) - f(x')| < \epsilon$). The *closure* of X, denoted $\overline{X}$, is defined to be the set of all points of R^n which can be written as limits of sequences of points of X. If f is uniformly continuous on X then the formula $f(\lim x_n) = \lim f(x_n)$ defines a function on the closure $\overline{X}$ of X which is uniformly continuous on $\overline{X}$ and which is the only uniformly continuous function on $\overline{X}$ which agrees with f on X.

All that is required to prove this theorem is a simple unravelling of the definitions of 'uniform continuity', 'limit', and 'real number', which will be omitted.

The definitions of the arithmetic operations $+$, $-$, $\times$, $\div$ for real numbers are special cases of this theorem, based on the fact that these operations are uniformly continuous functions on bounded intervals of *rational* numbers (excluding intervals which contain 0 in the case of $\div$). In the same way, any polynomial function $f: R^n \to R^m$ is uniformly continuous on any bounded subset X of R^n.

Loosely speaking, to say that a function is uniformly continuous on a set X means that the value is determined to an arbitrary degree of accuracy once the argument is determined with sufficient accuracy. From this point of view it is clear that any function which describes the result of a physical measurement must be uniformly continuous unless its value is in some sense indeterminate.

The notion of uniform continuity gives a very simple definition of the notion of 'differentiable', namely the following:

Definition

Let X be a subset of R^n and let $f: R^n \to R^m$ be a function which is defined at all points of R^n which are within δ of X; that is, a number $\delta > 0$ is given such that the domain of f includes all points y of R^n for which there is a point x of X satisfying $|x - y| \leq \delta$. Then

$$(1) \qquad M_{s,x}(h) = \frac{f(x + sh) - f(x)}{s}$$

is defined for all x in X, h an n-tuple satisfying $|h| \leq 1$, and s a real number satisfying $|s| \leq \delta$, $s \neq 0$. The function f is said to be *uniformly differentiable* on X if $M_{s,x}(h)$ is uniformly continuous on this set of (s, x, h) where it is defined. When this is the case $M_{s,x}(h)$ has (by the theorem above) a uniformly continuous extension to a set which includes the set $\{s = 0, x \text{ in } X, |h| \leq 1\}$. The function so defined is called the *derivative* of f at x evaluated on h. It is of the special form $(h_1, h_2, \ldots, h_n) \to \left(\sum_{j=1}^n a_{1j}h_j, \sum_{j=1}^n a_{2j}h_j, \ldots, \sum_{j=1}^n a_{mj}h_j\right)$ where the a_{ij} are uniformly continuous functions of $(x_1, x_2, \ldots, x_n)$. This formula is abbreviated by naming the coordinates on R^m, say

$y_1, y_2, \ldots, y_m$, and by writing it

$$dy_i = \sum_{j=1}^{n} \frac{\partial y_i}{\partial x_j} dx_j \qquad (i = 1, 2, \ldots, m)$$

where $\frac{\partial y_i}{\partial x_j}$ denotes a_{ij}. In the absence of names for the coordinates on $\mathbf{R}^m$ the functions a_{ij} are also denoted $a_{ij} = \frac{\partial f_i}{\partial x_j}$.

Intuitively, to say that f is uniformly differentiable means that the value of (1), which can be imagined as 'f at x under a microscope of power $1:s$', can be determined to within any prescribed ϵ by making s sufficiently small and by determining x, h with sufficient accuracy (see §5.3).

If f is a polynomial function and X is a bounded subset of R^n then f is uniformly differentiable on X for the simple reason that the function (1) is again a polynomial function (the terms in the numerator which contain no s cancel out so that the s in the denominator can be cancelled). The proof of the Implicit Function Theorem can be strengthened to show that whenever the given function f is uniformly differentiable, the solution functions g, h are also. Hence all algebraic functions are (locally) uniformly differentiable. Similarly, since the functions e^x, $\sin x$, $\cos x$ are (locally) uniformly differentiable (see Exercises 7 and 8) it follows that all elementary functions are (locally) uniformly differentiable.

As the terminology suggests, the notions of uniform continuity and uniform differentiability of a function on a set X are stronger forms of the notions of continuity and differentiability defined in the earlier chapters. For purposes of comparison these definitions are:

Let $f\colon R^n \to R^m$ be a function defined at all points of a subset X of R^n. The function f is said to be *continuous on* X if for every $\epsilon > 0$ and every x in X there is a $\delta > 0$ such that $|f(x) - f(x')| < \epsilon$ for all points x' of X satisfying $|x - x'| < \delta$. (The difference from *uniform* continuity is that δ may depend on the given x as well as on ϵ.) The function f is said to be *differentiable* on X if for every $x = (x_1, x_2, \ldots, x_n)$ in X

$$\frac{f_i(x_1, \ldots, x_j + s, \ldots, x_n) - f_i(x_1, \ldots, x_n)}{s}$$

$$(i = 1, 2, \ldots, m; \quad j = 1, 2, \ldots, n)$$

is defined for all sufficiently small $s \neq 0$, and if the limit as $s \to 0$ exists and is a continuous function on X for all mn values $i = 1, 2, \ldots, m; j = 1, 2, \ldots, n$.

Exercises

1 Prove directly that a polynomial function $f: R^n \to R$ is uniformly continuous on the set $\{|x| \leq K\}$ for any fixed K. [First show it is bounded, i.e. $|f(x)| < K'$ for some K'. Then consider $f(x + h) - f(x)$.]

2 Prove directly that a polynomial function $f: R^n \to R$ is uniformly differentiable on the set $\{|x| \leq K\}$ for any fixed K.

3 Give an explicit upper bound for the amount by which $4 \sum_{j=1}^{n} \left(\frac{j}{n}\right)^3 \cdot \frac{1}{n}$ differs from 1. [Do not use the formula for $\sum j^3$.] More generally, give an explicit upper bound for the amount by which an approximating sum

$$\sum(\alpha) = 4 \sum_j (x_j)^3 \, \Delta x_j$$

to $\int_0^1 4x^3 \, dx$ differs from 1 [where α denotes a subdivision of the interval $\{0 \leq x \leq 1\}$ into small subintervals indexed by j, together with a choice of a point x_j in the jth interval, and where Δx_j is the length of the jth interval]. Conclude that $\int_0^1 4x^3 \, dx$ converges and is equal to 1.

4 Generalizing 3, show that if $f(x)$ is uniformly differentiable on $\{0 \leq x \leq 1\}$ then $\int_0^1 f'(x) \, dx$ converges and is equal to $f(1) - f(0)$; that is, prove directly that the Fundamental Theorem of Calculus holds for uniformly differentiable functions.

5 Prove that $f(x) = x^{-1}$ is uniformly continuous (and uniformly differentiable) on any set $\{x > a\}$ for $a > 0$, but not for $a = 0$.

6 Prove that e^x (defined by its differential equation) is uniformly differentiable on any set $\{|x| \leq K\}$. [Use the formula for e^{x+h}.]

7 Prove that $\sin x$ and $\cos x$ are uniformly differentiable on the entire line $\{-\infty < x < \infty\}$. [Similar to 6.]

8 Prove that the function $e^{x \log 10}$ is identical with the function 10^x defined in the text, and hence that 10^x is uniformly differentiable on any interval $\{|x| \leq K\}$, with derivative $\log 10 \cdot 10^x$.

9 Prove that the function $f(x) = (\sin x)/x$ is uniformly differentiable on the entire line. [Rewrite $f(x)$ as $\int_0^1 \cos xt \, dt$. This extends f to $x = 0$.]

10 Prove that the function $f(x) = |x|^p$ is uniformly continuous on any interval $\{|x| \leq K\}$ for $p > 0$.

11 Let $f(x)$ be the function which is $x^2 \sin\left(\frac{1}{x^2}\right)$ when $x \neq 0$ and which is 0 when $x = 0$. Show that $\lim_{h\to 0} \frac{f(x+h) - f(x)}{h}$ exists for all x but that f is not differentiable.

12 Prove directly that if $f(x)$ is uniformly differentiable on an interval $\{a \leq x \leq b\}$ and if there is a $\delta > 0$ such that $f'(x) \geq \delta$ at all points x in the interval, then f does not assume a maximum value inside the interval.

13 Prove directly that if $f(x, y)$ is uniformly differentiable on the square $\{|x| \leq 1, |y| \leq 1\}$ and if there is a $\delta > 0$ such that at all points (x, y) of the square either $\left|\frac{\partial f}{\partial x}\right| \geq \delta$ or $\left|\frac{\partial f}{\partial y}\right| \geq \delta$ then f does not assume a maximum value inside the square.

14 Prove directly that if $A(x_1, x_2, \ldots, x_n)$ is uniformly continuous on an n-dimensional rectangle R in R^n then the integral $\int_R A\, dx_1\, dx_2 \ldots dx_n$ converges. [Using *uniform* continuity the statement $U(S) \to 0$ as $|S| \to 0$ of §2.3 is easily proved.]

15 Let X be the subset of R which consists of all numbers of the form $1/n$ where n is a positive integer. Prove that a function $f: R \to R$ defined at all points of X is uniformly continuous on X if and only if the sequence $f(1/n)$ is convergent.

16 *Differentiation under the integral sign.* Prove that if $f(x, y)$ is uniformly differentiable on $\{a \leq x \leq b, c \leq y \leq d\}$ then the function $F(y) = \int_a^b f(x, y)\, dx$ is uniformly differentiable on $\{c \leq y \leq d\}$ and its derivative is $F'(y) = \int_a^b \frac{\partial f}{\partial y}(x, y)\, dx$. [This is a very easy estimate.]

9.4

Compactness

A subset X of R^n is said to be *compact* if it is *closed* and *bounded*, that is, if every point which can be written as a limit of points of X is itself in X (closed) and if there is a real number K such that all coordinates of all points of X are less than K in absolute value (bounded).

Typically a compact set is a bounded set in R^n defined by a finite number of equations and inequalities involving $\geq$, for example the disk $x^2 + y^2 \leq 2$, the sphere $x^2 + y^2 + z^2 = 1$, the cube $\{|x| \leq 1, |y| \leq 1, |z| \leq 1\}$,

*Here and in the remainder of this section $|x - x'|$ denotes, for x, x' in R^n, the maximum absolute value of the difference of corresponding coordinates, i.e. if $x = (x_1, x_2, \ldots, x_n)$ and $x' = (x'_1, x'_2, \ldots, x'_n)$ then $|x - x'| = \max(|x_1 - x'_1|, |x_2 - x'_2|, \ldots, |x_n - x'_n|)$.

etc. Typically a non-compact set is a set which is not bounded (e.g. $x^2 + y^2 \geq 2$) or a set which is defined by inequalities involving $>$, for example $x^2 + y^2 < 2$ or $\{|x| < 1, |y| < 1, |z| < 1\}$. Such a set is not compact (in general) because a limit of points where an inequality $>$ is satisfied may not satisfy the inequality; for example, a point on the circle $x^2 + y^2 = 2$ can be written as a limit of points inside the disk $x^2 + y^2 < 2$. Of course when one speaks of subsets of R^n one has in mind very general sorts of sets, of which these subsets defined by equalities and inequalities are very simple special cases.

The usefulness of the notion of compactness is illustrated by the following four theorems.

Theorem 1

A continuous function on a compact set assumes a maximum value.

Let X be a compact subset of R^n and let $f: R^n \to R$ be a function which is defined and continuous at all points of X (for every x in X and for every $\epsilon > 0$ there is a $\delta > 0$ such that $|f(x) - f(x')| < \epsilon$ whenever x' is a point of X such that* $|x - x'| < \delta$). Then there is a point x of X such that $f(x) \geq f(x')$ for all x' in X.

Theorem 2

A continuous function on a compact set is uniformly continuous.

Let X be a compact subset of R^n and let $f: R^n \to R^m$ be a function which is defined and continuous at all points of X (for every x in X and for every $\epsilon > 0$ there is a $\delta > 0$ such that* $|f(x) - f(x')| < \epsilon$ whenever x' is a point of X such that $|x - x'| < \delta$). Then f is uniformly continuous on X (for every $\epsilon > 0$ there is a $\delta > 0$ such that $|f(x) - f(x')| < \epsilon$ whenever x, x' are points of X such that $|x - x'| < \delta$).

Bolzano-Weierstrass Theorem

A subset X of R^n is compact if and only if every infinite sequence of points of X has a point of accumulation in X.

If X is a compact subset of R^n and if $x^{(1)}, x^{(2)}, x^{(3)}, \ldots$ is an infinite sequence of points of X then there is a point $x^{(\infty)}$ of X with the property that for every $\epsilon > 0$ an infinite number of points $x^{(j)}$ of the sequence satisfy $|x^{(j)} - x^{(\infty)}| < \epsilon$. Conversely, if X is *not* compact then there is an infinite sequence $x^{(1)}, x^{(2)}, x^{(3)}, \ldots$ in X for which there is no such point $x^{(\infty)}$ in X.

Heine-Borel Theorem

Every (interior) covering of a compact set has a finite subcover.

Let X be a compact subset of R^n and let $\{U_\alpha\}$ be an infinite collection of subsets of R^n which 'cover' X in the sense that for every x in X there is a member U_α of the collection which contains x in its *interior*; that is, there is a $\delta > 0$ such that all points x' of X satisfying $|x' - x| < \delta$ lie in U_α. Then it is possible to select a *finite number* of the sets U_α in such a way that they 'cover' X in the same way.

The similarity of these theorems is less evident in their statements than it is in their proofs. In all four proofs it is useful to consider the subdivision of R^n by the planes

$$x_j = (i \text{ place decimal fraction}) \qquad (j = 1, 2, \ldots, n)$$

into cubes 10^{-i} on a side. If X is compact then for each i only a finite number of these cubes contain points of X (because X is bounded). In all four proofs one uses this observation to choose a nested sequence of cubes C_i, i.e. $C_0 \supseteq C_1 \supseteq C_2 \supseteq C_3 \supseteq \cdots$, such that C_i is 10^{-i} on a side and such that the points common to X and C_i have some desired property. By the completeness of the real number system, the sequence of centers of the cubes C_i has a limit, say $x^{(\infty)}$, which can also be described as the unique point of R^n which is contained in all cubes C_i of the nested sequence. For any $\delta > 0$ the cube $\{|x - x^{(\infty)}| < \delta\}$ contains all the cubes C_i for i sufficiently large; in particular, any such cube contains a point of X. Hence $x^{(\infty)}$ can be written as a limit of points of X and therefore $x^{(\infty)}$ is in X (because X is closed). The point $x^{(\infty)}$ is then shown to have the desired property.

Proof of the Bolzano-Weierstrass Theorem

At least one of the cubes of side 1 in the subdivision of R^n by planes $\{x_j = \text{integer}\}$ must contain an infinite number of points of the given sequence $x^{(1)}, x^{(2)}, x^{(3)}, \ldots$. Select such a cube and call it C_0. At least one of the 10^n cubes into which C_0 is divided by the planes $\{x_j = \text{integer}/10\}$ must contain an infinite number of points of the given sequence. Select such a cube and call it C_1. Continuing in this manner gives a nested sequence $C_0 \supseteq C_1 \supseteq C_2 \supseteq \cdots$ of cubes such that C_i has side 10^{-i} and contains an infinite number of points of the given sequence. Let $x^{(\infty)}$ be the unique point common to all the cubes C_i. Then $x^{(\infty)}$ is in X; and, given $\epsilon > 0$, the cube $\{|x - x^{(\infty)}| < \epsilon\}$ contains all the C_i beyond a certain point (take i so large that $10^{-i} < \epsilon$) hence the cube $\{|x - x^{(\infty)}| < \epsilon\}$ contains an infinite number of the points $x^{(j)}$ as desired. Conversely, if X is not compact then either X is not bounded, in which case there is a sequence $x^{(j)}$ of points in X such that $|x^{(j)}| \geq j$ or X is not closed, in which case there is a sequence $x^{(j)}$ of points of X converging to a point not in X, hence in either case there is a sequence $x^{(j)}$ of points in X for which there is no $x^{(\infty)}$ in X.

Proof of the Heine-Borel Theorem

If X cannot be covered by a finite number of the sets U_α then some cube C_0 of side 1 must have the property that the points of X which lie in C_0 cannot be covered by a finite number of the U_α. Subdividing C_0 into 10^n parts gives a cube C_1 of side 10^{-1} with the same property. Continuing this process *ad infinitum* gives a nested sequence of cubes with this property. Let $x^{(\infty)}$ be the point common to all the cubes C_i. By assumption there is a U_α which contains a cube of the form $\{|x - x^{(\infty)}| < \delta\}$, hence a U_α containing all of C_i for some i. This shows that the points of X in C_i can be covered by a *single* set U_α, contrary to assumption, and this contradiction shows that a finite number of the sets U_α must suffice to cover all of X.

Proof of Theorem 2

If f is not uniformly continuous then there is an ϵ for which no δ suffices; that is, there is an ϵ for which it is possible to find $x^{(j)}$, $\bar{x}^{(j)}$ in X such that $|x^{(j)} - \bar{x}^{(j)}| < 10^{-j}$ but such that $|f(x^{(j)}) - f(\bar{x}^{(j)})| \geq \epsilon$. One can then choose a nested sequence $C_0 \supseteq C_1 \supseteq C_2 \supseteq \cdots$ such that each C_i contains an infinite number of the points $x^{(j)}$ selected above. Let $x^{(\infty)}$ be the unique point common to the C_i. Then given $\delta > 0$ there is a j such that the cube $\{|x - x^{(\infty)}| < \delta\}$ contains both $x^{(j)}$, $\bar{x}^{(j)}$. Hence either $|f(x^{(\infty)}) - f(x^{(j)})| \geq \epsilon/2$ or $|f(x^{(\infty)}) - f(\bar{x}^{(j)})| \geq \epsilon/2$, which shows that f is not continuous at $x^{(\infty)}$. Therefore if f is continuous at all points of X it must be uniformly continuous.

Proof of Theorem 1

By Theorem 2 there is an i such that $|f(x) - f(x')| < 1$ whenever $|x - x'| < 10^{-i}$. Divide R^n into cubes of side 10^{-i} and evaluate f at one point of X in each cube which contains a point of X. This gives a finite number of values of f. Let K be the largest of these values. Then f assumes no value greater than $K + 1$ on X, i.e. f is bounded on

X. For each $j = 1, 2, 3, \ldots$, let q_j be the largest j place decimal fraction such that f assumes values greater than or equal to q_j on X and let $x^{(j)}$ be a point of X such that $f(x^{(j)}) \geq q_j$. The sequence q_j is contained in the interval $\{q_N \leq x \leq q_N + 10^{-N}\}$ for every N, hence converges to a real number, say r. Since f assumes no value on X greater than r (because such a value would be greater than $q_j + 10^{-j}$ for some j) it suffices to show that f assumes the value r. Let $C_0 \supseteq C_1 \supseteq C_2 \supseteq \cdots$ be a nested sequence of cubes such that C_i has side 10^{-i} and contains an infinite number of the points $x^{(j)}$, and let $x^{(\infty)}$ be the point common to all C_i. On any cube $\{|x - x^{(\infty)}| < \delta\}$ the function f assumes values (at the $x^{(j)}$) arbitrarily near r which, by the continuity of f, implies that $f(x^{(\infty)}) = r$, as was to be shown.

The proofs of many theorems about differentiable functions can be simplified using the following theorem:

Theorem 3

A (continuously) differentiable function on a compact set is uniformly differentiable. More precisely, a function $f: R^n \to R^m$ which is defined and continuously differentiable at all points within δ of a compact subset X of R^n (for some $\delta > 0$) is uniformly differentiable on X.

Proof

By assumption the function

$$M_{s,x}(h) = \frac{f(x + sh) - f(x)}{s}$$

is defined for x in X, s a real number $0 < |s| \leq \delta$, and h an n-tuple $|h| \leq 1$. On the basis of the assumption that the n functions $\frac{\partial f}{\partial x_i}: R^n \to R^m$ defined by

$$\frac{\partial f}{\partial x_i}(x) = \lim_{s \to 0} M_{s,x}(0, 0, \ldots, 0, 1, 0, \ldots, 0)$$

(1 in the ith place)

exist and are continuous at all points x within δ of X, it is to be shown that $M_{s,x}(h)$ is uniformly continuous on the set $\{0 < |s| \leq \delta, x \text{ in } X, |h| \leq 1\}$. The main step in the proof is, as in §5.3, to use the Fundamental Theorem of Calculus to write $M_{s,x}(h)$ as a sum of integrals of partial derivatives of f over n line segments parallel to the coordinate axes

$$(1)\quad M_{s,x}(h) = \sum_{i=1}^{n} \int_0^{h_i} \frac{\partial f}{\partial x_i}(x_1 + sh_1, \ldots, x_{i-1} + sh_{i-1}, x_i + sy, x_{i+1}, \ldots x_n)\, dy.$$

To say that $M_{s,x}(h)$ is uniformly continuous on the set $\{0 < |s| \leq \delta,\ x \text{ in } X,\ |h| \leq 1\}$ means that $|M_{s,x}(h) - M_{s',x'}(h')|$ can be made arbitrarily small (in all m coordinates) for (s, x, h) and (s', x', h') in this set by making $|s - s'|$, $|x - x'|$, and $|h - h'|$ sufficiently small. As usual, one makes the change from $(s, x_1, x_2, \ldots, x_n, h_1, h_2, \ldots, h_n)$ to $(s', x_1', x_2', \ldots, x_n', h_1', h_2', \ldots, h_n')$ one coordinate at a time and shows that for each such change the change in the value of $M_{s,x}(h)$ is small; then by the triangle inequality the total change from $M_{s,x}(h)$ to $M_{s',x'}(h')$ is small.

Specifically, if s is changed to s' then the integrands in (1) are changed from the values of $\frac{\partial f}{\partial x_i}$ at $(x_1 + sh_1, \ldots, x_n)$ to their values at another point $(x_1 + s'h_1, \ldots, x_n)$ which is a distance of at most $|(s - s')h_i| \leq |s - s'|$ away in any coordinate direction. The functions $\frac{\partial f}{\partial x_i}$ are continuous on the set of all points within δ of X and this set is compact (see Exercise 3) hence by Theorem 2 the change in the integrands can be made uniformly small by making $|s - s'|$ small. Since the change in the integrals is at most the change in the integrands times the length of the intervals of integration it follows that the change in (1) resulting from changing s to s' can be made arbitrarily small by making $|s - s'|$ sufficiently small. By the same argument, the change in (1) resulting from changing x_i to x_i' can be made arbitrarily small by making $|x_i - x_i'|$ sufficiently small. Finally, if h_i is changed to h_i' the first $i - 1$ integrals in (1) are not changed at all, and the last $n - i - 1$ are changed only slightly (by the same argument). In the ith integral, the upper limit of integration is changed from h_i to h_i' which changes its value by at most $|h_i - h_i'|$ times the largest value of the integrand $\frac{\partial f}{\partial x_i}$; since the integrand is bounded (Theorem 1) this change can be made arbitrarily small by making $|h_i - h_i'|$ small. This completes the proof of the theorem.

The use of the word 'compact' in the phrase 'compact,

oriented, differentiable, k-dimensional manifold-with-boundary' in Chapter 6 is easily seen to agree with its definition in this section. On the one hand, any set S in R^n which can be described by a finite number of charts as in Chapter 6 is easily seen to be compact (use the Bolzano-Weierstrass Theorem). On the other hand, any k-dimensional differentiable manifold in R^n (see §5.5) which is compact can be described by a finite number of charts (the Heine-Borel Theorem). If a k-dimensional 'manifold-with-boundary' is defined to be a set which can be parameterized locally by a k-dimensional rectangle and if an 'oriented, differentiable k-dimensional manifold' is defined to be a manifold for which a non-zero k-form is specified, then the separate meanings of all the terms in the phrase 'compact, oriented, differentiable, k-dimensional manifold-with-boundary' become clear.

The proofs of this section resemble very strongly the proof of the Fundamental Theorem of Calculus (§3.1) and the proof of the theorem which defines $\int_S \omega$ (§6.3) in that all these proofs involve a nested set of cubes or rectangles and an argument to a contradiction based on the fact that there is a limit point $x^{(\infty)}$ contained in them all. In fact, Theorems 2 and 3 can be used in the proofs of the Fundamental Theorem and of the definition of $\int_S \omega$ to conclude that all continuous functions are uniformly continuous and all differentiable functions are uniformly differentiable on the compact domain of integration, after which these theorems are easily proved (see Exercises 5 and 15 of §9.3) without recourse to the subdivision arguments given in Chapter 3 and Chapter 6.

Exercises

1 Let $C_0 \supseteq C_1 \supseteq C_2 \supseteq \cdots$ be a nested set of cubes in R^n, and let C_i have side 10^{-i}. On the basis of the completeness of the real number system, show that there is exactly one point $x^{(\infty)}$ contained in all the C_i.

2 Using the Heine-Borel Theorem, show that given a nested set of cubes $C_0 \supseteq C_1 \supseteq C_2 \supseteq \cdots$, there is at least one point which lies in them all. [Use the compactness of C_0.]

3 Prove that if X is a compact set in R^n and if X_δ is the set of all points which lie within δ of X, then X_δ is compact.

4 Show that if $f: R^n \to R^m$ is a continuous function and if X is a compact set in R^n then its image $f(X) = \{$all points of the form $f(x)$ where x is in $X\}$ is a compact set in R^m. [Use Bolzano-Weierstrass.]

5 *Differentiation under the integral sign.* Prove that if $f(x, y)$ is differentiable on $\{a \le x \le b, c \le y \le d\}$ then $F(y) = \int_a^b f(x, y)\,dx$ is differentiable on $\{c \le y \le d\}$ and its derivative is $\int_a^b \frac{\partial f}{\partial y}(x, y)\,dx$. [See Exercise 16, §9.3.]

9.5 Other Types of Limits

The limits discussed in the preceding chapters are limits of sequences, derivatives of functions, and integrals of forms over compact manifolds. For purposes of comparison the definitions of these limits will be reviewed before giving the analogous definitions for infinite series, infinite products, and improper integrals:

A sequence $x^{(1)}, x^{(2)}, x^{(3)}, \ldots$ of points in R^n is said to be *convergent* if for every $\epsilon > 0$ there is a natural number N such that* $|x^{(n)} - x^{(m)}| < \epsilon$ whenever $n, m \ge N$. When this is the case, the sequence determines a unique point $x^{(\infty)}$ in R^n, called the *limit* of the sequence, with the property that for every $\epsilon > 0$ there is a natural number N such that $|x^{(\infty)} - x^{(n)}| < \epsilon$ whenever $n \ge N$.

*Here and in the remainder of this section $|x - x'|$ denotes, for x, x' in R^n, the maximum absolute value of the difference of corresponding coordinates, i.e. if $x = (x_1, x_2, \ldots, x_n)$ and $x' = (x_1', x_2', \ldots, x_n')$ then $|x - x'| = \max(|x_1 - x_1'|, |x_2 - x_2'|, \ldots, |x_n - x_n'|)$.

A function $f: R^n \to R^m$ is said to be *uniformly differentiable* on a subset X of R^n if there is a $\delta > 0$ such that the function

$$M_{s,x}(h) = \frac{f(x + sh) - f(x)}{s}$$

is defined and uniformly continuous for all (s, x, h) such that s is a real number $0 < |s| \le \delta$, x is an n-tuple in X, and h is an n-tuple $|h| \le 1$. When this is the case, the formula

$$\begin{aligned} M_{0,x}(h) &= \lim_{s \to 0} M_{s,x}(h) \\ &= \lim_{s \to 0} \frac{f(x + sh) - f(x)}{s} \end{aligned}$$

defines an m-tuple $M_{0,x}(h)$ for all x in X and all n-tuples h. The function $M_{0,x}(h)$ is called the *derivative* of f.

The integral $\int_R A\,dx_1\,dx_2\ldots dx_k$ of a k-form over a k-dimensional rectangle is said to be *convergent* if for every $\epsilon > 0$ there is a $\delta > 0$ such that $|\sum(\alpha) - \sum(\alpha')| < \epsilon$ whenever $\sum(\alpha)$, $\sum(\alpha')$ are approximating sums to $\int_R A\,dx_1\,dx_2\ldots dx_k$ in which the mesh sizes $|\alpha|$, $|\alpha'|$ are less than δ. (See Chapter 2 for the definition of $\sum(\alpha)$.) When this is the case, there is a unique real number, denoted $\int_R A\,dx_1\,dx_2\ldots dx_k$ and called the *integral* of $A\,dx_1\,dx_2\ldots dx_k$ over R, with the property that for every $\epsilon > 0$ there is a $\delta > 0$ such that $|\int_R A\,dx_1\,dx_2\ldots dx_k - \sum(\alpha)| < \epsilon$ whenever $\sum(\alpha)$ is an approximating sum to $\int_R A\,dx_1\,dx_2\ldots dx_k$ in which the mesh size $|\alpha|$ is less than δ. The integral of a continuous k-form over a compact, oriented, differentiable, k-dimensional manifold-with-boundary was defined in §6.4 as a sum of a finite number of convergent integrals over rectangles.

The essence of the idea is that of a *process* for determining a number; the process is *convergent* if the number which results can be determined to within any margin for error $\epsilon > 0$ by carrying the process out with a sufficient (finite) degree of accuracy. When this is the case the process determines a real number which is called its *limit*. Processes other than those described above are infinite series, infinite products and improper integrals.

Infinite Series

In everyday speech the words 'series' and 'sequence' are more or less synonymous, but in mathematical terminology they are very sharply distinguished. A series is a sum, and a sequence is merely an infinite list (of numbers, points, functions, etc.).

An infinite series* $a_1 + a_2 + a_3 + \cdots$ in which the a_i are real numbers is said to be *convergent* if for every $\epsilon > 0$ there is a natural number N such that the sum of the first n terms differs by less than ϵ from the sum of the first m terms whenever $n, m \geq N$. Since $(a_1 + a_2 + \cdots + a_n) - (a_1 + a_2 + \cdots + a_m) = a_{m+1} + a_{m+2} + \cdots + a_n$ or $-a_{n+1} - a_{n+2} - \cdots - a_m$ depending on whether $n < n$ or $m > n$, this is the same as saying that $a_1 + a_2 + a_3 + \cdots$ is convergent if for every $\epsilon > 0$ there is an N such that $|a_{m+1} + a_{m+2} + \cdots + a_n| < \epsilon$ whenever $n > m \geq N$. When this is the case there is a unique real number, called the *sum* of the series and denoted $\sum_{i=1}^{\infty} a_i$, with the property that for every $\epsilon > 0$ there is an N such that $\left|\sum_{i=1}^{\infty} a_i - \sum_{i=1}^{n} a_i\right| < \epsilon$ whenever $n \geq N$ $\left(\text{where } \sum_{i=1}^{n} a_i \text{ denotes } a_1 + a_2 + \cdots + a_n\right)$.

Infinite Products

If none of the factors a_i are zero this is the same as saying that the product of the first n factors divided by the product of the first m factors differs from 1 by less than ϵ.

†*This statement requires proof. See Exercise 10.*

An infinite product $a_1a_2a_3\ldots$ in which the a_i are real numbers is said to be *convergent* if for every $\epsilon > 0$ there is an N such that* $|a_{m+1}a_{m+2}\ldots a_n - 1| < \epsilon$ whenever $n > m \geq N$. When this is the case there is† a unique real number, called the *product* of the a_i and denoted $\prod_{i=1}^{\infty} a_i$, with the property that for every $\epsilon > 0$ there is an N such that $\left|\prod_{i=1}^{\infty} a_i - \prod_{i=1}^{n} a_i\right| < \epsilon$ whenever $n \geq N$ $\left(\text{where } \prod_{i=1}^{n} a_i \text{ denotes } a_1a_2\ldots a_n\right)$.

Improper Integrals‡

‡*An integral is improper if the domain of integration is not compact.*

§*Here, once again, rectangles are used merely because they are the most simple compact domains. See §9.6 (p. 410) for a discussion of the possibility of using more general domains than rectangles.*

The integral $\int_{-\infty}^{\infty}\int_{-\infty}^{\infty} A(x, y)\,dx\,dy$ of a 2-form over the entire xy-plane is said to *converge* if the integrals of $A(x, y)\,dx\,dy$ over finite rectangles§ R converge and if for every $\epsilon > 0$ there is a number $K > 0$ such that $\int_R A(x, y)\,dx\,dy$ differs by less than ϵ from $\int_{R'} A(x, y)\,dx\,dy$ whenever R, R' are rectangles which contain the square $\{|x| \leq K, |y| \leq K\}$. When this is the case there is a unique real number, denoted $\int_{-\infty}^{\infty}\int_{-\infty}^{\infty} A(x, y)\,dx\,dy$, with the property that for every $\epsilon > 0$ there is a K such that $\left|\int_{-\infty}^{\infty}\int_{-\infty}^{\infty} A(x, y)\,dx\,dy - \int_R A(x, y)\,dx\,dy\right| < \epsilon$ whenever R is a rectangle containing the square $\{|x| \leq K, |y| \leq K\}$. Other improper integrals are defined analogously.

Examples and Applications

The usual tests for convergence of series are easy consequences of the above definition of convergence. For example the *ratio test* states that if $\lim_{n\to 0}\left|\frac{a_{n+1}}{a_n}\right|$ exists and is less than one then the series $a_1 + a_2 + a_3 + \cdots$ converges; this is proved by observing that

$$|a_{n+1}| \leq \rho|a_n| \qquad (\rho < 1)$$

for all sufficiently large n hence

$$\begin{aligned}
|a_{m+1} + a_{m+2} + \cdots + a_n| &\\
&\leq |a_{m+1}| + |a_{m+2}| + \cdots + |a_n| \\
&\leq \rho^{m-N+1}|a_N| + \cdots + \rho^{n-N}|a_N| \\
&= (\rho^{m-N} - \rho^{n-N})\frac{\rho}{1-\rho}|a_N| \leq \frac{\rho}{1-\rho}|a_N|.
\end{aligned}$$

Since $|a_N| \to 0$ this proves that the series converges. This test, applied to the series $x^{(1)} + (x^{(2)} - x^{(1)}) + (x^{(3)} - x^{(2)}) + \cdots$, was used repeatedly in Chapter 7 to prove, for example, that the series

$$e^x = 1 + x + \frac{x^2}{2!} + \frac{x^3}{3!} + \cdots$$

$$\sin x = x - \frac{x^3}{3!} + \frac{x^5}{5!} - \cdots$$

generated by Picard's iteration are convergent.

A series $a_1 + a_2 + a_3 + \cdots$ in which all terms are non-negative, $a_i \geq 0$, is either convergent or the 'partial sums' $\sum_{i=1}^{n} a_i$ can be made arbitrarily large, that is, given any real number K there is a natural number n such that $\sum_{i=1}^{n} a_i > K$. The proof of this fact is a simple reformulation of the denial of the statement '$\sum_{i=1}^{\infty} a_i$ is convergent'. (Exercise 10, §9.1.) This theorem is often stated as follows: A series $\sum_{i=1}^{\infty} a_i$ in which all terms are positive either converges to a finite sum or diverges to $+\infty$.

A very interesting infinite product is

$$\text{(1)} \quad \sin x = x\left(1 - \frac{x}{\pi}\right)\left(1 + \frac{x}{\pi}\right)\left(1 - \frac{x}{2\pi}\right)\left(1 + \frac{x}{2\pi}\right)\cdots = x\prod_{n=1}^{\infty}\left(1 - \frac{x^2}{n^2\pi^2}\right).$$

This formula was discovered by Euler, who was led to the discovery by the observation that $\sin x = x - \frac{x^3}{3!} + \frac{x^5}{5!} - \cdots$ is a 'polynomial of infinite degree' which has roots at $x = 0, \pm\pi, \pm 2\pi, \ldots$, and hence, by analogy with ordinary polynomials, that $\sin x$ should be the product of the factors on the right side of (1). A rigorous proof of (1) based on this idea is given in the next section.*

*For an excellent discussion of Euler's discovery see G. Polya, *Induction and Analogy in Mathematics*, Princeton University Press, 1954, pp. 17–22.

Euler had set out to find the sum of the series

$$1 + \frac{1}{4} + \frac{1}{9} + \frac{1}{16} + \frac{1}{25} + \cdots + \frac{1}{n^2} + \cdots.$$

It was known that this series was convergent

$$\left(\frac{1}{(m+1)^2} + \frac{1}{(m+2)^2} + \cdots + \frac{1}{n^2} < \frac{1}{m(m+1)} + \frac{1}{(m+1)(m+2)} + \cdots + \frac{1}{(n-1)n}\right.$$
$$\left. = \left[\frac{1}{m} - \frac{1}{m+1}\right] + \left[\frac{1}{m+1} - \frac{1}{m+2}\right] + \cdots + \left[\frac{1}{n-1} - \frac{1}{n}\right] = \frac{1}{m} - \frac{1}{n} < \frac{1}{m} < \frac{1}{N} \to 0\right)$$

and the sum had been found to several decimal places to be 1.644934 . . . , but the exact evaluation was a famous unsolved problem of the day. By equating the coefficients of x^3 in the formula (1) Euler concluded that

$$-\frac{1}{3!} = -\frac{1}{\pi^2} - \frac{1}{4\pi^2} - \frac{1}{9\pi^2} - \cdots$$

and hence gave the correct value

$$1 + \frac{1}{4} + \frac{1}{9} + \cdots + \frac{1}{n^2} + \cdots = \frac{\pi^2}{6}.$$

This example illustrates the vast difference between the statement 'the series $\sum a_i$ is convergent' and the statement 'the sum of the series $\sum a_i$ is such-and-such'. Even Euler was unable to find the sum of the convergent series

$$1 + \frac{1}{8} + \frac{1}{27} + \frac{1}{64} + \cdots + \frac{1}{n^3} + \cdots$$

(although, using (1), he was able to find $\sum \frac{1}{n^k}$ for all even values of k, e.g. $1 + \frac{1}{2^4} + \frac{1}{3^4} + \cdots = \frac{\pi^4}{90}$—see Exercise 9, §9.6.)

The fact that the infinite product (1) converges for all x is a special case of the following theorem: *If b_1, b_2, b_3, . . . is a sequence of non-negative* numbers $b_i \geq 0$ then either*

$$\prod_{i=1}^{\infty} (1 + b_i), \quad \prod_{i=1}^{\infty} (1 - b_i), \quad \sum_{i=1}^{\infty} b_i$$

all converge or none converge. Roughly speaking, none can converge unless $b_i \to 0$; if $b_i \to 0$ then products of two or more b's can be neglected relative to the b's themselves and

$$(1 \pm b_{m+1})(1 \pm b_{m+2}) \cdots (1 \pm b_n)$$
$$= 1 \pm (b_{m+1} + b_{m+2} + \cdots + b_n) + \text{small}$$

hence $\prod_{i=m+1}^{n} (1 \pm b_i)$ is near 1 if and only if $\sum_{i=m+1}^{n} b_i$ is near zero, which is the statement of the theorem. For a rigorous proof, see Exercises 12–14.

This theorem implies that the harmonic series $1 + \frac{1}{2} + \frac{1}{3} + \frac{1}{4} + \frac{1}{5} + \cdots$ is divergent because the product $(1 + 1)(1 + \frac{1}{2})(1 + \frac{1}{3})(1 + \frac{1}{4}) \cdots = \frac{2}{1} \cdot \frac{3}{2} \cdot \frac{4}{3} \cdot \frac{5}{4} \cdots$ is

*This condition is essential. See E. C. Titchmarsh, Theory of Functions, Oxford University Press, 1939, p. 17, for examples in which $\Pi(1 + b_i)$ converges but $\sum b_i$ diverges and vice versa.

obviously divergent. Euler used a similarly simple argument (Exercise 16) to prove that the series $\frac{1}{2} + \frac{1}{3} + \frac{1}{5} + \frac{1}{7} + \frac{1}{11} + \cdots + \frac{1}{p} + \cdots$ of reciprocal prime numbers is also divergent; this was a substantial strengthening of Euclid's theorem that there are infinitely many primes.

The improper integral $\int_{-\infty}^{\infty} \int_{-\infty}^{\infty} e^{-(x^2+y^2)}\, dx\, dy$ can be evaluated by converting to polar coordinates $x = r\cos\theta$, $y = r\sin\theta$ to find that the integral over the disk of radius R is

$$\int_0^{2\pi}\int_0^R e^{-r^2} r\, dr\, d\theta = \int_0^{2\pi}\int_0^R d[-\tfrac{1}{2}e^{-r^2}\, d\theta]$$
$$= \int_0^{2\pi} \tfrac{1}{2}[1 - e^{-R^2}]\, d\theta$$
$$= \pi[1 - e^{-R^2}].$$

This shows that the integral over any rectangle containing the square $\{|x| \le K, |y| \le K\}$ is at least $\pi[1 - e^{-K^2}]$ and at most π; hence the integral converges to π. On the other hand, the integral of $e^{-(x^2+y^2)}\, dx\, dy$ over the square $\{|x| \le K, |y| \le K\}$ is easily seen, by comparing approximating sums, to be $(\int_{-K}^{K} e^{-t^2}\, dt)^2$. This proves that the improper integral $\int_{-\infty}^{\infty} e^{-t^2}\, dt$ converges and has the value

$$\int_{-\infty}^{\infty} e^{-t^2}\, dt = \sqrt{\pi}.$$

Exercises

1 Prove that the series $1 + x + x^2/2! + x^3/3! + \cdots$ is convergent for all values of x.

2 Prove that the series $1 + x + x^2 + x^3 + \cdots$ is convergent if and only if $|x| < 1$. Show that the sum is $(1 - x)^{-1}$ (when $|x| < 1$).

3 Prove that the product $(1 + x)(1 + x^2)(1 + x^4) \times (1 + x^8) \ldots$ is convergent if and only if $|x| < 1$. Show that the product is $(1 - x)^{-1}$ (when $|x| < 1$).

4 Find $\pi^2/6$ to five decimal places. Estimate the number of terms of the series $1 + \frac{1}{4} + \frac{1}{9} + \frac{1}{16} + \cdots$ which would have to be added in order to obtain this accuracy. [Compare to the sum $\dfrac{1}{m(m+1)} + \dfrac{1}{(m+1)(m+2)} + \cdots$.]

5 Deduce Wallis' formula

$$\frac{\pi}{2} = \frac{2}{1} \cdot \frac{2}{3} \cdot \frac{4}{3} \cdot \frac{4}{5} \cdot \frac{6}{5} \cdot \frac{6}{7} \cdot \cdots$$

from the product formula (1).

6 The series $1 - 1 + \frac{1}{2} - \frac{1}{2} + \frac{1}{3} - \frac{1}{3} + \frac{1}{4} - \frac{1}{4} + \frac{1}{5} - \cdots$ converges to zero. Show that if it is rearranged so that k terms with the sign $+$ are taken for each term with the sign $-$; that is, if the series is rearranged

$$1 + \frac{1}{2} + \cdots + \frac{1}{k} - 1 + \frac{1}{k+1} + \frac{1}{k+2} + \cdots + \frac{1}{2k} - \frac{1}{2} + \frac{1}{2k+1} + \cdots$$
$$+ \frac{1}{3k} - \frac{1}{3} + \frac{1}{3k+1} + \cdots$$

then it converges to $\log k$. [Use

$$\log k = \int_1^k x^{-1}\,dx \sim \left(1 + \frac{1}{m}\right)^{-1}\left(\frac{1}{m}\right) + \left(1 + \frac{2}{m}\right)^{-1}\left(\frac{1}{m}\right) + \left(1 + \frac{3}{m}\right)^{-1}\left(\frac{1}{m}\right) + \cdots + k^{-1}\left(\frac{1}{m}\right)$$
$$= \sum_{n=1}^{mk} \frac{1}{n} - \sum_{n=1}^{m} \frac{1}{n}$$

to prove that the sum of the first N terms is within ϵ of $\log k$ for all sufficiently large N.] In particular conclude that

$$1 - \tfrac{1}{2} + \tfrac{1}{3} - \tfrac{1}{4} + \tfrac{1}{5} - \cdots = \log 2.$$

7 Describe a rearrangement of the series of Exercise 6 which converges to the sum 10. [Do not attempt to find a *formula* for the rearrangement, but describe the order in which positive and negative terms are to be taken.]

8 A series $b_1 + b_2 + b_3 + \cdots$ is said to be a *rearrangement* of a series $a_1 + a_2 + a_3 + \cdots$ if each a_i occurs exactly once among the b_i and vice versa (making the obvious provisions regarding numbers which may occur more than once among the a's). The preceding examples show that rearrangement of a convergent series may alter its sum. Show that if $a_1 + a_2 + a_3 + \cdots$ is *absolutely convergent*, that is, if the series $|a_1| + |a_2| + |a_3| + \cdots$ is convergent, then $a_1 + a_2 + a_3 + \cdots$ is convergent, any rearrangement of $a_1 + a_2 + a_3 + \cdots$ is also convergent, and the sum of any rearrangement is equal to the sum of the original series.

9 Show that if $a_1 + a_2 + a_3 + \cdots$ is a convergent series which is not absolutely convergent, then there is a rearrangement of the series which converges to 10 or, for that matter, to any sum whatsoever.

10 Prove that if an infinite product converges in the sense defined in the text then there is a number $\prod_{i=1}^{\infty} a_i$ as stated in the text. Show also that the number $\prod_{i=1}^{\infty} a_i$ is zero if and only if at least one of the factors a_i is zero. [First choose N such that $|a_{m+1}a_{m+2}\ldots a_n - 1| < \frac{1}{2}$ whenever $n > m \geq N$. Let $P = a_1a_2\ldots a_N$. If $P = 0$ then an a_i is zero and the products $\prod_{i=1}^{n} a_i$ are all zero for $n \geq N$. Otherwise $\left|\prod_{i=1}^{n} a_i\right| > \frac{1}{2}|P|$ for all $n \geq N$ and the limit, if it exists, is not zero. Finally, $\left|\prod_{i=1}^{n} a_i\right| < \frac{3}{2}|P|$ for all $n \geq N$ from which it follows that $\left|\prod_{i=1}^{n} a_i - \prod_{i=1}^{m} a_i\right|$ can be made small by making n, m large.]

11 Prove that an infinite product $a_1a_2a_3\ldots$ converges if and only if there is an N such that $\log a_i$ is defined for $i \geq N$ and $\sum_{i=N}^{\infty} \log a_i$ is a convergent series.

12 Prove that if $b_i \geq 0$ then $\prod_{i=1}^{\infty} (1 + b_i)$ converges if and only if $\sum b_i$ converges. [Use $b_1 + b_2 + \cdots + b_n \leq (1 + b_1)(1 + b_2)\ldots(1 + b_n) \leq e^{b_1+b_2+\cdots+b_n}$.]

13 Prove that if $a_1a_2a_3\ldots$ is a convergent product and if none of the a_i are zero, then $a_1^{-1}a_2^{-1}a_3^{-1}\ldots$ converges to the reciprocal of $a_1a_2a_3\ldots$.

14 Prove that if $b_i \geq 0$ then $\prod(1 + b_i)$ converges if and only if $\prod(1 - b_i)$ converges. [Show that $1 + b < (1 - b)^{-1} < 1 + 2b$ for small positive b.]

15 By analogy with (1) write $\cos x$ as an infinite product. Use this to guess the sum of the series

$$1 + \frac{1}{9} + \frac{1}{25} + \cdots + \frac{1}{(2n+1)^2} + \cdots.$$

Verify the result using

$$1 + \frac{1}{4} + \frac{1}{9} + \frac{1}{16} + \cdots = \frac{\pi^2}{6}.$$

16 If $\sum \frac{1}{p} = \frac{1}{2} + \frac{1}{3} + \frac{1}{5} + \frac{1}{7} + \frac{1}{11} + \cdots$ (p runs over the prime numbers) were a convergent series then the product $\prod\left(1 - \frac{1}{p}\right)^{-1}$ would converge, i.e.

$$\prod\left(1 + \frac{1}{p} + \frac{1}{p^2} + \frac{1}{p^3} + \cdots\right)$$

would converge. But, when expanded out, this product is $1 + \frac{1}{2} + \frac{1}{3} + \frac{1}{4} + \cdots$ (because every natural number can be written in just one way as a product of prime powers) which diverges. Hence $\sum \frac{1}{p}$ must diverge. Fill in the steps in this argument to prove that $\sum \frac{1}{p}$ diverges.

17 Evaluate $\int_{-\infty}^{\infty} e^{-\pi t^2}\, dt$.

18 The integral $\int_0^1 x^{-a}\, dx$ is improper if $a > 0$. Why? For which values of a does it converge?

19 The integral $\int_0^\infty e^{-t^2}\, dt$ is one half of the integral evaluated in the text, hence its value is $\frac{1}{2}\sqrt{\pi}$. This value can also be obtained from

$$\left(\int_0^\infty e^{-t^2}\, dt\right)^2 = \int_0^\infty \int_0^\infty e^{-(x^2+y^2)}\, dx\, dy$$

by making the change of variable $y = mx$. Sketch the 'new coordinates' (x, m), and evaluate the integral. As will be seen in the next section, the operation of changing the variable in an improper integral actually requires justification. However, in this example the integrand is positive, the improper integral converges absolutely, and the justification is immediate.

9.6 Interchange of Limits

Some of the most interesting and subtle facts in advanced calculus are related to the interchanging of two limit processes. This section contains several examples of such interchanges.

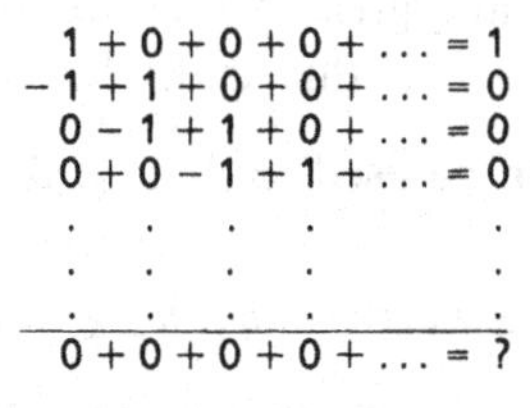

It is important first to understand that limits cannot always be interchanged, as is illustrated by the following example: For each pair m, n of natural numbers let $a_{m,n}$ be 1 if $m = n$, -1 if $m = n + 1$, and 0 in all other cases. Then for each n the series $\sum_{m=1}^{\infty} a_{m,n}$ has a 1, a -1, and the rest 0's, hence is convergent and has the sum zero. Thus $\sum_{n=1}^{\infty}\left(\sum_{m=1}^{\infty} a_{m,n}\right)$ exists and is zero because it is a series in which all terms are zero. On the other hand, the series $\sum_{n=1}^{\infty} a_{m,n}$ converges to 0 unless $m = 1$, in which case it converges to 1. Thus $\sum_{m=1}^{\infty}\left(\sum_{n=1}^{\infty} a_{m,n}\right) = 1 \neq$

$\sum_{n=1}^{\infty}\left(\sum_{m=1}^{\infty} a_{m,n}\right)$ and the two limits $\sum_{n=1}^{\infty}$ and $\sum_{m=1}^{\infty}$ cannot be interchanged.

An expression of the form $\sum_{m=1}^{\infty}\left(\sum_{n=1}^{\infty} a_{m,n}\right)$ is called a *double series*. A double series $\sum_{n=1}^{\infty}\left(\sum_{m=1}^{\infty} a_{m,n}\right)$ is said to converge to the sum S if each of the series $\sum_{m=1}^{\infty} a_{m,n}$ $(n = 1, 2, 3, \ldots)$ is convergent, say to b_n, and if the series $b_1 + b_2 + b_3 + \cdots$ converges to the sum S. It is easy to construct examples of convergent double series $\sum_{n=1}^{\infty}\left(\sum_{m=1}^{\infty} a_{m,n}\right)$ for which the double series $\sum_{m=1}^{\infty}\left(\sum_{n=1}^{\infty} a_{m,n}\right)$ obtained by summing in the reverse order is *not* convergent, and, as the above example shows, even if the reversed double series is convergent its sum need not be the same as that of the original series. Neither of these circumstances can occur, however, if all of the terms $a_{m,n}$ are positive, that is: *If* $\sum_{n=1}^{\infty}\left(\sum_{m=1}^{\infty} a_{m,n}\right)$ *is a convergent double series in which all terms* $a_{m,n}$ *are positive, then the double series* $\sum_{m=1}^{\infty}\left(\sum_{n=1}^{\infty} a_{m,n}\right)$ *is also convergent and the sums are equal,* $\sum_{n=1}^{\infty}\left(\sum_{m=1}^{\infty} a_{m,n}\right) = \sum_{m=1}^{\infty}\left(\sum_{n=1}^{\infty} a_{m,n}\right)$.

$$\begin{array}{l} 1 + \frac{1}{3} + \frac{1}{5} + \frac{1}{7} + \ldots = \text{div.} \\ 0 - \frac{1}{2} + 0 + 0 + \ldots = -\frac{1}{2} \\ 0 + 0 - \frac{1}{4} + 0 + \ldots = -\frac{1}{4} \\ 0 + 0 + 0 - \frac{1}{6} + \ldots = -\frac{1}{6} \\ \hline 1 - \frac{1}{2} + \frac{1}{3} - \frac{1}{4} + \frac{1}{5} - \frac{1}{6} + \frac{1}{7} - \ldots = \log 2 \end{array}$$

Briefly, the proof of this theorem is as follows: To prove that each of the series $\sum_{n=1}^{\infty} a_{m,n}$ is convergent it suffices to show that for every $\epsilon > 0$ there is an N such that *any* sum of a finite number of terms $a_{m,n}$ in which one or both indices are greater than N $(m \geq N$ or $n \geq N)$ is less than ϵ. To this end, the convergence of $\sum_{n=1}^{\infty}\left(\sum_{m=1}^{\infty} a_{m,n}\right)$ can be used to find an N_0 such that $\sum_{n=n_1}^{n_2}\left(\sum_{m=1}^{\infty} a_{m,n}\right) < \epsilon/2$ whenever $n_1, n_2 \geq N_0$. Then for each $n = 1, 2, 3, \ldots, N_0$ the convergence of $\sum_{m=1}^{\infty} a_{n,m}$ can be used to find an N_n $(n = 1, 2, \ldots, N_0)$ such that $\sum_{m=m_1}^{m_2} a_{m,n} < (\epsilon/2N_0)$ whenever $m_1, m_2 \geq N_n$. Let N be the largest of the integers N_0, N_n $(n = 1, 2, \ldots, N_0)$. Then any finite sum of terms $a_{m,n}$ in which $n \geq N$ or

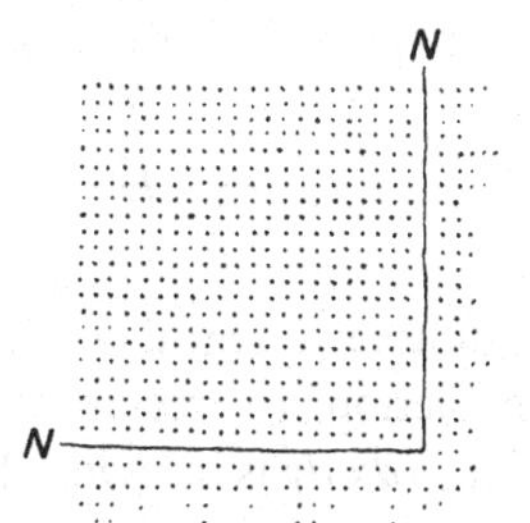

Any sum of a finite number of terms outside the box is less than ϵ.

$m \geq N$ is at most $\sum_{m=N}^{\infty} a_{m,1} + \sum_{m=N}^{\infty} a_{m,2} + \cdots + \sum_{m=N}^{\infty} a_{m,N_0} + \sum_{n=N_0}^{\infty}\left(\sum_{m=1}^{\infty} a_{m,n}\right) < \epsilon$. Thus $\sum_{n=1}^{\infty} a_{m,n}$ converges for $m = 1, 2, 3, \ldots$. Moreover, when N is chosen in this way any finite sum of terms $a_{m,n}$ which includes all terms $a_{m,n}$ for $m \leq N$, $n \leq N$, differs by at most ϵ from $\sum_{n=1}^{\infty}\left(\sum_{m=1}^{\infty} a_{m,n}\right)$. From this observation it is easy to conclude that $\sum_{m=1}^{\infty}\left(\sum_{n=1}^{\infty} a_{m,n}\right)$ converges to $\sum_{n=1}^{\infty}\left(\sum_{m=1}^{\infty} a_{m,n}\right)$ as desired.

More generally, a double series $\sum_{n=1}^{\infty}\left(\sum_{m=1}^{\infty} a_{m,n}\right)$ is said to *converge absolutely* if the double series $\sum_{n=1}^{\infty}\left(\sum_{m=1}^{\infty} |a_{m,n}|\right)$ of absolute values converges. When this is the case, a slight modification of the above proof shows that the double series itself converges and can be summed in either order $\sum_{n=1}^{\infty}\left(\sum_{m=1}^{\infty} a_{m,n}\right) = \sum_{m=1}^{\infty}\left(\sum_{n=1}^{\infty} a_{m,n}\right)$. Moreover, an absolutely convergent double series can be summed in any other order, e.g. $\sum_{N=1}^{\infty}\left(\sum_{m+n=N} a_{m,n}\right)$, and the result is the same. Similar results apply to *rearrangements* of ordinary series. (See Exercises 7, 8, 9 of §9.5.)

Similar considerations apply to improper double integrals. For example, if $A(x, y)$ is a function defined on the entire xy-plane then

$$\int_{-\infty}^{\infty}\left(\int_{-\infty}^{\infty} A(x, y)\, dx\right) dy, \qquad \int_{-\infty}^{\infty}\left(\int_{-\infty}^{\infty} A(x, y)\, dy\right) dx,$$
$$\int_{-\infty}^{\infty}\int_{-\infty}^{\infty} A(x, y)\, dx\, dy$$

are all defined in different ways and there is no guarantee *a priori* that if one converges then the others do or that if two converge then their values are equal. However, the technique applied to double series above is easily generalized to prove: *If A is continuous and if $\int_{-\infty}^{\infty}\int_{-\infty}^{\infty} A(x, y)\, dx\, dy$ is absolutely convergent*—i.e. if the improper integral $\int_{-\infty}^{\infty}\int_{-\infty}^{\infty} |A(x, y)|\, dx\, dy$ converges in the sense of §9.5—*then the improper integral $\int_{-\infty}^{\infty}\int_{-\infty}^{\infty} A(x, y)\, dx\, dy$ can be 'summed in any order' with the same result.* For example, the interated integrals above converge to the same value as the double integral, as does

$$\lim_{r\to\infty} \int_{\{x^2+y^2\le r^2\}} A(x, y)\,dx\,dy.$$

Thus, if an improper double integral is absolutely convergent then the restriction to rectangles in the definition of $\int_{\infty}^{-\infty}\int_{\infty}^{-\infty} A(x, y)\,dx\,dy$ is inessential and any other sort of domain could be used as well. On the other hand, if a double integral is convergent but not absolutely convergent*, then, as in the case of series, the order of summation is crucial and one must proceed with extreme care.

An integral (or series) which is convergent but not absolutely convergent is said to be conditionally convergent.

The formula

$$\begin{aligned}\log(1+x) &= \int_1^{1+x} \frac{1}{t}\,dt = \int_1^{1+x} [1 + (1-t) + (1-t)^2 + (1-t)^3 + \cdots]\,dt \\ &= \int_1^{1+x} dt + \int_1^{1+x} (1-t)\,dt + \int_1^{1+x} (1-t)^2\,dt + \cdots \\ &= x - \frac{x^2}{2} + \frac{x^3}{3} - \frac{x^4}{4} + \frac{x^5}{4} - \cdots\end{aligned}$$

is obtained by interchanging the limit $\int_1^{1+x}$ and the limit $t^{-1} = 1 + (1-t) + (1-t)^2 + \cdots$. Such an interchange must always be justified, but in this case the justification is quite easy: The formula

$$\frac{1}{t} = 1 + (1-t) + (1-t)^2 + \cdots + (1-t)^n + \frac{(1-t)^{n+1}}{t}$$

can be integrated from 1 to $1 + x$ (the integral of a *finite* sum is the sum of the integrals) to give

$$\log(1+x) = x - \frac{x^2}{2} + \frac{x^3}{3} - \cdots + (-1)^n \frac{x^{n+1}}{n+1} + \int_1^{1+x} \frac{(1-t)^{n+1}}{t}\,dt.$$

For any x in the range $|x| < 1$ the integrand $(1-t)^{n+1}/t$ is at most $|x|^{n+1}/(1-|x|)$ on the domain of integration, and the integral is therefore at most $|x|$ times this. If any $\epsilon > 0$ is given, then this integral can be made less than ϵ by making n large. Hence the desired formula

$$(1) \qquad \log(1+x) = x - \frac{x^2}{2} + \frac{x^3}{3} - \frac{x^4}{4} + \cdots$$

holds for $|x| < 1$.

This formula (1) cannot hold for $|x| > 1$ or for $x = -1$ because in these cases the series on the right does not converge. However, the alternating series $1 - \frac{1}{2} + \frac{1}{3} - \frac{1}{4} + \cdots$ obtained by setting $x = 1$ does converge and one would naturally expect that its sum is $\log(1 + 1) = \log 2$, even though the above proof is no longer valid. A slight refinement of this proof suffices to verify* that $\log 2 = 1 - \frac{1}{2} + \frac{1}{3} - \frac{1}{4} + \cdots$ but this conclusion can also be reached by appealing to an important theorem on the interchange of limits, namely *Abel's Theorem: If $a_1 + a_2 + a_3 + \cdots$ is a convergent series then the power series $a_1x + a_2x^2 + a_3x^3 + \cdots$ is convergent for all numbers x in the range*† $0 < x < 1$ *and*

$$\lim_{x\uparrow 1} [a_1x + a_2x^2 + \cdots] = a_1 + a_2 + a_3 + \cdots.$$

That is, given $\epsilon > 0$ there is a $\delta > 0$ such that for any x in the range $1 - \delta < x < 1$ the sum of the power series differs by less than ϵ from the sum of the original series.

The proof of Abel's Theorem is based on the formula‡

$$\begin{aligned}
a_mx^m &+ a_{m+1}x^{m+1} + \cdots + a_nx^n \\
&= a_m[(x^m - x^{m+1}) + (x^{m+1} - x^{m+2}) + \cdots + (x^{n-1} - x^n) + x^n] \\
&\quad + a_{m+1}[(x^{m+1} - x^{m+2}) + (x^{m+2} - x^{m+3}) + \cdots + (x^{n-1} - x^n) + x^n] \\
&\quad + \cdots + a_{n-1}[(x^{n-1} - x^n) + x^n] + a_nx^n \\
&= (x^m - x^{m+1})a_m + (x^{m+1} - x^{m+2})(a_m + a_{m+1}) \\
&\quad + (x^{m+3} - x^{m+2})(a_m + a_{m+1} + a_{m+2}) \\
&\quad + \cdots + (x^{n-1} - x^n)(a_m + a_{m+1} + \cdots + a_{n-1}) \\
&\quad + x^n(a_m + a_{m+1} + \cdots + a_n).
\end{aligned}$$

Given $\epsilon > 0$, let N be such that $|a_m + a_{m+1} + \cdots + a_n| < \epsilon$ whenever $m, n \geq N$. Taking absolute values in the above formula and using the triangle inequality gives

$$\begin{aligned}
|a_mx^m &+ a_{m+1}x^{m+1} + \cdots + a_nx^n| \\
&\leq |x^m - x^{m+1}|\epsilon + |x^{m+1} - x^{m+2}|\epsilon + \cdots + |x^{n-1} - x^n|\epsilon + |x^n|\epsilon \\
&= (x^m - x^{m+1} + x^{m+1} - x^{m+2} + \cdots + x^{n-1} - x^n + x^n)\epsilon = x^m\epsilon < \epsilon
\end{aligned}$$

for x in the range $0 < x < 1$. This proves not only that the power series is convergent but also that its sum differs by at most ϵ from the sum of the first N terms $a_1x + a_2x^2 + \cdots + a_Nx^N$. But $a_1 + a_2 + \cdots + a_N$ differs by at most ϵ from $\sum_{n=1}^{\infty} a_n$ and

$$\begin{aligned}
|(a_1 &+ a_2 + \cdots + a_N) - (a_1x + a_2x^2 + \cdots + a_Nx^N)| \\
&\leq |a_1(1 - x)| + |a_2(1 - x)(1 + x)| + \cdots + |a_N(1 - x)(1 + x + x^2 + \cdots + x^{N-1})| \\
&\leq |1 - x|(|a_1| + 2|a_2| + \cdots + N|a_N|)
\end{aligned}$$

*See Exercise 6, §9.5, for an alternative proof of this formula.

†In fact, comparison with the geometric series shows easily that it is convergent for $|x| < 1$. Only the values of x near 1 are of interest in the theorem.

‡This formula is analogous to integration by parts. Specifically, let A denote the sequence $A_i = a_m + a_{m+1} \cdots + a_i$ and let B denote the sequence x^i. Then the sum on the left is $\sum \Delta A \cdot B$ where ΔA represents the sequence of differences $A_i - A_{i-1} = a_i$. The formula is then the statement $\sum \Delta A \cdot B = -\sum A \cdot \Delta B + (A \cdot B)|_m^n$ analogous to integration by parts.

which (because N is fixed) is less than ϵ for x sufficiently near 1. Thus $\sum_{n=1}^{\infty} a_n x^n$ differs by at most 3ϵ from $\sum_{n=1}^{\infty} a_n$ for all $x < 1$ sufficiently near 1. Since ϵ was arbitrary this completes the proof of Abel's theorem.

An alternative statement of Abel's theorem which is important is the following: A series $a_1 + a_2 + a_3 + \cdots$ is said to be *Abel summable* if the power series $a_1 x + a_2 x^2 + a_3 x^3 + \cdots$ converges for $|x| < 1$ and if $\lim_{x \uparrow 1} (a_1 x + a_2 x^2 + \cdots)$ exists. When this is the case the *Abel sum* of the series is this limit. Then Abel's theorem states that a convergent series is Abel summable and that its Abel sum is its ordinary sum. The converse of this theorem is not true, that is, a series can be Abel summable without being convergent. For example, the series $1 - 1 + 1 - 1 + 1 - 1 + \cdots$ is not convergent, but

$$x - x^2 + x^3 - x^4 + \cdots = \frac{x}{1 + x}$$

is convergent for $|x| < 1$ and

$$\lim_{x \uparrow 1} \frac{x}{1 + x} = \frac{1}{2},$$

hence $1 - 1 + 1 - 1 + \cdots$ is Abel summable with Abel sum $\frac{1}{2}$. Note that any series $a_1 + a_2 + \cdots$ which is Abel summable but not convergent gives an example of limits which cannot be interchanged, namely,

$$\lim_{x \uparrow 1} \lim_{N \to \infty} \left(\sum_{n=1}^{N} a_n x^n \right) \neq \lim_{N \to \infty} \lim_{x \uparrow 1} \left(\sum_{n=1}^{N} a_n x^n \right)$$

because the limit on the right does not exist.

The formula of §9.2

$$\text{(2)} \qquad \int_{-\infty}^{\infty} \frac{\sin xy}{y}\, dy = \begin{cases} \pi & \text{if } x > 0 \\ 0 & \text{if } x = 0 \\ -\pi & \text{if } x < 0 \end{cases}$$

gives another example of a double limit which cannot be interchanged, namely

$$\lim_{n \to \infty} \int_{-\infty}^{\infty} \frac{\sin\left(\frac{y}{n}\right)}{y}\, dy = \pi$$

$$\int_{-\infty}^{\infty} \lim_{n \to \infty} \frac{\sin\left(\frac{y}{n}\right)}{y}\, dy = \int_{-\infty}^{\infty} 0\, dy = 0.$$

The formula (2) itself can be seen as the result of an inter· change of limits (this time a valid one) as follows: If $z > 0$ then

$$\int_0^\infty e^{-zt}\,dt = \lim_{K\to\infty} \int_0^K e^{-zt}\,dt$$

$$= \lim_{K\to\infty} \left[\frac{e^{-zK}}{-z} - \frac{e^{-0}}{-z}\right] = \frac{1}{z}\cdot$$

Assuming this formula is still valid for complex numbers $z = a + bi$ in which $a > 0$ gives

$$\int_0^\infty e^{-(a+bi)t}\,dt = \frac{1}{a+bi}$$

$$\int_0^\infty e^{-at}[\cos bt - i\sin bt]\,dt$$

$$= \frac{a - bi}{(a+bi)(a-bi)} = \frac{a-bi}{a^2+b^2}$$

and equating imaginary parts

$$\int_0^\infty e^{-at}\sin bt\,dt = \frac{b}{a^2+b^2}\cdot$$

Multiplying both sides by da and integrating from $a = 0$ to $a = \infty$ gives

$$\int_0^\infty \int_0^\infty e^{-at}\sin bt\,dt\,da = \int_0^\infty \frac{b\,da}{a^2+b^2}\cdot$$

The integral on the right is zero if $b = 0$, $\pi/2$ if $b > 0$ $\left(\text{take } u = a/b \text{ and use } \int_{-\infty}^\infty \frac{du}{u^2+1} = \pi\right)$, $-\pi/2$ if $b < 0$ (take $u = -a/b$). Interchanging the order of integration on the left gives

$$\int_0^\infty \left[\int_0^\infty e^{-at}\sin bt\,da\right] dt = \int_0^\infty \left[\frac{e^{-at}}{-t}\sin bt\Big|_{a=0}^{a=\infty}\right] dt$$

$$= \int_0^\infty \frac{\sin bt}{t}\,dt$$

and (2) follows. The steps in this argument can be justified to give a proof of (2) as follows: The essence of the argument is that the improper integral $\int_0^\infty \int_0^\infty e^{-at}$ $\sin bt\,dt\,da$ (b fixed) can be evaluated by using either the formula

$$d\left[\frac{e^{-at}}{t}\sin bt\,dt\right] = e^{-at}\sin bt\,dt\,da$$

or the formula

$$d\left[-\frac{e^{-at}(a\sin bt + b\cos bt)}{a^2+b^2}\,da\right] = e^{-at}\sin bt\,dt\,da$$

and applying Stokes' theorem. In fact, if the improper integral were absolutely convergent, i.e. if $\int_0^\infty \int_0^\infty e^{-at}$ $|\sin bt|\,dt\,da$ converged, then the analog of the argument proving $\sum_{m=1}^{\infty}\left(\sum_{n=1}^{\infty} a_{m,n}\right) = \sum_{n=1}^{\infty}\left(\sum_{m=1}^{\infty} a_{m,n}\right)$ for absolutely convergent double series would show that these two methods of computing $\int_0^\infty \int_0^\infty e^{-at}\sin bt\,dt\,da$ give the same result and hence prove (2). However, this improper integral is not absolutely convergent and a somewhat more careful analysis is necessary. Setting $K = 2\pi n/b$ for n a large positive integer, the formulas above give

$$\int_0^K \int_0^K e^{-at}\sin bt\,dt\,da$$

$$= \int_0^K \frac{\sin bt}{t}\,dt + \int_K^0 \frac{e^{-Kt}\sin bt}{t}\,dt$$

and

$$\int_0^K \int_0^K e^{-at}\sin bt\,dt\,da$$

$$= -\int_0^K \frac{e^{-Ka}b}{a^2+b^2}\,da - \int_K^0 \frac{b}{a^2+b^2}\,da.$$

Equating these two expressions for $\int_0^K \int_0^K e^{-at}\sin bt\,dt\,da$ and using the fact that the integrals

$$\int_0^\infty \frac{\sin bt}{t}\,dt, \quad \int_0^\infty \frac{b}{a^2+b^2}\,da$$

converge, their equality follows by taking the limit as $n \to \infty$ and showing that

$$\lim_{K\to\infty} \int_0^K e^{-Kt}\frac{\sin bt}{t}\,dt = 0$$

$$\lim_{K\to\infty} \int_0^K e^{-Ka}\frac{b}{a^2+b^2}\,da = 0.$$

To prove that the first of these integrals approaches zero as $K \to \infty$ let δ be a positive number and let the domain of integration $\{0 \le t \le K\}$ be broken into the intervals $\{0 \le t \le \delta\}$, $\{\delta \le t \le K\}$. This gives

$$\int_0^K e^{-Kt} \frac{\sin bt}{t} dt \le \int_0^\delta e^{-Kt} b\, dt + \int_\delta^K e^{-K\delta} \frac{\sin bt}{t} dt$$

(using the fact that $\sin bt/b$ assumes its maximum value b at $t = 0$)

$$\le \frac{1 - e^{-K\delta}}{K} b + e^{-K\delta} \int_0^\infty \frac{\sin bt}{t} dt$$

which approaches zero as $K \to \infty$. The proof that $\int_0^K e^{-Ka} \frac{b}{a^2 + b^2} da$ approaches zero as $K \to \infty$ is virtually identical and completes the proof of (2).

The product formula for the sine

$$\sin x = x \prod_{n=1}^\infty \left(1 - \frac{x^2}{n^2\pi^2}\right) \tag{3}$$

can be derived from the trigonometric identity

$$\sin x = p \sin\left(\frac{x}{p}\right) \prod_{n=1}^{\frac{p-1}{2}} \left(1 - \frac{\sin^2\left(\frac{x}{p}\right)}{\sin^2\left(\frac{n\pi}{p}\right)}\right) \tag{3'}$$

valid for all x and for all odd positive integers p. To derive the identity (3′) set $p = 2q + 1$, where q is a positive integer, and expand De Moivre's formula

$$\cos pA + i \sin pA = [\cos A + i \sin A]^p$$

by the binomial theorem to obtain

$$\cos pA + i \sin pA = \sum_{\nu=0}^p \binom{p}{\nu} i^\nu \cos^{p-\nu} A \sin^\nu A.$$

In taking the imaginary parts on both sides only the terms ν = odd appear on the right; setting $y = \sin A$, $\cos^2 A = 1 - y^2$, this gives an identity of the form

$$\begin{aligned} \sin pA &= p(1 - y^2)^q y + \cdots + (-1)^q y^p \\ &= py + \text{terms in } y^3, y^5, \ldots, y^p. \end{aligned}$$

The right-hand side is a polynomial of degree p in y which must be zero whenever $y = \sin A$ for $A = 0$, $\pm\frac{\pi}{p}, \pm\frac{2\pi}{p}, \pm\frac{q\pi}{p}$. Since these p numbers $\sin A$ are distinct, they account for all the roots of the polynomial and the polynomial must be

$$py \prod_{n=1}^{q} \left(1 - \frac{y}{\sin\left(\frac{n\pi}{p}\right)}\right)\left(1 + \frac{y}{\sin\left(\frac{n\pi}{p}\right)}\right).$$

Setting $x = pA$, $y = \sin A = \sin(x/p)$ then gives (3′).

Now since

$$\lim_{p\to\infty} p \sin\left(\frac{x}{p}\right) = \lim_{p\to\infty} x \cdot \frac{\sin\left(\frac{x}{p}\right) - \sin 0}{\frac{x}{p}} = x \cdot \cos 0 = x$$

and

$$\lim_{p\to\infty} \frac{\sin\left(\frac{x}{p}\right)}{\sin\left(\frac{n\pi}{p}\right)} = \lim_{p\to\infty} \frac{p\sin\left(\frac{x}{p}\right)}{p\sin\left(\frac{n\pi}{p}\right)} = \frac{x}{n\pi}$$

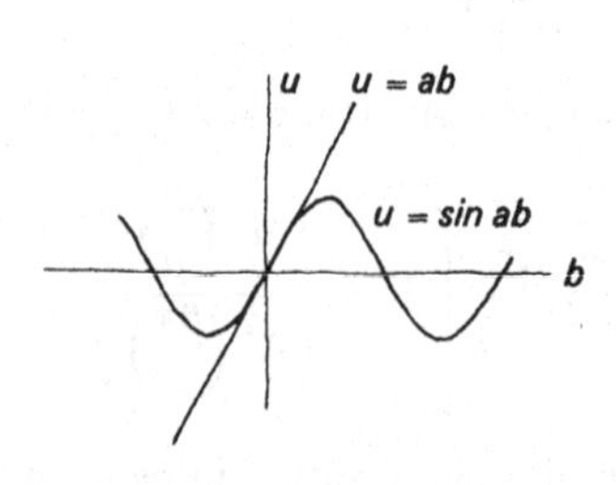

the formula (3) follows formally from (3′) by letting $p \to \infty$. This passage to the limit involves an interchange of limits (the product of the limits is equal to the limit of the products) which must be justified:

Let x be fixed. For each n let $a = \frac{n\pi}{p}$, $b = \frac{x}{n\pi}$ so that the term of (3′) corresponding to n is 1 if $a \geq \frac{\pi}{2}$ and $\sin ab/\sin a$ if $a < \frac{\pi}{2}$. Then the inequalities

$$|\sin ab| < a|b| \qquad (a > 0)$$

$$\frac{2}{\pi} a < \sin a \qquad \left(0 < a < \frac{\pi}{2}\right)$$

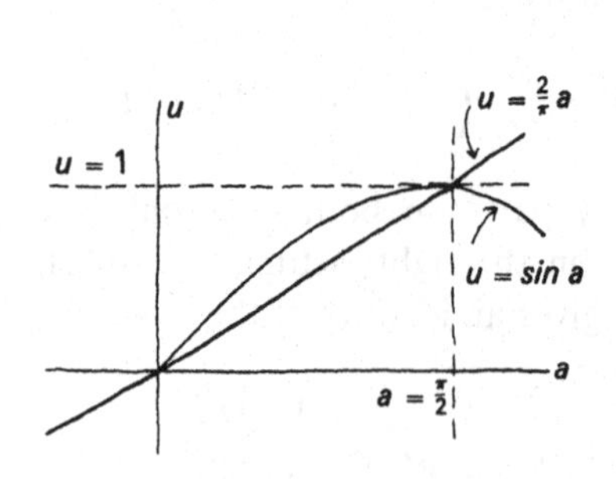

show that the nth factor of the product (3′) lies between 1 $\left(\text{if } n > \frac{p-1}{2}\right)$ and $1 - \left(\frac{a|b|}{\frac{2}{\pi}a}\right)^2 = 1 - \frac{x^2}{4n^2}$ for all p.

Now the product $\prod_{n=1}^{\infty}\left(1 - \frac{x^2}{4n^2}\right)$ converges by the theorem of §9.5, which means that for any ϵ there is an N such that any product of a finite number of factors of $\prod_{n=N}^{\infty}\left(1 - \frac{x^2}{4n^2}\right)$ is within ϵ of 1. It follows that the product of all factors past the Nth in (3′) (of which all but a finite number are 1) differs by less than ϵ from 1.

Hence for all (odd) p and for all sufficiently large N

$$\sin x = p \sin\left(\frac{x}{p}\right) \prod_{n=1}^{N} \left(1 - \frac{\sin^2\left(\frac{x}{p}\right)}{\sin^2\left(\frac{n\pi}{p}\right)}\right) \cdot (1 + \delta_1)$$

where $|\delta_1| < \epsilon$. On the other hand, for all sufficiently large N

$$x \prod_{n=1}^{\infty} \left(1 - \frac{x^2}{n^2\pi^2}\right) = x \prod_{n=1}^{N} \left(1 - \frac{x^2}{n^2\pi^2}\right)(1 + \delta_2)$$

where $|\delta_2| < \epsilon$ (by the definition of convergence of a product). Finally, for any fixed N

$$\lim_{p\to\infty} p \sin\left(\frac{x}{p}\right) \prod_{n=1}^{N} \left(1 - \frac{\sin^2\left(\frac{x}{p}\right)}{\sin^2\left(\frac{n\pi}{p}\right)}\right) = x \prod_{n=1}^{N} \left(1 - \frac{x^2}{n^2\pi^2}\right)$$

(in a *finite* product the limit of the product is the product of the limits) so that when p is sufficiently large the first N factors of (3′) can be obtained from $x \prod_{n=1}^{N} \left(1 - \frac{x^2}{n^2\pi^2}\right)$ by adding a number δ_3 with $|\delta_3| < \epsilon$. All together this gives

$$\left(\prod_{n=1}^{N} \left(1 - \frac{x^2}{n^2\pi^2}\right) + \delta_3\right)(1 + \delta_1)$$

$$= \left(\frac{x \cdot \prod_{n=1}^{\infty} \left(1 - \frac{x^2}{n^2\pi^2}\right)}{1 + \delta_2} + \delta_3\right)(1 + \delta_1).$$

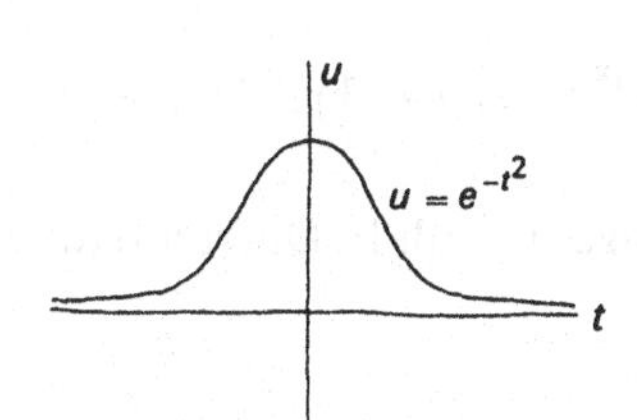

Since δ_1, δ_2, δ_3 are arbitrarily small, (3) follows.

As a final example, consider the evaluation of the improper integral

$$(4) \qquad f(x) = \int_{-\infty}^{\infty} e^{-t^2} \cos xt \, dt.$$

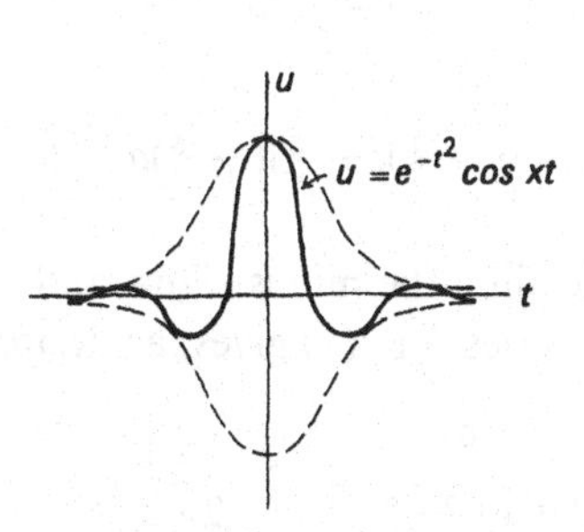

The first zero occurs when $xt = \frac{\pi}{2}$, that is, when $t = \frac{\pi}{2x}$.

(This problem is of central importance in the theory of diffusion and heat conduction.) For $x = 0$ this is the integral

$$(5) \qquad \int_{-\infty}^{\infty} e^{-t^2} \, dt = \sqrt{\pi}$$

which was evaluated in §9.5. Because the factor $\cos xt$ of (4) introduces cancellations into the integral (5), it can be seen that the given integral (4) converges for all x and that its value $f(x)$ is largest when $x = 0$, namely $f(0) = \sqrt{\pi} > f(x)$ for $x \neq 0$. Clearly $f(-x) = f(x)$. Plausible but less obvious are the facts that $\lim_{x\to\infty} f(x) = 0$ and $f(x) > 0$ (all x). Now $f(x)$ is in fact an elementary function (in the technical sense of §9.2) namely the function

(6) $$f(x) = \sqrt{\pi}\, e^{-(x^2/4)}.$$

This formula can be derived as follows:

Expanding $\cos xt$ as a power series $\cos xt = 1 - (xt)^2/2 + (xt)^4/4! - \cdots$ and interchanging the integral and the sum of the series gives

$$f(x) = \int_{-\infty}^{\infty} e^{-t^2}\, dt - \frac{x^2}{2}\int_{-\infty}^{\infty} t^2 e^{-t^2}\, dt + \frac{x^4}{4!}\int_{-\infty}^{\infty} t^4 e^{-t^2}\, dt - \cdots$$

assuming the interchange is valid. This will give a power series expansion of $f(x)$ if the integrals $\int_{-\infty}^{\infty} t^{2n} e^{-t^2}\, dt$ can be evaluated. Making the change of variable $t = \sqrt{a}\cdot u$ in (5) gives

(5′) $$\int_{-\infty}^{\infty} e^{-au^2}\, du = \sqrt{\frac{\pi}{a}}\cdot$$

Differentiating both sides of this equation with respect to a and interchanging the differentiation and integration gives

$$\int_{-\infty}^{\infty} (-u^2)e^{-au^2}\, du = (-\tfrac{1}{2})a^{-3/2}\sqrt{\pi}$$

assuming the interchange is valid. Differentiating repeatedly gives

$$\int_{-\infty}^{\infty} (-u^2)^2 e^{-au^2}\, du = (-\tfrac{1}{2})(-\tfrac{3}{2})a^{-5/2}\sqrt{\pi}$$

$$\int_{-\infty}^{\infty} (-u^2)^3 e^{-au^2}\, du = (-\tfrac{1}{2})(-\tfrac{3}{2})(-\tfrac{5}{2})a^{-7/2}\sqrt{\pi}$$

etc. Setting $a = 1$, cancelling the minus signs, and using the result in the power series for $f(x)$ gives, as desired,

$$\begin{aligned} f(x) &= \sqrt{\pi} - \frac{x^2}{2}\cdot\frac{1}{2}\sqrt{\pi} + \frac{x^4}{4\cdot 3\cdot 2\cdot 1}\cdot\frac{3}{2}\cdot\frac{1}{2}\cdot\sqrt{\pi} - \frac{x^6}{6\cdot 5\cdot 4\cdot 3\cdot 2\cdot 1}\cdot\frac{5\cdot 3\cdot 1}{2\cdot 2\cdot 2}\sqrt{\pi} + \cdots \\ &= \sqrt{\pi}\left[1 - \frac{x^2}{4} + \frac{x^4}{2^5} - \frac{x^6}{3!2^6} + \cdots + (-1)^n\frac{x^{2n}}{n!2^{2n}} + \cdots\right] \\ &= \sqrt{\pi}\, e^{-(x^2/4)}. \end{aligned}$$

In order to make this argument into a proof, one must justify the two operations of 'passing to the limit under an integral sign' which it involves, that is, one must prove that the integral of the infinite series is the sum of the integrals and that the integral of the derivative is the derivative of the integral. Since the main theorem on passing to the limit under an integral sign is given in the next section, these justifications are postponed until then.

There is no general theorem governing interchanges of limits, but rather many theorems, such as Abel's Theorem, covering various types of interchanges. In many cases it is easiest to justify an interchange directly, rather than by appeal to a general theorem; the comparison of the product (3′) to the product $\prod \left(1 - \frac{x^2}{4n^2}\right)$ is a standard technique for such proofs (see Exercise 5).

Exercises

1 Write $(1 + t^2)^{-1}$ as an infinite series, integrate, and apply Abel's Theorem to obtain Leibniz's formula

$$\frac{\pi}{4} = 1 - \frac{1}{3} + \frac{1}{5} - \frac{1}{7} + \cdots$$

justifying all steps fully.

2 Evaluate $\int_{-\infty}^{\infty} e^{-au^2} \cos (yu)\, du$ (y, a given, with $a > 0$) by performing a change of variables in the formula $\int_{-\infty}^{\infty} e^{-t^2} \cos xt\, dt = \sqrt{\pi}\, e^{-x^2/4}$.

3 Evaluate $f(x) = \displaystyle\int_{-\infty}^{\infty} \frac{\sin y}{y} \cos xy\, dy$. [Using a trigonometric identity for $\sin y \cos xy$ this can be reduced to two applications of formula (2) of the text.] Draw the graph of $f(x)$.

4 An example of a non-differentiable function. Let

$$f(x) = \sin x + \tfrac{1}{2} \sin 2x + \tfrac{1}{4} \sin 4x + \cdots + \frac{1}{2^n} \sin 2^n x + \cdots.$$

Show that $f(x)$ is defined for all x and is uniformly continuous. [As usual, break the step from $f(x)$ to $f(x')$ into three steps—from $f(x)$ to the sum of the first N terms evaluated at x, then to the sum of the first N terms at x', then to $f(x')$. Show that by making N large and then making $|x - x'|$ small, all three steps can be made small.] Sketch the graph of $\frac{1}{2^n} \sin 2^n x$ for $n = 1, 2, 3, 4$, and of their derivatives. These diagrams make it plausible that $f(x)$ is not differentiable at any point. It can in fact be shown that $\lim_{s\to 0} [f(x+s) - f(x)]/s$ does not exist at any point, and the proof is not difficult, although somewhat long.*

*See B. Sz.-Nagy, Introduction to Real Functions and Orthogonal Expansions, Oxford University Press, New York, 1965, pp. 101–103, for a different example with a complete proof of nondifferentiability.

5 *Weierstrass 'M-test'.* Generalizing the example of 4, show that if

$$f(x) = u_1(x) + u_2(x) + u_3(x) + \cdots$$

is an infinite series of *functions* of x, say on an interval $\{a \le x \le b\}$, such that (i) each of the functions $u_n(x)$ is uniformly continuous† on $\{a \le x \le b\}$ and such that (ii) there is a convergent series of positive constants $\sum_{n=1}^{\infty} M_n$ such that $|u_n(x)| \le M_n$ for all x in $\{a \le x \le b\}$ and for all $n = 1, 2, 3, \ldots$, then $f(x) = \sum_{n=1}^{\infty} u_n(x)$ is defined and uniformly continuous on $\{a \le x \le b\}$.

†Of course by Theorem 2, §9.4, every continuous function on $\{a \le x \le b\}$ is uniformly continuous.

6 Let $f(x) = a_0 + a_1x + a_2x^2 + a_3x^3 + \cdots$ be a function defined by a power series (for all values of x where this series converges). Show that if $\bar{x}$ is a point such that $f(\bar{x})$ is defined, then $f(x)$ is defined and uniformly continuous on any interval $\{-|\bar{x}| + \epsilon \le x \le |\bar{x}| - \epsilon\}$ for $\epsilon > 0$. [This is an application of the 'M-test'. If $\sum a_n\bar{x}^n$ converges then the absolute values $|a_n\bar{x}^n|$ must be bounded, from which $\sum |a_i|\rho^i$ can be shown to converge for $0 < \rho < |\bar{x}|$. Set $M_n = |a_n|\rho^n$ for ρ near $|\bar{x}|$.]

7 Show that if the power series for $1/(1+x)$ is multiplied formally by the power series for $1/(1-x)$ the result is the power series for $1/(1-x^2)$.

8 *Multiplication of power series.* Let $f(x) = a_0 + a_1x + a_2x^2 + \cdots$ and $g(x) = b_0 + b_1x + b_2x^2 + \cdots$ be two functions defined by power series. Show that if $\bar{x}$ is a point such that both $f(\bar{x})$ and $g(\bar{x})$ are defined then the power series

$$(*) \quad a_0b_0 + (a_0b_1 + a_1b_0)x + (a_0b_2 + a_1b_1 + a_2b_0)x^2 + \cdots + (a_nb_0 + a_{n-1}b_1 + \cdots + a_0b_n)x^n + \cdots$$

converges to $f(x)g(x)$ whenever $|x| < |\bar{x}|$. [Show that the double series $\sum_{m=0}^{\infty}\sum_{n=0}^{\infty} a_mb_nx^nx^m$ is *absolutely* convergent for $|x| < |\bar{x}|$, hence can be summed in any order and, in particular, in the order (*).]

9 Operating formally—i.e. without justifying all steps—write $\log\left(1 - \frac{x^2}{n^2\pi^2}\right)$ as a power series in x, then sum over n to obtain the power series for $\log(\sin x/x)$. A power series

$\log(\sin x/x) = a_2x^2 + a_4x^4 + a_6x^6 + \cdots$ can also be found by the following method: The power series

$$2a_2x + 4a_4x^3 + 6a_6x^5 + \cdots$$

represents

$$\frac{d}{dx}\log(\sin x/x)$$

and should therefore satisfy

$$\left(\frac{d}{dx}\log(\sin x/x)\right) x \sin x = x \cos x - \sin x.$$

Using the power series for $\sin x$, $\cos x$, multiplying, and equating coefficients gives equations which can be solved for $a_2, a_4, a_6, \ldots$ successively. Find a_2, a_4, a_6. The two formulas for a_2 give

$$1 + \frac{1}{2^2} + \frac{1}{3^2} + \cdots = \frac{\pi^2}{6}$$

and the two formulas for a_4 give

$$1 + \frac{1}{2^4} + \frac{1}{3^4} + \cdots = \frac{\pi^4}{90}.$$

Find

$$1 + \frac{1}{2^6} + \frac{1}{3^6} + \frac{1}{4^6} + \cdots.$$

10 *The factorial function.* Formally, the product formula for $\sin x$ can be written

$$\sin x = x \cdot \prod_{n=1}^{\infty}\left(1 + \frac{x}{n\pi}\right)\prod_{n=1}^{\infty}\left(1 - \frac{x}{n\pi}\right),$$

but this is in fact meaningless because these two infinite products diverge (unless $x = 0$). However, the limit

$$\lim_{N\to\infty}\prod_{n=1}^{N}\left(1 + \frac{x}{n\pi}\right)N^{-x/\pi}$$

does exist for all x, and defines a function $F(x)$ which satisfies $\sin x = x \cdot F(x) \cdot F(-x)$:

(a) Rather than the limit above, it is simpler to set $y = x/\pi$ and to consider

$$\lim_{N\to\infty}\prod_{n=1}^{N}\left(1 + \frac{y}{n}\right)\cdot N^{-y}.$$

By direct evaluation, show that for $y = 1, 2, 3, \ldots$ this limit exists and is equal to $1/y!$. Thus if one defines a function $\prod(y)$ by the equation

$$(*) \qquad \frac{1}{\prod(y)} = \lim_{N\to\infty}\prod_{n=1}^{N}\left(1 + \frac{y}{n}\right)N^{-y}$$

whenever this limit exists and is not zero, the function $\prod(y)$ is an *extension* of the factorial function.

(b) What value for 0! is indicated by the formula

$$\binom{n}{k} = \frac{n!}{k!(n-k)!} \,?$$

Does this agree with (*)?

(c) Show that the limit (*) exists for all values of y, and that it defines $\prod(y)$ for all values of y except $y = -1, -2, -3, \dots$. [Rewriting the limit (*) as

$$[1 + y]\left[\left(1 + \frac{y}{2}\right)\left(\frac{2}{1}\right)^{-y}\right] \cdots \left[\left(1 + \frac{y}{N}\right)\left(\frac{N}{N-1}\right)^{-y}\right]$$

the problem is to show that the product $a_1 a_2 a_3 \dots$ with

$$\begin{aligned} a_n &= \left(1 + \frac{y}{n}\right)\left(\frac{n-1}{n}\right)^y \\ &= \left(1 + \frac{y}{n}\right)\left(1 - \frac{1}{n}\right)^y \end{aligned}$$

converges. Now

$$\begin{aligned} a_n &= \left(1 + \frac{y}{n}\right)\left(1 - \frac{y}{n} + \text{terms in } \frac{1}{n^2}, \frac{1}{n^3}, \dots\right) \\ &= 1 + \text{terms in } \frac{1}{n^2}, \frac{1}{n^3}, \dots \end{aligned}$$

hence for large values of n the factors a_n are like $\left(1 + \frac{\text{const.}}{n^2}\right)$, which indicates that their product converges, as desired. To make this a rigorous proof, prove that there is a constant K such that

$$|\log(1 + x) - x| < Kx^2$$

for all sufficiently small x, hence that $|\log a_n| < K(y^2 + |y|)\frac{1}{n^2}$, hence that $\sum_{n=N}^{\infty} \log a_n$ converges, hence that $\prod_{n=1}^{\infty} a_n$ converges. The product is zero only if a factor is zero.]

(d) Prove the formulas

$$\sin \pi y = \frac{\pi y}{\prod(y)\prod(-y)}$$

and

$$\prod(y) = y\prod(y - 1).$$

(e) Combining the formulas of (d), prove that $\prod(-\frac{1}{2}) =$

$\sqrt{\pi}$ and derive from this the values of $\prod(\frac{1}{2}), \prod(\frac{3}{2}), \ldots \prod(n-\frac{1}{2})$, and $\prod(-\frac{3}{2}), \prod(-\frac{5}{2}), \ldots, \prod(-n-\frac{1}{2})$.

(f) Plot the values of $1/\prod(y)$ for $y = -5, -4\frac{1}{2}, -4, \ldots, 3\frac{1}{2}, 4, 4\frac{1}{2}, 5$. Make some guesses as to the other values and sketch the graph of $1/\prod(y)$.

(g) Prove that

$$\frac{2^{2x}\prod(x)\prod(x-\frac{1}{2})}{\prod(2x)}$$

is a constant independent of x. [It can be written as the limit of a quantity independent of x.] Evaluate the constant by setting $x = 0$.

(h) Prove that for any positive integer n

$$\frac{n^{nx}\prod(x)\prod\left(x-\frac{1}{n}\right)\prod\left(x-\frac{2}{n}\right)\cdots\prod\left(x-\frac{n-1}{n}\right)}{\prod(nx)}$$

is a constant independent of x.

(i) Let μ_n denote the constant of (h) for $n = 1, 2, 3, \ldots$. Use the formulas of (d) to write μ_n^2 in terms of π and $n-1$ values of $\sin x$. Find μ_n for $n = 2, 3, 4, 6$ and guess the general formula.

(j) Verify the guess of (i) for odd values of n using the formula for $\sin nA$ as a product (derived in the proof of formula (3)′ of the text). Using an analogous trigonometric identity, prove the formula for even values of n. The result is the *multiplication formula*

$$\frac{\prod(nx)}{n^{nx}} = \sqrt{n}\,(2\pi)^{(1-n)/2}\prod(x)\prod\left(x-\frac{1}{n}\right)\cdots\prod\left(x-\frac{n-1}{n}\right).$$

(k) Using a formula from the text and a change of variable, prove that

$$\prod\left(n-\frac{1}{2}\right) = \int_0^\infty t^{n-1/2}e^{-t}\,dt$$

(for $n = 0, 1, 2, \ldots$).

(l) Show that the improper integral $\int_0^\infty t^n e^{-t}\,dt$ converges for n a positive integer or zero, and use integration by parts to show that its value is $n! = \prod(n)$.

(m) For what values of x does the improper integral $\int_0^\infty t^x e^{-t}\,dt$ converge? Since its value is $\prod(x)$ for $x = -\frac{1}{2}, 0, \frac{1}{2}, 1, 1\frac{1}{2}, \ldots$ it is reasonable to guess that its value is $\prod(x)$ whenever it converges. This is proved in the following exercise.

(n) *Notation.* The gamma function $\Gamma(x)$ is defined by $\Gamma(x) = \prod(x-1)$. There is no apparent justification for this awkward notation other than tradition. Any equation involving the Γ-function can be immediately translated into an equation involving the $\prod$-function, and the result will usually be simpler. Translate the multiplication formula for $\prod$ (see (*j*)) into the multiplication formula of the Γ-function.

11 *Binomial coefficients.* The binomial coefficient $\binom{x}{n}$ is defined for all real numbers x and for all integers $n > 0$ to be

$$\binom{x}{n} = \frac{x(x-1)(x-2)\dots(x-n+1)}{n!}$$

$$= \frac{\prod(x)}{\prod(n)\prod(x-n)}.$$

[These numbers occur in the power series expansion of $(1+t)^x$, namely $(1+t)^x = 1 + xt + \frac{x(x-1)}{2}t^2 + \frac{x(x-1)(x-2)}{6}t^2 + \cdots = \sum_{n=0}^{\infty}\binom{x}{n}t^n$. See §8.4.]

(a) Show that $1/\prod(x) = \lim_{n\to\infty}\binom{x+n}{n}n^{-x}$. Thus $\prod(x)$ can be expressed in terms of a limit of *reciprocals* of binomial coefficients. The integral representation $\prod(x) = \int_0^\infty t^x e^{-t}\,dt$ can be obtained from an integral representation (Euler's first integral) of the reciprocal binomial coefficients, namely the integral

$$C(x,y) = (x+y+1)\int_0^1 t^x(1-t)^y\,dt.$$

(b) For what values of x, y does the integral $C(x,y)$ converge?

(c) Show that $C(x,y) = C(y,x)$ and that $C(x,0) = 1$ for all x.

(d) Integrating by parts [use the derivative of $t^{x+y+1}(t^{-1}-1)^y$] show that

$$C(x,y) = \frac{y}{x+y}C(x,y-1).$$

(e) Conclude from the above that

$$C(x,n) = \frac{1}{\binom{x+n}{n}}$$

hence $\prod(x) = \lim_{n\to 0} n^x C(x,n)$.

(f) Using the change of variable $u = nt$ in the integral $C(x,n)$, show that for very large integers n the integral $n^x C(x,n)$ is nearly $\int_0^\infty u^x e^{-u}\,du$.

(g) Formulate the equation $\prod(x) = \int_0^\infty u^x e^{-u}\, du$ as an interchange of limits. This interchange will be justified in §9.7.

(h) In the integral

$$\prod(x)\prod(y) = \int_0^\infty \int_0^\infty s^x e^{-s} t^y e^{-t}\, ds\, dt$$

perform the change of variables $u = s + t$, $v = s/(s + t)$ (hence $s = uv$, $t = u(1 - v)$) to obtain

$$\prod(x)\prod(y) = \frac{C(x, y)}{x + y + 1}\prod(x + y + 1) = C(x, y)\prod(x + y).$$

Thus $C(x, y)$ is the reciprocal binomial coefficient

$$C(x, y) = \frac{\prod(x)\prod(y)}{\prod(x + y)}$$

for all values of x, y for which it converges. [No elaborate discussion of the change of variables in the improper integral $\prod(x)\prod(y)$ is necessary since the integrand is positive and therefore absolutely convergent, hence can be 'summed' in any order.]

(i) Notation. The beta function is the function defined by the integral

$$B(x, y) = \int_0^1 t^{x-1}(1 - t)^{y-1}\, dt.$$

Express $B(x, y)$ in terms of the $\prod$-function and in terms of the Γ-function.

12 The 'volume' of the n-dimensional ball

$$B = \{x_1^2 + x_2^2 + \cdots + x_n^2 \le r^2\}$$

is

$$\frac{\pi^{n/2}}{\prod(n/2)} r^n = \int_B dx_1\, dx_2 \ldots dx_n.$$

(a) Verify this formula in the cases $n = 1, 2, 3$.

(b) Prove the formula as follows: The integral $\pi^{n/2} = \int_{-\infty}^{\infty}\int_{-\infty}^{\infty} \ldots \int_{-\infty}^{\infty} e^{-r^2}\, dx_1\, dx_2 \ldots dx_n$ where $r = \sqrt{x_1^2 + x_2^2 + \cdots + x_n^2}$ follows from a formula in the text. Write $dx_1\, dx_2 \ldots dx_n = r^{n-1}\, dr\omega$ where ω is a particular closed $(n - 1)$-form. (See Exercise 4, §8.3.) By Stokes' Theorem the 'volume' of the unit ball is n^{-1} times the integral of ω over any $(n - 1)$-dimensional sphere $r = $ const. Carrying out the integral of ω in the formula for $\pi^{n/2}$ it becomes a constant times $\int_0^\infty e^{-r^2} r^{n-1}\, dr$. Setting $u = r^2$, this number can be expressed in terms of the function $\prod$ and the desired formula follows.

13 Euler's formula

$$1 - 3 + 5 - 7 + 9 - 11 + \cdots = 0$$

is often held up to ridicule by people who are foolish enough to imagine that Euler had an inadequate grasp of the notion of convergence. Prove that the formula is true when it is interpreted as an Abel sum. [To sum the series set $y^2 = x$, factor out y^2, and integrate once. The function $y + \frac{1}{y}$ has a minimum at $y = 1$.]

14 The *Cesaro sum* of a series $a_1 + a_2 + a_3 + \cdots$ is defined to be the limit, if it exists, of the arithmetic means of the partial sums, that is,

$$\lim_{N\to\infty} \frac{1}{N}[S_1 + S_2 + S_3 + \cdots + S_N]$$

where

$$S_N = a_1 + a_2 + \cdots + a_N.$$

When this limit exists the series is said to be *Cesaro summable*.

(a) Show that $1 - 1 + 1 - 1 + 1 - 1 + \cdots$ is Cesaro summable and that its Cesaro sum is $\frac{1}{2}$.

(b) Prove that a convergent series is Cesaro summable and that its Cesaro sum is its ordinary sum. [Using $S = \frac{1}{N}[S + S + \cdots + S]$ this reduces to the statement that if x_n is a sequence such that $\lim x_n = 0$ then $\lim \frac{1}{N}[x_1 + x_2 + \cdots + x_N] = 0$, which is easily proved.] Show that the series of Exercise 13 is not Cesaro summable.

9.7 Lebesgue Integration

An ordinary integral $\int_a^b f(x)\,dx$ is defined as a limit of sums $\sum_i f(x_i)\,\Delta x_i$ formed by subdividing the (finite) interval $\{a \le x \le b\}$ into small intervals, letting Δx_i be the length of the ith interval of the subdivision, and letting x_i be a point of the ith interval; if it is true that for every $\epsilon > 0$ there is a $\delta > 0$ such that the resulting sum is determined to within ϵ whenever all Δx_i's are less than δ (regardless of the choice of subdivision and points x_i) the integral is said to converge and the real number it determines is denoted $\int_a^b f(x)\,dx$.

There are, however, cases in which an integral $\int_a^b f(x)\,dx$ does not converge but in which there is no doubt as to what the value of the integral 'should' be, for example the integral

$$\text{(1)} \qquad \int_{-1}^{1} \frac{1}{\sqrt{|x|}}\,dx.$$

(The integrand is not defined at zero so, for the sake of definiteness, the value at 0 will be defined to be 0.) Because the integrand is unbounded, it is easily seen that the integral (1) does not converge. But if a small interval $\{-\epsilon_1 \le x \le \epsilon_2\}$ (ϵ_1, ϵ_2 small positive numbers) is excluded then the integral converges and its value is found to be

$$\begin{aligned}\int_{-1}^{-\epsilon_1} \frac{d}{dx}[-2\sqrt{|x|}]\,dx + \int_{\epsilon_2}^{1} \frac{d}{dx}[2\sqrt{|x|}]\,dx \\ = -2\sqrt{\epsilon_1} + 2 + 2 - 2\sqrt{\epsilon_2} \\ = 4 - 2(\sqrt{\epsilon_1} + \sqrt{\epsilon_2})\end{aligned}$$

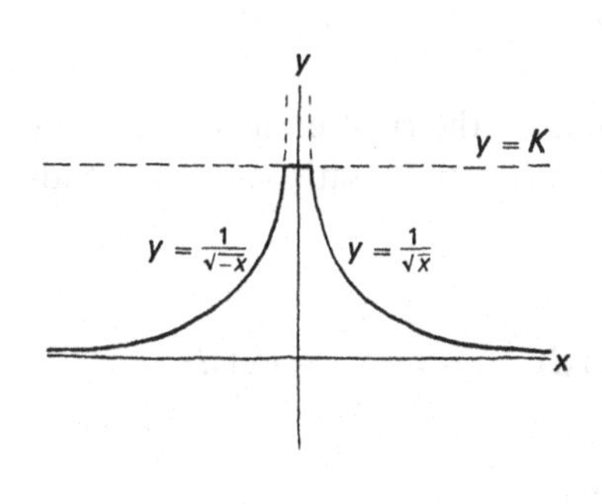

by the Fundamental Theorem. Thus the integral approaches the value 4 as the excluded interval is made small. Another way to 'evaluate' the non-convergent integral (1) would be to 'cut off the top', that is, to set $f_K(x) = \min(|x|^{-1/2}, K)$, to evaluate the convergent integral $\int_{-1}^{1} f_K(x)\,dx$, and to let $K \to \infty$. This gives

$$\begin{aligned}\int_{-1}^{-K^{-2}} \frac{1}{\sqrt{|x|}}\,dx + \int_{-K^{-2}}^{K^{-2}} K\,dx + \int_{K^{-2}}^{1} \frac{1}{\sqrt{|x|}}\,dx \\ = -2K^{-1} + 2 + K\cdot 2K^{-2} + 2 - 2K^{-1} \\ = 4 - \frac{4}{K} + \frac{2}{K} = 4 - \frac{2}{K}\end{aligned}$$

and again as $K \to \infty$ the limit is 4. Thus, even though the integral (1) is not convergent, its value 'should' be 4.

As this example indicates, there is an extended definition of $\int_a^b f(x)\,dx$ which coincides with the definition already given for all convergent integrals $\int_a^b f(x)\,dx$ but which is valid for other integrals, for example (1), which are not convergent in the original definition. This extended definition is called the *Lebesgue integral* in contradistinction to the Riemann integral which is the definition given above.

Theorem

*In accordance with the previous chapters, the integrand should be considered to be a 1-form $f(x)\,dx$, not a function $f(x)$. The distinction is of course purely pedantic and the common usage is to consider the integrand as a function and to let dx represent the 'measure' used in defining $\int_a^b f(x)\,dx$, namely 'oriented length'.

†But not 'only if'. See below.

Lebesgue integration. A function* $f(x)$ on an interval $\{a \le x \le b\}$ is said to be *Lebesgue integrable* iff† there exists a sequence $A_1(x)$, $A_2(x)$, $A_3(x)$, . . . of functions on $\{a \le x \le b\}$ such that:

(i) $\lim_{n\to\infty} A_n(x) = f(x)$ for all x in the interval $\{a \le x \le b\}$.

(ii) The (Riemann) integrals $\int_a^b A_n(x)\,dx$ converge $(n = 1, 2, 3, \ldots)$, that is, the approximating sums approach a limit in the sense defined above.

(iii) The Cauchy Criterion $\lim_{n,m\to\infty} \int_a^b |A_n(x) - A_m(x)|\,dx = 0$ is satisfied. More specifically, the integral $\int_a^b |A_n(x) - A_m(x)|\,dx$ (which converges by dint of (ii)) is less than any preassigned ϵ for all sufficiently large n, m.

‡Note the similarity of this theorem to the theorem of §6.3. The integral is defined by a process which involves choices, and it must be shown that the result is independent of these choices.

If $f(x)$ is Lebesgue integrable then the definition‡

$$\int_a^b f(x)\,dx = \lim_{n\to\infty} \int_a^b A_n(x)\,dx$$

is valid, that is, the limit on the right exists and if $B_n(x)$ is any other sequence of functions satisfying the conditions (i)–(iii) then

$$\lim_{n\to\infty} \int_a^b B_n(x)\,dx = \lim_{n\to\infty} \int_a^b A_n(x)\,dx.$$

Examples and Applications

a. The function of (1) is Lebesgue integrable (defining its value at 0 to be 0) as is shown by the sequence

$$A_n(x) = \begin{cases} \dfrac{1}{\sqrt{|x|}} & \dfrac{1}{n} \le |x| \le 1 \\ 0 & |x| < \dfrac{1}{n}. \end{cases}$$

The integral $\int_{-1}^{1} f(x)\,dx$ is therefore defined and is $\lim_{n\to\infty} \int_{-1}^{1} A_n(x)\,dx = 4$. If the sequence

$$B_n(x) = \begin{cases} \min\left(n, \dfrac{1}{\sqrt{|x|}}\right) & x \ne 0 \\ 0 & x = 0 \end{cases}$$

is used instead, the value

$$\int_{-1}^{1} f(x)\,dx = \lim_{n\to\infty} \int_{-1}^{1} B_n(x)\,dx = 4$$

is the same. The fact that the sequences $A_n(x)$, $B_n(x)$ satisfy the Cauchy Criterion (iii) is immediately verified using the Fundamental Theorem.

b. If $f(x)$ is Riemann integrable, that is, if $\int_a^b f(x)\,dx$ converges in the ordinary sense, then one can set $A_n(x) \equiv f(x)$ (all n). Thus $f(x)$ is Lebesgue integrable and the new definition of $\int_a^b f(x)\,dx$ coincides with the old one. This shows that the new definition is indeed an extension of the old one.

c. The Lebesgue integral $\int_{-1}^{1} |x|^{-a}\,dx$ is

$$\lim_{\epsilon_1,\epsilon_2\to 0}\left\{\int_{-1}^{-\epsilon_1} \frac{d}{dx}\left[-\frac{|x|^{1-a}}{1-a}\right]dx + \int_{\epsilon_2}^{1} \frac{d}{dx}\left[-\frac{|x|^{1-a}}{1-a}\right]dx\right\} = \frac{2}{1-a}$$

provided $a < 1$. This is a simple extension of the first example. If $a \le 0$ this integral converges as a Riemann integral. If $a \ge 1$ the function $|x|^{-a}$ is not Lebesgue integrable on $\{-1 \le x \le 1\}$ (see Exercise 6).

d. Let $f(x)$ be the function which is 1 if x is a rational number and 0 if x is irrational. Then $f(x)$ is Lebesgue integrable on $\{a \le x \le b\}$ and $\int_a^b f(x)\,dx = 0$. This is seen by defining

$$A_n(x) = \begin{cases} 1 \text{ if } x \text{ is a rational number with a} \\ \quad \text{denominator} \le n \\ 0 \text{ otherwise.} \end{cases}$$

Then A_n and $|A_n - A_m|$ are zero except at a finite number of points (where they are 1) which implies that $\int_a^b A_n(x)\,dx = 0$ as a Riemann integral and that the Cauchy Criterion is satisfied. For any fixed x, $\lim_{n\to\infty} A_n(x) = f(x)$, hence $\int_a^b f(x)\,dx = 0$.

e. Let $f(x)$ be the function which is zero if the decimal expansion of the real number x contains the digit 5 and which is one otherwise. (If x has two decimal expansions—that is, if x is a terminating decimal fraction—then $f(x)$ is ill-defined if one of these expansions has a 5 and the other does not, e.g. if $x = .6 = .5999\ldots$. For the sake of definiteness $f(x)$ will be defined to be one if either expansion has a 5, e.g. $f(.6) = 1$.) Then $\int_0^1 f(x)\,dx$ is defined (as a

Lebesgue integral) and $\int_0^1 f(x)\,dx = 0$. This can be proved by setting

$$A_n(x) = \begin{cases} 0 \text{ if the first } n \text{ places of the decimal} \\ \quad \text{expansion of } x \text{ contain a 5} \\ 1 \text{ if not} \end{cases}$$

(with $A_3(.260) = A_3(.2599\ldots) = 1$, etc., as for f). Then $A_1(x)$ is zero for $\{.5 < x < .6\}$ and one elsewhere in $\{0 \le x \le 1\}$. Thus $\int_0^1 A_1(x)\,dx = \frac{9}{10}$. Since $A_2(x)$ is zero on $\{.5 < x < .6\}$ and on the nine intervals $\{.05 < x < .06\}, \ldots, \{.95 < x < .96\}$ omitting $\{.55 < x < .56\}$ it follows in the same way that $\int_0^1 A_2(x)\,dx = \frac{81}{100}$. A similar argument shows that

$$\int_0^1 A_n(x)\,dx = \left(\frac{9}{10}\right)^n.$$

Thus $\int_0^1 A_n(x)\,dx \to 0$ as $n \to \infty$, and moreover

$$\int_0^1 |A_n(x) - A_m(x)|\,dx$$

$$\le \int_0^1 A_n(x)\,dx + \int_0^1 A_m(x)\,dx \to 0$$

as well. Therefore the conditions (i)–(iii) are satisfied and $\int_0^1 f(x)\,dx = 0$. In the theory of probability, this fact is stated, "The probability that a real number selected at random will have no 5's in its decimal expansion is zero." This accords with the common experience that the probability of rolling a die infinitely often without rolling a 5 is zero.

f. Lebesgue integrals combine in all the same ways as ordinary integrals, e.g.

$$\int_a^b (f(x) + g(x))\,dx = \int_a^b f(x)\,dx + \int_a^b g(x)\,dx$$

$$\int_a^b cf(x)\,dx = c\int_a^b f(x)\,dx \qquad (c = \text{const.})$$

$$\int_a^b |f(x)|\,dx \ge \left|\int_a^b f(x)\,dx\right|$$

$$\int_a^b \max(f, g)\,dx \ge \max\left(\int_a^b f(x)\,dx, \int_a^b g(x)\,dx\right)$$

etc., where the integrability of these functions is a consequence of the integrability of f and g. [The function $\max(f, g)$ assigns to each x the larger of the two numbers $f(x)$, $g(x)$.] These statements follow immediately from the analogous statements for ordinary Riemann integrals. The last identity above has the important consequence that

$$f(x) \geq g(x) \text{ on } \{a \leq x \leq b\} \quad \text{implies} \int_a^b f(x)\,dx \geq \int_a^b g(x)\,dx.$$

[$f \geq g$ means $\max(f, g) \equiv f$.]

Proof

Given a sequence of Riemann integrable functions A_n, the Riemann integrability of $|A_n - A_m|$ is easily proved. (The difference of Riemann integrable functions and the absolute value of a Riemann integrable function are Riemann integrable; see Exercise 1). If the Cauchy Criterion $\lim_{n,m\to\infty} \int_a^b |A_n - A_m|\,dx = 0$ is satisfied then the triangle inequality $|\int_a^b A_n\,dx - \int_a^b A_m\,dx| \leq \int_a^b |A_n - A_m|\,dx$ implies that $\lim_{n\to\infty} \int_a^b A_n\,dx$ exists. The difficult part of the proof is to show that if B_n is another sequence satisfying (i)–(iii) and if $\lim_{n\to\infty} B_n(x) = f(x) = \lim_{n\to\infty} A_n(x)$, then $\lim_{n\to\infty} \int_a^b B_n(x)\,dx = \lim_{n\to\infty} \int_a^b A_n(x)\,dx$. Setting

$$C_n(x) = B_n(x) - A_n(x)$$

the triangle inequality shows that $C_n(x)$ satisfies the Cauchy Criterion and the statement to be proved becomes: *If $C_n(x)$ is a sequence of Riemann integrable functions on $\{a \leq x \leq b\}$ such that (i) $\lim_{n\to\infty} C_n(x) = 0$ for all x in $\{a \leq x \leq b\}$, and such that (ii)*

$$\lim_{n,m\to\infty} \int_a^b |C_n(x) - C_m(x)|\,dx = 0,$$

then $\lim_{n\to\infty} \int_a^b C_n(x)\,dx = 0$. Seen in this light, the theorem simply states that the interchange of limits

$$\lim_{n\to\infty} \int_a^b C_n(x)\,dx = \int_a^b \lim_{n\to\infty} C_n(x)\,dx$$

is valid when $\lim_{n,m\to\infty} \int_a^b |C_n(x) - C_m(x)|\,dx = 0$ and $C_n(x) \to 0$ for all x in the interval.

Note first that it suffices to consider the case $C_n(x) \geq 0$ since the general case will then follow by considering* $|C_n(x)|$. Given a sequence $C_n(x)$ of non-negative Riemann integrable functions such that $\lim_{n\to\infty} C_n(x) = 0$ and such that $\lim_{n,m\to\infty} \int_a^b |C_n(x) - C_m(x)|\,dx = 0$, let $D_{n,K}$ denote the function $\min(C_n, K)$ for K a constant. (If $K \leq 0$ then $D_{n,K} \equiv K$ and only the cases $K > 0$ are of interest.) Intuitively, $D_{n,K}$ is 'C_n with the top cut off at K'. For each fixed $K > 0$ the sequence $D_{n,K}(x)$ $(n = 1, 2, 3, \ldots)$ is a sequence of Riemann integrable functions. Moreover, $\lim_{n\to\infty} D_{n,K}(x) = 0$ for all x and

*The inequality $|C_n - C_m| \geq ||C_n| - |C_m||$ shows that if C_n satisfies the Cauchy Criterion then $|C_n|$ does.

$$\lim_{n,m\to\infty} \int_a^b |D_{n,K} - D_{m,K}|\,dx = 0.$$

Hence $\lim_{n\to\infty} \int_a^b D_{n,K}(x)\,dx$ exists and, if the theorem is true, this limit is zero for all K.

Let $I_K = \lim_{n\to\infty} \int_a^b [C_n - D_{n,K}]\,dx$. Intuitively I_K measures the amount of C_n which, in the limit as $n \to \infty$, lies above K. If the theorem is true then

$$I_K = \lim_{n\to\infty} \int_a^b C_n\,dx - \lim_{n\to\infty} \int_a^b D_{n,K}\,dx = 0 - 0 = 0$$

for all $K > 0$. Conversely, if it can be shown that $I_K = 0$ for all $K > 0$, then $\lim_{n\to\infty} \int_a^b C_n\,dx = \lim_{n\to\infty} \int_a^b D_{n,K}\,dx \leq K(b - a)$ for all $K > 0$ and the desired conclusion $\lim_{n\to\infty} \int_a^b C_n\,dx = 0$ will follow.

If $K > K'$ then $I_K \leq I_{K'}$. Moreover, for every $\epsilon > 0$ there is a K such that $I_K \leq \epsilon$; this is proved by choosing N so large that $\int_a^b |C_n - C_N|\,dx < \epsilon/2$ whenever $n > N$ and by then choosing K so large that $C_N(x) \leq K$ for all x in $\{a \leq x \leq b\}$ (a Riemann integrable function is bounded). Then $C_N \equiv D_{N,K}$ hence

$$\begin{aligned}\int_a^b (C_n - D_{n,K})\,dx &= \int_a^b [(C_n - D_{n,K}) - (C_N - D_{N,K})]\,dx \\ &\leq \int_a^b |C_n - C_N|\,dx + \int_a^b |D_{n,K} - D_{N,K}|\,dx \\ &\leq 2\cdot(\epsilon/2) = \epsilon\end{aligned}$$

for all n, hence $I_K \le \epsilon$. It follows that if the desired conclusion ($I_K = 0$ for $K > 0$) is false then there must be positive numbers K, K', c such that $I_{K'} - I_K > c$ ($K > K'$). Now $I_{K'} - I_K = \lim_{n\to\infty} \int_a^b (D_{n,K} - D_{n,K'})\,dx$, hence if this is the case there is an N such that $\int_a^b (D_{n,K} - D_{n,K'})\,dx > c$ for $n \ge N$. The integral $\int_a^b (D_{n,K} - D_{n,K'})\,dx$ is a limit of approximating sums $\sum \tilde{C}_n(x_i)\,\Delta x_i$ where $\tilde{C}_n(x)$ is zero if $C_n(x) \le K'$, $C_n(x) - K'$ if $K' \le C_n(x) \le K$, and $K - K'$ if $C_n(x) \ge K$. Given $n \ge N$ let δ_n be so small that any approximating sum to the integral is greater than c provided that the subdivision is finer than δ_n, and let S_n be a subdivision finer than δ_n. Choose an x_i in each interval of S_n in such a way that $\tilde{C}(x_i) = 0$ if this is possible, i.e. if the interval contains any point x_i where $C(x_i) \le K'$. Let T_n denote the set of intervals of S_n where this is not possible, i.e. where all values of $C_n(x)$—including the values at the end points—are at least K'. Since

$$c < (K - K')(\text{total length of } T_n)$$

it follows that *if* $\int_a^b C_n(x)\,dx$ *does not approach zero then there are positive numbers* K' *and* $c' = c/(K - K')$ *and an integer* N *such that for every* $n \ge N$ *there is a finite collection of (closed) intervals* T_n *of total length greater than* c' *on which* $C_n(x)$ *assumes no value less than* K'. It is to be shown that this contradicts the assumptions $C_n(x) \to 0$ and $\int_a^b |C_n - C_m|\,dx \to 0$.

The sequence $C_n(x)$ can be 'weeded out' so that the Cauchy Criterion takes the form

$$(2) \qquad \int_a^b |C_n(x) - C_{n+k}(x)|\,dx < 10^{-n}.$$

One need only find an N such that $\int_a^b |C_n - C_N|\,dx < \frac{1}{10}$ for $n > N$, discard the first $N - 1$ functions in the sequence, and renumber $C_n = C_{n-N+1}$; then find an N such that $\int_a^b |C_n - C_N|\,dx < \frac{1}{100}$ for $n > N$, discard $C_2, C_3, \ldots, C_{N-1}$, and renumber. Continuing in this way, (2) is satisfied.

Now since $\int_a^b |C_2(x) - C_1(x)|\,dx < \frac{1}{10}$ there is a δ such that any approximating sum to this integral based on a subdivision finer than δ is also less than $\frac{1}{10}$. Consider the set where $|C_2(x) - C_1(x)| \ge \frac{1}{2}$. Roughly speaking, this set cannot be large because the integral is small. Quantitatively, let S be a subdivision of $\{a \le x \le b\}$

finer than δ, and let U_1 be the collection of intervals of S which contain points x where $|C_2(x) - C_1(x)| \geq \frac{1}{2}$. By forming an approximating sum to $\int_a^b |C_2(x) - C_1(x)|\,dx$ by choosing such points wherever possible, it follows that

$$\tfrac{1}{2} \cdot (\text{total length of intervals in } U_1) < \tfrac{1}{10}$$

hence the intervals of U_1 have total length less than $\frac{1}{5}$. Extending the intervals of U_1 slightly but keeping their total length less than $\frac{1}{5}$, they can be made to include all points x where $|C_2(x) - C_1(x)| \geq \frac{1}{2}$ in their *interiors.*

By exactly the same argument, the set of all points x for which $|C_3(x) - C_2(x)| \geq 2^{-2}$ can be contained in the interiors of a finite collection U_2 of intervals whose total length is less than $2^2 \cdot 10^{-2} = 5^{-2}$, and the set of all points x for which $|C_{n+1}(x) - C_n(x)| \geq 2^{-n}$ can be contained in the interiors of a finite collection U_n of intervals of total length less than 5^{-n}.

Consider now the set where $C_1(x) > 1$. Since $\lim_{n\to\infty} C_n(x) = 0$ there must be an N such that

$$|C_1(x) - C_N(x)| > 1$$
$$|C_1(x) - C_2(x)| + |C_2(x) - C_3(x)| + \cdots + |C_{N-1}(x) - C_N(x)| > 1$$

hence there must be at least one n such that

$$|C_{n+1}(x) - C_n(x)| \geq 2^{-n};$$

that is, if $C_1(x) > 1$ then x must be interior to an interval of U_n for some n. Therefore the set where $C_1(x) > 1$ can be contained in the interiors of an infinite collection of intervals whose total length* is less than $5^{-1} + 5^{-2} + 5^{-3} + \cdots = \frac{1}{4}$. In the same way, if $C_2(x) > \frac{1}{2}$ then x must be interior to one of the intervals of U_n for $n \geq 2$, and these intervals have total length less than $5^{-2} + 5^{-3} + 5^{-4} + \cdots = \frac{1}{20}$. Finally, *the set of all x for which $C_n(x) > 2^{-n}$ can be contained in the interiors of an infinite collection of intervals* (those of $U_n, U_{n+1}, U_{n+2}, \ldots$) *of total length*† *less than* $4^{-1} \cdot 5^{1-n}$.

*Because the lengths are positive, their total is well defined even though there are infinitely many of them. Very simply, to say that their total is less than 4^{-1} means that the total length of any finite number of the intervals is less than 4^{-1}.

†Again meaning that the total length of any finite number of them is less than $4^{-1} \cdot 5^{1-n}$.

It was shown above that if $\int_a^b C_n(x)\,dx$ does not approach zero as $n \to \infty$ then there exist positive numbers c', K' and an integer N such that for every $n \geq N$ there is a finite collection T_n of (closed) intervals of total length at least c' on which *all* values of $C_n(x)$ are $\geq K'$. Choose n so large that $c' > 4^{-1} \cdot 5^{1-n}$, $K' > 2^{-n}$, $n \geq N$, and let T_n be as above. Every point of T_n is interior to at least one of the intervals of U_n, U_{n+1}, $U_{n+2}, \ldots$. Since T_n is compact, the Heine-Borel Theorem states that it can be covered by a finite number of intervals of U_n, U_{n+1}, $U_{n+2}, \ldots$, and hence by a finite number of intervals whose total length is less than $4^{-1} \cdot 5^{1-n} < c'$. But a finite number of intervals of total length $> c'$ cannot be covered by a finite number of intervals of total length $< c'$, which shows that the assumption $\lim_{n\to\infty} \int_a^b C_n(x)\,dx \neq 0$ is untenable. Therefore $\lim_{n\to\infty} \int_a^b C_n(x)\,dx = 0$ and the theorem is proved.

An important strengthening of the theorem results from the observation that the proof just given is still valid even if there are points x in $\{a \le x \le b\}$ where $\lim_{n\to\infty} C_n(x) \neq 0$, provided that there are so 'few' of them that they cannot account for the length c' of the intervals T_n which result from the assumption that $\int_a^b C_n(x)\,dx$ does not approach zero. For this purpose it suffices to assume that for every $\epsilon > 0$ there is a (possibly infinite) collection of intervals of total length less than ϵ which contain in their interiors all points x where $\lim_{n\to\infty} C_n(x) = 0$ is false. This is what it means to say that '$C_n(x) \to 0$ for almost all x'.

Definition

A sequence $f_1(x), f_2(x), f_3(x), \ldots$ of functions on an interval $\{a \le x \le b\}$ is said to converge to a function $f(x)$ on $\{a \le x \le b\}$ 'for almost all x in $\{a \le x \le b\}$' if the set of values of x for which $\lim_{n\to\infty} f_n(x) = f(x)$ is false can be contained in the interior of a collection of intervals of arbitrarily small total length. More precisely, '$\lim_{n\to\infty} f_n(x) = f(x)$ for almost all x in $\{a \le x \le b\}$' means that for every $\epsilon > 0$ there is a collection of subintervals of $\{a \le x \le b\}$ such that (i) the total length of any finite number of intervals in the collection is less than ϵ, and such that (ii) for any point x not interior to an interval of the collection, $\lim_{n\to\infty} f_n(x) = f(x)$. Similarly, two functions $f(x)$, $g(x)$ are said to be equal 'for almost all x in $\{a \le x \le b\}$' if the set of all values where $f(x) \neq g(x)$ can be contained in the interiors of a collection of subintervals of arbitrarily small total length.

The proof above then shows that the theorem is still true when the word 'almost' is inserted in condition (i). One need only note that if $A_n(x)$, $B_n(x)$ satisfy conditions (i)–(iii) then $C_n = A_n - B_n$ satisfies $C_n(x) \to 0$ for almost all x. The proof then proceeds exactly as before except that one must add to the collection of intervals $U_n, U_{n+1}, U_{n+2}, \ldots$ a collection U_∞ of arbitrarily small total length, outside which $C_n(x) \to 0$. Then every point x of T_n is either interior to an interval of U_∞ or $\lim_{n\to\infty} C_n(x) = 0$, in which case it must be interior to an interval of $U_n, U_{n+1}, U_{n+2}, \ldots$. In this way T_n can still be covered by a collection of intervals of arbitrarily small total length and the assumption $\lim_{n\to\infty} \int_a^b C_n(x)\,dx \neq 0$ can still be contradicted. This proves the following theorem.

Theorem

Lebesgue integration, revised statement. A function $f(x)$ on the interval $\{a \le x \le b\}$ is said to be *Lebesgue integrable* if (and only if) there exists a sequence $A_1(x)$, $A_2(x), A_3(x), \ldots$ of functions on $\{a \le x \le b\}$ such that:

(i) $\lim_{n\to\infty} A_n(x) = f(x)$ for almost all x in $\{a \le x \le b\}$.

(ii) The Riemann integrals $\int_a^b A_n(x)\,dx$ converge $(n = 1, 2, 3, \ldots)$.

(iii) The Cauchy Criterion

$$\lim_{n,m\to\infty} \int_a^b |A_n(x) - A_m(x)|\,dx = 0$$

is satisfied.

When this is the case, the definition

$$\int_a^b f(x)\,dx = \lim_{n\to\infty} \int_a^b A_n(x)\,dx$$

is valid, that is, the limit on the right exists and depends only on f.

Examples and Applications

g. If $f(x)$ is a function which is integrable in the first definition then it is integrable in the revised definition and the two definitions of the integral agree. Thus the previous examples are all examples of Lebesgue integrable functions (revised definition).

Hereafter, 'Lebesgue integrable' will refer only to the revised definition.

h. There 'exist' Lebesgue integrable functions which are not Lebesgue integrable in the first definition. The proof of this fact is highly non-constructive, however, and no simple *example* of such a function can be given.

i. The revised definition simplifies the treatment of the integral (1). Setting $A_n(x) = \min(n, |x|^{-1/2})$ the conditions (i)–(iii) are satisfied (even though $\lim_{n\to\infty} A_n(0)$ does not exist) no matter what value is assigned to the integrand at $x = 0$.

j. If f, g are two functions on $\{a \leq x \leq b\}$, if $f(x) = g(x)$ for almost all x in the interval, and if $f(x)$ is Lebesgue integrable, then $g(x)$ is Lebesgue integrable and $\int_a^b g(x)\,dx = \int_a^b f(x)\,dx$. One can prove directly that the functions $f(x)$ of Examples 4 and 5 above are equal to zero for almost all x and hence conclude that their integrals are zero (see Exercise 3).

Apart from the desire to define $\int_a^b f(x)\,dx$ for as many functions $f(x)$ as possible, an important motivation of the theory of Lebesgue integration is the desire to establish the validity of the interchange of limits

$$(2) \qquad \lim_{n\to\infty} \int_a^b f_n(x)\,dx = \int_a^b (\lim_{n\to\infty} f_n(x))\,dx$$

under the weakest possible assumptions on the sequence of functions $f_n(x)$. As is shown by the examples a and d above, it is necessary to extend the definition of $\int_a^b f(x)\,dx$ beyond the ordinary (Riemann) definition to avoid cases where the integral on the right side of (2) is not defined even though it 'should' be. The theory of Lebesgue integration makes possible the following amazingly general theorem on the interchange (2).

Lebesgue Dominated Convergence Theorem

The interchange (2) is valid whenever there is an integrable function $F(x)$ on $\{a \leq x \leq b\}$ which 'dominates' the functions $f_n(x)$ in the sense that for each n the inequality $F(x) \geq |f_n(x)|$ holds for almost all x. More precisely, if $f_1(x), f_2(x), f_3(x), \ldots,$ and $f_\infty(x)$ are functions on $\{a \le x \le b\}$, if $\lim_{n\to\infty} f_n(x) = f_\infty(x)$ for almost all x, if there is a function $F(x)$ such that each $f_n(x)$ satisfies $|f_n(x)| \le F(x)$ for almost all x, and if $F(x)$ and $f_1(x), f_2(x), \ldots$ are all Lebesgue integrable on $\{a \le x \le b\}$, then $f_\infty(x)$ is Lebesgue integrable on $\{a \le x \le b\}$ and

$$\lim_{n\to\infty} \int_a^b f_n(x)\,dx = \int_a^b f_\infty(x)\,dx.$$

A proof of this theorem is given below. Note that the Lebesgue Dominated Convergence Theorem proves immediately that the interchange

$$\begin{aligned}
\log(1+x)& \\
&= \int_0^x \frac{1}{1+t}\,dt \\
&= \int_0^x [\lim_{n\to\infty} (1 - t + t^2 - t^3 \cdots + (-t)^n)]\,dt \\
&= \lim_{n\to\infty} \left[\int_0^x (1 - t + t^2 - \cdots + (-t)^n)\,dt\right] \\
&= x - \frac{x^2}{2} + \frac{x^3}{3} - \frac{x^4}{4} + \cdots
\end{aligned}$$

is valid for $-1 < x \le 1$; it suffices to note that all functions are dominated by $F(t) = \max\left(\frac{1}{1+t}, 1\right)$ and that this function is integrable on the interval in question.

The generalization of Lebesgue integration to higher dimensions and to improper integrals is straightforward. For example, a function $f(x, y)$ is said to be Lebesgue integrable on a rectangle R of the xy-plane if there is a sequence of functions $A_n(x, y)$ such that:

(i) $\lim_{n\to\infty} A_n(x, y) = f(x, y)$ for almost all points (x, y) in R. That is, for every $\epsilon > 0$ there is a (possibly infinite) collection of subrectangles of R whose total area is less than ϵ and whose interiors contain all points where $\lim_{n\to\infty} A_n(x, y) = f(x, y)$ is false.

(ii) The integrals $\int_R A_n(x, y)\,dx\,dy$ converge as ordinary (Riemann) integrals.

(iii) The Cauchy Criterion

$$\lim_{n,m\to\infty} \int_R |A_n(x, y) - A_m(x, y)|\, dx\, dy = 0$$

is satisfied.

When this is the case, the definition $\int_R f(x, y)\, dx\, dy = \lim_{n\to\infty} \int_R A_n(x, y)\, dx\, dy$ is valid, as is proved by the same argument as in the one-dimensional case.

A function $f(x, y)$ is Lebesgue integrable* on the entire xy-plane if it is Lebesgue integrable on every rectangle and if $\int_R f(x, y)\, dx\, dy$ approaches a limit as R becomes large; more precisely, the condition is that for every $\epsilon > 0$ there be a K such that the Lebesgue integrals $\int_R f(x, y)\, dx\, dy$, $\int_{R'} f(x, y)\, dx\, dy$ are defined and differ by less than ϵ whenever R, R' are rectangles containing the square $\{|x| \le K, |y| \le K\}$. The limiting value is in this case denoted $\int_{-\infty}^{\infty} \int_{-\infty}^{\infty} f(x, y)\, dx\, dy$.

*For reasons described in §9.6 this definition is usually used only when the integral is absolutely convergent, that is, only when $\int\int |f|\, dx\, dy$ converges as well. Improper integrals which converge but do not converge absolutely (called conditionally convergent improper integrals) must be handled very carefully, whether they be Riemann integrals or Lebesgue integrals.

The Lebesgue Dominated Convergence Theorem also holds for higher dimensions and for improper integrals. (The proofs are easy extensions of the proof below.) Thus the interchange

$$\lim_{n\to\infty} \int_0^n u^x \left(1 - \frac{u}{n}\right)^n du = \int_0^\infty u^x e^{-u}\, du \qquad (x > -1)$$

of Exercise 11 of §9.6 is justified merely by setting

$$f_n(u) = \begin{cases} u^x \left(1 - \dfrac{u}{n}\right)^n & \text{if } u \le n \\ 0 & \text{if } u \ge n \end{cases}$$

$f_\infty(u) = u^x e^{-u}$, showing that the improper integral $\int_0^\infty f_\infty(u)\, du$ converges (for $x > -1$), showing that $f_n(u) \le f_\infty(u)$ and showing that $\lim_{n\to\infty} f_n(u) = f_\infty(u)$, all of which are elementary. Similarly the interchange

$$\begin{aligned}
\int_{-\infty}^{\infty} e^{-t^2} \cos xt\, dt &= \int_{-\infty}^{\infty} \lim_{n\to\infty} e^{-t^2} \sum_{j=0}^{n} \frac{(-x^2t^2)^j}{(2j)!}\, dt \\
&= \lim_{n\to\infty} \int_{-\infty}^{\infty} e^{-t^2} \sum_{j=0}^{n} \frac{(-x^2t^2)^j}{(2j)!}\, dt \\
&= \sum_{j=0}^{\infty} \frac{(-x^2)^j}{(2j)!} \int_{-\infty}^{\infty} e^{-t^2} t^{2j}\, dt
\end{aligned}$$

of the last example of §9.6 is justified by showing that

the functions involved are dominated by the integrable function $F(t) = e^{-t^2}e^{|xt|}$. The differentiation under the integral sign of this example is also justified easily using the Lebesgue theorem (Exercise 5).

Proof of the Lebesgue Dominated Convergence Theorem. Replace f_n by $\bar{f}_n = \min(f_n, F)$. Then $f_n(x) = \bar{f}_n(x)$ for almost all x, hence $\lim_{n\to\infty} \bar{f}_n(x) = f_\infty(x)$ for almost all x (see Exercise 3). This shows that one can assume at the outset that $f_n(x) \le F(x)$ for *all* x. Similarly one can assume that $f_n(x) \ge -F(x)$.

Define a function $g_n(x)$ by

$$g_n = \lim_{j\to\infty} \max(f_n, f_{n+1}, \ldots, f_{n+j}). \tag{3}$$

[The limit on the right exists for all x because it is the limit of a non-decreasing sequence of real numbers bounded by $F(x)$.] The functions $\max(f_n, f_{n+1}, \ldots, f_{n+j})$ are Lebesgue integrable hence, assuming that the theorem is true, g_n is Lebesgue integrable and its integral is the limit of the integrals of $\max(f_n, f_{n+1}, \ldots, f_{n+j})$. On the other hand, it follows from the definition of $\lim_{n\to\infty} f_n(x)$ that

$$f_\infty(x) = \lim_{n\to\infty} g_n(x) \tag{4}$$

for almost all x. Thus, again assuming that the theorem is true, it follows that f_∞ is Lebesgue integrable and that $\int_a^b f_\infty(x)\,dx = \lim_{n\to\infty} \int_a^b g_n(x)\,dx$. In particular, for every $\epsilon > 0$ the inequality

$$\int_a^b f_\infty(x)\,dx + \epsilon \ge \int_a^b g_n(x)\,dx \ge \int_a^b f_n(x)\,dx$$

holds for all sufficiently large n.

Similarly the functions

$$h_n = \lim_{j\to\infty} \min(f_n, f_{n+1}, \ldots, f_{n+j}) \tag{3'}$$

are Lebesgue integrable,

$$f_\infty(x) = \lim_{n\to\infty} h_n(x) \tag{4'}$$

for almost all x, and therefore $f_\infty(x)$ is integrable and the inequality

$$\int_a^b f_\infty(x)\,dx - \epsilon \leq \int_a^b f_n(x)\,dx$$

holds for all sufficiently large n. Thus $\lim_{n\to\infty} \int_a^b f_n(x)\,dx$ exists and is equal to $\int_a^b f_\infty(x)\,dx$ as was to be shown.

This shows that the theorem will be proved if it is proved for the sequences (3), (4), (3′), (4′). These sequences have the added property of being *monotone*, that is

$$\max(f_1, f_2, \ldots, f_{n+j}) \leq \max(f_1, f_2, \ldots, f_{n+j+1})$$
$$g_n(x) \geq g_{n+1}(x), \text{ etc.}$$

for all x. Thus the general theorem is reduced to the case of monotone sequences.

It suffices, therefore, to consider the case where the given sequence $f_n(x)$ of Lebesgue integrable functions $|f_n(x)| \leq F(x)$ satisfies the additional condition $f_n(x) \leq f_{n+1}(x)$ for all n and x. (The case $f_n(x) \geq f_{n+1}(x)$ is reduced to this case by multiplying by -1.) In this case the integrals are a bounded, increasing sequence

$$\int_a^b f_1(x)\,dx \leq \int_a^b f_2(x)\,dx \leq \cdots \leq \int_a^b F(x)\,dx.$$

Such a sequence is necessarily convergent, hence for every ϵ there is an N such that

$$\left|\int_a^b f_n(x)\,dx - \int_a^b f_m(x)\,dx\right| < \epsilon$$

whenever $n, m \geq N$. But $f_n(x) \geq f_m(x)$ for $n \geq m$, hence

$$\int_a^b |f_n(x) - f_m(x)|\,dx = \left|\int_a^b (f_n(x) - f_m(x))\,dx\right|$$

and it follows that the sequence $f_n(x)$ satisfies the Cauchy Criterion

$$\lim_{m,n\to\infty} \int_a^b |f_n(x) - f_m(x)|\,dx = 0.$$

This reduces the theorem to the following theorem which is important in its own right.

Theorem

Completeness of the space of Lebesgue integrable functions. Let $f_1, f_2, f_3, \ldots$ be a sequence of Lebesgue integrable functions satisfying the Cauchy Criterion

$$\lim_{m,n\to\infty} \int_a^b |f_n - f_m|\, dx = 0.$$

Then there is a Lebesgue integrable function g such that $\lim_{n\to\infty} \int_a^b |f_n - g|\, dx = 0$. Moreover, if g is any Lebesgue integrable function with this property then $\lim_{n\to\infty} f_n(x) = g(x)$ for almost all x in $\{a \leq x \leq b\}$.

Using this theorem, the proof of the Lebesgue Dominated Convergence Theorem is completed by noting that the given monotone sequence $f_n(x)$ satisfies $\lim_{n\to\infty} f_n(x) = f_\infty(x)$ for almost all x (by assumption) and $\lim_{n\to\infty} f_n(x) = g(x)$ for almost all x (where $g(x)$ is the Lebesgue integrable function whose existence is asserted by the Completeness Theorem). Thus $f_\infty(x) = g(x)$ for almost all x, so $f_\infty(x)$ is Lebesgue integrable and $\int_a^b f_\infty(x)\, dx = \int_a^b g(x)\, dx$. Since

$$\lim_{n\to\infty} \left| \int_a^b f_n\, dx - \int_a^b g\, dx \right| \leq \lim_{n\to\infty} \int_a^b |f_n - g|\, dx = 0,$$

it follows that $\int_a^b f_\infty\, dx = \int_a^b g\, dx = \lim_{n\to\infty} \int_a^b f_n\, dx$ as desired.

The proof of the completeness theorem resembles, as is to be expected, the proof of the completeness of the real number system (§9.1). First let the given sequence $f_n(x)$ be 'weeded out' so that

$$\int_a^b |f_n(x) - f_{n+k}(x)|\, dx < 10^{-n}.$$

For each n let $A_{n,1}(x), A_{n,2}(x), A_{n,3}(x), \ldots$ be a sequence of Riemann integrable functions such that

$$\int_a^b |A_{n,j}(x) - A_{n,j+k}(x)|\, dx < 10^{-j}$$

and such that $\lim_{j\to\infty} A_{n,j}(x) = f_n(x)$ for almost all x. (Such sequences $A_{n,j}$ exist by the assumption that f_n is Lebesgue integrable.) The method of proof will be to show that the 'diagonal' sequence $A_{1,1}(x), A_{2,2}(x), A_{3,3}(x), \ldots$ converges for almost all x and thereby defines a function g with the desired properties.

To this end, note first that the diagonal sequence $A_{n,n}$ satisfies the Cauchy Criterion; in fact,

$$\begin{aligned}
\int_a^b &|A_{n,n}(x) - A_{n+k,n+k}(x)|\,dx \\
&\le \int_a^b |A_{n,n}(x) - A_{n,n+j}(x)|\,dx + \int_a^b |A_{n,n+j}(x) - A_{n+k,n+j}(x)|\,dx \\
&+ \int_a^b |A_{n+k,n+j}(x) - A_{n+k,n+k}(x)|\,dx \\
&< 10^{-n} + \int_a^b |A_{n,n+j}(x) - A_{n+k,n+j}(x)|dx + 10^{-(n+k)}
\end{aligned}$$

for $j > k$. As $j \to \infty$ the middle term approaches $\int_a^b |f_n(x) - f_{n+k}(x)|\,dx$ (by the definition of this integral), hence for large j it is less than $2 \cdot 10^{-n}$ and

$$\int_a^b |A_{n,n}(x) - A_{n+k,n+k}(x)|\,dx < 4 \cdot 10^{-n} \tag{5}$$

for all k.

To show that $\lim_{n\to\infty} A_{n,n}(x)$ exists for almost all x, consider the set where $|A_{n,n}(x) - A_{n+1,n+1}(x)| \ge 2^{-n}$. The argument used in the proof of the main theorem applied to the integral (5) shows that this set can be contained in the interior of a (finite) collection U_n of intervals of total length at most $4 \cdot 5^{-n}$. Let V_n denote the collection of all the intervals of $U_n, U_{n+1}, U_{n+2}, \ldots$. The total length of the intervals of V_n is at most $4 \cdot 5^{-n}(1 + 5^{-1} + 5^{-2} + \cdots) = 4 \cdot 5^{-n} \cdot (\frac{5}{4}) = 5 \cdot 5^{-n}$, and if x is any point not in V_n then

$$\begin{aligned}
|A_{n,n}(x) &- A_{n+k,n+k}(x)| \\
&\le |A_{n+1,n+1}(x) - A_{n,n}(x)| + \cdots + |A_{n+k,n+k}(x) - A_{n+k-1,n+k-1}(x)| \\
&\le 2^{-n}(1 + 2^{-1} + 2^{-2} + \cdots) = 2 \cdot 2^{-n}.
\end{aligned}$$

Thus the Cauchy Criterion is satisfied and $\lim_{n\to\infty} A_{n,n}(x)$ exists for all x not in V_n. Since V_n has arbitrarily small total length, this proves that $\lim_{n\to\infty} A_{n,n}(x)$ exists for almost all x. Let $g(x)$ be the function which is $\lim_{n\to\infty} A_{n,n}(x)$ when this limit exists and which is 0 otherwise. Then $|A_{n,n}(x) - g(x)| \leq 2 \cdot 2^{-n}$ whenever x is not in V_n.

By exactly the same argument applied to the sequence $A_{n,1}, A_{n,2}, \ldots, A_{n,j}, \ldots,$ it follows that $\lim_{j\to\infty} A_{n,j}(x)$ exists for almost all x and that there is a collection $W_{n,j}$ of intervals with total length less than $(\frac{5}{4}) \cdot 5^{-j}$ such that

$$|A_{n,j}(x) - \lim_{j\to\infty} A_{n,j}(x)| \leq 2 \cdot 2^{-j}$$

whenever x is not in $W_{n,j}$. Since $\lim_{j\to\infty} A_{n,j}(x) = f_n(x)$ for almost all x, the set where this fails can be contained in a collection of intervals of arbitrarily small total length, say $\leq (\frac{3}{4}) \cdot 5^{-j}$, which can be added to $W_{n,j}$ to give a collection $\overline{W_{n,j}}$ of total length $\leq 2 \cdot 5^{-j}$ such that

$$|A_{n,j}(x) - f_n(x)| \leq 2 \cdot 2^{-j}$$

whenever x is not in $\overline{W}_{n,j}$.

Finally, let Z_n be the collection of all intervals in V_n, $\overline{W}_{n,n}$, $\overline{W}_{n+1,n+1}$, $\overline{W}_{n+2,n+2}, \ldots$. Then Z_n has total length

$$\leq 5 \cdot 5^{-n} + 2 \cdot 5^{-n} + 2 \cdot 5^{-(n+1)} + 2 \cdot 5^{-(n+2)} + \cdots$$
$$= 5 \cdot 5^{-n} + 2 \cdot (\tfrac{5}{4}) \cdot 5^{-n} < 10 \cdot 5^{-n}.$$

If x is not in Z_n and if $m \geq n$ then x is not in V_m and

$$|f_m(x) - g(x)| \leq |f_m(x) - A_{m,m}(x)| + |A_{m,m}(x) - g(x)| \leq 2 \cdot 2^{-m} + 2 \cdot 2^{-m} = 4 \cdot 2^{-m}$$

so $\lim_{m\to\infty} f_m(x)$ exists and is equal to $g(x)$. This proves that $\lim_{n\to\infty} f_n(x) = g(x)$ for almost all x. Moreover, the sequence $A_{n,n}(x)$ shows that $g(x)$ is Lebesgue integrable and that

$$\int_a^b |A_{n,n+j} - A_{n+k,n+k}|\,dx \leq \int_a^b |A_{n,n+j} - A_{n+k,n+j}|\,dx + \int_a^b |A_{n+k,n+j} - A_{n+k,n+k}|\,dx$$
$$\leq \int_a^b |A_{n,n+j} - A_{n+k,n+j}|\,dx + 10^{-(n+k)}$$

for $j \geq k$. Hence as $j \to \infty$

$$\int_a^b |f_n - A_{n+k,n+k}|\,dx \leq \int_a^b |f_n - f_{n+k}|\,dx + 10^{-(n+k)}$$
$$\leq 10^{-n} + 10^{-(n+k)}.$$

Letting $k \to \infty$ gives

$$\int_a^b |f_n - g|\, dx \leq 10^{-n}$$

hence as $n \to \infty$

$$\lim_{n\to\infty} \int_a^b |f_n - g|\, dx = 0.$$

If $\bar{g}$ is any other function satisfying

$$\lim_{n\to\infty} \int_a^b |f_n - \bar{g}|\, dx = 0$$

then

$$\int_a^b |g - \bar{g}|\, dx \leq \int_a^b |g - f_n|\, dx + \int_a^b |f_n - \bar{g}|\, dx \to 0$$

hence $\int_a^b |g - \bar{g}|\, dx = 0$. It is to be shown that $\lim_{n\to\infty} f_n(x) = \bar{g}(x)$ for almost all x, that is, that $g(x) = \bar{g}(x)$ for almost all x. The proof of the theorem will therefore be completed by the following lemma.

Lemma

If h is a Lebesgue integrable function such that $h(x) \geq 0$ and such that $\int_a^b h(x)\, dx = 0$ then $h(x) = 0$ for almost all x in $\{a \leq x \leq b\}$.

Proof

Let $A_n(x)$ be a sequence of Riemann integrable functions satisfying the Cauchy Criterion such that $\lim_{n\to\infty} A_n(x) = h(x)$ for almost all x. By taking $|A_n|$ it can be assumed that $A_n \geq 0$. By assumption, $\lim_{n\to\infty} \int_a^b A_n(x)\, dx = 0$. By 'weeding out' the A's it can be assumed that

$$\int_a^b A_n(x)\, dx < 10^{-n}.$$

It follows that the set where $A_n(x) \geq 2^{-n}$ can be contained in the interiors of a finite collection U_n of intervals of total length $<5^{-n}$. Let V_n denote the collection of intervals in $U_n, U_{n+1}, U_{n+2}, \ldots$. If x is not in V_n then it is not in U_m for $m \geq n$ hence $0 \leq A_m(x) \leq 2^{-m}$ and

$\lim_{m\to\infty} A_m(x) = 0$. Thus $\lim_{m\to\infty} A_m(x) = 0$ for almost all x and $h(x) = 0$ for almost all x as was to be shown.

The theory of Lebesgue integration can be formulated in many ways. For an especially clear exposition of two formulations different from the one given here—including Lebesgue's original formulation in terms of the theory of measure—see B. Sz.-Nagy, *Introduction to Real Functions and Orthogonal Expansions*, Oxford University Press, 1965.

Exercises

1 Reviewing Chapter 2, show that a function $f(x)$ on $\{a \le x \le b\}$ is Riemann integrable—i.e. $\int_b^a f(x)\,dx$ converges in the ordinary sense—if and only if the following condition is satisfied: For every $\epsilon > 0$ there is a $\delta > 0$ such that

$$(*) \qquad \sum_i |f(x_i) - f(x_i')|\,\Delta x_i < \epsilon$$

whenever $\{a \le x \le b\}$ is divided into subintervals whose lengths Δx_i are all less than δ and whenever points x_i, x_i' are chosen in each interval to form the sum (*). Conclude that the relations listed under application f are valid for Riemann integrable functions.

2 Show that if $f(x) = g(x)$ for almost all x and if $g(x) = h(x)$ for almost all x then $f(x) = h(x)$ for almost all x, where $\{a \le x \le b\}$.

3 Show that if $f_n(x) \to f_\infty(x)$ for almost all x and if $f_n(x) = g_n(x)$ for almost all x then $g_n(x) \to f_\infty(x)$ for almost all x. [Use $\epsilon = \epsilon/2 + \epsilon/4 + \epsilon/8 + \cdots$.] Conclude that the function of Example d is zero for almost all x.

4 Find an example of a sequence of continuous functions $f_n(x)$ on $\{0 \le x \le 1\}$ which converge to zero $f_n(x) \to 0$ for all x in the interval but for which $\lim_{n\to\infty} \int_0^1 f_n(x)\,dx \ne 0$.

5 Use the Lebesgue Dominated Convergence Theorem to prove that the equations obtained by differentiation under the integral sign in the last example of §9.6 are valid. [The inequalities $-x^2 \le n(e^{-x^2/n} - 1) \le -x^2 + x^4$ can be used.]

6 Show that $|x|^{-a}$ is not Lebesgue integrable on

$$\{-1 \le x \le 1\}$$

for $a \ge 1$. [If it were, the Lebesgue Dominated Convergence Theorem could be contradicted.]

7 Let $r = \sqrt{x^2 + y^2}$. For what values of a is the function r^{-a} Lebesgue integrable on $\{-1 \le x \le 1, -1 \le y \le 1\}$?

8 Let $r = \sqrt{x^2 + y^2 + z^2}$. For what values of a is the function r^{-a} Lebesgue integrable on the cube

$$\{|x| \le 1, |y| \le 1, |z| \le 1\}?$$

9 Prove the Lebesgue Dominated Convergence Theorem for improper integrals

$$\lim_{n\to\infty} \int_{-\infty}^{\infty} f_n \, dx = \int_{-\infty}^{\infty} f_\infty \, dx.$$

[Use the theorem for finite intervals. If $\int_{-K}^{K} F\,dx$ is within ϵ of its limiting value then the same is true of any integral which it dominates.]

10 Prove *Beppo Levi's Theorem:* If $f_{n+1}(x) \ge f_n(x)$ is a monotone sequence of Lebesgue integrable functions on $\{a \le x \le b\}$ such that $\int_a^b f_n(x)\,dx$ is bounded for $n = 1, 2, 3, \ldots$ then $f_n(x)$ converges for almost all x to a Lebesgue integrable function $f_\infty(x)$.

11 Prove *Arzela's Theorem:* If a sequence of Riemann integrable functions is bounded on the interval $\{a \le x \le b\}$ and if it converges to a limit function f which is also Riemann integrable on $\{a \le x \le b\}$ then $\lim_{n\to\infty} \int_a^b f_n\,dx = \int_a^b f\,dx$. [Without the theory of Lebesgue integration this theorem is quite difficult; with it, it is trivial.]

12 Prove *Fatou's Lemma:* If $f_n \ge 0$ is a sequence of non-negative Lebesgue integrable functions which converge $f_n(x) \to f(x)$ for almost all x in $\{a \le x \le b\}$ to a limit $f(x)$, and if $\int_a^b f_n(x)\,dx \le K$ for all n, then f is Lebesgue integrable and $\int_a^b f(x)\,dx \le K$. [Use the method of the proof of the Lebesgue Dominated Convergence Theorem and the fact that $f_n \ge 0$.]

9.8

Banach Spaces

A *norm* on a vector space* is a function which assigns real numbers $|x|$ to elements x of the vector space in such a way that:

*See §4.5 for the definition of 'vector space'.

(i) $|ax| = |a|\,|x|$ where $|a|$ denotes the absolute value of the real number a and $|x|$, $|ax|$ denote the norms of the elements x, ax of the vector space.

(ii) *Triangle inequality*. The inequality $|x + y| \leq |x| + |y|$ holds for all elements x, y of the vector space.

(iii) $|x - y| = 0$ implies $x = y$.

Intuitively the norm $|x|$ is thought of as the length or size of x, and the norm $|x - y|$ is thought of as the distance from x to y or the size of their difference.

A *Banach space* is a vector space for which a norm is given, with respect to which the space is *complete*, that is, with respect to which every sequence $x_1, x_2, x_3, \ldots$ of elements of the space satisfying the Cauchy Criterion $\lim_{m,n\to\infty} |x_n - x_m| = 0$ has a limit x_∞ satisfying $\lim_{n\to\infty} |x_n - x_\infty| = 0$.

For x an n-tuple of real numbers $x = (x_1, x_2, \ldots, x_n)$ the notation $|x|$ has often been used above to denote the maximum of the absolute values

$$(1) \qquad |x| = \max(|x_1|, |x_2|, \ldots, |x_n|).$$

The properties (i)–(iii) and the completeness axiom hold, so that* R^n with this norm is a Banach space. This Banach space has formed the basis of most of the theorems and proofs of this book.

*The vector space of n-tuples of real numbers was denoted V_n in §4.5. Actually there is a useful distinction between R^n and V_n, which is the distinction between an affine space and a vector space; briefly this distinction is that V_n has the 'origin' or 'zero vector' (0, 0, . . . , 0) as a special point but R^n has no special points. This distinction will be ignored here and R^n will denote the vector space.

For any real number $p \geq 1$ the 'p-norm'

$$|x|_p = (|x_1|^p + |x_2|^p + \cdots + |x_n|^p)^{1/p}$$

also defines a Banach space structure on R^n. The triangle inequality $|x + y|_p \leq |x|_p + |y|_p$ is *Minkowski's Inequality* proved in Chapter 5 by the method of Lagrange multipliers (§5.4, Exercise 9). For $p = 2$ the p-norm is the Euclidean distance and $|x + y|_2 \leq |x|_2 + |y|_2$ is the ordinary triangle inequality. For any fixed x in R^n, $\lim_{p\to\infty} |x|_p = |x|$ where $|x|$ is the norm (1) (Exercise 7, §5.4). For this reason the norm (1) is also denoted $|x|_\infty$ to distinguish it from the other possible norms on R^n. The proof that R^n is complete in the norm $|x|_p$ is based on the simple inequalities

$$(2) \qquad |x|_\infty \leq |x|_p \leq n^{1/p}|x|_\infty.$$

The first inequality shows that if $\lim |x_n - x_m|_p = 0$ then $\lim |x_n - x_m|_\infty = 0$ hence there is an x_∞ satisfying $\lim |x_n - x_\infty|_\infty = 0$ which implies, by the second inequality, that $\lim |x_n - x_\infty|_p = 0$.

An example of an infinite-dimensional Banach space is provided by the space of all continuous functions $x(t)$ defined on the interval $\{a \leq t \leq b\}$ with the norm

$$|x|_\infty = \max_{a \leq t \leq b} |x(t)|. \tag{3}$$

To say that x is 'small' in this norm, say $|x|_\infty < \epsilon$, means that *all* values $x(t)$ of x are less than ϵ absolute value, i.e. x is *uniformly* small for all t in $\{a \leq t \leq b\}$. For this reason the norm (3) is called the 'uniform norm'. The fact that the space of all continuous functions is complete with respect to the uniform norm (3) is the important theorem: *A uniform limit of continuous functions is a continuous function.* More specifically, if $x_1, x_2, x_3, \ldots$ is a sequence of continuous functions such that for every $\epsilon > 0$ there is an N for which

$$|x_n - x_m|_\infty = \max_{a \leq t \leq b} |x_n(t) - x_m(t)| < \epsilon$$

whenever $n, m \geq N$, then the limit function $x_\infty(t)$ (which clearly exists and satisfies $\lim_{n \to \infty} |x_n - x_\infty|_\infty = 0$) is continuous. This is proved simply by writing

$$|x_\infty(t + h) - x_\infty(t)| \leq |x_\infty(t + h) - x_N(t + h)| + |x_N(t + h) - x_N(t)| + |x_N(t) - x_\infty(t)|$$

and noting that the first and last terms can be made small by making N large, after which the middle term (with N and t fixed) can be made small by making $|h|$ small; thus $|x_\infty(t + h) - x(t)|$ can be made small for fixed t by making $|h|$ small, which is the definition of continuity.

The analogous generalization of the 1-norm $|x|_1 = |x_1| + |x_2| + \cdots + |x_n|$ to functions $x(t)$ would be

$$|x|_1 = \int_a^b |x(t)|\, dt. \tag{4}$$

It is easily shown that this is a norm on the vector space of continuous functions, but the continuous functions are not complete in this norm. On the vector space of Riemann integrable functions the function (4) is no longer a norm (condition (iii) is violated) and the space is still not complete. However, the function (4) is defined on the vector space of Lebesgue integrable functions on $\{a \leq t \leq b\}$ and is a norm provided that two Lebesgue integrable functions $x(t)$, $y(t)$ are considered to be 'the same' whenever $\int_a^b |x(t) - y(t)|\, dt = 0$, that is, whenever $x(t) = y(t)$ for almost all t. The completeness theorem of §9.7 shows that the Lebesgue integrable functions are complete in the norm (4) and hence are a Banach space. This Banach space is denoted $L[a, b]$.

In a similar fashion, one can define the Banach space $L[-\infty, \infty]$ (=space of functions $x(t)$ for which the Lebesgue integral $\int_{-\infty}^{\infty} |x(t)|\, dt$ is defined) and Banach spaces of Lebesgue integrable functions of several variables. The p-norm

$$|x|_p = \left(\int_a^b |x(t)|^p\, dt\right)^{1/p}$$

($1 \le p \le \infty$) on the space of all functions x for which the Lebesgue integral $\int_a^b |x(t)|^p\, dt$ is defined can also be shown to satisfy the axioms of a Banach space. This Banach space is denoted $L^p[a, b]$.

All definitions and theorems concerning the Banach space R^n with the norm $|x|_\infty = \max(|x_1|, |x_2|, \ldots, |x_n|)$ can be extended immediately to arbitrary Banach spaces, provided that they depend only on the Banach space structure of R^n, that is, provided that they depend only on the vector space operations,* on the properties (i)–(iii) of the norm, and on completeness. For example: A function $f\colon E \to F$ which assigns elements of a Banach space F to elements of a Banach space E is said to be *uniformly continuous* on a subset X of E if it is defined for all x in X and if for every $\epsilon > 0$ there is a $\delta > 0$ such that $|f(x') - f(x)| < \epsilon$ whenever x', x are elements of X for which $|x' - x| < \delta$. (Here $|f(x') - f(x)|$ denotes the norm of the element $f(x') - f(x)$ of the Banach space F, and $|x' - x|$ the norm of $x' - x$ in E.) The *closure* $\overline{X}$ of a subset X of a Banach space E is the set of all elements of E which can be written as limits of sequences in X. If $f\colon E \to F$ is uniformly continuous on X, then the formula $f(\lim x_n) = \lim f(x_n)$ defines a uniformly continuous extension of f to $\overline{X}$.

*Actually the theorems listed below depend on even less; they depend only on the affine space structure of R^n. That is $(0, 0, \ldots, 0)$ plays no special role, except that a 'k-form on E with values in F' uses the vector space structure of F.

A function $f\colon E \to F$ from a Banach space E to a Banach space F is said to be *uniformly differentiable* on a subset X of E if there is a $\delta > 0$ such that the function

$$M_{s,x}(h) = \frac{f(x + sh) - f(x)}{s}$$

is defined and uniformly continuous on the set of triples (s, x, h) in which s is a real number $0 < |s| < \delta$, in which

x is an element of X, and in which h is an element of E with $|h| < 1$. (The norm of a triple (s, x, h) is $\max(|s|, |x|, |h|)$. See Exercise 7.) When this is the case the extension of $M_{s,x}(h)$ to the set $s = 0$ defines a function

$$L_x(h) = \lim_{s\to 0} \frac{f(x + sh) - f(x)}{s} \tag{5}$$

with values in F. The function $L_x(h)$ is linear in h (see Exercise 4) and uniformly continuous in (x, h) relative to the norm $\max(|x|, |h|)$ for x in X and for all h.

A function $f\colon E \to F$ is said to be *continuous on a set* X if it is defined for all x in X and if for every x in X and $\epsilon > 0$ there is a $\delta > 0$ such that $|f(x') - f(x)| < \epsilon$ whenever x' is in X and $|x' - x| < \delta$. A function $f\colon E \to F$ is said to be *continuous at a point* x if for every $\epsilon > 0$ there is a $\delta > 0$ such that $f(x')$ is defined and satisfies $|f(x') - f(x)| < \epsilon$ whenever x' is in E and $|x' - x| < \delta$.

A function $f\colon E \to F$ is said to be *differentiable at a point* x if there is a continuous* linear map $L_x\colon E \to F$ with the property that for every $\epsilon > 0$ there is a $\delta > 0$ such that $f(x + sh)$ is defined and satisfies

*If E is a finite-dimensional Banach space then a linear map $L\colon E \to F$ is necessarily continuous (Exercise 20). If E is not finite-dimensional this is no longer the case.

$$\left| \frac{f(x + sh) - f(x)}{s} - L_x(h) \right| < \epsilon$$

whenever $0 < |s| < \delta$ and $|h| \le 1$. It is *differentiable on a set* X if it is differentiable at every point x of X. It is *continuously differentiable on* X if it is differentiable on X and if the derivative L_x is a continuous function of x with values in the Banach space $L(E, F)$ of Exercise 9.

The notion of the 'rank' of a mapping $f\colon E \to F$ is not defined in general and the statement of the Implicit Function Theorem must be revised and weakened somewhat:

Implicit Function Theorem

Let $f\colon E_1 \times E_2 \to F$ be a mapping which assigns elements $y = f(x_1, x_2)$ in a Banach space F to pairs of elements (x_1, x_2) with x_1 in a Banach space E_1 and x_2 in a Banach space E_2. Let f be continuously differentiable on a 'cube' $\{|x_1 - \bar{x}_1| \le K, |x_2 - \bar{x}_2| \le K\}$ containing $(\bar{x}_1, \bar{x}_2)$ and let $L\colon E_1 \times E_2 \to F$ be its derivative at $(\bar{x}_1, \bar{x}_2)$. Then $y = f(x_1, x_2)$ can be solved locally for

$x_1 = g(y, x_2)$ if and only if $k = L(h, 0)$ can be solved for $h = M(k)$. That is, if there is a continuous linear function $M: F \to E_1$ such that the relations $k = L(h, 0)$ and $h = M(k)$ are equivalent then there are an $\epsilon > 0$ and a continuously differentiable map $g: F \times E_2 \to E_1$ such that the relations $y = f(x_1, x_2)$ and $x_1 = g(y, x_2)$ are defined and equivalent for all (x_1, x_2, y) satisfying $|x_1 - \bar{x}_1| < \epsilon$, $|x_2 - \bar{x}_2| < \epsilon$, $|y - f(\bar{x}_1, \bar{x}_2)| < \epsilon$. Conversely, if such ϵ, g exist so must such an M.

The proof is by successive approximations as before.

A 'continuous k-form $\omega = A\,dx_1\,dx_2 \ldots dx_k + \cdots$ on R^n with values in a Banach space F' is simply a k-form in which the coefficient functions A are continuous functions $A: R^n \to F$ rather than $A: R^n \to R$. Given such a k-form and given a compact, oriented, differentiable k-manifold S in R^n, an element $\int_S \omega$ of F is defined, exactly as before, as a limit of approximating sums. Similarly, one can define the notions of a 'continuous k-form ω on a Banach space E with values in a Banach space F' and of a 'compact oriented k-manifold S in a Banach space E' in such a way that $\int_S \omega$ can be defined as before (Exercises 16–19). However, the domain of integration in these integrals is always a k-dimensional manifold—that is, essentially R^k—and integration over a general Banach space is not defined (because integration over domains of R^k depends on more than just the vector space structure, the norm, and the completeness of R^k).

In summary, the abstract notion of a 'Banach space' gives a simple *vocabulary* for formulating many of the basic definitions and theorems of calculus in such a way that they are applicable under very general circumstances.

Exercises

1 Prove that a composition of continuous functions is continuous. That is, prove that if $f: E \to F$, $g: F \to G$ are functions (where E, F, G are Banach spaces) if f is continuous at $\bar{x}$ and if g is continuous at $f(\bar{x})$ then $g \circ f: E \to G$ is continuous at $\bar{x}$.

2 Prove that if $f: E \to F$ and $g: F \to G$ are uniformly continuous on subsets X of E and Y of F, and if $f(x)$ is in Y whenever x is in X, then $g \circ f: E \to G$ is uniformly continuous on X.

3 Prove the *Chain Rule:* If $f: E \to F$ is differentiable at a point $\bar{x}$ of E and if $g: F \to G$ is differentiable at $f(\bar{x})$ then

$g \circ f: E \to G$ is differentiable and its derivative mapping $E \to G$ is the composition of the derivatives of f and g.

4 Prove that if $f: E \to F$ is uniformly differentiable on a set X then its derivative $L_x(h)$ is linear in h for each x. [The proof that $L_x(ah) = aL_x(h)$ is easy. To prove that $L(h + h') = L(h) + L(h')$ show that

$$\lim_{s,t\to 0} \frac{f(x + sh + sh' + tk) - f(x + sh + tk) - f(x + sh' + tk) + f(x + tk)}{s}$$

exists and is independent of k. Then show that it is zero for all h, h' and hence that $L(h + h') = L(h) + L(h')$.]

5 Prove that a linear map $L: E \to F$ is continuous if and only if there is a constant B such that $|L(x)| < B|x|$, in which case it is uniformly continuous.

6 Use 5 to prove that if E is the Banach space R^n with the norm $|x|_\infty$ then any linear map $L: E \to F$ is continuous.

7 The *product* $E_1 \times E_2 \times \cdots \times E_n$ of a finite number of Banach spaces $E_1, E_2, \ldots, E_n$ is defined to be the set of n-tuples $(x_1, x_2, \ldots, x_n)$ in which x_i is an element of E_i $(i = 1, 2, \ldots, n)$ with the vector space operations defined componentwise and with the norm defined by

$$|(x_1, x_2, \ldots, x_n)| = \max(|x_1|, |x_2|, \ldots, |x_n|).$$

Show that $E_1 \times E_2 \times \cdots \times E_n$ is a Banach space. What is the Banach space $R \times R \times \cdots \times R$? [Note that there is only one norm on the Banach space R.]

8 Let $L(R^n, R^m)$ denote the set of all linear maps $R^n \to R^m$, i.e. the set of all $m \times n$ matrices M. Show that $L(R^n, R^m)$ is a vector space and that the norm

$$|M| = \max_{|x|_\infty = 1} |Mx|_\infty$$

makes it a Banach space. Find the norm $|M|$ as an explicit function of the matrix M.

9 Let $L(E, F)$ denote the set of all continuous linear maps $M: E \to F$ where E, F are Banach spaces. Show that $L(E, F)$ is a vector space, and show that

$$|M| = \underset{|x| \le 1}{\text{l.u.b.}}\, |Mx|$$
$= $ least number B which fulfills the property of Exercise 5

defines a norm on $L(E, F)$ with respect to which $L(E, F)$ is complete. Hence $L(E, F)$ is a Banach space.

10 Let E be the space R^n with the p-norm $|(x_1, \ldots, x_n)| = (|x_1|^p + \cdots + |x_n|^p)^{1/p}$ $(p > 1)$. Show that the Banach space $L(E, R)$ can be identified with R^n in the q-norm where $\frac{1}{p} + \frac{1}{q} = 1$.

11 In the Banach space $L(E, E)$ of continuous linear maps of a Banach space E to itself there is an operation of composition assigning to a pair of maps $M_1\colon E \to E$, $M_2\colon E \to E$ a new map $M_2 \circ M_1\colon E \to E$. Show that $|M_2 \circ M_1| \leq |M_2|\,|M_1|$.

12 An element L of the Banach space $L(E, E)$ is said to be *invertible* if there is an element M of $L(E, E)$ such that* $ML = LM = I$ where I denotes the identity map. When this is the case, M is called the *inverse* of L and is denoted L^{-1}. Show that if L is given and if M, N are given such that $|ML - I| < 1$, $|LN - I| < 1$, then L is invertible.

**If E is finite-dimensional then $ML = I$ implies $LM = I$. This is no longer true if E is infinite-dimensional (see Exercise 21).*

13 Refine the argument of 12 to show that the function 'inverse' from $L(E, E)$ to itself is continuous at all points where it is defined. That is, show that if L is invertible and if L_1 is near L then L_1 is invertible and L_1^{-1} is near L^{-1}. This fact is used in the following proof of the Inverse Function Theorem.

14 Write a complete proof of the *Inverse Function Theorem:* Let $f\colon E \to F$ be a map from a Banach space E to a Banach space F, let $\bar{x}$ be a point of E, let K be a real number such that f is continuously differentiable on the 'cube' $\{|x - \bar{x}| \leq K\}$ and let $L_{\bar{x}}\colon E \to F$ be the derivative of f at $\bar{x}$. If $L_{\bar{x}}(h) = k$ can be solved for h as a continuous function of k then $f(x) = y$ can be solved for x as a continuously differentiable function of y for all y sufficiently near $f(\bar{x})$. More precisely, if there is a continuous map $M\colon F \to E$ such that $M(k) = h$ is equivalent to $k = L_{\bar{x}}(h)$ (all h, k) then there is an $\epsilon > 0$ and a map $g\colon F \to E$, defined and continuously differentiable on $\{|y - f(\bar{x})| < \epsilon\}$, such that $y = f(x)$ is equivalent to $g(y) = x$ for all x, y satisfying $|x - \bar{x}| < \epsilon$, $|y - f(\bar{x})| < \epsilon$. [Use the method of successive approximation to show that $f(x) = y$ has a unique solution x for all y sufficiently near $f(\bar{x}) = \bar{y}$. Call this function g. Show that g is differentiable at $\bar{y}$ with derivative M. Conclude that g is differentiable at all points where it is defined, hence that g is continuous. Finally, the derivative $(L_{g(y)})^{-1}$, being a composition of continuous functions, is continuous.]

15 Deduce the Implicit Function Theorem stated in the text from the Inverse Function Theorem proved above.

16 A 'constant k-form on E with values in F' is a function $\phi\colon E \times E \times \cdots \times E \to F$ of k-tuples of elements of E to elements $\phi(v_1, v_2, \ldots, v_k)$ of F such that: (i) ϕ is continuous. (ii) ϕ is linear in each of its k variables, i.e. $\phi(av_1 + bv_1', v_2, v_3, \ldots, v_k) = a\phi(v_1, v_2, \ldots, v_k) + b\phi(v_1', v_2, \ldots, v_k)$, etc. (iii) ϕ is *alternating*, that is, interchanging two of the k variables in ϕ changes the sign of its value in F. Show that a 'constant k-form on R^n with values in R' in the sense of §4.2 gives rise to a 'constant k-form on R^n with values in R' in the sense just defined.

17 Let $A_k(E; F)$ denote the set of all constant k-forms on E with values in F. Given a continuous linear map $L: E_1 \to E_2$ define the pullback map $L^*: A_k(E_2; F) \to A_k(E_1; F)$ in a manner consistent with the previous definition, and prove the Chain Rule $(L \circ M)^* = M^* \circ L^*$.

18 A 'continuous k-form on E with values in F' is a continuous map $E \to A_k(E; F)$. Outline a definition of $\int_S \omega$ for S a compact, oriented, k-dimensional manifold in E (defining this concept) and for ω a continuous k-form on E.

19 Show that if $|x|$ is any norm on a vector space V then $|x| \geq 0$ for all x, i.e. the axioms (i)–(iii) imply $|x| \geq 0$.

20 Show that if $|x|$ is any norm on R^n then there exist positive constants c, C such that $c|x|_\infty \leq |x| \leq C|x|_\infty$. Conclude that any linear map $L: E \to F$ in which the domain E is finite dimensional must be continuous. [The inequality $|x| \leq C|x|_\infty$ is easy. The inequality $c|x|_\infty \leq |x|$ can be proved by considering $(R^n)^* = L(R^n, R)$ (see Exercise 10, §4.5). The last statement then follows from Exercise 6.]

21 Let E be the Banach space of all absolutely convergent series; that is, let E be the vector space of all infinite sequences $x = (x_1, x_2, x_3, \ldots)$ of real numbers such that $\sum_{i=1}^{\infty} |x_i|$ converges and let $|x|$ denote the norm $\sum |x_i|$ on E. Let L denote the operator 'shift right' defined by $L(x_1, x_2, x_3, \ldots) = (0, x_1, x_2, \ldots)$. Show that L is in $L(E, E)$, and that there is an M in $L(E, E)$ such that $ML = I$, but that L is not invertible.

the Cauchy Criterion

appendix 1

The central idea of calculus is the idea of the convergence of an infinite process. Because the Cauchy Convergence Criterion is the precise formulation of this idea, a clear understanding of the Cauchy Criterion is fundamental to an understanding of calculus. The Cauchy Criterion is emphasized at several points in this book: in the definition of the definite integral (§2.3, §6.2, §9.5), in the discussion of the convergence of a method of successive approximations (Chapter 7), in the definition of real numbers (§9.1), in the definition of infinite products (§9.5), and in the definition of Lebesgue integrals (§9.7). In this appendix the idea of convergence is reviewed from a picturesque and non-technical point of view.

Imagine that a number is to be determined by an experiment which involves some sort of apparatus. Imagine also that the apparatus is such that it can be set up with varying degrees of care and that different settings of the apparatus normally produce different results. The experiment can then be thought of as an infinite process, not because any single performance of the experiment is infinite, but because the apparatus can be set up in infinitely many different ways. The infinite process represented by the experiment is *convergent* if it is true that the resulting number can be determined to within any arbitrarily small margin for error by setting up the apparatus with sufficient care. This is the Cauchy Criterion.

More fully, one should imagine that there is some way of describing the amount of care which has been used in setting up the apparatus in any one performance of the experiment. Convergence of the experiment then means that for any given margin for error there is a degree of care such that any two results of the experiment will differ by less than the given margin for error provided only that the apparatus is set up with at least the specified degree of care.

Consider, for example, the determination of the number π (see §7.5). The formula

$$\begin{aligned} \pi = {} & 16[\tfrac{1}{5} - \tfrac{1}{3}(\tfrac{1}{5})^3 + \tfrac{1}{5}(\tfrac{1}{5})^5 - \tfrac{1}{7}(\tfrac{1}{5})^7 + \cdots] \\ & -4[\tfrac{1}{239} - \tfrac{1}{3}(\tfrac{1}{239})^3 + \tfrac{1}{5}(\tfrac{1}{239})^5 - \tfrac{1}{7}(\tfrac{1}{239})^7 + \cdots] \end{aligned} \tag{1}$$

gives a simple experiment for finding π. The number of decimal places which are retained in the calculations and the number of terms of the infinite series which are used represent the degree of care with which the experiment is performed. To say that this method converges means simply that if more and more decimal places and terms are retained then more and more decimal places of the answer will remain the same. This is the sense in which the formula (1) determines the real number π as the limit of a convergent infinite process.

The definition

$$\pi = \int_D dx\,dy$$

where D is the disc $\{x^2 + y^2 \leq 1\}$ oriented $dx\,dy$, describes another experiment (convergent infinite process) for determining π. This experiment is examined in detail in Exercise 1, §2.3.

It is worthwhile to observe that no physical experiment is convergent. That is, beyond a certain point any physically defined number becomes indeterminate. Examples of physical constants which can be determined to ten significant digits are rare, let alone constants which can be determined to a hundred significant digits or a thousand. Thus real numbers such as $\sqrt{2}$ and π are not at all 'real' in the fundamental sense of the word but, on the contrary, are prototypically ideal—that is, 'real' in the Platonic sense.

the Leibniz notation

appendix 2

The notation of differential forms is a generalization of the Liebniz notation for calculus. Although there is a great deal of controversy as to whether Newton or Leibniz should be credited with the invention of calculus, there is no question that it was Leibniz who introduced the notations $\frac{dy}{dx}$ and $\int_a^b y(x)\,dx$ for the derivative and the integral, and there is no question that this notation is superior to that of Newton.* Many historians of mathematics believe, in fact, that British mathematics was greatly impeded by the insistence of British mathematicians on using the Newtonian notation while German mathematics flourished using the notation of Leibniz. Nonetheless, the Leibniz notation has certain disadvantages which have caused it to be somewhat neglected in recent years—particularly on the level of intermediate mathematics.†

See F. Cajori, A History of Mathematical Notations, Vol. II, The Open Court Publishing Co., Chicago.

†On the level of elementary calculus the notation $\frac{dy}{dx}$ is so useful that only the most stubbornly 'modern' textbooks reject it completely. At the other end of the spectrum, on the level of differential geometry and advanced analysis, differential forms are very widely accepted.

Perhaps the most serious defect of the Leibniz notation is the fact that it leaves so much unsaid. For example, in writing

$$\frac{dy}{dx} = 2x$$

one is actually saying, "y represents a dependent variable which depends in some unspecified or understood way on an independent variable x, and the derivative of this functional relation between y and x is the function $2x$." The general tendency of contemporary mathematics is to make all statements as explicit as possible and to avoid such modes of terminology as this in which a great deal of what is being said is left implicit.

However, if one is concerned with computations and applications rather than with the abstract theory, then one is quickly led to the conclusion that the Leibniz notation is incomparably more efficient and that what it leaves unsaid is invariably either unimportant or is easily understood from the context. For example, the chain rule for functions of one variable

$$\frac{dz}{dx} = \frac{dz}{dy} \cdot \frac{dy}{dx} \tag{1}$$

could only mean: "z is a function of y, and y is a function of x. When z is considered as a function of x (the composed function) its derivative is the product of the derivatives $\frac{dz}{dy}$ and $\frac{dy}{dx}$. This derivative is a function of x, and $\frac{dz}{dy}$ is therefore to be considered as a function of x (by composition)." An alternative statement of (1) which is commonly found in contemporary calculus books is

$$(f \circ g)'(x) = f'[g(x)] \cdot g'(x). \tag{2}$$

This is a somewhat fuller statement of the chain rule than is (1), but the greater detail is bought at the cost of a great loss of clarity. One of the main objectives of this book has been to point out once again the superiority of the Leibniz notation. See, for example, the implicit differentiations in §5.2, the method of Lagrange multipliers in §5.4, and the integrability conditions of §8.6; each of these topics would be significantly more complicated without the Leibniz notation.

A second disadvantage of the Leibniz notation which has caused it to be neglected in recent years is the fact that it expresses the derivative $\frac{dy}{dx}$ as a quotient of two terms dx, dy which have not been defined. Is it indeed a quotient, and, if so, what are dx and dy? This question was central to the philosophical objections which were raised to the new calculus in the 17th and 18th centuries. In the 19th century the foundations of mathematics were profoundly reworked in order to establish calculus on a firm logical basis independent of vague notions of infinitesimals. This was done essentially by emphasizing the functional relationship, by writing $y = f(x)$, and by writing the derivative as a new function $f'(x)$. This relegated the Leibniz notation $\frac{dy}{dx}$ to the status of a convenient mnemonic device whose convenience was largely accidental and whose validity was highly suspect.

The viewpoint adopted in this book is that the pullback operation is of central importance and that $\frac{dy}{dx}$ is *not* a quotient but rather is the coefficient of dx in the pullback of dy under a function $y = f(x)$. That is, $f'(x) = \frac{dy}{dx}$ means that the pullback of dy is $f'(x)\,dx$. At the cost of some precision but without grave danger of misunderstanding one can write

(3) $$dy = f'(x)\,dx$$

meaning "the pullback of dy is $f'(x)\,dx$." This is the meaning of the equation $f'(x) = \frac{dy}{dx}$.* More generally, if y is a function of several variables $y = f(x_1, x_2, \ldots, x_n)$ then (3) becomes

(3′) $$dy = \frac{\partial f}{\partial x_1}dx_1 + \frac{\partial f}{\partial x_2}dx_2 + \cdots + \frac{\partial f}{\partial x_n}dx_n$$

or, as it has usually been written in this book,

$$dy = \frac{\partial y}{\partial x_1}dx_1 + \frac{\partial y}{\partial x_2}dx_2 + \cdots + \frac{\partial y}{\partial x_n}dx_n.$$

Here the "=" sign is not precise and one actually means that the right side is the pullback of dy under an understood function from $(x_1, x_2, \ldots, x_n)$ to y. The meaning of the pullback operation, and hence the meaning of the derivative $\frac{dy}{dx}$, is discussed in detail in §5.3 (see in particular formulas (3) and (4) of §5.3).

*Many writers refer to the derivative $\frac{dy}{dx}$ as the 'differential coefficient' of y with respect to x. This terminology agrees perfectly with the viewpoint adopted here, namely that $\frac{dy}{dx}$ is the coefficient of dx in (3).

on the foundations of mathematics

appendix ***3***

Many readers will be surprised by the fact that the discussion of the real number system in §9.1 contains no definition of 'number'—not even of 'natural number'. However, there should actually be nothing at all surprising about this. In defining any term, one must make use of other terms; if the definition is to be useful, then the terms used in the definition must be more simple and more familiar than the term being defined. But what could be more simple and more familiar than the idea of 'natural number'? Hence, how could 'natural number' possibly be defined?

There are various schools of thought concerning the foundations of mathematics and many questions are vigorously debated. In particular, the point of view that 'natural number' cannot be defined would be contested by many mathematicians who would maintain that the concept of 'set' is more primitive than that of 'number' and who would use it to define 'number'. Others would contend that the idea of 'set' is not at all intuitive and would contend that, in particular, the idea of an *infinite* set is very nebulous. They would consider a definition of 'number' in terms of sets to be an absurdity because it uses a difficult and perhaps meaningless concept to define a simple one. In short, they would contend that anyone who does not understand the meaning of 'natural number' has no chance of understanding the meaning of 'set'.

The point of view adopted in §9.1 (and adopted in §4.2 and §5.2 where analogous questions arise) is that mathematics is active, deductive, and computational, and that the important thing in mathematics is to define the logical and computational relationship between the terms employed. The terms themselves are inert and devoid of meaning other than the meaning imparted to them by the way in which they are used in active relationship with other terms.

constructive mathematics

appendix 4

In the opinion of many mathematicians, the theorems of §9.4 are logically unacceptable because they are non-constructive existence theorems, that is, theorems which assert that something or other 'exists' without telling how to find it explicitly. These mathematicians hold that it is pointless to say that something 'exists' if there is no way of finding it. For example, they hold that it is pointless to assert that an infinite sequence in a compact set has a point of accumulation (the Bolzano-Weierstrass Theorem) because there may be no way whatsoever of finding a point of accumulation. Either a point of accumulation can actually be found (in which case the theorem can be improved on) or there is no way to find a point of accumulation (in which case the theorem is futile).

If one adopts this constructive view of mathematical existence then several of the theorems of this book must be modified (and the theorems of §9.4 must be rejected altogether). However, the modifications are not as extensive as one might at first imagine, and the useful theorems of calculus survive it intact. In fact, a careful, constructive restatement of the theorems of calculus clarifies them and heightens their usefulness.

The proof that "the integral $\int_R A(x, y)\, dx\, dy$ of a continuous 2-form $A\, dx\, dy$ over a rectangle R of the xy-plane converges" (§2.3 and §6.3) used the theorem that a continuous function on R is necessarily uniformly continuous on R (Theorem 1, §9.4). From the constructivist point of view it is necessary to *assume* not only that the integrand $A\,dx\,dy$ is uniformly continuous but also that for any given ϵ one can explicitly find a δ such that $|A(x, y) - A(\bar{x}, \bar{y})| < \epsilon$ whenever $|x - \bar{x}| < \delta$, $|y - \bar{y}| < \delta$. This assumption is necessary in order to be able to state explicitly how $\int_R A\,dx\,dy$ can be computed with any prescribed degree of accuracy in a finite number of steps. This additional assumption on $A(x, y)$ is perfectly natural from the constructivist point of view because the very evaluation of $A(x, y)$ virtually requires that given ϵ a corresponding δ can be found (see §9.3). That is, if $A(x, y)$ does not satisfy the additional assumption, then from the constructivist point of view the integrand $A(x, y)$ is not even a well-defined function; hence the integral $\int_R A\,dx\,dy$ is without meaning and there is no point in discussing its convergence.

Similarly, from the constructivist point of view the Fundamental Theorem $F(b) - F(a) = \int_a^b F'(t)\,dt$ would be considered only in the case where $F(t)$ is uniformly differentiable on the interval $\{a \le t \le b\}$.* In this case it is easily shown that for every $\epsilon > 0$ there is an (explicitly constructible) $\delta > 0$ such that any approximating sum $\sum F'(t_i)\Delta t_i$ to $\int_a^b F'(t)\,dt$ differs by less than ϵ from $F(b) - F(a)$ whenever all Δt_i are less than δ (see Exercise 4, §9.3).

*In the constructive sense that given ϵ a corresponding δ can be found explicitly which satisfies the condition which defines uniform differentiability in §9.3.

Similarly, the statement of Stokes' Theorem can be amplified so that it becomes constructively true. The proof of the Implicit Function Theorem given in §7.1 is constructive,† so that this theorem is perfectly acceptable from the constructivist point of view. However, not every theorem can be interpreted constructively. A very surprising exception is the trichotomy law of §9.1, that is, the 'law' that *every real number is either positive, negative, or zero.* The following example shows that this 'law' is not entirely self-evident:

†However, the alternative proof given in the exercise of §5.3 is non-constructive.

The so-called Goldbach conjecture states that every even number greater than 4 can be written as the sum of two prime numbers. So far as is known, this conjecture is true. For example,

$$\begin{aligned} 6 &= 3 + 3 & 16 &= 3 + 13 = 5 + 11 \\ 8 &= 3 + 5 & 18 &= 5 + 13 = 7 + 11 \\ 10 &= 3 + 7 = 5 + 5 & 20 &= 3 + 17 = 7 + 13 \\ 12 &= 5 + 7 & 22 &= 3 + 19 = 5 + 17 = 11 + 11 \\ 14 &= 3 + 11 = 7 + 7 & 24 &= 5 + 19 = 7 + 17 = 11 + 13 \end{aligned}$$

For larger numbers there are more possibilities so that on a purely empirical basis the Goldbach conjecture appears very, very likely once it has been tested for low numbers. It has in fact been tested extensively by computer and no exception has ever been found. Consider now the real number $r = .a_1a_2a_3a_4\cdots = \frac{a_1}{10} + \frac{a_2}{100} + \frac{a_3}{1000} + \cdots$ defined by

$$a_n = \begin{cases} 0 & \text{if } 2n + 4 \text{ can be written as the sum of two primes,} \\ 1 & \text{otherwise.} \end{cases}$$

The Goldbach conjecture is that $r = 0$.

Millions of decimal places of r are known, and they are all zero. However, in order to prove $r = 0$ it is necessary to prove the Goldbach conjecture, and in order to prove $r > 0$ it is necessary to disprove the Goldbach conjecture. But it is quite conceivable that human (or inhuman) intelligence will never succeed either in proving or in disproving the Goldbach conjecture. Thus it may be that neither the statement $r = 0$ nor the statement $r > 0$ will ever be proved. The constructivist position is that it is pointless to assert, as the trichotomy law does, that either $r = 0$ or $r > 0$. What one means is simply that the statements $r = 0$ and $r > 0$ are contradictory, that is, that both cannot be true. To put this statement in the form of the trichotomy law gives the mistaken impression that the Goldbach conjecture necessarily can be resolved one way or the other.

What is involved is the so-called law of the excluded middle. If one proves that the denial of a statement is false, is one justified in concluding that the statement is true? Surprisingly enough, the answer is "no" if all statements are interpreted constructively. One might conceivably prove, for example, that the assumption $r = 0$ leads to a contradiction without being able to prove *constructively* that $r > 0$. Similarly, if $x_0, x_1, x_2, x_3, \ldots$ is some given sequence of points in the interval $\{0 \le x \le 1\}$ it is quite conceivable that one could succeed in proving that at most a finite number of the x_n lie in $\{\frac{1}{2} \le x \le 1\}$ without being able to prove *constructively* that an infinite number of the x_n lie in $\{0 \le x \le \frac{1}{2}\}$. In §9.4 the statement "there exist at most a finite number of the x_n in $\{\frac{1}{2} \le x \le 1\}$" was taken to imply "there exist an infinite number of the x_n in $\{0 \leq x \leq \frac{1}{2}\}$." In the constructivist view, however, mathematical existence means constructibility and this implication is invalid. It is for this reason that the proofs of §9.4 are not acceptable from the constructivist point of view.

the parable of the logician and the carpenter

appendix 5

The following tale was told, without further explanation, by a man who had spent many years contemplating the real number system:

Once upon a time, a Logician, seeking respite from the hurly-burly of academic life, came to pass a period of time in the country. Being in need of a desk to continue his researches, he sought out a Carpenter in the neighboring village and asked whether the Carpenter might build him a desk.

"You give me the proper specifications," said the Carpenter, "and I'll build you any kind of a desk you want."

"Heavens!" exclaimed the Logician, "do you realize that you just made a statement about the *set of all possible desks?* Do you have any conception of the logical complexities inherent in such a statement?"

"Nonsense," said the Carpenter. "I have no idea what a set is." He paused. "Come to think of it, I'm not sure just what you mean by a desk. I'm talking about carpentry. You give me the specifications and I'll build you the desk."

"How quaint," the Logician said to Mrs. Logician that night at home. "The chap was quite ignorant of even the rudiments of formal thought. Didn't know what a set was. However, I expect he'll build me a passable desk."

And indeed he did.

answers to exercises

§1.1 *pages 4–5*

1 (a) 2 (b) -33 (c) -14

2 (a) -13 (b) $8\,dx - 11\,dy$

3 (a) The flow is zero across any segment PQ of the form $P = (x, y)$, $Q = (x + 2t, y + 3t)$ where t is an arbitrary number. Therefore the flow lines are lines of the form $x = \text{const.} + 2t$, $y = \text{const.} + 3t$ or, what is the same, lines of the form $3x - 2y = \text{const.}$ Since the flow across the segment from $(0, 0)$ to $(0, 1)$ is -2 the flow must be from upper right to lower left when the axes are drawn in the usual way. (b) Flow is along lines $x + y = \text{const.}$ from upper left to lower right. (c) Flow is along lines $Ax + By = \text{const.}$ A particle at $(0, 0)$ passes across points $(Bt, -At)$ where t is a positive number.

4 $\frac{3}{4}\,dx + dy + 1\frac{3}{8}\,dz$

5 Points on the plane $3x + 4y - z = 0$ can be reached without work. The direction of the force is perpendicular to this plane. The directed line segment (vector) from $(0, 0, 0)$ to $(-3, -4, 1)$ indicates the direction of the force.

6 (a) Draw the arrow from $(0, 0)$ to $(2, -3)$ or any translate of this arrow. (b) Draw the arrow from $(0, 0)$ to $(-3, -2)$ or any translate.

7 Set $P_i = (x_i, y_i)$. The desired formula is then $\sum_{i=1}^{n} [A(x_i - x_{i-1}) + B(y_i - y_{i-1}) + C(z_i - z_{i-1})] = A(x_n - x_0) + B(y_n - y_0) + C(z_n - z_0)$ which follows from simple algebraic cancellation.

§1.2 *page 7*

1 $-3\frac{1}{2}$

2 -9. The oriented area of the projection on the xy-plane can be found by observing that it is the same as the oriented area of the triangle $(1, 1)$, $(3, 5)$, $(3, 0)$. The other two can be found by a similar method.

3 The oriented area is positive if and only if $x_1y_2 - x_2y_1 > 0$. This can be seen from the fact that the oriented area of $(0, 0)$, (x_1, y_1), $(-y_1, x_1)$ is positive.

4 The oriented area is positive if and only if $x_1y_2 + x_2y_3 + x_3y_1 - x_2y_1 - x_3y_2 - x_1y_3 > 0$. Interchanging two vertices interchanges positive and negative terms.

§1.3 *pages 13–15*

1 The pullback of $dx\,dy + 3\,dx\,dz$ under $x = u + v$, $y = 2u + 4v$, $z = 3u$ is $-7\,du\,dv$. The pullback of $dy\,dz + dz\,dx + dx\,dy$ under $x = 1 + 2u + 3v$, $y = 1 + 4u + v$, $z = 2 - 3u - v$ is $-18\,du\,dv$. The pullback of $dx\,dy$ under $x = x_1u + x_2v$, $y = y_1u + y_2v$ is $(x_1y_2 - x_2y_1)\,du\,dv$.

2 Self-checking

3 (a) -15 (b) $-4\frac{1}{2}$ (c) -35

4 (a) $\frac{1}{2}(x_1y_2 - x_2y_1)$ (b) $\frac{1}{2}(x_0y_1 - x_1y_0) + \frac{1}{2}(x_1y_2 - x_2y_1) + \frac{1}{2}(x_2y_0 - x_0y_2)$ (c) Total flow is zero. The three terms of (b) correspond to the three sides which touch S, and the original triangle PQR is the fourth side. (d) 'No' for ordinary area, 'yes' for oriented area.

5 (a) $\frac{1}{2}(x_0y_1 - x_1y_0) + \frac{1}{2}(x_1y_2 - x_2y_1) + \frac{1}{2}(x_2y_3 - x_3y_2) + \frac{1}{2}(x_3y_0 - x_0y_3)$. (b) If the vertices are (x_i, y_i) for $i = 1, 2, \ldots, n$ (in that order) then the oriented area is $\frac{1}{2}\sum_{i=1}^{n} (x_{i-1}y_i - x_iy_{i-1})$ where $(x_n, y_n) = (x_0, y_0)$ by definition. (c) If $P = (x_P, y_P, z_P)$, $Q = (x_Q, y_Q, z_Q)$ then the edge PQ contributes a term $\frac{1}{2}(x_Py_Q - x_Qy_P)$ to the value of $dx\,dy$ on one polygon and minus this amount to the value on another. Hence all terms cancel. (d) $dy\,dz$, $dz\,dx$ are just like $dx\,dy$. (e) Given two surfaces with the same boundary, reversing the orientation of one gives a *closed* surface and the desired result follows from (d).

6 Flow is in the direction of the line segment from $(0, 0, 0)$ to $(3, -7, 11)$. The pullback of the given 2-form under a map of the form $x = 3u + x_2v, y = -7u + y_2v, z = 11u + z_2v$ is zero.

7 (e) The pullback is $-dx\,dy$. Hence the map reverses orientations. This is easily shown by drawings.

8 The general case of this argument occurs in Exercise 3, §4.3, where it is simplified by matrix notation.

§1.4 *pages 18–19*

1 The composed map is $x = 2(r + 3s) + (2s + t) = 2r + 8s + t, y = 3r + 11s + t, z = 2r + 4s + 2t$ hence $dx\,dy\,dz = (2\,dr + 8\,ds + dt)(3\,dr + 11\,ds + dt)(2\,dr + 4\,ds + 2\,dt) = [(22 - 24)\,dr\,ds + (2 - 3)\,dr\,dt + (8 - 11)\,ds\,dt](2\,dr + 4\,ds + 2\,dt) = [-4 + 4 - 6]\,dr\,ds\,dt = -6\,dr\,ds\,dt$. The pullback of $du\,dv\,dw$ under the first map is $6\,dr\,ds\,dt$, and the pullback of $dx\,dy\,dz$ under the second map is $-du\,dv\,dw$, hence the pullback of the pullback is $-6\,dr\,ds\,dt$, as was to be shown.

2 Self-checking

3 (a) Counterclockwise (b) Clockwise (c) Collinear (d) Counterclockwise

4 (a) Right-handed (b) Right-handed (c) Coplanar

5 The pullback of $dx\,dy\,dz$ under $x = u + v + w$, $y = v + w$, $z = w$ is $du\,dv\,dw$. The tetrahedron with vertices $(0, 0, 0)$, $(1, 0, 0)$, $(1, 1, 0)$, $(1, 1, 1)$ is described by the inequalities $\{0 \le x \le y \le z \le 1\}$. The unit cube is divided into six such tetrahedra, one for each order of the coordinates x, y, z.

6 (a) $3\frac{5}{6}$ (b) 1 (c) 0

7 By giving two non-coincident points P_0P_1.

§1.5 *page 21*

1 $3\,du + 4\,dv$, $23\,du + 2\,dv$, $3\,dv$, $-51\,du\,dv$, $-183\,du\,dv$.

2 $-3\,dx + 11\,dy + 27\,dz$, $7\,dy + 15\,dz$, $-8\,dy\,dz + 15\,dz\,dx - 7\,dx\,dy$, $-32\,dy\,dz + 60\,dz\,dx - 28\,dx\,dy$, $-24\,dy\,dz + 45\,dz\,dx - 21\,dx\,dy$.

3 The work done between $t = 0$ and $t = 3$ is the work required to go from $(0, 0, 4)$ to $(9, 3, 13)$ which is $(3 \cdot 9) - (2 \cdot 3) + (2 \cdot 9) = 39$. In general the work required is $3\,dx - 2\,dy + 2\,dz = (3 \cdot 3\,dt) - (2 \cdot dt) + (2 \cdot 3\,dt) = 13\,dt$ (the pullback), that is, the work required is 26 in the second time interval, and 26 in the third.

4 A k-form in n variables has $\binom{n}{k}$ components, where $\binom{n}{k}$ denotes the binomial coefficient (see §4.2).

5 The amount of fluid which crosses the segment $(x_0, y_0), (x_1, y_1)$ in unit time is the amount of fluid in the parallelogram $(x_0, y_0), (x_0 + A, y_0 + B), (x_1 + A, y_1 + B), (x_1, y_1)$. The flow is $+$ if this parallelogram is oriented counterclockwise, $-$ otherwise. Hence the flow across the segment from (x_0, y_0) to (x_1, y_1) is the oriented area of this parallelogram, which is $A(y_1 - y_0) - B(x_1 - x_0)$. Thus 'flow across' is the 1-form $A\,dy - B\,dx$.

6 The flow across a parallelogram (x_0, y_0, z_0), (x_1, y_1, z_1), $(x_2 + x_1 - x_0, y_2 + y_1 - y_0, z_2 + z_1 - z_0)$, (x_2, y_2, z_2) is equal in magnitude to the volume of the parallelepiped generated by the parallelogram and the segment (x_0, y_0, z_0), $(x_0 + A, y_0 + B, z_0 + C)$ as in 5. Its sign is $+$ or $-$ depending on whether the orientation $(x_0, y_0, z_0), (x_1, y_1, z_1), (x_2, y_2, z_2), (x_0 + A, y_0 + B, z_0 + C)$ is right- or left-handed. Thus it is the coefficient of $du\,dv\,dw$ in the pullback of $dx\,dy\,dz$ under

$$x = x_0 + (x_1 - x_0)u + (x_2 - x_0)v + Aw$$
$$y = y_0 + (y_1 - y_0)u + (y_2 - y_0)v + Bw$$
$$z = z_0 + (z_1 - z_0)u + (z_2 - z_0)v + Cw.$$

This is $A[(y_1 - y_0)(z_2 - z_0) - (y_2 - y_0) \times (z_1 - z_0)] + \cdots$ which is the value of $A\,dy\,dz + B\,dz\,dx + C\,dx\,dy$ on the given parallelogram. Thus this 2-form describes the flow.

§2.1 *pages 23–24*

1 If the constant 1-form $A\,dx + B\,dy + C\,dz$ describes 'work' then the magnitude of the force is $\sqrt{A^2 + B^2 + C^2}$ because a unit displacement in the direction opposing the force is from $(0, 0, 0)$ to $(A/\sqrt{A^2 + B^2 + C^2}, B/\sqrt{A^2 + B^2 + C^2}, C/\sqrt{A^2 + B^2 + C^2})$ which requires an amount of work equal to $(A^2 + B^2 + C^2)/\sqrt{A^2 + B^2 + C^2} = \sqrt{A^2 + B^2 + C^2}$. To say that the force at (x, y, z) is radially outward means that it is of the form $cx\,dx + cy\,dy + cz\,dz$ where $c > 0$.

To say that its magnitude is proportional to $1/r^2$ means $\sqrt{(cx)^2 + (cy)^2 + (cz)^2} = k/r^2$ for some positive constant k, hence $c = k/r^3$, q. e. d.

2 If the constant 1-form $A\,dy - B\,dx$ describes a planar flow then the magnitude of the flow is $\sqrt{A^2 + B^2}$. To say that the flow at (x, y) is radially outward means it is of the form $cx\,dy - cy\,dx$. To say its magnitude is k/r means $\sqrt{(cx)^2 + (cy)^2} = k/r$, $c = k/r^2$ and the 1-form is $k(x\,dy - y\,dx)/r^2$.

3 By Exercise 6, §1.5 the flow is $c[x\,dy\,dz + y\,dz\,dx + z\,dx\,dy]$. The constant c is to be determined by the condition that the magnitude of the flow is proportional to $1/r^2$, say k/r^2. If $y = z = 0$ the magnitude of the flow is cx, hence $c = k/xr^2 = k/r^3$ in this case. By analogy with Exercise 2 this leads to the enlightened guess $k(x\,dy\,dz + y\,dz\,dx + z\,dx\,dy)/r^3$ (see Exercise 2, §8.3).

4 Flow from a source at 0 is the function $\operatorname{sgn} x = x/\sqrt{x^2}$. A 0-form is a function, and a constant 0-form is a number.

§2.2 *pages 27–28*

1 $\sum(n) = -4n[(2n+1)^{-2} + (2n+3)^{-2} + \cdots + (4n-1)^{-2}]$. The term $(25/20)^{-2}(1/10)$ of $\sum(10)$ corresponds to the interval $\{12/10 \le x \le 13/10\}$ and hence to the two terms $(49/40)^{-2}(1/20) + (51/40)^{-2}(1/20)$ of $\sum(20)$. The difference is estimated, using the given inequality, to be at most

$$\begin{aligned}&|(25/20)^{-2}(1/20) - (49/40)^{-2}(1/20)|\\ &\quad + |(25/20)^{-2}(1/20) - (51/40)^{-2}(1/20)|\\ &\quad \le (1/20)\cdot 2\cdot(1/40) + (1/20)\cdot 2\cdot(1/40)\\ &\quad = 1/200.\end{aligned}$$

Applying the same estimate on each of the ten intervals of $\sum(10)$ gives $|\sum(10) - \sum(20)| \le 1/20 = .05$. The resulting numbers differ by at most .05, i.e., by ± 5 in the last place. A similar argument $(10 = N, m = 2)$ gives $|\sum(N) - \sum(mN)| < N\cdot m\cdot(1/mN)\cdot 2\cdot(1/2N) = 1/N$. Thus $N = 200$ has the desired property. If $n \ge 200$ then $|\sum(n) - \sum(200)| \le |\sum(n) - \sum(200n)| + |\sum(200n) - \sum(200)| < (1/n) + (1/200) \le .01$. Therefore $\sum(n)$ rounded to two places differs from $\sum(200)$ rounded to two places by at most ± 1 in the last place.

2 (a) By the definition of rounding, $|a - a_3| \le .0005$. Thus $|a_3 - t_3| \le |a_3 - a| + |a - t| + |t - t_3| < .0005 + .001 + .0005 = .002$. But a_3, t_3 are integer multiples of .001 and two integers which differ by less than 2 differ by at most ± 1. (b) Take $a = .0004999999$ and $t = .0005000001$.

3 The argument of 1 gives $(x_1^{-1} - x_2^{-1})/(x_1 - x_2) > -1$, $|1/x_1 - 1/x_2| < |x_1 - x_2|$. Then $|\sum(N) - \sum(mN)| < N\cdot m\cdot(1/mN)\cdot(1/2N) = (1/2N)$. Therefore $N = 1{,}000$ has the desired property.

4 The vertices of the n-gon lie at $(\cos 2\pi j/n, \sin 2\pi j/n)$ $(j = 1, 2, \ldots, n)$. Between the jth and $(j+1)$st vertices the term in the approximating sum is $\cos(2\pi(j+\frac{1}{2})/n)[\sin 2\pi(j+1)/n - \sin(2\pi j/n)] - \sin(2\pi(j+\frac{1}{2})/n) \times [\cos(2\pi(j+1)/n) - \cos(2\pi j/n)] = \cos(2\pi(j+\frac{1}{2})/n)\cdot 2\cos(2\pi(j+\frac{1}{2})/n \sin(\pi/n) - \sin(2\pi(j+\frac{1}{2})/n)\cdot(-2)\cdot\sin(2\pi(j+\frac{1}{2})/n)\sin(\pi/n) = 2\sin(\pi/n)$. There are n such terms, hence $2n\sin(\pi/n)$ in all. As $n \to \infty$ the limit is 2π.

5 As in 1 and 3, one must estimate $|x_1^3y_1^2 - x_2^3y_2^3|$. For example the estimate $|x_1^3y_1^2 - x_2^3y_1^2 + x_2^3y_1^2 - x_2^3y_2^3| \le y_1^2|x_1^3 - x_2^3| + x_2^3|y_2^2 - y_1^2| = y_1^2|x_1 - x_2|\,|x_1^2 + x_1x_2 + x_2^2| + x_2^3|y_2 - y_1|\,|y_2 + y_1| \le 12\,|x_1 - x_2| + 4|y_1 - y_2|$ can be used. Then $|\sum(Nn, Mm) - \sum(n, m)| \le 2\cdot[12\cdot 2n^{-1} + 4\cdot m^{-1}] = (12/n) + (8/m)$. From this it can be shown that $N = 400{,}000$ has the desired property.

§2.3 *pages 34–38*

1 $A_{10} = 3.14$, $A_{20} = 3.135$. Further subdivision would at worst cause all uncertain squares to be omitted. This guarantees that A_{10} is within .38 (less than one decimal place) and that A_{20} is within .195 (still not one decimal place) of π. Consulting an accurate value of π (see §7.5) it is seen that A_{10} is correct to 2 decimal places and A_{20} is correct to almost 2 places. In general $U_n = 2n - 1$ because the circle crosses $n - 1$ lines $x =$ const. and $n - 1$ lines $y =$ const. in going from $(n, 0)$ to $(0, n)$. To guarantee $2n^{-2}U_n < .005$ one would therefore need $n \ge 800$. This is much too large because all uncertain squares are not omitted but, rather, only about half of them, which is allowed for in the formula for A_n.

2 (a) $N_5 = 81$, $N_{\sqrt{7}} = 21$. (b) Definition of approximating sum. (c) In the first quadrant the circle crosses at most r lines $x =$ const. and at most r lines $y =$ const., hence there are at most $2r + 1$ squares on the boundary in the first quadrant or at most $8r + 4$ in all. (d) In refining only the squares in (c) are uncertain.

3 There are $8n - 4$ squares on the boundary. If each is enlarged δ in all 4 directions their total area is $(8n - 4)(n^{-1} + 2\delta)^2$. If S is a subdivision with mesh size $|S| \le \delta$ then $U(S) = U_2(S) < (8n - 4)(n^{-1} + 2\delta)^2$. Let $\delta = n^{-1}$. Then $U(S) < (9(8n - 4)/n^2)$. Given $\epsilon > 0$ choose n so large that $9(8n - 4)n^{-2} < \epsilon$. Then $U(S) < \epsilon$ whenever $|S| \le n^{-1}$ as was to be shown.

4 Immediate

5 $|L - \sum(\alpha)| \le |L - \sum(\alpha')| + |\sum(\alpha') - \sum(\alpha)| \le \epsilon + U(S)$.

6 For the first reduction take a third rectangle containing both given rectangles. For the second note that if A_D is zero outside D then it is zero outside R'. In the last part it suffices to note that if $\sum(\alpha)$ is an approximating sum to $\int_R$ then there is an approximating sum $\sum'(\alpha')$ to $\int_{R'}$ which differs from $\sum(\alpha)$ by at most $M \cdot |\alpha| \cdot$ (perimeter of R') where $|\alpha|$ is the mesh size and M is a bound on A. Thus $\sum'(\alpha') \to L$ as $|\alpha'| \to 0$ implies $\sum(\alpha)$ converges (to L) as $|\alpha| \to 0$.

7 'Only if' is immediate. To prove 'if,' split $U(S)$ as directed. The first part is at most $\sigma \cdot$ area (R). Set $\sigma = \epsilon/[2 \text{ area } (R)]$. Let M be a bound for A on R. Choose δ so small that $s(S, \sigma) < \epsilon/(2M)$ whenever $|S| < \delta$. Then $U(S) < \sigma \cdot \text{area } R + 2M \cdot s(S, \sigma) \le \epsilon/2 + \epsilon/2 = \epsilon$.

8 The denial of the Cauchy Criterion is: There is an $\epsilon > 0$ such that for every $\delta > 0$ there exist approximating sums $\sum(\alpha)$, $\sum(\alpha')$ such that $|\alpha| < \delta$, $|\alpha'| < \delta$, and such that $|\sum(\alpha) - \sum(\alpha')| \ge \epsilon$.

9 Approximating sums are formed as in the case of rectangular subdivisions except that area $(R_{ij}) = (x_i - x_{i-1})(y_j - y_{j-1})$ must be replaced by the formula for the area of a parallelogram (§1.3). The proof of convergence is exactly the same as in the previous case. The proof that $U_2(S) \to 0$ as $|S| \to 0$ can be made rigorous by giving an explicit bound on the number of parallelograms which lie on the boundary of a rectangle.

10 Polygonal regions are used because only for them can 'area' be defined algebraically. Since arbitrary sums converge (by the same argument as before) and since rectangular sums converge to $\int_D A\,dx\,dy$, it follows that the limiting value of arbitrary sums is $\int_D A\,dx\,dy$.

§2.4 *pages 43–44*

1 (a) $uv^2 \cos^2 uv\,du\,dv$
(b) $-(u + v)u^2 \cos v\,du\,dv$ (c) $-16e^u\,du\,dv$

2 $(3/2)t^{1/2}$. The integral is thus $\int_1^4 (3/2)t^{1/2}\,dt$.

3 The line through $(u, v, 0)$ and $(0, 0, 1)$ can be parameterized $x = tu$, $y = tv$, $z = 1 - t$. Set $x^2 + y^2 + z^2 = 1$ and solve for t. The pullback is $-4(u^2 + v^2 + 1)^{-2}\,du\,dv$.

4 $\cos\varphi\,d\theta\,d\varphi$.

5 Set $x = \cos\theta$, $y = \sin\theta$ so that $x\,dy - y\,dx = \cos^2\theta\,d\theta + \sin^2\theta\,d\theta = d\theta$ and the integral is $\int_0^{2\pi} d\theta$.

6 $dx\,dy = r\,dr\,d\theta$. The area is approximately $2\pi r$ times Δr where r is the radius of the ring and Δr its width. If $r < 0$ the map $(r, \theta) \to (x, y)$ reverses orientations.

§2.5 *pages 48–49*

1 $x = u$, $y = v$, $z = 1 - u - v$ oriented $du\,dv$.

2 Spherical coordinates:
$dy\,dz = \cos\theta\cos^2\varphi\,d\theta\,d\varphi$,
$dz\,dx = \sin\theta\cos^2\varphi\,d\theta\,d\varphi$,
$dx\,dy = \sin\varphi\cos\varphi\,d\theta\,d\varphi$. Projection on xy-plane. To avoid confusion let (u, v) denote coordinates on the plane, (x, y, z) coordinates on space. Then $dy\,dz = \pm u(1 - u^2 - v^2)^{-1/2}\,du\,dv$, $dz\,dx = \pm v(1 - u^2 - v^2)^{-1/2}\,du\,dv$, $dx\,dy = du\,dv$. Stereographic projection: $dy\,dz = 8u(u^2 + v^2 + 1)^{-3}\,dv\,du$, $dz\,dx = 8v(u^2 + v^2 + 1)^{-3}\,dv\,du$, $dx\,dy = 4(u^2 + v^2 - 1) \times (u^2 + v^2 + 1)^{-3}\,dv\,du$.

3 $\int_b^a f(x)\,dx = \int_\alpha^\beta f(x(u))(dx/du)\,du$ if the map $u \to x$ preserves the orientation of the interval, i.e. carries α to a and β to b. If it reverses the orientation then there is a minus sign. In either case the formula $\int_{x(\alpha)}^{x(\beta)} f(x)\,dx = \int_\alpha^\beta f(x(u))(dx/du)\,du$ holds. Thus $\int_a^b x^n\,dx = \int_{\log a}^{\log b} e^{nu} \cdot e^u\,du = \int_{\log a}^{\log b} e^{(n+1)u}\,du$.

4 At $(\theta, \varphi) = (0, 0)$ the pullbacks are $dx = -\sin\theta\cos\varphi\,d\theta - \cos\theta\sin\varphi\,d\varphi = 0$, $dy = d\theta$, $dz = d\varphi$, $x\,dy\,dz + y\,dz\,dx + z\,dx\,dy = d\theta\,d\varphi$.

5 At $(u, v) = (0, 0)$ the x and y coordinates are 0 so only the term $z\,dx\,dy$ need be computed. Since $dx = 2\,du/(1 + u^2 + v^2) + 0 = 2\,du$ and $dy = 2\,dv$ the answer is $z\,dx\,dy = -4\,du\,dv$.

6 The top end is parameterized by $x = u$, $y = v$, $z = 1$ on $\{u^2 + v^2 \le 1\}$ oriented $du\,dv$, the bottom end by $x = u$, $y = v$, $z = -1$ on $\{u^2 + v^2 \le 1\}$ oriented $dv\,du$, and the sleeve by $x = \cos\theta$, $y = \sin\theta$, $z = h$ on $\{0 \le \theta \le 2\pi, -1 \le h \le 1\}$ oriented $d\theta\, dh$.

7 $x = (2 + \cos u)\cos v$, $y = (2 + \cos u)\sin v$, $z = \sin u$ on $\{0 \le u \le 2\pi, 0 \le v \le 2\pi\}$ oriented $dv\, du$.

§2.6 *page 51*

1 (i) is part of the definition of $\int_R A\, dx\, dy$. Since A is assumed to be continuous the integrals in (ii) all converge and the desired equation follows from the fact that arbitrarily fine approximating sums to $\int_{R_1+R_2} A\, dx\, dy$ split into the sum of an approximating sum to $\int_{R_1} A\, dx\, dy$ and one to $\int_{R_2} A\, dx\, dy$. (iii) and (iv) follow from the fact that approximating sums to the two sides are identical. (v) is easily proved using the continuity of A and the fact that if $m \le A(x, y) \le M$ throughout R then m area $(R) \le \int_R A\, dx\, dy \le M$ area (R), which in turn follows from the fact that any approximating sum to $\int_R A\, dx\, dy$ satisfies these inequalities.

2 Let $\sum(\alpha) = \sum_i [\int_a^b A(x, \hat{y}_i)\, dx]\, \Delta y_i$. Choose a subdivision of $\{a \le x \le b\}$ into intervals shorter than δ, and choose points $\hat{x}_j$ in each interval. By Exercise 5, §2.3, it follows both that $\int_a^b A(x, \hat{y}_i)\, dx$ differs by less than $\epsilon(b - a)$ from $\sum_j A(\hat{x}_j, \hat{y}_i)\, \Delta x_j$ and that $\int_R A\, dx\, dy$ differs by less than $\epsilon(b - a)(d - c)$ from

$$\textstyle\sum_i \sum_j A(\hat{x}_j, \hat{y}_i)\, \Delta x_j\, \Delta y_i.$$

Thus $\sum(\alpha)$ differs by less than $2\epsilon(b - a)(d - c)$ from $\int_R A\, dx\, dy$ whenever $|\alpha| < \delta$, hence $\sum(\alpha) \to \int_R A\, dx\, dy$ as $|\alpha| \to 0$, q.e.d.

§3.1 *pages 56–58*

1 F is the potential function so that $F'(t)\, dt$ gives the amount of work required for small (infinitesimal) displacements (see §8.8). The total amount of work required is expressed by (1) as a sum of small amounts.

2 $F(b) - F(a)$ is the total flow out of the interval $\{a \le x \le b\}$ across the ends a and b. Thus $F'(t)\, dt$ gives the amount of fluid emanating from small (infinitesimal) intervals. This is called the divergence of the flow $F(t)$. Equation (1) gives the total divergence as a sum of small divergences.

3 $F'(t)$ is velocity, $F'(t)\, dt$ gives the displacement which occurs during small (infinitesimal) time intervals, and equation (1) says that the total displacement can be written as a sum of small displacements.

4 Since $F(t + \Delta t) - F(t)$ is the area under the curve which lies over the interval $(t, t + \Delta t)$, it is approximately equal to $f(t)\,\Delta t$. As $\Delta t \to 0$ the approximation improves and $F'(t) = f(t)$. Hence $F'(t)\, dt = f(t)\, dt$ gives the amount of area lying over small (infinitesimal) intervals of the t-axis and (1) says that the total area can be expressed as a sum of small areas.

5 $F(n) - F(n - 1) = C \sin(nA + D)$ where $C = -2\sin(A/2)$ and $D = B - (A/2)$. Thus $\sin D + \sin(A + D) + \cdots + \sin(nA + D) = [F(n) - F(-1)]/C$ which gives

$$\sin\beta + \sin(\alpha + \beta) + \cdots + \sin(n\alpha + \beta) = \frac{\cos(n\alpha + \beta + (\alpha/2)) - \cos(\beta - (\alpha/2))}{-2\sin(\alpha/2)}.$$

6 $\dfrac{\sin(n\alpha + \beta + (\alpha/2)) - \sin(\beta - (\alpha/2))}{2\sin(\alpha/2)}$

7 $F(n) - F(n - 1) = (r - 1)r^{n-1}$. Therefore $F(n) - F(0) = (r - 1)[1 + r + r^2 + \cdots + r^{n-1}]$ and the desired sum is $[F(n) - F(0)]/(r - 1) = (r^n - 1)/(r - 1)$.

8 $n^2 = F(n) - F(0) = [1 + 3 + 5 + \cdots + (2n - 1)] = -n + 2[1 + 2 + 3 + \cdots + n]$ so the desired sum is $\frac{1}{2}(n^2 + n)$.

9 $n^3 = F(n) - F(0) = [1 + 7 + 19 + \cdots + (3n^2 - 3n + 1)] = n - 3 \cdot \frac{1}{2}(n^2 + n) + 3[1 + 2^2 + 3^2 + \cdots + n^2]$ so the desired sum is $\frac{1}{3}[n^3 - n + 3 \cdot \frac{1}{2} \cdot (n^2 + n)] = \frac{1}{3}n^3 + \frac{1}{2}n^2 + \frac{1}{6}n = (2n^3 + 3n^2 + n)/6 = ((2n + 1) \times (n + 1)n/6)$.

10 $n^4 = 4[1 + 2^3 + \cdots + n^3] - 6[1 + 2^2 + \cdots + n^2] + 4[1 + 2^1 + \cdots + n^1] - 1[1 + 2^0 + \cdots + n^0]$ which gives, when the formulas of Exercises 8, 9 are used, the formula $1 + 2^3 + \cdots + n^3 = n^2(n + 1)^2/4$.

11 As in 10 one obtains $n^{k+1} = (k + 1)[1 + 2^k + \cdots + n^k] - \binom{k}{2}[1 + 2^{k-1} + \cdots + n^{k-1}] + \cdots$. By induction, the sums on the right can be written, with the exception of the first, as polynomials in n of degree $\le k$ and the desired conclusion follows.

12 The formula of 5 gives $\sum \sin(j/n)N^{-1} = (\cos B - \cos A)/(-2N \sin(2N^{-1}))$ where B is near b and A is near a. As $N \to \infty$ the limit is $-[\cos b - \cos a]$. The others are similar.

13 By independence of parameter $F(b) - F(a) = \int_{F(a)}^{F(b)} dy = \int_a^b (dy/dx)\, dx = \int_a^b F'(x)\, dx$.

§3.2 *pages 62–65*

1 (a) A constant downward flow. No divergence. (b) A downward flow in the upper half plane and an upward flow in the lower half plane. Flow is converging at all points. (c) A downward flow in the right half plane, upward in the left. A shear with no divergence. (d) A flow from left to right at all points, converging in the left half plane, diverging in the right. (e) A vortical flow, no divergence. (f) A radial flow diverging at all points.

2 The divergence is identically zero as is easily checked. Cutting a small hole $\{x^2 + y^2 \leq \epsilon^2\}$ out of a domain containing the origin and applying the divergence theorem to the remaining domain shows that the flow across the boundary of the original domain is equal to the flow across a small circle $\{x^2 + y^2 = \epsilon^2\}$. This flow is easily seen, by parameterizing the circle $x = \epsilon \cos\theta$, $y = \epsilon \sin\theta$, to be 2π.

3 $-dF$ is the 1-form which describes the central force field.

4 If S is the line segment $\{a \leq x \leq b, y = c\}$ and if $G(x) = F(x, c)$ then $G'(x) = (\partial F/\partial x)(x, c)$ and $\int_S ((\partial F/\partial x)\,dx + (\partial F/\partial y)\,dy) = \int_a^b G'(x)\,dx = G(b) - G(a) = F(b, c) - F(a, c)$ as was to be shown.

5 The integral is $\frac{1}{2}(x_0^2 + y_0^2)$ in all three cases. The flow has no divergence so the total flow across any barrier from $(0, 0)$ to (x_0, y_0) is the same as the total flow across any other.

6 If $P_0P_1P_2P_3$ are the vertices of the rectangle then the integrals over the sides are $F(P_1) - F(P_0)$, $F(P_2) - F(P_1)$, $F(P_3) - F(P_2)$, $F(P_0) - F(P_3)$ for a total of zero. Thus by (2) the integral of the 2-form

$$[(\partial/\partial x)(\partial F/\partial y) - (\partial/\partial y)(\partial F/\partial x)]\,dx\,dy$$

over any rectangle is zero. By (v) of §2.6 this implies that $[(\partial/\partial x)(\partial F/\partial y) - (\partial/\partial y)(\partial F/\partial x)]$ is identically zero.

7 $F_1 = F_2$ because their difference is the integral of $A\,dx + B\,dy$ around the boundary of a rectangle which, by (2), is zero. It is easily shown that $(\partial F_1/\partial y) = B$ and that $(\partial F_2/\partial x) = A$.

8 (a) Exact. $F = \frac{1}{3}(x + y)^3$. (b) Exact. $F = xy$. (c) Exact. $F = \frac{1}{2}(x^2 + y^2)$. (d) Exact. $F = e^{xy}\cos y$. (e) Exact. $F = x\log(xy)$. (*f*) Exact. $F = \frac{1}{2}\log(x^2 + y^2)$. (g) Closed but not exact.

9 The first statement is proved by the method of Exercise 6. No, the force field described by the 1-form 8(*g*) has this property but is not conservative. Prove the last statement by the method of Exercise 7.

10 $dx = \cos\theta\,dr - r\sin\theta\,d\theta$, $dy = \sin\theta\,dr + r\cos\theta\,d\theta$, $x\,dy = r\cos\theta\sin\theta\,dr + r^2\cos^2\theta\,d\theta$, $x\,dy - y\,dx = r^2\,d\theta$, $(x\,dx + y\,dy)/(x^2 + y^2) = dr/r$, $(x\,dy - y\,dx)/(x^2 + y^2) = d\theta$. The first two are exact, the next two are not closed, the fifth is exact, and the sixth is closed and not exact. The same is true of the pullbacks except that the pullback of the last one is exact.

11 $\int_{\partial D} A\,dx + B\,dy = \int [(A\cos\theta + B\sin\theta)\,dr + (-rA\sin\theta + rB\cos\theta)\,d\theta]$. This can be considered as an integral around the boundary of the rectangle $\{0 \leq r \leq 1, 0 \leq \theta \leq 2\pi\}$. Applying formula (2) and simplifying gives $\int((\partial B/\partial x) - (\partial A/\partial y))r\,dr\,d\theta$ over the rectangle, which is (by definition) $\int_D ((\partial B/\partial x) - (\partial A/\partial y))\,dx\,dy$.

12 Let R be the region bounded by the curve S, the lines $x = a$, $x = b$, and the x-axis oriented so that S is part of ∂R. Then $\int_S y\,dx = \int_{\partial R} y\,dx$ because dx is zero on $x =$ const. and because y is zero on the x-axis. Thus $\int_S y\,dx = \int_R dy\,dx =$ area of R if S lies above the axis and $= -$area of R if S lies below the axis. Of course $\int_S y\,dx = \int_a^b f(x)\,dx$.

13 $\frac{1}{2}(x_{i-1}y_i - x_iy_{i-1})$

$$= \frac{1}{2}\left[\left(\frac{x_i + x_{i-1}}{2}\right)(y_i - y_{i-1}) - \left(\frac{y_i + y_{i-1}}{2}\right)(x_i - x_{i-1})\right]$$

$$= \tfrac{1}{2}[x\,(\Delta y) - y\,(\Delta x)].$$

14 By 13, $\pi = \int_{\partial D} \frac{1}{2}(x\,dy - y\,dx)$ where D is the disk $\{x^2 + y^2 \leq 1\}$ oriented $dx\,dy$. The parameterization of the circle by $x = 2u/(u^2 + 1)$, $y = (u^2 - 1)/(u^2 + 1)$ converts this integral to the given integral (see §7.5).

§3.3 *pages 70–72*

1 This is flow from a source at $(0, 0, 0)$, hence the divergence should be zero.

2 (a) $4\pi/3$ (b) $4\pi/3$ (c) $4\pi^2$

3 The force is $dy\,dz$, hence its total is $\int_{\partial D} dy\,dz = \int_D d(dy\,dz) = \int_D 0 = 0$.

4 $d(A\,dx + B\,dy) = d[A((\partial x/\partial r)\,dr + (\partial x/\partial\theta)\,d\theta) + B((\partial y/\partial r)\,dr + (\partial y/\partial\theta)\,d\theta)]$
$= [(\partial/\partial r)(A(\partial x/\partial\theta) + B(\partial y/\partial\theta)) - (\partial/\partial\theta)(A(\partial x/\partial r) + B(\partial y/\partial r))]\,dr\,d\theta$
$= [(\partial A/\partial r)(\partial x/\partial\theta) + (\partial B/\partial r)(\partial y/\partial\theta) - (\partial A/\partial\theta)(\partial x/\partial r) - (\partial B/\partial\theta)(\partial y/\partial r)]\,dr\,d\theta.$
$dA\,dx + dB\,dy = ((\partial A/\partial r)\,dr + (\partial A/\partial\theta)\,d\theta) \times ((\partial x/\partial r)\,dr + (\partial x/\partial\theta)\,d\theta) + ((\partial B/\partial r)\,dr + (\partial B/\partial\theta)\,d\theta)((\partial y/\partial r)\,dr + (\partial y/\partial\theta)\,d\theta)$
$= [(\partial A/\partial r)(\partial x/\partial\theta) + \cdots]\,dr\,d\theta.$

5 $(\partial H/\partial t) = k\,dE$, i.e. $(\partial H_1/\partial t) = k \times (\partial E_3/\partial y) - (\partial E_2/\partial t)), (\partial H_2/\partial t) = k \times ((\partial E_1/\partial z) - (\partial E_3/\partial x)),$ $(\partial H_3/\partial t) = k(\partial E_2/\partial x) - (\partial E_1/\partial y)).$

6 Radial lines emanating from the origin, of equal density in all directions. The derived 2-form is the 3-form which assigns to each portion of space the number of lines of force which end in it. [The number which enter minus the number which leave equal the number which terminate inside minus the number which originate inside.] $dH = 0$ means that magnetic lines of force never terminate. Electrical lines of force terminate on charges.

7 (a) is a simple computation. (b) depends on the observation that a boundary has no boundary.

8 Same as 7

9 The 1-form $(x\,dy - y\,dx)/(x^2 + y^2)$ is closed (where defined, i.e. except on the z-axis) but not exact. The 2-form of Exercise 1 is closed but not exact.

10 If $P = (x_P, y_P, z_P)$, $Q = (x_Q, y_Q, z_Q)$ then the value of $dy\,dz$ on the triangle OPQ is $\frac{1}{2}(y_P z_Q - y_Q z_P)$. If P, Q lie on a line $x =$ const., $z =$ const. this is $\frac{1}{2}z(y_P - y_Q) = -\frac{1}{2}z\,dy$. Similarly if P, Q lie on a line $x =$ const., $y =$ const. the value is $\frac{1}{2}y\,dz$. Thus the proposed 1-form is $\frac{1}{2}(y\,dz - z\,dy)$.

§3.4 *pages 74–75*

1 (a) $d\omega = 0$ (b) $d\omega = 3\,dx\,dy\,dz$ (c) $d\omega = e^{xyz}[yz\,dx + zx\,dy + xy\,dz]$ (d) $d\omega = \sin x\,dy\,dx + \cos x\,dx\,dz$ (e) $d\omega = 2(x + y) \times [dx\,dy + dx\,dz + dy\,dz]$ (f) $d\omega = (dx/x)$ (g) $d\omega = \cos x\,dx$ (h) $d\omega = 2x\,dx$ (i) $d\omega = dx$

2 The equation $\int_{\partial R} x\,dy\,dz = \int_R dx\,dy\,dz$ dictates that the face $x =$ const. with the larger x-coordinate must be oriented by $dy\,dz$.

3 See Exercises 7 and 8 of §3.3. $\int_R d\,(d\omega) = \int_{\partial R} d\omega = \int_{\partial(\partial R)} \omega = 0$ for all R, hence $d\,(d\omega) = 0$.

4 Simple differentiation gives $(\partial F/\partial x) = (2y(x^4 + 4x^2y^2 - y^4)/(x^2 + y^2)^2)$. This is continuous by the same argument which proves that F is continuous. On the line $x = 0$ it is $-2y$, hence $(\partial/\partial y)(\partial F/\partial x)$ exists at $(0, 0)$ and is -2. In the same way $(\partial/\partial x)(\partial F/\partial y)$ exists at $(0, 0)$ and is 2. The second partials have not been proved to be continuous, so equality of second partials is not contradicted.

5 The function must be constant because $F(Q) - F(P) = \int dF = 0$. If the domain consists of two pieces then F can have one constant value on one piece and another on the other so that $dF = 0$ without $F =$ const.

6 The method of Exercise 7, §3.2 proves that $\omega = dF$ for some function F. The 1-form $\omega = (x\,dy - y\,dx)/(x^2 + y^2)$ is the usual counterexample. See §8.6 for a continuation of this topic.

7 Straightforward

8 $\int_S d(A\,dx + B\,dy) = \int_S ((\partial B/\partial x) - (\partial A/\partial y))\,dx\,dy = \int_0^1 \int_0^{1-y} (\partial B/\partial x)\,dx\,dy - \int_0^1 \int_0^{1-x} (\partial A/\partial y)\,dy\,dx = \int_0^1 B(1 - y, y)\,dy - \int_0^1 B(0, y)\,dx + \int_0^1 A(x, 0)\,dx - \int_0^1 A(x, 1 - x)\,dx = \int_{\partial S} (A\,dx + B\,dy).$

9 The pullback of $d\omega$ is d of the pullback of ω; hence by Exercise 8 the integral of the pullback of $d\omega$ is the integral of the pullback of ω around the boundary.

10 Subdivide the polygonal surface into triangles, apply Exercise 9 to each triangle, and use cancellation on interior boundaries. The proof is necessarily sketchy because the notion of 'polygonal surface' has not been made precise.

§4.1 *pages 84–86*

1 (a) $x = -u + 2v + 7$
$y = 2u - 3v - 11$
$(r = 2)$

(b) $t = y - 2$
$x = 2y - 3$
$z = 4y - 11$

($r = 1$ and any of the three variables x, y, z can be moved to the right side.)

(c) $x = V - 2y + z - 7t - 4$

($r = 1$ and any of the four variables x, y, z, t can be moved to the left side.)

(d) $y = \frac{1}{5}v + \frac{2}{5}u \quad + \frac{4}{5}$
$z = \frac{1}{5}v - \frac{3}{5}u + 2x + \frac{4}{5}$

($r = 2$ and (x, y) can be moved to the left, but (x, z) cannot.)

(e) $p = -\frac{1}{2}c + \frac{1}{2}a - \frac{5}{2}$
$q = \frac{3}{2}c - \frac{1}{2}a + \frac{7}{2}$
$b = -\frac{5}{2}c + \frac{3}{2}a - \frac{13}{2}$

($r = 2$ and any two of the three variables a, b, c can be moved to the right.)

(f) $x = \frac{3}{10}u - \frac{1}{10}v + \frac{1}{2}z - \quad \frac{1}{5}$
$y = \frac{2}{5}u + \frac{1}{5}v \quad + 4t - \frac{3}{5}$

($r = 2$. The pairs (x, t), (y, z), (z, t) can be moved left, but the pairs (x, z), (y, t) cannot.)

(g) $p = 2x - 5y - z + 16$
$q = 9x - 23y - 4z + 71$
$r = -4x + 11y + 2z - 32$
$(r = 3)$

(h) $x = \frac{1}{13}v + \frac{3}{13}w - 2z$
$y = -\frac{4}{13}v + \frac{1}{13}w - z$
$u = -\frac{5}{13}v + \frac{11}{13}w$
($r = 2$. Any pair on the right can be moved left and vice versa.)

2 $x = -u - v + 3$
$y = -4u - 3v + 11$

3 $x = 4u - 6v + 4$
$y = 4u - 8v + 4$
$z = 6u - 9v - 3$
$t = 2u - 4v + 4$
The a_{ij}, b_i can be chosen in 6 ways, depending on which of the points $(0, 0)$, $(1, 0)$, $(1, 1)$ is assigned to each of the vertices of the triangle. $dx\,dy = -8\,du\,dv$, $dx\,dz = 0$, $dx\,dt = -4\,du\,dv$, $dy\,dz = 12\,du\,dv$, $dy\,dt = 0$, $dz\,dt = -6\,du\,dv$. Thus uv can be eliminated using any pair but (x, z) or (y, t). This is evidenced by the fact that the triangle projects to a line in the xz-plane and in the yt-plane.

4 $x = -3u + v + w + 2$
$y = -3u - v + 3w + 4$
$z = 3u - 2v + w + 1$
$t = 3u - 3v + 4w - 1$
This can be done in 24 ways, depending on the order of the vertices. $dx\,dy\,dz = 6\,du\,dv\,dw$, $dx\,dy\,dt = 18\,du\,dv\,dw$, $dx\,dz\,dt = 3\,du\,dv\,dw$, $dy\,dz\,dt = 15\,du\,dv\,dw$. Any triple.

5 If $a_{11}a_{22} - a_{12}a_{21} = 0$ then $a_{22}u - a_{12}v$ is constant for fixed z, hence not all values of (u, v) correspond to values of (x, y).

6 $u = (a_{22}/D)x - (a_{12}/D)y + (a_{12}b_2 - a_{22}b_1/D)$
$v = -(a_{21}/D)x + (a_{11}/D)y + (a_{21}b_1 - a_{11}b_2/D)$
$z = (a_{31}a_{22} - a_{32}a_{21})/D)x + (a_{32}a_{11} - a_{31}a_{12}/D)y + (a_{32}a_{21} - a_{31}a_{22}/D)b_1 + (a_{31}a_{12} - a_{32}a_{11}/D)b_2 + b_3$
where $D = a_{11}a_{22} - a_{12}a_{21}$. The condition $D \neq 0$ is necessary and sufficient for there to be a solution.

7 If and only if $a_{11}a_{22} - a_{12}a_{21}$, $a_{11}a_{23} - a_{13}a_{21}$, and $a_{12}a_{23} - a_{13}a_{22}$ are all zero. Geometrically this means that the map collapses xyz-space (the domain) to a line (the image) in the uv-plane (the range). The level surfaces are planes.

8 If and only if $a_{11}a_{22} - a_{12}a_{21}$, $a_{11}a_{32} - a_{12}a_{31}$, and $a_{21}a_{32} - a_{22}a_{31}$ are all zero. Then the map collapses the uv-plane (the domain) to a line (the image) in xyz-space (the range). The level surfaces are lines.

9 $dp\,dr = 2\frac{2}{3}\,dp\,dq$, $dq\,dr = -\frac{2}{3}\,dp\,dq$. This follows immediately from $dr = \frac{2}{3}\,dp + 2\frac{2}{3}\,dq$.

§4.2 *pages 93–94*

1 (a) $7\,du\,dv\,dw\,dx - 2\,du\,dw\,dy\,dz$ plus thirteen terms with the coefficient zero. (b) $-3\,du\,dv\,dw\,dy - 3\,du\,dv\,dw\,dz$ plus thirteen zero terms. (c) $2\,du\,dv\,dx\,dy$ plus fourteen zero terms. (d) $4\,du\,dv\,dw\,dy - du\,dv\,dx\,dz - 4\,du\,dx\,dy\,dz + dv\,dw\,dx\,dz - dv\,dx\,dy\,dz - 4\,dw\,dx\,dy\,dz$ (plus 9 zero terms).

2, 3, 4 Self-checking

5 There are 24 orders; half of them represent $dx\,dy\,dz\,dw$ and half $-dx\,dy\,dz\,dw$, e.g. $dx\,dy\,dz\,dw = dy\,dz\,dx\,dw = dz\,dx\,dy\,dw = dx\,dw\,dy\,dz = dw\,dy\,dx\,dz = \cdots$, $-dx\,dy\,dz\,dw = dy\,dx\,dz\,dw = dx\,dz\,dy\,dw = \cdots$.

§4.3 *pages 103–105*

1 (a) $M_a = \begin{pmatrix} 3 & 2 & 1 \\ 2 & 1 & -3 \end{pmatrix}$ $\quad M_b = \begin{pmatrix} 2 \\ 1 \\ 4 \end{pmatrix}$

$M_c = (1 \quad 2 \quad -1 \quad 7)$ $\quad M_d = \begin{pmatrix} 2 & 1 & -1 \\ -4 & 3 & 2 \end{pmatrix}$

$M_e = \begin{pmatrix} 3 & 1 \\ 2 & -1 \\ 1 & 1 \end{pmatrix}$ $\quad M_f = \begin{pmatrix} 2 & 1 & -1 & -4 \\ -4 & 3 & 2 & -12 \end{pmatrix}$

$M_g = \begin{pmatrix} 2 & 1 & 3 \\ 2 & 0 & 1 \\ -7 & 2 & 1 \end{pmatrix}$ $\quad M_h = \begin{pmatrix} 3 & 2 & 8 \\ 1 & -3 & -1 \\ 4 & 1 & 9 \end{pmatrix}$

(b), (c) $M_aM_b = \begin{pmatrix} 12 \\ -7 \end{pmatrix}$ $\quad M_aM_e = \begin{pmatrix} 14 & 2 \\ 5 & -2 \end{pmatrix}$

$M_aM_g = \begin{pmatrix} 3 & 5 & 12 \\ -15 & -4 & 4 \end{pmatrix}$

$M_aM_h = \begin{pmatrix} 15 & 1 & 31 \\ -5 & -2 & -12 \end{pmatrix}$

$M_eM_a = \begin{pmatrix} 11 & 7 & 0 \\ 4 & 3 & 5 \\ 5 & 3 & -2 \end{pmatrix}$

$M_bM_c = \begin{pmatrix} 2 & 4 & -2 & 14 \\ 1 & 2 & -1 & 7 \\ 4 & 8 & -4 & 28 \end{pmatrix}$

$M_dM_b = \begin{pmatrix} 1 \\ 3 \end{pmatrix}$ $\quad M_gM_b = \begin{pmatrix} 17 \\ 8 \\ -8 \end{pmatrix}$

$$M_hM_b = \begin{pmatrix} 40 \\ -5 \\ 45 \end{pmatrix} \qquad M_dM_e = \begin{pmatrix} 7 & 0 \\ -4 & -5 \end{pmatrix}$$

$$M_dM_g = \begin{pmatrix} 13 & 0 & 6 \\ -16 & 0 & -7 \end{pmatrix}$$

$$M_dM_h = \begin{pmatrix} 3 & 0 & 6 \\ -1 & -15 & -17 \end{pmatrix}$$

$$M_eM_d = \begin{pmatrix} 2 & 6 & -1 \\ 8 & -1 & -4 \\ -2 & 4 & 1 \end{pmatrix}$$

$$M_eM_f = \begin{pmatrix} 2 & 6 & -1 & -24 \\ 8 & -1 & -4 & 4 \\ -2 & 4 & 1 & -16 \end{pmatrix}$$

$$M_gM_e = \begin{pmatrix} 11 & 4 \\ 7 & 3 \\ -16 & -8 \end{pmatrix} \qquad M_hM_e = \begin{pmatrix} 21 & 9 \\ -4 & 3 \\ 23 & 12 \end{pmatrix}$$

$$M_gM_h = \begin{pmatrix} 19 & 4 & 42 \\ 10 & 5 & 25 \\ -15 & -19 & -49 \end{pmatrix}$$

$$M_hM_g = \begin{pmatrix} -46 & 19 & 19 \\ 3 & -1 & -1 \\ -53 & 22 & 22 \end{pmatrix}$$

(d) $$M_a^{(1)} = \begin{pmatrix} 3 & 2 \\ 2 & 1 \\ 1 & -3 \end{pmatrix} \qquad M_a^{(2)} = \begin{pmatrix} -1 \\ -11 \\ -7 \end{pmatrix}$$

$$M_b^{(1)} = (2 \quad 1 \quad 4)$$

$$M_c^{(1)} = \begin{pmatrix} 1 \\ 2 \\ -1 \\ 7 \end{pmatrix} \qquad M_d^{(1)} = \begin{pmatrix} 2 & -4 \\ 1 & 3 \\ -1 & 2 \end{pmatrix}$$

$$M_d^{(2)} = \begin{pmatrix} 10 \\ 0 \\ 5 \end{pmatrix} \qquad M_e^{(1)} = \begin{pmatrix} 3 & 2 & 1 \\ 1 & -1 & 1 \end{pmatrix}$$

$$M_e^{(2)} = (-5 \quad 2 \quad 3) \qquad M_f^{(1)} = \begin{pmatrix} 2 & -4 \\ 1 & 3 \\ -1 & 2 \\ -4 & -12 \end{pmatrix}$$

$$M_f^{(2)} = \begin{pmatrix} 10 \\ 0 \\ -40 \\ 5 \\ 0 \\ 20 \end{pmatrix} \qquad M_g^{(1)} = \begin{pmatrix} 2 & 2 & -7 \\ 1 & 0 & 2 \\ 3 & 1 & 1 \end{pmatrix}$$

$$M_g^{(2)} = \begin{pmatrix} -2 & 11 & 4 \\ -4 & 23 & 9 \\ 1 & -5 & -2 \end{pmatrix} \quad M_g^{(3)} = (-1)$$

$$M_h^{(1)} = \begin{pmatrix} 3 & 1 & 4 \\ 2 & -3 & 1 \\ 8 & -1 & 9 \end{pmatrix}$$

$$M_h^{(2)} = \begin{pmatrix} -11 & -5 & 13 \\ -11 & -5 & 13 \\ 22 & 10 & -26 \end{pmatrix} \quad M_h^{(3)} = (0).$$

(e) Self-checking. For example,

$$(M_gM_h)^{(2)} = \begin{pmatrix} 55 & -301 & -115 \\ 55 & -301 & -115 \\ -110 & 602 & 230 \end{pmatrix} = M_h^{(2)}M_g^{(2)}.$$

2 Interchanging the first two rows of a 3×3 matrix is the same as multiplying on the left by

$$\begin{pmatrix} 0 & 1 & 0 \\ 1 & 0 & 0 \\ 0 & 0 & 1 \end{pmatrix}, \text{ i.e.}$$

$$\begin{pmatrix} 0 & 1 & 0 \\ 1 & 0 & 0 \\ 0 & 0 & 1 \end{pmatrix}\begin{pmatrix} a & b & c \\ d & e & f \\ g & h & i \end{pmatrix} = \begin{pmatrix} d & e & f \\ a & b & c \\ g & h & i \end{pmatrix}.$$

Since the determinant of this matrix is -1 (it corresponds to $x = v$, $y = u$, $z = w$ hence $dx\,dy\,dz = -du\,dv\,dw$) the rule (a) follows for this case. Multiplying on the right by this matrix interchanges the first two columns. All cases of rule (a) can be proved in this way. The operation in (b) is the matrix form of the operation of composition with a map like $x = u + av$, $y = v$, $z = w$ for which $dx\,dy\,dz = du\,dv\,dw$. The operation in (c) is composition with a map like $x = cu$, $y = v$, $z = w$ for which $dx\,dy\,dz = c\,du\,dv\,dw$. If the first row is $(1, 0, 0)$ then using (b) the first column can be made to be $(1, 0, 0)$ without changing the other columns and without changing the determinant. The matrix then corresponds to an affine map of the form $x = u$, $y = av + bw$, $z = a'v + b'w$. In computing $dx\,dy\,dz = D\,du\,dv\,dw$ it suffices to find $dy\,dz = D\,dv\,dw$ and multiply by $dx = du$. The determinants are -42, 5, 45, 1, and -2242.

3 A rotation of 90° in a coordinate plane corresponds to the operation of interchanging two rows (or columns) and changing the sign of one of them. The corresponding matrix is a matrix which can be obtained from the identity matrix (1's in the ith row of the ith column for $i = 1, 2, \ldots, n$ and 0's elsewhere) by applying such an operation. A shear corresponds to the operation of adding a multiple of one row (or column) to another row (or column) and the corresponding matrix is a matrix which can be obtained from the identity matrix by applying such an operation. A scale factor corresponds to the operation of multiplying all entries of one row (or one column) by a given factor and the corresponding matrix is a matrix which can be obtained from the identity matrix by such an operation. By composing with a translation it can be assumed that the constants in the given affine map are zero. By composing with rotations it can be assumed that the entry in the upper left hand corner of the matrix of coefficients is not zero (unless it is identically zero, in which case it is a scale factor of 0 in all coordinate directions). Applying a scale factor it can be assumed that it has a 1 in the upper left hand corner. Composing with shears it can be assumed that all other entries in the first row and column of its matrix of coefficients are 0. Then use induction. Very simply, the statement is that every determinant can be evaluated by the method of Exercise 2, a fact of which one is easily convinced by a few examples.

4 By 3 every $n \times n$ matrix can be written as a composition (product) of matrices of three simple types (rotations, shears, scale factors). For each of these types the determinant of the transpose is immediately seen to be the determinant of the matrix itself. The transpose of a product is the product of the transposes in reverse order, and the determinant of a product is the product of the determinants. Hence if $M = M_1 M_2 \cdots M_n$ then $M^{(1)} = M_n^{(1)} \cdots M_2^{(1)} M_1^{(1)}$, $\det(M^{(1)}) = \det(M_n^{(1)}) \cdots \det(M_2^{(1)}) \det(M_1^{(1)}) = \det(M_1)\det(M_2) \cdots \det(M_n) = \det(M)$ as desired.

§4.4 *page 112*

1 $x = \frac{5}{3}u - \frac{2}{3}v - \frac{4}{3}w$, $y = \frac{11}{3}u - \frac{5}{3}v - \frac{7}{3}w$, $z = -4u + 2v + 3w$

2 Set $x = \frac{5}{3}u - \frac{2}{3}v + \frac{4}{3}w + b_1$, $y = \frac{11}{3}u - \frac{5}{3}v - \frac{7}{3}w + b_2$, $z = -4u + 2v + 3w + b_3$. The point $(x, y, z) = (0, 0, 0)$ corresponds to $(u, v, w) = (7, -2, 1)$. Substituting in the first equation gives $0 = \frac{5}{3}7 - \frac{2}{3}(-2) - \frac{4}{3} \cdot 1 + b_1$, hence $b_1 = -35/3$. Similarly $b_2 = -80/3$, $b_3 = 29$.

3 (a) Reduce the system to the form (2). Since $r \leq m = 2 < n$ the equations (2a) contain at least one of the variables (x, y, z) on the right side. Choosing two different values of this variable (or these variables) and fixing values of (u, v), the equations (2a) give two points on the same level surface. (b) Reduce the system to the form (2). Since $r \leq n = 2 < m$ the equations (2b) contain at least one equation. Fixing the values of (u, v, w) on the right side of (2b) gives values of those on the left. Choosing different values for those on the left gives points (u, v, w) not in the image.

4 As in 3, if $n > m$ then equations (2a) imply that the map is not one-to-one, and if $n < m$ then the equations (2b) imply that the map is not onto.

5 (a) The 'only if' half, because $1 = \det(MM^{-1}) = \det(M)\det(M^{-1})$ implies, if $\det(M^{-1})$ is an integer, that $\det(M) = \pm 1$. Hence M^{-1} has integer entries only if $\det(M) = \pm 1$. (b) The formula for M^{-1} gives its entries as integers divided by $\det(M)$, hence M^{-1} has integer entries if $\det(M) = \pm 1$.

§4.5 *pages 124–127*

1 (a) $(1, -1, 0), (0, 1, -1)$, dim $= 2$
(b) $(1, 1, 1), (-1, 0, 1)$, dim $= 2$
(c) $(1, 1, 1)$, dim $= 1$
(d) $(1, 3, 5)$, dim $= 1$
(e) dim $= 0$

2 (a) $(1, -1, 0, 0, 0, 0), (0, 1, -1, 0, 0, 0), \ldots, (0, 0, 0, 0, 1, -1)$, dim $= 5$ (b) $(1, 1, 1, 1, 1, 1), (0, 1, 2, 3, 4, 5)$, dim $= 2$

3 (a) Given an arrow A and an arrow B, place the beginning point of B at the ending point of A and take $A + B$ to be the arrow from the beginning point of A to the ending point of B. (b) Multiply the length of the arrow by the number, leaving the direction unchanged; if the number is negative, reverse the direction of the arrow. (c) The dimension is 2 and a basis is any two non-collinear arrows. (d) Displacements are described by arrows (directed line segments). The sum in (a) is the composition of two displacements. Forces are described by arrows. The sum of (a) tells how two forces give a resultant force. Velocities are described by arrows. The sum of (a) tells how the velocities of two motions (an airplane in the wind) add to give a resultant velocity.

4 Three arrows in space are linearly dependent if and only if they are coplanar.

5 $1 \cdot (1 \cdot v) = (1 \cdot 1)v = 1 \cdot v$. Thus $1 \cdot v$ and v satisfy the same equation and are therefore equal. This proves V. Similarly, $[v + 0 \cdot w] + w = v + [0 \cdot w + 1 \cdot w] = v + (0 + 1)w = v + w$. Thus $v + 0 \cdot w$ and v satisfy the same equation and are therefore equal. This proves VI. Using V and VI it is easily seen that $(1/a_1) \times [v - a_2 v_2 - \cdots - a_n v_n]$ satisfies the given equation, thus it is the solution guaranteed by IV.

6 By IV, if such a vector exists it is unique. If w is any vector then $0 \cdot w$ has the property which defines the zero vector (by II and VI). Hence the zero vector exists and is unique. If f is any function then $0 \cdot f$ is the zero vector and is the function whose values are all zero.

7 If $f(v) = 0$, $f(w) = 0$ then $f(v + w) = f(v) + f(w) = 0 + 0 = 0$, hence $v + w$ is in the kernel. Also $f(av) = af(v) = a \cdot 0 = 0$ so av is in the kernel, hence the kernel is a subspace. The subspace of 1(a) is the kernel of the linear map $(f_1, f_2, f_3) \to f_1 + f_2 + f_3$ of V_3 to V_1. Similarly the subspaces of 1(b) – (e) are kernels.

8 It shows that the solution v_1 asserted by IV lies in the subspace provided that $v, v_2, v_3, \ldots, v_n$ do; hence the elements of the subspace satisfy IV. I–III are immediate.

9 (a) See (b). (b) If S is any set and if W is any vector space then the rule $(f + g)(x) = f(x) + g(x)$ defines an addition operation on the set of functions $\{f : S \to W\}$ and $(af)(x) = a[f(x)]$ defines an operation of multiplication by numbers. Axioms I–IV are satisfied; thus $\{f : S \to W\}$ is a vector space. If $S = V$ is a vector space then the set of all *linear* maps $\{f : V \to W\}$ is a subspace of the set of all maps (the sum of linear maps is linear, and any multiple of a linear map is linear). Thus the space of linear maps $\mathrm{Hom}(V, W)$ is itself a vector space. (c) Let δ_i be the element of V_n which is 1 on i and 0 on all other integers $1, 2, 3, \ldots, n$. Similarly let ϵ_i be the element of V_m which is 1 on i and 0 on other integers. Then $\epsilon_1, \epsilon_2, \ldots, \epsilon_m$ are a basis of V_m and $\delta_1, \delta_2, \ldots, \delta_n$ a basis of V_n. Given a linear map $f: V \to W$, define numbers a_{ij} by $f(\delta_j) = \sum_{i=1}^m a_{ij}\epsilon_i$. Then $f(x_1, x_2, \ldots, x_n)$

$$\begin{aligned} &= f(x_1\delta_1 + x_2\delta_2 + \cdots + x_n\delta_n) \\ &= x_1 f(\delta_1) + x_2 f(\delta_2) + \cdots + x_n f(\delta_n) \\ &= x_1 \textstyle\sum_{i=1}^m a_{i1}\epsilon_i + \cdots + x_n \sum_{i=1}^m a_{in}\epsilon_i \\ &= y_1\epsilon_1 + \cdots + y_m\epsilon_m \\ &= (y_1, y_2, \ldots, y_m) \end{aligned}$$

where

$$y_1 = x_1 a_{11} + \cdots + x_n a_{1n} = \textstyle\sum_{j=1}^n a_{1j}x_j,$$

etc. Addition of elements of $\mathrm{Hom}(V_n, W_m)$ is the operation of adding matrices by adding corresponding components. Multiplication by numbers in $\mathrm{Hom}(V_n, V_m)$ is the operation of multiplying all entries of a matrix by a given number. (d) dim $= nm$. A basis is given by the matrices which are 1 in one position and 0 in all others.

10 (a) An element ϕ of W^* is a linear map $\phi: W \to V_1$. If $f: V \to W$ is linear then the composition $\phi \circ f: V \to V_1$ is an element of V^*, denoted $f^*(\phi)$. This map $f^*: W^* \to V^*$ is linear. (b) $\dim(V^*) = \dim(V)$. (c) Given a basis $v_1, v_2, \ldots, v_n$ of V, define $\phi_1: V \to V_1$ by $\phi_1(x_1v_1 + x_2v_2 + \cdots + x_nv_n) = x_1$. Defining $\phi_2, \phi_3, \ldots, \phi_n$ analogously gives a basis of V^*.

11 (a) Elements of $(V_n)^*$ are row matrices, i.e. $1 \times n$ matrices. (b) By the transposed matrix.

12 If $r < n$ then the zero vector must be the image of a non-zero vector (v_n), hence the first alternative does not hold. Any element of V_n which is in the image is the image of more than one element (add any multiple of v_n) hence the second alternative holds. If $r = n$ then the map is one-to-one and onto, i.e. every element of V_n is the image of just one element of V_n under the given map; hence the first alternative holds.

13 $$(1 \;\; 1 \;\; 1)\begin{pmatrix} 1 & -1 & -1 \\ 0 & 1 & 0 \\ 0 & 0 & 1 \end{pmatrix} = (1 \;\; 0 \;\; 0).$$

$$(1 \;\; -2 \;\; 1)\begin{pmatrix} 1 & 2 & -1 \\ 0 & 1 & 0 \\ 0 & 0 & 1 \end{pmatrix} = (1 \;\; 0 \;\; 0).$$

$$\begin{pmatrix} 1 & 0 & 0 \\ -\frac{7}{9} & \frac{2}{9} & \frac{13}{9} \\ \frac{1}{9} & \frac{1}{9} & \frac{2}{9} \end{pmatrix}\begin{pmatrix} 4 & -3 & 1 \\ 1 & 1 & -3 \\ 2 & 1 & 1 \end{pmatrix}\begin{pmatrix} 0 & 0 & 1 \\ 0 & \frac{1}{4} & 0 \\ 1 & \frac{3}{4} & -4 \end{pmatrix} = \begin{pmatrix} 1 & 0 & 0 \\ 0 & 1 & 0 \\ 0 & 0 & 1 \end{pmatrix}.$$

Of course there are many solutions, for example

$$\begin{pmatrix} 4 & -3 & 1 \\ 1 & 1 & -3 \\ 2 & 1 & 1 \end{pmatrix}\begin{pmatrix} \frac{4}{36} & \frac{4}{36} & \frac{8}{36} \\ -\frac{7}{36} & \frac{2}{36} & \frac{13}{36} \\ -\frac{1}{36} & -\frac{10}{36} & \frac{7}{36} \end{pmatrix} = \begin{pmatrix} 1 & 0 & 0 \\ 0 & 1 & 0 \\ 0 & 0 & 1 \end{pmatrix}.$$

$$(1 \;\; 1 \;\; 1 \;\; 1 \;\; 1 \;\; 1) \times \begin{pmatrix} 1 & -1 & -1 & -1 & -1 & -1 \\ 0 & 1 & 0 & 0 & 0 & 0 \\ 0 & 0 & 1 & 0 & 0 & 0 \\ 0 & 0 & 0 & 1 & 0 & 0 \\ 0 & 0 & 0 & 0 & 1 & 0 \\ 0 & 0 & 0 & 0 & 0 & 1 \end{pmatrix} = (1 \;\; 0 \;\; 0 \;\; 0 \;\; 0 \;\; 0).$$

$$\begin{pmatrix} 1 & 0 & 0 \\ 0 & 0 & 1 \\ 1 & 1 & 1 \end{pmatrix}\begin{pmatrix} 1 & -1 & 0 \\ 0 & 1 & -1 \\ -1 & 0 & 1 \end{pmatrix}\begin{pmatrix} 1 & 0 & 1 \\ 0 & 0 & 1 \\ 1 & 1 & 1 \end{pmatrix} = \begin{pmatrix} 1 & 0 & 0 \\ 0 & 1 & 0 \\ 0 & 0 & 0 \end{pmatrix}$$

$$\begin{pmatrix} 0 & 1 \\ \frac{1}{5} & -\frac{2}{5} \end{pmatrix}\begin{pmatrix} 2 & 1 & -1 \\ 1 & -2 & 1 \end{pmatrix}\begin{pmatrix} 1 & 2 & \frac{1}{5} \\ 0 & 1 & \frac{3}{5} \\ 0 & 0 & 1 \end{pmatrix} = \begin{pmatrix} 1 & 0 & 0 \\ 0 & 1 & 0 \end{pmatrix}.$$

$$\begin{pmatrix} 1 & -2 & 1 & 0 & 0 & 0 \\ 0 & 1 & -2 & 1 & 0 & 0 \\ 0 & 0 & 1 & -2 & 1 & 0 \\ 0 & 0 & 0 & 1 & -2 & 1 \end{pmatrix}$$

$$\times \begin{pmatrix} 1 & 2 & 3 & 4 & 5 & -4 \\ 0 & 1 & 2 & 3 & 4 & -3 \\ 0 & 0 & 1 & 2 & 3 & -2 \\ 0 & 0 & 0 & 1 & 2 & -1 \\ 0 & 0 & 0 & 0 & 1 & 0 \\ 0 & 0 & 0 & 0 & 0 & 1 \end{pmatrix}$$

$$= \begin{pmatrix} 1 & 0 & 0 & 0 & 0 & 0 \\ 0 & 1 & 0 & 0 & 0 & 0 \\ 0 & 0 & 1 & 0 & 0 & 0 \\ 0 & 0 & 0 & 1 & 0 & 0 \end{pmatrix}$$

Set $v_i = Q(\delta_i) = i$th column of Q, and set $w_i = P^{-1}(\delta_i) = i$th column of P^{-1}. Since Q and P^{-1} are invertible these are bases. $M(v_i) = P^{-1}PMQ(\delta_i) = P^{-1}C_r(\delta_i) = w_i$ if $i \leq r$ and $= 0$ if $i > r$.

§4.6 *pages 130–131*

1 If the map (1) in the Implicit Function Theorem is onto then equations (2b) are absent and the equations (2a) give a parameterization of each level surface $y =$ const. by $x_{m+1}, x_{m+2}, \ldots, x_n$. If the map (1) is one-to-one then the equations (2b) give $m - n$ equations which define the image.

2 If the pullbacks of all k-forms under M_2 are zero then the pullbacks of all k-forms under $M_1 = P_1^{-1}M_2Q_1^{-1}$ are zero (by the Chain Rule) and vice versa. This shows that rank $M_1 =$ rank M_2. Since $M^{(k)} = Q^{(k)}C_r^{(k)}P^{(k)}$ and since $Q^{(k)}$, $P^{(k)}$ are invertible, it follows that rank $M^{(k)} =$ rank $C_r^{(k)} = \binom{r}{k}$.

3 (a) In this case $f_P \circ f_Q^{-1}$ is of the form $f_Q \circ A \circ f_Q^{-1}$ where $A: \mathbf{R}^n \to \mathbf{R}^n$ is the map $y_1 = 1 - x_1 - x_2 - \cdots - x_n$, $y_2 = x_2$, $y_3 = x_3, \ldots, y_n = x_n$ which interchanges $(0, 0, \ldots, 0) \leftrightarrow (1, 0, \ldots, 0)$ and leaves $(0, 1, 0, \ldots, 0), \ldots, (0, \ldots, 0, 1)$ fixed. Since $dy_1\, dy_2 \cdots dy_n = -dx_1\, dx_2 \cdots dx_n$ this map has Jacobian

$$\partial(y_1, y_2, \ldots, y_n)/\partial(x_1, x_2, \ldots, x_n) = -1$$

and the result follows. (b) The Chain Rule implies that the composition of two maps with negative Jacobian has positive Jacobian. (c) $\begin{pmatrix} \cos\theta & \sin\theta \\ -\sin\theta & \cos\theta \end{pmatrix}$ for $0 \leq \theta \leq \pi/2$ gives a continuous deformation of $\begin{pmatrix} 1 & 0 \\ 0 & 1 \end{pmatrix}$ to $\begin{pmatrix} 0 & 1 \\ -1 & 0 \end{pmatrix}$. To do this for shears, translations, and positive scale factors is easy. Writing f_Q as a product of such factors and deforming the factors to the identity one by one deforms f_Q to the identity, hence deforms Q to P. (d) Clear (e) The given determinant is the Jacobian of the map

$$\begin{aligned} x &= x_0 + (x_1 - x_0)u \\ &\qquad + (x_2 - x_0)v + (x_3 - x_0)w \\ y &= y_0 + (y_1 - y_0)u \\ &\qquad + (y_2 - y_0)v + (y_3 - y_0)w \\ z &= z_0 + (z_1 - z_0)u \\ &\qquad + (z_2 - z_0)v + (z_3 - z_0)w \end{aligned}$$

carrying $(0, 0, 0)$, $(1, 0, 0)$, $(0, 1, 0)$, $(0, 0, 1)$ to the given quadruple, hence its sign tells whether it agrees or disagrees with the standard quadruple.

4 Let r_i be the rank of f_i. Then $n_i - r_{i+1} =$ dimension of level surface of $f_{i+1} =$ dimension of image of $f_i = r_i$. Hence $n_i = r_i + r_{i+1}$. Thus $n_1 - n_2 + n_3 - \cdots \pm n_\nu = (r_1 + r_2) - (r_2 + r_3) + (r_3 + r_4) - \cdots \pm (r_\nu + r_{\nu+1}) = r_1 \pm r_{\nu+1} = 0 \pm 0 = 0$.

§5.1 *pages 140–142*

1 Any point where $\bar{u} \neq 0$. Any point where $\bar{v} \neq 0$. Explicit solutions are $u = \pm\sqrt{y + v^2}$, $v = \pm\sqrt{-y + u^2}$, with the sign chosen in accord with $(\bar{u}, \bar{v})$. The equation $y = u^2 - v^2$ can be solved by

$$(u, v) = \begin{cases} (\sqrt{y}, 0) & \text{if } y > 0 \\ (0, 0) & \text{if } y = 0 \\ (0, \sqrt{-y}) & \text{if } y < 0. \end{cases}$$

Hence for all y there is a solution. Among other things, the level curves $y =$ const. are very different near $(0, 0)$ in the three cases $y > 0$, $y = 0$, $y < 0$.

2 $y + 2x = (u + v)^2$, $y - 2x = (u - v)^2$; hence $u = \frac{1}{2}[\pm\sqrt{y + 2x} \pm \sqrt{y - 2x}]$, $v = \frac{1}{2}[\pm\sqrt{y + 2x} \mp \sqrt{y - 2x}]$. Points of the image must satisfy $y + 2x \geq 0$, $y - 2x \geq 0$. The lines $u + v = 0$, $u - v = 0$ divide the uv-plane into 4 parts, each of which is mapped one-to-one onto the wedge $y \geq -2x$, $y \geq 2x$. The signs under the radical signs in (6′) are necessarily opposite, which leaves 8 possibilities. One choice determines the sign of u, one the sign of v, and the sign under the radical sign determines whether $|u| > |v|$ or $|v| > |u|$. Hence each of the 8 choices is valid in one of the 8 regions into which the uv-plane is divided by the lines $u = 0$, $v = 0$, $u = v$, $u = -v$.

3 Solution is possible provided that the line $v = \bar{v}$, $w = \bar{w}$ is not tangent to the sphere $u^2 + v^2 + w^2 = \bar{u}^2 + \bar{v}^2 + \bar{w}^2$, i.e. provided $\bar{u} \neq 0$.

4 This map is non-singular of rank 2 at all points other than $(u, v) = (0, 0)$. The curves $x =$ const. and $y =$ const. are rectangular hyperbolae intersecting at right angles. For each $(x, y) \neq (0, 0)$ there are two solutions (u, v) given by $u = \pm\sqrt{\frac{1}{2}(x + \sqrt{x^2 + y^2})}$, $v = y/2u = \pm y[2(x + \sqrt{x^2 + y^2})]^{-1/2}$.

5 Non-singular of rank 2 at all points. $x^2 + y^2 = e^{2u}$ gives $u = \frac{1}{2}\log(x^2 + y^2)$ and $\sin v/\cos v = y/x$ gives $v = \arctan(y/x)$. Giving $(x, y) \neq (0, 0)$ determines u and determines infinitely many values of v, any two of which differ by a multiple of 2π. The point $(x, y) = (0, 0)$ is not in the image.

6 The map of 5 is everywhere non-singular of rank 2 but not onto ($(0, 0)$ is omitted) or one-to-one ($(0, 0)$ and $(0, 2\pi)$ have the same image). The sign of the derivative of a non-singular map $f: \mathbf{R} \to \mathbf{R}$ does not change, which implies that the function is increasing or decreasing, hence one-to-one. The map $y = e^x$ is non-singular of rank 1 but not onto.

7 The map $t = u + v$ has as its level surfaces the lines $u + v =$ const. The map $x = \cos t$, $y = \sin t$ wraps the t-line around the circle $x^2 + y^2 = 1$.

8 See §7.1. Arcsin x, defined for $|x| \leq 1$, is the unique number y in the interval $\{-\pi/2 \leq y \leq \pi/2\}$ such that $x = \sin y$.

9 The curve intersects $y = \alpha x + \beta$ if and only if x satisfies a cubic equation, hence in 1 or 3 points, unless $\alpha = -1$, in which case the equation has degree 2 in x. It intersects $x + y = C$ in 1 point (where $t = 0$) if $C = 1$, in 2 points for $0 < C < 1$, in 1 point for $C = 0$, in two points for $-\frac{1}{3} < C < 0$ and in no points for $C \leq -\frac{1}{3}$ or $C > 1$; these conclusions all follow from the fact that $t^2 = \alpha$ has two solutions if $\alpha > 0$, one if $\alpha = 0$, none if $\alpha < 0$. The curve has $x + y = -\frac{1}{3}$ as an asymptote and makes a loop (or 'leaf' = folium) in the first quadrant. F is singular where $3x^2 - y = 0$, $3y^2 - x = 0$, which gives $(x, y) = (0, 0)$ or $(x, y) = (\frac{1}{3}, \frac{1}{3})$. F is negative inside the folium and to the southwest of the folium, positive to the northeast. The singularity at $(\frac{1}{3}, \frac{1}{3})$ is a local minimum of F. The curves $F =$ const. consist of two pieces when the constant is negative, one if positive. Solving for y as a function of (F, x) is possible provided $3y^2 - x \neq 0$. Assuming $F = 0$ this is true provided $(x, y) \neq (0, 0)$ and $(x, y) \neq (\sqrt[3]{4}/3, \sqrt[3]{2}/3)$. These points cut the curve into 4 pieces of the form $(x, f(x))$.

10 If $\partial F/\partial z \neq 0$ then z can be written as a function of x, y, and F. Setting $F = 0$ gives z as a function of x and y.

11 $dF \neq 0$ by assumption. If $dF\,dG \equiv 0$ the map $(x, y, z) \to (F, G)$ is non-singular of rank 1 and the equation (2b) gives a functional relation $G - g(F) \equiv 0$. Conversely, if $dF\,dG \neq 0$, then F, G cannot be functionally related because $f(F, G) \equiv 0$ implies $df = (\partial f/\partial F)\,dF + (\partial f/\partial G)\,dG$ is identically zero as a 1-form in (x, y, z); hence $0 = df\,dG = (\partial f/\partial F)\,dF\,dG$, $(\partial f/\partial F) \equiv 0$ as a function of (x, y, z). Similarly $(\partial f/\partial G) \equiv 0$. Since $(x, y, z) \to (F, G)$ covers all points near $(F(\bar{x}, \bar{y}, \bar{z}), G(\bar{x}, \bar{y}, \bar{z}))$ this implies both partials of f are identically zero near this point, hence $f \equiv 0$ and $f(F, G) \equiv 0$ is not a relation between F and G. Thus 'functionally related' implies $dF\,dG = 0$.

12 $f'(x)$ exists and is equal to $2x \sin(1/x^2) - 2x^{-1}\cos(1/x^2)$ for $x \neq 0$ by the usual rules of differentiation. For $x = 0$ it is the limit of $h \sin(1/h^2)$ as $h \to 0$ which is $\leq |h|$ in absolute value and hence $f'(0) = 0$. But $\lim_{x\to 0} f'(x)$ does not exist so $f'(x)$ is not a continuous function.

13 A map is non-singular of rank r if and only if the local rank and the infinitesimal rank are both r. It is singular if and only if the infinitesimal rank is strictly less than the local rank.

§5.2 *pages 147–151*

1 (a) $ya + xb$ (c) 1, 2, -9, -6, $-2\frac{1}{2}$ (d) $-3a + 2b = 0$ (e) $-3(x - 2) + 2(y + 3) = 0$ (f) $\bar{y}(x - \bar{x}) + \bar{x}(y - \bar{y}) = 0$ (g) At $(0, 0)$ this is not the equation of a line. The curve $xy = 0$ is singular at $(0, 0)$ and has no tangent line. (h) The line has the equation $(y - \bar{y}) = -(\bar{y}/\bar{x})(x - \bar{x}) = -\bar{x}^{-2}(x - \bar{x})$, hence its slope is $-\bar{x}^{-2}$. (i) $y\,dx + x\,dy = 0$, $dy = -(y/x)\,dx$; hence $dy/dx = -y/x = -1/x^2$.

2 (a) $4\bar{x}a + 2\bar{y}b$ (b) $F > 12$ away from $(0, 0)$ (e) $4\bar{x}(x - \bar{x}) + 2\bar{y}(y - \bar{y}) = 0$ (f) The tangent can also be written $(y - \bar{y}) = (-2\bar{x}/\bar{y})(x - \bar{x})$, hence its slope is $-2\bar{x}/\bar{y}$. (g) $dz = 4x\,dx + 2y\,dy$, $dy = (1/2y)\,dz - (2x/y)\,dx$; hence when z is constant $dy/dx = -2x/y = 2x(1 - 2x^2)^{-1/2}$.

3 (a) $(3\bar{x}^2 - \bar{y})(x - \bar{x}) + (3\bar{y}^2 - \bar{x})(y - \bar{y}) = 0$. (b) $3\bar{y}^2 - \bar{x} = 0$, which combines with $\bar{x}^3 + \bar{y}^3 - \bar{x}\bar{y} = 0$ to give $(\bar{x}, \bar{y}) = (\sqrt[3]{4}/3, \sqrt[3]{2}/3)$. The point $(\bar{x}, \bar{y}) = (0, 0)$ is a singularity at which one branch of the curve has a vertical tangent. (c) $(\sqrt[3]{2}/3, \sqrt[3]{4}/3)$ and one branch at $(0, 0)$ (d) $3\bar{x}^2 - \bar{y} = 3\bar{y}^2 - \bar{x}$ and $\bar{x}^3 + \bar{y}^3 = \bar{x}\bar{y}$ imply $\bar{x} = \bar{y} = \frac{1}{2}$ (or $\bar{x} = \bar{y} = 0$).

4 (a) $(\partial F/\partial x)(\bar{x}, \bar{y})(x - \bar{x}) + (\partial F/\partial y)(\bar{x}, \bar{y}) \times (y - \bar{y}) = 0$ (b) $f'(x) = \text{slope} = -(\partial F/\partial x)/(\partial F/\partial y)$ (c) Set $z = F(x, y)$, $dz = (\partial F/\partial x)\,dx + (\partial F/\partial y)\,dy$ hence when $dz = 0$, $dy/dx = -(\partial F/\partial x)/(\partial F/\partial y)$.

5 $y = f(x)$, $dy = f'(x)\,dx$. If the equation is solved for x as a function of y, $x = g(y)$, then by implicit differentiation $g'(y)$ is found by solving $dx = (1/f'(x))\,dy$. If $y = \log x$ then $x = e^y, dx = e^y\,dy, dy = (1/e^y)\,dx = (1/x)dx$, hence $d[\log x]/dx = 1/x$.

6 (a) Since $x^p = y^q = t^{pq}$ it suffices to show that every positive number has a unique positive qth root. This is obvious from the fact that y^q increases from 0 at $y = 0$ to arbitrarily large numbers for large y. See §7.5 for a complete proof. (b) If q is odd then every number has a unique qth root. (c) $dy = pt^{p-1}\,dt$, $dx = qt^{q-1}\,dt$ hence $dy/dx = (p/q)t^{p-q} = (p/q)(y/x) = (p/q)x^{(p/q)-1}$. (e) Only $x = 0$ is at issue. Here it is a question of evaluating $\lim_{h\to 0} |h|^r/h = \lim_{h\to 0} (\text{sign } h)|h|^{r-1}$. If $r > 1$ this exists and is 0 which is also the limit of the derivative as $x \to 0$. If $r \le 1$ the limit doesn't exist. (f) $rx|x|^{r-2}$ or $r(\text{sign } x)|x|^{r-1}$

7 (a) $12a - 12b + c$ (b) Into, if $12a - 12b + c < 0$; out of, if > 0; tangent, if $= 0$ (c) $6\bar{x}a + 4\bar{y}b + c$

8 (a) If $F(f(t), g(t), h(t))$ is constant then its derivative $(\partial F/\partial x)a + (\partial F/\partial y)b + (\partial F/\partial z)c$ is zero. (b) $12(x - 2) - 12(y + 3) + (z - 1) = 0$

9 (a) $z - \bar{z} = (\partial f/\partial x)(x - \bar{x}) + (\partial f/\partial y) \times (y - \bar{y})$ (b) $(\partial F/\partial x)(x - \bar{x}) + (\partial F/\partial y) \times (y - \bar{y}) + (\partial F/\partial z)(z - \bar{z}) = 0$ is the same plane as $(\partial f/\partial x)(x - \bar{x}) + (\partial f/\partial y)(y - \bar{y}) - (z - \bar{z}) = 0$ where all derivatives are evaluated at $(\bar{x}, \bar{y}, \bar{z})$. Therefore

$$(\partial f/\partial x)\colon -1 = (\partial F/\partial x)\colon (\partial F/\partial z) \text{ and}$$
$$(\partial f/\partial y)\colon -1 = (\partial F/\partial y)\colon (\partial F/\partial z).$$

10 (a) $r^2 \cos\varphi$ (b) The planes $r = 0$ and $\varphi = (2n + 1)\pi/2$ are singularities. (c) $\{r > 0,\ 0 < \theta < 2\pi,\ -\pi/2 < \varphi < \pi/2\}$. (d) $-\sin\theta\,dx + \cos\theta\,dy = r\cos\varphi\,d\varphi$. Since $y/(x^2 + y^2) = r\sin\theta\cos\varphi/r^2\cos^2\varphi$ and similarly for $x/(x^2 + y^2)$, this gives the derivatives of θ. The equation $\cos\theta\,dx + \sin\theta\,dy = \cos\varphi\,dr - r\sin\varphi\,d\varphi$ combines with $dz = \sin\varphi\,dr + r\cos\varphi\,d\varphi$ to give $dr, d\varphi$ in terms of dx, dy, dz.

§5.4 *pages 183–190*

1 $(7, 2, -3)$ is the nearest point. In general, one can set $x = \lambda_1 A_1 + \lambda_2 A_2$, $y = \lambda_1 B_1 + \lambda_2 B_2$, $z = \lambda_1 C_1 + \lambda_2 C_2$ in (*) and solve for λ_1, λ_2, hence for (x, y, z). This procedure applies if and only if the planes (*) are not parallel or coincident.

2 The problem is $v_1^{-1}\sqrt{a_1^2 + b_1^2} + v_1^{-1} \times \sqrt{a_2^2 + b_2^2} = \text{min.}$, subject to $b_1 + b_2 = \text{const.}$, and the result follows immediately by the method of Lagrange multipliers.

3 width = length = $2\sqrt{2}$, height = $\sqrt{2}$.

4 Critical points satisfy $yz = \lambda x$, $zx = \lambda y$, $xy = \lambda z$ for some λ. Then $3xyz = \lambda(x^2 + y^2 + z^2) = \lambda$ so the value at the critical point is $\lambda/3$. This excludes the possibility that $\lambda = 0$ at a maximum or a minimum. Hence $x \ne 0$, $y \ne 0$, $z \ne 0$ at max. or min. Dividing the first two equations gives $y/x = x/y$ hence $x = \pm y = \pm z = \pm\lambda = \pm 3^{-1/2}$ and the maximum and minimum values are $1/3\sqrt{3}$ and $-1/3\sqrt{3}$, assumed at the points $(\pm 1/\sqrt{3}, \pm 1/\sqrt{3}, \pm 1/\sqrt{3})$.

5 The critical points satisfy $2\lambda_1 u = u - x$, $2\lambda_1 v = v - y$, $\lambda_2(\partial f/\partial x) = x - u$, $\lambda_2(\partial f/\partial y) = y - v$, $u^2 + v^2 = 1$, $f(x, y) = 0$. From $x = (1 - 2\lambda_1)u$, $y = (1 - 2\lambda_1)v$ it follows that $1 - 2\lambda_1 \ne 0$ (the origin does not lie on $f = 0$). Thus $(x - u, y - v)$ is a multiple of (x, y) and the desired conclusions follow.

6 (a) $\frac{1}{4}$ at $(x, y) = (\frac{1}{2}, \frac{1}{2})$ (b) No, $(x, y, z) = (2K + 1, -K, -K)$ for large values of K gives large values of xyz. (c) $\lambda = xy = yz = zx$ gives $\lambda x = xyz = \lambda y = \lambda z$. If $\lambda \ne 0$ then $x = y = z = \frac{1}{3}$. If $\lambda = 0$ then two of the 3 must be zero and $(x, y, z) = (1, 0, 0)$ or $(0, 1, 0)$ or $(0, 0, 1)$. (d) From $\lambda Ax = Ax^Ay^Bx^C$, $\lambda By = Bx^Ay^Bz^C$, $\lambda Cz = Cx^Ay^Bz^C$ it follows that $\lambda \ne 0$, hence that $x = y = z = (A + B + C)^{-1}$ and the only critical point is a point where the value is $(A + B + C)^{-(A+B+C)}$. (e) A triangle on whose sides the function is identically zero, which is a minimum. (f) As before $x_1 = x_2 = \cdots = x_n = K$ and the maximum is K. If $\sum A_i = 1$ and if $x_i > 0$ then

$x_1^{A}x_2^{A_2}\cdots x_n^{A_n} \le \sum A_ix_i$. Hence

$$(x_1^{A_1}\cdots x_n^{A_n})^{(1/\Sigma A_i)} \le \sum A_ix_i/\sum A_i.$$

(g) geometric mean $\le$ arithmetic mean (h) Only when the x's are all equal.

7 (a) For $p = 2$ it is a circle. For large p it is nearly the square with vertices $(\pm 1, \pm 1)$. For p slightly larger than 1 it is nearly the square with vertices $(\pm 1, 0)$, $(0, \pm 1)$, which is what it is when $p = 1$. (b) As max $(|x|, |y|)$ (c) 13 at $(x, y) = (1, 1)$ (d) 9 at $(1, -1)$ (e) $|A| + |B|$ at $(x, y) =$ (sign A, sign B). If $A = 0$ it is near the largest value all along one side of the 'square' $|x|^p + |y|^p = 1$.

8 (a) $|A_1| + |A_2| + \cdots + |A_n|$ (b) At points where $x_i =$ sign A_i for those i for which $A_i \ne 0$. (c) $|A_1x_1 + \cdots + A_nx_n| \le |A|_1|x|_\infty$ (d) max $(|A_1|, |A_2|, \ldots, |A_n|)$.

9 Fix (x_1, x_2) and solve $(x_1 + y_1)^4 + (x_2 + y_2)^4 =$ max. or min. on $y_1^4 + y_2^4 = K$. It is easily seen that $(x_1 + y_1, x_2 + y_2)$ is a multiple of (y_1, y_2) hence (y_1, y_2) and (x_1, x_2) are collinear; this means there are just two critical points, which must be the max. and min. Similarly in (c) it is easily shown that the max. and min. occur when $(x_1, x_2, \ldots, x_n)$ and $(y_1, y_2, \ldots, y_n)$ are collinear, which is when equality occurs.

10 Elliptic. The minor axis is $(1 - \sqrt{5})x + 2y = 0$ and the max. on $x^2 + y^2 = 1$ occurs at points of the minor axis. The major axis is of course perpendicular.

11 Hyperbolic. The minimum value on $x^2 + y^2 = 1$ is at the intersections with the line $(1 + \sqrt{5})x + 2y = 0$ where it is $2 - \sqrt{5}$.

12 There are essentially two configurations. Concentric ellipsoids is one and the other is an elliptical double cone surrounded by hyperboloids of one sheet and containing hyperboloids of two sheets. Otherwise there are the 'degenerate' cases where the level surfaces are elliptical cylinders, or hyperbolic cylinders, or planes.

13 If not, then (x, y, z) is a multiple of (x_1, y_1, z_1), which implies, since $x_1x + y_1y + z_1z = 0$, that the first equation is not satisfied.

14 (a) $u^2 + v^2 + w^2 = 1$ (b) $\lambda_1u^2 + \lambda_2v^2 + \lambda_3w^2$ (c) The seven points $(0, 0, 0)$, $(\pm 1, 0, 0)$, $(0 \pm 1, 0)$, $(0, 0, \pm 1)$ are the only critical points if $0, \lambda_1, \lambda_2, \lambda_3$ are distinct. Otherwise there can be disks, line segments, circles, or spheres as before.

15 (a) The polynomial has leading term $(-\lambda)^n$, hence it is positive for $\lambda = -K$ (K large). If its constant term is negative it must change sign, and hence have a root, between 0 and $-K$.

16 Let $A = (A_{ij})$ be the symmetric matrix corresponding to Q. As in the text there is a matrix M such that $AM = M\Lambda$ where Λ is a matrix with λ_i in the ith column of the ith row $(i = 1, 2, \ldots, n)$ and 0 elsewhere, and where the transpose $M^{(1)}$ of M is its inverse, i.e. $MM^{(1)} = I$. Since det $(A) =$ det (Λ), two of the λ_i must be negative—unless Q is positive definite. But then Q would be negative definite on a 2-dimensional affine manifold through the origin. Since this manifold must intersect $x_n = 0$, the assumption that $Q(x_1, x_2, \ldots, x_{n-1}, 0) \ge 0$ would then be contradicted.

17 The determinants are ≥ 0.

18 (a) Use estimates such as those on p. 154. (b) Choose ϵ smaller than the least value of Q on $u^2 + v^2 + w^2 = 1$ and choose S as in (a) with $B = 1$. Then for (x, y, z) satisfying $(x - \bar{x})^2 + (y - \bar{y})^2 + (z - \bar{z})^2 \le S$, setting $s = [(x - \bar{x})^2 + (y - \bar{y})^2 + (z - \bar{z})^2]^{1/2}$ shows that $|s^{-2}[F(x, y, z) - F(\bar{x}, \bar{y}, \bar{z})] - Q(u, v, w)| < \epsilon$ where $u^2 + v^2 + w^2 = 1$. Thus $F(x, y, z) > F(\bar{x}, \bar{y}, \bar{z})$ as desired. (c) If Q assumed negative values then an argument like (b) would show that there must be points arbitrarily near $(\bar{x}, \bar{y}, \bar{z})$ where $F(x, y, z) < F(\bar{x}, \bar{y}, \bar{z})$. (d) If all are positive there is a local minimum. If one is negative there is not a local minimum. Otherwise neither conclusion can be drawn.

19 If two consecutive points coincided there would be an $(n - 1)$-gon with the same L and A, hence $A \le (1/4(n - 1))(\cot(\pi/n - 1))L^2 < (1/4n)(\cot(\pi/n))L^2$ and the given polygon would not be a solution of $A =$ max., $L =$ const., because it is less than a regular n-gon. L is differentiable because $\sqrt{x}$ is differentiable for $x > 0$. The equations $dA = \lambda\, dL$ are $v_{i+1} + v_i/2 = \lambda(u_i - u_{i+1})/l_i, -(u_{i+1} + u_i)/2 = \lambda(v_i - v_{i+1})/l_i$. $dL \ne 0$ because $dL = 0$ would imply all u's are equal and all v's are equal, hence, since $\sum_{i=1}^n u_i = 0, \sum_{i=1}^n v_i = 0$, would imply all u's and v's were zero. Since $l_{i+1}^2 - l_i^2 = (u_{i+1} + {} + u_i)(u_{i+1} - u_i) + (v_{i+1} + v_i)(v_{i+1} - v_i) = (v_{i+1} - v_i)(u_{i+1} - u_i)(2\lambda - 2\lambda)l_i^{-1} = 0$ it follows that $l_i = l_{i+1} = L/n$. Then

$$\begin{aligned} u_{i+1} - \lambda' v_{i+1} &= -u_i - \lambda' v_i \\ \lambda' u_{i+1} + v_{i+1} &= \lambda' u_i - v_i \end{aligned}$$

where $\lambda' = 2\lambda/nL$, hence

$$\begin{pmatrix} u_{i+1} \\ v_{i+1} \end{pmatrix} = -\begin{pmatrix} 1 & -\lambda' \\ \lambda' & 1 \end{pmatrix}^{-1} \begin{pmatrix} 1 & \lambda' \\ -\lambda' & 1 \end{pmatrix} \begin{pmatrix} u_i \\ v_i \end{pmatrix}$$

The matrix $M(\lambda)$ is (see §7.5) of the form

$$\begin{pmatrix} a & -b \\ b & a \end{pmatrix}$$

where $a^2 + b^2 = 1$ and where $a \neq 1$. Thus $(a, b) = (\cos\theta, \sin\theta)$ for $0 < \theta < 2\pi$ and

$$[M(\lambda)]^n = \begin{pmatrix} \cos n\theta & -\sin n\theta \\ \sin n\theta & \cos n\theta \end{pmatrix}$$

and the fact that $[M(\lambda)]^n - I$ is not one-to-one gives $\cos n\theta = 1$, $n\theta = 2\pi j$ for some j.

§5.5 *pages 193–195*

1 If $n = k$ in (1) and if the map is non-singular of rank k then the image is defined by the $m - k$ equations (2b), i.e. by the equations $z_i = 0$ where $z_i = y_i - h_i(y_1, \ldots, y_r)$. These equations are independent because $(\partial(z_{k+1}, \ldots, z_m)/\partial(y_{k+1}, \ldots, y_m)) = 1$. If $m = n - k$ and if (1) is nonsingular of rank m then the equations (2a) parameterize the level surfaces using the k independent variables $x_{n-k+1}, \ldots, x_n$.

2 Set $y = x_1^2 + \cdots + x_n^2$. Then $dy = 0$ only at the origin so the equation $y =$ const. defines a differentiable manifold near any point other than $(0, 0, \ldots, 0)$. This point is the unique point (zero-dimensional manifold) where $y = 0$.

3 No. Near $(0, 0)$ it consists of two branches and cannot be described by an equation $F =$ const.

4 Such a matrix is invertible if and only if $x_1x_4 - x_2x_3 \neq 0$. The set of such matrices is a 4-dimensional manifold parameterized by x_1, x_2, x_3, x_4.

5 Set $a_1 = x_1^2 + x_2^2$, $a_2 = x_3^2 + x_4^2$, $b = x_1x_3 + x_2x_4$. Then orthogonal matrices are defined by the relations $a_1 = 1$, $a_2 = 1$, $b = 0$. These relations are independent because $da_1\,da_2\,db = [-4x_1x_3x_4 + 4x_2x_3^2]\,dx_1\,dx_2\,dx_3 + \cdots = 4\{x_2\,dx_1\,dx_2\,dx_3 - x_1\,dx_1\,dx_2\,dx_4 - x_4\,dx_1\,dx_3\,dx_4 + x_3\,dx_2\,dx_3\,dx_4\}$ is not zero at any point (x_1, x_2, x_3, x_4) satisfying $a_1 = 1$, $a_2 = 1$, $b = 0$. Thus these relations define a 1-dimensional manifold. The parametric curves

$$\begin{pmatrix} \sqrt{1 - t^2} & t \\ -t & \sqrt{1 - t^2} \end{pmatrix}, \begin{pmatrix} \cos t & -\sin t \\ \sin t & \cos t \end{pmatrix},$$

$$\begin{pmatrix} \dfrac{1 - t^2}{1 + t^2} & \dfrac{-2t}{1 + t^2} \\ \dfrac{2t}{1 + t^2} & \dfrac{1 - t^2}{1 + t^2} \end{pmatrix}$$

for t near zero all parameterize the orthogonal matrices near the identity matrix.

6 Let x_{ij} $(i, j = 1, 2, \ldots, n)$ denote the entries of a typical $n \times n$ matrix, and let $a_i = \sum_{j=1}^n (x_{ij})^2$, $b_{ij} = \sum_{\nu=1}^n x_{i\nu}x_{j\nu}$. A matrix is orthogonal if and only if it satisfies the $n + n(n - 1)/2$ relations $a_i = 1$ $(i = 1, 2, \ldots, n)$, $b_{ij} = 0$ $(1 \leq i < j \leq n)$. If it is shown that these relations are independent it will follow that orthogonal matrices are an $n^2 - n - n(n - 1)/2 = n(n - 1)/2$-dimensional manifold. They are independent near the identity matrix ($x_{ij} = 1$ if $i = j$ and 0 otherwise) because at this point $da_i = 2\,dx_{ii}$, $db_{ij} = dx_{ij} + dx_{ji}$ so that $da_1 \cdots da_n\,db_{12}\,db_{13} \cdots db_{n-1,n} \neq 0$. Near any orthogonal matrix M_1 the orthogonal matrices are those for which $M_1^{-1}M$ is orthogonal (because the inverse of an orthogonal matrix is orthogonal and the product of two orthogonal matrices is orthogonal). Hence multiplication by M_1^{-1} composed with the functions $a_1, \ldots, b_{12}, \ldots$ gives $n(n + 1)/2$ independent relations defining the orthogonal matrices near M_1.

7 (a) $(x - \alpha)^2 + y^2 = 1$ and $-2(x - \alpha) = 0$ gives $y^2 = 1$, $y = \pm 1$. These two lines are obviously envelopes of the circles (*). (b) The equations $f =$ const., $f_\alpha = 0$ define a 1-dimensional differentiable manifold near $(\bar{x}, \bar{y}, \bar{\alpha})$. Let $(x(t), y(t), \alpha(t))$ be a parameterization of this curve. Since $(x(t), y(t))$ satisfies $g(x, y) =$ const. it suffices to show that this curve $(x(t), y(t))$ is tangent to the curve $f(x, y, \alpha(t)) =$ const. in the xy-plane. Using $f =$ const. and $f_\alpha = 0$ gives $0 = ((\partial f/\partial x)(dx/dt)) + ((\partial f/\partial y) \times (dy/dt)) + ((\partial f/\partial \alpha)(d\alpha/dt)) = ((\partial f/\partial x) \times (dx/dt)) + ((\partial f/\partial y)(dy/dt))$ and the desired result follows. (c) This example illustrates very well the difference between the theoretical solutions of (b) and actual solutions by formulas. The formulas here can get very complicated. A fairly easy solution is the following: Set $v_x = v\cos\alpha$, $v_y = v\sin\alpha$. Eliminating t, clearing denominators and using $2\cos^2\alpha =$

$\cos 2\alpha + 1$, $2\sin\alpha\cos\alpha = \sin 2\alpha$, puts the equation of the trajectory in the form $yv^2 + gx^2 + yv^2\cos 2\alpha - xv^2\sin 2\alpha = 0$. Differentiating with respect to α and setting equal to zero gives $yv^2\sin 2\alpha + xv^2\cos 2\alpha = 0$ which leads to $\sin 2\alpha = (\pm x/\sqrt{x^2+y^2})$, $\cos 2\alpha = (\mp y/\sqrt{x^2+y^2})$. Using these to eliminate α from the original equation gives $yv^2 + gx^2 = \pm v^2\sqrt{x^2+y^2}$. Squaring and cancelling gives then $2gv^2y + g^2x^2 = v^4$ as an envelope of the trajectories.

§6.2 *page 200*

1 πab.

2 The absolute value of

$$\frac{1}{6}\begin{vmatrix} 1 & x_0 & y_0 & z_0 \\ 1 & x_1 & y_1 & x_1 \\ 1 & x_2 & y_2 & z_2 \\ 1 & x_3 & y_3 & z_3 \end{vmatrix}$$

[See Ex. 3(e), §4.6 and Ex. 5, §1.4.]

3 Adopting the notation of §2.3, if $|S| < \delta$ then $V(S)$, which is the total volume of all rectangular parallelepipeds of S which lie partly inside D and partly outside D, is at most $(1+\delta)^3 - (1-\delta)^3 = 6\delta + 2\delta^3 \to 0$.

4 The matrix equation

$$\begin{pmatrix} c & 0 \\ 0 & 1 \end{pmatrix}\begin{pmatrix} 1 & 1 \\ 0 & 1 \end{pmatrix}\begin{pmatrix} c^{-1} & 0 \\ 0 & 1 \end{pmatrix} = \begin{pmatrix} 1 & c \\ 0 & 1 \end{pmatrix}$$

shows that an arbitrary shear can be written as a composition of the types (i)–(iv). Then apply Ex. 3, §4.3.

5 (a) As in Ex. 3, give an explicit estimate of the total volume of the rectangles of S which lie partly inside and partly outside $f(D)$. (b) For each δ give a *particular* approximating sum $\sum(\alpha)$ to $\int_{f(D)} dy$ such that $|\alpha| < \delta$ and such that $\sum(\alpha) = \int_D dx$. (c) If the formula holds for disjoint (non-overlapping) sets $D_1, D_2, \ldots, D_N$ it holds for their union. (d) Follows from the definition of $V = \int_D dx$. (e) To prove (a)–(e), note that any approximating sum to $\int_{f(D)} dy$ is less than the corresponding approximating sum to $\int_{f(\bar{D})} dy$ and that the latter converges to $\int_{\bar{D}} dx$ which is less than $V + \epsilon$.

§6.3 *pages 213–214*

1 Let $g(x)$ denote the square of the suggested function $f(x)$. Then $g(x)$ is differentiable. Given $P = (\bar{x}, \bar{y}, \bar{z})$ set $c(x, y, z) = g((x-\bar{x})/\epsilon) \times g((\bar{x}-x)/\epsilon)g((y-\bar{y})/\epsilon)g((\bar{y}-y)/\epsilon) \times g((z-z)/\epsilon)g((\bar{z}-z)/\epsilon)$ where ϵ is a small positive number. Then $c(x, y, z)$ is differentiable, $c(\bar{x}, \bar{y}, \bar{z}) = 1$, and $c(x, y, z) = 0$ unless $|x-\bar{x}| < \epsilon$, $|y-\bar{y}| < \epsilon$, $|z-\bar{z}| < \epsilon$. For ϵ sufficiently small this function will serve as c_P.

2 S near P can be parameterized by (x, y) or (x, z) or (y, z), hence F_i and F_j can be expressed as an invertible relation between these two coordinates and (u, v). By composition $F_j^{-1}\circ F_i$ is an invertible relation $(u, v) \to (u, v)$. Set $g = F_j^{-1}\circ F_i$. Then g has positive Jacobian if and only if $g^*(\sigma)$ is a positive multiple of σ for every 2-form σ in uv. Since every 2-form σ in uv can be written in the form $\sigma = F_j^*(\omega)$ where ω is a 2-form in xyz, it follows that this is true if and only if $F_i^*(\omega) = (F_j\circ F_j^{-1}\circ F_i)^*(\omega) = (F_j^{-1}\circ F_i)^*(F_j^*(\omega)] = g^*(\sigma)$ is a positive multiple of $F_j^*(\omega) = \sigma$ for all ω, q.e.d.

3 Let $x = f_1(u, v)$, $y = f_2(u, v)$, $z = f_3(u, v)$ be the parameterization of the torus given in Ex. 7, §2.5. Then any square $\{\bar{u} \le u \le \bar{u} + 2\pi, \bar{v} \le v \le \bar{v} + 2\pi\}$ in the uv-plane parameterizes the entire torus. The only problem is to include every point of the torus in the *interior* of the image of such a square. The 3 squares obtained by setting $(\bar{u}, \bar{v}) = (0, 0)$, $(2\pi/3, 2\pi/3)$, $(4\pi/3, 4\pi/3)$ accomplish this.

§6.4 *page 218*

1 If the map is affine

$$x = au + bv + c$$
$$y = a'u + b'v + c'$$

then one can assume, by rotating the xy-plane if necessary, that $b' > 0$. Then $y = c'$ is not a boundary of R because points where $y = \pm b'\epsilon + c'$ are in R. Therefore $x = c$ must be a boundary of R because R is only allowed two kinds of boundaries — $x =$ const. or $y =$ const. — and the point $(a\epsilon + c, a'\epsilon + c')$ must lie outside R while (c, c') must lie inside. Therefore $b = 0$. Therefore $a > 0$. The remaining statements follow easily.

2 If f has continuous second partials then the functions $\partial y_i/\partial x_j$ are differentiable; thus if $\omega = A\,dy + \cdots$ is a differentiable form then $f^*(\omega) = A(\partial y/\partial x)\,dx + \cdots$ and all terms in $f^*(\omega)$ are differentiable because sums, products, and compositions of differentiable functions are differentiable. Conversely, since dy_i is differentiable, the assumption that the pullback of a differentiable k-form is differentiable implies that $f^*(dy_i) = (\partial y_i/\partial x_1)\,dx_1 + \cdots +$

$(\partial y_i/\partial x_n)\,dx_n$ is differentiable, hence that the functions $\partial y_i/\partial x_j$ are differentiable, hence that the functions y_i are twice differentiable.

3 One possibility is the following. Let $F_1: (u, v) \to (x, y, z)$ be the affine map which carries (0, 0), (0, 2), (2, 0) to P_0, P_1, P_2 respectively and let $R_1 = \{0 \le u \le 1, 0 \le v \le 1\}$. Similarly let F_2, F_3 be the affine maps which carry (0, 0), (0, 2), (2, 0) to $P_1P_2P_0$, $P_2P_0P_1$, and let $R_2 = R_3 = R_1$. This is almost a set of charts for $P_0P_1P_2$ except that the midpoint of each side is not in the interior of any chart. Let $F_4: (u, v) \to (x, y, z)$ be the affine map which carries (−2, 0), (2, 0), (0, 2) to $P_0P_1P_2$ and let $R_4 = \{0 \le u \le 1, -1 \le v \le 1\}$. Defining F_5, F_6 analogously gives a set of charts.

§6.5 *page 223*

1 If R_i is all of $\{|u_j| \le 1\}$ then no points of ∂S are involved. If R_i has just one side inside $\{|u_j| \le 1\}$, then, by rotating $u_1u_2 \cdots u_k$-space if necessary, it can be assumed that this side is of the form u_1 = const. and that R_i lies on the side $u_1 \le$ const. Map the square $\{|v_1| \le 1, \ldots, |v_{k-1}| \le 1\}$ to this side of R_i by $u_j = v_{j-1}$. Composing with the F_i, the resulting maps $(v_1, \ldots, v_{k-1}) \to (x_1, x_2, \ldots, x_n)$ give a set of charts for ∂S, as can be proved by the method of Ex. 1, §6.4. [The case $k = 0$ must be handled separately.]

2 Let $R = \{a_j \le u_j \le b_j\}$ where $a_j < b_j$; $j = 1, 2, \ldots, k$. If $\omega = A_1\,du_2\,du_3 \cdots du_k$ the formula is $\int_{\partial R} \omega = \int_{|u_j| \le 1} [A_1(b_1, u_2, u_3, \ldots, u_k) - A_1(a_1, u_2, \ldots, u_k)]\,du_2\,du_3 \cdots du_k$. The other terms of the general formula are accompanied by the sign $(-1)^{j-1}$ where u_j is the variable held constant in the integration.

§6.6 *page 225*

1 A function (assigning numbers to points). A finite collection of points, for each of which a sign (±) is designated.

2 If material is continuously distributed in xyz-space then 'mass' assigns to each compact 3-dimensional region of xyz-space the mass of the material contained in the region. 'Density' is the ratio of mass to volume ($= dx\,dy\,dz$) for infinitesimal cubes in xyz-space.

§7.1 *pages 234–235*

1 The formula

$$u^{(N+1)} = u^{(N)} + (1 - [(u^{(N)})^2 + \epsilon^2])/2$$

with the initial approximation $u^{(0)} = 1$ gives easily $u^{(3)} = 1 - \epsilon^2/2 - \epsilon^4/8 - \epsilon^6/16 - \epsilon^8/128$ whereas the binomial series $(1 + y)^n = 1 + ny + n(n-1)y/2 + \cdots$ (see §8.4) gives $(1 - x^2)^{1/2} = 1 - (1/2)x^2 - (1/8)x^4 - (1/16)x^6 - (5/128)x^8 - \cdots$.

2 The line through $(x^{(N)}, f(x^{(N)}))$ parallel to the original tangent line is the line $(y - f(x^{(N)}))/(x - x^{(N)}) = a$. This line intersects the line $y = \bar{\bar{y}}$ at $(x^{(N+1)}, \bar{\bar{y}})$ where $x^{(N+1)} = x^{(N)} + (\bar{\bar{y}} - f(x^{(N)}))/a$. The process of solving the equation $y = bx + c$ when y is given, and when the approximate slope a is used, leads to $x^{(N+1)} = x^{(N)} + (y - bx^{(N)} - c)/a$. This gives $x^{(N)} = (1 + r + r^2 + \cdots + r^{N-1})(y - c)/a + r^N x^{(0)} = (1 - r^N)(1 - r)^{-1}(y - c)a^{-1} + r^N x^{(0)}$ where $r = 1 - (b/a)$. This converges if and only if $|r| < 1$. When this is the case $x^{(\infty)} - x^{(N)} = r^N((y - c)/b - x^{(0)})$ so the error decreases by a factor of $r = 1 - [f'(x^{(\infty)})/f'(\bar{x})]$ with each step.

§7.2 *pages 240–242*

1 $x_i^{(N+1)} = (1/a_{ii})[y_i - \sum_{j \ne i} a_{ij}x_j^{(N)}] = x_i^{(N)} + (1/a_{ii})[y_i - \sum_{j=1}^{n} a_{ij}x_j^{(N)}]$ hence $x^{(N+1)} = x^{(N)} + M[y - Lx^{(N)}]$ where M is the matrix which is $(1/a_{ii})$ in the ith row of the ith column ($i = 1, 2, \ldots, n$) and which is zero elsewhere. (5) is satisfied if and only if $\sum_{j \ne i} |a_{ij}| < |a_{ii}|$ for $i = 1, 2, \ldots, n$; in words, the 'diagonal term' a_{ii} in each row must be larger in absolute value than the sum of the absolute values of the other terms in that row.

2 The terms on the diagonal (a_{ii}) should be large relative to the terms not on the diagonal (a_{ij} where $i \ne j$). The Gauss-Seidel method would be expected to converge more quickly because it uses a more recent approximation to $x_j^{(N)}$ ($j < i$) in computing $x_i^{(N)}$.

3 Self-checking

4 Check by using the formula for the inverse of a matrix (§4.4) to express the entries of L^{-1} as explicit rational numbers.

6 Set $ML = (c_{ij})$. Then (1′) is $x_i^{(N+1)} = x_i^{(N)} + (My)_i - \sum_{j=1}^{i-1} c_{ij}x_j^{(N+1)} - \sum_{j=i}^{n} c_{ij}x_j^{(N)}$. Let $ML = T + U$ where T is c_{ij} if $j < i$, 0 if $j \ge i$ and where U is c_{ij} if $j \ge i$ and 0 if $j < i$. Then $x^{(N+1)} = x^{(N)} + My - Tx^{(N+1)} - Ux^{(N)}$, $x^{(N+1)} + Tx^{(N+1)} = x^{(N)} + Tx^{(N)} + My - Tx^{(N)} - Ux^{(N)}$, $(I + T)x^{(N+1)} = (I + T)x^{(N)} + My - (T + U)x^{(N)}$, $x^{(N+1)} = x^{(N)} + M'(y - Lx^{(N)})$ where $M' = (I + T)^{-1}M$. The product of $I + T$ and $I - T + T^2 - T^3 + \cdots + (-T)^{n-1}$

is $I - (-T)^n$, so it suffices to show that T^n is 0. It is easily shown that T^k has the property that the entry in the ith row and the jth column is zero unless $j \le i - k$, which implies $T^n = 0$.

7 It suffices to assume that $\sum_{i=1}^n \max \{|b_{i1}|, |b_{i2}|, \ldots, |b_{in}|\} < 1$ because $|MLx - x| = \sum_{i=1}^n |\sum_{j=1}^n b_{ij}x_j| \le \sum_{i=1}^n [\mu_i \sum_{j=1}^n |x_j|] = \rho|x|$ where $\mu_i = \max \{|b_{i1}|, \ldots, |b_{in}|\}$ and where $\rho = \sum_{i=1}^n \mu_i$.

8 Exactly as in 7 it suffices to assume that the p-norm (see §9.8) of the q-norms of the rows is less than 1.

9 $|x^{(N)} - x^{(M)}| \le \sum_{i=N}^{M-1} |x^{(i)} - x^{(i+1)}| \le \sum_{i=N}^{M-1} \rho^i |x^{(1)} - x^{(0)}| = (\rho^N - \rho^M)(1-\rho)^{-1} \times |x^{(1)} - x^{(0)}| < \rho^N(1-\rho)^{-1}|x^{(1)} - x^{(0)}|$ and the desired result follows by letting $M \to \infty$. This estimate is more useful because $\bar{x}$ is not known, whereas $x^{(0)}$, $x^{(1)}$ are known.

10 If $Lx = 0$ then $|MLx - x| = |-x| = |x|$ and (3) implies $|x| \le \rho|x|$ with $\rho < 1$, hence $|x| = 0$. Thus (3) implies L is one-to-one, hence by dimensionality L is onto. The method of 9 shows that (3) implies (1) converges, and passage to the limit in (1) gives $x^{(\infty)} = x^{(\infty)} + M(y - Lx^{(\infty)})$, $M(y - Lx^{(\infty)}) = 0$. Set $z = y - Lx^{(\infty)}$. Since L is onto, there is a w such that $z = Lw$. Then $\rho|w| \ge |MLw - w| = |Mz - w| = |-w| = |w|$ so $w = 0$, $y - Lw^{(\infty)} = z = Lw = 0$, $y = Lx^{(\infty)}$, q.e.d.

11 Let x be the n-tuple $(x_1, x_2, \ldots, x_n)$ which is 1 in the ith position and 0 in the other positions. Then the n-tuple $A_N x$ is equal to the ith column of A_N, so the equation $\lim_{N\to\infty} A_N x = Bx$ implies that the limit of the ith column of A_N is the ith column of B. Since this holds for all i, the desired conclusion follows.

13 The solution $\bar{x}$ exists, since otherwise L would not be onto and consequently Q would not be positive definite. At the step in the Gauss-Seidel iteration, where x_i is being 'corrected', the other x's are held constant and x_i is moved to the point on the line $x_j =$ const. $(j \ne i)$ where $Q(x - \bar{x})$ is a minimum; this follows from the observation that at the minimum the partial derivative of $Q(x - \bar{x})$ with respect to x_i must be zero, $\sum_{j=1}^n 2a_{ij}(x_j - \bar{x}_j) = 0$, $\sum_{j=1}^n a_{ij}x_j = y_i$, and that the Gauss-Seidel iteration consists in using this equation $\sum_{j=1}^n a_{ij}x_j = y_i$ and the fixed values of x_j $(j \ne i)$ to determine x_i. If n consecutive steps do not change x then $\sum a_{ij}x_j = y_i$, for all i and x is the solution $\bar{x}$. Otherwise, since Gauss-Seidel is (1′) when M is as in Ex. 1, Ex. 6 implies it can be written $x^{(N+1)} = x^{(N)} + M'(y - Lx^{(N)})$, $x^{(N+1)} - \bar{x} = x^{(N)} - \bar{x} + M'L(\bar{x} - x^{(N)})$, $x^{(N+1)} = \bar{x} + N(x^{(N)} - \bar{x})$ where $N = I - M'L$. Let $\rho < 1$ be the maximum value of $Q(Ny)$ subject to the constraint $Q(y) = 1$. Then $Q(y) = k$ implies $Q(Ny) \le \rho k$, hence $Q(x^{(N)} - \bar{x}) \le \rho^N Q(x^{(0)} - \bar{x})$. Let $m > 0$ be the minimum value of $Q(x)$ subject to the constraint $|x| = 1$. Then $Q(x) < km$ implies $|x| < k$. Hence $|x^{(N)} - \bar{x}| < \rho^N Q(x^{(0)} - \bar{x})/m \to 0$ and $x^{(N)} \to \bar{x}$ as desired regardless of the choice of $x^{(0)}$.

§7.3 *pages 244–245*

1 The formula is an immediate consequence of (3). When $y = 2$ and $x^{(0)} = 1$ this gives $x^{(1)} = 3/2$, $x^{(2)} = 17/12$, $x^{(3)} = 577/408$, $x^{(4)} = 665{,}857/470{,}832$. Now $x^{(1)} - x^{(2)} = 1/2 \cdot 2 \cdot 3$, $x^{(2)} - x^{(3)} = 1/2 \cdot 12 \cdot 17$, $x^{(3)} - x^{(4)} = 1/2 \cdot 408 \cdot 577$ and, by an easy calculation, $x^{(4)} - x^{(5)} = 1/2 \cdot 665857 \cdot 470832$. Thus the step from $x^{(4)}$ to $x^{(5)}$ affects only the 12th decimal place, and division of 665,857 by 470,832 gives $\sqrt{2}$ to eleven places. If decimals are used throughout the calculation then the first eleven places of $x^{(4)}$ are correct. The answer is $\sqrt{2} = 1.41421\ 35623\ 73095$ to fifteen places.

2 $x^{(N+1)} = \frac{1}{3}[2x^{(N)} + (y/[x^{(N)}]^2)]$. The cube root of 2 is 1.25992 10498 to ten decimal places.

3 If x is a positive number such that $x^2 = \pi$ with six place accuracy, and if $\delta = x - \sqrt{\pi}$, then $x^2 = (\sqrt{\pi} + \delta)^2 = \pi + 2\delta\sqrt{\pi} + \delta^2 = \pi + \delta\sqrt{\pi} + \delta(\sqrt{\pi} + \delta)$, $\delta = (x^2 - \pi)/(\sqrt{\pi} + x)$. Since $x > 1$, $\sqrt{\pi} > 1$ this gives $|\delta| < \frac{1}{2}|x^2 - \pi|$. Thus it suffices to retain 6 places of π. The square root of π is 1.77245 38509 to ten places.

4 Newton's method gives the formula $x_{N+1} = -\frac{1}{12}((1 + 16x_N^3)/(1 - 2x_N^2))$. Taking $x_0 = 0$ gives $x_1 = -1/12$, $x_2 = -(1/12)(214/213)$. Now $(214/213) = 1.004695\ldots$ from which $x_2 = -.0837246\ldots$. Take $x_2 = -.08372$. Then $(1 + 16x_2^3)(1 - 2x_2^2)^{-1}$ can be found without too much calculation to be about 1.004697, from which $x_3 = -.0837247\ldots$. One can safely say then that $-.083725$ represents the root to 6 places.

5 The diagram is changed in that the line through $(x^{(N)}, f(x^{(N)}))$ parallel to the tangent at $(\bar{x}, f(\bar{x}))$ is replaced by the tangent at $(x^{(N)}, f(x^{(N)}))$.

6 $x^{(N+1)} = x^{(N)} + (y - (1/x^{(N)})/(-(x^{(N)})^{-2})$

$= 2x^{(N)} - x^{(N)}yx^{(N)}$. If $y = 1 - \epsilon$, $x^{(0)} = 1$ then $x^{(1)} = 2 - (1 - \epsilon) = 1 + \epsilon$, $x^{(2)} = 2(1 + \epsilon) - (1 + \epsilon)^2(1 - \epsilon) = (1 + \epsilon)[2 - (1 - \epsilon^2)] = (1 + \epsilon)(1 + \epsilon^2)$, $x^{(3)} = x^{(2)}[2 - (1 - \epsilon)x^{(2)}] = x^{(2)}[2 - (1 - \epsilon^4)] = (1 + \epsilon)(1 + \epsilon^2)(1 + \epsilon^4)$, $x^{(4)} = (1 + \epsilon)(1 + \epsilon^2)(1 + \epsilon^4)(1 + \epsilon^8)$, etc.

§7.4 *pages 251–257*

1 (a) The arrows form a vortex rotating counterclockwise. (b) The derivative of this function of t is zero, hence the function is constant. Geometrically, all points $(x(t), y(t))$ of the solution curve lie on a circle. (c) Adopting a slightly different notation from that of the text,

$$\begin{pmatrix} x_0(t) \\ y_0(t) \end{pmatrix} = \begin{pmatrix} 1 \\ 0 \end{pmatrix}$$

$$\begin{pmatrix} x_1(t) \\ y_1(t) \end{pmatrix} = \begin{pmatrix} 1 - \int_0^t y_0(u)\,du \\ 0 + \int_0^t x_0(u)\,du \end{pmatrix} = \begin{pmatrix} 1 \\ t \end{pmatrix}$$

$$\begin{pmatrix} x_2(t) \\ y_2(t) \end{pmatrix} = \begin{pmatrix} 1 - \int_0^t y_1(u)\,du \\ 0 + \int_0^t x_1(u)\,du \end{pmatrix} = \begin{pmatrix} 1 - \frac{t^2}{2} \\ t \end{pmatrix}$$

$$\begin{pmatrix} x_3(t) \\ y_3(t) \end{pmatrix} = \begin{pmatrix} 1 - \int_0^t y_2(u)\,du \\ 0 + \int_0^t x_2(u)\,du \end{pmatrix} = \begin{pmatrix} 1 - \frac{t^2}{2} \\ t - \frac{t^3}{6} \end{pmatrix}$$

etc. In general $x_{2N}(t) = x_{2N+1}(t) = 1 - (t^2/2) + (t^4/24) - \cdots + (-1)^N(t^{2N}/(2N)!)$ and $y_{2N+1}(t) = y_{2N+2}(t) = t - (t^3/6) + \cdots + (-1)^N(t^{2N+1}/(2N + 1)!)$. The method of p. 248 (choose J such that $J > |t|$ and set $\rho = |t|/J < 1$) can be used to show that $\lim_{N\to\infty} x_N(t)$ and $\lim_{N\to\infty} y_N(t)$ exist for all t. The fact that the functions so defined are continuous and satisfy the differential equation follows, as before, by passing to the limit under the integral sign (see Ex. 7). (d) The functions $(x(t), y(t)) = (\bar{x} \cos t - \bar{y} \sin t, \bar{x} \sin t + \bar{y} \cos t)$ begin at $(\bar{x}, \bar{y})$ when $t = 0$ and satisfy the equation. (e) The functions $(\cos (a + t), \sin (a + t))$ and $(\cos a \cos t - \sin a \sin t, \cos a \sin t + \sin a \cos t)$ both start at $(\cos a, \sin a)$ and both satisfy the equation, therefore they must be identical. This gives the addition formulas.

(f) $\begin{pmatrix} \cos a & \sin a \\ -\sin a & \cos a \end{pmatrix}\begin{pmatrix} \cos b & \sin b \\ -\sin b & \cos b \end{pmatrix} = \begin{pmatrix} \cos (a + b) & \sin (a + b) \\ -\sin (a + b) & \cos (a + b) \end{pmatrix}$.

(g) Multiplying formally, setting $i^2 = 1$ and using "$a + ib = c + id$ implies $a = c, b = d$" the given equation implies the addition formulas. (h) $\cos t > 0$ for small t, hence $\sin t$ is increasing for small t, hence $\sin t$ is positive for small positive t, hence $\cos t$ is decreasing for small positive t. It can stop decreasing only if $\sin t = 0$, which implies $\cos t = \pm 1$. Since $\cos t$ starts at 1 and decreases, this implies $\cos t$ decreases until it reaches -1. Then an analogous argument shows that it increases until $\cos t = 1$ again. (i) 2π is defined to be the smallest positive solution t of the equation $\cos t = 1$. Then $\sin 2\pi = 0$ and by the addition formulas $\cos (t + 2\pi) \equiv \cos t$, $\sin (t + 2\pi) \equiv \sin t$ for all t. The functions $(\cos t, \sin t)$ for $\{0 \le t \le 2\pi\}$ parameterize the boundary of the disk $D = \{x^2 + y^2 \le 1\}$, hence $\int_D dx\,dy = \int_D d[\frac{1}{2}(x\,dy - y\,dx)] = \int_{\partial D} \frac{1}{2}(x\,dy - y\,dx) = \frac{1}{2}\int_0^{2\pi} [\cos^2 t\,dt + \sin^2 t\,dt] = \frac{1}{2}\int_0^{2\pi} dt = \pi$. (j) The advantage lies in the *periodicity* of the functions $\cos (t + 2\pi) = \cos t$, $\sin (t + 2\pi) = \sin t$. Once the values in the range $\{0 \le t \le 2\pi\}$ are known—and it is most convenient to space t evenly over this interval—all other values can be found by periodicity. (k)–(n) Compare to a trig table.

2 (a) It is the solution of $dx/dt = x$ which satisfies $x(0) = 1$. (b) Both e^{t+y} and $e^t e^y$ satisfy the differential equation $dx/dt = x$ and both start at e^y. Hence they are identical. (c) $x_0(t) = 1$. $x_1(t) = 1 + \int_0^t 1\,du = 1 + t$. $x_2(t) = 1 + \int_0^t (1 + u)\,du = 1 + t + (t^2/2)$, etc., $x_N(t) = 1 + t + \cdots + (t^N/N!)$. $\lim_{N\to\infty} x_N(t)$ exists and satisfies the differential equation for all t as in 1(c). (d) $\exp (t) = [\exp (t/2)]^2 \ge 0$ hence the function $\exp (t)$ is always increasing, hence $\exp (t) \ge 1$ for $t \ge 0$, hence $\exp (1/n) - 1 = \int_0^{1/n} \exp (t)\,dt \ge \int_0^{1/n} dt = (1/n)$; $\exp (1) = [\exp (1/n)]^n \ge (1 + 1/n)^n$. In the same way $1 - \exp(-1/n) = \int_{-1/n}^0 \exp (t)\,dt \le \int_{-1/n}^0 dt = 1/n$, $(n - 1)/n \le \exp (-1/n)$. Multiplying by the positive number $(n/(n - 1)) \exp (1/n)$ gives $\exp (1/n) \le (n/(n - 1))$, $\exp (1/(n + 1)) \le (n + 1)/n = 1 + 1/n$, $\exp (1) \le (1 + 1/n)^{n+1}$. Thus the difference between $(1 + 1/n)^n$ and e is less than $(1 + 1/n)^{n+1} - (1 + 1/n)^n = (1/n)(1 + 1/n)^n \le (1/n) \exp (1) \to 0$, q.e.d. (e) Such a table is a table of 10^x for equally spaced rational values of x. It suffices to construct such a table for $\{0 \le x \le 1\}$ since the formula $10^{x+1} = 10 \cdot 10^x$ enables one to find all other values easily. (f) Use the power series of (c). Check by consulting a table of anti-logarithms to base 10. $10^{1.001}$ is $\exp((1001/1000)\alpha)$ because the thousandth power of this number is $\exp(1001\,\alpha)$, which is the thousand-and-first power of $\exp(\alpha) = 10$.

3 $\cosh^2 t - \sinh^2 t \equiv 1$, $\cosh(a+b) = \cosh a \times \cosh b + \sinh a \sinh b$, $\sinh(a+b) = \sinh a \cosh b + \cosh a \sinh b$. $\cosh t = 1 + t^2/2 + \cdots + t^{2n}/(2n)! + \cdots$, $\sinh t = t + t^3/6 + \cdots + t^{2n+1}/(2n+1)! + \cdots$.

4

$$\begin{pmatrix} x_0(t) \\ y_0(t) \end{pmatrix} = \begin{pmatrix} \bar{x} \\ \bar{y} \end{pmatrix}$$

$$\begin{pmatrix} x_1(t) \\ y_1(t) \end{pmatrix} = \begin{pmatrix} \bar{x} \\ \bar{y} \end{pmatrix} + \int_0^t M \begin{pmatrix} \bar{x} \\ \bar{y} \end{pmatrix} du = (I + tM) \begin{pmatrix} \bar{x} \\ \bar{y} \end{pmatrix}$$

and by induction

$$\begin{pmatrix} x_{N+1}(t) \\ y_{N+1}(t) \end{pmatrix} = \begin{pmatrix} \bar{x} \\ \bar{y} \end{pmatrix} + \int_0^t M \begin{pmatrix} x_N(u) \\ y_N(u) \end{pmatrix} du = \begin{pmatrix} \bar{x} \\ \bar{y} \end{pmatrix} + \int_0^t M\left(I + uM + \frac{u^2M^2}{2} + \cdots + \frac{u^N M^N}{N!}\right)\begin{pmatrix} \bar{x} \\ \bar{y} \end{pmatrix} du = \left(I + tM + \frac{t^2M^2}{2} + \cdots + \frac{t^{N+1}M^{N+1}}{(N+1)!}\right)\begin{pmatrix} \bar{x} \\ \bar{y} \end{pmatrix}.$$

Defining $\exp(tM) = \sum_{n=0}^{\infty} (t^n M^n/n!)$ then gives the desired formula.

$$\exp\begin{pmatrix} a & -b \\ b & a \end{pmatrix} = \begin{pmatrix} e^a \cos b & -e^a \sin b \\ e^a \sin b & e^a \cos b \end{pmatrix}.$$

The identity $\exp(z_1 + z_2) = \exp(z_1)\exp(z_2)$ follows by direct multiplication (using 1(*f*)) or by the last part of this problem. For the matrices M_1, M_2 that are given

$$\exp(M_1 + M_2) = \begin{pmatrix} \cosh 1 & \sinh 1 \\ \sinh 1 & \cosh 1 \end{pmatrix}$$

$$\exp M_1 \exp M_2 = \begin{pmatrix} 1 & 1 \\ 0 & 1 \end{pmatrix}\begin{pmatrix} 1 & 0 \\ 1 & 1 \end{pmatrix} = \begin{pmatrix} 2 & 1 \\ 1 & 1 \end{pmatrix}.$$

The multiplication formula $\exp(M_1 + M_2) = \exp(M_1)\exp(M_2)$ (when $M_1M_2 = M_2M_1$) can be deduced from the uniqueness of the solution of a differential equation as in Ex. 1 and Ex. 2. Specifically, $\exp(M_1 + tM_2)(\bar{x})$ (where $\bar{x}$ is a fixed n-tuple) and $\exp(M_1)\exp(tM_2)(\bar{x})$ start at the same point $x = \exp(M_1)(\bar{x})$ and satisfy the same differential equation $dx/dt = M_2 x$, hence they are identical and the desired result follows from setting $t = 1$. Algebraically the multiplication formula $\exp(M_1 + M_2) = \exp M_1 \exp M_2$ is the formula $I + (M_1 + M_2) + (M_1 + M_2)^2/2! + \cdots = (I + M_1 + M_1^2/2! + \cdots)(I + M_2 + M_2^2/2! + \cdots)$ which, upon equating terms of like degree, becomes the algebraic identity

$$(M_1 + M_2)^n/n! = \textstyle\sum_{i+j=n} (M_1^i M_2^j / i!j!)$$

which is essentially the binomial theorem. This leads to an alternative proof that $\exp(M_1 + M_2) = \exp(M_1)\exp(M_2)$ (provided $M_1M_2 = M_2M_1$).

5 (a) The vertices of the polygon are $(i/10, (11/10)^i)$ for $i = -10, -9, \ldots, 9, 10$. (b) The 21 points are found by the formula

$$\begin{pmatrix} 1 & -\frac{1}{10} \\ \frac{1}{10} & 1 \end{pmatrix}^i \begin{pmatrix} 1 \\ 0 \end{pmatrix}$$

for $i = -10, -9, \ldots, 9, 10$. E.g. for $i = 3$ the point is $\begin{pmatrix} 1 & -\frac{1}{10} \\ \frac{1}{10} & 1 \end{pmatrix}^3 \begin{pmatrix} 1 \\ 0 \end{pmatrix} = \begin{pmatrix} .970 \\ .299 \end{pmatrix}$. (c) For each N let $P_N(t)$ denote the function whose graph is the polygon. Let t be given. For each N there is an integer j such that $j/N \le t < (j+1)/N$. By definition of P_N, $(1 + 1/N)^j \le P_N(t) < (1 + 1/N)^{j+1}$. As $N \to \infty$ this interval becomes arbitrarily short, so it will suffice to show that $(1 + 1/N)^j \to e^t$ as $N \to \infty$. Raising $(1 + 1/N)^j$ to the power $1/t$ and using $N - 1 < j/t \le N$ gives $(1 + 1/N)^{N-1} \le [(1 + 1/N)^j]^{1/t} \le (1 + 1/N)^N$. As $N \to \infty$ this gives $[(1 + 1/N)^j]^{1/t} \to e$, $(1 + 1/N)^j \to e^t$, q. e. d.

6 The theorem is not contradicted because $x^{2/3}$ is not differentiable at $x = 0$.

7 For each fixed t the Cauchy criterion is satisfied by the numbers $x^{(n)}(t)$, hence they determine a limiting value $x^{(\infty)}(t)$ for each t. Moreover, $|x^{(\infty)}(t) - x^{(n)}(t)| < \epsilon$ for all $n > N$ and for all $|t| \le \delta$ when ϵ, N are as before. Fixing n and $\bar{t}$, the continuity of $x^{(n)}(t)$ implies that there is a δ such that $|x^{(n)}(t) - x^{(n)}(\bar{t})| < \epsilon$ whenever $|t - \bar{t}| < \delta$. Hence $|x^{(\infty)}(t) - x^{(\infty)}(\bar{t})| \le |x^{(\infty)}(t) - x^{(n)}(t)| + |x^{(n)}(t) - x^{(n)}(\bar{t})| + |x^{(n)}(\bar{t}) - x^{(\infty)}(\bar{t})| \le 3\epsilon$ whenever $|t - \bar{t}| < \delta$. Since ϵ was arbitrary, this shows that $x^{(\infty)}(t)$ is a continuous function (see §9.8). Finally, $f(x^{(\infty)}(t))$ is a continuous function, hence $\int_0^t f[x^{(\infty)}(s)]\,ds$ exists and, by (4),

$$\left|\int_0^t f[x^{(\infty)}(s)]\,ds - \int_0^t f[x^{(N)}(s)]\,ds\right| \le K \int_0^t |x^{(\infty)}(s) - x^{(N)}(s)|\,ds \le K\epsilon t \to 0.$$

8 Define function $s(t)$, $x(t)$, $y_1(t)$, $y_2(t)$, ..., $y_k(t)$ by applying the theorem of the text to the differential equations

$$\frac{ds}{dt} = 1 \qquad s(0) = 0$$
$$\frac{dx}{dt} = y_1 \qquad x(0) = \text{given}$$
$$\frac{dy_1}{dt} = y_2 \qquad y_1(0) = \frac{dx}{dt}(0) = \text{given}$$
$$\vdots \qquad \vdots$$
$$\frac{dy_{k-1}}{dt} = y_k \qquad y_{k-1}(0) = \frac{d^{k-1}x}{dt^{k-1}}(0) = \text{given}$$
$$\frac{dy_k}{dt} = f(y_{k-1}, \ldots, y_1, x, s)$$
$$y_k(0) = \frac{d^k x}{dt^k}(0) = \text{given}$$

where f is the solution of (*) for d^kx/dt^k as a function of the remaining variables. Then $x(t)$ is a solution of (*) and is the only solution with the required initial values.

9 Setting $y = dx/dt$ converts this equation to the equation of Ex. 1. Hence the solution is $\bar{x}\cos t + \bar{y}\sin t$ where $\bar{x}$ is the intial value of x and $\bar{y}$ is its initial derivative.

10 Identical to 4

11 Let M be the $k \times k$ matrix

$$M = \begin{pmatrix} 0 & 1 & 0 & 0 & \cdots & 0 \\ 0 & 0 & 1 & 0 & \cdots & 0 \\ 0 & 0 & 0 & 1 & \cdots & 0 \\ \vdots & \vdots & \vdots & \vdots & & \vdots \\ 0 & 0 & 0 & 0 & \cdots & 1 \\ -a_0 & -a_1 & -a_2 & -a_3 & \cdots & -a_k \end{pmatrix}.$$

Then for any k-tuple $\bar{x} = (\bar{x}_1, \bar{x}_2, \ldots, \bar{x}_k)$ the first entry of $\exp(tM)(\bar{x})$, considered as a function of t, is a solution of the given differential equation and every solution is of this form.

§7.5 *page 264*

1 $\pi = 3.14159\ 26535\ 89793$ to fifteen places.

2 $\log 2 = .69314\ 71805$ to ten places.

3 $\log 3 = 1.09861\ 22886$ to ten places.

4 $\log i = \log(i(1-i)/(1-i)) = \log((1+i)/(1-i)) = 2[i + i^3/3 + i^5/5 + i^7/7 + \cdots] = 2i[1 - 1/3 + 1/5 - 1/7 + \cdots] = 2i\int_0^1 (1+t^2)^{-1}dt = 2i \times$ (area subtended by the arc from (1, 0) to (0, 1)) $= 2i \cdot \pi/4 = i\pi/2$. The exponential of the matrix $i\frac{\pi}{2} = \begin{pmatrix} 0 & -\frac{\pi}{2} \\ \frac{\pi}{2} & 0 \end{pmatrix}$ is

$$\begin{pmatrix} \cos\frac{\pi}{2} & -\sin\frac{\pi}{2} \\ \sin\frac{\pi}{2} & \cos\frac{\pi}{2} \end{pmatrix} = \begin{pmatrix} 0 & -1 \\ 1 & 0 \end{pmatrix} = i.$$

§8.1 *pages 269–270*

1 (a) $\text{grad } f = 2x\mathbf{i} + 2y\mathbf{j} + 2z\mathbf{k}$ (b) $\text{grad } f = -r^{-3}(x\mathbf{i} + y\mathbf{j} + z\mathbf{k})$ (c) $\text{grad } f = 2(x^2 + y^2)^{-1}(x\mathbf{i} + y\mathbf{j})$

2 The divergence of (a) is 6, and the divergence of (b) and (c) are zero.

3 For the first one div $\mathbf{X} = 3$, curl $\mathbf{X} = 0$. For the second one div $\mathbf{X} = 2x + xz$, curl $\mathbf{X} = [\cos(x+y) - xy]\mathbf{i} - \cos(x+y)\mathbf{j} + yz\mathbf{k}$.

4 As in Exercise 6, §1.5, the value of $A\,dy\,dz + B\,dz\,dx + C\,dx\,dy$ on a parallelogram is equal to the oriented volume of the parallelepiped generated by the parallelogram and by a line segment parallel to the segment from (0, 0, 0) to (A, B, C). This is zero if the line segment is in the plane of the parallelogram and is the length of the line segment times the area of the parallelogram if they are perpendicular. The same is true of $\mathbf{X} \cdot \mathbf{n}\,d\sigma$ (the dot product of perpendicular vectors is zero and of parallel vectors is the product of their lengths) in these cases, so the two are equal in all cases.

5 Both numbers are equal to the 3×3 determinant whose rows are (A, B, C), (a_1, a_2, a_3), and (b_1, b_2, b_3).

6 $\mathbf{X} \cdot \mathbf{n}\,d\sigma = \mathbf{X} \cdot (\mathbf{a} \times \mathbf{b})$ for all $\mathbf{X}$, therefore $\mathbf{n}\,d\sigma = \mathbf{a} \times \mathbf{b}$.

7 A small rectangle with sides du, dv in the uv-plane parameterizes a small piece of S which is nearly a parallelogram. Denoting the sides of this parallelogram by $\mathbf{du}$, $\mathbf{dv}$ the value of the integrand on the infinitesimal parallelogram is $\mathbf{X} \cdot (\mathbf{du} \times \mathbf{dv})$ by Exercises 4 and 6.

§8.2 *pages 276–277*

1 (a) $F(x, y) = x^2 + y^2$ (b) $F(x, y) = \frac{1}{2}\log(1 + x^2) + \text{Arctan } y$ (c) $(1/yx^2)$ is an integrating factor. $F(x, y) = \log y - (y/x)$ (d) $F(x, y) = x^2 + xy + y^2$ (e) $F(x, y) = \log|y| + \sin x$ (f) $e^x(x^2 + y^2)$ (g) e^{x^2} is an integrating factor, as is shown by the last example of the text. Thus the equation is $e^{x^2}\,dy + 2xe^{x^2}y = 2x^3e^{x^2}$, $d[e^{x^2}y] = d[x^2e^{x^2}] - 2xe^{x^2}$, $d[e^{x^2}y] = d[x^2e^{x^2} - e^{x^2}]$, $e^{x^2}y = (x^2 - 1)e^{x^2} + \text{const.}$ so $F(x, y) = e^{x^2}[y + 1 - x^2]$ has the desired property.

2 (a) $x\,dx - y\,dy = 0$ (b) $x\,dy + y\,dx = 0$ (c) $(x - 1)\,dx + y\,dy = 0$ (d) $x\,dx + 4y\,dy = 0$ (e) The curves are $x^2 + y^2 + Cy = 1$ which gives, when one solves for C and differentiates, $(1 + y^2 - x^2)\,dy + 2xy\,dx = 0$.

3 (a) and (b) are orthogonal trajectories of each other. The radial lines $y = C(x - 1)$ are orthogonal to (c). The curves $y = \text{const} \cdot x^4$ are orthogonal to (d).

4 The orthogonal trajectories are the ellipses $x^2 + py^2 = \text{const.}$

§8.3 *pages 288–289*

1 A function $u(x)$ of one variable is harmonic $(2\delta)^{-1} \int_{\bar{x}-\delta}^{\bar{x}+\delta} u(x)\,dx = u(\bar{x})$ if and only if it is affine $u = Ax + B$, because the analog of Laplace's equation is $(d^2u/dx^2) = 0$. The analog of the Poisson Integral Formula is the formula $u(x) = (x - a)(b - a)^{-1}u(b) + (b - x)(b - a)^{-1}u(a)$ giving $u(x)$ for $a \le x \le b$ in terms of $u(a)$, $u(b)$.

2 From $r\,dr = x\,dx + y\,dy + z\,dz$ the identity $r^2\,dr\omega = dx\,dy\,dz$ follows easily. The identity $d\omega = 0$ can be proved by simple differentiation. [A more satisfying, if not simpler, method is to observe that ω is unchanged by the substitution $x \to cx$, $y \to cy$, $z \to cz$; this can be shown to imply that if ω is expressed in terms of the new (local) coordinates $r = \sqrt{x^2 + y^2 + z^2}$, $u = x/r$, $v = y/r$ it does not involve r, hence is a 2-form in 2 variables and must be closed.] The proof that the double integral $\int_D \omega\,dr$ can be written as an iterated integral will be omitted. By (v) of §2.6,

$$\frac{d}{dr}\int_0^r f(r)\,dr = \lim_{\Delta r \to 0} \frac{1}{\Delta r}\int_r^{r+\Delta r} f(r)\,dr = f(r).$$

The left side is constant because $d\omega = 0$. For the last step, one must observe that on the sphere $a^2 + b^2 + c^2 = 1$ the identity $a\,da + b\,db + c\,dc = 0$ is satisfied, which gives $a^2db\,dc + ab\,dc\,da + ac\,da\,db = a^2db\,dc + b\,dc(-b\,db - c\,dc) + c(-b\,db - c\,dc)db = a^2db\,dc - b^2\,dc\,db - c^2dc\,db = db\,dc$ and analogous identities for $dc\,da$, $da\,db$.

3 The difference of two harmonic functions with the same values on ∂D is a harmonic function which is zero on ∂D. It is to be shown that such a function is identically zero. If not, then its absolute value has a non-zero maximum at some point P inside D. Its average over a ball with center P is equal to its maximum value, which means it is constant on such a ball. Letting the ball expand until it touches ∂D, this implies a contradiction.

4 $d(r^{-1}) = -r^{-2}\,dr = -\frac{1}{2}r^{-3}\,d(r^2) = -(x\,dx + y\,dy + z\,dz)r^{-3}$. Hence $\omega = -((\partial(r^{-1})/\partial x)\,dy\,dz + (\partial(r^{-1})/\partial y)\,dz\,dx + (\partial(r^{-1})/\partial z)\,dx\,dy)$ and $d\omega = 0$ is Laplace's equation for r^{-1}. Similarly, in n dimensions $d(r^{2-n}) = (2 - n)r^{1-n}\,dr = (2 - n)(x_1\,dx_1 + \cdots + x_n dx_n)/r^n$. Thus $(\partial(r^{2-n})/\partial x_1)dx_2\,dx_3 \cdots dx_n + \cdots$ is $(2 - n)$ times $(1/r^n)[x_1\,dx_2\,dx_3 \cdots dx_n - \cdots + (-1)^{n-1}x_n\,dx_1\,dx_2 \cdots dx_{n-1}]$ which is the n-dimensional analog of ω and can be shown to be closed in the same way.

5 Following Exercise 2 the average of $u(x, y, z)$ over $\{(x - \bar{x})^2 + (y - \bar{y})^2 + (z - \bar{z})^2 = r^2\}$ should be defined to be

$$\frac{\displaystyle\int_{a^2+b^2+c^2=1} u(\bar{x} + ra, \bar{y} + rb, \bar{z} + rc) \times (a\,db\,dc + b\,dc\,da + c\,da\,db)}{\displaystyle\int_{a^2+b^2+c^2=1} a\,db\,dc + b\,dc\,da + c\,da\,db}$$

which can also be written

$$\frac{1}{4\pi r^3}\int_{(x-\bar{x})^2+(y-\bar{y})^2+(z-\bar{z})^2=r^2} u(x, y, z)[(x - \bar{x}) \times dy\,dz + (y - \bar{y})\,dz\,dx + (z - \bar{z})\,dx\,dy].$$

6, 7 Straightforward differentiation. The matrix formulation is that columns of

$$\begin{pmatrix} u & -v \\ v & u \end{pmatrix}\begin{pmatrix} r & -s \\ s & r \end{pmatrix}$$

are conformal coordinates if (u, v) and (r, s) are.

§8.4 *pages 310–313*

1 It suffices to show that $|z_1 + z_2|^2 \le (|z_1| + |z_2|)^2$. Expanding the right side and using the definition of $|z|^2$ this becomes the Schwartz inequality $x_1x_2 + y_1y_2 \le \sqrt{x_1^2 + y_1^2}\sqrt{x_2^2 + y_2^2}$.

2 The sequence $z_n = x_n + iy_n$ converges to $z_\infty = x_\infty + iy_\infty$ if and only if both of the real sequences converge $x_n \to x_\infty$, $y_n \to y_\infty$.

3 Choose an N such that $|z_n - z_N| < 1$ for $n > N$. Let K' be the largest of the numbers

$|z_1|, |z_2|, \ldots, |z_N|$ and set $K = K' + 1$. Then $|z_N| < K$ for all n.

4 Every complex number can be written in the form

$$\begin{pmatrix} r\cos\theta & -r\sin\theta \\ r\sin\theta & r\cos\theta \end{pmatrix}.$$

Geometrically this matrix represents a transformation $\mathbf{R}^2 \to \mathbf{R}^2$ which is a rotation through an angle θ and a scale factor of r.

5 The vertices are 1, $a + bi$, $a^2 - b^2 + 2abi$, $a^3 - 3ab^2 + 3a^2bi - b^3i$, $a^4 - 6a^2b^2 + b^4 + 4a^3bi - 4ab^3i$. In short, they are A^0, A^1, A^2, A^3, A^4 where $A = a + bi$. The basic relation is $A^5 = 1$, which gives $a^5 - 10a^3b^2 + 5ab^4 = 1$, $5a^4b - 10a^2b^3 + b^5 = 0$. Cancelling b in the last relation and using $b^2 = 1 - a^2$ gives $a^2 = (3 \pm \sqrt{5})/8$. Since $a < \frac{1}{2}$ this gives $a^2 = (3 - \sqrt{5})/8$ which has the solution $a = (\sqrt{5} - 1)/4$ as desired. Then $b = \sqrt{1 - a^2}$ gives b (see also Ex. 15, §9.1).

6 If $z^2 + 1 = z$ then $(z^2 - z + 1)(z + 1)$ $(z^3 - 1) = 0$, $z^6 - 1 = 0$, $z^6 = 1$. The solutions of this equation are the vertices of a regular hexagon $\cos(2\pi j/6) + i\sin(2\pi j/6)$ $(j = 1, 2, \ldots, 6)$. Of these 6 possibilities, only the points $j = 1, 5$ satisfy the given equation, i.e. the solutions are $(1/2) \pm i(\sqrt{3}/2)$.

7 $(d/d\theta)e^{i\theta} = ie^{i\theta}$, $(d/d\theta)\cos\theta + i\,(d/d\theta)\sin\theta = i\cos\theta - \sin\theta$, $(d/d\theta)\cos\theta = -\sin\theta$, $(d/d\theta)\sin\theta = \cos\theta$.

8 $(d/dz)\log(1 + z) = 1/(1 + z)$. When $z = 0$ this gives $\lim_{n\to\infty}(\log(1 + z/n)/(z/n)) = 1$, $\lim_{n\to\infty} n\log(1 + z/n) = z$, $\lim_{n\to\infty} e^{n\log(1+z/n)} = e^z$, $\lim_{n\to\infty}(1 + z/n)^n = e^z$.

9 By the theorem, c_n is the average value of $z^{-n}f(z)$ on the circle $|z| = \rho$. Since the average of numbers of modulus $\le K\rho^{-n}$ also has modulus $\le K\rho^{-n}$ it follows that $|c_n| \le K\rho^{-n}$ for all ρ, hence $c_n = 0$ for $n > 0$.

10 The basic estimate $(s^{-1}[f(x + sh) - f(x)]$ is within ϵ of $f'(x)\cdot h$ for all sufficiently small s) is proved, as usual, by writing $f(x + sh) - f(x) = \int_x^{x+sh} f'(t)\,dt = \int_x^{x+sh} f'(x)\,dt + \int_x^{x+sh}[f'(t) - f'(x)]\,dt = f'(x)\cdot sh + \text{small}\cdot sh$. In short, the entire proof is the same as in §5.3.

11 $x^{3/2}$ is one example.

12 Immediate. Differentiation under the integral sign requires justification, of course. In the present case it suffices to show that z^{-n} is uniformly differentiable on $0 < r \le |z| \le R$, which is easily proved by estimating $(z + h)^{-n} - z^{-n}$ directly.

13 Elementary

14 It is easy to show that there is a constant K such that $|A^{-1}w| \le K|w|$ where A^{-1} is the inverse matrix. Let $\alpha = K^{-1}$ and $w = Av$. Let $K' = \max(|c_0|, |c_1|, \ldots, |c_{n-1}|)$. Then by induction $\max(|c_k|, |c_{k+1}|, \ldots, |c_{k+n-1}|) \ge K'\alpha^k$, so $K = K'\alpha$ has the desired property. Since the terms of the series $\sum c_k\alpha^{-k}$ do not approach zero, the series can't converge.

15 As in the text, it suffices to show that $c_k\alpha^k$ is bounded for some α, which follows by the method of Exercise 14. Let $g(x)$ be the function defined by the power series $c_0 + c_1x + c_2x^2 + \cdots$. Let $F(x) = f(x)g(x)$. Then $F(0) = a_0c_0 = 1$, $F'(0) = f'(0)g(0) + f(0)g'(0) = a_1c_0 + a_0c_1 = 0$ and, similarly, all derivatives of F are 0 at 0. Therefore the Taylor expansion of F is $F(x) \equiv 1$, $g(x) \equiv [f(x)]^{-1}$

§8.5 *pages 319–320*

1 (a) Set $u = A(x - \bar{x}) + B(y - \bar{y}) + C(z - \bar{z})$ and $v = D(x - \bar{x}) + E(y - \bar{y}) + F(z - \bar{z})$. If $du\,dv \ne 0$ then by the Implicit Function Theorem these equations can be solved for two of the variables in terms of u, v and the third variable. Setting $u = v = 0$ then gives the line in parametric form. If $du\,dv = 0$ then $u = 0$, $v = 0$ is a plane or all of xyz-space, hence not a line. (b) The coefficient of ω_3 in ω_1' is the coefficient of $\omega_1\omega_2\omega_3$ in $\omega_1'\omega_1\omega_2$. If $\omega_1\omega_2$ is a multiple of $\omega_1'\omega_2'$ this is zero. (c) This is essentially the same as (b), only with $(x - \bar{x})$ in place of dx, $(y - \bar{y})$ in place of dy and $(z - \bar{z})$ in place of dz.

2 Choose $\omega_{k+1}, \ldots, \omega_n$ such that $\omega_1\omega_2\cdots\omega_n \ne 0$. Then $\omega' = a_1\omega_1 + a_2\omega_2 + \cdots + a_n\omega_n$ and a_j is the coefficient of $\omega_j\omega_1\cdots\omega_{j-1} \times \omega_{j+1}\cdots\omega_n$ in $\omega'\omega_1\cdots\omega_{j-1}\omega_{j+1}\cdots\omega_n$. If $\omega_1'\omega_2'\cdots\omega_k'$ is a multiple of $\omega_1\omega_2\cdots\omega_k$ then the expression of $\omega_i'(i = 1, 2, \ldots, k)$ has no term in ω_j for $j > k$.

3 $(AF - BE)(\partial D/\partial z - \partial C/\partial t) + (AG - CE)(\partial B/\partial t - \partial D/\partial y) + (AH - DE)(\partial C/\partial y - \partial B/\partial z) + (BG - CF)(\partial D/\partial x - \partial A/\partial t) + (BH - DF)(\partial A/\partial z - \partial C/\partial x) + (CH - DG)$ $(\partial B/\partial x - \partial A/\partial y) = 0$ and the analogous condition obtained by interchanging $A \leftrightarrow E$, $B \leftrightarrow F$, $C \leftrightarrow G$, $D \leftrightarrow H$.

4 By the theorem there is an F such that the curves $F = \text{const.}$ are solutions of the differential equation. This means that dF is a multiple of $A\,dx + B\,dy$, i.e. a multiple of $A\,dx + B\,dy$ is exact.

5 Given $f: \mathbf{R}^n \to \mathbf{R}^n$ as in the theorem, consider the differential equations $dx_1 - f_1(x)\,dt = 0, dx_2 - f_2(x)\,dt = 0, \ldots, dx_n - f_n(x)\,dt = 0$ in the $(n+1)$ variables $x_1, x_2, \ldots, x_n, t$. By the theorem of this section there is a nonsingular map $F: R^{n+1} \to R^n$ of rank n such that the curves $F = $ const. satisfy the differential equations. The pullback of $dF_1 dF_2 \cdots dF_n$ is $dx_1 dx_2 \cdots dx_n +$ terms in dt, so the equations $F =$ const. can be solved for $x_1, x_2, \ldots, x_n$ as functions of $F_1, F_2, \ldots, F_n, t$. On the other hand, $F_1, F_2, \ldots, F_n$ can be written as functions of $x_1, x_2, \ldots, x_n$ for $t = 0$.

6 Since each dx_i can be written as a combination of the ω_i, each 2-form $dx_i dx_j$ can be written as a combination of the $\omega_i\omega_j$, hence any 2-form can be written as a combination of the $\omega_i\omega_j$. Moreover, the coefficient of $\omega_1\omega_2$, say, in the 2-form ω is determined by the fact that it is the coefficient of $\omega_1\omega_2\omega_3\cdots\omega_n$ in $\omega\omega_3\omega_4\cdots\omega_n$. In this way the integrability conditions can be reduced to the condition that the 2-forms $d\omega_i$ $(i = 1, 2, \ldots, k)$, when expressed as combinations of $\omega_i\omega_j$, contain no terms in which both $i > k$ and $j > k$. This imposes $\binom{n-k}{2}$ conditions on each $d\omega_i$.

§8.6 *pages 326–327*

1 $d\theta = (x\,dy - y\,dx)/(x^2 + y^2)$ has the property that its integral over γ_1 is 2π, and that its integral over γ_2 is zero. The closed 1-form

$$\frac{(r-2)\,dz - z\,dr}{(r-2)^2 + z^2} = \frac{\sqrt{x^2+y^2}(\sqrt{x^2+y^2} - 2)\,dz - z(x\,dx + y\,dy)}{\sqrt{x^2+y^2}[(\sqrt{x^2+y^2} - 2)^2 + z^2]}$$

has the property that its integral over γ_1 is zero and that its integral over γ_2 is 2π.

2 (a) Let $P_0 = (1, 1), P_1 = (-1, 1), P_2 = (-1, -1), P_3 = (1, -1)$. Then the integral of $d\sigma$ over the boundary of the square is $[\sigma(P_0) - \sigma(P_3)] + [\sigma(P_1) - \sigma(P_0)] + [\sigma(P_2) - \sigma(P_1)] + [\sigma(P_3) - \sigma(P_2)] = 0$ by $\int_P^Q d\sigma = \sigma(Q) - \sigma(P)$. (b) Parameterizing the sides $\{x = 1, y = t, -1 \le t \le 1\}, \{y = 1, x = -t, -1 \le t \le 1\}$, etc., the integral is $4\int_{-1}^{1} (1 + t^2)^{-1}\,dt$. On the other hand it is the integral of $d\theta$ over the boundary of a domain containing $(0, 0)$ and hence it is 2π. Therefore

$$\pi/2 = \int_{-1}^{1} (1 + t^2)^{-1}\,dt.$$

3 Instead of going around the circle to $(x/\sqrt{x^2+y^2}, y/\sqrt{x^2+y^2})$, go around the square to $(x/\max(|x|, |y|), y/\max(|x|, |y|))$.

4 Given a 1-form ω with $d\omega = 0$ define $F(P)$ to be the integral of ω over a smooth curve from $(1, 0, 0)$ to P which does not pass through $(0, 0, 0)$. The problem is to show that $F(P)$ is well-defined, i.e. independent of the choice of the curve. This is hard to do rigorously, but not hard to imagine.

5 See Exercise 10, §3.3.

6, 7 Hard to do rigorously. Good exercises for the imagination.

8 The circles $\{(x-1)^2 + y^2 = \epsilon^2\}, \{(x+1)^2 + y^2 = \epsilon^2\}$ oriented counterclockwise are a basis $(0 < \epsilon < 1)$.

9 Assuming $f(x)$ is differentiable, Poincaré's Lemma says that since $d[f(x)\,dx] = 0$ there must be a function $F(x)$ such that $dF = f(x)\,dx$. The two proofs are the same except that in Chapter 3 the end point a was used rather than the center point $(a + b)/2$ of the interval $\{a \le x \le b\}$.

10 The integrals of $\omega - \sum b_j\omega_j$ over $\gamma_1, \gamma_2, \ldots$, and γ_ν are zero, therefore, since it is closed, it is exact by definition of 'homology basis.' Therefore its integrals over $\gamma_1', \gamma_2', \ldots$, and γ_μ' are zero, which gives the desired equations.

§8.7 *page 333*

1 Follows immediately from the continuity equation.

2 (a) The lines $x =$ const. equally spaced. (b) Radial lines equally spaced. (c) Circles, centered at $(0, 0)$ drawn with a density proportional to r^{-1}, i.e. denser near $(0, 0)$, thinner far away. (d) The lines $x =$ const., $y =$ const. equally spaced. (e) Radial lines equally spaced.

3 Since $\omega \neq 0$, one can assume that the $dx_1\,dx_2 \cdots dx_{n-1}$ term is not zero, hence that $\omega = A(dx_1\,dx_2 \cdots dx_{n-1} - u_1\,dx_n\,dx_2 \cdots dx_{n-1} - u_2\,dx_1\,dx_n\,dx_3 \cdots dx_{n-1} - \cdots) = A(dx_1 - u_1\,dx_n)(dx_2 - u_2\,dx_n) \cdots (dx_{n-1} - u_{n-1}\,dx_n)$. By the theorem of §8.6 there exist functions $y_1, y_2, \ldots, y_{n-1}$ such that $dx_1 - u_1\,dx_n, \ldots, dx_{n-1} - u_{n-1}\,dx_n$ are combinations of $dy_1, dy_2, \ldots, dy_{n-1}$ and vice versa. Thus $\omega = A\, dy_1\, dy_2 \cdots dy_{n-1}$. Setting $\rho = A^{-1}$ then gives $\rho\omega = dy_1\, dy_2 \cdots dy_{n-1}$. If A can be expressed as a function of $y_1, y_2, \ldots, y_{n-1}$ then dA can be expressed in terms of $dy_1, dy_2, \ldots, dy_{n-1}$ and $d\omega = dA\, dy_1\, dy_2 \cdots dy_{n-1} = 0$. Conversely, if $d\omega = 0$ then the map $(x_1, x_2, \ldots, x_n) \to (y_1, y_2, \ldots, y_{n-1}, A)$ is of rank $n-1$ and the Implicit Function Theorem gives A as a function of $y_1, y_2, \ldots, y_{n-1}$. If $z_1, z_2, \ldots, z_{n-1}$ is another set of functions satisfying $dz_1\, dz_2 \cdots dz_{n-1} = \rho'\omega$ for some ρ', then $dz_1\, dz_2 \cdots dz_{n-1}$ is a multiple of $dy_1\, dy_2 \cdots dy_{n-1}$ and therefore $dz_1, dz_2, \ldots, dz_{n-1}$ are combinations of $dy_1, dy_2, \ldots, dy_{n-1}$ and vice versa. Hence the differential equations $dz_i = 0$ ($z_i =$ const.) and $dy_i = 0$ ($y_i =$ const.) define the same curves in $x_1x_2 \cdots x_n$-space. Drawing points P in $y_1y_2 \cdots y_{n-1}$-space with density $A(y_1, y_2, \ldots, y_{n-1})$ and then drawing the curves $y = P$ in $x_1x_2 \cdots x_n$-space gives curves such that $\int_S \omega$ is proportional to the number of curves which cross the $(n-1)$-dimension manifold S, counting the sign of an intersection as $+$ or $-$ depending on whether ω is $+$ or $-$ on the oriented surface S at the point of intersection.

§8.8 *pages 354–356*

1 Because r^{-1} is harmonic, the 2-form

$$\frac{\partial}{\partial x}(u - Ar^{-1})\, dy\, dz + \frac{\partial}{\partial y}(u - Ar^{-1})\, dz\, dx + \frac{\partial}{\partial z}(u - Ar^{-1})\, dx\, dy$$

is closed and its integral over any two large spheres is the same. Fix a sphere and choose A so that this integral is zero. Since u is radially symmetric it is constant on spheres $r =$ const. and by the Implicit Function Theorem u can be expressed (locally) as a function of r. Hence $du = u'(r)\, dr = f(r)(x\, dx + y\, dy + z\, dz)$. The corresponding 2-form is therefore $f(r)[x\, dy\, dz + y\, dz\, dx + z\, dx\, dy]$ and its integral over a large sphere S, which is zero, is also equal to $f(r) \times \int_S (x\, dy\, dz + y\, dz\, dx + z\, dx\, dy) = f(r) \cdot 4\pi r^3$ because $f(r)$ is constant on S. Hence $f(r) = 0$, $f(r) \equiv 0$, $du \equiv 0$, $u =$ const.

2 Let $x' = \Upsilon(x - vt)$, $y' = y$, $z' = z$, $t' = \Upsilon(-\epsilon\mu v x + t)$ where $\Upsilon = (1 - \epsilon\mu v^2)^{-1/2}$. Then the charged particle is stationary at $(0, 0, 0, 0)$ relative to the coordinates $x'z'y't'$. In $x'y'z't'$-coordinates B is zero and the electromagnetic field is given by Coulomb's law as

$$d\left(\frac{e}{4\pi\epsilon r}\right) dt' = -\frac{e}{4\pi\epsilon r^3}(x'\, dx'\, dt' + y'\, dy'\, dt' + z'\, dz'\, dt')$$

where e is the charge and $r = (x'^2 + y'^2 + z'^2)^{1/2}$. Expressing this in terms of x, y, z, t it becomes $(-e\Upsilon/4\pi r^3)[((x - vt)/\epsilon)dx\,dt + (y/\epsilon) \times dy\, dt + (z/\epsilon)\, dz\, dt + 0\, dy\, dz - \mu v z\, dz\, dx + \mu v y\, dx\, dy]$ where $\Upsilon = (1 - \epsilon\mu v^2)^{-1/2}$ and where $r = \Upsilon^2(x - vt)^2 + y^2 + z^2$.

3 (a) The laws are expressed entirely in terms of these correspondences and in terms of the operation of differentiation of forms which is the same in any coordinate system (not just Lorentz transformations of coordinates). (b) If, for example, one starts with the 2-form $dz'\, dx'$, forms the pullback $\Upsilon\, dz\, dx - \Upsilon v\, dz\, dt$, and takes the corresponding 2-form $(1/\mu)[\epsilon\mu \times (-\Upsilon v)\, dx\, dy - \Upsilon\, dy\, dt]$ the result is the same as if one starts with $dz'\, dx'$, takes the corresponding 2-form $(1/\mu)[-dy'\, dt']$, and forms the pullback $(1/\mu)[-\Upsilon\, dy\, dt + \epsilon\mu\Upsilon v\, dy\, dx]$. The proof of the other cases is the same. (e) This is the so-called 'star operation' {1-forms} $\leftrightarrow$ {3-forms}, {2-forms} $\leftrightarrow$ {2-forms} determined by the 'Lorentz metric' $x^2 + y^2 + z^2 - c^2t^2$.

§9.1 *pages 375–380*

1 $(8, 2, 5) + (5, 7, 1) \equiv (40, 10 + 56, 25 + 8) \equiv (40, 66, 33) \equiv (40, 34, 1)$. $(40, 34, 1) + (7, 4, 9) \equiv (280, 238 + 160, 7 + 360) \equiv (280, 38, 7)$. $(3/8) + (-6/5) + (5/7) = (105 - 336 + 200)/280 = -31/280$. $(8, 2, 5)(5, 7, 1) \equiv (40, 2 + 35, 5 + 14) \equiv (40, 32, 14) \equiv (20, 16, 7)$. $(20, 16, 7)(7, 4, 9) \equiv (140, 144 + 28, 63 + 64) \equiv (140, 80 + 28, 63) \equiv (140, 110, 65) \equiv (28, 22, 13)$. $(3/8)(-6/5)(5/7) = (3/4)(-3)(1/7) = -9/28$.

2 $(a, b, b + d)(d, 1, a + 1) \equiv (1, 1, 1 + 1)$. $(a, c + d, c)(d, a + 1, 1) \equiv (1, 1, 1 + 1)$. If $b = c$ then (a, b, c) is zero and q^{-1} does not exist.

3 Reflexive and symmetric are obvious. If $a + 6p = b + 6q$ and if $b + 6p' = c + 6q'$ then $a + 6(p + p') = (a + 6p) + 6p' = (b + 6q) + 6p' = (b + 6p') + 6q = c + 6(q' + q)$ hence $a \equiv c \pmod 6$ and the relation is transitive. If $a \equiv b \pmod 6$ then $ac \equiv bc \pmod 6$ because $(a + 6p)c = (b + 6q)c$. Thus $a \equiv b$ and $c \equiv d \pmod 6$ implies $ac \equiv bc \equiv cb \equiv db \equiv bd$, i.e. $ac \equiv bd \pmod 6$. The proof that $a + c \equiv b + d \pmod 6$ is even easier. By division by 6, every natural number a can be written in the form

$a = 6q + r$ where $r = 0, 1, 2, 3, 4, 5$. Thus $a \equiv 1$ or $a \equiv 2$ or $\cdots$ or $a \equiv 6 \pmod 6$. Given a, b, in solving $a + x \equiv b \pmod 6$ one can assume $b > a$, hence $b = a + d$ and the equation is solved by $x \equiv d \pmod 6$. $3 \cdot 2 \equiv 6 \pmod 6$ and $3 \cdot 6 \equiv 6 \pmod 6$, but $2 \not\equiv 6 \pmod 6$. Further, $3x \equiv 1 \pmod 6$ has no solution x. $1^2 \equiv 1$, $2^2 \equiv 4$, $3^2 \equiv 3$, $4^2 \equiv 4$, $5^2 \equiv 1$, $6^2 \equiv 6$; therefore the squares are 1, 3, 4, 6.

4 $ax \equiv b \pmod n$ has a unique solution x if and only if a is invertible mod n. This is true if and only if a and n have no common factors other than 1. This is true for all $a < n$ if and only if n is prime.

5 $ab \equiv 1 \pmod n$ means $ab + pn = 1 + qn$, which means that b steps of size a plus p steps of size n to the right followed by q steps of size n to the left results in one step to the right all told. Repeating the process one can take any number of steps to the right. To take a step to the left, take $n - 1$ steps to the right and a step of size n to the left. Conversely, if one can go from 1 to 2 in steps of size a and n then, adding steps to the right and left separately, there exist natural numbers p, p', q, q' such that $1 + pa + p'n = 2 + qa + q'n$. Hence $pa \equiv 1 + qa \pmod n$. Choose x such that $p \equiv q + x \pmod n$. Then $ax \equiv 1 \pmod n$, i.e. a is invertible mod n. Given a, n let d be the smallest natural number such that it is possible to go from 1 to $1 + d$ in steps of size a and n. Then it is possible to take steps of size d in either direction. Thus the points $1 + 2d$, $1 + 3d, \ldots$ can all be reached from 1, and if any point between them could be reached then a point between 1 and $1 + d$ could be reached, contrary to assumption. Since $1 + a$ and $1 + n$ can be reached it follows that a and n are multiples of d. If c is any factor common to a and n then all steps can be broken into steps of size c, the step from 1 to $1 + d$ can be accomplished in steps of size c, hence c divides d. This proves: *Theorem.* a is invertible mod n if and only if the greatest common factor of a and n is 1.

6 $49 = 32 + 17$, $32 = 17 + 15$, $17 = 15 + 2$, $15 = 7 \cdot 2 + 1$, $1 = 15 - 7 \cdot 2 = 15 - 7(17 - 15) = 8 \cdot 15 - 7 \cdot 17 = 8(32 - 17) - 7 \cdot 17 = 8 \cdot 32 - 15 \cdot 17 = 8 \cdot 32 - 15(49 - 32) = 23 \cdot 32 - 15 \cdot 49$. Hence $49 \cdot 15 = -1 + (32 \cdot 23)$, $49 \cdot 15 \cdot 31 = -31 + 32 \cdot 23 \cdot 31 = 1 + 32(23 \cdot 31 - 1)$, $49 \cdot 465 = 1 + 32 \cdot 712$. Similarly, $48 \cdot 11 = 1 + 31 \cdot 17$, $63 \cdot 7 = 1 + 40 \cdot 11$.

7 In other words, are 1953, 5115 without common factors, or, in other words, can steps of size 1 be taken using steps of size 1953, 5115? From the equations $5115 = 2 \cdot 1953 + 1209$, $1953 = 1 \cdot 1209 + 744$, $1209 = 1 \cdot 744 + 465$, $744 = 1 \cdot 465 + 279$, $465 = 1 \cdot 279 + 186$, $279 = 1 \cdot 186 + 93$, $186 = 2 \cdot 93 + 0$ it follows that 93 divides 186, hence divides 279, hence divides 465, etc. Thus one is led to the reduction $(1953/5115) = (93/93)(21/55) = 21/55$.

8 The $(j + 1)$st term in the expansion of $(1 + 1/n)^n$ is $(n \cdot (n-1) \cdots (n - j + 1))/j!n^j$. This is less than $1/(j!)$ but approaches $1/(j!)$ as $n \to \infty$. The desired equation follows easily from this observation. It is easier to sum the series $1 + 1 + (1/2) + (1/6) + (1/24) + \cdots$ than to find the limit directly. The first 7 terms give three place accuracy.

9 The interval between s_n and s_{n+1} contains the sequence and has length $a_{n+1} \to 0$.

10 'Not convergent' means there is an ϵ such that for every N there is an $n > N$ such that $|q_n - q_N| \geq \epsilon$. Applying this again gives an $m > n$ such that $|q_m - q_n| \geq \epsilon$, from which $q_m - q_N = |q_m - q_n| + |q_n - q_N| \geq 2\epsilon$. Applying it a third time gives a $p > m$ such that $q_p - q_N \geq 3\epsilon$. Repeating *ad infinitum* gives $q_n - q_N \geq p\epsilon$ for all p. Therefore q_n is arbitrarily large.

11 $(b + 1)a = ba + a \geq b + a > b$. If a, b are positive rational numbers then there is a natural number d such that da, db are natural numbers. Hence there is a natural number n such that $n\,da > db$, $na > b$. By the binomial theorem (or by an elementary induction on n) $(1 + a)^n > 1 + na$. Thus $(1 + a)^n > b$ (if $a > 0, b > 0$) for n sufficiently large. If $|q_1| < 1$ then either $q_1 = 0$ or $|q_1|^{-1} = 1 + a$ for $a > 0$. In the latter case there is an n such that $|q_1|^{-n} > q_2^{-1}$, hence $|q_1|^n < q_2$. In the former case $|q_1| < q_2$.

12 $\begin{pmatrix}1037\\2379\end{pmatrix}$

$$= \begin{pmatrix}0&1\\1&2\end{pmatrix}\begin{pmatrix}0&1\\1&3\end{pmatrix}\begin{pmatrix}0&1\\1&2\end{pmatrix}\begin{pmatrix}0&1\\1&2\end{pmatrix}\begin{pmatrix}0\\61\end{pmatrix}$$

$$= \begin{pmatrix}7&17\\16&39\end{pmatrix}\begin{pmatrix}0\\61\end{pmatrix}$$

which gives $1037/2379 = 17/39$. Written as a continued fraction it is 1 over 2 plus 1 over 3 plus 1 over 2 plus 1 over 2.

13 Because the determinant of a product is the product of the determinants it follows that $p_{n-1}q_n - p_nq_{n-1} = (-1)^n$. If p_n and q_n are both divisible by c then c divides $(-1)^n$ and therefore $c = \pm 1$, i.e. p_n/q_n is in lowest terms. The equations

$$\begin{pmatrix} r_n \\ r_{n-1} \end{pmatrix} = \begin{pmatrix} 0 & 1 \\ 1 & a_n \end{pmatrix} \begin{pmatrix} r_{n+1} \\ r_n \end{pmatrix}$$

imply

$$\begin{pmatrix} r_1 \\ r_0 \end{pmatrix} = \begin{pmatrix} p_{n-1} & p_n \\ q_{n-1} & q_n \end{pmatrix} \begin{pmatrix} r_{n+1} \\ r_n \end{pmatrix}$$

and hence the desired equation if $r_{n+1} = 0$. If $r_1/r_0 = m_1/m_0$ where m_1, m_0 are natural numbers, set $c = m_0 r_0^{-1}$. Then $cr_0 = m_0$, $cr_1 = m_1$. Set $m_n = cr_n$. Then $m_{n-1} = a_n m_n + m_{n+1}$. Thus all the numbers $m_0 > m_1 > m_2 > m_3 > \cdots$ are natural numbers or zero. It follows then that within m_0 steps one must arrive at $m_n = 0$, $r_n = 0$. If r_n is never zero the assumption $r_1/r_0 = m_1/m_0$ must be false, i.e. r_1/r_0 is irrational. Dividing $p_{n-1}q_n - p_n q_{n-1} = (-1)^n$ by $q_{n-1}q_n$ gives the stated relation, from which $(p_n/q_n) = (p_n/q_n) - (p_{n-1}/q_{n-1}) + (p_{n-1}/q_{n-1}) - (p_{n-2}/q_{n-2}) + \cdots + (p_1/q_1) - (p_0/q_0) + (p_0/q_0) = 0 + (1/q_0 q_1) - (1/q_1 q_2) + (1/q_2 q_3) - (1/q_3 q_4) + \cdots + (-1)^n (1/q_{n-1}q_n)$. Since $q_n q_{n+1} = q_n(q_{n-1} + a_{n+1}q_n) > q_n q_{n-1}$ the convergence of p_n/q_n follows by the alternating series test (Exercise 9) and it remains only to show that the limit is r_1/r_0. Since $(r_1/r_0) = (p_{n-1}r_{n+1} + p_n r_n)/(q_{n-1}r_{n+1} + q_n r_n)$ it follows that r_1/r_0 lies between p_{n-1}/q_{n-1} and p_n/q_n for all n and hence that $p_n/q_n \to r_1/r_0$.

14 Since p/q lies between p_{n-1}/q_{n-1} and $p_n/q_n = (p_{n-1}/q_{n-1}) \pm (1/q_{n-1}q_n)$ multiplication by q_{n-1} shows that $(pq_{n-1})/q$ lies between an integer and an integer $\pm(1/q_n)$. This clearly implies that the fraction $(pq_{n-1})/q$ has a denominator larger than q_n.

15 It is easily shown that $p_{n+1} = q_n$, $q_{n+1} = p_n + q_n$. Hence $p_{n+1}/q_{n+1} = q_n/(p_n + q_n) = [(p_n/q_n) + 1]^{-1}$. Passing to the limit as $n \to \infty$ gives $x = (x + 1)^{-1}$ and hence the desired equation. By the quadratic formula $r_1/r_0 = \frac{1}{2}(\sqrt{5} - 1)$. By the margin diagram, the chord subtended by an angle of $\pi/5$ satisfies $1:x = x + 1:1$. To compute this number, square the matrix $\begin{pmatrix} 0 & 1 \\ 1 & 1 \end{pmatrix}$ several times until $q_n q_{n-1} > 1{,}000$. Squaring the matrix four times gives its sixteenth power $\begin{pmatrix} 610 & 987 \\ 987 & 1597 \end{pmatrix}$ from which it follows that 610/987 represents the limit with about six place accuracy. Hence it is .618 to three places.

16 $2\sin(\pi/720) = 2[(\pi/720) - (1/6)(\pi/720)^3 + \cdots] = (10\pi/60^2) - (1/3)(\pi/720)^3 + \cdots$. The second term is much smaller than 60^{-3} and all terms other than the first can be ignored. Since $10\pi = 31.4159\ldots$ this term can be written $31 \cdot 60^{-2} + [.4159\ldots]60^{-2} = 31 \cdot 60^{-2} + [24.95\ldots]60^{-3}$, which gives Ptolemy's result.

17 (a) Multiplying by 10^n this becomes the statement that for every real number x there is an integer i such that $|x - i| < 1$. This is obviously true. (b) Two integers which lie within 1 of the same real number lie within 2 of each other and hence differ by at most ± 1.

18 The error is at most half of 60^{-6}, which is just short of eleven decimal places.

19 (a) Let r_n be the largest decimal fraction less than r. Then the sequence $\{r_n\}$ and the real number r lie in the interval $\{r_n \le x \le r_n + 10^{-n}\}$ for all n, hence $\{r_n\}$ converges and its limit is r. (b) Let q_n be another sequence such that $q_n \to r$ and such that q_n is the first n places of an infinite decimal fraction. Assume the fraction is > 0. Then the sequence q_n is *non-decreasing*. Hence q_n is $\le r$ with equal only if $q_n = q_{n+1} = q_{n+2} = \cdots$. Since $q_n + 10^{-n} > q_m$ for $m > n$ it follows that $q_n + 10^{-n} \ge r$ and q_n is the greatest n place decimal less than r, hence $q_n = r_n$ (all n), unless r is a decimal fraction.

§9.2 *pages 386–387*

1 Polynomial functions whose coefficients are natural numbers.

2 Implicit differentiation shows that the derivative of $\log x$ is x^{-1}. Since $\log 1 = 0$, the first equation is the Fundamental Theorem applied to $F(x) = \log x$. The second equation follows from the fact that d/dh of $x^h = e^{h \log x}$ at $h = 0$ is $\log x$, i.e. $\lim_{n\to\infty} [x^{1/n} - x^0]/(1/n) = \log x$.

3 Arcsin 0 = 0 and $dx = \cos y\, dy$, $\cos y = \sqrt{1 - \sin^2 y}$ (cosine is ≥ 0 for $-(\pi/2) \le y \le (\pi/2)$) gives $dy/dx = (1 - x^2)^{-1/2}$ and therefore Arcsin $x = \int_0^x (1 - t^2)^{-1/2}\, dt$ for $|x| < 1$. As y varies from $-\pi/2$ to $\pi/2$, $x = \tan y$ varies from $-\infty$ to ∞. Since $dx = (1 + x^2)\, dy$ and $0 = \tan 0$ it follows that Arctan $x = \int_0^x (1 + x^2)^{-1}\, dx$ for all x.

4 Let $(x(t), y(t))$ be a differentiable closed curve which does not pass through $(0, 0)$. Without loss of generality one may assume that the range of t is $0 \le t \le 1$. Choose (r_0, θ_0)

such that $(x(0), y(0)) = (r_0 \cos\theta_0, r_0 \sin\theta_0)$. Define $r(t) = \sqrt{x(t)^2 + y(t)^2}$ and $\theta(t)$ by $\theta(t) = \theta_0 + \int_0^t d\theta$ where $d\theta$ denotes the pullback of $(x\,dy - y\,dx)/(x^2 + y^2)$. Then $(x(t), y(t)) \equiv (r(t) \cos\theta(t), r(t) \sin\theta(t))$ because they are equal for $t = 0$ and have the same derivative for all t. Since $(x(1), y(1)) = (x(0), y(0))$ it follows that $\cos\theta(1) = \cos\theta(0)$, $\sin\theta(1) = \sin\theta(0)$. This implies $\theta(1)$ differs from $\theta(0)$ by a multiple of 2π.

5 If $z = a_n w^n + \cdots$ then $dz = [na_n w^{n-1} + (n-1)a_{n-1}w^{n-2} + \cdots]\,dw$. Now $dz = dx + i\,dy$, $dw = du + i\,dv$. The expression in parentheses can be expanded algebraically and put in the form $p(u, v) + iq(u, v)$. Then $dx = p(u, v)\,du - q(u, v)\,dv$, $dy = q(u, v)\,du + p(u, v)\,dv$, $dx\,dy = (p^2 + q^2)\,du\,dv$ and $dx\,dy \neq 0$ except when $p = q = 0$.

6 Since $e^{-t^2} \cos xt < e^{-t}$ for $t > 1$ the integral converges for all x by comparison with $\int_1^\infty e^{-t}\,dt = e^{-1}$. It is largest when $x = 0$ because then $\cos xt \equiv 1$ and e^{-t^2} is undiminished when multiplied by $\cos xt$. Since $\cos(-xt) = \cos xt$ it follows that $f(-x) = f(x)$. If x is very large then the alternating series test shows that $f(x)$ is at most the area of the central bump of the curve $y = e^{-t^2} \cos xt$, which is small. Thus $\lim_{x\to\infty} f(x) = 0$.

§9.3 *pages 391–392*

1 The triangle inequality suffices to prove boundedness. Since $f(x + h) - f(x)$ is a polynomial in $x_1, x_2, \ldots, x_n, h_1, h_2, \ldots, h_n$ in which every term contains an h, it follows that $|f(x + h) - f(x)| \leq |h_1|\,|P_1| + |h_2|\,|P_2| + \cdots + |h_n|\,|P_n|$ where P_i is a polynomial in $2n$ variables. Since $|P_i| < K_i$ whenever $|x| \leq K$, $|h| \leq 1$ it follows that $|f(x + h) - f(x)|$ can be made small by making $|h| = \max(|h_1|, |h_2|, \ldots, |h_n|)$ small.

2 $s^{-1}[f(x + sh) - f(x)]$ is a polynomial in $x_1, x_2, \ldots, x_n, h_1, h_2, \ldots, h_n, s$, hence it is uniformly continuous by Exercise 1.

3 $1 = \sum_{j=1}^n [(j/n)^4 - ((j-1)/n)^4] = \sum_{j=1}^n n^{-4}[j^4 - j^4 + 4j^3 - 6j^2 + 4j - 1]$ which differs from the proposed approximation by $n^{-2}\sum_{j=1}^n [6(j/n)^2 - 4(j/n)n^{-1} - n^{-2}]$. The terms of the sum are easily less than 11 and there are n of them, hence $11/n$ is a bound on the difference. More generally, $(x + h)^4 - x^4 = 4x^3h + h^2[6x^2 + 4xh + h^2]$ which shows that $(x + h)^4 - x^4$ differs from $4x^3h$ by at most $11h^2$ when $0 \leq x \leq 1$, $|h| \leq 1$. Thus over an interval $\{x_j - k \leq x \leq x_j + h\}$ the difference $(x_j + h)^4 - (x_j - k)^4 = [(x_j + h)^4 - x_j^4] - [(x_j - k)^4 - x_j^4]$ differs from $4x_j^3(\Delta x) = 4x_j^3[h + k]$ by at most $22(\Delta x)^2$. Summing over all intervals of a subdivision of $\{0 \leq x \leq 1\}$ shows that $1^4 - 0^4$ differs from $\sum(\alpha)$ by at most $\sum 22(\Delta x)^2 \leq \sum 22(\Delta x)\epsilon = 22\epsilon$ provided all Δx's are less than ϵ. Thus $\sum(\alpha)$ can be made to differ arbitrarily little from 1 by making $|\alpha|$ small.

4 Given ϵ, choose δ such that $|h^{-1}[f(x + h) - f(x)] - f'(x)| < \epsilon$ when $|h| < \delta$, $0 \leq x \leq 1$. Then on any interval $\{x_j - k \leq x \leq x_j + h\}$ the change in f, $f(x_j + h) - f(x_j - k)$, differs from $f'(x_j)(\Delta x) = f'(x_j)(h + k)$ by at most $\epsilon h + \epsilon k = \epsilon\Delta x$. Summing shows that $f(1) - f(0)$ differs from $\sum(\alpha) = \sum f'(x_i)\,\Delta x_i$ by at most $\epsilon\sum \Delta x = \epsilon$.

5 $(x + h)^{-1} - x^{-1} = -h/x(x + h)$. If $|x| > a$ and $|h| < a/2$ this is less than $2|h|a^{-2}$ in absolute value, hence it can be made small by making $|h|$ small. On the other hand, given any $\delta > 0$ it is easy to find numbers x, x' such that $|x - x'| < \delta$ but $|x^{-1} - (x')^{-1}| \geq 1$.

6 The fact that the derivative of e^x at $x = 0$ is 1 means that given $\epsilon > 0$ there is a δ such that $(e^h - 1)/h$ differs from 1 by less than ϵ whenever $|h| < \delta$. Hence $|e^h - 1| < (1 + \epsilon)|h|$. Since $|e^{x+h} - e^x| = e^x|e^h - 1|$, to prove e^x is uniformly continuous on $\{|x| \leq K\}$ it suffices to prove that e^x is bounded on $\{|x| \leq K\}$. From the power series $e^x = 1 + x + (x^2/2) + \cdots + (x^n/n!) + \cdots$ it follows easily that $|e^x| \leq e^K$ when $|x| \leq K$. Since $(e^{x+h} - e^x)/h = e^x(e^h - 1)/h$ and since e^x is uniformly continuous, to prove that e^x is uniformly differentiable it suffices to prove that $(e^h - 1)/h$ is uniformly continuous for $0 < |h| \leq 1$. Let $f(h)$ denote this function $(e^h - 1)/h$. Given $\epsilon > 0$ there is a $\delta > 0$ such that $|f(h) - 1| < \epsilon/2$ whenever $|h| < \delta$. It is easily shown that $f(h)$ is uniformly continuous on $\{\delta/2 \leq |h| \leq 1\}$; hence there is a δ_0 such that if $|h|$ and $|h'|$ both lie between $\delta/2$ and 1, and if $|h - h'| < \delta_0$, then $|f(h) - f(h')| < \epsilon$. Let $\delta_1 = \min(\delta_0, \delta/2)$. Then $|h| \leq 1$, $|h'| \leq 1$, $|h - h'| < \delta_1$ imply $|f(h) - f(h')| < \epsilon$. Since ϵ was arbitrary the result follows.

7 Because sine, cosine are perodic $[\sin(x + 2\pi) = \sin x$, $\cos(x + 2\pi) = \cos x]$ it suffices to prove uniform differentiability on the interval $\{|x| \leq \pi\}$.

8 $e^{x \log 10}$ is uniformly continuous and agrees with 10^x whenever x is rational.

9 $h^{-1}[f(x+h) - f(x)] = \int_0^1 h^{-1}[\cos(x+h)t - \cos xt]\,dt$. For small h the integrand differs by less than ϵ (in fact less than ϵt) from $t \sin xt$, hence the integral differs by less than ϵ from $\int_0^1 t \sin xt\,dt$. This is a uniformly continuous function of x and the result follows.

10 $|x|^p = e^{p \log|x|}$ by definition. Prove that $\log|x| = \int_1^{|x|} (dt/t)$ is uniformly continuous on $\{\delta/2 \le |x| \le K\}$ for any δ, hence $|x|^p$ is. Following 6 it suffices to show then that $\lim_{x\to 0} |x|^p = 0$, which is easy.

11 For $x \neq 0$ the limit exists because $f(x)$ is the product of the differentiable functions x^2 and $\sin(1/x^2)$. For $x = 0$ the limit is $\lim_{h\to 0} h^{-1}[h^2 \sin(h^{-2})]$. Since $|\sin(h^{-2})| \le 1$ this number has absolute value $\le |h|$ and the limit is 0. For $x \neq 0$ the derivative is $2x \sin x^{-2} - 2x^{-1} \cos x^{-2}$ which does not approach 0 as $x \to 0$.

12 For any ϵ, $|h^{-1}(f(x+h) - f(x)) - f'(x)| < \epsilon$ for h sufficiently small. If $\epsilon < \delta$ this implies $h^{-1}(f(x+h) - f(x)) > 0$, hence if $h > 0$ it implies $f(x+h) > f(x)$ so $f(x)$ is not a maximum.

13 Same as 12

14 See §2.3

15 Uniformly continuous means that for every $\epsilon >$ there is a $\delta > 0$ such that $|f(n^{-1}) - f(m^{-1})| < \epsilon$ whenever $|n^{-1} - m^{-1}| < \delta$. If $N > 2\delta^{-1}$ then $n, m > N$ implies $|n^{-1} - m^{-1}| < \delta$, hence the sequence $f(n^{-1})$ is convergent. Conversely, if $f(n^{-1})$ is convergent then $|f(n^{-1}) - f(m^{-1})| < \epsilon$ for $n, m > N$. Take $\delta = (N-1)^{-1} - N^{-1}$. Then $|n^{-1} - m^{-1}| < \delta$ implies $n, m > N$ and uniform continuity follows.

16 $h^{-1}[F(y+h) - F(y)] = \int_a^b h^{-1}[f(x, y+h) - f(x,y)]\,dx$. When h is sufficiently small the integrand differs by at most ϵ from $\partial f/\partial y$ for all x, y, hence the integral differs by at most $\epsilon(b-a)$ from $\int_a^b (\partial f/\partial y)\,dx$ for all y. It follows that F is uniformly differentiable.

§9.4 *pages 398–399*

1 Let P_i be the midpoint of C_i. If $j > i$ then $|P_i - P_j| \le 2^{-1}10^{-i}$ because P_j lies in C_i. Therefore the sequence $P_0, P_1, P_2, \ldots$ converges. Let P_∞ be its limit. Then passing to the limit as $j \to \infty$ gives $|P_i - P_\infty| < 2^{-1}10^{-i}$. Hence P_∞ lies in C_i for all sufficiently large i. Therefore P_∞ lies in all C_i. If Q also lies in C_i for all i then $|Q - P_\infty| \le 10^{-i}$ for all i, hence $Q = P_\infty$.

2 Let $U_i = \{$all points not in $C_i\}$. If no point lies in all the C_i then every point of C_0 lies in at least one of the U_i, hence U_i is an interior cover of C_0. Thus a finite number of the U_i cover C_0. Thus there is a finite number of the C's such that no point lies in all of them. This contradicts the assumption that the C's are nested unless the assumption that no point lies in all C_i is false. Hence some point lies in all the C_i.

3 It suffices to show that if $x^{(1)}, x^{(2)}, x^{(3)}, \ldots$ is an infinite sequence of points of X_δ then there is an $x^{(\infty)}$ in X_δ satisfying the condition of the Bolzano-Weierstrass Theorem. Since the sequence $x^{(i)}$ is bounded it follows from the Bolzano-Weierstrass Theorem applied to a cube $\{|x| \le K\}$ that there is a point $x^{(\infty)}$ and it suffices to show that $x^{(\infty)}$ is in X_δ. Let U_j be all points whose distance from $x^{(\infty)}$ is strictly greater than $\delta + j^{-1}$. Then $U_0 \subset U_1 \subset \cdots$ and for each j there is a point of X not in U_j; hence no finite number of the U's cover X. Hence by Heine-Borel the U's do not cover X and some point of X lies in no U. Hence $x^{(\infty)}$ is in X_δ.

4 Let $y^{(1)}, y^{(2)}, \ldots$ be an infinite sequence of points in $f(X)$. It suffices to show that there is a point $y^{(\infty)}$ as in Bolzano-Weierstrass. For each i let $x^{(i)}$ be a point of X such that $f(x^{(i)}) = y^{(i)}$. Let $x^{(\infty)}$ be as in Bolzano-Weierstrass, and set $y^{(\infty)} = f(x^{(\infty)})$. Given $\epsilon > 0$ there is, because f is continuous, a $\delta > 0$ such that $|x' - x^{(\infty)}| < \delta$ implies $|f(x') - f(x^{(\infty)})| < \epsilon$. By assumption there are an infinite number of $x^{(i)}$ such that $|x^{(i)} - x^{(\infty)}| < \delta$, hence an infinite number of $y^{(i)}$ such that $|y^{(i)} - y^{(\infty)}| < \epsilon$.

5 Combine Theorem 3 with Exercise 16, §9.3.

§9.5 *pages 404–407*

1 If $a_n = x^n/n!$ then $|a_{n+1}/a_n| = |x|/(n+1) \to 0$ and the ratio test implies convergence.

2 If $|x| \ge 1$ then $|x|^n \ge 1$ and the definition of convergence is not fulfilled (take $n = m + 1$). Since $(1-x)(1 + x + x^2 + \cdots + x^n) = 1 - x^{n+1}$ and since $x^{n+1} \to 0$ the desired result follows as $n \to \infty$, because $A(\lim x_n) = \lim(Ax_n)$ for any constant A.

3 If $|x| \ge 1$ then the factors do not approach 1 and the definition of convergence is not fulfilled (take $n = m + 1$). The identities $(1-x) \times (1+x) = 1 - x^2$, $(1-x)(1+x)(1+x^2) =$

$1 - x^4$, $(1 - x)(1 + x)(1 + x^2)(1 + x^4) = 1 - x^8$, etc., show that $1 - x$ times the infinite product is 1.

4 $\pi^2/6 = 1.64493\ldots$. To obtain this accuracy it would suffice to take the sum of the first 100,000 terms. Needless to say, this is not the method by which Euler estimated $1 + \frac{1}{4} + \frac{1}{9} + \frac{1}{16} + \cdots$.

5 Set $x = \pi/2$ and use $1 - (2n)^{-2} = (2n - 1)(2n + 1)(2n)^{-2}$.

6 The suggested method shows that for every $\epsilon > 0$ there is an M such that the sum of the first $m(k + 1)$ terms differs from $\log k$ by less than $\epsilon/2$ whenever $m \geq M$. The sums between the $m(k + 1)$st and the $(m + 1)(k + 1)$st differ from the $m(k + 1)$st by at most $(1/(mk + 1)) + (1/(mk + 2)) + \cdots + (1/(mk + k)) \leq (k/mk) = (1/m)$ which is less than $\epsilon/2$ for m sufficiently large.

7 Add positive terms until a total greater than 10 is reached, then add negative terms until a total less than 10 is reached (one term will do it), then add positive terms until a total greater than 10 is reached, then add negative terms, etc. It is easily shown that all negative (and positive) terms are eventually used and that the sum approaches 10.

8 Given $\epsilon > 0$ use the convergence of $\sum_n |a_n|$ to find an N such that the sum of any finite collection of $|a_n|$ for which $n > N$ is $<\epsilon$. It follows that the sum of the first M terms of any rearrangement of $\sum a_n$ differs from $a_1 + a_2 + \cdots + a_N$ by at most ϵ provided only that M is large enough to include all the terms $a_1, a_2, \ldots, a_N$. Thus the sum of the first M terms differs from $\sum_{n=1}^{\infty} a_n$ by at most 2ϵ.

9 Let $\sum p_n$ be the series of positive terms obtained by striking all negative terms from $\sum a_n$, and let $\sum q_n$ be the series of negative terms obtained by striking the positive terms. Either $\sum p_n$ diverges or $\sum q_n$ diverges since otherwise $\sum a_n$ converges absolutely. Thus both $\sum p_n$, $\sum q_n$ must diverge, since otherwise $\sum a_n$ would diverge. Add terms p_n until 10 is passed, then add terms q_n until 10 is passed, etc. This gives a rearrangement of $\sum a_n$ converging to 10.

10 $|\prod_{i=1}^n a_i - \prod_{i=1}^m a_i| = |\prod_{i=1}^m a_i| |a_{m+1}a_{m+2}\cdots a_n - 1| \leq \frac{3}{2}|P|\,\epsilon$ for $n > m \geq N$. Therefore the sequence $x_n = \prod_{i=1}^n a_i$ is convergent.

11 Since $|1 - a_i|$ is small for large i (by convergence) $\log a_i$ is defined for large i. Then $a_{m+1}a_{m+2}\cdots a_n$ is near 1 if and only if $\log(a_{m+1}a_{m+2}\cdots a_n)$ is near $\log 1$ (by continuity of log).

12 If $\prod(1 + b_i)$ converges then its value is a bound on the increasing sequence $b_1, b_1 + b_2, b_1 + b_2 + b_3, \ldots$ which therefore converges (Exercise 10, §9.1). Similarly, if $\sum b_i$ converges then $(1 + b_1), (1 + b_1)(1 + b_2), \ldots$ is increasing and bounded.

13 A number is near 1 if and only if its inverse is near 1. Hence $a_{m+1}^{-1}a_{m+2}^{-1}\cdots a_n^{-1}$ is near 1 if and only if $a_{m+1}a_{m+2}\cdots a_n$ is. Since $(\prod_{i=1}^n a_i)(\prod_{i=1}^n a_i^{-1}) \equiv 1$ the desired formula follows by letting $n \to \infty$.

14 By the method of 12 it follows that $\prod(1 - b_i)^{-1}$ converges if and only if $\prod(1 + b_i)$ converges. Then apply 13.

15 $\cos x$ is zero for $x = \pm\pi/2, \pm 3\pi/2, \ldots$ and is 1 when $x = 0$ which leads to the conjecture that $\cos x = (1 - (2x/\pi)^2)(1 - (2x/3\pi)^2)\cdots$. Since $\cos x = 1 - (x^2/2) + \cdots$, equating the terms in x^2 gives $-(1/2) = -(4/\pi^2) - (4/9\pi^2) - (4/25\pi^2) - \cdots$, $(\pi^2/8) = 1 + (1/9) + (1/25) + \cdots$. On the other hand $1 + (1/9) + (1/25) + \cdots = (1 + (1/4) + (1/9) + \cdots) - (1/4)(1 + (1/4) + (1/9) + \cdots) = (\pi^2/6) - (1/4) \times (\pi^2/6) = (\pi^2/8)$.

16 Using 12–14, in order to prove that $\sum(1/p)$ diverges, it suffices to show that $\prod(1 - p^{-1})^{-1}$ diverges. For this it suffices to show that this product can be made arbitrarily large by including enough terms. Let N be a given integer. For each prime p let $S_p = 1 + p^{-1} + p^{-2} + \cdots + p^{-k}$ where p^k is the largest power of p less than N. Then $S_p < (1 - p^{-1})^{-1}$ and $S_p = 1$ for all primes $p \geq N$. Thus $\prod S_p$ is a finite product. It is easily seen that $\prod S_p > 1 + 1/2 + 1/3 + \cdots + 1/(N - 1)$. Since the series $\sum(1/n)$ diverges this can be made large, hence $\prod(1 - p^{-1})^{-1}$ can be made large.

17 Set $u = t\sqrt{\pi}$. Then $\int_{-\infty}^{\infty} \exp(-\pi t^2)\,dt = \pi^{-1/2}\int_{-\infty}^{\infty} \exp(-u^2)\,du = 1$.

18 Because the integrand is not defined at $x = 0$, hence the domain of integration must be taken to be the non-compact set $\{0 < x \leq 1\}$. Since $\lim_{\epsilon\to 0}\int_\epsilon^1 x^{-a}\,dx = \lim_{\epsilon\to 0}[x^{1-a}/(1 - a)]\,|_\epsilon^1 = (1/(1 - a)) - \lim_{\epsilon\to 0}(\epsilon^{1-a}/(1 - a))$ it exists and is $(1 - a)^{-1}$ provided $a < 1$.

19 Set $x = u$, $y = uv$. Then the pullback of $\exp -(x^2 + y^2)\,dx\,dy$ is $\exp(-u^2(1 + v^2))\,u\,du\,dv$. Writing the double integral as an iterated integral and using $\int_0^\infty \exp(-u^2k)\,u\,du = (1/2k)$ and $\int_0^\infty (1+v^2)^{-1}\,dv = \pi/2$ gives the value $\pi/4$.

§9.6 *pages 419–426*

1 $(1+t^2)(1-t^2+t^4-t^6+\cdots+(-t^2)^n) = 1-(-t^2)^{n+1}$ hence $(1+t^2)^{-1} = 1-t^2+t^4-\cdots+(-t^2)^n+((-t^2)^{n+1}/(1+t^2))$. The last term is uniformly small for $0 \le t \le x < 1$ when n is large, from which $\int_0^x (1+t^2)^{-1}\,dt = \sum_{n=0}^\infty \int_0^x (-t^2)^n\,dt = \sum_{n=0}^\infty (-1)^n(x^{2n+1}/(2n+1))$. Thus the series $1+0-\frac{1}{3}+0+\frac{1}{5}+0-\frac{1}{7}+\cdots$ is Abel summable and its Abel sum is $\lim_{x\to 1}\int_0^x (1+t^2)^{-1}\,dt = \int_0^1 (1+t^2)^{-1}\,dt = \pi/4$ (see §7.5). Since $1-\frac{1}{3}+\frac{1}{5}-\frac{1}{7}+\cdots$ is convergent, its sum is $\pi/4$ by Abel's Theorem.

2 $(\pi/a)^{1/2}\,e^{-(y^2/4a)}$.

3 The given integral is equal to $(1/2)\int_{-\infty}^\infty (\sin(x+1)y/y)\,dy - (1/2)\int_{-\infty}^\infty (\sin(x-1)y/y)\,dy$ which is 0 if $|x| > 1$, π if $|x| < 1$ and $\pi/2$ if $|x| = 1$.

4 Let $f_N(x)$ denote the sum of the first N terms. Then for $M > N$, $|f_N(x) - f_M(x)| \le 2^{-(N+1)} + 2^{-(N+2)} + \cdots = 2^{-N}(\frac{1}{2}+\frac{1}{4}+\frac{1}{8}+\cdots) = 2^{-N}$. Thus $f(x) = \lim_{N\to\infty} f_N(x)$ exists and satisfies $|f(x) - f_N(x)| \le 2^{-N}$ for all x. The function $f_N(x)$ is the sum of a finite number of uniformly continuous functions and can therefore be shown to be uniformly continuous. Given $\epsilon > 0$ choose N so large that $2^{-N} < \epsilon/3$ and choose δ so small that $|x' - x| < \delta$ implies $|f_N(x') - f_N(x)| < \epsilon/3$. Then $|x'-x| < \delta$ implies $|f(x') - f(x)| < \epsilon$ so f is uniformly continuous.

5 Set $f_N(x) = u_1(x) + \cdots + u_N(x)$. Then $|f_N(x) - f_K(x)| \le M_{N+1} + \cdots + M_K$ for all x. Since $\sum M_n$ converges it follows easily that $\lim_{N\to\infty} f_N(x) = f(x)$ exists and that for every $\epsilon > 0$ there is an N such that $|f_N(x) - f(x)| < \epsilon/3$ for all x. Since each $f_N(x)$ is uniformly continuous, the uniform continuity of $f(x)$ follows from $|f(x') - f(x)| \le |f(x') - f_N(x')| + |f_N(x') - f_N(x)| + |f_N(x) - f(x)| \le (\epsilon/3) + (\epsilon/3) + (\epsilon/3)$ as in 4.

6 If $\sum a_n\bar{x}^n$ converges then $a_n\bar{x}^n \to 0$ as $n\to\infty$, from which it is easily shown that there is a K such that $|a_n\bar{x}^n| \le K$ for all n. Set $\rho = |\bar{x}| - (\epsilon/2)$, and $M_n = |a_n|\rho^n \le K|\rho/\bar{x}|^n$. Since $|\rho/\bar{x}| < 1$ the series $\sum M_n$ converges.

7 $(1-x+x^2-x^3+x^4-\cdots)(1+x+x^2+x^3+x^4+\cdots) = 1+(1-1)x+(1-1+1)x^2+(1-1+1-1)x^3+\cdots = 1+x^2+x^4+\cdots$.

8 As in 6 there is a K_1 such that $|a_n\bar{x}^n| \le K_1$ for all n and a K_2 such that $|b_n\bar{x}^n| \le K_2$. Let r be slightly less than 1, and set $M_{n,m} = K_1K_2r^{n+m}$. For $|x| < r\bar{x}$ it follows that $|a_nb_mx^nx^m| \le M_{n,m}$. The double series $\sum_{n,m} M_{n,m} = \sum_{n=0}^\infty [K_1K_2r^n\sum_{m=0}^\infty r^m] = \sum_{n=0}^\infty K_1K_2(1-r)^{-1}r^n = K_1K_2(1-r)^{-2}$ converges absolutely. Therefore $\sum a_nb_mx^nx^m$ can be summed either in the order (*) or in the order $\sum_{n=0}^\infty a_nx^n[\sum_{m=0}^\infty b_mx^m] = g(x)\sum_{n=0}^\infty a_nx^n = g(x)f(x)$.

9 Setting $y = -x^2n^{-2}\pi^{-2}$ in $\log(1+y) = y - \frac{1}{2}y^2 + \frac{1}{3}y^3 - \frac{1}{4}y^4 + \cdots$ and summing over $n = 1, 2, 3, \ldots$ gives $\log(\sin x/x) = a_2x^2 + a_4x^4 + a_6x^6 + \cdots$ where $a_{2n} = -n^{-1}\pi^{-2n} \times [1 + 2^{-2n} + 3^{-2n} + 4^{-2n} + \cdots]$. On the other hand, the equation $(2a_2x + 4a_4x^3 + 6a_6x^5 + \cdots)[x^2 - (1/6)x^4 + (1/120)x^6 - \cdots] = (x - (1/2)x^3 + (1/24)x^5 - (1/720)x^7 + \cdots) - (x - (1/6)x^3 + (1/120)x^5 - (1/5040)\,x^7 + \cdots)$ yields $a_2 = -1/6$, $a_4 = -1/180$, $a_6 = -(1/2835)$. This verifies the first two formulas and indicates that $1 + 2^{-6} + 3^{-6} + 4^{-6} + \cdots = \pi^6/945$.

10 (a) If y is an integer the expression can be written as $(y+1)(y+2)\cdots(y+N)$ divided by $N!N^y$. Taking N very large and cancelling $(y+1)(y+2)\cdots N$ from numerator and denominator leaves $(N+1)(N+2)\cdots(N+y)$ in the numerator and $y!N^y$ in the denominator. Hence it is $(1/y!)$ times $(1+(1/N))(1+(2/N))\cdots(1+(y/N))$, and the second factor approaches 1 as $N\to\infty$. (b) 0! should be 1 and taking $y = 0$ in (a) gives this answer. (c) All steps are easy. The inequality follows from the formula $\log(1+x) - x = -\frac{1}{2}x^2 + \frac{1}{3}x^3 - \cdots$ which gives for $|x| < \frac{1}{2}$ the desired inequality with $K = \frac{1}{2}+\frac{1}{4}+\frac{1}{8}+\cdots = 1$. (d) $\pi y/\prod(y)\prod(-y)$ is the limit as $N\to\infty$ of $\pi y\prod_{n=1}^N (1+(y/n))N^{-y}\prod_{n=1}^N (1-(y/n))N^y = \pi y\prod_{n=1}^N (1-(y^2/n^2))$ which is $\sin\pi y$. Using $N-1$ in the formula for $\prod(y)$ shows that $y\prod(y-1)/\prod(y)$ is the limit as $N\to\infty$ of $y\cdot(y+1)(y+2)\cdot\cdots\cdot(y+N-1)\cdot 1^{-1}\cdot 2^{-1}\cdot\cdots\cdot(N-1)^{-1}(N-1)^{-y}\cdot 1\cdot 2\cdot\cdots\cdot N\cdot y^{-1}\cdot(y+1)^{-1}\cdot\cdots\cdot(y-1+N)^{-1}N^{y-1} = (1-(1/N))^{-y}$ which is 1. (e) From (d) $\sin(\pi/2) = \pi/\prod(-1/2)\prod(-1/2)$. Since $\prod(-1/2) > 0$ this gives $\prod(-1/2) = \sqrt{\pi}$. Thus $\prod(n-(1/2)) = ((2n-1)/2)\prod(n-(3/2)) = \cdots = ((2n-1)/2)((2n-3)/2)\cdots(3/2)\times(1/2)\sqrt{\pi}$ and $\prod(-n-(1/2)) = (-2)(-2/3)\times$

$(-2/5)\cdots(-2/(2n-1))\sqrt{\pi}$. (*f*) At the integers $-5, -4, \ldots, 4, 5$ the value of the function is 0, 0, 0, 0, 0, 1, 1, 1/2, 1/6, 1/24, 1/120. The values at $-\frac{1}{2}, \frac{1}{2}, 1\frac{1}{2}, 2\frac{1}{2}, 3\frac{1}{2}, 4\frac{1}{2}$, can be guessed (approximately) by filling in a smooth curve through these points, and verified by (e). The oscillations of $1/\prod(y)$ between the negative integers increase greatly as $y \to -\infty$. (E.g. $1/\prod(-4\frac{1}{2}) \sim 3.7$.) (g) $1/\prod(2x)$ is the limit as $N \to \infty$ of $((1+2x)/1)((2+2x)/2) \times ((3+2x)/3)\cdots((2N+2x)/2N) \times (2N)^{-2x}$, while $1/\prod(x)$ is the limit of $((1+x)/1)((2+x)/2)\cdots((N+x)/N)N^{-x}$ and $1/\prod(x-\frac{1}{2})$ is the limit of $((1+2x)/2)((3+2x)/4)\cdots((2N-1+2x)/2N)N^{-x+1/2}$. Thus the expression in question is the limit of an expression which does not involve x. The value of the limit is $\prod(-1/2) = \sqrt{\pi}$. (h) Similar to (g). Use N in the expressions of $\prod(x), \prod(x-1/n), \ldots$ and nN in the expression of $\prod(nx)$. (i) $\mu_n = \prod(-1/n)\prod(-2/n)\cdots\prod(-(n-1)/n)$. By (d) $\prod(y-1)\prod(-y) = \pi/\sin(\pi y)$. Setting $y = 1/n, 2/n, \ldots, (n-1)/n$ and multiplying gives

$$\mu_n^2 = \pi^{n-1}/(\sin(\pi/n)\sin(2\pi/n)\cdots \sin((n-1)\pi/n))$$

$\mu_2^2 = \pi$, $\mu_3^2 = 4\pi^2/3$, $\mu_4^2 = 2\pi^3$, $\mu_6^2 = 2^4\pi^5/3$ which leads to the guess $\mu_n^2 = (2\pi)^{n-1}/n$. (j) For n odd this is essentially the problem of finding the coefficient of y^p in the expression of $\sin pA$ as a polynomial in $y = \sin A$. The formula

$$\binom{p}{1} + \binom{p}{3} + \cdots + \binom{p}{p} = \frac{1}{2}\left[\binom{p}{0} + \binom{p}{1} + \cdots + \binom{p}{p}\right] = \frac{1}{2}(1+1)^p = 2^{p-1}$$

is used. (k) In the formulas obtained by differentiating (5′) set $a = 1$ and cancel minus signs to obtain $\prod((2n-1)/2) = \int_{-\infty}^{\infty}(u^2)^n e^{-u^2}\,du$. Write this as twice the integral from 0 to ∞ and substitute $t = u^2$. (l) The integral is proper at $t = 0$. Since $e^t > t^n/n!$ for all $t > 0$ and all n it follows (using $n+2$ in place of n) that $t^n e^{-t} < \text{const.}\ t^{-2}$. Since $\int_1^\infty t^{-2}\,dt$ converges so does $\int_1^\infty t^n e^{-t}\,dt$. The desired formula follows from $\int_0^\infty t^n e^{-t}\,dt = \lim_{K\to\infty}\int_0^K \{(d/dt)(-t^n e^{-t}) + nt^{n-1}e^{-t}\}\,dt = n\int_0^\infty t^{n-1}e^{-t}\,dt$ and $\int_0^\infty e^{-t}\,dt = 1$. (m) The integral converges as $t \to \infty$ for all x. As $t \to 0$ the term e^{-t} is like 1 and the integral is like $\int_0^1 t^x\,dt$. By Exercise 18 of §9.5 this converges for $x > -1$. (n) Set $\prod(nx) = \Gamma(nx+1)$, $\prod(x) = \Gamma(x+1)$, etc. and $y = nx+1$ to obtain $\Gamma(y/n)\Gamma((y+1)/n)\cdots\Gamma((y+n-1)/n) = n^{(1/2)-y}(2\pi)^{(n-1)/2}\Gamma(y)$.

11 (a) The formula (*) of 10 (b) The integral converges near 0 if $x > -1$ and near 1 if $y > -1$, hence converges if both $x > -1$ and $y > -1$. (c) Set $u = 1 - t$. Then $du = -dt$ and the orientation is reversed. $\int_0^1 t^x\,dt = (x+1)^{-1}$. (d) Elementary. Note $y > 0$. (e) $C(x,n) = (n/(x+n))((n-1)/(x+n-1))\cdots(1/(x+1))\cdot 1$. (*f*) $n^x C(x,n) = (1+(x+1)/n)\int_0^n u^x(1-(u/n))^n\,du$. For very large n the factor in front is nearly 1, the domain of integration is nearly $\{0 < u < \infty\}$ and the integrand is nearly $u^x e^{-u}$. (g) Let $f_n(u)$ be the function which is $(1-(u/n))^n$ for $u \le n$ and which is 0 for $u \ge n$. It suffices to prove then that $\lim_{n\to\infty}\int_0^\infty f_n(u)\,du = \int_0^\infty [\lim_{n\to\infty} f_n(u)]\,du$. (h) Elementary (i) $B(x,y) = \prod(x-1) \times \prod(y-1)/\prod(x+y-1) = \Gamma(x)\Gamma(y)/\Gamma(x+y)$

12 (a) Both are $2r$, πr^2, $4\pi r^3/3$. (b) $\omega = r^{-n}(x_1\,dx_2\,dx_3\cdots dx_n - x_2\,dx_1\,dx_3\cdots dx_n + \cdots + (-1)^{n-1}x_n\,dx_1\,dx_2\cdots dx_{n-1}]$. ω is closed (Ex. 4, §8.3). The integral of ω over $r =$ const. is r^{-n} times the integral of $x_1\,dx_2\cdots dx_n - \cdots$ which is the integral of $n\,dx_1\,dx_2\cdots dx_n$ over the interior which is $n\sigma_n r^n$ where σ_n is the 'volume' of the unit sphere. Thus $\pi^{n/2} = n\sigma_n\int_0^\infty e^{-r^2}r^{n-1}\,dr = (n/2)\sigma_n\int_0^\infty e^{-u}u^{(n/2)-1}\,du = \sigma_n\prod(n/2)$.

13 Since $y + y^{-1}$ has a minimum at $y = 1$ its inverse $(y+y^{-1})^{-1}$ has a maximum at $y = 1$. Thus $\lim_{y\to 1}(d/dy)[y/(1+y^2)] = 0$. For $|y| < 1$ this derivative can be written as the sum of the series $(d/dy)[y - y^3 + y^5 - y^7 + \cdots] = 1 - 3y^2 + 5y^4 - 7y^6 + \cdots = x^{-1}[x - 3x^2 + 5x^3 - 7x^4 + \cdots]$ where $x = y^2$. Thus $\lim_{x\to 1}$ of $[x - 3x^2 + \cdots] = 0$.

14 (a) S_N is 1 or 0 depending on whether N is odd or even. (b) Let S_∞ be the sum of the series, and $x_n = S_n - S_\infty$. Then $x_n \to 0$. Given $\epsilon > 0$ choose M so large that $|x_n| < \epsilon/2$ for $n \ge M$. Then for $N > M$, $N^{-1}[x_1 + x_2 + \cdots + x_N] \le (|x_1|/N) + \cdots + (|x_M|/N) + (\epsilon/2)((N-M)/N) \to (\epsilon/2)$ as $N \to \infty$. Thus $N^{-1}[x_1 + x_2 + \cdots + x_N] \to 0$, $N^{-1}[S_1 + S_2\cdots + S_N] \to S_\infty$.

§9.7 *pages 446–447*

1 See the proof in §2.3 that an integral converges if and only if $U(S) \to 0$ as $|S| \to 0$. This implies that if f, g are Riemann integrable then $f + g$, cf, $|f|$ and $\max(f, g)$ are Riemann integrable.

2 Given $\epsilon > 0$ let V be a collection of intervals of total length less than $\epsilon/2$ which contain all points where $f(x) \neq g(x)$ and W a collection of total length less than $\epsilon/2$ which contain all points where $g(x) \neq h(x)$. Then V and W together give a collection of total length less than ϵ containing all points where $f(x) \neq h(x)$.

3 Given $\epsilon > 0$ choose a collection of intervals V_n of total length less than $\epsilon/2^{n+1}$ such that $f_n(x) = g_n(x)$ for x outside V_n $(n = 1, 2, 3, \ldots)$. Let V_∞ be a collection of total length $\epsilon/2$ such that $f_n(x) \to f_\infty(x)$ for x outside V_∞. Then the collection of all intervals $V_\infty, V_1, V_2, \ldots$ has total length less than ϵ and outside these intervals $g_n(x) = f_n(x)$ (all n), $\lim_{n\to\infty} g_n(x) = \lim_{n\to\infty} f_n(x) = f_\infty(x)$.

4 One example is

$$f_n(x) = \begin{cases} 0 & \text{if } x \leq 0 \\ n^2x & \text{if } 0 \leq x \leq 1/n \\ 2n - n^2x & \text{if } 1/n \leq x \leq 2/n \\ 0 & \text{if } 2/n \leq x. \end{cases}$$

Then $\lim_{n\to\infty} f_n(x) = 0$ for all x, but $\int_0^1 f_n(x)\,dx = 1$ for all n.

5 Set $f_n(x) = n[x^{2k}\exp(-(a + n^{-1})x^2) - x^{2k}\exp(-ax^2)]$. Then $f_n(x)$ lies between the integrable (see Ex. 10(m), §9.6) functions $x^{2k}\exp(-ax^2)(-x^2)$ and $x^{2k}\exp(-ax^2) \times (-x^2 + x^4)$. Since $F \leq f \leq G$ implies $|f| \leq \max(|F|, |G|)$ the sequence f_n is dominated by an integrable function, hence $\lim_{n\to\infty} \int_{-\infty}^{\infty} f_n\,dx = \int_{-\infty}^{\infty} (\lim_{n\to\infty} f_n)\,dx$, which gives the desired formula.

6 Let $f_n(x)$ be 0 for $\{|x| < (1/n)\}$ and $|x|^{-a}$ elsewhere. Then $\int_{-1}^{1} f_n(x)\,dx = (a-1)^{-1} \times (2n^{a-1} - 2)$ if $a > 1$ and $2\log n$ if $a = 1$. In either case $\lim_{n\to\infty}$ does not exist. If $|x|^{-a}$ were integrable then this limit would have to exist (and be $\int_{-1}^{1} |x|^{-a}\,dx$).

7 Let $A_n(x, y)$ be 0 if $x^2 + y^2 < (1/n)$ and r^{-a} if $x^2 + y^2 \geq (1/n)$. If $m > n$ then the integral of $A_m - A_n$ is the integral of r^{-a} over the annulus $\{m^{-1} \leq x^2 + y^2 \leq n^{-1}\}$ which is $\iint r^{-a}\,r\,dr\,d\theta = 2\pi(2-a)^{-1}[n^{a-2} - m^{a-2}]$ which proves convergence for $a < 2$. The proof of divergence for $a \geq 2$ is similar to 6.

8 Integrable if and only if $a < 3$

9 Given $\epsilon > 0$, choose K so large that $\int_{-K}^{K} F(x)\,dx$ differs from $\int_{-\infty}^{\infty} F(x)\,dx$ by less than $\epsilon/3$. Choose N so large that $\int_{-K}^{K} f_n(x)\,dx$ differs from $\int_{-K}^{K} f_\infty(x)\,dx$ by at most $\epsilon/3$ whenever $n \geq N$ (by the Dominated Convergence Theorem by proper integrals). Then $\int_{-\infty}^{\infty} f_n(x)\,dx$ differs from $\int_{-\infty}^{\infty} f_m(x)\,dx$ by less than ϵ whenever $n, m \geq N$. Thus $\lim_{n\to\infty} \int_{-\infty}^{\infty} f_n(x)\,dx$ exists. Moreover if $K' > K$ then $\int_{-K'}^{K'} f_\infty(x)\,dx$ differs from this limit by at most ϵ and the theorem follows.

10 $\int_a^b |f_m(x) - f_n(x)|\,dx = \int_a^b [f_m(x) - f_n(x)]\,dx = \int_a^b f_m(x)\,dx - \int_a^b f_n(x)\,dx$ whenever $m > n$. By assumption $\int_a^b f_n(x)\,dx$ is a bounded increasing sequence. It therefore converges and $\int_a^b |f_m(x) - f_n(x)|\,dx < \epsilon$ for $n, m > N$. The result then follows from the completeness theorem.

11 Since Riemann integrable functions are Lebesgue integrable this is a special case of the Dominated Convergence Theorem.

12 Let $g_n(x) = \lim_{j\to\infty} \min(f_n, f_{n+1}, \ldots, f_{n+j})$. This limit exists for all x because it is a decreasing sequence of non-negative numbers. Each g_n is the dominated (by f_n) limit of integrable functions, hence $\int_a^b g_n\,dx$ exists and is $\leq K$. The sequence $g_1, g_2, \ldots$ is increasing, hence as in 10 it satisfies the Cauchy Criterion and there is an integrable g_∞ such that $\int_a^b |g_\infty - g_n|\,dx \to \infty$ and $g_n(x) \to g_\infty(x)$ for almost all x. Since $g_n(x) \to f(x)$ for almost all x, it follows that $g_\infty(x) = f(x)$ for almost all x and that $\int_a^b f(x)\,dx = \lim_{n\to\infty} \int_a^b g_n(x)\,dx \leq K$.

§9.8 *pages 452–455*

1 Given $\epsilon > 0$ there is a δ such that $|y - f(\bar{x})| < \delta$ implies $|g(y) - g(f(\bar{x}))| < \epsilon$. But there is a δ_1 such that $|x - \bar{x}| < \delta_1$ implies $|f(x) - f(\bar{x})| < \delta$. Thus $|x - \bar{x}| < \delta_1$ implies $|g(f(x)) - g(f(\bar{x}))| < \epsilon$.

2 Given $\epsilon > 0$ there is a $\delta > 0$ such that y, y' in Y and $|y - y'| < \delta$ implies $|g(y) - g(y')| < \epsilon$. But there is a δ_1 such that x, x' in X and $|x - x'| < \delta_1$ implies $|f(x) - f(x')| < \delta$. Thus $|x - x'| < \delta_1$ implies $|g(f(x)) - g(f(x'))| < \epsilon$.

3 Let $L: E \to F$ denote the derivative of f at $\bar{x}$ and $M: F \to G$ denote the derivative of g at $f(\bar{x})$. Then $((g(f(\bar{x} + sh)) - g(f(\bar{x})))/s) - M(L(h)) = ((g(f(\bar{x}) + sy) - g(f(\bar{x})))/s) - M(L(h))$ where $y = ((f(\bar{x} + sh) - f(\bar{x}))/s)$. Since $y \to L(h)$ as $S \to 0$, y is in a bounded set and for every $\epsilon > 0$ there is a δ such that $s^{-1}[g(f(\bar{x}) + sy) - g(f(\bar{x}))]$ differs from $M(y)$ by less than $\epsilon/2$ whenever $|s| < \delta$. The continuity of M implies $M(y)$ differs by less than $\epsilon/2$ from $M(L(h))$ for s small and the result follows.

4 $L_x(ah) = \lim_{s\to 0} ((f(x+sah) - f(x))/s) = \lim_{s\to 0} a((f(x+sah) - f(x))/sa) = a \lim_{t\to 0} (f(x+th) - f(x))/t) = aL_x(h)$. Since the functions $s^{-1}[f(x+uh+sh'+tk) - f(x+uh)]$ and $s^{-1}[f(x+sh') - f(x)]$ are uniformly continuous in s, u, t, x, h, h', k, their difference is uniformly continuous on the set $s = u$. Therefore it has an extension to $s = t = 0$ which is the limit. The function is defined and independent of k for $t = 0$, hence the same is true of the limit. Take $k = \frac{1}{2}(h+h')$ and $t = -\frac{1}{2}s$. Adding and subtracting $f(x)$ in each of the 4 terms shows that the limit is $L_x(\frac{1}{2}h + \frac{1}{2}h') + L_x(-\frac{1}{2}h - \frac{1}{2}h') + L_x(\frac{1}{2}h - \frac{1}{2}h') + L_x(-\frac{1}{2}h + \frac{1}{2}h') = 0$ by $L_x(ah) = aL_x(h)$. Then setting $t = 0$ and adding and subtracting $f(x)$ gives $L(h+h') - L(h) - L(h') = 0$ as desired.

5 If the condition is fulfilled then the map is uniformly continuous because $|L(x) - L(x')| = |L(x-x')| < B|x-x'|$ can be made $<\epsilon$ by making $|x-x'| < \epsilon/B$. If the map is continuous then there is a δ such that $|x-0| \le \delta$ implies $|L(x) - L(0)| < 1$, i.e. $|x| \le \delta$ implies $|L(x)| < 1$. Then for any x', $|\delta x'/|x'|| = \delta$, $|L(\delta x'/|x'|)| < 1$, $(\delta/|x'|)\,|L(x')| < 1$, $|L(x')| < \delta^{-1}|x'|$ and the condition is fulfilled with $B = \delta^{-1}$.

6 Let $\delta_1 = (1, 0, 0, \ldots, 0)$, $\delta_2 = (0, 1, 0, \ldots, 0)$, $\ldots$, $\delta_n = (0, 0, 0, \ldots, 1)$ and let B be a number such that $B/n \ge |L(\delta_i)|$ for $i = 1, 2, \ldots, n$. Then, for $x = (x_1, x_2, \ldots, x_n)$, $|L(x)| = |L(\sum x_i\delta_i)| = |\sum x_i L(\delta_i)| \le \sum |x_i| \times |L(\delta_i)| \le B|(x)|_\infty$, hence L is continuous.

7 Elementary. $R \times R \times \cdots \times R$ is R^n with the norm $|x|_\infty$.

8 If $|x|_\infty = 1$ then $|M_1x + M_2x|_\infty \le |M_1x|_\infty + |M_2x|_\infty \le \max|M_1x| + \max|M_2x| = |M_1| + |M_2|$, hence $|M_1 + M_2| \le |M_1| + |M_2|$. The other axioms are easily proved. If $Mx = \sum_j a_{ij}x_j$ then $|M| = \max_i (|a_{i1}| + |a_{i2}| + \cdots + |a_{in}|)$.

9 $|M|$ is easily seen to be a norm. Let $M_1, M_2, \ldots$ be a sequence of elements of $L(E, F)$ which satisfies the Cauchy Criterion $|M_n - M_m| \to 0$. For each x in E, $|M_nx - M_mx| = |(M_n - M_m)x| \le |M_n - M_m|\,|x| \to 0$. Therefore M_nx is a convergent sequence in F. Define $M_\infty x$ to be its limit. Then $M_\infty: E \to F$ is a well-defined function. It is clearly linear, so it suffices to show that there is a B such that $|M_\infty x| \le B|x|$. For this it suffices to choose B such that $|M_n| \le B$ for all n.

10 The Hölder inequality (see §5.4)

11 $|M_2(M_1x)| = |M_1x|\,|M_2y|$ where $y = |M_1x|^{-1}(M_1x)$ so that $|y| = 1$. (If $|M_1x| = 0$ then $M_2(M_1x) = 0$.) Thus $|M_2(M_1x)| \le |M_1|\,|M_2|$ for all x satisfying $|x| = 1$, and $|M_2 \circ M_1| \le |M_1|\,|M_2|$ follows.

12 Given M such that $|ML - I| < 1$, define a sequence M_n in $L(E, E)$ by $M_0 = M$, $M_{n+1} = 2M_n - M_nLM_n$. Then $(I - M_{n+1}L) = (I - M_nL)^2 = P^{2^{n+1}}$ where $P = I - ML$. Consequently $M_{n+1} = M_n + (I - M_nL)M_n = (I + P^{2^n})M_n = (I + P^{2^n})\cdots(I + P^4)(I + P^2)(I + P)M$. The existence of the limit $M_\infty = \lim_{n\to\infty} M_n$ is easily proved by proving the convergence of the product $(I + P^{2^n})\cdots(I + P^2)(I + P)$ or, what is the same, of the series $I + P + P^2 + P^3 + \cdots + P^{2^n}$. Letting $n \to \infty$ in $M_nL = I - P^{2^n}$ gives $M_\infty L = I$. An analogous argument shows that if there is an N such that $|LN - I| < 1$ then there is an N_∞ such that $LN_\infty = I$. Finally, $M_\infty = M_\infty I = M_\infty LN_\infty = IN_\infty = N_\infty$, hence L is invertible with inverse $L^{-1} = M_\infty = N_\infty$.

13 Given L_1 near L, let $\delta = |L - L_1|$, $\rho = |I - L^{-1}L_1| \le |L^{-1}|\delta$, $M_0 = L^{-1}$, $M_{n+1} = 2M_n - M_nL_1M_n$. Then, as in 12, the sequence M_n converges to $M_\infty = L_1^{-1}$ and satisfies $|M_n - M_0| = |(P + P^2 + \cdots + P^{2^n})M_0| \le (\rho/(1-\rho))\,|M_0|$, hence $|L_1^{-1} - L^{-1}| < (\rho/(1-\rho))\,|L^{-1}|$. This can be made small by making δ small, q.e.d.

14 For any given y the sequence of successive approximations is $x_0 = \bar{x}$, $x_{n+1} = x_n + M(y - f(x_n))$, i.e. $\Delta x = M(\text{desired } \Delta y)$. The proof of convergence turns on the inequality $|f(x') - f(x) - L(x'-x)| < \epsilon\,|x'-x|$, i.e. $|\Delta y - L(\Delta x)| < \epsilon\,|\Delta x|$ for x', x near $\bar{x}$. This can be proved by parameterizing the line segment from x to x' by $x(t) = x + t(x'-x)$ on the interval $0 \le t \le 1$. By the Fundamental Theorem of Calculus $\Delta y = \int_0^1 (dy/dt)\,dt = \int_0^1 L_{x(t)}(\Delta x)\,dt$ where the integral is the integral of a 1-form on $\{0 \le t \le 1\}$ with values in F. Given $\epsilon > 0$ there is a $\delta > 0$ such that $|x - \bar{x}| < \delta$ implies $|L_x - L_{\bar{x}}| < \epsilon$. Therefore $|L_{\bar{x}}(\Delta x) - L_x(\Delta x)| < \epsilon\,|\Delta x|$, hence $|\Delta y - L_{\bar{x}}(\Delta x)| < \epsilon|\Delta x|$ as desired provided $|x' - \bar{x}| < \delta$ and $|x - \bar{x}| < \delta$. Let $\rho < 1$ be given. Choose δ such that $|\Delta y - L_{\bar{x}}(\Delta x)| < \rho\,|M|^{-1}|\Delta x|$ whenever x', x are within δ of $\bar{x}$. Then for any y satisfying $|M(y - \bar{y})| < (1-\rho)\delta$ the successive approximations satisfy $x_{n+1} - x_n = x_n - x_{n-1} - M(f(x_n) - f(x_{n-1})) = \Delta x -$

$M(\Delta y) = M(L(\Delta x) - \Delta y)$ which gives $|x_{n+1} - x_n| < \rho|x_n - x_{n-1}|$ and hence $|x_{n+1} - \bar{x}| = |x_{n+1} - x_n| + \cdots + |x_1 - x_0| < (\rho^n + \rho^{n-1} + \cdots + \rho + 1)|M(y - \bar{y})| < \delta$ as in §7.1. Thus the sequence x_n satisfies the Cauchy Criterion for all y sufficiently near $\bar{y}$. Its limit satisfies $x_\infty = x_\infty + M(y - f(x_\infty))$, hence $y = f(x_\infty)$. Set $x_\infty = g(y)$. For ρ small $g(y)$ differs from $x_1 = \bar{x} + M(y - \bar{y})$ by a constant times $\rho(1 - \rho)$ times $|\Delta y|$. This implies that g is differentiable with derivative M. For x near $\bar{x}$ the derivative L_x is near $L_{\bar{x}}$, hence by 13 the inverse $M_x = L_x^{-1}$ exists and is continuous. Therefore g is differentiable at $f(x)$ with derivative M_x by the proof above. Therefore g is continuous. Thus its derivative $L_{g(y)}^{-1}$ is continuous, q.e.d.

15 Apply the Inverse Function theorem to the map $\hat{f}$: $E_1 \times E_2 \to F \times E_2$ defined by $\hat{f}(x_1, x_2) = (f(x_1, x_2), x_2)$. Alternatively, follow the proof of §7.1.

16 Given a k-form $\sum A\, dx_{i_1} \ldots dx_{i_k}$ (in the sense of §4.2) and given a k-tuple $(v_1, v_2, \ldots, v_k)$ of elements of R^n, define the 'value' of the given k-form on the given k-tuple to be the coefficient of $du_1\, du_2 \ldots du_k$ in the pullback of the given k-form under the map $(u_1, u_2, \ldots, u_k) \to u_1v_1 + u_2v_2 + \cdots + u_kv_k$ of R^k to R^n.

17 $L^*(\phi)$ for ϕ in $A_k(E_2, F)$ is the function which assigns to a k-tuple $(v_1, v_2, \ldots, v_k)$ of elements of E_1 the element $\phi\,(Lv_1, Lv_2, \ldots, Lv_k)$ of F. The Chain Rule is immediate from this definition.

18 S is a subset of E which can be described by a finite number of charts $R^k \to E$ satisfying the conditions of Chap. 6. The approximating sums give elements of F, hence the limit is an element of F.

19 By (i), $|0 \cdot x| = 0$, $|-x| = |x|$. Thus by (iii) $0 = |x + (-x)| \leq |x| + |-x| = 2|x|$ and division by 2 gives $|x| \geq 0$ as desired.

20 Let δ_i be the element of R^n which is 1 in the ith coordinate and 0 in the others. Choose B so that $|\delta_i| \leq B$ for $i = 1, 2, \ldots, n$. Then $|x| = |(x_1, x_2, \ldots, x_n)| = |\sum x_i\delta_i| \leq \sum |x_i|\, B \leq nB \max\{|x_i|\} = C|x|_\infty$ where $C = nB$. Algebraically the space of all linear maps ϕ: $R^n \to R$ is the same as R^n. Let $|\phi|^*$ be the norm defined by Exercise 9 and the given norm $|x|$, i.e. define $|\phi|^*$ to be the least real number such that $|\phi(x)| \leq |\phi|^*|x|$ for all x in R^n. Then $|\phi|^*$ is a norm so, by the first part, there is a C such that $|\phi|^* \leq C\,|\phi|_\infty$. Thus $|\phi(x)| \leq C\,|\phi|_\infty|x|$ for all ϕ, x. But $|x|_1$ is the least real number such that $|\phi(x)| \leq |\phi|_\infty|x|_1$ for all ϕ. Hence $|x|_1 \leq C|x|$, so $|x| \geq C^{-1}\,|x|_1 \geq$ const. $|x|_\infty$ as desired. This proves that if E is any finite dimensional Banach space and if $R^n \leftrightarrow E$ is any basis of E then a map f: $E \to F$ is continuous if and only if the corresponding map $R^n \to F$ is continuous in the norm $|x|_\infty$. Hence the last statement follows from Exercise 6.

21 L is continuous by Ex. 5 because $|Lx| = |x|$. Let M be the operator 'shift left', i.e. the operator $M(x_1, x_2, x_3, \ldots) = (x_2, x_3, x_4, \ldots)$. Then M is continuous because $|Mx| \leq |x|$. Also $ML = I$. However, L is not invertible because the equation $Lx = (1, 0, 0, \ldots)$ has no solution x in E.

index